Introductory Plant Biology

Introductory Plant Biology

edition seven

KINGSLEY R. STERN

California State University ~ Chico

WCB **Wm. C. Brown Publishers**

Dubuque, IA   Bogotá   Buenos Aires   Caracas   Chicago   Guilford, CT   London
Madrid   Mexico City   Seoul   Singapore   Sydney   Taipei   Tokyo   Toronto

Project Team

Editor *Marge Kemp*
Developmental Editor *Kathy Loewenberg*
Production Editor *Kay J. Brimeyer*
Marketing Manager *Tom Lyon*
Designer *Kaye Farmer*
Art Editor *Jodi K. Banowetz*
Photo Editor *Carrie Burger*
Advertising Coordinator *Heather Wagner*

 Wm. C. Brown Publishers

 **A Times Mirror Company**

Copyedited by *Sarah Lane*
Interior & cover design by *Jamie O'Neal*
Cover photo: © Lois Ellen Frank/West Light

Copyright © 1997 Times Mirror Higher Education Group, Inc.
All rights reserved

Library of Congress Catalog Card Number: 96–60094

ISBN 0–697–25773–8 (paper)
ISBN 0–697–25772–X (case)

Printed in the United States of America by Times Mirror Higher Education Group, Inc.,
2460 Kerper Boulevard, Dubuque, IA 52001

10 9 8 7 6 5 4 3 2 1

To the memory of Franklin Charles Lane

1928–1971

Botanist, Teacher, Friend

Table of Contents

7 Leaves 101

8 Flowers, Fruits, and Seeds 120

9 Water in Plants 143

10 Plant Metabolism 159

11 Growth 180

12 Meiosis and Alternation of Generations 203

19　Kingdom Fungi and Lichens　319

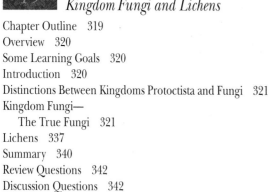

20　Introduction to the Plant Kingdom: Byrophytes　343

21　*Introduction to Vascular Plants: Ferns and Their Relatives*　358

22　*Introduction to Seed Plants: Gymnosperms*　382

23　*Flowering Plants*　401

Preface

Introductory Plant Biology is designed as an introductory text in botany. It assumes little knowledge of the sciences on the part of the student. The text includes sufficient information for some shorter introductory botany courses open to both majors and nonmajors, but it is arranged so that certain sections—for example, "Early History and Development of Plant Study," "Soils," "Division Psilotophyta"—can be omitted without disrupting the overall continuity of the course.

Botany instructors vary greatly in their opinions concerning the depth of coverage needed for the topics of photosynthesis and respiration in a text of this type. Some feel that nonmajors, in particular, should have a brief introduction only, while others consider a more detailed discussion essential. In this text, photosynthesis and respiration are discussed at three levels. Some may find one or two levels sufficient, and others may wish their students to become familiar with the processes at all three levels.

Despite eye-catching chapter titles and headings, many texts for majors and nonmajors relegate the current interests of a significant number of students to comparative obscurity. This text emphasizes current interests without giving short shrift to botanical principles. Present interests of students include subjects such as global warming, ozone layer depletion, acid rain, genetic engineering, organic gardening, Native American and pioneer uses of plants, pollution and recycling, houseplants, backyard vegetable gardens, natural dye plants, poisonous and hallucinogenic plants, and the nutritional values of edible plants. The rather perfunctory coverage or absence of such topics in many botany texts has occurred partly because botanists previously have tended to believe that some of the topics are more appropriately covered in anthropology and horticulture courses. I have found, however, that both majors and nonmajors in botany who may be initially disinterested in the subject matter of a required course frequently become engrossed if the material is repeatedly related to such topics. Accordingly, a considerable amount of ecological and ethnobotanical material has been included with traditional botany throughout the book—without, however, resorting to excessive use of technical terms.

ORGANIZATION OF THE TEXT

A relatively conventional sequence of botanical subjects is included. Chapters 1 and 2 cover introductory and background information; chapters 3 through 11 deal with structure and function; chapters 12 and 13 introduce meiosis and genetics. Chapter 14 discusses plant biotechnology; chapter 15 introduces evolution. Chapter 16 presents a five-kingdom system of classification; chapters 17 through 23 stress, in phylogenetic sequence, the diversity of organisms traditionally regarded as plants, and chapter 24 deals with ethnobotanical aspects and information of general interest pertaining to sixteen major families of flowering plants. Chapter 25 is an overview of the vast topic of ecology, although ecological topics and applied botany are included in most of the preceding chapters as well. Some of these subjects are broached in anecdotes that introduce the chapters, while others are mentioned in the human and ecological relevance sections (with which most of the chapters in the latter half of the book conclude).

AIDS TO THE READER

Review questions, discussion questions, and additional reading lists are provided for each chapter. New terms are defined as they are introduced, and those used more than once are boldfaced and included in a pronunciation glossary. The use of scientific names throughout the body of the text has been held to a minimum, but a list of the scientific names of all organisms mentioned is given in Appendix 1. Appendix 2 deals with biological controls and companion planting; Appendix 3 lists wild edible plants, poisonous plants, hallucinogenic plants, spices, and natural dye plants. Appendix 4 discusses pruning and grafting, and gives horticultural information on houseplants; information on the cultivation and nutritional value of vegetables is included. Appendix 5 gives some metric equivalents.

NEW TO THIS EDITION

More than 100 new or revised illustrations have been added to this edition, and information throughout the text has been updated or augmented, particularly in the area of plant physiology. Boxed inserts by Dr. Daniel Scheirer about interesting recent specific events and discoveries have been added to about half of the chapters; his contributions are gratefully acknowledged. Discussion of major types of grafting has been moved from the Plant Biotechnology chapter to Appendix 4. The metric conversion table in Appendix 5 has been expanded.

ADDITIONAL LEARNING AIDS

Instructor's Manual/Test Item File

The Instructor's Manual/Test Item File is available with *Introductory Plant Biology* and offers a variety of course schedules while providing overviews, goals, suggested answers, film sources, and examination questions for each text chapter.

Laboratory Manual

The Laboratory Manual that accompanies *Introductory Plant Biology* has been revised for the seventh edition. It is written for the student entering the study of botany for the first time. The exercises utilize plants to introduce biological principles and the scientific method, and are written to allow for maximum flexibility in sequencing.

Student Study Guide

A Student Study Guide, prepared by Daniel Scheirer, Northeastern University, is also available. The guide provides students an opportunity to study at their own pace. It contains learning objectives, chapter outlines, key terms/concepts (referenced to the text), and a set of objective questions for each chapter.

Transparencies

The transparency package includes 100 two- and four-color acetate overlays that are available free to adopters. These figures represent key illustrations from the text that merit additional visual review and discussion.

Micro Test III

Micro Test III is a computerized testing program that offers you the most effective and flexible software to date. And there are several convenient format options available to all adopters: Use your MacIntosh or IBM PC to pick and choose your test questions, edit them, and add your own questions.

If you do not have a microcomputer, you can still pick and choose your questions via our call-in/mail-in service. Within two working days of your request, we'll put a test master, a student answer sheet, and an answer key in the mail to you. Call-in hours are 8:30-5:00 CST, Monday through Friday.

Readings

Critical Thinking: A Collection of Readings, by David J. Stroup and Robert D. Allen is available to instructors who are working to integrate critical thinking into their curriculum. This inexpensive text will help in the planning and implementation of programs intended to develop students' abilities to think logically and analytically. The reader is a collection of articles that provide instruction and examples of current programs. The authors have included descriptions and evaluations of their personal experience with incorporating critical thinking study in their coursework. (ISBN 14556)

In addition to helping students learn about botany through writing, *Writing to Learn Botany,* by Randy Moore, will help improve communication skills—crucial to any profession. (ISBN 17455)

Videotapes

Tapes One, Two, and Five in the *Life Science Animations Videotape Series* will provide you with more than 30 animations of the most difficult-to-learn concepts found in a botany course.

ACKNOWLEDGMENTS

The contributions of many individuals to the development of this book are gratefully acknowledged. Critical reviewers, who provided many valuable suggestions for improving and updating the text, include:

Reviewers

Holly Adrian, *St. Johns University*
Steve K. Alexander, *University of Mary Hardin-Baylor*
Dale Benham, *Nebraska Wesleyan University*
George S. Ellmore, *Tufts University*
John Green, *Nicholls State University*
Helen G. Kiss, *Miami University, Oxford, Ohio*
John Z. Kiss, *Miami University, Oxford, Ohio*
Jerry McClure, *Miami University, Oxford, Ohio*
H. Gordon Morris, *University of Tennessee—Martin*
Alison M. Mostrom, *Philadelphia College of Pharmacy and Science*
Jim Nelson, *Southwest Texas State University*
Daniel C. Scheirer, *Northeastern University*
C. Gerald Van Dyke, *North Carolina State University*

Additional persons who read parts of the manuscripts of various editions and made many helpful criticisms and suggestions include Robert I. Ediger, Richard S. Demaree, Jr., Robert B. McNairn, Donald T. Kowalski, Larry Hanne, Patricia Parker, and Robert A. Schlising. Others whose encouragement and contributions are deeply appreciated include Timothy Devine, W. T. Stearn, Lorraine Wiley, Isabella A. Abbott, Paul C. Silva, Donald E. Brink, Jr., William F. Derr, Beverly Marcum, Robert McNulty, the faculty and staff of the Department of Biological Sciences, California State University, Chico, my many inspiring students, the Lyon Arboretum of the University of Hawaii, the editorial, production, and design staffs of the Wm. C. Brown Company Publishers, and most of all my wife, Janet, and my children, Kevin and Sharon. Special thanks are due to artists Denise Robertson Devine, Janet Monelo, and Sharon Stern.

Kingsley R. Stern
Chico, California

Chapter Outline

Steershead (Dicentra uniflora), *a diminutive relative of bleeding hearts, native to the mountains of the western United States and Canada.*

The Development
of Plant Study

1

Overview ─────────────────────────────

This chapter introduces you to botany: what it is, how it developed, how it relates to our everyday lives, and what its potential is for the future. The discussion includes a brief introduction to the beginning of plant study in ancient civilizations, an examination of the scientific method, and a look at the development of botany after the invention of the microscope. It concludes with a brief survey of the major disciplines within the field of botany.

Some Learning Goals

1. Explain briefly what the scientific method is and what hypotheses are.
2. Name or identify a contribution to the development of botany as a science made by each of the following: Theophrastus, Leeuwenhoek, Malpighi, Grew, van Helmont, Linnaeus.
3. Know what the Doctrine of Signatures and associated herbals are.
4. Understand the major botanical disciplines and indicate briefly the particular aspect of botany with which each is concerned.

O
n one occasion, I was dining on seafood in a dimly lit restaurant and, after I had begun to eat, I noticed what looked like a faint light coming from the surface of a cooked shrimp. I cupped my hands to shield what room light there was from the plate and found the shrimp was, indeed, glowing in the dark. I called the waiter, who was startled to see the phenomenon for himself. He immediately removed my plate and in due course returned with "ordinary" shrimp. He also reported to me later that they had discovered other shrimp glowing in the restaurant refrigerator.

The glow on the surface of the shrimp came from harmless luminescent bacteria, which produce light instead of heat as part of the chemical activities that occur within their cells. This is similar to the bioluminescence found in fireflies, glowworms, and certain marine algae (discussed in chapters 11 and 18). The incident reminded me of one small facet of current genetic engineering research (discussed in Chapter 14) involving the insertion of the gene for firefly bioluminescence into tobacco plants, which then produce light whenever the gene is activated. Scientists hope, eventually, to insert the gene in the self-defense mechanisms of a variety of crop plants, which as a result would glow when attacked by disease or pests, alerting the growers to the problem.

Genetic Engineering and Plant Breeding

Genetic engineering, which involves the introduction of desirable genes from one organism into another, holds enormous potential for improvements in the quality and yield of crop plants and in the suppression or elimination of human defects and diseases.

Recent developments in the field of genetic engineering include the development of tomatoes that can remain in good condition on a shelf for weeks and others that produce thicker catsups and pastes. Bacteria that can prevent frost damage in plants down to 27°F (–3°C) have also been developed. In addition, other bacteria that can break up crude-oil spills and break down refinery, sewage, food, and paper pulping wastes and still others that can remove grease and fat clogs in drains and sewers are now in production.

In the near future, we will probably see plants with properties that were completely unknown to our grandparents. Plant breeders have already given us many hybrid plants with greater vigor, disease resistance, and yield than those of their parents. Now the breeders are trying to develop plants that, among other things, will inhibit weeds or grow in areas that are inhospitable to crop plants.

Other plants that naturally repel insects are being investigated and improved. Plant breeders at Cornell University, for example, are developing a potato plant with sticky hairs that will trap insects, while others are trying to improve the protein and vitamin content of crop plants. Plant breeders also are developing varieties that can thrive in relatively dry or salty soil and others they call "green glue" that can bind and stabilize soils and even reclaim land that has become desert.

Dependence of Humans and All Animal Life on Plants

The efforts of genetic engineers and plant breeders are particularly significant when one considers that all of us have been totally dependent on green organisms, such as plants and algae, since before we were born. If that sounds like an exaggeration, consider that all green organisms are capable of transforming the sun's energy into forms usable by animal life and that such energy is vital to the very existence of animal life. Consider further that virtually all the oxygen in the air we breathe is produced by green organisms and that they alone can remove the carbon dioxide waste we give off into the atmosphere. If some major disease were to kill off all or most of the green organisms on earth and in the oceans, all the animals on land, in the sea, and in the air would soon starve. Even if some alternative source of energy were available, animal life would suffocate within 11 years—the time estimated for all the earth's oxygen to be completely used up if it were not replaced.

Apart from these aspects, green organisms and plant products are so much a part of human society that we largely take them for granted. We know, of course, that rice, corn, potatoes, and other vegetables are plants (Fig. 1.1); but all foods, including meat, fish, poultry, eggs, cheese, and milk, to mention but a few, owe their existence to plants. Condiments such as spices, mustard, and pepper (Fig. 1.2) and luxuries such as perfumes are produced by plants, as are dyes,

FIGURE 1.1A   Rice cakes being manufactured. Unprocessed rice is poured into small ovens where the kernels are expanded. The kernels are then compressed into cakes, which are conveyed by belt to a packaging area.

FIGURE 1.1B   Part of a produce section in a supermarket.

adhesives, digestible surgical stitching fiber, food stabilizers, beverages (Fig. 1.3), and emulsifiers.

Our houses are constructed with lumber from trees, which also furnish the cellulose for paper, cardboard, and

FIGURE 1.2   Some of the spices derived from plants.

synthetic fibers. Our clothing, camping equipment, bedding, draperies, and other textile goods are made from fibers of many different plant families (Fig. 1.4). Coal is fossilized plant material, and oil may have come from microscopic green organisms or animals that either directly or indirectly were plant consumers. All medicines and drugs at one time came from plants, fungi, or bacteria, and many important ones, including many of the antibiotics, still do. Microscopic organisms are responsible for recycling both plant and animal wastes and for aiding in the building of healthy soils. They are responsible for human diseases and allergies.

Whether or not gluts or shortages of oil and other fossil fuels are politically or economically manipulated, there is no question that these fuels are finite and eventually will disappear. Accordingly, geothermal, atomic, and other alternative energy sources are receiving increased attention.

Methane gas, which can be used as a substitute for natural gas, has been produced from animal manures and decomposed plants in numerous villages in India and elsewhere for many years, and after several years of trial on a small scale in the United States, plans are now under way for production on a larger scale of methane from human sewage. Potatoes, grains, and other sources of carbohydrates are currently used in the manufacture of alcohols, some of which are being blended with gasoline ("gasohol"), and such uses may increase in the future. In fact, automobiles that can run on propane and either methanol (wood alcohol) or gasoline, or a mixture of both, are now in use in the United States and other parts of the world.

What of plants and the future? By the time you read this, the population of the earth may have already exceeded 6 billion persons, every one of whom needs food, clothing, and shelter in order to survive. To ensure survival, a majority of us eventually may need to learn not only how to cultivate food plants but also how to use plants in removing pollutants from water (Fig. 1.5), in making land productive again, and in renewing urban areas. In addition, many more of us may need to be involved in helping halt the destruction

FIGURE 1.3A Ripening coffee berries. They are picked by hand when they are red. The seeds are extracted for roasting after the berries have been fermented.

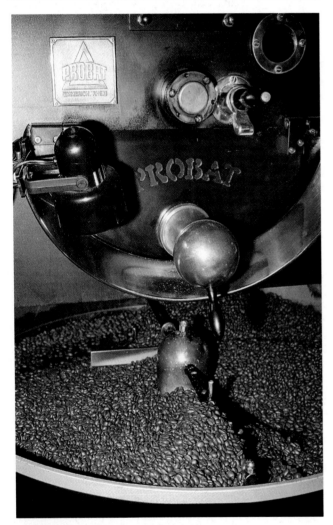

FIGURE 1.3B Coffee beans being roasted.

of plant habitats caused primarily by the huge increase in the number of earth inhabitants. This subject and related matters are further discussed in Chapter 25.

Some have suggested that the odds are against humanity saving itself from itself and have even indicated that it might become necessary to emigrate to another planet. If so, microscopic algae could play a vital role in space exploration. Experiments with portable oxygen generators have been in progress for many years. Tanks of water teeming with tiny green algae are taken aboard a spacecraft and installed so that they are exposed to light for at least part of the time. The algae not only produce oxygen, which the spacecraft inhabitants can breathe, but they also utilize the waste carbon dioxide end product of respiration. In addition, as they multiply, excess algae can be fed to a special kind of shrimp, which in turn multiply and become food for the space travelers. Other wastes are recycled by different microscopic organisms. When this self-supporting arrangement, called a *closed system,* is perfected, the range of spacecraft should greatly increase because heavy oxygen tanks will not be necessary, and fewer food reserves will be needed.

Inhabitants of undeveloped areas still use plants not only for food, shelter, clothing, and medicine but also in hunting and fishing. For example, plant poisons are applied to the tips of blow gun darts used in hunting, and the bulbs and seeds of certain plants, when thrown in dammed streams, stupefy fish so that they float to the surface. Today, small teams of botanists, anthropologists, and medical doctors are interviewing medical practitioners and herbal healers in remote tropical regions about various uses of plants by primitive peoples. These scientists are doing so in the hope of preserving at least some plants with potential uses for modern civilization before disruption of their habitats results in their extinction.

EARLY HISTORY AND DEVELOPMENT OF PLANT STUDY

On a number of occasions, I have received visits from anxious mothers or from pediatricians wanting to know if part of backyard plants that young children have consumed are poisonous. I have also been consulted by various law enforcement officials, landscape architects, pollen collectors, students interested in wildflowers, vegetarians, organic and traditional gardeners, and a variety of other professional and amateur persons with a wide range of interest in plant life. Some have wanted plants identified. Others have wanted suggestions for treating diseased plants. Still others have wanted to know about grafting techniques, the effect of "gray water" (water that has been used for bathing or washing dishes) on plants, the suitability of plants for specific locations, the preservation of plants, the edibility of wild plants, and a host of other plant-related subjects.

FIGURE 1.4 Cotton plants. The white fibers, in which seeds are embedded, are the source of textiles and fabrics. The seeds are the source of vegetable oils used in margarine and shortening. After the oils have been extracted, the remaining "cotton cake" is used for cattle feed.

FIGURE 1.5 A polluted waterway in an urban area.

Knowledge about plant life throughout the world has now become so vast that it is impossible for any person to be an authority on more than a tiny fraction of it. Our libraries contain thousands of books dealing with virtually every facet of botanical investigation, and research journals publish the latest discoveries from around the world on an almost daily basis. Why and how has all this knowledge accumulated? To answer this question we need to take a brief look at the early history and development of plant study.

Plants and Primitive Peoples

Archaeological evidence indicates that between 15,000 and 35,000 years ago humans migrated to the Americas and Australia. Migrations to the Americas evidently occurred between Siberia and Alaska across the Bering Strait, which was a land bridge during parts of the Pleistocene era. These early humans were primarily hunters. Indeed, the extinction of many large land mammals in North America coincides with the appearance and activities of humans between 13,000 and 9000 B.C., although major changes in climate and vegetation also occurred at that time. If you are a hunter or a fisherman, however, you are well aware that success in hunting or fishing can vary considerably, depending on various environmental and other circumstances.

By 8000 B.C., our ancestors had begun to develop more reliable sources of food through primitive forms of agriculture. Archaeological evidence obtained from the walls of tombs, mummy wrappings, hieroglyphics, cave paintings, and carvings indicates the cultivation of grains legumes and certain fruits (e.g., figs, olives, pomegranates, dates) was well established in the Near East by 6501 B.C. The Near Eastern Center and other major centers of origin of cultivated plants are discussed in Chapter 24.

By 4000 B.C., the date had become one of the most important crops to the Assyrians and Egyptians. It was eaten fresh or dried, and the sap of date palms was fermented for wine. Although the Assyrians knew nothing of the details, they were aware of sexuality in the date palm and pollinated the female trees by hand (Fig. 1.6). By the seventh century B.C., they had produced a systematically arranged list of medicinal plants, which suggests that the physicians and pharmacists of the day had a noteworthy knowledge of plants and their uses.

The Chinese have been cultivating medicinal and other useful plants for at least 4,500 years. Records from that far back in history are fragmentary, and it is often impossible to separate fact from legend. Some authorities agree, however, that the founder of Chinese agriculture was an emperor by the name of Shen Nung, who was born in 2737 B.C. Shen Nung is said to have invented the plow. He is also believed to have established an annual ceremony during which seeds of soybeans, wheat, rice, millet, and sorghum were sown or planted by royalty. He appears to have been an authority on poisons and antidotes. He also wrote a book on drugs and medicines that was incorporated into the Pun-tsao, a Chinese *pharmacopoeia* (an officially recognized book describing

FIGURE 1.6   A date palm.

and how they were put together. This inquisitiveness led to plant study becoming a **science,** which broadly defined is simply "a search for knowledge of the natural world." A science may be distinguished from other fields of intellectual endeavor by several features. It involves the observation, recording, organization, and classification of facts, and more importantly, it involves what is done with the facts. Scientific procedure involves experimentation, observation, and the verifying or discarding of information, chiefly through inductive reasoning from known samples. There is no universal agreement on the precise details of the process. A few decades ago, scientific procedure was considered to involve a routine series of steps. This series of steps came to be known as the *scientific method,* and there are still instances where such a structured approach works well. In general, however, the scientific method now describes the procedures of assuming and testing *hypotheses.*

Hypotheses

A **hypothesis** is simply a tentative, unproven explanation for something that has been observed. It may not be the correct explanation—testing will determine whether it is correct or incorrect. To be accepted by scientists, the results of any experiments designed to test a hypothesis must be repeatable and capable of being duplicated by others. The nature of the testing will vary according to the circumstances and materials. We may, for example, *observe* that apples are red fruits that taste sweet. We may then *hypothesize* that all red fruits taste sweet. We may *test* the hypothesis by tasting red fruits, and as a result of our testing (since red crab apples are bitter), we may *modify* the *hypothesis* to state that only some red fruits are sweet.

When a hypothesis is tested, *data* (bits of information) are accumulated and may lead to the formulation of a useful generalization called a *principle.* Several related principles may lend themselves to grouping into a *theory,* which is not simply a guess. A theory is a group of generalizations (principles) that help us understand something. We reject or modify theories only when new principles increase our understanding of a phenomenon.

Plant Science and Ancient Greece

However one defines *science,* it is clear that plant science existed in ancient Greece. As in even older cultures, the study of plants in Greece started when people developed a practical interest in food and drug plants, and it grew as they also became curious about the structure and function of plants. As the physicians and pharmacists of the era gathered and used medicinal plants, they studied the variations and forms and came to recognize apparent relationships among them.

One of these Greek herbal physicians, in 384 B.C., had a son who became one of the most renowned philosophers of all time—Aristotle. Although Aristotle is perhaps better known for his philosophical works, he was an accomplished mathematician and he also acquired extensive knowledge in

drugs and medicines) of 40 volumes, published during the 17th century. During the Han dynasties, which lasted from about 200 B.C. until the birth of Christ, gardens became very extensive in China and many ornamental plants were cultivated. Plants such as primroses, poppies, and chrysanthemums were brought to the Western world from China over 2,000 years ago.

The Egyptians cultivated primitive forms of wheat and barley from about 5000 to 3400 B.C., although some authorities place the cultivation of these two cereals as far back as 10,000 to 15,000 B.C. More modern forms of cereals, such as six-rowed barley, may have been under cultivation by 2000 B.C. By the fifth century B.C., the Egyptians apparently were brewing *booza,* a beer, from barley.

Botany as a Science

The study of plants, called **botany**—from the French word *botanique* (botanical) and the three Greek words *botanikos* (botanical), *botane* (plant or herb), and *boskein* (to feed)—appears to have had its origins with Stone Age peoples who sought to modify their surroundings and feed themselves. Initially, the primary interest in plants was practical, centering around how plants might provide food, fibers, fuel, and medicine. Eventually, however, an intellectual interest arose. Individuals became curious about how plants reproduced

nearly all aspects of natural history. In fact, he combined philosophical and scientific interests as few other philosophers have done.

At the age of 17, Aristotle went to Athens, where he met and became a pupil of Plato. He left Athens after Plato's death in 347 B.C. and studied marine animals at a coastal area for several years, eventually returning to Athens to found the first botanical garden of which there is any record.

When Aristotle died, he willed the botanical garden and its associated library to his pupil and assistant, Theophrastus (Fig. 1.7). Theophrastus was an extraordinary man who not only learned virtually all Aristotle knew about plants but also added prodigiously to Aristotle's knowledge from his own observations. It is said that he had 2,000 disciples and wrote 200 treatises. The most important of the latter to have survived are two books entitled *History of Plants* and *Causes of Plants*. So great were his contributions to botany as a science that the famous 18th-century Swedish botanist Linnaeus gave him the title "Father of Botany." Few, if any, dispute his right to the honor.

Herbals Appear

Two books that had a significant influence on botanical studies appeared during the second century A.D. Pliny's *Historia Naturalis* contained lists of food or medicinal plants. Dioscorides' *Materia Medica* was the first book to contain illustrations of plants, all laboriously copied by hand. Many of the common names used by Dioscorides are still used today. European scholars who followed Dioscorides continued, by hand, to copy these books, which became known as **herbals,** and held them in such high esteem that it was considered heresy to question anything in their contents; consequently, few new ideas were added during the Dark and Middle Ages that lasted from 400 to 1400 A.D.

When the printing press appeared in the middle of the 15th century, the number of herbals mushroomed, and the period from about 1500 to 1700 A.D. became known as the Age of Herbals. These botanical works were primarily the products of German botanists, although some Italian and English botanists made their own contributions between 1470 and 1670.

The *herbalists,* as they were called, were mostly concerned with medicinal plants, which they studied in the botanical gardens that had become numerous and extensive in Europe by this time. They produced elaborate and intriguing illustrations for the herbals, occasionally accompanied by outlandish stories and descriptions. Some of the stories became legends and developed into the *Doctrine of Signatures.* According to this doctrine, if a part of a plant had the shape of a part of the human body, it would be useful in treating a disease of the human part it most closely resembled. For example, the meat of a walnut, which somewhat resembles a miniature brain, was used in treating brain diseases, and hepatica leaves, which have lobes reminiscent of those of the liver, were used to treat ailments of that organ (Fig. 1.8). One of the more famous herbalists was Otto Brunfels,

FIGURE 1.7 Theophrastus.
(Courtesy National Library of Medicine)

FIGURE 1.8 *Hepatica* plants.

who published a three-volume herbal in 1530. His work had excellent illustrations and is considered to be a link between ancient and modern botany (Fig. 1.9).

THE FIRST MICROSCOPES

The microscope had and continues to have a profound effect not only on plant studies but also on the biological sciences and related fields as a whole. In 1590, Zacharias and Francis

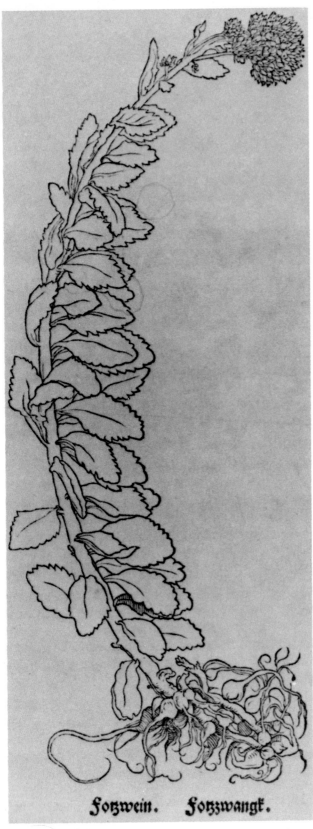

FIGURE 1.9 An illustration from Otto Brunfel's herbal (1530 A.D.).

(Courtesy National Library of Medicine)

Janssen, Dutch brothers who were spectacle makers, drew on the experience of their father, Hans, who was famous for his optical work. They discovered how to combine two convex lenses in the interior of a tube and produced the first instrument for magnifying minute objects. Because of this work, Zacharias Janssen, in particular, is often referred to as the inventor of the compound microscope, although it was Faber of Bamberg, a physician serving Pope Urban VIII, who originally applied the term microscope to the instrument during the first half of the 17th century. A Dutch draper by the name of Anton van Leeuwenhoek (1632–1723), who ground lenses and made microscopes in his spare time, is best known for his development of primitive microscopes. Leeuwenhoek was the first to describe bacteria, sperms, and other tiny cells he observed with his microscopes, some of which could magnify as much as 200 times. In his will, he left 26 of his 400 handmade microscopes to the Royal Society of London. Pictures of both primitive and modern microscopes are found on pages 28 and 30.

Before the invention of the microscope, plant study had been dominated by investigations based primarily on the external features of plants. The magnification of the early microscopes was not very great by present standards, but these instruments nevertheless led to the discovery of *cells* (discussed in Chapter 3) and opened up whole new areas of study.

DIVERSIFICATION OF PLANT STUDY

Plant anatomy, which is concerned chiefly with the internal structure of plants, was established through the efforts of several scientific pioneers. Early plant anatomists of note included Marcello Malpighi (1628–1694) of Italy, who discovered various tissues in stems and roots, and Nehemiah Grew (1628–1711) of England, who described the structure of wood more precisely than any of his predecessors (Fig. 1.10).

Plant physiology, which is concerned with plant function, was established by J. B. van Helmont (1577–1644), a Flemish physician and chemist, who was the first to demonstrate that plants do not have the same nutritional needs as animals. In a classic experiment, van Helmont planted a willow branch weighing 5 pounds in an earthenware tub filled with 74.4 kilograms (200 pounds) of dry soil. He covered the soil to prevent dust settling on it from the air, and after five years he reweighed the willow and the soil. He found that the soil weighed 56.7 grams (2 ounces) less than it had at the beginning of the experiment, but that the willow had gained 164 pounds. He concluded that the tree had added to its bulk and size from the water it had absorbed. We know now that most of the weight came as a result of photosynthetic activity (discussed in Chapter 10), but van Helmont deserves credit for landmark experimentation in plant physiology.

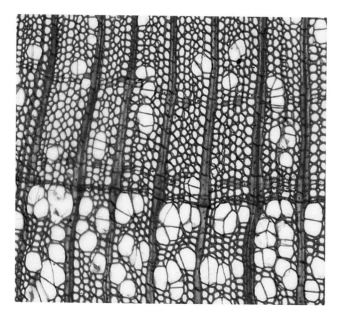

FIGURE 1.10 A thin section of wood as seen through a microscope.

FIGURE 1.11 Ecologists, geographers, and other biologists recognize large communities of plants and animals that occur in areas with distinctive combinations of environmental features. These areas, called *biomes,* are represented here by the Tropical Rain Forest, which, although occupying less than 5% of the earth's surface, is home to more than half of the world's species of organisms.

In the 15th century when Columbus visited Cuba, he found local Indian tribes cultivating corn (maize). This important food plant had apparently been in use by the pre-Incas of Peru some 5,000 years earlier. By the time explorers ventured into the Americas in the 1600s, they found that maize culture had spread from Argentina in the South to the St. Lawrence River area in the North. American Indians had also domesticated the white potato, and Indians were cultivating flowers and medicinal plants in Mexico.

The 17th century saw a marked increase in botanical explorations to various parts of the globe. The explorers took large numbers of plants back to Europe with them, and it soon became clear to those working with the plants that some sort of formalized system was necessary just to keep the collections straight. Several *plant taxonomists* (botanists who specialize in the identifying, naming, and classifying of plants) proposed ways of accomplishing this, but we owe our present system of naming and classifying plants to the Swedish botanist Carolus Linnaeus (1707–1778) (see Fig. 16.2).

Plant taxonomy (also called *plant systematics),* which is the oldest branch of plant study, began in antiquity, but Linnaeus did more for the field than any other person in history. Thousands of plant names in use today are those originally recorded in Linnaeus's book *Species Plantarum,* published in 1753. An expanded account of Linnaeus and his system of classification is given in Chapter 16.

Theophrastus (fourth century B.C.) was the first person on record to have systematically discussed the relationship of plants to their surroundings, but the discipline of **plant geography,** the study of how and why plants are distributed where they are, did not develop until the 19th century (Fig. 1.11). The allied field of **plant ecology,** which is the study of the interaction of plants with one another and with their environment, also developed in the 19th century.

Public awareness of the field of ecology as a whole increased considerably after 1962, following publication of a book entitled *Silent Spring* by Rachel Carson. In this best-seller, based on more than four years of literature research, the author called attention to the fact that more than 500 new toxic chemicals annually are put to use as pesticides in the United States alone, and she detailed the insidious impact of these chemicals and other pollutants on all facets of human life and the environment.

Among the most noteworthy of the early plant geographers were two natives of Berlin, Germany, Carl Willdenow (1765–1812) and Alexander von Humboldt (1769–1859), who published books on the relationship of seed dispersal to plant distribution and on the associations of various plants with one another in tropical and temperate climates. These studies were brought to a climax by Sir Joseph D. Hooker (1817–1911), who eventually became director of the Royal Botanic Gardens in Kew, England. Hooker traveled widely, studying plant life in both the Northern and Southern Hemispheres, and he also published floras (accounts of the plants of a specific region) of India and Antarctica. Charles Darwin, whose books (particularly his *On the Origin of Species*) revolutionized some basic biological concepts of the adaptation of organisms to their environment, said of Hooker's *Flora Antarctica,* "It is by far the grandest and most interesting essay on subjects of nature I have ever read."

The study of the form and structure of plants, **plant morphology,** was developed during the 19th century, and by the beginning of the 20th century, much of the basic information incorporated in the plant sciences today had been

Plant Biology and the Web

The **World Wide Web** (WWW) is a rich area of cyberspace that contains formatted text documents, color graphics, maps, audio clips, video images, and other interesting items. It contains a virtual storehouse of scientific knowledge just waiting to be explored. If you have never experienced the **Internet** and the **World Wide Web,** you are about to experience virtual biology.

What is the Internet? A technical answer would include a description of the historical origins of the Internet in national defense, research, and education, as well as the physical connection of computers to one another. However, the Internet has come to mean much more than this. It is frequently described as the *Information Superhighway,* the *Infobahn,* or *Cyberspace.* What do people do on the Internet? There are several components to the Internet such as exchanging e-mail, following newsgroups, and downloading data files, images, and sound files.

One aspect of the Internet is that it is international. Because it is a global network, one minute you may be retrieving a file from France or Japan, and the next minute you are tapped into a computer at your local university or college. The interesting part about this is that you frequently do not even know that you have crossed national boundaries.

Access to the Internet begins with a connection that can be supplied by numerous Internet providers or some large commercial on-line services like America Online or CompuServe. These services charge a fee for access to their computer, but once connected, you have the full global capabilities of exploring the vast amounts of information and entertainment features on the Internet.

The *client* and *server* are terms that are used to explain the information flow from remote computers (server) to your computer (client). Your personal computer has software that controls what you see on the screen (user interface menus) and responds to your interactions. This is the *client.* When you request a file, the client software program sends a message to a *server* (on another computer) to retrieve the file. The server then returns the file to the client software, which interprets and displays the information in the file. The diagram below summarizes this interaction.

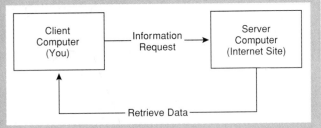

The Internet has several information servers that provide different ways to access information. They range from the easy-to-use to the more complex and arcane. The World Wide Web is similar to the other information servers (FTP, Gopher, WAIS, Veronica) but has several distinct advantages that make it a very popular way of browsing for information. First, it offers formatted text and graphics in the form of pages instead of menu lists. These pages begin with a *home page* (a central navigational point) and are read much like the pages of a book. Additionally, documents are linked together using hypertext formatting that allows users to browse from one linked document to another, not in an hierarchical tree, but in a true web of interrelated topics.

The first thing you need to start browsing on the web is web browser software like *Netscape Navigator* or *Mosaic.* You are then ready to type in an address, called a URL (universal resource locator) and start exploring. If you type in a URL address, the client software interprets the URL and initiates communication with the specified server. The parts of a URL consists of the following:

protocol://server/directory/filename

This is the URL for the web site at Wm. C. Brown Publishers (the publishers of this text).

---

discovered and elucidated. The number of scientists engaged in investigating plants had also increased conspicuously. **Genetics,** the science of heredity, had been founded by the Austrian monk Gregor Mendel (1822–1884) through his classic experiments with peas. **Cytology,** the science of cell structure and function (now called **cell biology**), had received great impetus from the discovery of how cells multiply and function in sexual reproduction. The mid-20th-century development of *electron microscopes* (see Chapter 3) further spurred cell research and led to vast new insights into cells and new forms of cell research that continue to the present. Many other significant 20th-century discoveries are discussed at appropriate junctures throughout the chapters to follow.

http://www.wcbp.com/wcbprintf.html#plants

1. **http** is the protocol called hypertext transfer protocol and is used by the client and server to communicate with each other.

2. **www.wcbp.com** is the name of the server

3. **wcbprintf.html#plants** is the directory and filename (pathname)

What if you want to search the Web for a specific topic? There are several *search engines* available that allow you to search by key word(s). The search software scans the numerous Web servers for your key word(s) and returns to you any number of hits, or positive matches. You can go directly to any of the matches by clicking the mouse pointer on the hyperlinked search results. One widely used search engine is called *WebCrawler (http://www.Webcrawler.com/)*. Another one that offers several different search engines is **Quarterdeck** *(http://www.qdeck.com/qdeck/search/)*.

What botanical information is available on the Web? You will be surprised at the variety and amount of information accessible. The following are some interesting web sites that I have explored and their URLs. Try them out sometime! Maybe you'll find a good idea for a research paper.

Some plant biology Web sites:

Australian National Botanic Gardens WWW Server provides a wealth of botanical and biological information about Australia.

http://www.anbg.gov.au/

California Flora Database contains geographic and ecological distribution information for 6,717 California vascular plant taxa, as well as additional habitat information for rare taxa and species of the Sierra Nevada.

http://s27w007.pswfs.gov/calflora/index.html

Carnivorous Plants Database includes over 3,000 entries giving an exhaustive nomenclatural synopsis of all carnivorous plants.

http://www.hpl.hp.com/bot/cp_home

Common Conifers of the Pacific Northwest provides information about the conifers of Oregon, including a dichotomous key for their identification.

http://www.orst.edu/instruct/for241/con/

Florida Wildflower Page provides hundreds of photos of Florida wildflowers.

http://www-wane-leon.scri.fsu.edu/~mikems

GardenNet is an information center for garden and gardening enthusiasts.

http://trine.com/gardennet/

National Wildflower Research Center is a nonprofit research and educational organization committed to the conservation and reestablishment of native wildflowers, grasses, shrubs, and trees.

http://www.onr.com/wildflower.html

Orchid Greenhouse provides information, photographs, and methods of growing orchids.

http://yakko.cs.wmich.edu/~charles/orchids/

Poisonous Plant Database is a set of working files of scientific information about the animal and human toxicology of vascular plants and herbal products of the world.

http://vm.cfsan.fda.gov/~djw/readme.html

Tropical Rainforest in Surinam provides a virtual tour of the rain forests of Surinam, complete with many fine photographs.

http://www.euronet.nl/users/mbleeker/suri_eng.html

PLANT SCIENCES AND THE FUTURE

There is still a vast amount of information to be discovered, and new discoveries continue to be made daily. Back in 1938, for example, 11,000 papers on botanical subjects were published in that year alone; the number in recent years is many times greater. Further, it appears probable that at least one-third of all the organisms traditionally regarded as plants (particularly algae and fungi) have yet to be named, let alone thoroughly investigated.

Wild plants and animals are becoming extinct at a rapidly accelerating rate as their natural habitats are destroyed through development and pollution; in fact, it is evident that many undescribed organisms are becoming extinct before we have learned anything about them. Efforts must be upgraded to educate the general public on the necessity of preserving natural habitats so that the numerous tangible and aesthetic benefits of doing so will be available to succeeding generations. Also, both basic and applied research in botany need additional support if the earth's burgeoning human population is to continue to be fed, clothed, and housed.

Summary

1. Genetic engineering, which involves the introduction of desirable genes from one organism into another, holds much potential for the future in many areas including crop improvement, frost damage control, pollution control, weed inhibition, insect repulsion, and soil reclamation and binding.

2. We are totally dependent on green plants because they alone can convert the sun's energy into forms that are usable by and vital to the very existence of animal life.

3. We largely take plants and plant products for granted. Animals, animal products, many luxuries and condiments, and other useful substances, such as fibers, lumber, coal, medicines, and drugs, either depend on plants or are produced by them.

4. To ensure human survival, all persons soon may need to acquire some knowledge of plants and how to use them. Plants will undoubtedly play a vital role in space exploration as portable oxygen generators.

5. Teams of scientists are interviewing medical practitioners and herbal healers in the tropics to locate little-known plants used by primitive peoples before the plants become extinct.

6. There are thousands of books dealing with botanical subjects, the knowledge having accumulated over the span of human existence.

7. Humans migrated to the Americas between 10,000 and 35,000 years ago; they originally were hunters, but gradually they learned how to cultivate crops. The Assyrians and Egyptians were cultivating fruits and cereals by 4000 B.C., and the Chinese practiced primitive agriculture at least 4,500 years ago.

8. Botany, the study of plants, apparently began with Stone Age peoples' practical uses of plants. Eventually, botany became a science, as intellectual curiosity about plants arose.

9. A science involves observation, recording, organization, and classification of facts. The verifying or discarding of facts is done chiefly from known samples through inductive reasoning. The scientific method involves specifically following a routine series of steps and generally assuming and testing hypotheses.

10. Plant science existed in ancient Greece. A fourth-century B.C. Greek, Theophrastus, contributed much to the field of botany.

11. During the second century A.D., Pliny and Dioscorides produced books on food or medicinal plants. These books were copied by hand and became known as *herbals*. During the 15th century, the number of herbals increased greatly with the advent of the printing press. The herbals were illustrated and sometimes contained bizarre stories about plants; these stories resulted in the development of the *Doctrine of Signatures*.

12. The compound microscope had a profound effect on studies in the biological sciences and led to the discovery of cells.

13. Plant anatomy and plant physiology developed during the 17th century. J. B. van Helmont was the first to demonstrate that plants have nutritional needs different from those of animals. During the 17th century, Europeans engaged in botanical exploration on other continents and took plants back to Europe.

14. During the 18th century, Linnaeus produced the elements of our present system of naming and classifying plants.

15. During the 19th century, plant ecology, plant geography, and plant morphology developed, and by the beginning of the 20th century, genetics and cell biology became established. Much remains yet to be discovered and investigated.

Review Questions

1. Briefly indicate contributions to plant science made by the following: Shen Nung, the ancient Egyptians, Theophrastus, Grew, and Linnaeus.

2. What is meant by the *scientific method?*

3. Distinguish among *herb, herbal,* and *herbalist.*

4. What was the significance of van Helmont's experiment with the willow tree?

5. What is the oldest branch of botany, and why did it precede other branches?

6. What is the thrust of each of the other branches of botany?

7. What is the *Doctrine of Signatures?*

Discussion Questions

1. Since humans survived on wild plants for thousands of years, might it be desirable to return to that practice?

2. On the basis of what you have read, would you single out any one individual as having contributed the most to the development of plant study? Why?

3. How would you guess that Stone Age peoples discovered medicinal uses for plants?

4. Many of the early botanists were also doctors. Why do you suppose this is no longer so?

5. Consider the following hypothesis: "The majority of mushrooms that grow in grassy areas are not poisonous." How could you go about testing this hypothesis scientifically?

Additional Reading

Anderson, F. J. 1985. *An illustrated history of the herbals*. New York: Columbia University Press.

Carson, R. 1994. *Silent spring*. Boston: Houghton Mifflin Co.

Ewan, J. (Ed.). 1969. *A short history of botany in the United States*. Forestburgh, NY: Lubrecht & Cramer.

Greene, E. L. 1983. In F. N. Egerton (Ed.), *Landmarks of botanical history* (2 vols.). Palo Alto, CA: Stanford University Press.

Harvey-Gibson, R. J. 1981. In I. B. Cohen (Ed.), *Outlines of the history of botany*. Salem, NH: Ayer Company Publishers.

Morton, A. G. 1981. *History of botanical science: An account of the development of botany from the ancient time to the present*. San Diego, CA: Academic Press.

Nordenskiold, E. 1988. *The history of biology*. Irvine, CA: Reprint Services Corp. (Reprint of 1935 edition)

Swift, L. H. 1974. *Botanical bibliographies: A guide to bibliographic materials applicable to botany*. Ann Arbor, MI: University Microfilms International. (Reprint of 1970 edition)

A *flower of brownstain collinsia* (Collinsia tinctoria). *The clear sap of this plant, on contact with human skin, produces a yellowish-brown stain; hence the plant's name.*

2 The Nature of Life

Overview———————

This chapter begins with a discussion of the attributes of living organisms. These include growth, reproduction, response to stimuli, metabolism, movement, complexity of organization, and adaptation to the environment. Then the chapter examines the chemical and physical bases of life. A brief look at the elements and their atoms is followed by a discussion of compounds, molecules, bonds, ions, valence, mixtures, acids, bases, and salts. Forms of energy and the chemical components of protoplasm are examined next. The chapter concludes with an introduction to macromolecules: carbohydrates, lipids, proteins, and nucleic acids.

Some Learning Goals

1. Learn the attributes of living organisms.
2. Define *matter*; describe its basic state.
3. Differentiate compounds from mixtures and describe acids, bases, and salts.
4. Know the various forms of energy.
5. Learn the elements found in protoplasm.
6. Understand the nature of carbohydrates, lipids, and proteins.

H ave you ever dropped a pellet of dry ice (frozen carbon dioxide) into a pan of water and watched what happens? As the solid pellet is rapidly converted to a gas due to its contact with the warmer water, it darts randomly about the surface looking like a highly energetic bug waterskiing. Does all that motion make the dry ice alive? Hardly; yet one of the attributes of living things is movement. But if living things move, what about plants? Is a tree not alive because it does not crawl down the sidewalk? Again the answer is no, but these questions do serve to point out some of the difficulties encountered in defining *life*. In fact, some contend that there is no such thing as life—only living organisms—and that life is a concept based on the collective attributes of living organisms.

ATTRIBUTES OF LIVING ORGANISMS

Composition

The activities of living organisms emanate from *protoplasm,* the physical basis or the "stuff of life," discussed in the next chapter.

Structure

The protoplasm is contained in tiny structural units called *cells,* which are unique to living things. Cells are discussed in the next chapter.

Growth

The complex phenomenon of **growth** has been described simply as an increase in mass (a body of matter), which is usually also correlated with an increase in volume. Growth, which results primarily from the production of new protoplasm, includes variations in *form*—some the result of inheritance, some the result of environmental response. As an example of environmental response, consider what might happen if you were to plant two apple seeds of the same variety in poor soil and subsequently give them unequal treatment. If you were to give one just barely enough water to allow it to germinate and grow, while you not only gave the other an ample water supply but also worked fertilizers and conditioners into the soil around it, you might expect the second one to grow larger and be more productive than the first. In other words, although your two apple trees grew from the same variety of apple seed, they would differ in form, following patterns of growth dictated by the protoplasm and the environment. Various aspects of growth are discussed in Chapter 11.

Reproduction

Dinosaurs were abundant 160 million years ago, but none exist today. Numerous mammals, birds, reptiles, plants, and other organisms are now on lists of endangered or threatened species, and many species are doomed to extinction within the next decade or two. All these once-living or living things have one feature in common: It became impossible or it has become difficult for them to reproduce. **Reproduction** is such an obvious feature of living organisms that we take it for granted—until it is lost.

When reproduction occurs, the offspring are always similar to the parents: Guppies never have puppies—just more guppies—and a petunia seed, when planted, will not develop into a pineapple plant. In addition, offspring of one kind tend to resemble their parents more than they do other individuals of the same kind. The laws governing these aspects of inheritance are discussed in Chapter 13.

Response to Stimuli

If you were to stick a pin into a pillow, you certainly would not expect any reaction from the pillow, but if you were to stick the same pin into a friend, you know the reaction would be instantaneous (assuming the friend was conscious) because response to stimuli is a major characteristic of all living things. You might argue, however, that when you stuck a pin into your houseplant nothing happened, even though you were fairly certain the plant was alive. What you might not have been aware of is that the houseplant did indeed respond but in a manner very different from that of a human. Plant responses to stimuli are generally much slower than those of animals and usually are of a different nature. If the houseplant's food-conducting tissue was pierced, the plant probably responded by producing a plugging substance called callose in the affected cells. Some studies have shown that callose may form within as little as 5 seconds after wounding. In addition, an unorganized tissue called **callus,** which forms much more slowly, may be produced at the site of the wound. Responses of plants to injury and to other stimuli, such as light, temperature, and gravity, are discussed in Chapters 9 through 11.

Metabolism

Metabolism has been defined as the "collective product of all the biochemical reactions taking place within an organism." All living organisms undergo various metabolic activities, which include the production of new protoplasm, the repair of damage, and normal maintenance. The most important activities include **respiration,** an energy-releasing process that takes place in all living things; **photosynthesis,** an energy-harnessing process in green cells that is, in turn, associated with energy storage; **digestion,** the conversion of large or insoluble food molecules to smaller soluble ones; and **assimilation,** the conversion of raw materials into protoplasm and other cell substances. These topics are discussed in Chapters 9 through 11.

Movement

As observed at the beginning of this chapter, plants generally do not move from one place to another (although their reproductive cells may do so). This does not mean, however, that plants do not exhibit movement, a universal characteristic of living things. The leaves of sensitive plants (*Mimosa pudica*) fold within a few seconds after being disturbed or subjected to sudden environmental changes, and the tiny underwater traps of bladderworts (*Utricularia*) snap shut in less than one 100th of a second. But most plant movements, when compared with those of animals, are slow and imperceptible and are primarily related to growth phenomena. They become obvious only when demonstrated experimentally or when shown by time-lapse photography. The latter often reveals many types and directions of motion, particularly in young organs. Movement is not confined to the organism as a whole but occurs down to the cellular level. *Cyclosis,* a streaming motion of protoplasm, occurs constantly in living cells. The streaming tends to resemble a river flowing clockwise or counterclockwise within the boundaries of each cell, but movement may actually be in various directions.

Complexity of Organization

The cells of living organisms are composed of large numbers of **molecules** (the smallest unit of an element or compound retaining its own identity). There are typically more than 1 trillion molecules in a single cell. The molecules are not simply mixed, like the ingredients of a cake or the concrete in a sidewalk, but are organized into "compartments," membranes, and other structures within cells and tissues. Even the most complex nonliving object has only a tiny fraction of the types of molecules of the simplest living organism, and in the living organism, the arrangements of these molecules are highly structured and complex. Bacteria, for example, are considered to have the simplest cells known, yet each cell contains a minimum of 600 different kinds of protein in addition to hundreds of other substances, and each component has a specific place or structure within the cell. When larger living objects, such as flowering plants, are examined, the complexity of organization is overwhelming, and the number of molecule types can run into the millions.

Adaptation to the Environment

Assume that you skip a flat stone across a body of water and it lands on the opposite shore. The stone is not affected by the change from air to water to land during its quick journey; it does not respond to its environment. Living organisms, however, do respond to the air, light, water, and soil of their environment, as will be explained in later chapters. In addition, they are in many subtle ways genetically adapted to their environment, after countless generations of natural selection (as discussed in Chapter 15). Some weeds (e.g., dandelions) can thrive in a wide variety of soils and climates, while many species now threatened with extinction have adaptations to their environment that are so specific they cannot tolerate even relatively minor changes.

CHEMICAL AND PHYSICAL BASES OF LIFE

The Elements: Units of Matter

The basic "stuff of the universe" is called matter. On earth, matter occurs in three states: *solid, liquid,* and *gas.* In simple terms, matter's characteristics are as follows:

1. It occupies space.

2. It has mass (with which we commonly associate weight).

3. It is composed of **elements,** of which there are at present 109 known (92 occur naturally on our planet; the others have been produced artificially). Only a few of the natural elements (e.g., nitrogen, oxygen, gold, silver, copper) occur in pure form; the others are found combined together chemically in various ways. Each element has a designated symbol, often derived from its Latin name. The symbol for copper, for example, is **Cu** (from the Latin *cuprum*); and for sodium it is **Na** (from the Latin *natrium*). The symbols for carbon, hydrogen, and oxygen are **C, H,** and **O,** respectively.

The smallest stable subdivision of an element that can exist is called an **atom.** Atoms are so minute that individual atoms were not rendered directly visible to us until the mid-1980s

Table 2.1

ATOMIC NUMBERS AND MASSES OF SOME ELEMENTS FOUND IN PLANTS

ELEMENT	ATOMIC NUMBER	USUAL ATOMIC MASS
Hydrogen (H)	1	1
Boron (B)	5	11
Carbon (C)	6	12
Nitrogen (N)	7	14
Oxygen (O)	8	16
Magnesium (Mg)	12	24
Phosphorus (P)	15	31
Sulphur (S)	16	32
Chlorine (Cl)	17	35
Potassium (K)	19	39
Calcium (Ca)	20	40
Iron (Fe)	26	56

by even the most powerful electron microscopes. We do know from experimental evidence, however, that each atom has a tiny **nucleus** consisting of particles called **protons,** which have positive electrical charges, and other particles called **neutrons,** which have no electrical charges. If the nucleus, which contains nearly all of the atom's mass, were enlarged so that it were as big as a beach ball, the atom, which is mostly space, would be larger than an average-sized professional football stadium. Each kind of atom has a specific number of protons in its nucleus, ranging from 1 in the lightest element, hydrogen, to 92 in the heaviest element, uranium. Each element has an *atomic number* that is based on the number of protons present in a single atom. The atomic mass of an element is determined by the number of protons and neutrons present in a single atom (Table 2.1). An atom's protons and neutrons equal each other in mass.

Whirling around the nucleus are associated negative electric charges called **electrons.** Electron masses are about 1,840 times lighter than those of both protons and neutrons and are so infinitesimal that they are generally disregarded.

The region in which electrons whirl around the nucleus is called an **orbital.** Orbitals each have an imaginary axis and are somewhat cloudlike but are without precise boundaries, so we cannot be certain of an electron's position within an orbital at any time. This has led to an orbital being defined as a *volume of space in which a given electron occurs 90% of the time.*

An important feature of orbitals is that each is limited to two electrons; an orbital with only one electron can attract another electron to fill the available space. The innermost orbital is more or less spherical and is so close to the nucleus that it is usually not shown on diagrams of atoms. The one to several additional orbitals, which are shaped mostly like the tips of cotton swabs, generally occupy much more space.

The electrons of each orbital tend to repel those of other orbitals, so the axes of all the orbitals of an atom are oriented as far apart from each other as possible, although the outer parts of the orbitals actually overlap more than shown in diagrams of them. Orbitals usually have diameters thousands of times more extensive than that of an atomic nucleus (Fig. 2.1).

Electrons usually equal the protons in number, so the positive electric charges of the protons balance the negative charges of the electrons, making the atom electrically neutral. The number of neutrons in the atoms of an element can vary slightly, so the element may occur in forms having different weights but with all forms behaving alike chemically. Such variations of an element are called **isotopes.** The element oxygen (Fig. 2.2), for example, has seven known isotopes. The nucleus of one of these isotopes contains eight protons and eight neutrons; the nucleus of another isotope holds eight protons and ten neutrons, and the nucleus of a third isotope consists of eight protons and nine neutrons. If the number of neutrons in an isotope of a particular element varies too greatly from the average number of neutrons for its atoms, the isotope may be unstable and split into smaller parts, with the release of a great deal of energy. Such an isotope is said to be *radioactive.*

Molecules: Combinations of Elements

The atoms of most elements have the property of binding to other atoms of the same or different elements and forming new combinations; in fact, most elements do not exist independently as single atoms. When two or more elements are united in a definite ratio by chemical bonds, the substance is

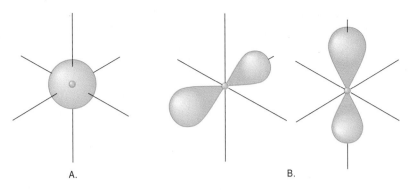

A.   B.

FIGURE 2.1   Models of orbitals. *A*. The two electrons closest to the atom's nucleus occupy a single spherical orbital. *B*. Additional orbitals are dumbbell-shaped, with axes that are perpendicular to one another. The atom's nucleus is at the intersection of the axes.

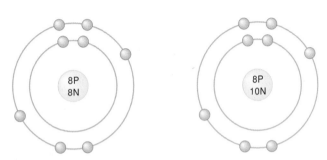

FIGURE 2.2   Isotopes of oxygen portrayed two-dimensionally.

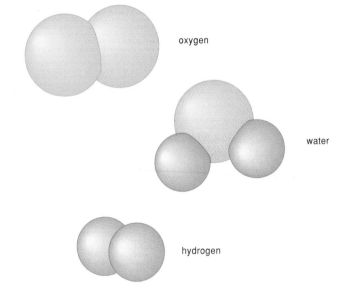

FIGURE 2.3   Models of oxygen, water, and hydrogen molecules. A water molecule is 0.6 nanometer in diameter.

called a **compound**. A **molecule** consists of two or more atoms bound together and is the smallest independently existing particle of a compound or element. The molecules of the gases oxygen and hydrogen, for example, exist in nature as combinations of two atoms of oxygen or two atoms of hydrogen, respectively. Water molecules consist of two atoms of hydrogen and one atom of oxygen (Fig. 2.3).

Molecules are in constant motion, the motion speeding up or slowing down with an increase or decrease in temperature. The more molecular movement there is, the greater the chances are that some molecules will collide with each other. Also, the chances of random collisions increase in proportion to the density of the molecules (i.e., the number of molecules present in a given space).

Random collisions between molecules capable of sharing electrons are the basis for all chemical reactions. The reactions often result in new molecules being formed. Each chemical reaction in a cell usually takes place in a watery fluid and is controlled by a specific *enzyme*. Enzymes are organic *catalysts*. (A catalyst speeds up a chemical reaction without being used up in the reaction. See the discussion of enzymes on page 23.)

When a water molecule is formed, two hydrogen atoms become attached to an oxygen atom at an angle averaging 105° in liquid water (for ice, the angle is precisely 105°). The electrons of the three atoms are shared and form an electron cloud around the core, giving the molecule an asymmetrical shape. Although the electron and proton charges balance each other, the asymmetrical shape and unequal sharing of the electrons in the bond between oxygen and hydrogen cause one side of the water molecule to have a slight positive charge and the other a slight negative charge. Such molecules are said to be *polar*. Since negative charges attract positive charges, polarity affects the way in which molecules become aligned toward each other; polarity also causes molecules other than water to be water soluble.

Water molecules form a cohesive network as their slightly positive hydrogen atoms are attracted to the slightly negative oxygen atoms of other water molecules. The cohesion between water molecules is partly responsible for their movement through fine (capillary) tubes, such as those present in the wood and other parts of plants. The attraction between the hydrogen atoms of water and other negatively charged molecules, such as those of fibers, also causes *adhesion* (attraction of dissimilar molecules to each other) and is the basis for substances that can be wet by water. When there is no attraction between water and other substances (e.g., between water and the waxy surface of a cabbage leaf), the cohesion between the water molecules results in droplets beading in the same way that raindrops bead on a freshly waxed automobile.

Valence

The combining capacity of an atom or ion is called valence. Atoms of the element calcium, for example, have a valence of two, while those of the element chlorine have a valence of one. In order for the atoms of these two elements to combine, there must be a balance between electrons lost or gained (i.e., the valences must balance); for example, it takes two chlorine atoms to combine with *one* calcium atom. The compound formed by the union of calcium and chlorine is called *calcium chloride*. It is customary to use standard abbreviations taken from the Latin names of the elements when giving chemical formulas or equations. Calcium chloride, for example, is $CaCl_2$, indicating that one atom of calcium (Ca^{++}) requires two atoms of chorine (Cl^-) to form a calcium chloride molecule.

Bonds and Ions

Bonds that hold atoms together in molecules form in several different ways. There are three types of chemical bonds that are of major importance:

1. *Hydrogen bonds* form as a result of attraction between positively charged hydrogen atoms and other negatively charged atoms. Negatively charged oxygen and/or nitrogen atoms of one molecule may attract positively but weakly charged hydrogen atoms of other molecules, forming a weak bond. Hydrogen bonds are very important in nature because of their abundance in many biologically significant molecules. They have, however, only about 5% of the strength of covalent bonds.

2. *Covalent bonds* involve the sharing of a pair of electrons occurring in the outermost orbital; they hold together two or more atomic nuclei and travel between them, keeping them at a stable distance from each other. For example, the single orbital of a hydrogen atom, which has just one electron, is usually filled by attracting an electron from another hydrogen atom. As a result, two hydrogen atoms share their single electrons, making a combined orbital with two electrons. The combined orbital, with its two hydrogen atoms, makes a molecule of hydrogen gas, shown as H_2.

    Covalent bonds are the strongest of the three types of bonds discussed here and the principal force binding together atoms that make up some important biological molecules discussed later in this chapter (Fig. 2.4).

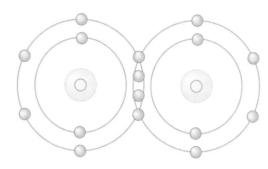

FIGURE 2.4   A covalent bond between two oxygen atoms. In a covalent bond, electrons are shared as outer shells of atoms overlap. In this instance, two pairs of electrons are shared between the two atoms, and the shared electrons are counted as belonging to each atom.

3. *Ionic bonds* form whenever one or more electrons are donated to another atom and result whenever two oppositely charged ions come in contact. In nature, some electrons in the outermost orbital are not really shared but instead are completely removed from one atom and transferred to another, particularly between elements that can strongly attract or easily give up an electron. Molecules that lose or gain electrons become positively or negatively charged particles called ions. Ions are shown with their charges as superscripts. For example, table salt (sodium chloride) is formed by ionic bonding between an ion of sodium (Na^+) and an ion of chlorine (Cl^-). The sodium becomes a positively charged ion when it loses one of its electrons, which is gained by an atom of chlorine. This extra electron makes the chlorine ion negatively charged, and the sodium ion and chlorine ion become bonded together by the force of the opposite charge.

Some ions such as those of magnesium (Mg^{++}), for example, give up two electrons and therefore have two positive charges. Such ions can form ionic bonds with two single negatively charged ions such as those of chlorine (Cl^-), forming magnesium chloride ($MgCl_2$). Most biologically important molecules exist as ions in living matter. Types of bonds other than hydrogen, covalent, and ionic need not be considered here.

Mixtures

A **mixture** differs from a compound in that not all of its molecules or atoms are united in definite ratios. For example, granite is composed of several different materials that vary in proportion to one another throughout the rock; likewise, a cake consists of ingredients that can vary in proportion to one another. Accordingly, granite, cakes, and a myriad of other substances with variable proportions of molecules are mixtures.

Acids, Bases, and Salts

As previously indicated, water molecules are held together by weak hydrogen bonds (see also the cohesion-tension theory in Chapter 9). In pure water, however, a few molecules sometimes separate into hydrogen (H^+) and hydroxyl (OH^-) ions, with the number of H^+ ions precisely equaling the number of OH^- ions. **Acids,** which taste sour like cranberry or lemon juice, are defined as substances that release hydrogen (H^+) ions when dissolved in water, with the result that there are proportionately more hydrogen than hydroxyl ions present. Conversely, **bases** (also referred to as *alkaline compounds*), which usually feel slippery or soapy, are defined as compounds that release negatively charged hydroxyl (OH^-) ions when dissolved in water. Bases may also be defined as compounds that accept H^+ ions.

The pH Scale

The concentration of H^+ ions present is used to define degrees of acidity or alkalinity on a specific scale, called the **pH scale.** The scale ranges from 0 to 14, with each unit representing a tenfold change in H^+ concentration. Pure water has a pH of 7—the point on the scale where the number of H^+ and OH^- ions is exactly the same, or the neutral point.[1] The lower a number is below 7, the higher the degree of acidity, and conversely, the higher a number is above 7, the higher the degree of alkalinity. Vinegar, for example, has a pH of 3, tomato juice has a pH of 4.3, and egg white has a pH of 8.

When an acid and a base are mixed, the H^+ ions of the acid bond with the OH^- ions of the base—forming water (H_2O). The remaining ions bond together, forming a **salt.** If hydrochloric acid (HCl) is mixed with a base—for example, sodium hydroxide (NaOH)—water (H_2O) and sodium chloride (NaCl), a salt, are formed. The reaction is represented by symbols in an equation that shows what occurs:

$$HCl + NaOH \longrightarrow H_2O + NaCl$$

Energy

Energy, which can be defined as "the ability to do work" or "the ability to produce a change in motion or matter," is required for all the activities of living things to take place, whether at the level of whole organisms, cells, or molecules. The ultimate source on earth of that energy is the sun.

Scientists have characterized energy with laws of thermodynamics. The *first law of thermodynamics* states that energy is constant—it cannot be increased or diminished—but it can be converted from one form to another. Among its forms are *chemical, electrical, heat,* and *light* energy. The *second law of thermodynamics* states that when energy is converted from one form to another in a given system in which no energy enters or leaves, it flows in one direction and the amount of useful energy remaining after the conversion will always be less than before the conversion. For example, heat will always flow from a hot iron to cold clothing but never from the cold clothing to the hot iron. Such energy-yielding reactions are vital to the normal functions of cells and provide the energy needed for other cell reactions. Both types of reactions are discussed in Chapter 10. Forms of energy include *kinetic* (motion) and *potential* energy. Potential energy is defined as the "capacity to do work owing to the position or state of a particle." For example, a ball resting at the top of a hill possesses potential energy that is converted to kinetic energy if the ball rolls down the hill. Some chemical reactions release energy and others require an input of energy.

Although all electrons have the same weight and electrical charge, they vary in the amount of potential energy they possess. Electrons with the least potential energy are located within a single spherical orbital closest to the atom's nucleus. At the next highest energy level, there are up to four orbitals, each with two electrons. Depending on the kind of atom, there can be higher numbers of orbitals at still higher energy levels, but the outermost level is limited to four orbitals and a maximum of four electrons. Each energy level is usually referred to as an *electron shell.* The outermost electron shell determines how or if an atom reacts with another atom. Atoms with eight electrons in the outer shell have no reaction with other atoms.

The various energy levels are due to the attraction between the positive charges of the protons and negative charges of the electrons. The farther away from the nucleus an electron is, the greater the amount of energy required to keep it there (Fig. 2.5). Some of the numerous energy exchanges and carriers occurring in living cells are discussed in later chapters.

Chemical Components of Protoplasm

The living substance of all cells is called **protoplasm.** Protoplasm is organized into numerous bodies of various sizes, most of which are discussed in Chapter 3. About 96% of protoplasm is composed of the elements carbon, hydrogen, oxygen, and nitrogen; 3% consists of phosphorus, potassium, and sulphur. The remaining 1% includes calcium, iron, magnesium, sodium, chlorine, copper, manganese, cobalt, zinc, and minute quantities of other elements. When a plant first absorbs these elements from the soil or atmosphere (see the section on essential elements in Chapter 9) or when it utilizes breakdown products within the cell, the elements are in the form of simple molecules or ions. These simple forms may be converted to very large, complex molecules through the metabolism of the cells.

The large molecules invariably have "backbones" of carbon atoms within them and are said to be **organic.** Other molecules that contain no carbon atoms are called **inorganic.**

---

1. Note that although distilled water is theoretically "pure," its pH is always less than 7 because carbon dioxide from the air with which it is in contact dissolves in it, forming carbonic acid (H_2CO_3); the actual pH of distilled water is usually approximately 5.7.

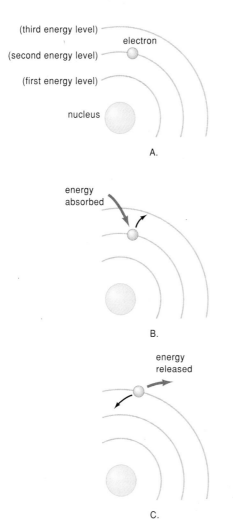

FIGURE 2.5 Energy levels of electrons. The closer electrons are to the nucleus, the less energy they possess and vice versa. The energy levels are referred to as electron shells. *A.* An electron at a second energy level. *B.* An electron can absorb energy from sunlight or some other source and be boosted to a higher energy level. *C.* The absorbed energy can be released, with the electron dropping back to its original level (see Fig. 10.8).

Exceptions include carbon dioxide (CO_2) and sodium bicarbonate ($NaHCO_3$).

The name **organic** was given to most of the chemicals of living things when it was believed that only living organisms could produce molecules containing carbon. Today, many organic compounds can be produced artificially in the laboratory, and scientists sometimes hesitate to classify as either organic or inorganic some of the 4 million carbon-containing compounds thus far identified. Most scientists, nevertheless, agree that inorganic compounds usually do not contain carbon.

Macromolecules

The large molecules making up the majority of cell components are called *macromolecules* or **polymers.** Polymers are formed when two or more small units called **monomers**

(third energy level)

electron

(second energy level)

(first energy level)

nucleus

energy absorbed

energy released

bond together. The bonding between monomers occurs when a hydrogen (H^+) is removed from one and a hydroxyl (OH^-) is removed from another, creating an electrical attraction between them. Since the components of water (H^+ and OH^-) are removed (*dehydration*) in the formation (*synthesis*) of a bond, the process is referred to as *dehydration synthesis.* Dehydration synthesis is controlled by an enzyme (see page 23).

Hydrolysis, which is essentially the opposite of dehydration synthesis, occurs when a hydrogen from water becomes attached to one monomer and a hydroxyl group to the other. Energy is released when a bond is broken by hydrolysis. This energy may be stored temporarily or used in the manufacture or renewal of cell components.

Four of the most important classes of polymers found in protoplasm are *carbohydrates, lipids, proteins,* and *nucleic acids.*

Carbohydrates *Carbohydrates* are the most abundant organic compounds in nature. They contain C, H, and O in or close to a ratio of 1C:2H:1O. Carbohydrates include *monosaccharides,* which are simple sugars with backbones consisting of three to seven carbon atoms. Among the most common monosaccharides are glucose and fructose, each of which has six carbon atoms. Glucose is a primary source of energy in cells. *Fructose,* which is found in fruits, is an *isomer* of glucose. Isomers are molecules with identical numbers and kinds of atoms but with different structures and shapes (Fig. 2.6).

Disaccharides are formed when two monosaccharides become bonded together by dehydration synthesis. The common table sugar **sucrose** is a disaccharide formed from a molecule of glucose and a molecule of fructose. Sucrose is the form in which sugar is usually transported throughout plants.

Polysaccharides are formed when several monosaccharides are bonded together. Polysaccharide polymers sometimes consist of thousands of simple sugars attached to one another in long chains or coils. *Starches,* for example, are polysaccharides that usually consist of several hundred to several thousand coiled glucose units. When numerous glucose molecules become a starch molecule, each glucose gives up a molecule of water. The formula for starch is $(C_6H_{10}O_5)\boldsymbol{n},$ the $\boldsymbol{n}$ representing many units. *Cellulose,* a principal substance in plant cell walls, is a polysaccharide consisting of 3,000 to 10,000 unbranched chains of glucose molecules. In order for a starch molecule to become available as an energy source in cells, it has to be hydrolyzed; that is, it has to be broken up into individual glucose molecules through the restoration of a water molecule for each unit.

Lipids *Lipids* are fatty or oily substances that are mostly insoluble in water because they have no polarized components. They store more energy than carbohydrates and play an important role in the longer term energy reserves and structural components of cells. Like carbohydrates, their molecules contain principally carbon, hydrogen, and oxygen, but there is proportionately much less oxygen present.

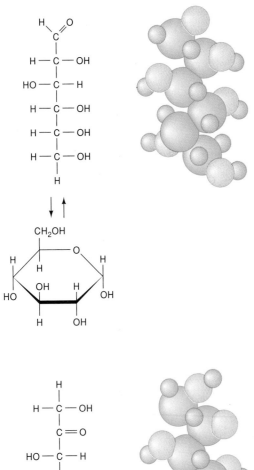

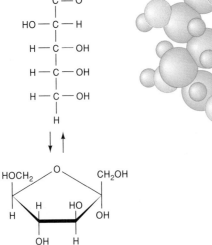

FIGURE 2.6 Structural formulas and models of glucose and fructose molecules. The molecules can exist either as straight chains or in the form of rings. Each chain is 3 nanometers long. The ring forms are more common in cells.

Examples of lipids include **fats** (Fig. 2.7) and **oils,** whose molecules are manufactured from sugars and are composed of a unit of glycerol (an alcohol) with three fatty acids attached. The fatty acids have carbon atoms to which hydrogen atoms can become attached.

Most fatty acid molecules consist of a chain with 16 to 18 carbon atoms. If hydrogen atoms are attached to every available attachment point of these fatty acid carbon atoms,

the fat is said to be *saturated*. However, if fewer hydrogen atoms are attached, as is the case with some vegetable oils that are currently receiving publicity, the fat is said to be *polyunsaturated*. Like polysaccharides and proteins (discussed in the next section), lipids are broken down by hydrolysis.

Waxes are lipids consisting of very long-chain fatty acids bonded to a very long-chain alcohol other than glycerol. Waxes, which are solid at room temperature, are found on the surfaces of aboveground plant organs. They are usually embedded in a matrix of *cutin* or *suberin,* which are also insoluble lipid polymers. The combinations of wax and cutin or wax and suberin function in waterproofing, reduction of water loss, and protection against microorganisms and small insects.

Phospholipids are constructed like fats, but one of the three fatty acids is usually replaced by a phosphate group; this can cause the molecule to become a polarized ion. When phospholipids are placed in water, they form a double-layered sheet resembling a membrane. Indeed, phospholipids are important components of all membranes found in living organisms.

Amino Acids and Proteins The cells of living organisms contain from several hundred to many thousand different kinds of **proteins.** These important molecules are usually very large and are composed of monomers called **amino acids** (Fig. 2.8). Proteins consist of carbon, hydrogen, oxygen, and nitrogen atoms and sometimes also sulphur atoms. Aside from water, proteins form the bulk of protoplasm.

There are 20 different kinds of amino acids, and from 50 to 50,000 or more of them are present in various combinations in each protein molecule. Every amino acid has two special functional groups of atoms plus a remainder called the *R group*. One functional group is called the *amino group* (NH₂); the other, which is acid, is called the *carboxyl group* (COOH). The makeup of the R group is distinctive for each of the 20 amino acids. Some R groups are polar while others are not.

Peptides A *peptide* consists of two or more amino acids bonded together. Bonds between amino acids are called *peptide bonds;* many amino acids may form long *polypeptide* chains. Each polypeptide usually coils, bends, and folds in a specific fashion within a protein, which characteristically has three levels of structure and sometimes four:

1. A sequence of amino acids fastened together by peptide bonds forms the *primary structure* of a protein.

2. As hydrogen bonds form between oxygen and nitrogen atoms of different amino acids, the polypeptide chain coils like a spiral staircase, and the *secondary structure* develops. Some secondary structures include polypeptide chains that double back and form hydrogen bonds between the two lengths in what is referred to as a *beta sheet,* or *pleated sheet.*

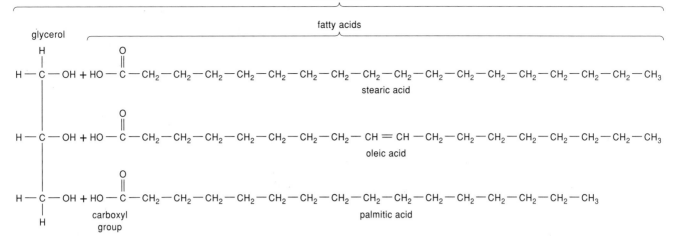

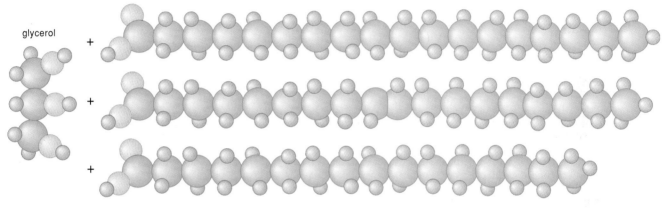

FIGURE 2.7 Structural formula and model of a fat molecule. H = hydrogen, C = carbon, O = oxygen. A typical fatty acid is 4 nanometers long.

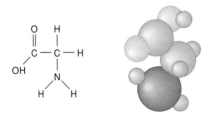

FIGURE 2.8 Structural formula and model of the amino acid glycine.

3. *Tertiary structure* develops as the polypeptide further coils and folds. The tertiary structure is maintained by bonds between R groups.

4. If a protein happens to have more than one kind of polypeptide, a fourth or *quaternary structure* may form (Fig. 2.9).

Anything that disturbs the normal pattern of bonds between parts of the protein molecule and thereby alters the characteristic coiling and folding, will *denature* the protein (adversely affect its function or properties). Denaturing, which

is often brought about by high temperatures or chemicals, may kill the cell of which the protein is a part.

Enzymes *Enzymes* are large, complex proteins (or in a few instances ribonucleic acid molecules—see the section on nucleic acids that follows). Enzymes function as organic catalysts under specific conditions of pH and temperature. They facilitate cellular chemical reactions, even at very low concentrations, and they can increase the rate of reaction as much as a billion times. Enzymes are essential, for none of the 2,000 or more chemical reactions in cells can take place unless the enzyme specific for each one is present and functional in the cell in which it is produced. In addition, enzymes do not usually break down in the reactions they accelerate and are often used repeatedly.

The names of enzymes normally end in *-ase*. One of the most common is maltase, which catalyzes the hydrolysis of maltose to glucose; maltose is a disaccharide composed of two glucose monomers. Enzymes lower the *energy of activation,* which is the energy needed to cause molecules to react with one another. An enzyme brings about its effect by bonding with potentially reactive molecules at a surface site. The reactive molecules temporarily fit into the active site, where a short-lived complex is formed. The reaction then occurs

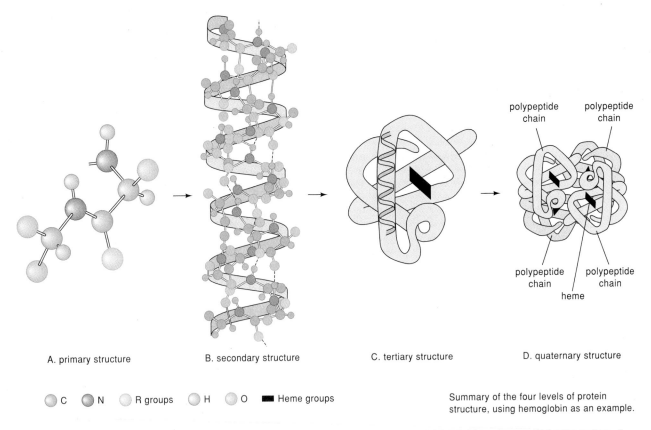

A. primary structure   B. secondary structure   C. tertiary structure   D. quaternary structure

○ C ● N ○ R groups ○ H ○ O ■ Heme groups

Summary of the four levels of protein structure, using hemoglobin as an example.

FIGURE 2.9   The four levels of protein structure. The example shown is for hemoglobin. *A.* The primary structure consists of a chain of amino acids bonded together. *B.* As the amino acid chain grows it coils and often doubles back, with hydrogen bonds forming between the two lengths and creating a *beta-* or *pleated sheet. C.* The coil or helix folds further, forming a somewhat globular structure. *D.* Several chains combine into a single functional protein molecule.

After Caret, R. L., K. J. Denniston, and J. J. Topping. 1993. *Inorganic, Organic & Biological Chemistry.* Copyright 1993 by Wm C. Brown Publishers, Dubuque, IA.

rapidly, often at rates exceeding 500,000 times per second. The complex then breaks down as the products of the reaction are released, with the enzyme remaining unchanged and capable of once more facilitating the reaction (Fig. 2.10).

Many enzymes, derived mostly from bacteria and fungi, have very important industrial uses. For example, waste treatment plants, the dairy industry, and manufacturers of detergents all use enzymes that have been mass-produced by microorganisms in large vats. One such commercially marketed enzyme produced by the activities of *Aspergillus,* a mold, breaks down complex sugars found in beans, broccoli, and many other vegetables consumed by humans. A few drops of the enzyme placed on these foods while they are being consumed effectively reduces the gas produced when enzymes in human digestive tracts are otherwise unable to accomplish the breakdown.

Nucleic Acids   *Nucleic acids* are exceptionally large complex polymers originally thought to be confined to the nuclei of cells but now known also to be associated with other cell parts. They are vital to the normal internal communication and functioning of all living cells. The two types of nucleic acids—deoxyribonucleic acid (DNA) and ribonucleic acid (RNA)—are briefly introduced here and discussed in more detail in Chapter 13.

Deoxyribonucleic acid (DNA) molecules consist of double helical (spiral) coils of repeating subunits called **nucleotides,** each composed of a base, a sugar, and a phosphate. Four kinds of nucleotides, each with a unique nitrogenous base, occur in DNA. DNA molecules contain, in units known as **genes,** the coded information that precisely determines the nature and proportions of the myriad substances found in cells and also the ultimate form and structure of the organism itself. If this coded information were written out, it would fill over 1,000 books of 300 pages each—at least for the more complex organisms. DNA molecules can replicate (duplicate themselves) in precise fashion. When a cell divides, the hereditary information contained in the DNA of the new cells is an exact copy of the original and can be passed on from generation to generation without change, except in the event of a *mutation* (discussed in Chapter 13).

Ribonucleic acid (RNA) is similar to DNA but differs in its sugar and one of its nucleotide components. It usually occurs as a single strand. Forms of RNA facilitate protein synthesis.

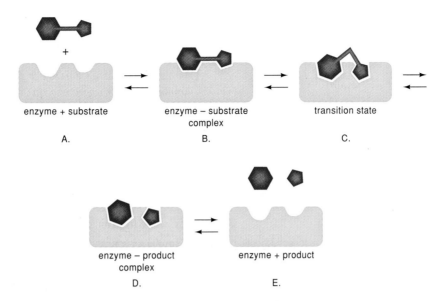

FIGURE 2.10 How an enzyme facilitates a reaction. *A.* An enzyme and the raw material (substrate) for which it is specific. *B.* The substrate fits into the active site on the enzyme. *C.* The enzyme then changes shape, putting stress on the linkage between parts of the substrate. *D.* The bonds (linkage) are broken. *E.* The enzyme returns to its original shape and the products are released. When an enzyme is combining substrates the events shown proceed in reverse.

Summary

1. Activities of living organisms stem from protoplasm, the physical basis or the "stuff of life." Structure and growth are among the attributes of living organisms. Growth has been described as an increase in volume; it results primarily from the production of new protoplasm. Variations in form may be inherited or result from response to the environment.

2. Reproduction involves offspring that are always similar in form to their parents; if reproduction ceases, the organism becomes extinct. Plants generally respond to stimuli more slowly than animals and in a different fashion.

3. All living organisms exhibit metabolic activities, including production of new protoplasm, respiration, digestion, and assimilation (green organisms can, in addition, carry on photosynthesis); they also all exhibit movement to varying degrees. Cyclosis is the stream-ing motion of protoplasm within living cells. Living organisms have a much more complex structure than nonliving objects and are adapted to their individual environments.

4. The basic "stuff of the universe" is called *matter;* it occurs in three states: solid, liquid, and gas. It is composed of elements, the smallest stable subdivision of which is an atom. Atoms contain, in a tiny nucleus, positively charged protons and uncharged neutrons, surrounded by much larger orbitals or regions of whirling, negatively charged electrons. Isotopes are forms of elements that have slight variations in the number of neutrons in their atoms.

5. The combining capacities of atoms or ions are called *valence.* Atoms of elements can bond to other atoms, and those of most elements do not exist independently; compounds are substances composed of two or more elements combined in a definite ratio by chemical bonds. Molecules are the smallest independently existing particles. If a molecule loses or gains electrons, it becomes an ion, which may form an ionic bond with another ion. In a covalent bond, pairs of electrons link two or more atomic nuclei; nitrogen and/or oxygen atoms of one molecule may form weak hydrogen bonds with hydrogen atoms of other molecules.

6. Water molecules are asymmetrical in shape, causing them to have slight electrical charges on each side (i.e., they are polar). Water molecules cohere to each other and adhere to other molecules.

7. The atoms of mixtures are not all chemically united in definite ratios. Acids release positively charged hydrogen ions when dissolved in water. Bases release negatively charged hydroxyl ions when dissolved in water. The pH scale is used to measure degrees of acidity or alkalinity. Salts and water are formed when acids and bases are mixed.

8. Energy can be defined as "ability to produce a change in motion or matter" or simply as "ability to do work." Its forms include chemical, electrical, heat, light, kinetic, and potential. The farther away from the nucleus an electron is, the greater the amount of energy required to keep it there.

9. Protoplasm is composed primarily of carbon, hydrogen, oxygen, and nitrogen, with a little phosphorus and potassium, plus small amounts of other elements. A plant may convert the simple molecules or ions it recycles or absorbs from the soil to very large, complex molecules. Organic molecules are usually large polymers that have a "backbone" of carbon atoms.

10. Carbohydrates contain carbon, hydrogen, and oxygen in a ratio of 1C:2H:1O. Carbohydrates occur as monosaccharides (simple sugars such as glucose) and disaccharides (two simple sugars joined together such as sucrose). Some polysaccharides, (e.g., starch, cellulose) consist of many simple sugars condensed together, and others (e.g., pectin) are a little more complex. Lignins are often associated with structural polysaccharides. Simple sugars, when they are attached to one another, each give up a molecule of water, forming starch. Hydrolysis is the process of restoring a water molecule to each simple sugar when starch is broken down during digestion.

11. Lipids (e.g., fats, oils, and waxes) consist of a unit of glycerol or other alcohol with three fatty acids attached. They are insoluble in water and contain carbon, hydrogen, and oxygen, with proportionately much less oxygen than found in carbohydrates. Saturated fats are those that have hydrogen atoms attached to every available attachment point of their carbon atoms; if there are very few places for hydrogen atoms to attach, the fat is said to be polyunsaturated. Phospholipids have a phosphate group replacing one fatty acid.

12. Proteins are usually large molecules composed of subunits called *amino acids*. Each amino acid has two special groups of atoms: an amino group (NH_2) and a carboxyl group (COOH). These groups bond amino acids together, forming polypeptide chains; the bonds are called peptide bonds. Enzymes are large protein molecules that function as organic catalysts. Their names end in -ase. Some have important industrial uses.

13. There are two nucleic acids (DNA and RNA) associated primarily with cell nuclei. DNA and RNA molecules consist of chains of building blocks called *nucleotides,* each of which has a nitrogenous base, a 5-carbon sugar, and a phosphate group. Four kinds of nucleotides, each with a unique nitrogenous base, occur in DNA. Helical coils of DNA contain coded information determining the nature and proportions of substances in cells and the ultimate form and structure of the organism. RNA has a different sugar and nucleotide.

Review Questions

1. What distinguishes a living organism from a nonliving object, such as a rock or a tin can?

2. What is meant by the term *organic?*

3. How are acids, bases, and salts distinguished from one another?

4. Differentiate among carbohydrates, lipids, and proteins.

5. What is energy and what forms does it take?

6. How are macromolecules formed?

7. How is a protein molecule different from a nucleic acid molecule?

Discussion Questions

1. Can part of an organism be alive while another part is dead? Explain.

2. What is the difference between inherited form and form resulting from response to the environment?

3. What might happen if all enzymes were to work at half their usual speed?

Additional Reading

Day, W. 1984. *Genesis on planet earth.* New Haven, CT: Yale University Press.

Dickerson, R. E. 1981. Chemical evolution and the origin of life. *Scientific American* 239(3): 70.

Lehninger, A. L., D. L. Nelson, and M. M. Cox. 1992. *Principles of biochemistry,* 2d ed. New York: Worth Publishers.

Margulis, L. 1992. *Diversity of life: The five kingdoms.* Hillside, NJ: Enslow Publications.

Oparin, A. I. 1953. *Origin of life,* 2d ed. Mineola, NY: Dover Publications.

Raven, P. H., R. F. Evert, and S. E. Eichhorn. 1992. *Biology of plants,* 5th ed. New York: Worth Publishers.

Sackheim, G. 1991. *Introduction to chemistry for biology students,* 4th ed. Redwood City, CA: Benjamin/Cummings.

Smith, C. A., and E. J. Wood (Eds.). 1991. *Biological molecules.* New York: Chapman and Hall.

Chapter Outline

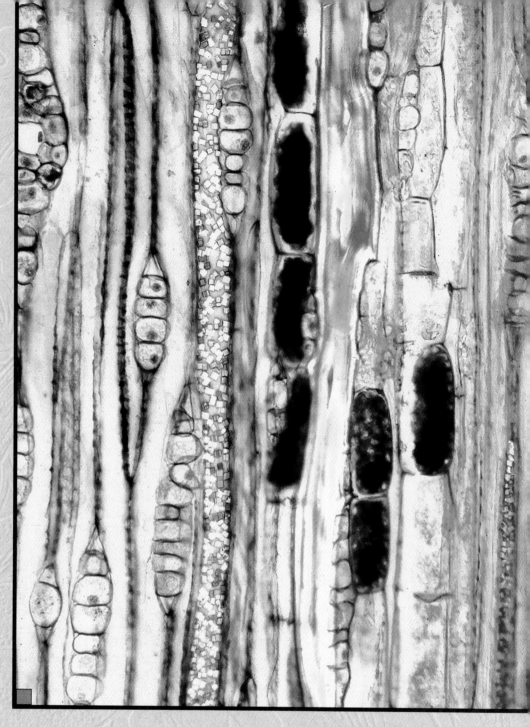

Phloem of Douglas fir (Pseudotsuga menziesii), *×500. (Polarized light photomicrograph by G. S. Ellmore.)*

Cells

3

O v e r v i e w

This chapter gives a brief review of the history of the discovery of cells and the development of cell theory. Differences between prokaryotic and eukaryotic cells are mentioned, and observations on cell size and structure follow. Each of a cell's particulates are discussed, beginning with the cell wall. Included are the plasma membrane, endoplasmic reticulum, ribosomes, Golgi bodies, mitochondria, plastids, microtubules, microfilaments, the nucleus, other organelles, vacuoles, and vacuolar membranes. Distinctions between plant and animal cells are then given. The chapter next discusses mitosis and cytokinesis and concludes with a brief review of intercellular communication.

Some Learning Goals

1. Trace the development of modern cell theory and show how the advances of early researchers have led us to our current understanding.
2. Know the following cell structures and organelles and indicate the function of each: plasma membrane, mitochondria, plastids, ribosomes, endoplasmic reticulum, Golgi bodies, vacuoles.
3. Describe the components of a nucleus and understand the function of each component.
4. Contrast plant cells with animal cells.
5. Understand the cell cycle and the events that take place in each phase of mitosis.

FIGURE 3.1   Robert Hooke's microscope, as illustrated in one of his works.
(Courtesy National Library of Medicine)

All living organisms, from aardvarks and almond trees to zebras and zinnias, are composed of cells, and all living organisms, including each of us, also generally begin life as a single cell. This single cell divides repeatedly until it develops into an organism consisting of perhaps billions of cells. During the first few hours of any organism's development, the cells all resemble each other, but changes soon occur, not only in the appearance of the cells but also in their function. The modifications of some, for example, permit them to serve as conduits for food and water, while others come to function in secretion or support. Some cells live and function for many years; others mature and degenerate in just a few days. Even as you read this, millions of new cells are being produced in your body. Some cells add to your total body mass (if you have not yet stopped growing), but most replace the millions of older cells that are destroyed every second you remain alive. The variety and form of cells seem almost infinite, but certain features are shared by most of them. A discussion of these features forms the body of this chapter.

CELLS

History

The discovery of cells is associated with the development of the microscope in the 17th century (see Chapter 1). In 1665,

the English physicist Robert Hooke, using a primitive microscope (Fig. 3.1), examined thin slices of cork he had cut with a sharp penknife. Hooke compared the boxlike compartments he saw to the surface of a honeycomb and is credited with applying the term *cell* to those compartments. He also estimated that a cubic inch of cork would contain approximately 1,259 million such cells. What Hooke saw in the cork were really only the walls of dead cells, but he also observed "juices" in living cells of elderberry plants and thought he had found something similar to the veins and arteries of animals.

Two physicians, Marcello Malpighi in Italy and Hooke's compatriot Nehemiah Grew in England, along with Anton van Leeuwenhoek reported for 50 years on the organization of cells in a variety of plant tissues. In the 1670s, they also reported on the form and structure of single-celled organisms, which they referred to as "animalcules."

After this period, little more was reported on cells until the early 1800s. This lack of progress was due in large part to the imperfections of the primitive microscopes and also to the crude methods of tissue preparation used. But both microscopes and tissue preparations slowly improved, and by 1809, the famous French biologist Jean Baptiste de Lamarck had seen a wide enough variety of cells and tissues to conclude that "no body can have life if its constituent parts are not cellular tissue or are not formed by cellular tissue." In 1824, René J. H. Dutrochet, also of France, reinforced Lamarck's conclusions that all animal and plant tissues are composed of cells of various kinds. Neither of them, however, realized that each cell could, in most cases, reproduce itself and exist independently.

In 1831, the English botanist Robert Brown discovered that all cells contain a relatively large body that he called the *nucleus.* Shortly thereafter, the German botanist Matthias Schleiden observed a smaller body within the nucleus that he called the *nucleolus.* Schleiden and a German zoologist, Theodor Schwann, were not the first to grasp the significance of cells, but they explained them with greater clarity and perception than others before them had done. They are generally credited with developing the *cell theory,* beginning with their publications of 1838 to 1839. In essence, this theory holds that all living organisms are composed of cells and that cells form a unifying structural basis of organization.

In 1858, an important augmentation of the cell theory appeared in a classic textbook by another German scientist, Rudolf Virchow. He argued cogently that every cell comes from a preexisting cell ("*omnis cellula e cellula*") and that there is no spontaneous generation of cells. Virchow's publication stirred up a great controversy, because prior to this time there was a widespread belief among scientists and nonscientists alike that animals could originate spontaneously from dust. Many who had microscopes were thoroughly convinced they could see "animalcules" appearing in decomposing substances.

The controversy became so heated that in 1860, the Paris Academy of Sciences offered a prize to anyone who could, through experiments, shed light on the matter. Just two years later, the brilliant French scientist Louis Pasteur was awarded the prize. Pasteur, using swan-necked flasks, demonstrated conclusively that boiled media remained sterile indefinitely if microorganisms from the air were excluded from the media.

In 1871, Pasteur proved that natural alcoholic fermentation always involves the activity of yeast cells. In 1897, the German scientist Eduard Buchner accidentally discovered that the yeast cells did not need to be alive for fermentation to occur. He found that extracts from the yeast cells would convert sugar to alcohol. This discovery was a major surprise to the biologists of the time and quickly led to the identification and description of enzymes (discussed in Chapter 2), the organic catalysts (substances that aid chemical reactions without themselves being changed) found in all living cells; it also led to the belief that cells were little more than miniature packets of enzymes. During the first half of the 20th century, however, great advances were made in the refinement of microscopes and in tissue preparation techniques. Numerous structures and bodies, in addition to the nucleus, were observed in cells, and the relationship between structure and function came to be realized and understood on a much broader scale than previously had been possible.

Modern Microscopes

Without microscopes, very little would be known about cells. Our present vast knowledge of cells and all aspects of biological investigations associated with them is directly related to the development of these instruments.

Light microscopes increase magnification as light passes through a series of transparent lenses, currently made of various types of glass or calcium fluoride crystals. The curvatures of the lens materials and their composition are designed to minimize distortion of image shapes and colors.

Light microscopes are of two basic types: *compound microscopes,* which require the material being examined to be sliced thinly enough for light to pass through, and *dissecting microscopes (stereomicroscopes),* which permit three-dimensional viewing of opaque objects. The best compound microscopes in use today can produce useful magnifications of up to 1,500 times under ideal conditions. Most dissecting microscopes used in teaching laboratories magnify up to 30 times, but higher magnifications are possible with both types of microscopes. Magnifications of more than 1,500 times, however, are considered "empty" because *resolution* (the capacity of lenses to aid in separating closely adjacent tiny objects) does not improve with magnification beyond a certain point. Light microscopes will continue to be useful, particularly for observing living cells, into the foreseeable future (Fig. 3.2).

Since the 1950s, the production and development of high resolution modern electron microscopes has resulted in observation of much greater detail than is possible with light microscopes. Instead of using light, electron microscopes use a beam of electrons produced when electricity of high voltage is passed through a wire. This electron beam is directed through a vacuum in a large tube or column. When the beam passes through a specimen, an image is formed on a plate. Magnification is controlled by powerful electromagnetic lenses located on the column.

Like light microscopes, electron microscopes are of two basic types. *Transmission electron microscopes* (Fig. 3.3 *A*) permit magnifications of 200,000 or more times, but the material to be viewed must be sliced extremely thinly and introduced into the column's vacuum, so living objects cannot be observed. *Scanning electron microscopes* (Fig. 3.3 *B*) usually do not achieve such high magnifications (3,000 to 10,000 times is the usual range), but opaque objects can be observed as a scanner renders the object visible on a cathode tube like a television screen. The techniques for such observation have become so refined that even preserved material can appear exceptionally lifelike, and high resolution three-dimensional images can be obtained.

In 1986, the Nobel Prize in physics was awarded to two International Business Machine scientists, Gerd Binnig and Heinrich Rohrer, for their invention in 1982 of a *scanning tunneling microscope.* This microscope uses a minute probe rather than electrons or light to scan across a surface and then reproduces an image of it down to the atomic level, doing so without damaging the probed area. The probe can scan areas barely twice the width of an atom and theoretically could be used to print on the head of an ordinary pin the words contained in more than 50,000 single-spaced pages of books.

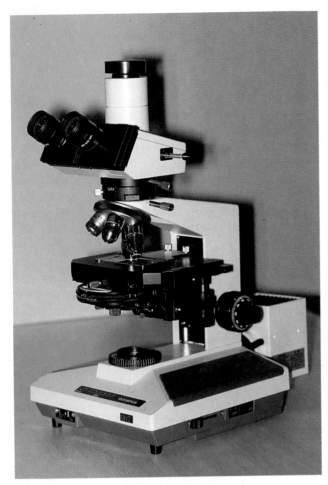

FIGURE 3.2A   A compound light microscope.

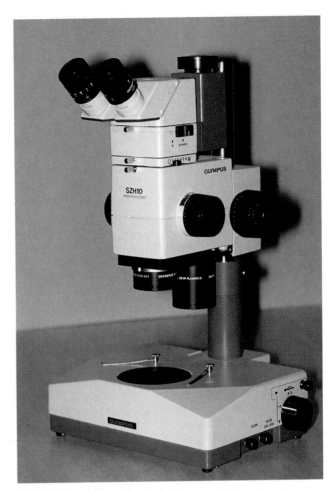

FIGURE 3.2B   A stereomicroscope (dissecting microscope).

Early in 1989, the first picture of a segment of DNA showing its helical structure was taken with a scanning tunneling microscope by an undergraduate student associated with the Lawrence Laboratories in northern California. Several variations of this microscope, each using a slightly different type of probe, have now been produced. Significant new discoveries by cell biologists using one or more of all three types of microscopes in their research have now become frequent events.

EUKARYOTIC VERSUS PROKARYOTIC CELLS

Nearly all higher plant and animal cells exhibit most of the various features that are discussed in the sections that follow. There are some very primitive organisms, however, whose cells lack a number of these features (e.g., true nuclei and other bodies bound by membranes). Such cells, called **prokaryotic** to distinguish them from the typical **eukaryotic** cells discussed here, may have been the origin of several cell components almost universally found in cells of less

primitive organisms. These and other aspects of prokaryotic cells are covered in Chapter 17.

CELL SIZE AND STRUCTURE

Cell Size

Most plant cells, and the vast majority of animal cells, are so tiny they are invisible to the unaided eye. Cells of higher plants generally vary in length between 10 and 100 micrometers.[1] Since there are roughly 25,000 micrometers to the inch, it would take about 500 average-sized cells to extend across 2.54 centimeters (1 inch) of space; 30 of them could easily be placed across the head of a pin. Some bacterial cells are less than 0.5 micrometer in diameter, while cells of the green alga mermaid's wineglass (*Acetabularia*)

1. See Appendix 5 for metric conversion tables.

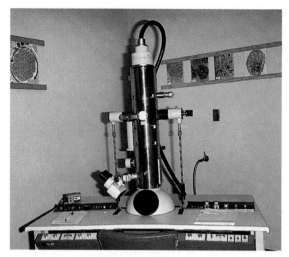

FIGURE 3.3A A modern transmission electron microscope.

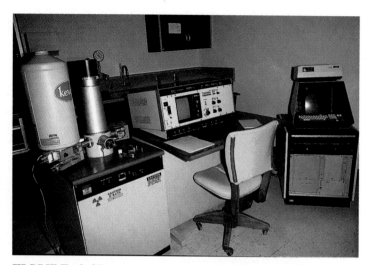

FIGURE 3.3B A scanning electron microscope.

are mostly between 2 and 5 centimeters in length, and fiber cells of some nettles are about 20 centimeters long.

Because cells are so minute, the numbers occurring in full-grown organisms are astronomical. For example, it has been calculated that a single mature leaf of a pear tree contains 50 million cells and that the total number of cells in the roots, stem, branches, leaves, and fruit of a full-grown pear tree exceeds 15 trillion. Can you imagine how many cells there are in a 3,000-year-old redwood tree of California, which may reach heights of 90 meters (300 feet) and measure up to 4.5 meters (15 feet) in diameter near the base?

Some cells are boxlike with six walls, but others assume a wide variety of shapes, depending on their location and function. The most abundant cells in the younger parts of plants and fruits may be more or less spherical when they are first formed, but they are packed together in such a way that they commonly have 14 sides by the time they are mature. These are discussed in the next chapter.

As indicated at the beginning of Chapter 2, the living part of the cell within the wall is called *protoplasm.* Two main components of protoplasm are readily discernible: the *nucleus,* which controls the cell's activities, and the **cytoplasm,** which is a souplike fluid containing water, dissolved substances, and many small **organelles** (persistent structures of various shapes and sizes with specialized functions in the cell; most, but not all, are bound by membranes). The organelles are the sites of many different activities that take place within the cell. A brief examination of each of the various cell components follows (Figs. 3.4 and 3.5 show these various components).

The Cell Wall

A popular novelty song of 50 or more years ago had several verses listing food items the author purportedly disliked,

with each verse ending, "But I like bananas because they have no bones!" Indeed, bananas and all parts of plants differ from animals in that they have no bones or similar internal skeletal structures. Yet large trees support branches and leaves weighing many tons. They are able to do this because most plant cells have semirigid or rigid walls that perform the functions of bones; that is, they provide strength and support for the plants (and also protect the delicate cell contents within). When millions of these cells function together as a tissue, their collective strength is enormous. The largest trees alive today, the redwoods, exceed the mass, or volume, of the largest land animals, the elephants, by more than a hundred times. The wood of one tree could support the combined weight of a thousand elephants.

Cell walls are composed of *cellulose,* a substance having long molecules made up of as many as 10,000 simple glucose molecules attached end to end, the polysaccharides **pectin** (the complex organic material that gives stiffness to fruit jellies) and *hemicellulose* (a gluelike substance unrelated to cellulose that holds cellulose fibrils together), and *glycoproteins* (proteins that have sugars associated with their molecules).

When new cell walls are first formed, a **middle lamella,** consisting primarily of pectin, appears. This middle lamella is normally shared by two adjacent cells and is so thin that it may not be visible with an ordinary light microscope unless it is specially stained. A fine network of cellulose is laid down on either side of the middle lamella (Fig. 3.6). The long cellulose molecules are grouped together in bundles known as *microfibrils,* which, in turn, are twisted together in ropelike fashion, forming larger bundles. The larger bundles are held together by pectin and related substances and make up the bulk of the cell wall.

Sometimes, the cellulose is deposited in two stages, forming a *primary cell wall* and then a *secondary cell wall* inside the primary wall. When this happens, the secondary

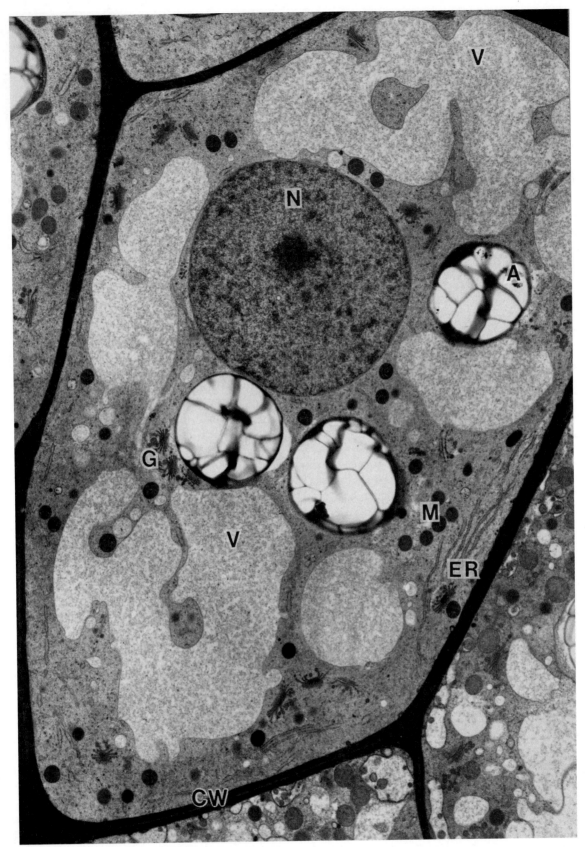

FIGURE 3.4   A cross section of a root cap cell of tobacco. V = vacuole; N = nucleus; A = amyloplast; G = Golgi bodies; M = mitochondria; ER = endoplasmic reticulum; CW = cell wall. The cell wall has a thickness of about 0.1–0.4 micrometer; the nucleus is 3 micrometers in diameter, and the cell itself is about 8 micrometers long.

(Transmission electron micrograph courtesy John Z. Kiss)

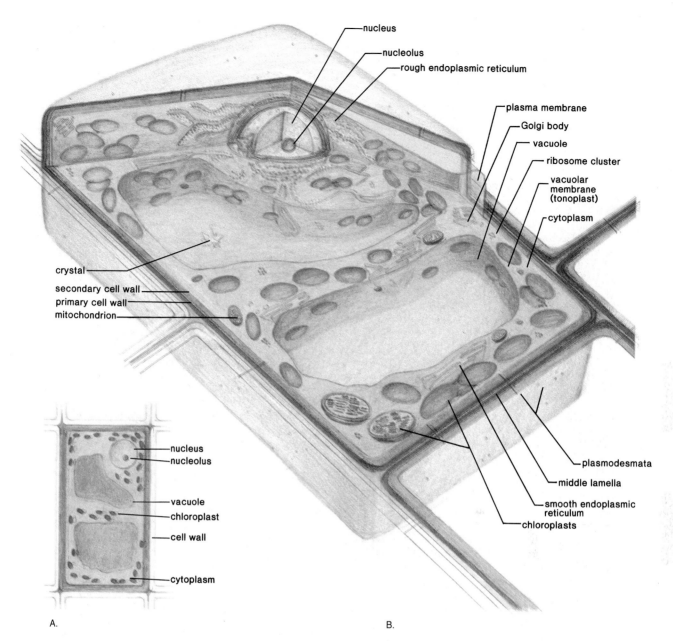

labels (clockwise from top):
nucleus
nucleolus
rough endoplasmic reticulum
plasma membrane
Golgi body
vacuole
ribosome cluster
vacuolar membrane (tonoplast)
cytoplasm
plasmodesmata
middle lamella
smooth endoplasmic reticulum
chloroplasts
mitochondrion
primary cell wall
secondary cell wall
crystal

A.
nucleus
nucleolus
vacuole
chloroplast
cell wall
cytoplasm

A.

B.

FIGURE 3.5   *A.* A leaf cell diagrammed with the aid of a light microscope. *B.* The same cell greatly enlarged to show submicroscopic features. The nucleus of the enlarged cell would be about 10 micrometers in diameter and the cell itself would be about 60 micrometers long.

cell wall is usually the more extensive of the two structures. Depending on the type of cell involved, other substances, such as sugars and **lignin** (a complex organic substance that adds mechanical strength to cell walls), may be impregnated into the wall.

The thickness of the wall can also vary, occupying more than 95% of the volume of the cell in some instances and as little as 5% in others. Cells that function in food storage or manufacture usually have thin walls, while those primarily involved in support usually have walls of moderate to extensive thickness. Fluids and dissolved substances usually can pass relatively rapidly through most cell walls

via **plasmodesmata** (singular: **plasmodesma**), which are tiny strands of protoplasm that extend between adjacent cells through minute holes in the walls.

Protoplasm

The Plasma Membrane   The outer boundary of the living part of the cell, the **plasma membrane,** is roughly eight-millionths of a millimeter thick. To get an idea of how incredibly thin that is, consider that it would take 12,500 such membranes neatly stacked in a pile to achieve the thickness of an ordinary piece of writing paper. Yet this delicate structure is of vital importance in regulating what substances enter and leave the cell and in producing and assembling cellulose for cell walls. Evidence obtained since the early 1970s indicates that this and other cell membranes are mosaics composed of lipids, with proteins interspersed throughout (Fig. 3.7). Covalent bonds link carbohydrates to both the lipids and the proteins on the outer surfaces of the membranes. Some proteins extend across the entire width of the membrane, while others are embedded in or apparently loosely bound to the outer surface.

The remainder of the cell contents usually push the plasma membrane up against the cell wall because of pressures developed by osmosis (see Chapter 9), but the membrane is quite flexible and often forms folds, which may in turn become little hollow spheres or vesicles that float off into the cell. In fact, experiments have shown that adding detergents to a unit membrane can break up and disperse the membrane, yet it can re-form (albeit imperfectly) when the detergents are removed. The membrane may even shrink away from the wall temporarily, but if it ever ruptures, the cell soon dies.

The Endoplasmic Reticulum The **endoplasmic reticulum** (often referred to simply as **ER**) is a network of flattened sacs and tubes that form channels throughout the cytoplasm, the amount and form varying considerably from cell to cell. It appears, in section, as a series of parallel membranes that resemble long, narrow bags, tubes, or sacs that create subcompartments within the cell.

The nucleus, which directs the various activities of the cell, is connected to the endoplasmic reticulum, and many of the most important activities occur either on the surface of the endoplasmic reticulum, where various enzymes are attached, or between its compartments.

Ribosomes (discussed in the section that follows) may line the outer surfaces of the endoplasmic reticulum. Such endoplasmic reticulum is said to be "rough" and is primarily associated with the synthesis, secretion, or storage of proteins (Fig. 3.8; see also Chapter 13). This contrasts with "smooth" endoplasmic reticulum, which has few, if any, ribosomes lining the surface and which is associated with lipid secretion. Both types of endoplasmic reticulum can

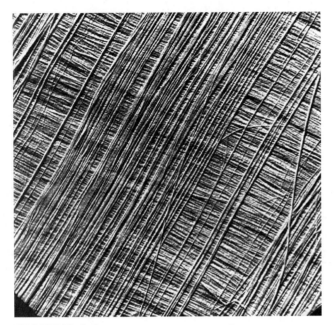

FIGURE 3.6   A small portion of a cell wall of the green alga *Chaetomorpha melagonium*, showing how cellulose microfibrils are laid down. Each microfibril is composed of numerous molecules of cellulose, ×24,000.

(Electron micrograph courtesy Eva Frei and R. D. Preston)

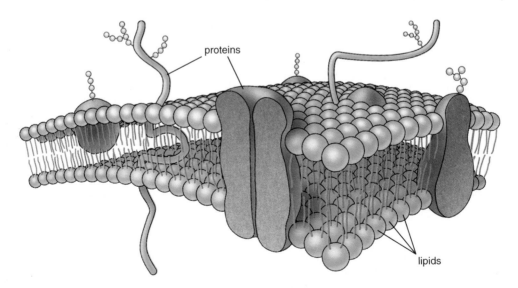

FIGURE 3.7   A model of a small portion of a plasma membrane, showing its fluid-mosaic nature. The proteins shown here are actually coiled chains of polypeptides. They occur either on the surfaces or are embedded. Some of the embedded proteins extend all the way through the continuous part of the membrane, which is a double layer of lipids. The pairs of vertical lines extending inward from each lipid represent long chain fatty acids. The membrane is about 8 nanometers thick.

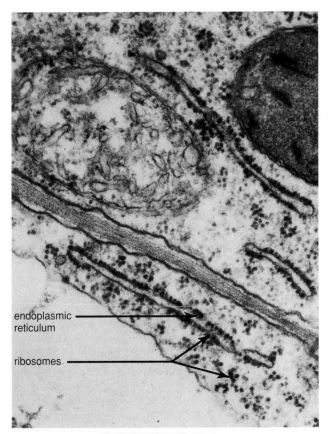

FIGURE 3.8 A small portion of the endoplasmic reticulum and ribosomes in a young leaf cell of corn (*Zea mays*). The ribosomes are 20 nanometers in diameter.

(Electron micrograph courtesy Jean Whatley)

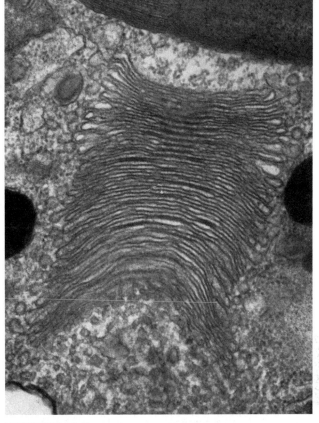

FIGURE 3.9 A Golgi body from the alga *Euglena*, ×40,000.

(Electron micrograph courtesy John Z. Kiss)

occur in the same cell. Many enzymes involved in the process of respiration are synthesized on the endoplasmic reticulum. However, they enter other organelles (primarily mitochondria) without passing through the endoplasmic reticulum. The endoplasmic reticulum also appears to be the primary site of membrane synthesis within the cell.

Ribosomes As indicated, **ribosomes** may line the endoplasmic reticulum (see Fig. 3.8), but they may also occur unattached in the cytoplasm, nucleus, chloroplasts, or other organelles. They are tiny, averaging about 20 nanometers in diameter in most plant cells. The unattached ribosomes often occur in clusters of 5 to 100, particularly when they are involved in performing their function of linking amino acids together to form the large, complex protein molecules that are a basic part of all living organisms.

Ribosomes are roughly ellipsoidal in shape, although recent evidence suggests the surface topography is varied and complex. Each ribosome is composed of two subunits, which in turn are made up of RNA and proteins. About 55 kinds of protein are found in each ribosome of prokaryotic cells and a slightly higher number in those of eukaryotic cells (see the discussion of various types of RNA in Chapter 13). Unlike other organelles, ribosomes are not bounded by membranes.

The Golgi Apparatus (Dictyosomes) **Golgi apparatus** is a term applied collectively to all of the **Golgi bodies** (**Golgi stacks** or **dictyosomes**) of a cell (Fig. 3.9), the latter terms more often being applied to plant cells, which may contain up to several hundred of them. They appear as groups of flat, roundish sacs, frequently bounded by branching tubules that originate from the endoplasmic reticulum but are not directly connected to it. A number of them are usually scattered throughout the cytoplasm of a living cell. The sacs are often organized into stacks of 5 to 8, but up to 30 or more are not uncommon in simpler organisms.

Complex carbohydrates (polysaccharides) are assembled within the Golgi apparatus and collect in small vesicles (blisterlike bodies) that are pinched off from the margins. These vesicles migrate to the plasma membrane or other parts of the cell. There they discharge their contents to the outside, where they become a part of the cell wall. The enzymes needed for the "packaging" process are produced or contained within the dictyosomes or Golgi stacks themselves. One might describe the components of the Golgi apparatus as collecting, packaging, and delivery centers.

inner membrane

outer membrane

cristae

matrix (fluid)

FIGURE 3.10 A mitochondrion greatly enlarged and cut away to show the cristae (folds of the interior membrane). A mitochondrion is about 2 micrometers long.

Mitochondria **Mitochondria** are often referred to as the "powerhouses" of the cell, for it is within them that energy is released from organic molecules by the process of respiration (the role of mitochondria in respiration is further discussed in Chapter 10). This energy is needed to keep the individual cells and the plant functioning as a whole. Mitochondria are numerous and tiny, typically measuring from 1 to 3 or more micrometers in length and having a width of roughly one-half micrometer; they are barely visible with light microscopes. In living cells, they are in constant motion and tend to accumulate in groups where energy is needed. They often divide in two; in fact, all originate from the division of existing mitochondria.

Mitochondria assume various shapes, such as those of gherkins or paddles. A sectioned mitochondrion resembles a scooped-out watermelon with inward extensions of the rind forming mostly incomplete partitions perpendicular to the surface (Fig. 3.10). The appearance of incomplete partitions results from the fact that each mitochondrion is bounded by two membranes, with the inner membrane forming numerous platelike folds called *cristae,* which greatly increase the surface area available to the enzymes contained in a matrix fluid. This fluid also contains DNA, RNA, ribosomes, proteins, and dissolved substances.

Plastids Several kinds of **plastids** are generally found in living cells, with the **chloroplasts** (Fig. 3.11 *A*) of green organisms usually being the most conspicuous. They occur in a variety of shapes and sizes, such as the beautiful corkscrewlike ribbons found in cells of the green alga *Spirogyra* (see Fig. 18.12) and the bracelet-shaped chloroplasts of other green algae, such as *Ulothrix* (see Figs. 18.8 *D* and 18.11). The chloroplasts of higher plants, however, tend to be shaped somewhat like two Frisbees glued together along their edges, and when they are sliced in median section, they resemble the outline of a football.

Although a number of the algae and a few other plants have only one or two chloroplasts per cell, the number of chloroplasts is usually much greater in a green cell of higher plants. Seventy-five to 125 is quite common, with up to several hundred occurring in certain green cells of a few plants. The chloroplasts may be from 2 to 10 micrometers in diameter, and each is bounded by an envelope consisting of two delicate membranes. The outer membrane apparently is derived from endoplasmic reticulum, while the inner membrane is believed to have originated from the cell membrane of a blue-green bacterium (discussed in Chapter 17). Within is a colorless fluid matrix, the **stroma,** that contains enzymes. Most of the activities of chloroplasts are dictated by genes in the nucleus, but each chloroplast contains a small circular molecule of DNA that encodes a few of the many photosynthetic and other activities within the chloroplast itself.

Grana (singular: **granum**), which are stacks of coin-shaped double membranes called **thylakoids** (Fig. 3.11 *B, C*), are suspended in the stroma. The membranes of the thylakoids contain green **chlorophyll** and other pigments. These "coin stacks" of grana are vital to life as we know it on our planet today, for it is within the thylakoids that the first steps of the important process of *photosynthesis* (see Chapter 10) occur. In photosynthesis, green plants convert water and carbon dioxide (from the air) to simple food substances, harnessing energy from the sun in the process. The existence of human and all other animal life depends on the activities of the chloroplasts.

In each chloroplast, there are usually about 40 to 60 grana linked together by arms, and each granum may contain from 2 or 3 to more than 100 stacked thylakoids (Fig. 3.11). There are usually four or five starch grains in the stroma, as well as oil droplets and enzymes. Some plastids (e.g., those of tobacco) store proteins. In addition to containing the DNA molecule, the stroma, like the matrix fluid in mitochondria, contains ribosomes.

A second type of plastid found in some cells of more complex plants is the **chromoplast.** Although chromoplasts are similar to chloroplasts in size, they vary considerably in shape, often being somewhat angular. They sometimes

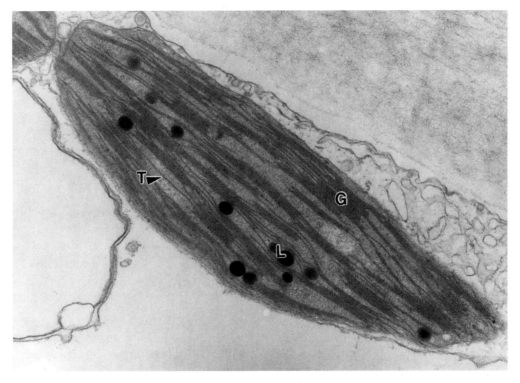

A.

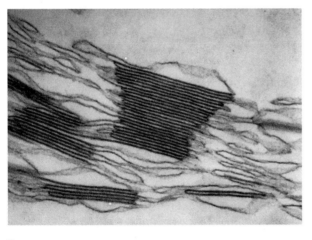

B.

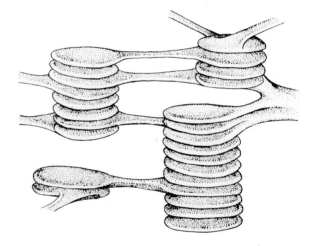

C.

FIGURE 3.11   *A*. A chloroplast ×10,000. G = granum; T = A thylakoid; L = lipid bodies. *B*. Grana ×40,000. *C*. Thylakoids. Each thylakoid is about 25 nanometers wide.

(*A*. Electron micrograph courtesy John Z. Kiss; *B*. Electron micrograph courtesy Blake Rowe)

develop from chloroplasts through internal changes, including the disappearance of chlorophyll. Chromoplasts are yellow, orange, or red in color due to the presence of carotenoid pigments, which they synthesize and accumulate. They are most abundant in the yellow, orange, or some red parts of plants such as ripe tomatoes, carrots, or red peppers. These carotenoid pigments are not, however, the predominant pigments in red flower petals.

Leucoplasts are a third type of plastid common to cells of higher plants. They are essentially colorless and include

amyloplasts, which are known to synthesize starches, and *elaioplasts*, which synthesize oils. If exposed to light, some leucoplasts will develop into chloroplasts, and vice versa.

Plastids of all types develop from **proplastids**, which are small pale green or colorless organelles having roughly the size and form of mitochondria. They are simpler in internal structure than plastids and have fewer thylakoids, the thylakoids not being arranged in grana stacks. Proplastids frequently divide and become distributed throughout the cell; after a cell itself divides, each daughter cell has a

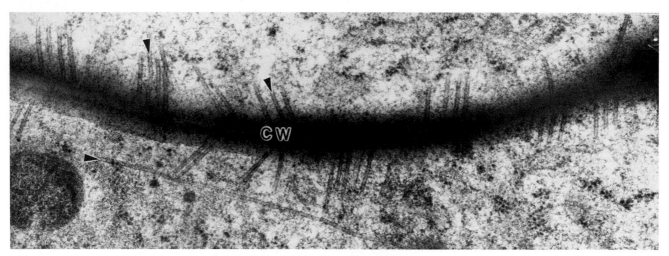

FIGURE 3.12 A small portion of a plant cell wall with microtubules more or less perpendicular to it, ×100,000. (Electron micrograph courtesy John Z. Kiss)

proportionate share. Plastids also arise through the division of existing mature plastids.

Microtubules and Microfilaments **Microtubules** are unbranched, thin, hollow, tubelike structures that resemble tiny straws. They are composed of proteins called *tubulins* and are of varying lengths, most being between 15 and 25 nanometers in diameter. They are most commonly found just inside the plasma membrane, where they apparently control the addition of cellulose to the cell wall (Fig. 3.12). Other functions of microtubules include the steering of vesicles containing cell wall components synthesized by dictyosomes (Golgi bodies) to the cell wall and the aiding of movement of the tiny whiplike *flagella* and *cilia* possessed by some cells (see the section on plant movements in Chapter 11). Microtubules are also found in the special fibers that form the *spindles* and *cell plates* of dividing cells discussed later in this chapter.

Microfilaments occur in nearly all cells as long protein filaments with an average diameter of 6 nanometers. They are often in bundles and appear to play a role in *cytoplasmic streaming* (discussed on p. 39). They play a major role in the contraction and movement of cells in multicellular animals. Microtubules and microfilaments, as well as filaments of intermediate length, form a flexible framework within the cytoplasm.

The Nucleus The **nucleus** is frequently the most conspicuous organelle in a living cell, although it may be obscured by chloroplasts in green cells when they are observed with the aid of a light microscope. In cells without chloroplasts, the nucleus may appear as a grayish lump, somewhat spherical or ellipsoidal in shape and often lying against the plasma membrane to one side of the cell or in a corner. Some nuclei are irregular in form and, like most other organelles, they can vary greatly in size. They are, however, generally from 2 to 15 micrometers or larger

in diameter. Certain fungi and algae have numerous nuclei within a single extensively branched cell, but more complex plants usually have a single nucleus located within the cytoplasm of each living cell.

The nucleus is the control center of the cell. In some ways it functions like a combination of a DNA-programmed computer and a dispatcher. The rest of the cell receives coded messages or "blueprints" from the nucleus and assembles items called for from the raw materials available to it. These raw materials are either absorbed by the plant from the soil or recycled from other areas. The nucleus not only directs the myriad activities of the complex cell "factory" but also stores hereditary information, which is passed from cell to cell as new cells are formed. Each nucleus is bounded by two membranes, which together constitute the **nuclear envelope.**

The nuclear envelope, which originates from the endoplasmic reticulum after a nucleus divides, appears to be essentially a specialized part of the endoplasmic reticulum and is continuous with it. Structurally complex pores, about 50 to 75 nanometers apart, occupy up to one-third of the total surface area of the nuclear envelope (Fig. 3.13). Proteins, which act as channels for molecules, are embedded within the pores, which apparently permit only certain kinds of molecules (for example, proteins being carried into the nucleus and RNA being carried out) to pass between the nucleus and the cytoplasm.

Within the nuclear envelope is a granular-appearing fluid called the *nucleoplasm*. It differs from the cytoplasm around it in that it contains no ribosomes. The nucleoplasm is packed with short fibers that are about 10 nanometers in diameter, and several different larger bodies are suspended within it. The most noticeable of these larger bodies are **nucleoli** (singular: **nucleolus**), which superficially resemble miniature nuclei. There may be present in each nucleus from one to several nucleoli, each of which is composed primarily of protein and is not membrane-bound. They are involved

FIGURE 3.13 The nuclear envelope of a barley seed embryo cell nucleus showing the nuclear pores. Magnification ca. ×3,000.

(Electron micrograph courtesy K. A. Platt-Aloia)

with the synthesis of ribosomal RNA (discussed in Chapter 13), which is exported outside the nucleus where, with proteins, it forms ribosomes.

Other important nuclear structures that are not apparent with light microscopy unless the cell is stained or is in the process of dividing include thin strands of **chromatin.** When a nucleus divides, the chromatin strands coil, becoming shorter and thicker, and in their condensed condition, they are called **chromosomes.** Chromatin is composed of protein and DNA (discussed in Chapters 2 and 13). Each species of plant or animal has its own fixed number and composition of chromosomes in each of its cells; the cells involved in sexual reproduction have half the number found in other cells of the same organism. The number of chromosomes present in a nucleus normally bears no relation to the size and complexity of the organism. Each body cell of a radish, for example, has 18 chromosomes in its nucleus, while a cell of one species of goldenweed has 4, and a cell of a tropical adder's tongue fern has over 1,000. Humans have 46 chromosomes in each body cell.

Other Organelles Various small bodies distributed throughout the cytoplasm tend to give it a granular appearance. Examples of such components are types of small spherical organelles called *microbodies,* which contain enzymes and are bounded by a single membrane. One microbody contains enzymes involved in a phase of photosynthesis and photorespiration (discussed in Chapter 10); another contains enzymes that aid in the conversion of fats to carbohydrates. At one time, lipid, fat, or wax droplets, which are common in cytoplasm, were believed to be bounded by a membrane; recent evidence suggests no membrane is present, and some cell biologists, therefore, do not consider them true organelles. One organelle, called a *lysosome,* stores enzymes that digest proteins and certain other large molecules but is apparently confined to animal cells. The digestive activities of lysosomes are similar to those of the *vacuoles* of plant cells.

Vacuoles In a mature living plant cell, as much as 95% or more of the volume may be taken up by one or two large central **vacuoles** that are bounded by **vacuolar membranes (tonoplasts)** (Fig. 3.14). The vacuolar membranes, which constitute the inner boundaries of the living part of the cell, are similar in structure and function to plasma membranes.

The vacuole was evidently so called because of a belief that it was just an empty space; hence its name has the same Latin root as the word *vacuum* (from vacuus, meaning "empty"). Vacuoles, however, are filled with a watery fluid called **cell sap,** which is slightly to significantly acidic and plays a role in maintaining pressures within the cell (see the discussion of *osmosis* in Chapter 9). Cell sap contains dissolved substances, such as salts, sugars, organic acids, and small quantities of soluble proteins. It also frequently contains water-soluble pigments. These pigments, called **anthocyanins,** are responsible for many of the red, blue, or purple colors of flowers and some reddish leaves. In some instances, anthocyanins accumulate to a greater extent in response to cold temperatures in the fall. They should not be confused, however, with the red and orange carotenoid pigments confined to the chromoplasts. Carotenoid pigments are not soluble in water at all, but the yellow carotenes play a role in fall leaf coloration (discussed in Chapter 7).

Sometimes, large crystals of waste products form within the cell sap after certain ions have become concentrated there. Vacuoles in newly formed cells are usually tiny and numerous. They increase in size and unite as the cell matures. In addition to accumulating the various substances and ions already mentioned, vacuoles are apparently also involved in the recycling of certain materials within the cell and even aid in the breakdown and digestion of organelles, such as mitochondria and plastids.

CYCLOSIS

Cyclosis, or **cytoplasmic streaming,** occurs in living cells. When a living cell is examined with the light microscope, the organelles may appear to be moving as a current within the protoplasm carries them around within the walls. This streaming probably facilitates exchanges of materials within the cell and plays a role in the movement of substances from

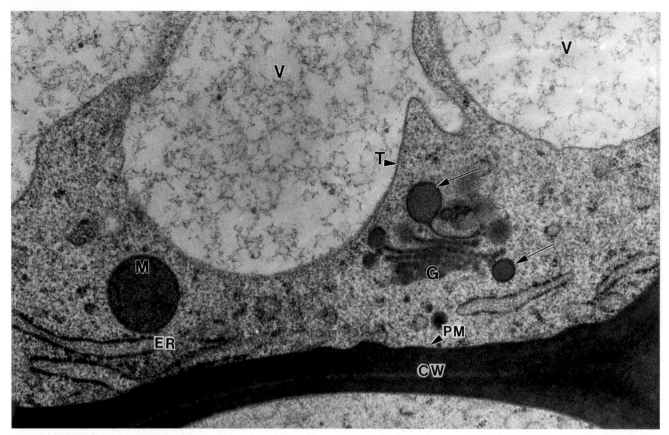

FIGURE 3.14 A small portion of a root cap cell of tobacco ×100,000. V = vacuole; T = vacuolar membrane (tonoplast); G = Golgi body with vesicles (arrows); M = mitochondrion; ER = endoplasmic reticulum; PM = plasma membrane; CW = cell wall. (Electron micrograph courtesy John Z. Kiss)

cell to cell. The precise nature and origin of cyclosis is still not known, but there is evidence that bundles of microfilaments may be responsible for it. Other evidence suggests that it may be related to the transport of cellular substances by microtubules.

CELLULAR REPRODUCTION

The Cell Cycle

When most cells divide, they go through an orderly series of events known as the *cell cycle* (Fig. 3.15). This cycle is usually divided into *interphase* and *mitosis,* mitosis itself being subdivided into four phases. The length of the cell cycle varies with the kind of organism involved, the type of cell within the organism, and temperature and other environmental factors. In most instances, however, interphase may occupy up to 90% or more of the time it takes to complete the cycle.

Interphase

Living cells that are not dividing are said to be in *interphase,* a period during which chromosomes are not visible with light microscopes. It is such cells that have been discussed up to this point.

For many years, immature cells were considered to be "resting" when they were not actually dividing, but we know now that three consecutive periods of intense activity take place during interphase. These intervals are designated as *gap* (or *growth*) *1, synthesis,* and *gap* (or *growth*) *2* periods, usually referred to as G_1, S, and G_2, respectively.

The G_1 period, which is relatively lengthy, begins immediately after a nucleus has divided. During this period, the cell increases in size. Also, ribosomes, RNA, and substances that either inhibit or stimulate the S period that follows are produced. During the S period, the unique process of DNA replication (duplication) takes place. Details of this process and of DNA structure are discussed in Chapter 13. In the G_2 period, mitochondria and other organelles divide, and microtubules and other substances directly involved in mitosis are produced. Coiling and condensation of chromosomes also begin during G_2.

Mitosis

All organisms begin life as a single cell. Almost immediately, this cell usually begins to divide, producing two new cells. These two cells, in turn, divide with each of them

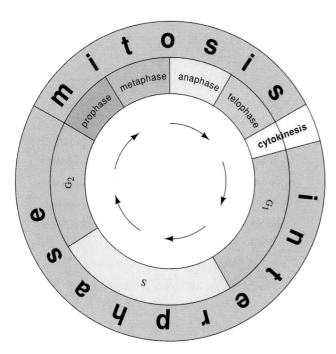

FIGURE 3.15   A diagram of a cell cycle.

membrane, is usually divided into four arbitrary phases, primarily for convenience. Descriptions of the phases follow.

Prophase

The main features of *prophase* (Fig. 3.16 *A*) are (1) the chromosomes become shorter and thicker, and their two-stranded nature becomes apparent; and (2) the nuclear envelope and nucleolus disassociate.

Prophase takes up about as much time as the remaining three phases combined. The beginning of this phase is marked by the appearance of the chromosomes as faint threads in the nucleus. These chromosomes gradually coil or fold into thicker and shorter structures, and soon two strands, or *chromatids*, can be distinguished for each chromosome. The chromatids, which are identical to each other, are themselves independently coiled. These coils appear to tighten and condense until the chromosomes have become relatively short, thick, and rodlike, with areas called **centromeres** holding each pair of chromatids together.

The centromere is located at a constriction on the chromosome (Fig. 3.17). A dense granule called a **kinetochore,** to which spindle fibers become attached, is located near each centromere. When examined with the aid of a light microscope, the centromeres appear to be single structures, but they actually have doubled by the G_2 stage of interphase and simply function as a single unit at this point. They may be located almost anywhere on a chromosome but tend to be toward the middle. Sometimes, other constrictions may appear on individual chromosomes, usually toward one end, giving them the appearance of having extra knobs; these knobs are referred to as *satellites*. The constrictions at the base of the satellites have no known function, but the satellites themselves are useful in helping to distinguish certain chromosomes from others in a nucleus.

As prophase progresses, the nucleolus gradually becomes less distinct and eventually disassociates. By the end of prophase, *spindle fibers,* which consist of microtubules, have developed and extend in arcs between two invisible *poles* located toward the ends of the cell. The tips of the spindle fibers become anchored at the poles. An additional spindle fiber appears to grow out from opposite sides of each centromere until the tips reach the poles. At the conclusion of prophase, the nuclear envelope has been reabsorbed into the endoplasmic reticulum and has totally disassociated.

In certain simpler organisms, such as fungi and algae, and in virtually all animal cells, the cytoplasm just outside the nucleolus contains dual pairs of tiny keg-shaped organelles called *centrioles*. The centrioles are surrounded by microtubules, which radiate out from them and arrange cytoplasmic particles in the vicinity into starlike rays, collectively called an *aster.* At the beginning of prophase, the aster divides into two parts; one part remains at its original location, while the other part migrates around the nuclear envelope to the opposite side. Centrioles and asters have not been detected in the cells of most of the more complex members of the Plant Kingdom.

producing two more cells. This process, called *mitosis* (shown in Fig. 3.16), ensures that the two new cells (*daughter cells*) resulting from each cell undergoing mitosis have precisely equal amounts of DNA and certain other substances duplicated during interphase. Mitosis occurs in an organism until it dies. Strictly speaking, mitosis refers to the division of the nucleus alone, but with a few exceptions seen in algae and fungi (discussed in Chapters 18 and 19), the division of the remainder of the cell, called **cytokinesis,** normally accompanies or follows mitosis. Both processes will be considered together here.

In higher plants, such as conifers and flowering plants, mitosis occurs in specific regions, or tissues, called **meristems** (shown in Fig. 4.1). Meristems are found in the root and stem tips and also in a thin, perforated, and branching cylinder of tissue called the **vascular cambium** (often referred to simply as the **cambium**), located in the interior of some stems and roots a short distance from the surface. In some herbaceous and most woody plants, a second meristem similar in form to the cambium lies between the cambium and the outer bark. This second meristem is called the **cork cambium.** These specific tissues are discussed in chapters 4, 5, and 6.

When mitosis takes place, it makes no difference how many chromosomes are in the nucleus. The daughter cells that result from the process each have exactly the same number of chromosomes and DNA distribution as the parent cell. It is a continuous process, which may take as little as 5 minutes or as long as several hours from start to finish. Typically, however, it takes from 30 minutes to 2 or 3 hours. The process, which is initiated with the appearance of a ringlike *preprophase band* of microtubules just beneath the plasma

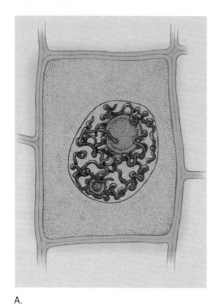

A.

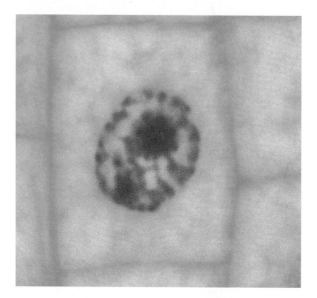

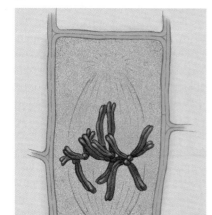

B.

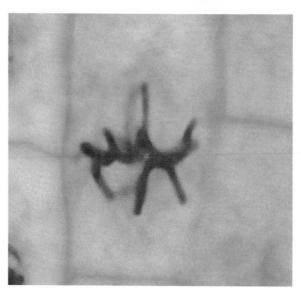

FIGURE 3.16 The phases of mitosis as seen in onion root-tip cells. These chromosomes are about 4 micrometers long. *A.* Cell (center) in prophase. *B.* Cell (center) in metaphase.

Metaphase

The main feature of *metaphase* (Fig. 3.16 *B*) is the alignment of the chromosomes in a ring around the circumference of the cell. The ring is more or less perpendicular to the axis of the spindle fibers, like the equator of a globe.

As indicated in our discussion of prophase, spindle fibers can be seen in the area previously occupied by the nucleus after the nuclear envelope has disassociated. They form a structure that looks like an old-fashioned spinning top made of fine threads. Collectively, the spindle fibers are referred to as the **spindle.** The chromosomes become aligned so that their centromeres are in a plane roughly in the center of the cell. This invisible circular plate, called the equator, is analogous to the equator of the earth. At the end of metaphase, the centromeres holding the two strands (*sister chromatids*) of each chromosome together split lengthwise.

Anaphase

The main feature of *anaphase*—the briefest of the phases—is that the sister chromatids of each chromosome separate and are pulled to opposite poles (Fig. 3.16 *C*).

Until the end of metaphase, the sister chromatids of each chromosome have been united at their centromeres. Anaphase begins with all the sister chromatids separating in unison and moving toward the poles. The chromatids, which after separation at their centromeres are called *daughter chromosomes,* are pulled toward the poles as their spindle fibers gradually shorten. The shortening occurs as a result of material continuously being removed from the polar ends of the spindle fibers. If the centromere is in the center of the daughter chromosome, it leads the way, with the rest of the chromosome assuming a V shape as it appears to drag in the cytoplasm.

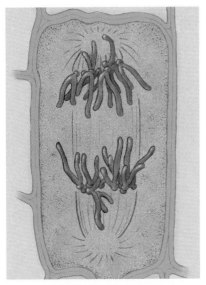

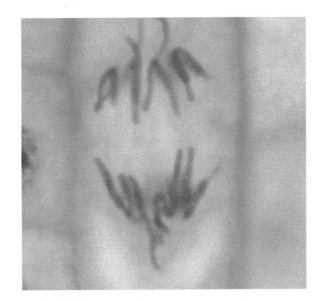

C.

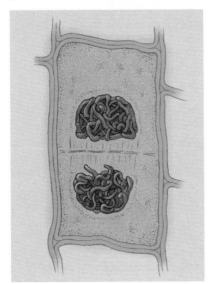

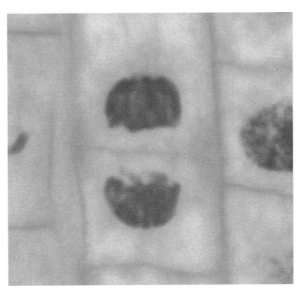

D.

FIGURE 3.16
continued. *C.* Cell (center) in anaphase. *D.* Cell (center) in telophase.

All of the chromosomes separate and move at the same time. Although experiments have shown that a chromosome will not migrate to a pole if the fiber attached to its centromere is severed, other experiments have shown that the chromosomes will separate from one another but not move to the poles, even if no spindle is present. The force or forces involved in the latter phenomenon have not yet been identified.

Telophase

The five main features of *telophase* (Fig. 3.16 *D*) are (1) each group of daughter chromosomes becomes surrounded by a reassociated nuclear envelope; (2) the daughter chromosomes become longer and thinner and are finally indistinguishable; (3) reassociated nucleoli appear; (4) many of the spindle fibers disassociate; and (5) a cell plate forms.

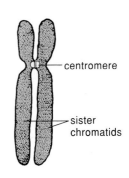

centromere

sister chromatids

FIGURE 3.17 Parts of a chromosome.

Microscapes

Scientists use the scanning electron microscope (SEM) to study the details of many different types of surfaces. Unlike the light microscope or even a transmission electron microscope that forms images by passing either a beam of light or electrons through a thin slice of fixed tissue, the SEM's great advantage is its ability to look at surfaces of specimens and provide topographical detail not possible with other types of microscopy.

The basic concept of a scanning electron microscope is that a finely focused beam of electrons is scanned across the surface of the specimen. The high velocity electrons from the beam create an energetic interaction with the surface layers. These electron-specimen interactions generate particles that are emitted from the specimen and can be collected with a detector and sent to a TV screen (cathode ray tube). Particles that form the typical scanning electron image are called secondary electrons because they come from the electrons in the specimen itself. The more electrons a particular region emits, the brighter the image will be on the TV screen. The end result, therefore, is brightness associated with surface characteristics and an image that looks very much like a normally illuminated subject. SEM images typically contain a good deal of topographical detail because the electrons that are emitted and produced on the TV screen represent a one-for-one correspondence with the contours of the specimen.

All scanning electron images have one very distinctive characteristic because of this feature of electron emission and display—the images are three dimensional

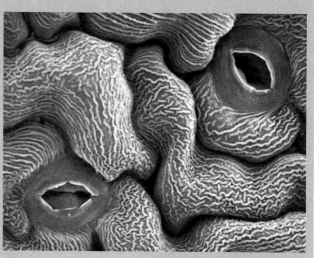

BOX FIGURE 3.1A.   Scanning electron micrograph of the surface of a sepal (modified leaf) from a flower of the mouse-ear cress, *Arabidopsis thaliana*, ×2,000.
(Electron micrograph by Daniel Scheirer)

rather than the flat 2-dimensional images obtained from other types of microscopes. The images can be understood even by the lay person because the eye is accustomed to interpreting objects that are in three dimensions.

Take, for instance, a leaf surface, which looks smooth with an ordinary light microscope. But with a scanning electron microscope, the leaf surface is a rich composition of undulating cell walls, cells joined together like pieces of a jigsaw puzzle, squiggly ridges of waxes that look like frosting decorations on a cake, and lens-shaped stomatal pores (Box Figure 3.1 *A*). The

---

The transition from anaphase to telophase is not distinct, but telophase is definitely in progress when elements of new nuclear envelopes appear around each group of daughter chromosomes at the poles. These elements gradually form intact envelopes as the daughter chromosomes return to the diffuse, indistinct threads seen at the onset of prophase. The new nucleoli appear on specific regions of certain chromosomes.

During telophase, the spindle gradually disappears and a set of shorter fibers (fibrils), composed of micro-tubules, develops in the region of the equator between the daughter nuclei. This set of fibrils, which appears somewhat keg-shaped, is called a **phragmoplast.** Golgi bodies produce small vesicles containing raw materials for the cell wall and membranes. Some of these vesicles, which resemble tiny droplets of fluid when viewed with a light microscope, are directed toward the center of the spindle (equator) by the remaining spindle fibers.

The microtubules apparently trap the Golgi body vesicles, which then fuse together into one large, flattened but

stomatal pore even provides a window into the interior of the leaf where deeper cellular layers are visible. Or look at the tentacles seen in Box Figure 3.1 *B* that remind us of some sinister sea creature. No stinging tentacles here but rather the surface of a small flower of a common weed called mouse-ear cress (*Arabidopsis*). The "tentacles" are actually stigmatic papillae that serve to trap pollen grains that are released from the pollen sacs of the flower. (*See* Chapter 8 for details of flower structure.)

While biologists utilize the SEM extensively, other types of scientists put it to work in diverse ways as well, whether looking at "moon rocks" brought to Earth by the Apollo astronauts or studying the impact craters created by micrometeorite projectiles striking the space shuttle's heat resistant tiles. Recently, a textile technologist in England examined a piece of the frayed linen tunic of King Tut, the ancient Egyptian boy Pharaoh whose tomb was discovered in 1922. Apparently, the tunic had either been washed about 40 times in water or had been washed less frequently in a solution of sodium carbonate, a chemical used to whiten as it cleans. Additionally, unlike the clothing of ordinary people, King Tut's tunic had few mends in it—not surprising considering the wealth of the deceased. The tomb was filled with golden treasures, as well as wooden chests containing his clothes and footwear.

Whether used by biologists or material scientists, the scanning electron microscope provides a stunning view of the previously unseen, but nevertheless real, world. As the beauty of nature is seen for the first time in startling detail, micrographs do indeed become "microscapes."

BOX FIGURE 3.1B.   Scanning electron micrograph of the surface of the stigma from a flower of the mouse-ear cress, *Arabidopsis thaliana*, ×200.

(Electron micrograph by Daniel Scheirer)

hollow structure called a **cell plate** (Fig. 3.18). Carbohydrates in the vesicles are synthesized into two new primary cell walls and a *middle lamella.* The middle lamella is shared by what now have become two new *daughter cells.* The cell plate grows outward until it contacts and unites with the plasma membrane of the mother cell. *Plasmodesmata* (tiny strands of protoplasm that extend through the walls between cells—see Fig. 3.5) are apparently formed as portions of the endoplasmic reticulum are trapped between fusing vesicles of the cell plate.

New plasma membranes develop on either side of the cell plate as it forms, and new cell wall materials are deposited between the middle lamella and the plasma membranes. These new walls are relatively flexible and remain so until the cells increase to their mature size. At that time, additional cellulose and other substances may be added, forming a secondary cell wall interior to the primary wall. In some instances, cell plate formation does not accompany division of the nucleus.

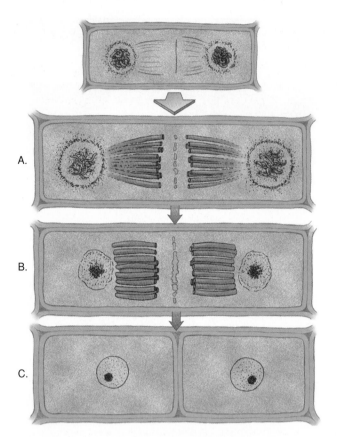

FIGURE 3.18   How a cell plate is formed. *A*. During telophase a *phragmoplast* (a complex of microtubules and endoplasmic reticulum) develops between the two daughter nuclei. The microtubules trap Golgi body vesicles along a central plane. *B*. The vesicles fuse into a flattened hollow structure that becomes a *cell plate*. *C*. The cell plate grows outward; two primary cell walls and two plasma membranes form. When the cell plate reaches the mother cell walls, the plasma membranes unite with the existing plasma membrane and the production of two daughter cells is complete.

COMMUNICATION BETWEEN CELLS

Although each living cell is capable of independently carrying on complex activities, it is essential that these activities be coordinated through some means of communication among all the living cells of an organism. Living plant cells are in contact with each other via the plasmodesmata. A plasmodesma extends through a minute hole in the walls of two adjacent cells. The hole is located within a pair of more or less circular depressions (one in each wall) where the adjacent secondary walls are very thin or nonexistent. These paired areas, where little more than the middle lamella and a small amount of primary wall separates abutting cells, are called pits (shown in Fig. 4.7). Although most pit pairs are relatively simple in structure, some, called *bordered pits,* have blisterlike, arched secondary wall "covers" over the

thin areas, giving them the appearance of tiny doughnuts in surface view. The translocation of sugars, amino acids, ions, and other substances occurs through the plasmodesmata.

HIGHER PLANT CELLS VERSUS ANIMAL CELLS

All animals have either internal or external support for their tissues from a skeleton of some kind. Animal cells do not have cell walls; instead, the plasma membrane, called the *cell membrane* by most zoologists (animal scientists), forms the outer boundary of animal cells. Higher plant cells have walls that are thickened and rigid to varying degrees, with a framework of cellulose fibrils. Higher plant cells also have plasmodesmata connecting the protoplasts with each other through microscopic holes in the walls. Animal cells lack plasmodesmata since they have no walls. When higher plant cells divide, a cell plate is formed during the telophase of mitosis, but cell plates do not form in animal cells, which divide by pinching in two.

Other differences pertain to the presence or absence of certain organelles. Centrioles, for example, the tiny, paired, keg-shaped organelles found just outside the nucleus, occur in all animal cells but are generally absent from higher plant cells. Plastids, common in plant cells, are not found in animal cells. Vacuoles, which are often large in plant cells, are either small or absent in animal cells.

Summary

1. All living organisms are composed of cells. Cells are modified according to the functions they perform; some live for a few days while others live for many years.

2. The discovery of cells is associated with the development of the microscope. Robert Hooke was the first to use the word *cells* for boxlike compartments he saw in cork in 1665 with the aid of a primitive microscope. Leeuwenhoek and Grew reported frequently during the next 50 years on the existence of cells in a variety of tissues.

3. Lamarck concluded in 1809 that all living tissue is composed of cells, and Dutrochet, in 1824, reinforced Lamarck's conclusions. Brown discovered in 1833 that all cells contain a nucleus, and shortly thereafter Schleiden observed a nucleolus within a nucleus. Schleiden and Schwann are generally credited with developing the cell theory during 1838 and 1839. The theory holds that all living organisms are composed of cells and that cells form a unifying structural basis of organization.

4. In 1858, Virchow contended that every cell comes from a preexisting cell and that there is no spontaneous generation of cells from dust. Pasteur confirmed Virchow's contentions in 1862, with experiments involving microorganisms in swan-necked flasks. Pasteur later proved that fermentation involves activity of yeast cells. Buchner, in 1897, found that yeast cells do not need to be alive for fermentation to occur. This led to the discovery of organic catalysts called *enzymes.*

5. Light microscopes, which magnify up to 1,500 times, utilize light and glass or calcium fluoride lenses to achieve magnification. Compound microscopes require material being viewed to be sliced thinly enough for the light to pass through. Stereomicroscopes can be used to view opaque objects; most magnify up to 30 times.

6. Electron microscopes utilize electromagnetic lenses and a beam of electrons within a vacuum to achieve magnification. Transmission electron microscopes magnify up to 200,000 or more times. Scanning electron microscopes, which can be used with opaque objects, usually magnify up to 10,000 times.

7. Eukaryotic cells are the subject of this chapter. Prokaryotic cells, which lack some of the features of eukaryotic cells, are discussed in Chapter 17.

8. Scanning tunneling microscopes use a minute probe to scan surfaces at a width as narrow as that of two atoms.

9. Cells are minute; most vary in diameter from 10 to 100 micrometers. They number into the billions in larger organisms, such as trees. Plant cells are bounded by walls. The living part of the cell within the walls, protoplasm, has two main components: the nucleus, which functions as a control center, and cytoplasm, a fluid in which various membrane-bounded organelles are suspended.

10. A pectic middle lamella is sandwiched between the primary cell walls of adjacent cells. The primary wall and also the secondary cell wall, often added inside the primary wall, are composed mostly of cellulose, but secondary cell walls may also contain lignin and other substances.

11. A flexible plasma membrane, which is sandwichlike and often forms folds, constitutes the outer boundary of the protoplasm. It regulates the substances that enter and leave the cell.

12. The endoplasmic reticulum is a system of flattened sacs and tubes associated with the storing and transporting of protein and other cell products. Granular particles called *ribosomes,* which function in protein synthesis, may line the outer surfaces of the endoplasmic reticulum.

13. Dictyosomes, or Golgi bodies, which constitute the Golgi apparatus, are organelles that appear as groups of sacs and function as collecting and packaging centers for the cell.

14. Mitochondria are tiny, numerous organelles that are bounded by two membranes with inner platelike folds called *cristae;* they are associated with respiration.

15. Plastids are larger organelles that are green (chloroplasts), orange or red (chromoplasts), or colorless (leucoplasts). Chloroplasts contain enzymes, in a matrix called the *stroma,* and grana, which are stacks of coin-shaped membranes (thylakoids) containing green chlorophyll pigments; photosynthesis occurs in the thylakoids. Plastids develop from proplastids, which divide frequently, and also arise from the division of mature plastids.

16. Microtubules are tiny tubelike structures that are commonly found just inside the plasma membrane. They apparently control the addition of cellulose to the cell wall. Microfilaments are associated with cytoplasmic streaming.

17. The nucleus is bounded by a nuclear envelope consisting of two membranes that are perforated by numerous pores. Within the nucleus are a fluid called *nucleoplasm,* one or more spherical nucleoli, and thin strands of chromatin, which condense and become chromosomes when nuclei divide. Each species of organism has a specific number of chromosomes in each cell.

18. As much as 90% of the volume of a mature cell may be taken up with a vacuole or vacuoles that are bounded by a vacuolar membrane (tonoplast) and contain a watery fluid called *cell sap.* Cell sap contains dissolved substances and frequently also water-soluble pigments called *anthocyanins,* which are red or blue. Movement of cytoplasm within a cell in a streaming motion is called *cyclosis.*

19. In the cell cycle, cells that are not dividing are said to be in interphase, which is subdivided into three periods of intense activity that precede mitosis, or division of the nucleus. Mitosis is usually accompanied by division of the rest of the cell and takes place in plant regions known as *meristems.*

20. Mitosis is arbitrarily divided into four phases: (1) prophase, in which the chromosomes and their two-stranded nature become apparent and the nuclear envelope breaks down; (2) metaphase, in which the chromosomes become aligned at the equator of the cell and a spindle composed of spindle fibers is fully developed, with some spindle fibers being attached to the chromosomes at their centromeres; (3) anaphase, in which the sister chromatids of each chromosome (now called *daughter chromosomes*) separate lengthwise, with each group of daughter chromosomes migrating to opposite poles of the cell; and (4) telophase, in which each group of daughter chromosomes becomes surrounded by a nuclear envelope, thus becoming new

nuclei, and a wall dividing the daughter nuclei forms, creating two daughter cells.

21. The development of the dividing wall is initiated by the appearance of a set of short fibrils constituting the phragmoplast. Droplets, or vesicles, of pectin merge, forming a cell plate that grows to become the middle lamella of the new cell wall.

22. Living cells are in contact with one another via fine strands of cytoplasm called *plasmodesmata,* which often extend through minute holes in paired wall depressions called *pits.*

23. Animal cells differ from those of higher plants in not having a wall, plastids, or large vacuoles. Also, they have keg-shaped centrioles in pairs just outside the nucleus and pinch in two instead of forming a cell plate when they divide.

Review Questions

1. How is cellular structure beneficial to plants and animals?

2. Of what importance to the cell is the wall?

3. What is the difference between protoplasm and cytoplasm?

4. How can you distinguish between cytoplasm and vacuoles?

5. Of what are chloroplasts composed? What is the function of each component?

6. What is the function of a cell nucleus? How does it perform its function?

7. What are plasmodesmata, and what is their importance to living plants?

8. What are pits? Where are they located?

9. What are the differences and similarities between plant and animal cells?

10. Are prophase and telophase of mitosis exactly the reverse of one another? Explain.

Discussion Questions

1. Would you consider any one type of cell more useful than another? Why?

2. After you have completed your introductory plant science course, do you believe you will be able to determine the function of each of a cell's organelles in a laboratory? Explain.

Additional Reading

Alberts, B., et al. 1994. *Molecular biology of the cell*, 3d ed. New York: Garland Publishing, Inc.

Baker, N. R., and J. Barber (Eds.). 1984. *Chloroplast biogenesis.* New York: Elsevier Science Publishing Co.

Buvat, R. 1989. *Ontogeny, cell differentiation and structure of vascular plants.* New York: Springer-Verlag.

Cross, P. C., and K. L. Mercer. 1993. *Cell and tissue ultrastructure: A unique perspective.* New York: W. H. Freeman.

DeRobertis, E. D., and E. M. DeRobertis Jr. 1987. *Cell and molecular biology*, 8th ed. Baltimore, MD: Williams and Wilkins.

Loewy, A. G., et al. 1991. *Cell structure and function*, 3d ed. Philadelphia, PA: Saunders College Publishing.

Murray, A., and T. Hunt. 1993. *Cell cycle: An introduction.* New York: Oxford University Press.

Ohki, S. (Ed.). 1992. *Cell and model membrane interactions.* New York: Plenum.

Sadava, D. E. 1993. *Cell biology: Organelle structure.* Boston, MA: Jones & Bartlett.

Sheeler, P., and D. E. Bianchi. 1987. *Cell and molecular biology,* 3d ed. New York: John Wiley & Sons, Inc.

Smith, C. A., and E. J. Wood (Eds.). 1991. *Cell biology.* New York: Chapman and Hall.

Stein, W. D., and F. Bronner (Eds.). 1989. *Cell shape: Determinants, regulation, and regulatory role.* San Diego, CA: Academic Press.

Sussex, I., et al. (Eds.). 1985. *Plant cell interactions.* New York: Cold Spring Harbor.

Wolfe, S. L. 1993. *Molecular and cell biology.* Belmont, CA: Wadsworth Publishing Co.

Chapter Outline

A hanging heliconia (Heloconia rostrata), native to Peru and Ecuador. There are about 200 known species of these striking plants, many of which are becoming increasingly popular as ornamentals in tropical and subtropical areas.

Tissues

4

O v e r v i e w————————————

A discussion of meristems (apical meristems, vascular cambium, cork cambium, intercalary meristems) and nonmeristematic tissues (parenchyma, collenchyma, sclerenchyma, secretory tissues, xylem, phloem, epidermis, periderm) forms the body of this chapter.

S o m e   L e a r n i n g   G o a l s

1. Know the meristems present in plants and where they are found.
2. Learn the conducting tissues of plants and the function of each cell component.
3. Learn tissues of plants that neither are meristematic nor function in conduction at maturity.

I once was privileged to have a new house constructed for me on a vacant lot. I followed the stages of construction with considerable interest. First, a foundation was laid; then trucks arrived with various building materials, and construction of a frame began. This was followed by the installation of plumbing, electrical wiring, windows, heating and air-conditioning units, vents, and various other devices. Finally, waterproof walls and a roof were added and, upon occupation, food and other materials were stored in their appropriate niches.

In a sense, the growth of a plant from a seed is something like the construction of a house. Using raw materials from the soil and a superb manufacturing process, each plant develops a framework; "plumbing"; and a waterproof covering that includes "windows," vents, means of disposing of wastes, and food storage areas. Even a form of air-conditioning, which enables plants to survive and thrive in the hottest summer sun, is included in each mature plant package. The building components of the framework, plumbing, and related features of plants form the body of this chapter.

There are many interesting modifications of higher plants discussed in the three chapters that follow this one but, regardless of their outer form, most plants have three major groups of organs: **roots, stems,** and **leaves.** Each of these organs is composed of **tissues,** which are defined as "groups of cells performing a common function." Any plant organ may be composed of a number of different tissues, each of which is classified according to structure, origin, or function. In addition, three basic tissue patterns occur in roots and stems (see *woody dicots, herbaceous dicots,* and *monocots,* discussed in chapters 5 and 6). The following are major kinds of tissues found in higher plants. The specific types of cells associated with each tissue, as well as illustrations of them, are included in the discussions that follow the classification.

MERISTEMATIC TISSUES

Meristems or *meristematic tissues,* are tissues in which cells actively divide (Fig. 4.1). As new cells are produced, they typically are small, six-sided, boxlike structures, each with a proportionately large nucleus in the center and tiny vacuoles, or no vacuoles at all. As they mature, however, they assume many different shapes and sizes, each related to the cell's ultimate function, and the vacuoles may become more than 95% of the volume of the cell.

Apical Meristems

Apical meristems are found at, or near, the tips of roots and shoots, which increase in length as the apical meristems produce new cells. This type of growth is known as *primary growth.* Three *primary meristems,* as well as embryo leaves and buds, develop from each apical meristem. These primary meristems are called **protoderm, ground meristem,** and **procambium.** The tissues they produce are called **primary tissues.** Note their locations in Figure 4.1; they are discussed in chapters 5 and 6.

Lateral Meristems

As just indicated, apical meristems are involved in *primary growth* in the plant body. The vascular cambium and cork cambium, discussed next, are *lateral meristems* involved in *secondary growth.*

Vascular Cambium

The **vascular cambium,** often referred to simply as the **cambium,** forms a thin, frequently branching cylinder that, except at the tips, runs the length of the roots and stems of most perennial plants and many herbaceous annuals. It is primarily responsible for the production of tissues that increase the girth of a plant. The individual remaining cells of the cambium are referred to as *initials,* while their sister cells are called *derivatives.* Both the cambium and its initials are discussed in chapters 5 and 6.

Cork Cambium

The **cork cambium,** like the vascular cambium, forms a thin cylinder that runs the length of roots and stems of woody plants. It is located to the exterior of the vascular cambium. The cork cambium is primarily responsible for producing the outer bark of woody plants; it is discussed in chapters 5 and 6. The tissues laid down by the vascular cambium and the cork cambium are produced *after* the primary tissues have matured and are called *secondary tissues.*

Intercalary Meristems

Grasses and related plants have neither a vascular cambium nor a cork cambium. They do, however, have apical meristems and, in the vicinity of **nodes** (leaf attachment areas—discussed in chapter 7), they have other meristematic tissues called *intercalary meristems,* which occur at intervals along stems. Intercalary meristems, like apical meristems, produce increases in stem length.

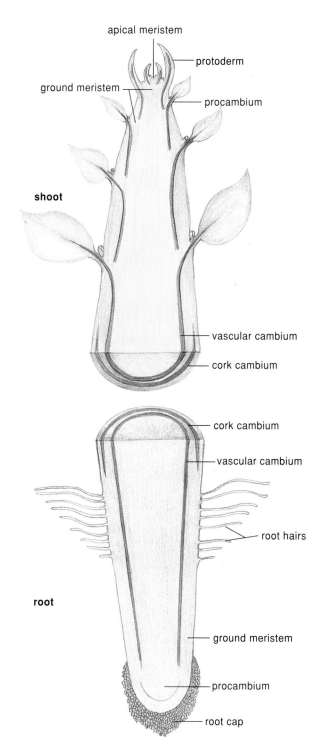

FIGURE 4.1   Diagram of the longitudinal axis of a plant, showing the location of meristems.

NONMERISTEMATIC TISSUES

Nonmeristematic tissues are composed of cells that, after being produced by meristems, assumed various shapes and sizes related to their functions as they developed and matured. Some of these tissues consist of only one kind of cell, while others may have two to several kinds of cells. Simpler

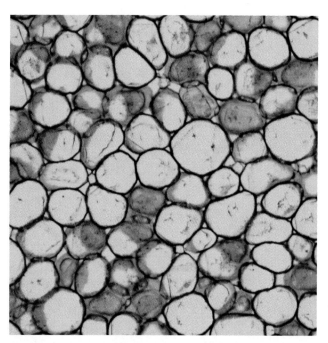

FIGURE 4.2   Parenchyma cells. They are more or less spherical when first formed but end up with an average of 14 sides at maturity, ×100.

(Photomicrograph by G.S. Ellmore)

basic types of nonmeristematic tissues are discussed first, followed by those that are more complex.

Simple Tissues

Parenchyma

Parenchyma tissue is composed of *parenchyma cells* (Fig. 4.2), which are the most abundant of the cell types; they are found in almost all major parts of higher plants. They are more or less spherical in shape when they are first produced, but when all the parenchyma cells push up against one another, their thin, pliable walls are flattened at the points of contact. As a result, parenchyma cells assume various shapes and sizes, with the majority having 14 sides. They tend to have large vacuoles and may contain starch grains, oils, tannins (tanning or dyeing substances), crystals, and various other secretions.

Spaces commonly occur between parenchyma cells; in fact, in water lilies and other aquatic plants, the intercellular spaces are quite extensive and form a network throughout the entire plant. This type of parenchyma tissue—with extensive connected air spaces—is referred to as *aerenchyma*.

Parenchyma cells containing numerous chloroplasts (as found in leaves) form **chlorenchyma** tissue. The chief function of chlorenchyma tissue is photosynthesis, while parenchyma tissues without chloroplasts function primarily in food or water storage. The edible parts of most fruits and vegetables consist largely of parenchyma. Some parenchyma cells develop irregular extensions of the inner wall

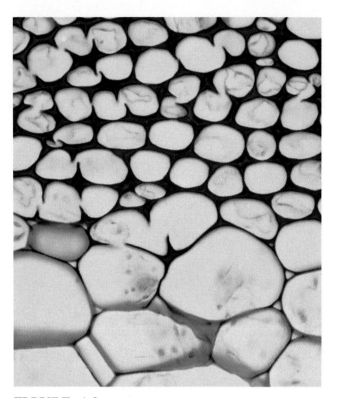

FIGURE 4.3   Collenchyma cells. Note the unevenly thickened walls, ×100.

(Photomicrograph by G.S. Ellmore)

that greatly increase the surface area of the plasma membrane. Such cells, called *transfer cells,* apparently play a role in transferring dissolved substances between adjacent cells. Many parenchyma cells live a long time; in some cacti, for example, they may live to be over 100 years old.

Mature parenchyma cells can divide. When a plant is damaged or wounded, the capacity of parenchyma cells to multiply is especially important in repair of tissues.

Collenchyma

Collenchyma cells (Fig. 4.3), like parenchyma cells, have living protoplasm and may remain alive a long time. They are distinguished from parenchyma cells primarily by the thicker walls, the thickness usually varying enough that in cross section the walls appear uneven. Collenchyma cells often occur just beneath the epidermis; typically, they are longer than they are wide and their walls are pliable as well as strong. They provide flexible support for both growing organs and mature organs, such as leaves and floral parts. The "strings" of celery that get stuck in our teeth, for example, are composed of collenchyma cells.

Sclerenchyma

Sclerenchyma tissue consists of cells that have thick, tough, secondary walls, normally impregnated with **lignin.** Most sclerenchyma cells are dead at maturity and function in support. Two forms of sclerenchyma occur: **sclereids** and

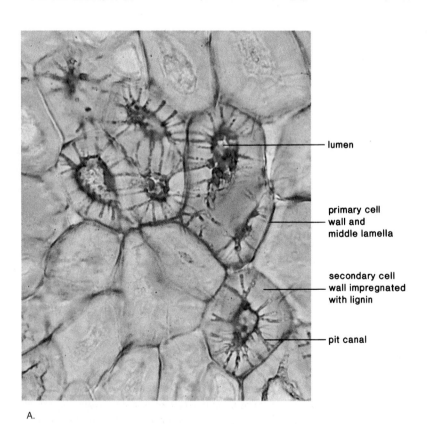

— lumen

— primary cell wall and middle lamella

— secondary cell wall impregnated with lignin

— pit canal

A.

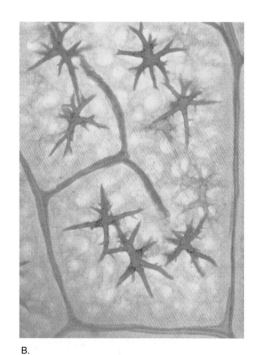

B.

FIGURE 4.4   *A.* Sclereids (stone cells) of a pear in cross section. *B.* Sclereids in the leaf of a wheel tree. Both photomicrographs ca. ×1,000.

(*B.* by G.S. Ellmore)

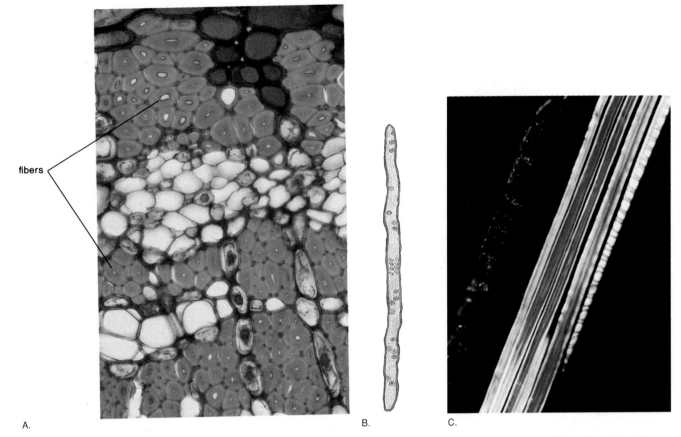

fibers

A.

B.

C.

FIGURE 4.5 Fibers. *A*. A cross section of a portion of stem tissue from a linden tree. Note the thickness of the walls of the darker fibers. *B*. A single fiber in longitudinal section. *C*. A longitudinal section through fibers in a *Welwitschia* leaf. Photomicrographs ca. ×1,000.

(*C*. by G.S. Ellmore)

fibers. Sclereids (Fig. 4.4) may be randomly distributed in other tissues. For example, the slightly gritty texture of pears is due to the presence of groups of sclereids, or *stone cells,* as they are sometimes called. They tend to be about as long as they are wide and sometimes occur in specific zones (e.g., the margins of camellia leaves) rather than being scattered within other tissues.

Fibers (Fig. 4.5) may be found in association with a number of different tissues in roots, stems, leaves, and fruits. They are usually much longer than they are broad and have a proportionately tiny cavity, or *lumen,* in the center of the cell. At the present time, fibers from more than 40 different families of plants are in commercial use in the manufacture of textile goods, ropes, string, canvas, and similar products. Archaeological evidence indicates that humans have been using plant fibers for at least 10,000 years.

Secretory Cells and Tissues

All cells secrete certain substances that, if allowed to accumulate internally, can damage the protoplasm. Such materials must be either isolated from the protoplasm of the cells in which they originate or moved outside of the plant body. Often, the substances consist of waste products that are of no further use to the plant, but some substances, such as nectar,

perfumes, and *plant hormones* (discussed in Chapter 11), are vital to normal plant functions.

Secretory cells may function individually or as part of a **secretory tissue.** Secretory cells or tissues, which often are derived from parenchyma, can occur in a wide variety of places in a plant. Among the most common secretory tissues are those that secrete nectar in flowers; oils in citrus, mint, and many other leaves; mucilage in the glandular hairs of sundews and other insect-trapping plants; latex in members of several plant families, such as the Spurge Family; and resins in coniferous plants, such as pine trees. Latex and resins are usually secreted by cells lining tubelike ducts that form networks throughout certain plant species (see Fig. 6.11). Some plant secretions, such as pine resin, rubber, mint oil, and opium, have considerable commercial value.

Complex Tissues

Most of the tissues thus far discussed consist of one kind of cell, but a few important tissues are always composed of two or more kinds of cells and are sometimes referred to in a general sense as *complex tissues.* The two most important complex tissues in plants, *xylem* and *phloem,* function primarily in the transport of water, ions, and soluble food substances

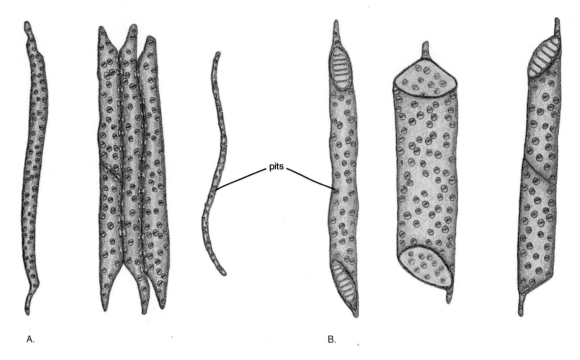

FIGURE 4.6   Lengthwise views of water-conducting cells of xylem. *A.* Tracheids. *B.* Vessel members.

throughout the plant. Some complex tissues are produced by apical meristems, but the bulk of such tissues in woody plants is produced by the vascular cambium and is often referred to as *vascular tissue.*

The *epidermis,* which forms a protective layer covering all plant organs, consists primarily of parenchyma or parenchymalike cells, but it also includes specialized cells involved in the movement of water and gases in and out of plants; secretory glands, various hairs, cells in which crystals are isolated, and others that greatly increase absorptive parts of roots are all frequent components of an epidermis. Accordingly, the epidermis is discussed in this section.

Periderm, which comprises the outer bark of woody plants, consists mostly of *cork* cells, but it is included in this discussion because it contains pockets of parenchyma cells.

Xylem

Xylem tissue, which is an important component of the "plumbing" system of a plant, conducts water and dissolved substances throughout all organs; it consists of a combination of parenchyma cells, fibers, *vessels, tracheids,* and *ray cells* (Fig. 4.6). **Vessels** are long tubes made up of individual cells called **vessel members** that are open at each end, with bars of wall material extending across the open areas in some instances. The cells are joined end to end, forming the tubes.

Tracheids, which, like vessel members, are dead at maturity and possess thick secondary cell walls, are tapered at each end and have no openings similar to those of vessels. The ends overlap with those of other tracheids, and wherever two tracheids are in contact with one another, pairs of *pits* (Fig. 4.7) are usually present. Pits are areas in which no secondary wall material has been deposited, and, as indicated in

Chapter 3, pit pairs permit water to pass from cell to cell. How pit pairs function in permitting water to pass between adjacent cells is shown in Figure 4.8.

In certain kinds of plants, such as cone-bearing trees, the xylem is composed almost entirely of tracheids. The walls of many tracheids have spiral thickenings on them that are easily seen with the light microscope. Most conduction is up and down, but some lateral (sideways) conduction also occurs. The lateral conduction takes place in the **rays.** Ray cells, which also function in food storage, are actually long-lived parenchyma cells that are produced in horizontal rows by special *ray initials* of the vascular cambium. In woody plants, the rays radiate out from the center of stems and roots like the spokes of a wheel (see figs. 6.6 and 6.9).

Phloem

Phloem tissue (Fig. 4.9), which functions primarily in the conduction throughout the plant of dissolved food substances, is mostly composed of **sieve-tube members** and **companion cells,** which have no secondary walls. Phloem is derived from the parent cells of the cambium, which also produce xylem cells; it often also includes fibers, parenchyma, and ray cells. Sieve-tube members, like vessel members, are laid end to end, forming **sieve tubes.** Unlike vessel members, however, the end walls have no large openings; instead, the walls are full of small pores through which the cytoplasm extends from cell to cell. The porous regions of sieve members are called **sieve plates.** Sieve-tube members have no nuclei at maturity, despite the fact that their cytoplasm is very active in the conduction of food materials in solution throughout the plant. Apparently, the adjacent companion cells form a very close relationship with the sieve tubes next to them and function in some manner that brings about the conduction of the food.

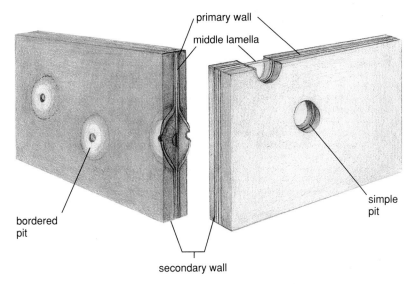

primary wall

middle lamella

bordered pit

secondary wall

simple pit

FIGURE 4.7 Pits. Pits are depressions or cavities in cell walls where the secondary wall does not form. There may be from one or two to several thousand in a cell. They always occur in pairs, with one on each side of the middle lamella. Some, called *bordered pits* (*left*), bulge out from the wall and resemble doughnuts in surface view, while others, called *simple pits* (*right*), do not bulge.

Living sieve-tube members contain a polymer called *callose* that stays in solution as long as the cell contents are under pressure. If an insect such as an aphid injures a cell, however, the pressure drops and the callose precipitates. The callose and a phloem protein are then carried to the nearest sieve plate where they form a plug that prevents leaking of the sieve tube contents.

Sieve cells, which are found in ferns and cone-bearing trees, are similar to sieve-tube members but tend to overlap at their ends rather than form continuous tubes. Like sieve-tube members, they have no nuclei at maturity, and they also have no adjacent companion cells. They do have equivalent adjacent parenchyma cells, called *albuminous cells,* which apparently function in the same manner as companion cells.

Epidermis

The outermost layer of cells of all young plant organs is called the **epidermis.** Since it is in direct contact with the environment, it is subject to modification by environmental factors and often includes several different kinds of cells. The epidermis is usually one cell thick, but a few plants produce aerial roots called **velamen** roots (e.g., orchids) in which the epidermis may be several cells thick, with the outer cells functioning something like a sponge. Such a multiple-layered epidermis also occurs in the leaves of some tropical figs and members of the Pepper Family, where it protects a plant from desiccation.

Most epidermal cells secrete a fatty substance called **cutin** within and on the surface of the outer walls. Cutin forms a protective layer called the **cuticle** (Fig. 4.10). The thickness of the cuticle (or, more importantly, wax secreted on top of the cuticle by the epidermis) to a large extent determines how much water is lost through the cell walls by evaporation. The cuticle is also exceptionally resistant to bacteria and other disease organisms and has been recovered from fossil plants millions of years old. The waxes deposited

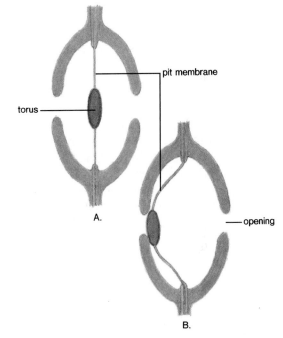

pit membrane

torus

opening

A.

B.

FIGURE 4.8 How water flow is controlled in adjacent pairs of bordered pits. The pits are separated by a *pit membrane* consisting of the middle lamella and two thin layers of primary walls. *A.* Water moves relatively freely through the pit openings and pit membrane when the *torus* (a thickened region of the pit membrane) is in the center. *B.* If the flexible pit membrane swings to one side so that the torus blocks an opening, water movement through the pit pair is restricted.

on the cuticle in a number of plants apparently reach the surface through microscopic channels in the cell walls. The susceptibility of a plant to herbicides may depend on the thickness of these wax layers. Some wax deposits are extensive enough to have commercial value. Carnauba wax, for

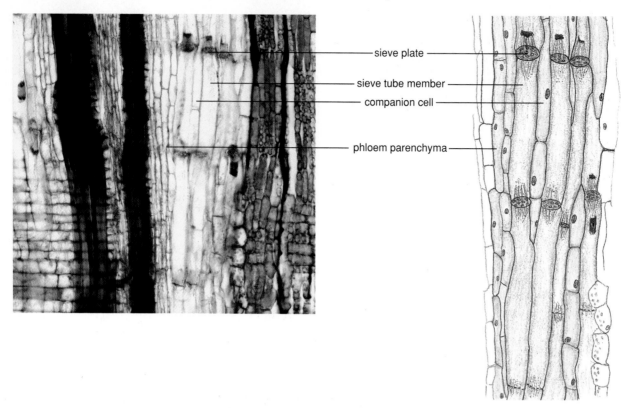

sieve plate

sieve tube member

companion cell

phloem parenchyma

FIGURE 4.9   Longitudinal view of part of the phloem of a black locust tree, ×1,000.

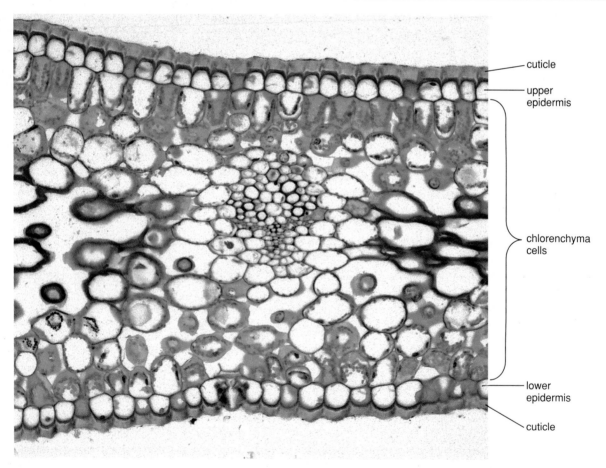

cuticle

upper epidermis

chlorenchyma cells

lower epidermis

cuticle

FIGURE 4.10   A portion of a cross section of a kaffir lily leaf, showing the thick cuticle secreted by the epidermis, ×1,000.

FIGURE 4.11 Hairs on the surface of an ornamental mint plant, ×50.

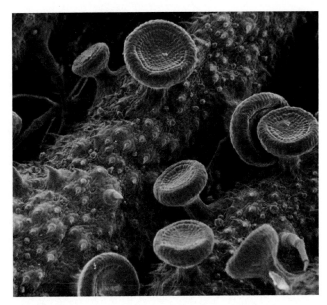

FIGURE 4.12A Tack-shaped glands and epidermal hairs of various sizes on the surface of flower bracts of a western tarweed. Scanning electron micrograph ca. ×200.

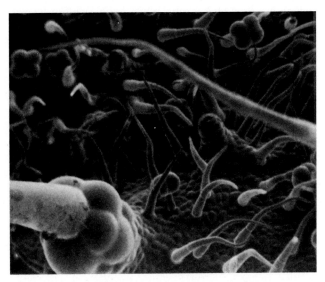

FIGURE 4.12B Hairs on the surface of a tomato plant stem. Note the raised stoma to the right of center. Scanning electron micrograph ca. ×300.

(*A.* courtesy Robert L. Carr; *B.* courtesy Dr. Tahany H.I. Sherif)

example, is deposited on the leaves of the wax palm. It and other waxes are harvested for use in polishes and phonograph records. In colonial times, a wax obtained from boiling leaves and fruits of the wax myrtle was used to make bayberry candles.

In leaves, the epidermal cell walls perpendicular to the surface often assume bizarre shapes that, under the microscope, give them the appearance of pieces of a jigsaw puzzle. Epidermal cells of roots produce tubular extensions called *root hairs* (shown in Fig. 5.4) a short distance behind the growing tips. The root hairs greatly increase the absorptive area of the surface.

Hairs of a different nature occur on the epidermis of aboveground parts of plants. These form outgrowths consisting of one to several cells (Fig. 4.11). Leaves also have numerous small pores, the **stomata,** bordered by pairs of specialized epidermal cells called **guard cells** (see figs. 7.7 and 9.15B). Guard cells differ in shape from other epidermal cells; they also differ in that chloroplasts are present within them. Stomata and guard cells are discussed in chapters 7 and 9. Some epidermal cells may be modified as **glands** that secrete protective or other substances, or modified as hairs that either reduce water loss or repel insects and animals that might otherwise consume them (Fig. 4.12).

Periderm

In woody plants, the epidermis is sloughed off and replaced by a **periderm** after the cork cambium begins producing new tissues that increase the girth of the stem or root. The periderm constitutes the outer bark and is primarily composed of somewhat rectangular and boxlike **cork** cells,

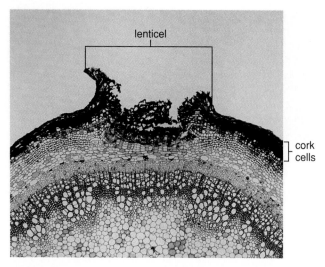

lenticel

cork cells

FIGURE 4.13 Periderm and a lenticel. A cross section through a small portion of elderberry periderm, showing a large lenticel, ca. ×250.

which are dead at maturity. While the protoplasm of cork cells is still functioning, it secretes a fatty substance, **suberin,** into the walls. This makes cork cells waterproof and helps them protect the tissues beneath the bark. Some cork tissues, such as those produced by the cork oak, are harvested commercially and are used for bottle corks and in the manufacture of linoleum and gaskets.

Some parts of a cork cambium form pockets of loosely arranged cork cells that are not impregnated with suberin. These pockets of tissue protrude through the surface of the periderm; they are called **lenticels** (Fig. 4.13) and function in gas exchange between the air and the interior of the stem. The fissures in the bark of trees have lenticels at their bases. The various tissues discussed are shown in Figure 6.6 as they occur in a woody stem.

Summary

1. A group of cells performing a common function is called a *tissue*.
2. Apical meristems are found in the vicinity of the tips of roots and stems; the vascular cambium and the cork cambium occur as lengthwise cylinders within roots and stems; intercalary meristems occur in the vicinity of nodes of grasses and related plants.
3. Nonmeristematic tissues are produced by meristems, and each tissue consists of one to several kinds of cells. Parenchyma, collenchyma, sclerenchyma, secretory tissues, epidermis, xylem, phloem, and periderm are common nonmeristematic tissues.
4. Parenchyma cells are thin walled, while collenchyma cells have unevenly thickened walls that provide flexible support for various plant organs.
5. There are two types of sclerenchyma—fibers (which are longer and tapering) and sclereids (which are short in length); both types have thick walls and are usually dead at maturity.
6. Secretory tissues occur in various places in plants; they secrete substances, such as nectar, oils, mucilage, latex, and resins.
7. Complex tissues have more than one kind of cell. The principal types are xylem, phloem, epidermis, and periderm.
8. Xylem conducts water and dissolved substances throughout the plant. It consists of a combination of parenchyma, fibers, vessels (tubular channels), tracheids (cells with tapering end walls that overlap), and ray cells (involved in lateral conduction).
9. Phloem conducts dissolved food materials throughout the plant. It is composed of sieve tubes (made up of cells called *sieve-tube members*), companion cells (which apparently regulate adjacent sieve-tube members), parenchyma, ray cells, and fibers. Callose aids in plugging injured sieve tubes. Sieve cells, which have overlapping end walls, and adjacent albuminous cells take the place of sieve-tube members and companion cells in ferns and cone-bearing trees.
10. The epidermis is usually one cell thick, with fatty cutin (forming the cuticle) within and on the surface of the outer walls. The epidermis may include guard cells that border pores called *stomata*; root hairs, which are tubular extensions of single cells; other hairs that consist of one to several cells; and glands that secrete protective substances.
11. Periderm, which consists of cork cells and loosely arranged groups of cells comprising lenticels involved in gas exchange, constitutes the outer bark of woody plants.

Review Questions

1. What is the function of meristems? Where are they located?
2. How are parenchyma, collenchyma, and sclerenchyma distinguished from one another?
3. Distinguish between epidermis and periderm.
4. What types of substances do secretory cells secrete?
5. What are the functions of xylem and phloem? What cells are involved in their normal activities?

Discussion Questions

1. Most plant meristems are located at the tips of shoots and roots and in cylindrical layers within stems and roots. What could happen if they were present in leaves?

2. The cambium produces xylem toward the center of a tree and phloem toward the outside. Do you think it would make any difference if the positions of the xylem and phloem were reversed? Why?

Additional Reading

Cutler, E.F., and K.L. Alvin. 1982. *The plant cuticle*. San Diego, CA: Academic Press.

Cutter, E.G. 1978. *Plant anatomy, Part I: Cells and tissues*, 2d ed. Reading, MA: Addison-Wesley Publishing Co.

Fahn, A. 1990. *Plant anatomy,* 4th ed. Elmsford, NY: Pergamon Press.

Foskett, D. E. 1994. *Plant growth and development*. San Diego, CA: Academic Press.

Lloyd, C.W. (Ed.). 1991. *The cytoskeletal basis of plant growth and form*. San Diego, CA: Academic Press.

Mauseth, J.D. 1988. *Plant anatomy*. Menlo Park, CA: Benjamin/Cummings Publishing Co., Inc.

Metcalfe, C.R. (Ed.). 1969–1982. *Anatomy of the monocotyledons*, 7 vols. Fair Lawn, NY: Oxford University Press.

Metcalfe, C.R., and L. Chalk (Eds.). 1988–1989. *Anatomy of the dicotyledons,* 2 vols. Fair Lawn, NY: Oxford University Press.

Romberger, J. A. 1993. *Plant structure: Function and development.* New York: Springer-Verlag.

Aerial roots of a tropical fig tree (Ficus sp.).

5 Roots and Soils

O v e r v i e w

This chapter discusses roots, beginning with the functions and continuing with the development of roots from a seed. It covers the function and structure of the root cap, region of cell division, region of elongation, and region of maturation (with its tissues). The endodermis and pericycle are also discussed.

Specialized roots (food-storage roots, water-storage roots, propagative roots, pneumatophores, aerial roots, contractile roots, buttress roots, parasitic roots, mycorrhizae) are given brief treatment. This is followed by some observations on the economic importance of roots.

After a brief examination of soil horizons, the chapter concludes with a discussion of the development of soil, its texture, its composition, its structure, and its water.

Some Learning Goals

1. Know the primary functions and forms of roots.
2. Learn the root regions, including the root cap, region of cell division, region of cell elongation, and region of maturation (including root hairs and all tissues), and know the function of each.
3. Discuss the specific functions of the endodermis and the pericycle.
4. Understand the differences among the various types of specialized roots.
5. Know at least 10 practical uses of roots.
6. Understand how a good agricultural soil is developed from raw materials.
7. Contrast the various forms of soil particles and soil water with regard to specific location and availability to plants.

You have probably seen pictures of the destruction caused by a tornado as it cut a swath through a village or a city, but have you ever seen what a twister can do to a forest? Large trees may be snapped off above the ground or knocked down, and branches may be stripped bare of leaves. Depending on the type of forest and the composition of the soil, however, you will probably see few trees completely torn up by the roots and blown elsewhere. And in the tropics it is indeed rare to find healthy palm trees uprooted after a hurricane.

Roots anchor trees firmly in the soil, usually by forming an extensive branching network that constitutes about one-third of the total dry weight of the plant. The roots of most plants do not usually extend down into the earth beyond a depth of 3 to 5 meters (10 to 16 feet); those of many herbaceous species are confined to the upper 0.6 to 0.9 meter (2 to 3 feet). The roots of a few plants, such as alfalfa, however, frequently extend more than 6 meters (20 feet) into the earth. When the Suez Canal was being built, workers encountered roots of tamarisk at depths of nearly 30 meters (100 feet), and mesquite roots have been seen 53.4 meters (175 feet) deep in a pit mine in the southwestern United States. Some plants, such as cacti, form very shallow root systems but still effectively achieve *anchorage* by means of

a densely branching mass of roots radiating out in all directions as far as 15 meters (50 feet) from the stem.

In addition to serving as anchors, roots function extensively in the *absorption of water and minerals in solution*, with the bulk of such "feeder" roots being confined to the upper meter (3.3 feet) of soil. The roots of some plants have specialized functions, such as food or water storage.

Although some aquatic plants (e.g., duckweeds) normally produce roots in water, and others (e.g., many orchids) produce aerial roots, the great majority of vascular plants develop their root systems in soils. Soils vary considerably in composition, texture, and other characteristics; they are discussed toward the end of this chapter.

Root Origins

When a seed germinates, a part of the **embryo** within it, the **radicle,** grows out and develops into the first root. This may develop either into a thickened *taproot*, from which thinner branch roots arise, or into numerous *adventitious roots*. A *fibrous* root system with numerous fine roots of similar diameter then develops from the adventitious roots (Fig. 5.1). Many mature plants have a combination of taproot and fibrous root systems. The number of roots produced by a single plant may be prodigious. For example, a single mature ryegrass plant may have as many as 15 million individual roots and branch roots, with a combined length of 644 kilometers (400 miles) and a total surface area larger than a volleyball court, all contained within 0.57 cubic meter (2 cubic feet) of soil. Significant additional surface area is provided by *root hairs* discussed in the section on the region of maturation.

Most *dicot* plants (plants such as peas and carrots, whose seeds have two "seed leaves") have taproot systems with one, or occasionally more, primary roots from which secondary roots develop. *Monocots* (plants such as corn and rice, whose seeds have one "seed leaf"), on the other hand, have fibrous root systems. Other types of roots, such as adventitious roots (discussed in Chapter 6), may develop in both dicots and monocots. In English and other ivies, lateral adventitious roots that aid in climbing appear along the aerial stems, and in certain plants with specialized stems (e.g., rhizomes, corms, and bulbs; see Fig. 6.14), adventitious roots are the only kind produced.

ROOT STRUCTURE

Botanists have traditionally recognized four regions or zones in developing young roots. Three of the regions are not sharply defined at their boundaries. The cells of each region gradually develop the form of those of the next region, and the extent of each region varies considerably, depending on the species involved. These regions are called (1) the *root cap*, (2) the *region of cell division*, (3) the *region of elongation*, and (4) the *region of maturation* (Fig. 5.2).

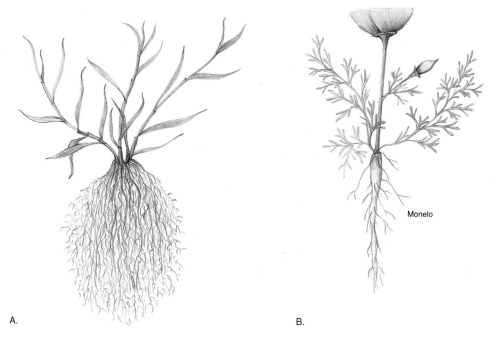

A.

B.

Monelo

FIGURE 5.1   Root systems. *A.* A fibrous root system of a grass. *B.* A taproot system of a poppy.

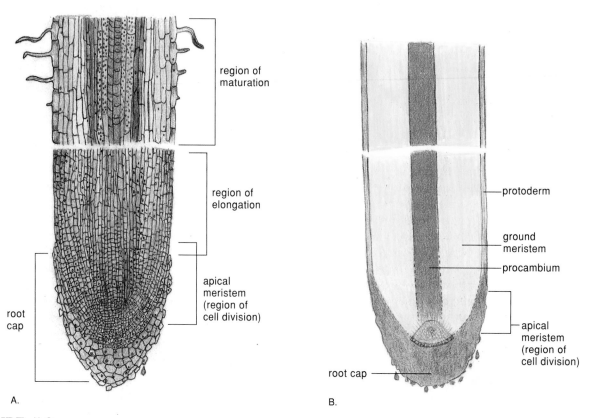

region of
maturation

region of
elongation

apical
meristem
(region of
cell division)

root
cap

A.

protoderm

ground
meristem

procambium

apical
meristem
(region of
cell division)

root cap

B.

FIGURE 5.2   A longitudinal section through a dicot root tip. *A.* Regions of the root. *B.* Locations of the primary meristems of the root.

The Root Cap

The **root cap** is composed of a thimble-shaped mass of parenchyma cells covering the tip of each root. It is quite large and obvious in some plants, while in others it is nearly invisible. One of its functions is to *protect* from damage the delicate tissues behind it as the young root tip pushes through often angular and abrasive soil particles.

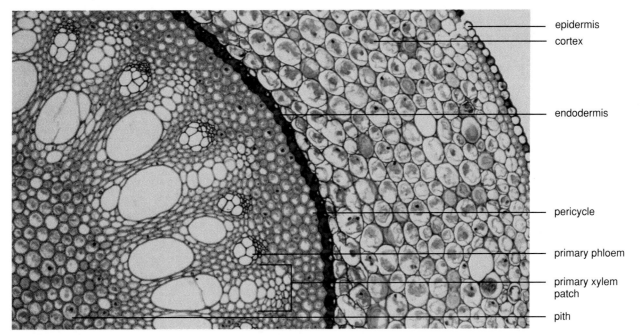

epidermis
cortex
endodermis
pericycle
primary phloem
primary xylem patch
pith

FIGURE 5.3   A cross section of a portion of a root of greenbrier (*Smilax*), a monocot, ×500. (Photomicrograph by G.S. Ellmore)

The root cap has no equivalent in stems. The Golgi bodies of the root cap's outer cells secrete and release a slimy substance that lodges in the walls and eventually passes to the outside. The cells, which are replaced from the inside, constantly slough off, forming a slimy lubricant that facilitates movement through the soil. This mucilaginous lubricant also provides a medium favorable to the growth of beneficial bacteria.

The root cap, whose cells have an average life of less than a week, can be slipped off or cut from a living root, and when this is done, a new root cap is produced. Until the root cap has been renewed, however, the root seems to grow randomly instead of downward, suggesting that the root cap has another function, namely, the *perception of gravity* (see *gravitropism* in Chapter 11). It is known that *amyloplasts* (plastids containing starch grains) collect on the sides of root cap cells facing the direction of gravitational force. When a root that has been growing vertically is artificially tipped horizontally, the amyloplasts tumble or float down to the "bottom" of the cells in which they occur. Within 30 minutes to a few hours, the root begins growing downward again. The exact nature of this gravitational response is not known, but there is some evidence that calcium ions known to be present in the amyloplasts influence the distribution of growth hormones in the cells.

The Region of Cell Division

The root cap is produced by cells in the region of *cell division,* which is in the center of the root tip and surrounded by the root cap. It is composed of an **apical meristem.** Most of the cell divisions take place at the edges of an inverted, cup-shaped zone a short distance behind the actual base of the meristem just adjacent to the root cap. Here the cells divide every 12 to 36 hours, while at the base of the meristem, they may divide only once in every 200 to 500 hours. The divisions are often rhythmic, reaching a peak once or twice each day, usually toward noon and midnight, with relatively quiescent intermediate periods. Cells in this region are mostly cube shaped, with relatively large centrally located nuclei and few, if any, small vacuoles. In both roots and stems, the apical meristem soon subdivides into three meristematic areas: (1) the **protoderm** gives rise to an outer layer of cells, the *epidermis;* (2) the **ground meristem,** to the inside of the protoderm, produces parenchyma cells of the *cortex;* and (3) the **procambium,** which appears as a solid cylinder in the center of the root, produces *primary xylem* and *phloem* (Fig. 5.3). *Pith* tissue, which is seen in most stems, is absent in most dicot roots but is found in those of many monocots, such as grasses.

The Region of Elongation

The *region of elongation,* which merges with the apical meristem, usually extends about 1 centimeter (0.4 inch) or less from the tip of the root. Here the cells become several times their original length and also somewhat wider. At the same time, the tiny vacuoles merge and grow until one or two large vacuoles, occupying up to 90% or more of the volume of each cell, have been formed. Only the root cap and apical meristem are actually pushing through the soil, since no further increase in cell size takes place above the region of elongation. The usually extensive remainder of each root remains stationary for the life of the plant. If a cambium is present, however, there normally is a gradual increase in girth through the addition of **secondary tissues.**

The Region of Maturation

Most of the cells mature into the various distinctive cell types of the primary tissues in this region, which is sometimes called the *region of differentiation*, or *root-hair zone*. It is given this last name because of the numerous, hairlike, delicate protuberances that develop from many of the epidermal cells. The **root hairs,** as they are called, adhere tightly to soil particles (Fig. 5.4) with the aid of microscopic fibers they produce.

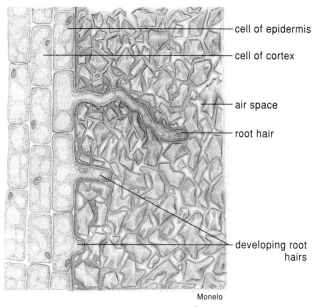

cell of epidermis

cell of cortex

air space

root hair

developing root hairs

Monelo

FIGURE 5.4   Root hairs in contact with soil particles.

The root hairs are not separate cells; in fact, the nucleus of the epidermal cell to which each is attached often moves out into the protuberance. They are so numerous that they appear as fine down to the naked eye, typically numbering more than 38,000 per square centimeter (250,000 per square inch) of surface area in roots of plants such as corn, and they seldom exceed 1 centimeter (0.4 inch) in length. A single ryegrass plant occupying less than 0.6 cubic meter (2 cubic feet) of soil has been found to have more than 14 billion root hairs, with a total surface area almost the size of a football field. Obviously, the root hairs greatly increase the absorptive surface of the root.

When a seedling or plant is moved, many of the delicate root hairs are torn off or die within seconds if exposed to the sun, thus greatly reducing the plant's capacity to absorb water and minerals in solution. This is why plants should be watered, shaded, and pruned after transplanting until new root hairs have formed. In any growing root, the extent of the root-hair zone remains fairly constant, with new root hairs being formed toward the root cap and older root hairs dying back in the more mature regions. The life of the average root hair is usually not more than a few days, although a few live for a maximum of perhaps three weeks.

An examination of the region of maturation (Fig. 5.5) reveals that the cuticle (shown in Fig. 4.10), which may be relatively thick on the epidermal cells of stems and leaves, is very thin on the root hairs and epidermal cells of roots. Presumably, any significant amount of fatty substance present in or on the walls of water-absorbing cells would interfere with their function.

The **cortex,** a tissue composed of parenchyma cells adjacent to the epidermis, functions primarily in food storage. It may be many cells thick and is similar to the cortex of

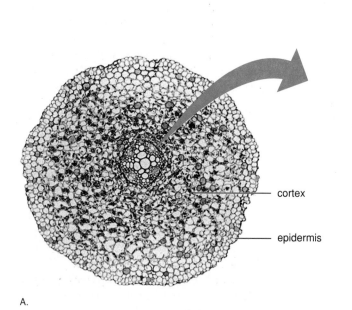

cortex

epidermis

A.

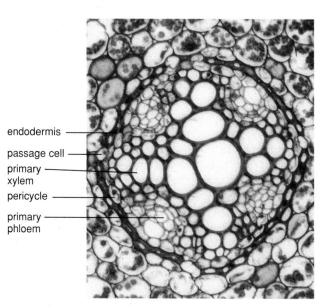

endodermis

passage cell

primary xylem

pericycle

primary phloem

B.

FIGURE 5.5   A cross section through the region of maturation of a root of buttercup (*Ranunculus*), a dicot, ×500.

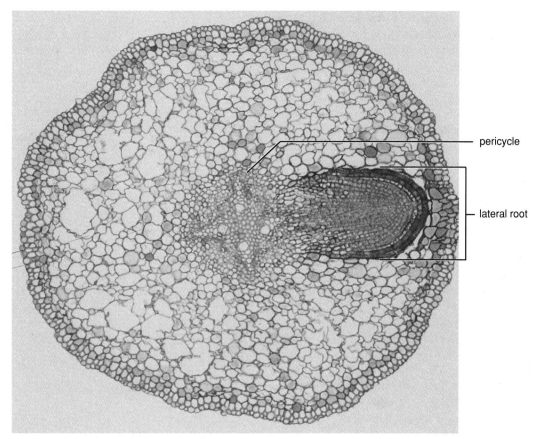

pericycle

lateral root

FIGURE 5.6   A cross section through a willow (*Salix*—a dicot) root showing the origin of a lateral (branch) root. (Photomicrograph by G.S. Ellmore)

stems except for the presence of an **endodermis** at its inner boundary. An endodermis is rare in stems but so universal in roots that only three species of plants are known to lack a root endodermis. It is a cylinder formed by a single layer of compactly arranged cells whose primary walls are impregnated with suberin.

The suberin forms bands around the walls perpendicular to the root's surface (i.e., the radial and transverse walls). These suberin bands are called **Casparian strips.** The plasma membranes of the endodermal cells are fused to the Casparian strips, which prevent water from passing through the otherwise permeable (porous) cell walls. The Casparian strip barrier forces all water and dissolved substances entering and leaving the central core of tissues to pass through the plasma membranes of the endodermal cells or their plasmodesmata. This has a regulatory effect on the types of minerals absorbed and transported by the root to the stems, leaves, and seeds. Harmful minerals may be excluded by the plasma membranes while useful ones generally are retained.

In some roots, the epidermis, cortex, and endodermis are sloughed off as their girth increases, but in those roots where the endodermis is retained, the inner walls of the endodermal cells eventually become thickened by the addition of alternating layers of suberin and wax. Later, cellulose and sometimes lignin are also deposited. Some endodermal

cells, called **passage cells,** may remain thin-walled and retain their Casparian strips for a while, but they, too, eventually tend to become suberized.

On the inside of the endodermis and immediately adjacent to it is a cylinder of parenchyma cells called the **pericycle.** This is usually one cell wide, although in some plants it may extend for several cells. It is a very important tissue, for the cells retain their capacity to divide even after they have matured, and it is within the pericycle that the *lateral (branch) roots* and part of the *vascular cambium* of dicots arise (Fig. 5.6).

The *primary xylem*, with its water-conducting cells such as tracheids, forms a solid core in the center of most dicot and conifer roots. Unlike stem xylem, however, root xylem of most dicots first forms with short spirelike ridges or arms, which point toward the pericycle. There are often four of these arms, but there can be from two to several. Branch roots usually arise in the pericycle opposite the xylem arms. In many monocot and a few dicot roots, the primary xylem forms a cylinder of tissue in which the arms may not be well-defined; this tissue surrounds the pith parenchyma cells. *Primary phloem*, which contains food-conducting cells, forms in discrete patches between the xylem arms of both dicot and monocot roots (as shown in Fig. 5.5).

In woody dicots, some herbaceous dicots, and conifers, parts of the pericycle and parenchyma cells between the xylem arms and the phloem patches become a vascular cambium. This cambium starts producing secondary phloem to the outside and secondary xylem to the inside. Soon, the cambium follows the outline of the primary xylem, and eventually, instead of appearing as patches and arms, the secondary conducting tissues appear as concentric cylinders. The primary tissues are crushed out of existence or sloughed off and lost as secondary tissues are added.

In woody plants, a cork cambium normally arises in the pericycle and gives rise to *cork* tissue (*periderm*) similar to that of stems. In fact, the cross section of a woody root several years old, with its annual rings of xylem along with its sapwood and heartwood, may be distinguished from that of a stem of the same species only by the relatively larger proportion of parenchyma cells present in the root. Although there are exceptions, monocot roots generally have no secondary meristems and therefore no secondary growth.

Natural grafting may occur between roots of different trees of the same species, particularly in the tropics. When the roots come in contact with one another, they unite through secondary growth, but the details of the uniting process are not yet known. One unfortunate aspect of this grafting is that if one tree becomes diseased, the disease can be transmitted through these grafts to all the other trees connected to it.

Branch or lateral roots arise internally in the pericycle, pushing their way out to the surface through the endodermis, cortex, and epidermis (Fig. 5.6). This process is in contrast with that in stems, where branches develop at the surface from axillary buds. Some specialized roots, which at first glance may appear similar to specialized stems, such as rhizomes, bulbs, and corms (discussed in Chapter 6), can be distinguished by their complete lack of leaves.

SPECIALIZED ROOTS

As indicated earlier, most plants produce either a fibrous root system, a taproot system, or, more commonly, a combination of the two types. Some plants, however, have roots with modifications that permit specific functions in addition to the absorption of water and minerals in solution.

Food-Storage Roots

Most roots and stems store some food, but in certain plants the roots are enlarged and store large quantities of starch and other carbohydrates (Fig. 5.7). In sweet potatoes and yams, for example, extra cambial cells develop in parts of the xylem of branch roots and produce numerous parenchyma cells. These cause the organs to swell, providing storage areas for considerable quantities of carbohydrates. Similar food-storage roots are found in the deadly poisonous water hemlocks, in dandelions, and in salsify. Some food-storage

FIGURE 5.7   A sweet potato plant. Note the food-storage roots.

organs, such as those of carrots, beets, turnips, and radishes, are actually a combination of root and stem. In carrots, for example, approximately 2 centimeters (0.8 inch) at the top are derived from stem tissue, yet a casual inspection will not reveal any differences of origin.

Water-Storage Roots

Some members of the Pumpkin Family produce huge water-storage roots. This is particularly characteristic of those that grow in arid regions or in those areas where there may be no precipitation for several months of the year. In certain man-roots, for example, roots weighing 30 kilograms (66 pounds) or more are frequently produced (Fig. 5.8), and a major root of one calabazilla plant was found to weigh 72.12 kilograms (159 pounds). The water in the roots is apparently used by the plants when the supply in the soil is inadequate.

Propagative Roots

Many plants produce **adventitious buds** (buds appearing in unusual places) along the roots that grow near the surface of the ground. The buds develop into aerial stems called *suckers,* which have additional rootlets at their bases. The rooted suckers can be separated from the original root and grown individually. A number of fruit trees, such as cherry and pear trees, frequently produce such roots. The adventitious roots of horseradish and rice-paper plants can become a nuisance in gardens, the latter often producing propagative roots within a radius of 10 meters (33 feet) or more from the

FIGURE 5.8 A manroot water-storage root weighing over 25.3 kilograms (60 pounds).

(Courtesy Robert A. Schlising)

FIGURE 5.9 Pneumatophores (foreground) of tropical mangroves rising above the sand at low tide. The pneumatophores are spongy outgrowths from the roots beneath the surface. Pneumatophores facilitate the exchange of oxygen and carbon dioxide for the roots, which grow in areas where little oxygen is otherwise available to them.

(Courtesy Lani Stemmerman)

parent plant. Tree of Heaven, Canada thistle, and some other weeds have a remarkable facility to reproduce in this fashion as well as by means of seeds. This ability made it difficult to control them in the past, but some **biological controls** now being investigated (see Appendix 2) may be an answer to the problem in the future.

Pneumatophores

Water, even after air has been bubbled through it, contains less than one 30th of the amount of free oxygen found in the air. Plants growing with their roots in water may therefore encounter less than the amount of oxygen necessary for normal respiration. Some swamp plants, such as the black mangrove and the yellow water weed, develop special "spongy" roots, called *pneumatophores*, which rise above the surface and facilitate gas exchange between the atmosphere and the subsurface roots to which they are connected (Fig. 5.9). The bald cypress, which occur in southern swamps produces "knees" (shown in Fig. 22.19) that were once thought to function as pneumatophores, but there is no conclusive evidence for this theory.

Aerial Roots

The various kinds of aerial roots produced by plants include the *velamen* roots of orchids, *prop* roots of corn and banyan trees (Fig. 5.10), *adventitious* roots of ivies, and *photosynthetic* roots of certain orchids. Velamen roots have an epidermis several cells thick. It was assumed that this thick epidermis aids in the absorption of rainwater, but it now appears it may function more in preventing loss of moisture from the root. Corn prop roots, produced toward the base of the stems, support the plants in a high wind. A number of tropical plants, including the screw pines and various mangroves, produce sizable prop roots extending for several feet above the surface of the ground or water. Debris collects between them and helps to create additional soil.

Many of the tropical figs or banyan trees produce roots that grow down from the branches until they contact the soil. Once they are established, they continue secondary growth and look just like additional trunks. Banyan trees may live for hundreds of years and can become very large. In India and southeast Asia, several such trees display almost 1,000 root-trunks and have circumferences approaching 450 meters (1,476 feet). The oldest is estimated to be about 2,000 years old.

FIGURE 5.10   A banyan tree, with numerous large prop roots that have developed from the branches.

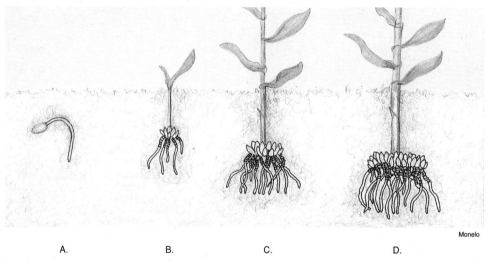

Monelo

A.   B.   C.   D.

FIGURE 5.11   How a lily bulb, over three seasons, is withdrawn deeper into the soil through the action of contractile roots. *A.* A seed germinates. *B.* Contractile roots pull the newly formed bulb down several millimeters during the first season. *C.* The bulb is pulled down farther the second season. *D.* The bulb is pulled down even farther the third season. The bulb will continue to be pulled down in succeeding seasons until it reaches an area of relatively stable soil temperatures.

The vanilla orchid, from which we obtain vanilla flavoring, produces chlorophyll in its aerial roots and thus, through photosynthesis, can manufacture food with them. The adventitious roots of English ivy, Boston ivy, and Virginia creeper appear along the stem and aid the plants in climbing.

Contractile Roots

A number of herbaceous dicots and monocots have *contractile* roots that pull the plant deeper into the soil. Many lily bulbs are pulled a little deeper into the soil each year as

FIGURE 5.12   Buttress roots of a tropical fig tree.

additional sets of contractile roots are developed (Fig. 5.11). The bulbs continue to be pulled down until an area of relatively stable temperatures is reached. The leaves of plants such as dandelions always seem to be coming out of the ground as the top of the stem is pulled down a small amount each year when the root contracts. The contractile part of the root may lose as much as two-thirds of its length within a few weeks. Mechanisms of contraction include the thickening and constriction of parenchyma cells, causing the xylem elements to spiral somewhat like a corkscrew.

Buttress Roots

Some tropical trees produce huge buttress roots toward the base of the trunk, giving them great stability (Fig. 5.12). Except for their angular appearance, these roots look like a part of the trunk.

Parasitic Roots

A number of plants, such as the dodders, broomrapes, and pinedrops, have no chlorophyll (necessary for photosynthesis) and have become dependent on plants with chlorophyll for their nutrition. They parasitize their host plants via somewhat rootlike projections called **haustoria** (singular: **haustorium**), which develop along the stem in contact with the host. The haustoria penetrate the outer tissues and establish connections with the water-conducting and food-conducting tissues (Fig. 5.13). Some green plants, such as

Indian warrior and the mistletoes, also form haustoria. These haustoria, however, apparently function primarily in obtaining water and dissolved minerals from the host plants, since the partially parasitic plants are capable of manufacturing at least some of their own food through photosynthesis.

Mycorrhizae

More than three quarters of all seed plant species have various fungi associated with their roots. The association is *mutualistic;* that is, both the fungus and the root benefit from it and are dependent upon the association for normal development. (Mutualism is a form of *symbiosis;* see page 267.)

The fungus is able to absorb and concentrate phosphorus much better than it can be absorbed by the root hairs. In fact, if mycorrhizal fungi have been killed by fumigation or are otherwise absent, many plants appear to have considerable difficulty absorbing phosphorus, even when the element is abundant in the soil. The phosphorus is stored in granular form until it is utilized by the plant. The fungus also often forms a mantle of millions of threadlike strands that facilitate the absorption of water and nutrients. The plant furnishes sugars and amino acids without which the fungus cannot survive.

These "fungus-roots," or **mycorrhizae** (Fig. 5.14), are essential to the normal growth and development of forest trees and many herbaceous plants. Orchid seeds, for example, will not germinate until mycorrhizal fungi invade their cells.

A.

B.

FIGURE 5.13   *A.* Pale stems of a dodder plant twining around other vegetation. *B.* A close-up view of dodder, showing the peglike *haustoria* that penetrate the tissues of the host plant, ×40.

In virtually all of the woody trees and shrubs found in forests, the fungal threads grow between the walls of the outer cells of the cortex but rarely penetrate into the cells themselves. If they should happen to penetrate, they are apparently broken down and digested by the host plants. In herbaceous plants,

the fungi do penetrate the cortex cells as far as the endodermis, but they cannot grow beyond the Casparian strips. Once inside the cells, the fungi branch repeatedly but do not break down the plasma or vacuolar membranes.

Some plants do not seem to need mycorrhizae unless essential elements in the soil are present in amounts barely sufficient for healthy growth. Plants with mycorrhizae develop few root hairs as compared with those growing without an associated fungus. Mycorrhizae have proved to be particularly susceptible to acid rain (discussed in Chapter 25); this may portend major problems for our coniferous forests in the future if the problem of acid rain is not solved.

Root Nodules

Although almost 80% of our atmosphere consists of nitrogen gas, plants are not capable of converting the nitrogen gas to forms they can use. Bacteria, however, produce enzymes that facilitate the transformation of nitrogen into nitrates and other substances that can readily be absorbed by plant roots. Members of the Legume Family, including peas and beans, and a few other plants such as alders form associations with soil bacteria that result in the production of numerous small swellings called *root nodules* that are clearly visible when such plants are uprooted (Fig. 5.15).

A substance exuded into the soil by plant roots stimulates *Rhizobium* bacteria which, in turn, respond with another substance that prompts root hairs to bend sharply. A bacterium may attach to the concave side of a bend and then invade the cell with a tubular *infection thread* that does not actually break the host cell wall and plasma membrane. The infection thread grows through to the cortex, which is stimulated to produce new cells that become a part of the root nodule; here the bacteria multiply and engage in nitrogen conversion. (See also the discussion of the nitrogen cycle in Chapter 25.)

Root nodules should not be confused with *root knots*, which are also swellings and may be seen in the roots of tomatoes and many other plants. Root knots develop in response to the invasion of tissue by small, parasitic roundworms (nematodes). Unlike bacterial nodules, root knots are not beneficial and the activities of the parasites within them can eventually lead to the premature death of the plant.

HUMAN RELEVANCE OF ROOTS

Roots have been important sources of food for humans since prehistoric times, and some, such as the carrot, have been in cultivation in Europe for at least 2,000 years. A number of cultivated root crops involve biennials (i.e., plants that complete their life cycles from seed to flowering and back to seed in two seasons). Such plants store food in a swollen taproot during the first year of growth and then utilize the stored food to produce flowers in the second season. Among the best-known biennial root crops are sugar beets, beets, turnips, rutabagas, parsnips, horseradishes, and carrots.

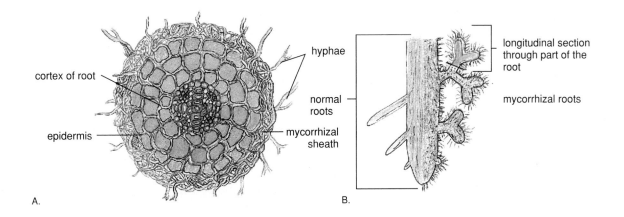

cortex of root

epidermis

hyphae

normal roots

mycorrhizal sheath

A.

longitudinal section through part of the root

mycorrhizal roots

B.

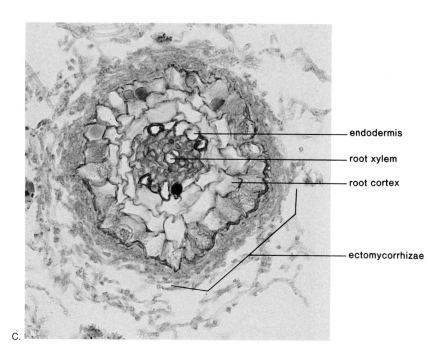

endodermis

root xylem

root cortex

ectomycorrhizae

C.

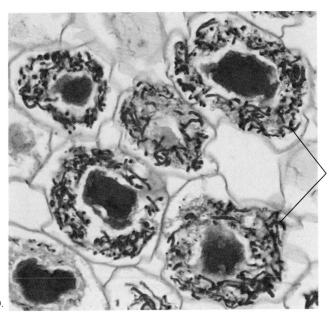

endomycorrhizae

D.

FIGURE 5.14 Mycorrhizae. *A.* A diagram of a cross section of a root with ectomycorrhizae. *B.* Comparison, in longitudinal section, of a root with ectomycorrhizae and one without mycorrhizae. *C.* Photomicrograph of a cross section of a root around which ectomycorrhizae have formed a mantle. The fungal cells have not penetrated beyond the outermost layers of root cells. *D.* A few root parenchyma cells with endomycorrhizae.

FIGURE 5.15   Root nodules on a bur clover plant. The nodules contain bacteria that convert nitrogen from the air into forms that can be used by plants, ×5.

Other important root crops include sweet potatoes, yams, and cassava. Cassava (Fig. 5.16), from which tapioca is made, forms a major part of the basic diet for millions of inhabitants of the tropics. It yields more starch per hectare (about 45 metric tons, the equivalent of 20 tons per acre) than any other cultivated crop, with a minimum of human labor. A number of minor root crops are cultivated in South America and other parts of the world. Among such plants are relatives of wild mustards, nasturtiums, and sorrel.

A number of spices, such as sassafras, sarsaparilla, licorice, and angelica, are obtained from roots. Sweet potatoes are used in the production of alcohol in Japan. Several important red to brownish dyes are obtained from roots of members of the Madder Family, to which coffee plants belong. Drugs obtained from roots include aconite, ipecac, gentian, and reserpine, a tranquilizer. A valuable insecticide, rotenone, is obtained from the barbasco plant, which has been cultivated for centuries as a fish poison by primitive South American tribes. When thrown into a dammed stream, the roots containing rotenone cause the fish to float but in no way poison them for human consumption. Other uses of roots are discussed in Chapter 24.

FIGURE 5.16   Cassava plants. Note the food-storage roots on the plant that has been dug up.
(Courtesy Monica E. Emerson)

SOILS

The soil is a dynamic, complex, constantly changing part of the earth's crust, which extends from a few centimeters deep in some places to hundreds of meters deep in others. It is essential not only to our existence but also to the existence of most living organisms. It has a pronounced effect on the plants that grow in it, and they have an effect on it.

If you dig up a shovelful of soil from your yard and examine it, you will probably find a mixture of ingredients, including several grades of sand, rocks and pebbles, powdery silt, clay, humus, dead leaves and twigs, clods consisting of soil particles held together by clay and organic matter, plant roots, and small animals, such as ants, pill bugs, millipedes, and earthworms. Also present, but not visible, would be millions of microorganisms, particularly bacteria, and, of course, air and water.

The soil became what it is today through the interaction of a number of factors: climate, parent material, topography of the area, vegetation, living organisms, and time. Because there are thousands of ways in which these factors may interact, there are many thousands of different soils. The solid portion of a soil consists of mineral matter and organic matter. Pore spaces, shared by variable amounts of water and air, occur between the solid particles. The smaller pores often contain water, and the larger ones usually contain air. The sizes of the pores and the connections between them largely determine how well the soil is aerated.

If we were to dig down 1 or 2 meters (3 to 6 feet) in an undisturbed area, we would probably expose a soil profile of three intergrading regions called *horizons* (Fig. 5.17). The horizons show the soil in different stages of development, and the composition varies accordingly. The upper layer, usually extending down 10 to 20 centimeters (4 to 8 inches), is called the *A horizon*, or *topsoil*. It is usually subdivided into a darker upper portion called the A_1 *horizon* and a lighter lower portion called the A_2 *horizon*. The A_1 portion contains more organic matter than do the layers below.

The next 0.3 to 0.6 meter (1 to 2 feet) is called the *B horizon*, or *subsoil*. It usually contains more clay and is lighter in color than the topsoil. The *C horizon* at the bottom may vary from about 10 centimeters (4 inches) to several meters (6 to 10 feet or more) in depth; it may even be absent. It is commonly referred to as the *soil parent material* and extends down to *bedrock*.

Parent Material

The first step in the development of soil is the formation of *parent material*. This material accumulates through the weathering of three types of rock that originate from various sources. These sources and types include volcanic activity (*igneous rocks*); fragments deposited by glaciers, water, or wind (*sedimentary rocks*); and changes in igneous or sedimentary rocks brought about by great pressures, heat, or both (*metamorphic rocks*).

Climate

Climate varies greatly throughout the globe, and its role in the weathering of rocks varies correspondingly. In desert

A_1 horizon (dark loam)

A_2 horizon (lighter-colored loam)

B horizon (clay loam)

C horizon (parent material)

FIGURE 5.17 A soil profile.

Reproduced from Marbut Memorial Slide Set, 1968, SSSA, by permission of Soil Sciences of America, Inc.

areas, for example, there is little weathering by rain, and soils are poorly developed. In areas of high rainfall, however, well-developed soils are common. In areas where there are great temperature ranges, rocks may split or crack as their outer surfaces expand or contract at rates different from those of the material beneath the surface. Low temperatures cause water in rock crevices to freeze, which also causes cracks and splits.

Living Organisms and Organic Composition

In addition to roots and other parts of plants, there are numerous organisms of many kinds in the soil. In the upper 30 centimeters (1 foot) of a good agricultural soil, living organisms constitute about one 1,000th of the total weight of the soil. This may not sound significant, but it amounts to approximately 6.73 metric tons per hectare (3 tons per acre). Bacteria and fungi present in the soil decompose all types of organic matter, which accumulates when leaves fall and roots and animals die. (This process is further discussed under the section on composting in Chapter 17.) Roots and all other living organisms produce carbon dioxide, which combines with water and forms an acid, thereby increasing the rate at which minerals dissolve. Ants and other insects, earthworms, burrowing animals, and birds all alter the soil through their activities and add to its organic increment either through wastes that they deposit or through the decomposition of their bodies when they die.

The total organic composition of a soil varies greatly. An average topsoil might consist of about 25% air, 25% water, 48% minerals, and 2% organic matter. Soils in low, wet areas, where a lack of oxygen keeps microorganisms from their normal activities, may contain as much as 90% organic matter. Except in legumes and a few other plants, almost all of the nitrogen utilized by growing plants, as well as much of the phosphorus and sulphur, comes from decomposing organic matter. In addition, as organic matter breaks down, it produces acids, which, in turn, decompose minerals. Other roles of organic matter in the soil are discussed in the section on soil structure.

Topography of the Area

If the topography (surface features) of an area is steep, soil may wash away or erode through the actions of wind, water, and ice as soon as the soil is weathered from the parent material. It has been estimated that more than 20 metric tons of topsoil per hectare (8.2 tons per acre) are washed away annually from some prime croplands in the central United States. However, if an area is flat and poorly drained, pools and ponds may appear in slight depressions when it rains. If these bodies of water cannot drain quickly, the activities of organisms in the soil are interrupted and the development of the soil is arrested. The ideal topography for the development of soil is one that permits drainage without erosion.

Soil Texture and Mineral Composition

Soil *texture* refers to the sizes of the individual soil particles (Table 5.1). The principal texture designations are *sand, silt,* and *clay.* Sands and gravels are usually composed of many small particles bound together chemically or by a cementing matrix material. Silt consists of particles that are mostly too small to be seen without a lens or a microscope.

Clay particles are so tiny that even a powerful light microscope will not render them visible, although they can be seen with an electron microscope. Individual clay particles are called *micelles.* Micelles are somewhat sheetlike, negatively charged, and held together by chemical bonds. They attract, exchange, or retain positively charged ions. Clay is plastic in nature because the water that adheres tightly to the surface of the particles acts both as a binding agent and as a lubricant. Physically, clay is a **colloid**—that is, a suspension of particles that are larger than molecules but that do not settle out of a fluid medium.

"Light" soils have high sand and low clay content. "Heavy" soils have high clay content. Coarse-textured soils don't retain much water, while clay soils have high water content and allow little water to pass through.

Over half of the composition by weight of mineral matter is oxygen. Other elements commonly present are hydrogen, silicon, aluminum, iron, potassium, calcium, magnesium, and sodium. However, soil inherits hundreds of different mineral combinations from parent material.

Compared to sand and silt, clay and organic matter store, in the form of ions, significantly more plant nutrients. This is because the smaller clay and organic particles have a greater total surface area to which ions can become attached than does an equal volume of larger sand and silt particles.

Soil Structure

Soil *structure* refers to the arrangement of the soil particles into groups called *aggregates.* Aggregates in sands and gravels show little cohesion, but most agricultural soils have aggregates that stick together. Structure develops when colloidal particles clump together primarily as a result of the activities of soil organisms, freezing, and thawing. If the individual granules do not become coated with organic matter, they may continue to clump until they become clods.

Productive agricultural soils are granular soils with pore spaces that occupy between 40% and 60% of the total volume of the soil. The pores contain air and water, and their sizes are more important than their total volume. Clay soils, for example, have more pore space than sandy soils, but the pores are so small that water and air are restricted in their movement through the soil. When the pores are full of water, air is kept out, and there is not enough oxygen for root growth. Sandy soils have large pores, which drain by gravity soon after they are filled. The water is replaced by air, but too much air speeds up nitrogen release by microorganisms. Plants can't use the nitrogen that quickly, and much of it is lost.

Water itself can be harmful. Too much of it leaches mineral nutrients and slows the mineralization process under the anaerobic (marked by the absence of free oxygen) conditions. Too much water also slows the release of nitrogen, interferes with plant growth, and accelerates the breakdown of nitrates to the extent that virtually all the nitrates present may be lost in as little as half an hour. Many gardeners and houseplant enthusiasts have stunted or killed the very plants

Table 5.1

SOIL MINERAL COMPONENTS AS CLASSIFIED BY THE U.S. DEPARTMENT OF AGRICULTURE

MINERAL	DIAMETER (RANGE IN MM)	COMMENTS
Stones	>76 mm	do not support plant growth but affect permeability and erosion of the soil
Gravel	76.0 mm–2.0 mm	
Very coarse sand	2.0 mm–1.0 mm	
Coarse sand	1.0 mm–0.5 mm	
Medium sand	0.5 mm–0.25 mm	makes soil feel gritty
Fine sand	0.25 mm–0.10 mm	
Very fine sand	0.10 mm–0.05 mm	
Silt	0.05 mm–0.002 mm	lens or microscope needed to see any but the coarsest silt particles
Clay	<0.002 mm	may absorb water, swell, and later shrink, causing the soil to crack as it dries

they were trying to foster by "drowning" them with excessive or too frequent watering. As a general rule, plants should not be watered until the soil surface feels dry.

Water in the Soil

Water in the soil occurs in three forms. **Hygroscopic water** is physically bound to the soil particles and is unavailable to plants. **Gravitational water** drains out of the pore spaces after a rain. If drainage is poor, it is this water that interferes with normal plant growth. Plants are mainly dependent on the third type, **capillary water,** which is water held against the force of gravity in pores of the soil for their needs. The structure and organic matter of the soil, which enable the soil to hold water against the force of gravity; the density and type of vegetational cover; and the location of underground water tables largely determine the amount of capillary water available to the plant.

The ancient Incas of Peru knew that water would rise just so far in some areas. Where the water table was close to the surface, they removed the upper 0.6 meter (2 feet) of soil and planted their crops down in the hollowed-out areas so that the roots would be able to reach the capillary water. They knew that in some areas having sandy soils and low annual precipitation, soils could be compacted by a heavy roller to create finer capillaries to raise water from below. This technique is effective only if the available water is within 1.5 to 3.0 meters (5 to 10 feet) of the surface and if the soils do not contain much silt or clay.

After rain or irrigation, water in the soil drains away by gravity. The water remaining after such draining is referred to as the *field capacity* of the soil. Field capacity is mainly governed by the texture of the soil, but the structure and organic content also influence it to a certain extent. Plants readily absorb water from the soil when it is at, or near, field capacity. As the soil dries, the film of water around each soil particle becomes thinner and more tightly bound to the soil particle and less likely to enter the root. If water is not added to the soil, eventually a point is reached at which the rate of absorption of water by the plant is insufficient for its needs, and the plant wilts permanently. The soil is then said to be at the *permanent wilting point*. In clay soils, the permanent wilting point is reached when the water content drops below 15%, while in sandy soils, the permanent wilting point may be as low as 4%. *Available water* is soil water between field capacity and the permanent wilting point.

The pH (acidity or alkalinity) of a soil affects both the soil and the plants growing in it in various ways. A soil that is unusually acid or alkaline may be toxic to the roots, but these conditions do not normally directly affect plants nearly as much as they affect nutrient availability. For example, alkalinity causes minerals, such as copper, iron, and manganese, to become less available to plants, while acidity, if high enough, inhibits the growth of nitrogen-fixing bacteria. Acid soils tend to be common in areas of high precipitation where significant amounts of bases are leached from the topsoil.

It is a common agricultural practice to counteract soil acidity by adding compounds of calcium or magnesium in a process known as *liming*. Alkaline soils can be made more acid by the addition of sulphur, which is converted by bacteria to sulphuric acid. The addition of some nitrogenous fertilizers may have the same effect.

Metal Munching Plants

Years ago prospectors looking for promising sites to mine silver or gold often noticed that certain plants grew on old mine tailings. They reasoned that these plants might be indicators of the precious metals entrapped in the soil. It turns out that these prospectors had it right.

Certain plants not only grow in heavy, metal-laden soils but are able to extract these metals through their root systems and accumulate them in their tissues without being damaged. Plants absorb metals because they require certain ones such as zinc and copper as components of their proteins and enzymes for normal growth and development. Some plants absorb heavy metals such as cadmium or selenium along with their required elements. In either case, these metals are absorbed by the plant's extensive root system, which may extend a meter or more in depth. Scientists have coined a new term to describe this process—**phytoremediation**—the use of plants to facilitate the removal of toxic compounds from ground water and soil. The plants that "munch metal" so well are called "hyperaccumulating" plants.

What makes a good hyperaccumulator? Researchers are attempting to answer this question, but so far, nobody knows for sure. One clue may be metal-binding polypeptides called **phytochelatins** that sequester and detoxify heavy metals in plant tissue. A survey of numerous plants has shown that phytochelatins are produced when these plants are exposed to heavy metals in soil. Interestingly, a wide range of plants from the most advanced flowering plants (even orchids) to red, green, and brown algae produce these detoxifying polypeptides. One possible method of accumulating heavy metals is for the plant to transport them into the cell's vacuoles, a sort of waste disposal dump.

Phytoremediation is attracting the increased attention not only of scientists, but also of regulators who see it as a low-cost alternative for cleaning up contaminated sites across the country. The conventional process of soil excavation and reburial in a landfill is very expensive, typically costing about $1.5 million per acre, depending on the pollutant. The price tag opens the door for lots of alternative ideas. These plants can be harvested and disposed of—and in certain instances the metals can even be recovered by sending the plants to a smelter. But can hyperaccumulators effectively get the job done?

Alpine pennycress (*Thlaspi caerulescens*) is a remarkable hyperaccumulator, able to accumulate 4% of its dry body weight in zinc. This translates into 10 metric tons of zinc per hectare. The problem, however, is that *Thlaspi* is a small and slow-growing plant. For phytoremediation to become practical, plants with metal uptake rates comparable to *Thlaspi* but with faster growth rates and larger tissue mass must be found. Screening of other plants is being done in several laboratories around the world. One plant that has been identified so far is Indian mustard, *Brassica juncea,* a relative of some highly nutritious vegetables—cabbage, cauliflower, broccoli, collard, kale, and mustard greens. Indian mustard can accumulate 3.5% of its dry body weight in lead. It also can absorb cadmium, chromium, nickel, selenium, zinc, and copper.

Despite the advantages of phytoremediation, one drawback is that multiple crops must be planted over several years to reduce contamination to acceptable levels, while removal of the soil provides an immediate resolution. Additionally, there is concern about increasing the accumulation of these metals in the food chain as wildlife and insects eat the plants, accumulating toxicity in their bodies. In this way toxic metals could work their way up the food chain and pose a new set of problems.

Phytoremediation is being put to a big test at Chernobyl, the site of the largest environmental disaster in modern history. In 1986 the meltdown of the nuclear reactor left radioactive wastes scattered over the Ukrainian countryside. A pilot study there suggests that Indian mustard could remove radioactive strontium from the soil over a five-year period. While scientists continue to test this promising natural process, environmental cleanups of the future could be as simple as letting the flowers grow.

Summary

1. Roots function in anchorage and in absorption of water and minerals in solution. A germinating seed radicle becomes the first root. Taproots with branch roots or adventitious roots that become a fibrous root system develop from the radicle. Many plants have a combination of both systems.

2. Four zones or regions of young roots are recognized: (1) a protective root cap that also aids in the perception of gravity; (2) a region of cell division (its apical meristem subdivides into a protoderm, which produces the epidermis); a ground meristem, which produces the cortex; and a procambium, which produces primary

xylem and primary phloem; (3) a region of elongation in which the cells produced by the apical meristem become considerably longer and slightly wider; and (4) a region of maturation in which the cells mature into the distinctive cell types of primary tissues.

3. Some of the epidermal cells in the region of maturation develop root hairs; the root hairs greatly increase the absorptive surface of the root. The tissues that mature in this region are similar to those of stem tips, but pith is absent in most dicot roots and originates from the procambium in monocot roots.

4. The cortex has an endodermis with suberized Casparian strips at its inner boundary.

5. Next to the endodermis toward the center of the root are parenchyma cells constituting the pericycle. Branch roots and the vascular cambium arise in the pericycle.

6. In dicot roots, the primary xylem usually first forms a solid core with two to several arms in the center of the root. A pith may be present in monocot roots.

7. Primary phloem first is produced in discrete patches between the primary xylem arms, but the tissues eventually appear as concentric cylinders. In woody plants, a cork cambium usually arises in the pericycle and produces cork tissues similar to those of stems. Roots may graft together naturally. There are no nodes or internodes in roots.

8. Specialized roots include those for food or water storage; pneumatophores; aerial roots (velamen roots, prop roots, photosynthetic roots, and adventitious roots); contractile roots; buttress roots; haustoria; and mycorrhizal roots. Some plants have nitrogen-fixing bacteria in nodules on their roots.

9. Root crops include sugar beets, beets, turnips, rutabagas, parsnips, carrots, sweet potatoes, yams, and cassava. Several spices are obtained from roots. Other uses of roots include the production of alcohol and the extraction of dyes, drugs, insecticides, and poisons.

10. Soils contain a mixture of ingredients, including sands, rocks, silt, clay, humus, dead organic matter, plant roots, small animals, microorganisms, and air and water within pore spaces of various sizes.

11. A vertical column of soil exhibits horizons; The A horizon (topsoil) is divided into an upper A_1 horizon and a lower A_2 horizon. The B horizon (subsoil) usually contains more clay and is lighter in color than the A horizon. The C horizon (bottom portion) constitutes the weathered soil parent material.

12. Living organisms in the soil decompose organic matter, the source of most important plant nutrients. Animals also cultivate the soil. Soil erosion is affected by the topography of an area.

13. Soil texture pertains to sizes of soil particles that include sands, gravels, silt, and clay. More than half of the composition by weight of mineral matter is oxygen.

14. Soil structure is the arrangement of the soil particles into aggregates. Good soils are highly granular and have pore spaces that constitute about half the total volume.

15. Water in the soil occurs as hygroscopic water, gravitational water, and capillary water.

16. The field capacity of the soil is the amount of water that remains after the rest of the water has drained away by gravity. Soil reaches the permanent wilting point when plants wilt permanently because they can no longer extract enough water from the soil for their needs. Available water is soil water between field capacity and the permanent wilting point.

Review Questions

1. Distinguish between a tiny root and a root hair. What is the function of a root hair?

2. What is the difference between parasitic roots and mycorrhizae?

3. If you were shown cross sections of a young root and a young stem from the same dicot plant, how could you tell them apart?

4. What is the function of the root cap and from which meristem does it originate?

5. How do endodermal cells differ from other types of cells?

6. Where do branch roots originate?

7. List some spices and drugs obtained from roots.

8. What is soil parent material? Where does it come from, and how does it become soil?

9. What is the difference between soil *texture* and soil *structure?*

10. What types of soil water are recognized? What is *available* soil water?

Discussion Questions

1. Japanese gardeners regularly trim away parts of the root system to assist in dwarfing a plant. A plant's food is obtained through photosynthesis in the leaves; can you suggest why trimming the roots can cause dwarfing?

2. It was suggested that roots perceive gravity through the root cap. Would it really matter if roots grew randomly in the soil instead of responding to gravity?

3. From the viewpoint of the plant, can you suggest a practical reason for branch roots originating internally instead of at the surface?

4. When you eat a yam or a sweet potato, what kinds of compounds and cells are you consuming?

5. Persons associated with commercial nurseries and greenhouses often sterilize their soil by heating it to get rid of pests, but then they have to wait for a short time after it has cooled to use it. Why?

Additional Reading

Allen, M.F. (Ed.). 1992. *Mycorrhizal functioning*. New York: Chapman & Hall.

Brady, N. C. 1989. *The nature and properties of soils*, 10th ed. New York: Macmillan.

Cutter, E.G. 1971. *Plant anatomy: Experiment and interpretation. Part 2: Organs*. Reading, MA: Addison-Wesley Publishing Co.

Davis, T. D., and B. E. Haissig (Eds.). 1994. *Biology of adventitious root formation*. New York: Plenum Publishing Corp.

Donahue, R. L., et al. 1990. *Our soils and their management*, 6th ed. Danville, IL: Interstate.

Epstein, E. 1973. "Roots." *Scientific American* 228: 48–58.

Fahn, A. 1990. *Plant anatomy*, 4th ed. Elmsford, NY: Pergamon Press.

Gregory, P.J., et al. (Eds.). 1987. *Root development and function: Effects of the physical environment*. Fair Lawn, NY: Cambridge University Press.

Hatfield, J. L., and B. A. Stewart (Eds.). 1992. *Limitations to plant root growth*. New York: Springer-Verlag.

Mauseth, J.D. 1988. *Plant anatomy*. Menlo Park, CA: Benjamin/Cummings Publishing Co., Inc.

Miller, R., and R. L. Donahue. 1990. *Soils: An introduction to soils and plant growth*, 6th ed. New York: Prentice-Hall.

Singer, M. J. 1991. *Soils: An introduction*. New York: Macmillan.

Torrey, J.G., and D. Clarkson (Eds.). 1975. *The development and function of roots*. San Diego, CA: Academic Press.

Waisel, Y., and E. Amram (Eds.). 1991. *Plant roots: The hidden half*. New York: Dekker, Marcel, Inc.

Chapter Outline

The trunk of a Mindanao gum tree (Eucalyptus deglupta) *of Indonesia and the Philippine Islands.*

Stems

Overview

After a brief introduction, this chapter discusses in general the origin and development of stems. Items such as the apical meristem and the tissues derived from it, leaf gaps, cambia, secondary tissues, and lenticels are included. This general discussion is followed by notes on the distinctions between herbaceous and woody dicot stems and on monocot stems. This section covers annual rings, rays, heartwood and sapwood, resin canals, bark, laticifers, and vascular bundles.

Next there is a survey of specialized stems (rhizomes, stolons, tubers, bulbs, corms, cladophylls, and others). The chapter concludes with a discussion of the economic importance of wood and stems.

Some Learning Goals

1. Know the tissues that develop from shoot apices and the meristems from which each tissue is derived. Distinguish between primary tissues and secondary tissues.
2. Learn and give the function of each of the following: vascular cambium, cork cambium, stomata, lenticels.
3. Contrast the stems of herbaceous and woody dicots with the stems of monocots.
4. Understand the composition of wood and its annual rings, sapwood, heartwood, and bark. Explain how a log is sawed for commercial use.
5. Distinguish among rhizomes, stolons, tubers, bulbs, corms, cladophylls, and tendrils.
6. Learn at least 10 human uses of wood and stems in general.

When you order a dozen long-stemmed roses for someone special, use a toothpick or chopsticks, build a wood-framed house, sit in a wooden chair and read a newspaper, or graft one variety of fruit tree to another, you may or may not be conscious of the fact that these and literally hundreds of other activities either directly or indirectly involve plant stems, and that these particular plant organs have been an integral part of human life since cave dwellers first used wooden clubs to kill for food.

One of the activities just mentioned—*grafting*—usually involves artificially uniting stems, or parts of stems, of different but related varieties of plants. The careful matching of certain tissues is critical to its success, as is seen in the discussion of grafting in Chapter 14. To understand how and why grafts may or may not be successful and to identify which parts of stems are useful, we first need to examine the structure of stems and learn the basic functions of the various tissues.

Unlike animals, some plants can grow indefinitely, with the meristems at their tips increasing their length and other meristems increasing their girth for hundreds or even thousands of years. In stems, the cells produced by the meristems usually become the familiar, erect, aerial *shoot system* with branches and leaves. In certain plants, such as ferns or perennial grasses, this shoot system may develop horizontally beneath or at the surface of the ground; in other plants, the stem may be so short and inconspicuous as to appear nonexistent. In a number of plants, modifications of the stem permit specialized functions, such as climbing or the storage of food or water.

EXTERNAL FORM OF A WOODY TWIG

A woody twig consists basically of an axis with leaves attached (Fig. 6.1). The leaves may be attached to the twig in a spiral around the stem (they are then said to be *alternate* or *alternately arranged*), they may occur in pairs (*opposite* or *oppositely arranged*), or they may occur in groups of three or more called *whorls* (*whorled arrangement*). The area, or region (*not* structure), of a stem where a leaf or leaves are attached is called a **node,** and a stem region between nodes is called an **internode.** A leaf usually has a flattened *blade* and a stalk, called the **petiole,** by which it is attached to the twig.

A bud occurs in the angle between a petiole and the stem. This angle is called an **axil,** and the bud located in the axil is an *axillary bud.* Axillary buds may become branches or they may contain tissues that will develop into the next season's flowers. Most, but not all, buds are protected by one to several *bud scales,* which fall off when bud tissue growth begins.

A *terminal bud* is often present at the tip of each twig. It usually resembles an axillary bud, but it is frequently a little larger. Unlike axillary buds, terminal buds do not form separate branches but, instead, normally produce tissues that extend the length of the twig during the growing season. The bud scales of a terminal bud leave tiny scars around the twig when they fall off in the spring. A twig's age can be determined by counting the number of groups of *bud scale scars* on it.

Sometimes, scars with a different origin also occur on a twig. These scars come from a leaf that has paired appendages called **stipules** at the base of the petiole. The stipules may remain throughout the life of the leaf, but in some plants they fall off as the buds expand in the spring, leaving tiny *stipule scars,* which may resemble a fine line encircling the twig or may be very inconspicuous small scars on either side of the petiole base.

Deciduous trees and shrubs (those that lose their leaves annually) have characteristic **leaf scars,** with dormant axillary buds above them after the leaves fall. Tiny **bundle scars,** which mark the location of the water-conducting and food-conducting tissues, can be seen within the leaf scars. There are frequently three bundle scars present, but the number can vary from one to many. The shape and size of the leaf scars and the arrangement and numbers of the bundle scars are characteristic of each species. It is often possible to determine the identity of a woody plant in its winter condition by means of these structures.

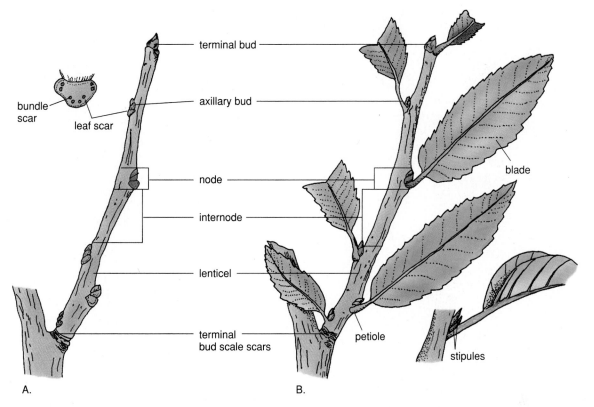

FIGURE 6.1 A woody twig. *A*. The twig in its winter condition. *B*. The same twig as it appeared the summer before.

ORIGIN AND DEVELOPMENT OF STEMS

As indicated in Chapter 4, there is an *apical meristem* at the tip of each stem, and it is this meristem that produces the tissues resulting in the stem's increase in length. Before the onset of the growing season, the apical meristem is dormant. It is protected by bud scales of the bud in which it is located and also to a certain extent by leaf **primordia** (singular: **primordium**), the tiny embryonic leaves that will develop into mature leaves after the bud scales drop off and growth begins. The apical meristem in the embryonic stem of a seed is also dormant until the seed begins to germinate.

When a bud begins to expand, or a seed germinates, the cells of the apical meristem undergo mitosis, and soon three primary meristems develop from it (see Fig. 4.1). The outermost of these primary meristems, the **protoderm,** gives rise to the epidermis. As noted in Chapter 4, the epidermis is typically one cell thick and usually becomes coated with a thin, fatty, protective layer, the *cuticle.* A cylinder of strands constituting the **procambium** appears to the interior of the protoderm. (The procambium produces water-conducting *primary xylem cells* and food-conducting *primary phloem cells.*)

The remainder of the meristematic tissue, called **ground meristem,** produces two tissues composed of parenchyma cells. The parenchyma tissue in the center of the stem is the **pith.** Pith cells tend to be very large and may break down shortly after they are formed, leaving a cylindrical hollow area. Even if they do not break down early, they may eventually be crushed as new tissues produced by other meristems add to the girth of the stem, particularly in woody plants. The other tissue produced by the ground meristem is the **cortex.** The cortex may become more extensive than the pith, but in woody plants it, too, eventually will be crushed and replaced by new tissues produced from within. The parenchyma of both the pith and the cortex function in storing food or, sometimes, if chloroplasts are present, in manufacturing it.

All five of the tissues (epidermis, primary xylem, primary phloem, pith, and cortex) produced by this apical meristem complex arise while the stem is increasing in length and are called *primary tissues*. As these primary tissues are produced, the leaf primordia and the *bud primordia* (embryonic buds in the axils of the leaf primordia) develop into mature leaves and buds (Fig. 6.2). As each leaf and each bud develop, a strand of xylem and phloem, called a *trace,* branches off from the cylinder of xylem and phloem extending up and down the stem and enters the leaf or the bud. As the traces branch from the main cylinder of xylem and phloem, each trace leaves a little thumbnail-shaped gap in the cylinder of tissue. These gaps are called **leaf gaps** and *bud gaps* (Fig. 6.3).

A narrow band of cells between the primary xylem and the primary phloem may retain its meristematic nature and become the *vascular cambium,* one of the two lateral meristems. The vascular cambium is often referred

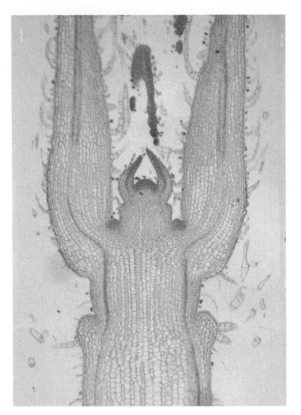

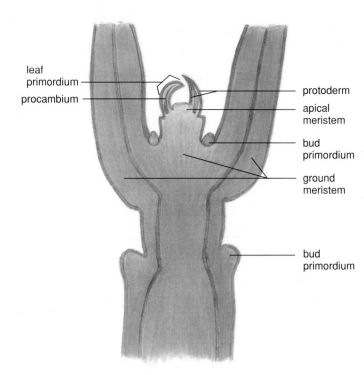

leaf
primordium

procambium

protoderm

apical
meristem

bud
primordium

ground
meristem

bud
primordium

FIGURE 6.2   A longitudinal section through the tip of a *Coleus* stem, ca. ×800.
(Photomicrograph by G.S. Ellmore)

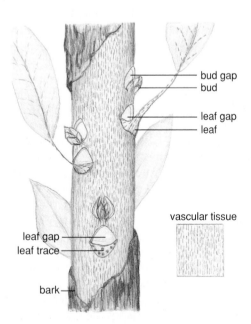

bud gap
bud

leaf gap
leaf

vascular tissue

leaf gap
leaf trace

bark

FIGURE 6.3   A portion of a young stem showing leaf gaps
and bud gaps in the cylinder of vascular tissue.

to simply as the *cambium.* The cells of the cambium continue to divide indefinitely, with the divisions taking place mostly in a plane parallel to the surface of the plant. The *secondary tissues* produced by the vascular cambium and the cork cambium thus add to the girth of the stem instead of to its length (Fig. 6.4).

Cells produced by the vascular cambium become *tracheids, vessel members, fibers,* or other components of *secondary xylem* (*inside* of the meristem, toward the center), or they become sieve-tube members, companion cells, or other components of *secondary phloem* (*outside* of the meristem, toward the surface). The functions of these secondary tissues are the same as those of their primary counterparts—secondary xylem conducts water and soluble nutrients, while secondary phloem conducts food, in soluble form, throughout the plant.

In many plants, especially woody species, a second cambium arises within the cortex or, in some instances, develops from the epidermis or phloem. This is called the **cork cambium,** or **phellogen.** The cork cambium produces box-like **cork cells,** which become impregnated with **suberin,** a fatty substance that makes the cells impervious to moisture. The cork cells, which are produced annually in cylindrical layers, die shortly after they are formed. The cork cambium may also produce parenchymalike *phelloderm* cells to the inside. Cork tissue makes up the outer bark of woody plants; it functions in reducing water loss and in protecting the stem

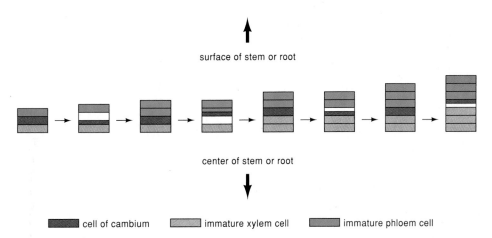

surface of stem or root

center of stem or root

■ cell of cambium   ▭ immature xylem cell   ▬ immature phloem cell

FIGURE 6.4 An illustration of how a cell of the vascular cambium produces new secondary phloem cells to the outside and new secondary xylem cells to the inside. Note, in cross section, that the cambium gradually becomes shifted away from the center as new cells are produced. Phloem is produced before xylem in secondary growth.

against mechanical injury (the periderm is discussed in Chapter 4).

Cork tissue cuts off water and food supplies to the epidermis, which soon dies and is sloughed off. In fact, if the cork were to be formed as a solid cylinder covering the entire stem, vital gas exchange with the interior of the stem would not be possible. In young stems, such gas exchange takes place through the *stomata* located in the epidermis (see figs. 7.7 and 9.15). As woody stems age, **lenticels** (see Fig. 4.13) develop beneath the stomata. As cork is produced, the unsuberized cells of the lenticels remain, so exchange of gases (e.g., oxygen, carbon dioxide) can continue through spaces between the cells. As indicated in Chapter 4, lenticels occur in the fissures of the bark of older trees and often appear as small bumps on younger bark. In birch and cherry trees, the lenticels form conspicuous horizontal lines.

Differences between the activities of the apical meristem and those of the cambium and cork cambium become apparent if one drives a nail into the side of a tree and observes it over a period of years. The nail may eventually become embedded as the stem increases in girth, but it will always remain at the same height above the ground, as the cells that increase the length of a stem are produced only at the tips.

TISSUE PATTERNS IN STEMS

Steles

Primary xylem, primary phloem, and the pith, if present, make up a central cylinder called the **stele** in at least most younger stems and roots. The simplest form of stele, called a *protostele,* consists of a solid core of conducting tissues in which the phloem usually surrounds the xylem. Protosteles were common in primitive seed plants that are now extinct, and they are also found in relatives of ferns, such as whisk ferns and club mosses (discussed in Chapter 21). *Siphono-*

steles, which are tubular with pith in the center, occur in most ferns.

Most present-day flowering plants and conifers have *eusteles* in which the primary xylem and primary phloem are in *vascular bundles,* as discussed later.

Flowering plants develop from seeds that have either one or two "seed leaves," called **cotyledons,** attached to the embryonic stem (see chapters 11 and 23), while the seeds of cone-bearing trees, such as pines, have several (usually eight) cotyledons. The cotyledons may function in storing food needed by the young seedling until its first true leaves can produce food themselves.

Flowering plants that develop from seeds having two cotyledons are called **dicotyledons** (often abbreviated to *dicot*), while those developing from seeds with a single cotyledon are called **monocotyledons** (abbreviated to *monocot*). Dicots and monocots differ from one another in several other respects; differences in stem structure are noted in the following sections, and a summary of these and other differences between these two classes of flowering plants is given in Table 8.1.

Herbaceous Dicotyledonous Stems

In general, plants that die after going from seed to maturity within one year (**annuals**) have green herbaceous (nonwoody) stems. Most monocots (discussed after herbaceous dicots) are annuals, but many dicots (discussed next) are also annuals.

The tissues of annual dicots are largely primary, although cambia (plural of cambium) may develop some secondary tissues. Herbaceous dicot stems (Fig. 6.5) have discrete patches of xylem and phloem called **vascular bundles,** which occur in a ring that separates the cortex from the pith, although in a few plants (e.g., foxgloves), the xylem and the phloem are produced as continuous rings.

As previously noted, the procambium produces only primary xylem and phloem, but later a vascular cambium

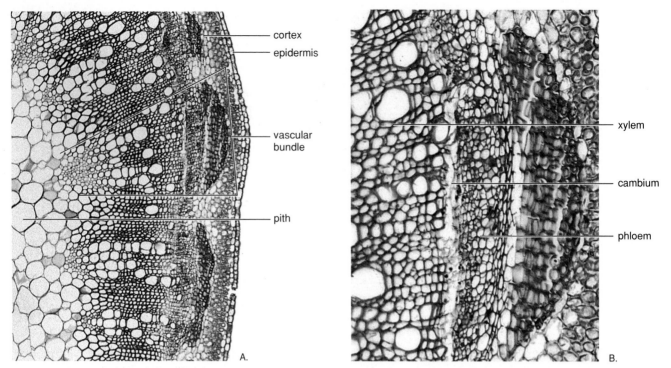

FIGURE 6.5   *A.* A cross section of an alfalfa (*Medicago*) stem, ×100. The tissue arrangement is typical of herbaceous dicot stems. *B.* An enlargement of a small portion of the outer part of the stem, ×500.

arises between these two primary tissues and adds secondary xylem and phloem to the vascular bundles. In some plants, the cambium extends between the vascular bundles, appearing as a narrow ring producing the conducting tissues within the bundles and the parenchyma cells between them. In other plants, the cambium is confined to the bundles, each of which has its own small band of this meristematic tissue between the xylem and phloem.

Woody Dicotyledonous Stems

The arrangement of primary tissues in woody dicot stems (and also in those of cone-bearing trees) is very similar to that found in herbaceous dicot stems during the early stages of growth. As soon as the vascular cambium and the cork cambium start functioning, however, obvious differences begin to appear, the most conspicuous of which involve the secondary xylem, or *wood,* as it is best known (Fig. 6.6). Some tropical trees (e.g., ebony), in which both the vascular cambium and the cork cambium are active all year, produce an ungrained, uniform wood. The wood of most trees, however, exhibits seasonal growth. In trees of temperate climates, virtually all growth takes place during the spring and summer and then ceases for the year.

When the vascular cambium of a typical broadleaf tree first becomes active in the spring, it usually produces relatively large vessel members of secondary xylem; such xylem is referred to as *spring wood*. As the season progresses, the vascular cambium may produce vessel members whose diameters become progressively smaller in each succeeding

series of cells produced, or there may be fewer vessel members in proportion to tracheids produced until tracheids (and sometimes fibers) predominate.

This xylem, which is produced after the spring wood, and which has smaller or fewer vessel members and larger numbers of tracheids, is referred to as *summer wood*. Over a period of years, the result of this type of switch between the early spring and the summer growth is a series of alternating concentric rings of light and dark cells. One year's growth of xylem is called an **annual ring.** In conifers (discussed in Chapter 22), vessels and fibers are absent, the wood consisting mostly of tracheids. Annual rings are still visible, however, since the first tracheids produced in the spring are considerably larger and lighter in color than those produced later in the growing season. Note that an annual ring normally may contain many layers of xylem cells, and it is all the layers produced in one growing season that constitute an annual ring, not just the dark layers.

The vascular cambium produces more secondary xylem than it does phloem. In addition, xylem cells have stronger, more rigid walls than do phloem cells and thus are less subject to collapse under pressure. As a result, the bulk of a tree trunk consists of annual rings of wood. The annual rings not only indicate the age of the tree (since normally only one is produced each year), but they can also tell something of the climate and other conditions occurring during the tree's lifetime (Fig. 6.7). For example, if the rainfall during a particular year is higher than normal, the annual ring for that year will be wider than usual. Sometimes, caterpillars or locusts will strip the leaves of a tree shortly after they have appeared. This usually results in two annual rings being very close together, since very little growth can occur under such conditions.

If a fire not resulting in the death of the tree occurs, it may be possible to determine the year of its occurrence, since the burn scar may appear among the rings. The most recent season's growth is directly adjacent to the vascular cambium, and one need only count the rings back from the cambium to determine the actual year of the fire.

It is not necessary to cut down a tree to determine its age. Botanists and foresters employ an *increment borer* for this purpose. This device, which resembles a piece of pipe with a handle on one end, removes from the tree a plug of wood perpendicular to the trunk. The annual rings can then be counted in the plug; the small hole left in the tree can be treated with a disinfectant and covered up to prevent disease and any harm to the tree.

A count of annual rings has produced some red faces on at least one occasion. The Hooker Oak, located in the community of Chico, California, and named in honor of a famous British botanist who once examined it, was a huge, much-visited tree until its demise in 1977. Beneath the tree was a plaque indicating the tree to be over 1,000 years old. A count of rings after its death, however, revealed it to be less than 300 years old.

When a tree trunk is examined in transverse, or cross section, lighter streaks or lines called **rays** can be seen radiating out from the center across the annual rings (see figs. 6.8 and 6.14). That part of a ray within the xylem is called a *xylem ray;* its extension through the phloem is called a *phloem ray.* In basswood trees, some of the phloem rays, when observed in cross section, flare out from a width of two or three cells near the cambium to many cells wide in the part next to the cortex. In longitudinal section, rays may be from two or three cells to 50 or more cells deep, but the majority of rays in both xylem and phloem are one or two cells wide.

Rays consist of parenchyma cells that may remain alive for 10 or more years. Their primary function is the lateral conduction of nutrients and water from the stele to the cortex, with some cells also functioning in food storage. Ray cells can be seen in cross section if a woody stem is cut or split lengthwise along a ray (Fig. 6.8). Another view of rays (in tangential section) is obtained when the stem is cut at a tangent (i.e., cut lengthwise and off center)(see Fig. 6.17).

As a tree ages, the protoplasts of some of the parenchyma cells that surround the vessels and tracheids grow through the pits in the walls of these conducting cells and balloon out into the cavities. The growth continues until much of the cavity of the vessel or tracheid has been filled. Such protrusions, called *tyloses* (singular: *tylosis*), prevent further conduction of water and dissolved substances. When this occurs, resins, gums, and tannins begin to accumulate, along with pigments that darken the color of the wood.

This older, darker wood at the center is called **heartwood,** while the lighter, still-functioning xylem closest to the cambium is called **sapwood** (Fig. 6.9). Except for giving strength and support, the heartwood is not of much use to the

tree since it can no longer conduct materials. A tree may live and function perfectly well after the heartwood has rotted away and left the interior hollow (Fig. 6.10). It is even possible to remove part of the sapwood and other tissues and apparently not affect the tree very much, as has been done with giant trees, such as the coastal redwoods of California, where holes big enough to drive a car through have been cut out without killing the trees.

Sapwood forms at approximately the same rate as heartwood develops, so there is always sufficient "plumbing" for the vital conducting functions. The relative widths of the two types of wood, however, vary considerably from species to species. For example, in the golden chain tree (a native of Europe and a member of the Legume Family), the sapwood is usually only one or two rings wide, while in several North American trees (e.g., maple, ash, and beech), the sapwood may be many rings wide.

As mentioned earlier, cone-bearing trees, such as pine trees (discussed in Chapter 22), have xylem that consists primarily of tracheids; no fibers or vessel elements are produced. The wood tends to be softer than that of trees with fibers and is commonly referred to as *softwood,* while the wood of woody dicot trees is called *hardwood.*

In many cone-bearing trees, **resin canals** are scattered not only through the xylem but also throughout other tissues. These canals are tubelike and may or may not be branched; they are lined with specialized cells that secrete resin (discussed in Chapter 22) into their cavities (Fig. 6.11). Although resin canals are commonly associated with cone-bearing trees, they are not confined to them. Tropical flowering plants, such as olibanum and myrrh trees, have in the bark resin ducts that produce the soft resins frankincense and myrrh of biblical note.

While the vascular cambium is producing secondary xylem to the inside, it is also producing secondary phloem to the outside. The term **bark** is usually applied to all the tissues outside the cambium, including the phloem. Some scientists distinguish between the *inner bark,* consisting of primary and secondary phloem, and the *outer bark* (periderm), consisting of cork tissue and cork cambium. Despite the presence of thick-walled fibers in the phloem, thin-walled conducting cells of this tissue are not usually able to withstand for many seasons the pressure of thousands of new cells added to their interior, and the older layers become crushed and functionless.

The parenchyma cells of the cortex to the outside of the phloem also function only briefly because they too become crushed or sloughed off. Before they disappear, however, the cork cambium begins its production of cork, and since new xylem and phloem tissues produced by the vascular cambium arise to the inside of the older phloem, the mature bark may consist of alternating layers of crushed phloem and cork.

The younger layers of phloem nearest to the cambium transport, via their sieve tubes, sugars and other substances in solution from the points of origin in the leaves to various parts of the plant, where they are either stored or used in the

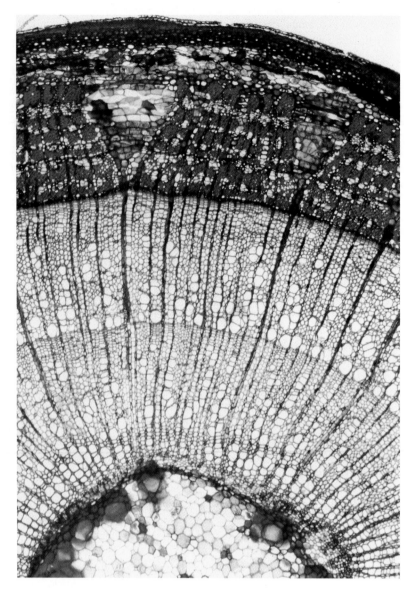

FIGURE 6.6 A cross section of a portion of a young linden (*Tilia*) stem, ca. ×300.

process of *respiration* (discussed in Chapter 10). This sugar content of the phloem was recognized by North American Indian tribes. Some stripped the young phloem and cambium from Douglas fir trees and used the dried strips as food for winter and in emergencies.

Specialized cells or ducts called **laticifers** are found in about 20 families of herbaceous and woody flowering plants. These cells are most common in the phloem but occur throughout all parts of the plants. The laticifers, which resemble vessels, form extensive branched networks of latex-secreting cells originating from rows of meristematic cells. Unlike vessels, however, the cells remain living and may have many nuclei.

Latex is a thick fluid that is white, yellow, orange, or red in color and consists of gums, proteins, sugars, oils, salts, alkaloidal drugs, enzymes, and other substances. Its function in the plant is not clear, although some believe it aids in closing wounds. Some forms of latex have considerable commercial value (see the discussion of the Spurge Family in

Chapter 24). Of these, rubber is the most important. Amazon Indians utilized rubber for making balls and containers hundreds of years before Pará rubber trees were cultivated for their latex. The chicle tree produces a latex used in the making of chewing gum. Several poppies, notably the opium poppy, produce a latex containing important medicinal drugs, such as morphine, and other drug complexes, such as heroin. Other well-known latex producers include milkweeds, dogbanes, and dandelions.

Monocotyledonous Stems

Most monocots (e.g., grasses, lilies) are herbaceous plants that do not attain great size. The stems have neither a vascular cambium nor a cork cambium and thus produce no secondary vascular tissues or cork. As in herbaceous dicots, the surfaces of the stems are covered by an epidermis, but the xylem and phloem tissues produced by the procambium appear in cross section as discrete vascular bundles,

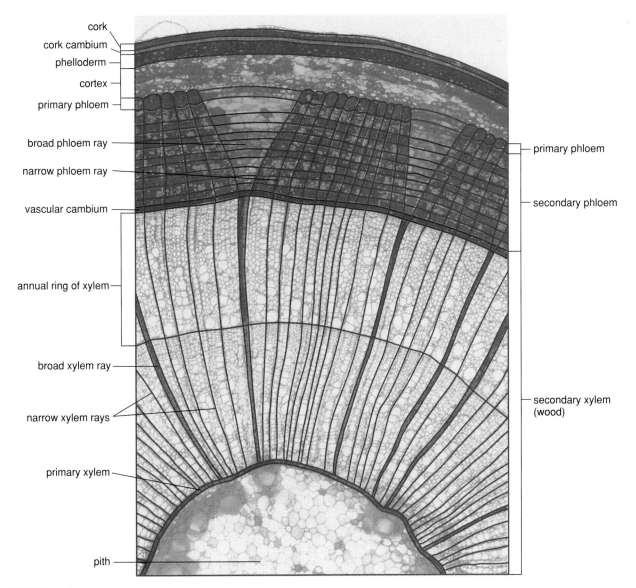

cork
cork cambium
phelloderm
cortex
primary phloem
broad phloem ray
narrow phloem ray
vascular cambium
annual ring of xylem
broad xylem ray
narrow xylem rays
primary xylem
pith

primary phloem
secondary phloem
secondary xylem
(wood)

FIGURE 6.6 continued

scattered throughout the stem instead of being arranged in a ring (Fig. 6.12).

Each bundle, regardless of its specific location, is oriented so that its xylem is closest to the center of the stem and its phloem is closest to the surface. In a typical monocot such as corn, a bundle's xylem usually contains two large vessels with several small vessels between them (Fig. 6.13). The first-formed xylem cells usually stretch and collapse under the stresses of early growth and leave an irregularly shaped air space toward the base of the bundle; the remnants of a vessel are often present in this air space. The phloem consists entirely of sieve tubes and companion cells, and the entire bundle is surrounded by a sheath of sclerenchyma cells. The parenchyma tissue between the vascular bundles is not separated into cortex and pith in monocots, although its function and appearance are the same as those of the parenchyma cells in cortex and pith.

The bundles of a corn stem are more numerous just beneath the surface than they are toward the center. Also, a band of sclerenchyma cells, usually two or three cells thick, occurs immediately beneath the epidermis, and parenchyma cells in the area develop thicker walls as the stem matures. The concentration of bundles, combined with the additional band of sclerenchyma cells beneath the epidermis and the thicker-walled parenchyma cells, all contribute to giving the stem the capacity to withstand stresses resulting from summer storms and the weight of the leaves and the ears of corn as they mature.

In wheat, rice, barley, oats, rye, and other grasses, there is an *intercalary meristem* (discussed in Chapter 4) at the base of each internode; like the apical meristem, it contributes to elongation in growth. The stems of such plants elongate rapidly during the growing season, but because there is no vascular cambium producing tissues that would add to the girth of the stems, growth is columnar, with little difference in diameter between the top and the bottom.

Palm trees differ from most monocots in that they attain considerable size, but they do so primarily as a result of their parenchyma cells continuing to divide and enlarge without a true cambium developing. Several popular houseplants (e.g., ti plants, *Dracaena, Sansevieria*) are monocots in which a secondary meristem develops as a cylinder that extends throughout the stem. Unlike the vascular cambium of dicots and conifers, this secondary meristem produces only parenchyma cells to the outside and secondary vascular bundles to the inside.

Several commercially important cordage fibers (e.g., broomcorn, Mauritius and Manila hemps, sisal) come from the stems and leaves of monocots, but the individual cells are not separated from one another by *retting* (a process that utilizes the rotting power of microorganisms thriving under moist conditions to break down thin-walled cells) as they are when fibers from dicots are obtained. Instead, during commercial preparation, entire vascular bundles are scraped free of the surrounding parenchyma cells by hand; the individual bundles then serve as unit "fibers." If such fibers are treated with chemicals or bleached, the cementing middle lamella between the cells breaks down. Monocot fibers are not as strong or as durable as most dicot fibers.

SPECIALIZED STEMS

While an erect shoot system is characteristic of most higher plants, many species have stems that perform specialized functions; these stems are modified accordingly (Fig. 6.14). Although the overall appearance of specialized stems may differ markedly from that of the stems discussed so far, all stems have *nodes, internodes,* and *axillary buds;* these features distinguish them from roots and leaves, which do not have them. The leaves at the nodes of these specialized stems are often small and scalelike. They are seldom green, but full-sized functioning leaves are also produced. Descriptions of some of the specialized stems follow.

Rhizomes

Rhizomes (Fig. 6.14) are horizontal stems that grow below ground, often near the surface of the soil. Superficially, they resemble roots, but a close examination will reveal scalelike leaves and axillary buds at each node, at least during some stage of development, with short to long internodes in between. **Adventitious** roots are produced all along the rhizome, mainly on the lower surface. As indicated in Chapter 5, the word *adventitious* refers to structures arising at unusual places, such as roots growing from stems or leaves, or buds appearing at places other than leaf axils and tips of stems. A rhizome may be a relatively thick, fleshy, food-storage organ, as in irises, or it may be quite slender, as in many perennial grasses and some ferns.

Runners and Stolons

Runners are horizontal stems that differ from rhizomes in that they grow above ground, generally along the surface; they also have long internodes (Fig. 6.14). In strawberries,

FIGURE 6.7   Climatic history illustrated by a cross section of a 62-year-old tree.

(Courtesy St. Regis Paper Company)

1914
When the tree was 6 years old, something pushed against it, making it lean. The rings are now wider on the lower side, as the tree builds "reaction wood" to help support it.

1924
The tree is growing straight again. But its neighbors are growing too, and their crowns and root systems take much of the water and sunshine the tree needs.

1927
The surrounding trees are harvested. The larger trees are removed and there is once again ample nourishment and sunlight. The tree can now grow rapidly again.

1930
A fire sweeps through the forest. Fortunately, the tree is only scarred, and year by year more and more of the scar is covered over by newly formed wood.

1942
These narrow rings may have been caused by a prolonged dry spell. One or two dry summers would not have dried the ground enough to slow the tree's growth this much.

1957
Another series of narrow rings may have been caused by an insect like the larva of the sawfly. It eats the leaves and leafbuds of many kinds of coniferous trees.

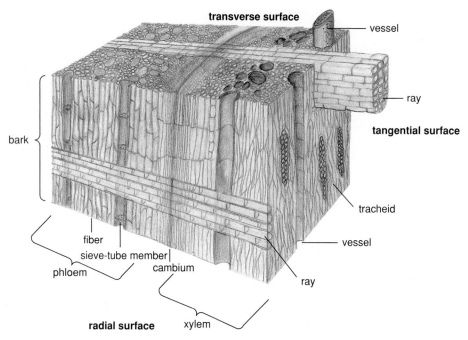

FIGURE 6.8   A three-dimensional, magnified view of a block of a woody dicot.

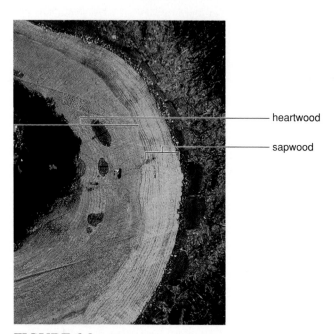

FIGURE 6.9   This tree was 100 years old when it was cut down. Note the proportion of *sapwood,* which consists of functional cells, to *heartwood,* in which the cells are no longer capable of conduction.

FIGURE 6.10   An African baobab tree.

runners are usually produced after the first flowering of the season has occurred. Several may radiate out from the parent plant, attaining lengths of up to 1 meter (3 feet) or more within a few weeks. Adventitious buds appear at alternate nodes along the runners and develop into new strawberry plants, which can be separated and grown independently. In some house plants, such as the saxifrages, runners may grow out and hang over the edge of the pot, producing new plants at intervals.

Stolons are similar to runners but usually grow more or less vertically beneath the surface of the ground. In Irish potato plants, *tubers* are produced at the tips of stolons.

Tubers

As food accumulates at the tips of stolons, such as those produced by Irish or white potato plants, several internodes swell, becoming **tubers** (Fig. 6.14). After the tuber is mature, the stolon dies, isolating it. The "eyes" of the potato are actually nodes formed in a spiral around the modified stem.

resin canals

FIGURE 6.11   Resin canals in a portion of a pine (*Pinus*) stem, ca. ×400.

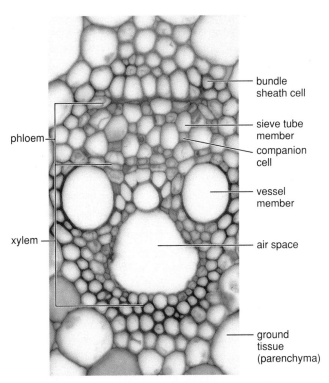

bundle sheath cell

phloem

sieve tube member

companion cell

vessel member

xylem

air space

ground tissue (parenchyma)

FIGURE 6.13   A single vascular bundle of corn (*Zea mays*) enlarged, ca. ×800.

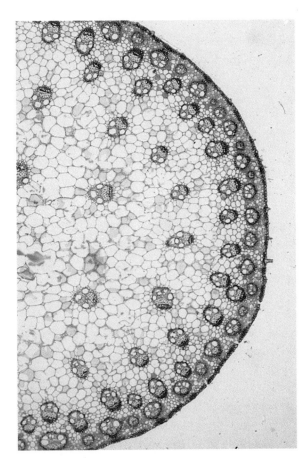

FIGURE 6.12   A portion of a cross section of a monocot (corn—*Zea mays*) stem, ca. ×100.

(Photomicrograph by G.S. Ellmore)

Each eye consists of an axillary bud in the axil of a scalelike leaf, although the latter is visible only in very young tubers, the small ridges seen on mature tubers being leaf scars.

Bulbs

Bulbs (Fig. 6.14) are actually large buds with a small stem at the lower end surrounded by numerous fleshy leaves. Adventitious roots grow from the bottom of the stem, but the fleshy leaves comprise the bulk of the bulb tissue, which functions in food storage. In onions, the fleshy leaves usually are surrounded by the scalelike leaf bases of long, green, aboveground leaves. Other plants producing bulbs include lilies, hyacinths, and tulips.

Corms

Corms, which superficially resemble bulbs, differ from them in being composed almost entirely of stem tissue except for the few papery scalelike leaves sparsely covering the outside (Fig. 6.14). Adventitious roots are produced at the base, and corms, like bulbs, function in food storage. Well-known plants producing corms include crocuses and gladioli.

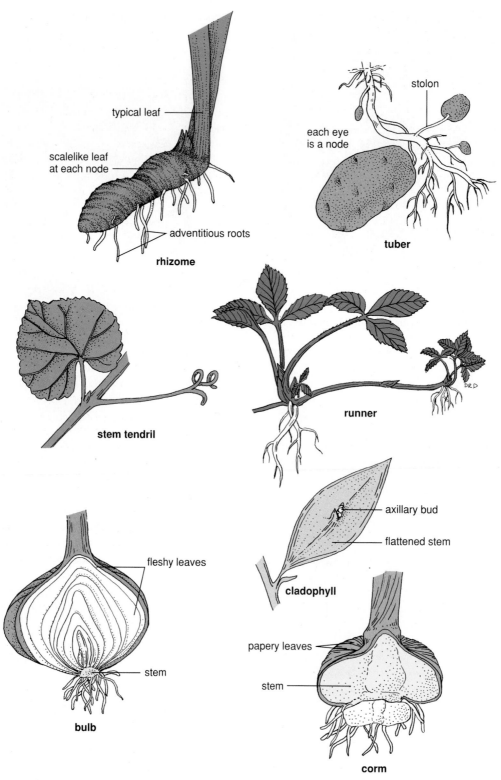

typical leaf

scalelike leaf
at each node

adventitious roots

rhizome

stolon

each eye
is a node

tuber

stem tendril

runner

axillary bud

flattened stem

cladophyll

fleshy leaves

stem

bulb

papery leaves

stem

corm

FIGURE 6.14   Types of specialized stems.

FIGURE 6.15   The flattened stems of prickly pear cacti (*Opuntia*) are cladophylls on which the leaves have been reduced to spines.

Cladophylls

In butcher's-broom plants, stems are flattened and leaflike in appearance. Such stems are called **cladophylls** (or *cladodes* or *phylloclades*) (Fig. 6.14). In the center of each butcher's-broom cladophyll is a node bearing very small scalelike leaves with axillary buds. The feathery appearance of asparagus is due to the presence of numerous small cladophylls. Cladophylls also occur in greenbriers, certain orchids, prickly pear cacti (Fig. 6.15), and several lesser-known plants.

Other Specialized Stems

In many cacti and in some of the spurges, the stems are stout and fleshy. Such stems are modified for water and food storage. Other stems may be modified in the form of *thorns,* as in the honey locust whose branched thorns may exceed 3 decimeters (1 foot) in length, but all thornlike objects are not necessarily modified stems. For example, the black locust has a pair of *spines* at the base of each petiole of most leaves (these are parts of the leaf called *stipules,*

mentioned in the discussion of twigs and discussed further in Chapter 7). The prickles of raspberries and roses, both of which originate from the epidermis, are neither thorns nor spines. Tiger lilies produce small aerial bulblets in the axils of their leaves.

Climbing plants have stems modified in various ways that adapt them for their growth habit. Some stems, called *ramblers,* simply rest on the tops of other plants, but many produce **tendrils** (Fig. 6.14). These are specialized stems in the grape and Boston ivy but are modified leaves or leaf parts in plants such as peas and cucumbers. In Boston ivy, the tendrils have adhesive disks. In English ivy, the stems climb with the aid of adventitious roots that arise along the sides of the stem and become embedded in the bark or other support material over which the plant is growing.

WOOD AND ITS USES

The use of wood by humans dates back into antiquity, and present uses are so numerous that it would be impossible to list in a work of this type more than the most important ones. Before looking at the economic importance of wood, a brief discussion of its properties is in order.

In a living tree, up to 50% of the weight of the wood comes from the water content. Before the wood can be used, the moisture content is reduced to 10% or less through *seasoning,* either by air-drying the wood in ventilated piles or stacks or by drying it in special ovens known as *kilns.* The seasoning has to be done gradually and under carefully controlled conditions or the timber may warp and split along the rays, making it unfit for most uses. The dry part of wood is composed of 60% to 75% *cellulose* and about 15% to 25% *lignin,* an organic substance that is deposited in the walls of xylem cells and makes the walls tough and hard. Other substances present in smaller amounts include resins, gums, oils, dyes, tannins, and starch. The proportions and amounts of these and other substances determine how various woods will be used (Fig. 6.16).

Density

Among the most important physical properties of wood is its *density.* Technically, the density is the weight per unit volume. The weight is compared with that of an equal volume of water and is stated as a fraction of 1.0. Because of the considerable air space within the cells, most woods have a *specific gravity,* as the comparative density is called, of less than 1.0. The range of specific gravities of known woods varies from 0.04 to 1.40, the lightest commercially used wood being balsa with a specific gravity of about 0.12. Woods with specific gravities of less than 0.50 are considered light; those with specific gravities of above 0.70 are considered heavy. Among the heaviest woods are the South

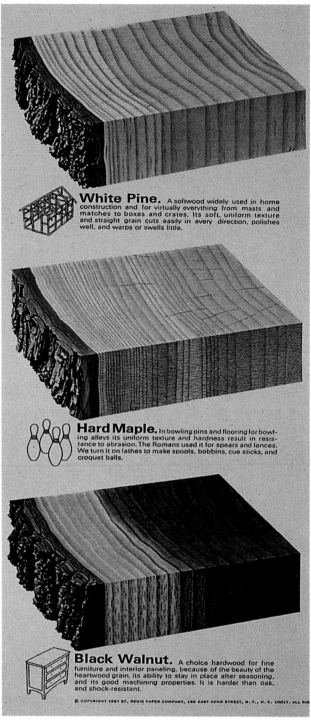

White Pine. A softwood widely used in home construction and for virtually everything from masts and matches to boxes and crates. Its soft, uniform texture and straight grain cuts easily in every direction, polishes well, and warps or swells little.

Hard Maple. In bowling pins and flooring for bowling alleys its uniform texture and hardness result in resistance to abrasion. The Romans used it for spears and lances. We turn it on lathes to make spools, bobbins, cue sticks, and croquet balls.

Black Walnut. A choice hardwood for fine furniture and interior paneling, because of the beauty of the heartwood grain, its ability to stay in place after seasoning, and its good machining properties. It is harder than oak, and shock-resistant.

White Oak. Makes good barrels because the wood is resilient, durable, and impermeable to liquids. This hardwood, which is about twice as dense as white pine, has many other uses ranging from flooring to fine cabinet work.

Baldcypress. Because it is weather-resistant without treatment, this wood was widely used for cross ties in the early days of railroading. Today it is used for water tanks and other applications requiring prolonged contact with water.

White Ash. Perfect for baseball bats, tennis racquets, oars and long tool handles. This hardwood's major virtues are straight grain, stiffness, strength, moderate weight, good bending qualities, and capacity for wearing smooth.

FIGURE 6.16 Uses of some common North American woods.
(Courtesy St. Regis Paper Company)

American ironwood and lignum vitae, with specific gravities of over 1.25. Lignum vitae, obtained from West Indian trees, is extremely hard wood and is used instead of metal in the manufacture of main bearings for drive shafts of submarines, because it is self-lubricating and less noisy.

Durability

A wood's ability to withstand decay organisms and insects is referred to as its *durability*. Moisture is needed for the enzymatic breakdown of cellulose and other wood

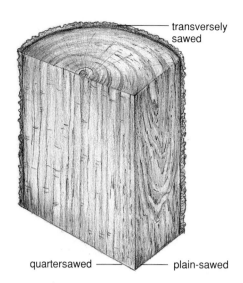

FIGURE 6.17 How the surfaces of plain-sawed, quartersawed, and transversely sawed wood appear (See also Fig. 6.8).

FIGURE 6.16 continued

or more may survive on a forest floor for many years after the tree has been felled by diseases of the phloem or other causes. Among the most durable of American woods are cedar, catalpa, black locust, red mulberry, and Osage orange. The least durable woods include cottonwood, willow, fir, and basswood.

Types of Sawing

Logs are usually cut longitudinally in one of two ways: along the radius or perpendicular to the rays (Fig. 6.17). Radially cut, or *quartersawed,* boards show the annual rings in side view; they appear as longitudinal streaks and are the most conspicuous feature of the wood. Only a few perfect quartersawed boards can be obtained from a log, making them quite expensive. Boards cut perpendicular to the rays (tangentially cut boards) are more common. In these, the annual rings appear as irregular bands of light and dark alternating streaks or patches, with the ends of the rays visible as narrower and less conspicuous vertical streaks. Lumber cut tangentially is referred to as being *plain-sawed* or *slab cut.* Slabs are the boards with rounded sides at the outside of the log; they are usually made into chips for pulping.

Knots

Knots are the bases of lost branches that have become covered, over a period of time, by new annual rings of wood produced by the cambium of the trunk. They are found in greater concentration in the older parts of the log toward the center, because in the forest, the lowermost branches of a tree (produced while the trunk was small in girth) often die from lack of sufficient light. When a branch dies and falls

substances by decay organisms, but the seasoning process usually reduces the moisture to a level below that necessary for the fungi and other decay organisms to survive. Other natural constituents of wood that repel decay organisms include tannins and oils. Wood with a tannin content of 15%

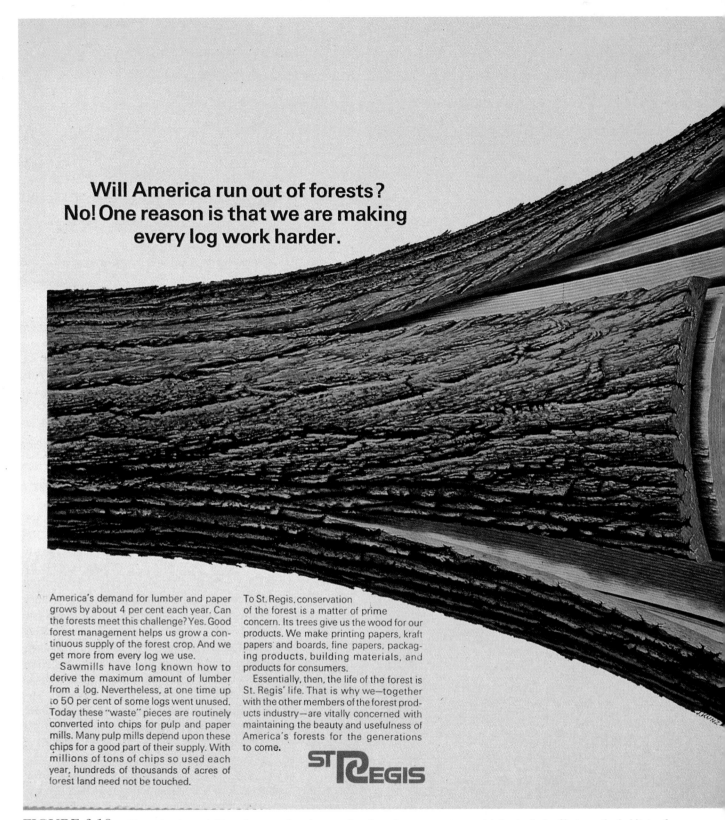

FIGURE 6.18 How a log is used. Note that some logging practices have become controversial due to their effects on the habitats of threatened species of living organisms, as discussed in Chapter 25.

(Courtesy St. Regis Paper Company)

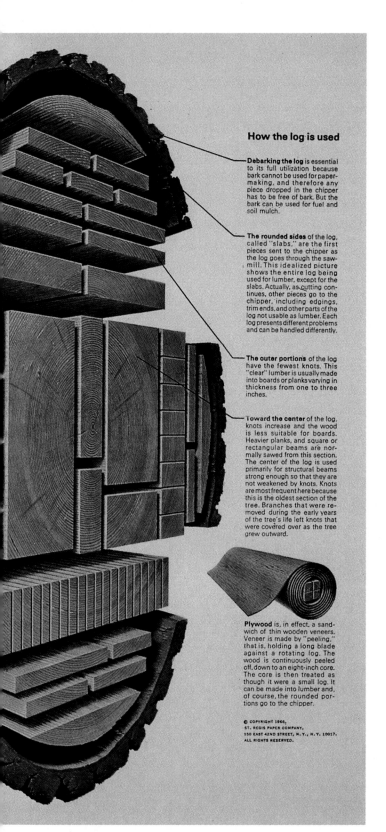

How the log is used

Debarking the log is essential to its full utilization because bark cannot be used for paper-making, and therefore any piece dropped in the chipper has to be free of bark. But the bark can be used for fuel and soil mulch.

The rounded sides of the log, called "slabs," are the first pieces sent to the chipper as the log goes through the saw-mill. This idealized picture shows the entire log being used for lumber, except for the slabs. Actually, as cutting continues, other pieces go to the chipper, including edgings, trim ends, and other parts of the log not usable as lumber. Each log presents different problems and can be handled differently.

The outer portions of the log have the fewest knots. This "clear" lumber is usually made into boards or planks varying in thickness from one to three inches.

Toward the center of the log, knots increase and the wood is less suitable for boards. Heavier planks, and square or rectangular beams are normally sawed from this section. The center of the log is used primarily for structural beams strong enough so that they are not weakened by knots. Knots are most frequent here because this is the oldest section of the tree. Branches that were removed during the early years of the tree's life left knots that were covered over as the tree grew outward.

Plywood is, in effect, a sandwich of thin wooden veneers. Veneer is made by "peeling," that is, holding a long blade against a rotating log. The wood is continuously peeled off, down to an eight-inch core. The core is then treated as though it were a small log. It can be made into lumber and, of course, the rounded portions go to the chipper.

off, the cambium at its base also dies, but the cambium of the trunk remains alive and increases the girth of the tree, slowly enveloping the dead tissue of the branch base until it may be completely buried and not visible from the surface. Knots usually weaken the boards in which they occur.

Wood Products

In the United States, about half of the wood produced is used as lumber, primarily for construction; the sawdust and other waste formed in processing the boards is converted to particle board and pulp. A considerable amount of lumber goes into the making of furniture, which may be constructed of solid wood or covered with a *veneer*. A veneer is a very thin sheet of desirable wood that is glued to cheaper lumber; it is carefully cut so as to produce the best possible view of the grain (Fig. 6.18).

The next most extensive use of wood is for *pulp*, which, among other things, is converted by various processes to paper, synthetic fibers, plastics, and linoleum. In recent years, it has been added as a filler to commercial ice cream and bread. Some hardwoods are treated chemically or heated under controlled conditions to yield a number of chemicals, such as wood alcohol and acetic acid, but other sources of these products are now usually considered more economical. Charcoal, excelsior, cooperage (kegs, casks, and barrels), railroad ties, boxes and crates, musical instruments, bowling pins, tool handles, pilings, cellophane, photographic film, and Christmas trees are but a few of the additional wood products worth billions of dollars annually on the world market (see Fig. 6.16).

In developing countries, approximately half of the timber cut is used for fuel, but in the United States, a little less than 10% is currently used for that purpose. In colonial times, wood was the almost exclusive source of heating energy. In the 1980s, Brazil's major cities were still using scrub timber from the surrounding forests to energize their utilities, but rapidly depleting supplies and problems related to the greenhouse effect (discussed in Chapter 25) have pointed to the need for alternate sources of energy. Many types of coal are wood that has been compressed for millions of years until nearly pure carbon remains. The formation of coal and other fossils is discussed in Chapter 21. Although the world's supply of coal is still plentiful, the rate at which this fossil fuel is being consumed makes it obvious that resources will eventually be exhausted unless our energy demands find renewable or less destructive alternatives.

Some of the vast array of secondary products from stems, including dyes, medicines, spices, and foods, are discussed in later chapters and in the appendices.

Standing in Fields of Stone

The combination of high altitude, dry cool air, low rainfall, high winds, and poor soil has provided sustenance for the oldest known living species on the earth! These ancient warriors, whose great age was unknown until 1953, are the **bristlecone pines** that flourish atop the arid mountains of the Great Basin from Colorado to California. The oldest—determined to be almost 5,000 years old—is located in the Ancient Bristlecone Pine Forest high in the White Mountains of California. Some of the trees standing today were seedlings when the great pyramids were built, were middle-aged trees during the time of Christ, and today are hoary patriarchs standing in fields of stone.

There are two species of bristlecones, one living in the westernmost regions (*Pinus longaeva*) and the other inhabiting the eastern regions (*Pinus aristata*). The trees do not grow very tall, with none over 60 feet, and many much shorter. Typical of their girth is the bristlecone named the "Patriarch" that is just over 36 1/2 feet around, but it is a relative youngster at 1,500 years. With the short summer growing season high in these mountains, bristlecones typically grow 1/100th of an inch or less in diameter in any given year. The trees stand in isolation, each appearing as sentinels, overlooking an otherwise barren rock-strewn landscape. Their needles are remarkable. While occurring five per bundle and about 1 to 1.5 inches in length, they can live for 20 to 30 years before being cast off. This extraordinary leaf longevity gives the trees a stable photosynthetic output and sustains the tree during years of unusual stress, when producing new leaf tissue would be difficult. The trees must generate new leaves and cones as well as produce enough reserves for the long winter months, all on scant annual precipitation of about 10 inches.

Ecologists have long noted the peculiar distribution and growth habits of certain plants. Why do plants grow where they do? What adaptations permit them to survive in life-threatening environments? The bristlecones age and habitat offer several insights. These trees grow in places on earth where no other plant wants to grow. One answer appears to be the type of soil in which they are anchored. Stands of brislecone pine grow on outcrops of dolomite, an alkaline limestone substrate of low nutrient but higher moisture content than the surrounding sandstone. The granite and sandstone formations surrounding the dolomite outcroppings support sagebrush and Limber pine, but not bristlecones. At these altitudes the radiant sunlight is extreme. Dolomite reflects more sunlight than other rocks and thus keeps the root zones cooler and moisture-laden during the important growing season.

Another survival tactic is revealed by taking small core samples of the wood. These trees have large amounts of die-back (deadwood that is no longer functional), reducing the amount of tissue that the leaves need to supply with food. For example, one bristlecone over 4,000 years old is nearly four feet in diameter, but it has only a 10 inch wide strip of bark to sustain the plant.

BOX FIGURE 6.1 Ancient bristlecone pine from the White Montains of California.

Other characteristics provide bristlecones with survival advantages. Because their dense resin-filled wood renders them inhospitable sites for colonization by pathogenic fungi or bacteria, they are relatively free from these attacks. These trees are often struck by lightning but with the absence of ground cover and decaying leaf litter, fire rarely spreads from tree to tree.

The age of these living trees is determined, like others, by an instrument called an **increment borer,** a type of drill that is inserted into the tree trunk at its widest girth. Using a hollow drill bit, a linear core of wood is obtained, which can be "read" for the number of **annual growth rings.** In this way the age of a tree can be obtained without cutting the tree down to examine the annual rings revealed in the cross-section of the stump.

Each year a tree will add a layer of wood to its trunk and these become the annual rings that can be observed in a cross-section of the trunk. During spring growth of wood, large-diametered water-conducting cells are formed; later in the summer, the water-conducting cells produced have a smaller diameter. This difference in appearance between early (spring) wood and late (summer) wood is sufficient to make each growth increment distinctive, and thus counting of the rings possible.

In 1957 "Methuselah" was discovered and determined to be 4,723 years old. Methuselah remains today as the world's oldest living organism. But what about tomorrow? After surviving nearly five millennia, Methuselah is being protected against a more insidious enemy. Standing in its field of stone without a marker because of fear from vandalism, Methuselah serves as a reminder that the human species can be just as destructive as any microbe.

Summary

1. To understand practical uses of stems, an examination of stem structure and function is needed.

2. The shoot system of plants is usually erect, but some stems may be horizontal or modified for climbing or food and water storage.

3. Leaves of woody twigs may be arranged alternately, oppositely, or in a whorl. Nodes are stem regions where leaves are attached; internodes occur between nodes. Most leaves have petioles and blades. Axillary buds occur in leaf axils. Most buds are protected by bud scales. Terminal buds occur at twig tips. Terminal bud scales, when they fall, leave bud scale scars that help determine the age of the twig.

4. Stipules are paired appendages present at the base of some leaves; when they fall off, they leave small scars on the twig. When whole leaves fall, they cause leaf scars on the twig, with tiny bundle scars within the leaf-scar surfaces.

5. Each stem has an apical meristem at its tip that produces tissues resulting in increase in length. Leaf primordia develop into mature leaves when growth begins. Three primary meristems develop from an apical meristem: The protoderm gives rise to the epidermis; the procambium produces primary xylem and primary phloem; and the ground meristem produces pith and cortex.

6. As leaves and buds develop from primordia, traces of xylem and phloem branch off from the main cylinder, leaving leaf gaps or bud gaps.

7. A vascular cambium, producing secondary tissues, may arise between primary xylem and phloem. Secondary xylem cells include tracheids, vessel members, and fibers. Secondary phloem cells include sieve-tube members and companion cells.

8. In many plants, a cork cambium producing cork and phelloderm cells develops near the surface of the stem. Cork cells, which are part of the outer bark (periderm), have suberin in their walls. Suberin is impervious to moisture and the outer bark therefore aids in protection. Lenticels in the bark permit gas exchange.

9. Primary vascular tissues and the pith, if present, constitute the stele. Protosteles have a solid core of xylem, usually surrounded by phloem; siphonosteles are tubular, with pith in the center; eusteles have the vascular tissues in discrete bundles.

10. Dicotyledons (dicots) are plants whose seeds have two seed leaves (cotyledons), while monocotyledons (monocots) have seeds with one seed leaf. Herbaceous dicots have vascular bundles arranged in a ring in the stem.

11. Woody dicots have most of their secondary tissues arranged in concentric layers. The most conspicuous tissue is wood (secondary xylem). In broadleaf trees, spring wood usually has relatively large vessel members, while summer wood has smaller vessels and/or a predominance of tracheids.

12. An annual ring is one year's growth of xylem. A tree's age and other aspects of its history can be determined from annual rings. Rays, which function in lateral conduction, radiate out from the center of the trunk. Older wood toward the center (heartwood) ceases to function when its cells become plugged with tyloses. Younger, functioning wood (sapwood) is closer to the surface. A tree's functions are not particularly affected by the rotting of its heartwood.

13. The wood of cone-bearing trees consists primarily of tracheids, and resin canals are often present. The wood of conifers has no fibers or vessels and is called *softwood*, while the wood of woody dicots is called *hardwood*. In woody plants, older tissues composed of thin-walled cells become crushed and functionless, and some are sloughed off.

14. Laticifers are latex-secreting cells or ducts found in various flowering plants. The latex of some plants has considerable commercial value.

15. Monocot stems have scattered vascular bundles and no cambia. The parenchyma tissue is not divided into pith and cortex. Each vascular bundle is surrounded by a sheath of sclerenchyma cells. Numerous bundles and a band of sclerenchyma cells and thicker-walled parenchyma cells just beneath the surface of monocot stems aid in withstanding stresses.

16. Palm trees are monocots that become large because their parenchyma cells continue to divide. Other monocots develop a secondary meristem that produces parenchyma cells and secondary vascular bundles. Grasses have at the base of each internode intercalary meristems that contribute to rapid increases in length. Several commercially important cordage fibers are obtained from monocots.

17. Specialized stems include rhizomes, stolons, tubers, bulbs, corms, cladophylls, and tendrils. Such stems may have adventitious roots.

18. The dry part of wood consists primarily of cellulose and lignin. Resins, gums, oils, dyes, tannins, and starch are also present. Properties of wood that play a role in its use include density, specific gravity, and durability.

19. Logs are usually cut longitudinally along the radius (quartersawed) or perpendicular to the rays (tangentially, plain-sawed, or slab cut). Knots are bases of lost branches that have become covered over by new wood; they usually weaken the boards in which they occur.

20. About half the timber produced in the United States is used as lumber. Sawdust and waste are converted to particle board and pulp for paper, synthetics, and linoleum. Other timber is used for cooperage, charcoal, railroad ties, boxes, tool handles, and so on. Developing countries use a greater proportion of their timber for fuel than do other countries.

Review Questions

1. What is the function of bud scales?

2. How can you tell the age of a twig?

3. Distinguish among procambium, vascular cambium, and cork cambium.

4. How can you tell, when you look at a cross section of a young stem, whether it is a dicot or a monocot?

5. What are laticifers?

6. An Irish or white potato is a stem, but a sweet potato is a root. How can you tell the difference?

7. Distinguish among corms, bulbs, and tubers.

8. If you were examining the top of a wooden desk, how could you tell if the wood had been radially or tangentially cut (quartersawed or plain-sawed)?

9. What differences are there between heartwood and sapwood?

10. What is meant by the specific gravity of wood?

Discussion Questions

1. If the cambium of a tropical tree were active all year long, how would its wood differ from that of a typical temperate climate tree?

2. It was mentioned that a nail driven into the side of a tree will remain at exactly the same distance from the ground for the life of the tree. Why?

3. Do climbing plants have any advantages over erect plants? Any disadvantages?

4. If two leaves are removed from a plant and one is coated with petroleum jelly while the other is not, the uncoated leaf will shrivel considerably sooner than will the coated one. Would it be helpful to coat the stems of young trees with petroleum jelly? Explain.

5. Suggest some reasons for heartwood being preferred to sapwood for making furniture.

Additional Reading

Core, H.A., W.A. Cote, and A.C. Day. 1979. *Wood structure and identification*, 2d ed. Syracuse, NY: Syracuse University Press.

Cutter, E.G. 1978. *Plant anatomy: Experiment and interpretation. Part I: Cells and tissues*, 2d ed. Reading, MA: Addison-Wesley Publishing Co.

Fahn, A. 1990. *Plant anatomy*, 4th ed. Elmsford, NY: Pergamon Press.

Flynn, J. H., Jr. (Ed.). 1994. *A guide to useful woods of the world.* Portland, ME: King Philip Publishing Co.

Hughs, M.K., et al. (Eds.). 1982. *Climate from tree rings.* Fair Lawn, NY: Cambridge University Press.

Lewin, M. 1991. *Wood structure and composition.* New York: Dekker, Marcel, Inc.

Mauseth, J.D. 1988. *Plant anatomy.* Menlo Park, CA: Benjamin/Cummings Publishing Co., Inc.

Metcalfe, C.R., and L. Chalk (Eds.). 1988–1989. *Anatomy of the dicotyledons*, 2 vols. Fair Lawn, NY: Oxford University Press.

Meylan, B.A., and B.G. Butterfield. 1972. *Three-dimensional structure of wood: A scanning electron microscope study.* Syracuse, NY: Syracuse University Press.

Schweingruber, F. H. 1993. *Trees and wood in dendrochronology.* New York: Springer-Verlag.

Steeves, T.A., and I.M. Sussex. 1989. *Patterns in plant development*, 2d ed. Englewood Cliffs, NJ: Prentice-Hall.

Chapter Outline

Leaves of Croton, *a colorful member of the Spurge Family. The intense red anthocyanin pigments mask the visibility of the green chlorophyll pigments, which are also present.*

Leaves

Overview————

This chapter introduces leaves by comparing them with solar panels and by discussing their general functions, morphology, and dimensions. This discussion is followed by descriptive information on basic leaf types and specific forms and arrangements. The chapter next discusses the internal structure of leaves, including epidermis and cuticle, stomata, glands, mesophyll, and veins.

Specialized leaves, including tendrils, spines, flower pot leaves, window leaves, reproductive leaves, floral leaves, and insectivorous leaves, are then examined. The chapter concludes with an explanation of autumnal color changes and abscission and some observations on the human and ecological relevance of leaves.

Some Learning Goals

1. Learn the external forms and parts of leaves. Know the functions of a typical leaf and the specific tissues and cells that contribute to those functions.
2. Understand the differences among pinnate, palmate, and dichotomous venation and also the differences between simple and compound leaves.
3. Contrast tendrils, spines, storage leaves, flower pot leaves, window leaves, reproductive leaves, floral leaves, and different types of insect-trapping leaves.
4. Explain why deciduous leaves turn various colors in the fall and how such leaves are shed.
5. Know at least 15 uses of leaves by humans.

FIGURE 7.1   English ivy on a tree trunk. Note how each leaf blade is oriented to receive the maximum amount of light.

The earliest records of glass being used by humans date back to about 2600 B.C., when the ancient Egyptians and Babylonians made beads from the material. The use of glass panes for windows, however, did not begin until the Roman Imperial period a little over 2,000 years ago. Since then, the use of glass windows for admitting light to buildings of all sizes and shapes has become almost universal.

A comparatively recent use of glass involves solar energy as an alternative to nonrenewable sources of energy such as fossil fuels. The construction industry, particularly in the southwestern United States, is building new houses with flat panels and windows inclined at angles that maximize the amount of energy captured from the sun's rays. Some buildings have mechanical devices that slowly move the solar panels so that they will follow the sun in its daily course across the sky. The use of such means of capturing solar energy is now spreading to other countries and could become commonplace during the 21st century.

Plants had a highly efficient form of solar panel that captured the sun's energy many aeons before modern civilization began to realize that fossil fuel supplies eventually would be exhausted. These remarkably constructed solar panels are the plant organs known to us as **leaves.**

The flattened surfaces of leaves, which are completely covered with a transparent protective layer of cells, the *epi-dermis,* admit light to all parts of the interior. Many leaves daily twist on their stalks, or *petioles,* so that their upper surfaces incline at right angles to the sun's rays throughout daylight hours (Fig. 7.1).

Green leaves capture the light energy available to them by means of the most important process for life on earth, at least life as we know it today. This process, called *photosynthesis* (discussed in Chapter 10), involves the trapping of energy in sugar molecules that are constructed from ordinary water and from carbon dioxide present in the atmosphere. All the energy needs of living organisms ultimately depend on photosynthesis, from the first day of their existence to the last.

The lower surfaces of leaves (and sometimes the upper surfaces as well) are dotted with tiny pores called *stomata* (singular: *stoma*), which not only permit air circulation but also are involved with cooling the interior of the leaf, as water evaporates from the moist cell surfaces and passes out of them into the air in vapor form.

Leaves also perform other functions. For example, all living cells *respire* (respiration is also discussed in Chapter

10), and in the process of this and other metabolic activities, waste products are produced. These wastes are deposited outside the plant when the leaves are shed, mostly in the fall, after being sealed off at the bases of their petioles (see the discussion of leaf colors and *abscission* later in this chapter). The following season, the discarded leaves are replaced with new ones.

Leaves are involved in the movement of the water absorbed by the roots and transported throughout the plant. Most of the water reaching the leaves evaporates into the atmosphere by a process known as **transpiration** (discussed in Chapter 9). In some plants, *root pressure* (see page 150) forces water out of *hydathodes,* which are special openings at the tips of leaf veins, usually at night when transpiration is not occurring (see page 153). The water thus expelled may contain ions secreted by root cells. Other functions of leaves are discussed throughout this chapter.

LEAF ARRANGEMENTS AND TYPES

Many of the roughly 275,000 different kinds of plants that produce leaves can be distinguished from one another by their leaves alone. In addition, many plants, such as lilies, ferns, and pine trees, produce on the same plant different forms of leaves, including tiny papery scales. The variety of shapes, sizes, and textures of leaves seems to be almost infinite. Some of the smaller duckweeds have leaves less than 1 millimeter (0.04 inch) wide. The leaves of the Seychelles

Island palm attain spans of 6 meters (20 feet), and the floating leaves of a giant water lily, which reach 2 meters (6.5 feet) in diameter (Fig. 7.2), can support, without sinking, weights of more than 45 kilograms (100 pounds) distributed over their surface.

In addition to flattened, variously shaped leaves, there are others that are tubular, feathery, cup shaped, spinelike, or needlelike; in fact, leaves may assume virtually any form (Fig. 7.3). They may be smooth or hairy, slippery or sticky, waxy or glossy, pleasantly fragrant or foul smelling, edible or poisonous. They also may be of almost every color of the rainbow and of exquisite beauty, especially when viewed with a microscope.

As indicated in Chapter 6, all leaves originate as *primordia* in the buds, regardless of their ultimate size or form. At maturity, most leaves consist of a stalk, called the **petiole,** and a flattened **blade,** or *lamina,* which has a network of *veins (vascular bundles).* A pair of appendages, called **stipules,** are sometimes present at the base of the petiole. Occasionally, leaves may lack petioles; when they do, they are said to be *sessile.* Leaves of deciduous trees normally live through only one growing season, and even those of evergreen trees rarely are functional for more than two to seven years. Leaves of flowering plants are associated with *leaf gaps* (see Fig. 6.3 and associated text) and may be *simple* or *compound.*

The arrangement of leaves on a stem (*phyllotaxy*) may be one in which the leaves are in a *spiral* (*alternate*) pattern, or two leaves may be attached at the same node, providing an *opposite* arrangement. When three or more leaves occur at a node, they are said to be **whorled.** The leaf itself may be

FIGURE 7.2 Floating leaves of a giant water lily, which sometimes attain a diameter of 2 meters (6.5 feet). The larger leaves are capable of supporting, without sinking, the weight of a child.

FIGURE 7.3 Types of leaves and leaf arrangements. *A.* Palmately compound leaf of a buckeye. *B.* Pinnately compound leaf of a black walnut. *C.* Alternate, simple but lobed leaves of a tulip tree. *D.* Opposite, simple leaves of a dogwood. *E.* Palmately veined leaf of a maple. *F.* Globe-shaped succulent leaves of string-of-pearls. *G.* Pinnately veined, lobed leaf of an oak. *H.* Parallel-veined leaf of a grass. *I.* Whorled leaves of a bedstraw. *J.* Linear leaves of a yew. *K.* Fan-shaped leaf of a *Ginkgo* tree, showing dichotomous venation.

a **simple leaf,** with an undivided blade, or it may be a **compound leaf,** with the blade divided into *leaflets* in various ways (see Fig. 7.3). **Pinnately compound** leaves have the leaflets in pairs along a central stalklike **rachis,** while **palmately compound** leaves have all the leaflets attached at

the same point at the end of the petiole. Sometimes, the leaflets of a pinnately compound leaf may be subdivided into still smaller leaflets, forming a *bipinnately compound* leaf.

The arrangement of veins in a leaf or leaflet blade (*venation*) may also be either pinnate or palmate. In **pinnately**

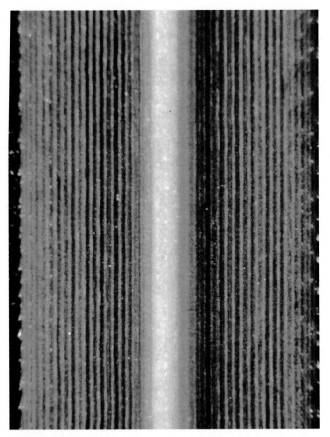

FIGURE 7.4 A portion of a monocot leaf showing the parallel veins.

veined leaves, there is a main vein, called the **midrib,** with secondary veins branching from it, but in **palmately veined** leaves, several veins fan out from the base of the blade. The larger veins parallel one another in monocots (Fig. 7.4) and diverge from one another in various ways in dicots (see Fig. 7.8). In a few leaves (e.g., those of *Ginkgo*), no midrib or other large veins are present. Instead, the veins fork evenly and progressively from the base of the blade to the opposite margin. This is called *dichotomous venation* (Fig. 7.3 K).

INTERNAL STRUCTURE OF LEAVES

If a typical leaf is cut transversely and examined with the aid of a microscope, three regions stand out: *epidermis, mesophyll,* and *veins* (referred to as *vascular bundles* in our discussion of roots and stems) (Fig. 7.5). The epidermis is a single layer of cells covering the entire surface of the leaf. The epidermis on the lower surface of the blade can sometimes be distinguished from the upper epidermis by the presence of *stomata,* which are discussed in the section that follows.

When seen from the top, the wavy, undulating epidermal cells often resemble pieces of a jigsaw puzzle fitted together. Except for *guard cells,* the upper epidermal cells contain no chloroplasts, their function being primarily protection of the delicate tissues to the interior. A thin coating of waxy **cutin** (the **cuticle**—see Fig. 4.10) is normally present,

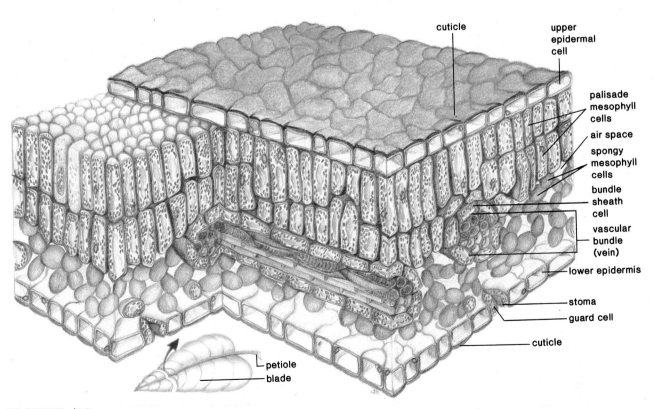

FIGURE 7.5 A stereoscopic view of a portion of a typical leaf.

FIGURE 7.6 Waxes, in addition to those of the cuticle, are sometimes produced on the surfaces of leaves, stems, and fruits, giving them a whitish appearance, as seen on this black raspberry cane.

although it may not be visible with ordinary light microscopes without being specially stained. Many plants produce, in addition to the cuticle, other waxy substances on their surfaces (Fig. 7.6).

In studies of the effects on plants of smog and auto exhaust fumes, it was found that these waxes may be produced in abnormal fashion on beet leaves within as little as 24 hours of exposure to the pollutants. Beet leaves also respond to aphid damage by producing wax around each tiny puncture. Occasionally, epidermal cells contain crystals of waste materials. Different types of **glands** may also be present in the epidermis. Glands occur in the form of depressions, protuberances, or appendages either directly on the leaf surface or on the ends of hairs (see Fig. 4.11). Glands often secrete sticky substances.

STOMATA

The lower epidermis of most plants generally resembles the upper epidermis, but it is perforated by numerous tiny pores, the **stomata** (Fig. 7.7). They also occur in both leaf surfaces of some plants (e.g., alfalfa, corn), exclusively on the upper epidermis of other plants such as water lilies where the lower epidermis is in contact with water, and are absent altogether from the submerged leaves of aquatic plants. Stomata are very numerous, ranging from about 1,000 to more than 1.2 million per square centimeter (6,300 to 8 million per square inch) of surface. An average-sized sunflower leaf has about two million of these pores throughout its lower epidermis. Each stoma is bordered by two sausage- or dumbbell-shaped cells that usually are smaller than most of the neighboring epidermal cells. These are called **guard cells,** and unlike other cells of either epidermis, they contain chloroplasts.

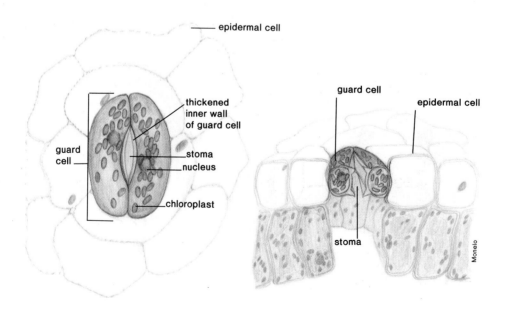

FIGURE 7.7 A typical dicot stoma. *A*. Surface view. *B*. View in transverse section.

The photosynthesis that takes place in the guard cells aids in the functioning of the cells. The primary functions include permitting gas exchange between the interior of the leaf and the atmosphere and regulation of evaporation of most of the water entering the plant at the roots.

The guard cell walls are distinctly thickened but quite flexible on the side adjacent to the pore. As the guard cells expand or contract with changes in the amount of water within the cells, their unique construction causes the stomata to open or close. When the guard cells are expanded, the stomata are open; when the water content of the guard cells decreases, the cells contract and the stomata close. (For more detailed discussions of this stomatal mechanism see the sections on regulation of transpiration in Chapter 9 and turgor movements in Chapter 11.)

MESOPHYLL AND VEINS

Most photosynthesis takes place in the **mesophyll** between the two epidermal layers. When two regions of the mesophyll are distinguishable, as is often the case, the uppermost region consists of compactly stacked parenchyma cells; the cells are mostly shaped like short posts and are commonly in two rows. This region is called the **palisade mesophyll** and may contain more than 80% of the leaf's chloroplasts. The lower region, consisting of loosely arranged parenchyma cells with abundant air spaces between them, is called the **spongy mesophyll.** Its cells also contain numerous chloroplasts.

Parenchyma tissue containing numerous chloroplasts is referred to as *chlorenchyma* tissue. Besides being present in leaves, chlorenchyma tissue is found in the outer parts of the cortex in the stems of herbaceous plants. Inside the leaf, the surfaces of mesophyll cells in contact with the air are moist. If moisture decreases below a certain level, the stomata close, thereby significantly reducing further drying.

Veins (*vascular bundles*) of various sizes are scattered throughout the mesophyll (Fig. 7.8). They consist of xylem and phloem tissues surrounded by a jacket of fibers called the **bundle sheath.** The veins give the leaf its "skeleton." The carbohydrates produced in the mesophyll cells are transported in solution throughout the plant by the phloem. Water, sometimes located more than 100 meters (330 feet) away in the ground below, is brought up to the leaf by the xylem, which, like the phloem, is part of a vast network of "plumbing" throughout the plant. Since veins run in all directions in the network, particularly in dicots, it is common, when examining a cross section of a leaf under the microscope, to see veins cut transversely, lengthwise, and at a tangent, all in the same section.

Monocot leaves, in addition to having *parallel veins,* usually do not have the mesophyll differentiated into palisade and spongy layers. Some monocot leaves (e.g., those of grasses) have large, thin-walled *bulliform cells* on either side of the main central vein (midrib) toward the upper surface (Fig. 7.9). The bulliform cells partly collapse under dry

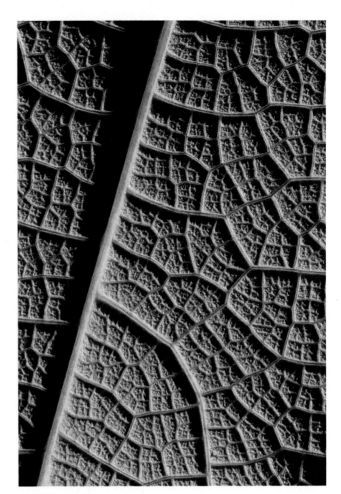

FIGURE 7.8   Net venation in a leaf, ×50.

conditions, causing the leaf blade to fold or roll, thus reducing transpiration (see Fig. 11.16).

SPECIALIZED LEAVES

If the leaves of all plants could function normally under any environmental condition, various leaf modifications would provide no special benefits to a plant. But the form and structure of plants of tropical rain forests do not permit them to thrive in a desert and cacti soon die if planted in a creek because their structure, form, and life cycles are attuned to specific combinations of environmental factors, such as temperature, humidity, light, water, and soil conditions. In addition, the modifications of leaves occupying any single ecological niche may be very diverse, resulting in such a rich variety of leaf forms and specializations throughout the Plant Kingdom that only a few may be mentioned here.

Shade Leaves

Some leaves with inconspicuous modifications may occur along with unmodified leaves on the same plant. For example, leaves in the shade tend to be thinner and have fewer

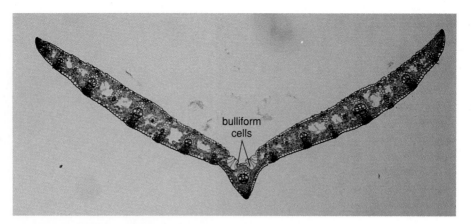

FIGURE 7.9   A cross section of a grass leaf.

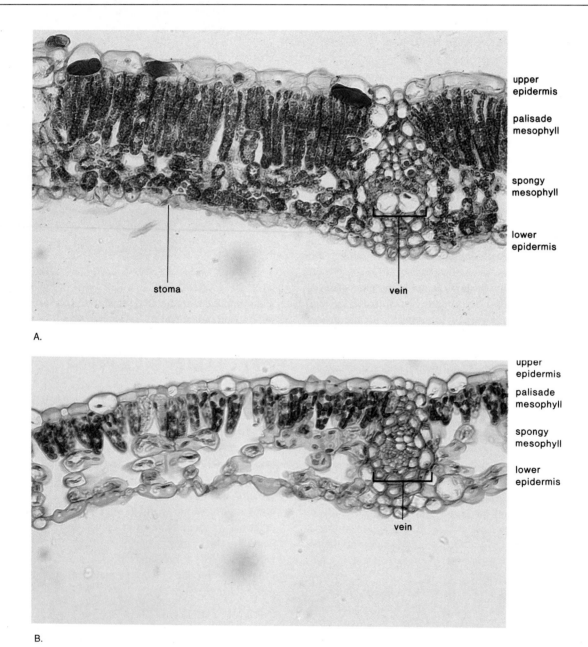

upper
epidermis

palisade
mesophyll

spongy
mesophyll

lower
epidermis

stoma

vein

A.

upper
epidermis

palisade
mesophyll

spongy
mesophyll

lower
epidermis

vein

B.

FIGURE 7.10   Portions of cross sections of maple (*Acer*) leaves. The chloroplasts are stained red. *A*. A leaf exposed to full sun. *B*. A leaf exposed to shade. Note the reduction in mesophyll cells and chloroplasts in the shade leaf.

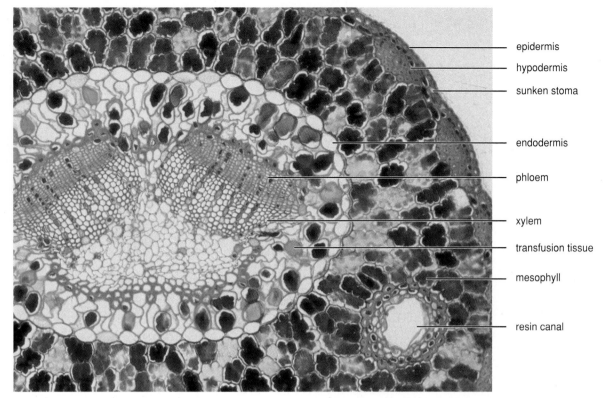

epidermis
hypodermis
sunken stoma
endodermis
phloem
xylem
transfusion tissue
mesophyll
resin canal

FIGURE 7.11 A pine needle in cross section.
(Photomicrograph by G.S. Ellmore)

hairs than leaves on the same tree that are exposed to direct light. They also tend to be larger and to have fewer well-defined mesophyll layers and fewer chloroplasts than their counterparts in the sun (Fig. 7.10).

Leaves of Arid Regions

In different climatic zones or habitats, leaf modifications are generally more pronounced. Plants growing in arid regions may have thick and leathery leaves and have stomata that are fewer in number or sunken below the surface in special depressions. They also may have succulent water-retaining leaves or no leaves at all (with the stems taking over the function of photosynthesis), or they may have dense, hairy coverings. Pine trees, whose water supply may be severely restricted in the winter when the soil is frozen, have some leaf modifications similar to those of desert plants. The modifications include sunken stomata, a thick cuticle, and a layer of thick-walled cells (the **hypodermis**) beneath the epidermis (Fig. 7.11).

The leaves of the compass plant face the East and West, with the blades perpendicular to the ground, so when the sun is overhead it strikes only the thin edge of the leaf, minimizing moisture loss (Fig. 7.12). In plants that grow in water, the submerged leaves usually have considerably less xylem than phloem. Large spaces are found in the mesophyll, which is not differentiated into palisade and spongy layers. Other modifications are described in the following sections.

Tendrils

There are many plants whose leaves are partly or completely modified as **tendrils.** These leaves, when curled tightly around more rigid objects, help the plant in climbing or in supporting weak stems. In garden peas, the leaves are compound (Fig. 7.13) and the upper leaflets are reduced to whip-like strands that, like all tendrils, are very sensitive to contact. In fact, if you lightly stroke a healthy tendril, there is a sudden rapid growth of cells on the opposite side, and it starts curling in the direction of the contact within a minute or two. If the contact is very brief, the tendril undergoes a reverse movement and straightens out again. If, however, the tendril encounters a suitable solid support (e.g., a twig), the stimulation is continuous, and the tendril coils tightly around the support as it grows.

In the yellow vetchling, the whole leaf is modified as a tendril, and photosynthesis is carried on by the leaflike stipules at the base. In the potato vine and the garden nasturtium, the petioles serve as tendrils, while in some greenbriers, stipules are modified as tendrils. In *Clematis,* the rachises of some of the compound leaves serve very effectively as tendrils. Members of the Pumpkin Family, which include

FIGURE 7.12 A compass plant. The leaves, which have their blades parallel to the sun, are oriented toward the East and West.

FIGURE 7.13 Tendrils at the tip of a garden pea plant leaf.

mesquite, black locust), the stipules at the bases of the leaves are modified as short, paired spines. Most *spines* are modifications of the whole leaf, in which much of the normal leaf tissue is replaced with sclerenchyma. As is the case with tendrils, many spinelike objects arising in the axils of leaves are modified stems rather than modified leaves. Such modifications should be referred to as *thorns* to distinguish them from true spines. The *prickles* of roses and raspberries, however, are neither leaves nor stems but outgrowths from the epidermis or cortex just beneath it (Fig. 7.14).

Storage Leaves

As previously mentioned, desert plants may have *succulent leaves* (i.e., leaves that are modified so that they retain water). The modifications that permit water storage involve large, thin-walled parenchyma cells without chloroplasts to the interior of chlorenchyma tissue just beneath the epidermis. These nonphotosynthetic cells contain large vacuoles that can store proportionately substantial quantities of water. If removed from the plant and set aside, the leaves will often retain much of the water for several weeks. Many plants with succulent leaves carry on a special form of photosynthesis discussed in Chapter 10.

Other fleshy leaves, such as those that comprise the bulk of onion and lily bulbs, store large amounts of carbohydrates, which are utilized by the plant in the subsequent growing season.

Flower Pot Leaves

Some of the leaves of *Dischidia* (Fig. 7.15), a greenhouse curiosity from tropical Australasia, develop into urnlike pouches that become the homes of ant colonies. The ants carry in soil and add nitrogenous wastes, while moisture collects in the leaves through condensation of the water vapor

squashes, melons, and cucumbers, produce tendrils that may be up to 3 decimeters (1 foot) in length.

As the tendrils develop, they become coiled like a spring. When contact with a support is made, the tip not only curls around it, but the direction of the coil reverses (see Fig. 11.8); sclerenchyma cells then develop in the vicinity of contact. This makes a very strong but flexible attachment that protects the plant from damage during high winds. It should be noted that the tendrils of many other plants (e.g., grapes) are not modified leaves but have developed from stems.

Spines, Thorns, and Prickles

Many desert plants have their leaves modified as **spines.** This reduction in leaf surface correspondingly reduces water loss from the plants, and the spines also apparently protect the plants from browsing animals. Photosynthesis in such desert plants, which would otherwise take place in leaves, occurs in the green stems. In a number of woody plants (e.g.,

FIGURE 7.14 A. The *spines* of this barberry are modified leaves.

FIGURE 7.14 C. *Prickles* of raspberry stems. The prickles are neither leaves nor stems; they are outgrowths from the epidermis or the cortex just beneath it.

FIGURE 7.14 B. *Thorns* (modified stems) produced in the axils of leaves.

coming through stomata from the mesophyll. This situation creates a good growing medium for roots, which develop adventitiously from the same node as the leaf and grow down into the soil contained in the urnlike pouch. In other words, this extraordinary plant not only reproduces itself by conventional means but also, with the aid of ants, provides its own fertilized growing medium and flower pots and then produces special roots, which "take advantage" of the situation.

Window Leaves

In the Kalahari desert of South Africa, there are at least three plants belonging to the Carpetweed Family that have unique adaptations to living in dry, sandy areas. Their leaves, which are shaped like ice cream cones, are about 3.75 centimeters (1.5 inches) long (Fig. 7.16) and are buried in the sand; only the dime-sized wide end of a leaf is exposed at the surface. This exposed end is covered with a thick epidermis and cuticle. It has a few stomata and is relatively transparent. Below the exposed end is a mass of tightly packed water-storage cells; these allow light coming through the "windows" to penetrate to the chloroplasts in the mesophyll, located all around the inside of the shell of the leaf. This arrangement,

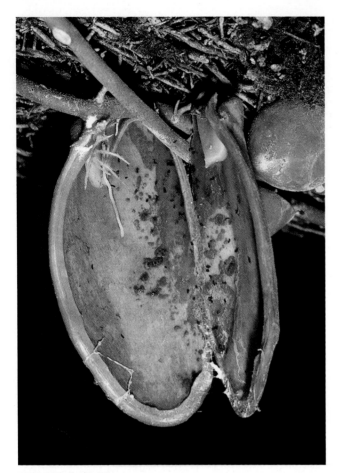

FIGURE 7.15 A flower pot leaf of *Dischidia*. The leaf has been sliced lengthwise to reveal the roots inside that have started to grow down from the top.

FIGURE 7.16 A window plant. Note the transparent tips of the window leaves.

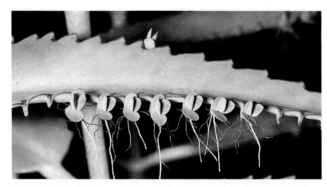

FIGURE 7.17 A leaf of an air plant, showing plantlets being produced along the margins.

which keeps most of the plant buried and away from drying winds, allows the plant to thrive under circumstances that most other plants could not tolerate. Window leaves also occur in succulent plants of a few other families.

Reproductive Leaves

Some of the leaves of the walking fern are most unusual in that they produce new plants at their tips. Occasionally, three generations of plants may be found linked together. The succulent leaves of air plants (Fig. 7.17) have little notches along the leaf margins in which tiny plantlets are produced, complete with roots and leaves, even after a leaf has been removed from the parent plant and pinned to a curtain or left on a counter. Each of the plantlets can develop into a mature plant if given the opportunity to do so.

Floral Leaves (Bracts)

Specialized leaves known as **bracts** are found at the bases of flowers or flower stalks. In some plants, such as the Christ-

mas flower (poinsettia), the flowers themselves have no petals, but the brightly colored floral bracts that surround the small flowers make up for the absence of petals (Fig. 7.18). In other plants, such as the dogwood, the tiny flowers in their buttonlike clusters have inconspicuous petals, and the large white-to-pink bracts that surround the flower clusters, although they appear to the casual observer to be petals, are actually modified leaves.

Insect-Trapping Leaves

Highly specialized *insect-trapping leaves* have intrigued humans for hundreds of years. Almost 200 species of flowering plants are known to have these leaves. Such plants occur mainly in swampy areas and bogs of tropical and temperate regions. In such environments, certain needed elements may be deficient in the soil, or they may be in a form not readily available to the plants. Some of these elements are furnished

FIGURE 7.18   A poinsettia "flower." There are several flowers without petals in the center. The most conspicuous parts of the "flower" are the colored bracts (modified leaves) surrounding the true flowers.

when the soft parts of insects and other small organisms trapped by the specialized leaves are broken down and digested. All the plants have chlorophyll and are able to make their own food. It has been demonstrated that they apparently can develop in normal fashion without insects if they are given the nutrients they need. The following plants represent four types of insect-trapping mechanisms.

Pitcher Plants

Some of the leaves of *pitcher plants* (Fig. 7.19) have flattened blades and function like any other leaves, but others are formed like vases of various sizes and shapes, with or without umbrellalike flaps over the open ends. Some Asian pitcher plants are vines whose long petioles are twisted around branches for support and whose pitchers are formed at the tips of leaves. The plants give off a distinctive odor that attracts insects; these, upon arrival, seek out nectar-secreting glands located at the rim of the pitcher and usually fall into a pool of fluid at the bottom. If they try to climb out, they find the walls highly polished and slippery. In fact, in some species of pitcher plants the walls are coated with wax, and as the insects struggle up the surface, their feet become coated with the wax, which builds up until the victims seem to have acquired heavy clodlike boots. Most insects never make it up the walls, but even if they do, they still face a formidable barricade of stiff downward-pointing hairs near the rim. Eventually they drown, and their soft parts are digested

FIGURE 7.19   Insect-trapping leaves of pitcher plants.

FIGURE 7.20   Sundew leaves.

by bacteria and by enzymes secreted by digestive glands near the bottom of the plant.

Malaysian tree frogs lay their eggs in the pitchers of pitcher plants. The eggs, presumably, contain something that neutralizes digestive enzymes. In North America, the pitcher plants produce their pitchers in erect clusters on the ground, but their mechanisms for trapping insects are similar to those of Asian species.

Merchandisers have been shameless in their wholesale collection of these plants for sale, and pitcher plants may become extinct in the wild. The cobra plant, a pitcher plant restricted to a few swampy areas in California and Oregon, has been placed on official threatened species lists.

Sundews

The tiny plants called *sundews* (Fig. 7.20) often do not measure more than 2.5 to 5.0 centimeters (1 to 2 inches) in diameter. The roundish to oval leaves are covered with about 150 to 200 upright glandular hairs that look like miniature clubs. There is a clear glistening drop of sticky fluid containing digestive enzymes at the tip of each hair. As the droplets sparkle in the sun, they may attract insects, which find themselves stuck if they alight. The hairs are exceptionally sensitive to contact, responding to weights of less than one 1,000th of a milligram, and bend inward, surrounding any trapped insect

within the space of a few minutes. The digestive enzymes break down the soft parts of the insects, and after digestion has been completed (within a few days), the glandular hairs return to their original positions. If bits of nonliving debris happen to catch in the sticky fluid, the hairs barely respond, showing they can distinguish between protein and something "inedible." Some sundew owners regularly feed their plants tiny bits of hamburger and boiled egg white.

Portuguese peasants use relatives of sundews with less specialized leaves in their homes as an effective substitute for flypaper. In response to contact by living insects, the edges of specialized leaves of similar plants called *butterworts* rapidly curl over and trap unwary victims.

Venus Flytraps

The *Venus flytrap* (Fig. 7.21), which has leaves constructed along the lines of an old-fashioned steel trap, is found in nature only in wet areas of North Carolina and South Carolina. The two halves of the blade have the appearance of being hinged along the midrib, with stiff hairlike projections located along their margins. There are three tiny trigger hairs on the inner surface of each half. If two trigger hairs are touched simultaneously or if any one of them is touched twice within a few seconds, the blade halves suddenly snap together, trapping the insect or other small animal. As the

FIGURE 7.21   A Venus flytrap plant.

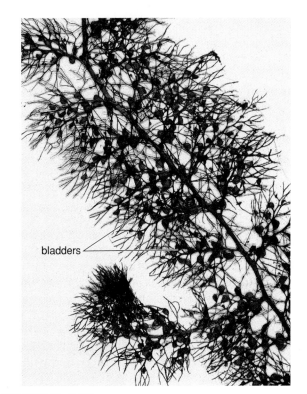

bladders

FIGURE 7.22   A bladderwort.

organism struggles, the trap closes even more tightly. Digestive enzymes secreted by the leaf break down the soft parts of the insect, which are then absorbed. After digestion has been completed, the trap reopens, ready to repeat the process. As is the case with sundews, the traps do not normally close for bits of debris that might accidentally fall on the leaf, presumably because nonliving material does not move about and stimulate the trigger hairs.

Bladderworts

Bladderworts (Fig. 7.22) are plants with finely dissected leaves that are found submerged and floating in the shallow water along the margins of lakes and streams. Toward the bases of many of the leaves are tiny stomach-shaped bladders, each with a trapdoor over the opening at one end. The bladders, which are between 0.3 and 0.6 centimeter (0.125 to 0.250 inch) in diameter, trap aquatic insects and other small animals through a complex mechanism. Four curled but stiff hairs located toward one end of the trapdoor act as triggers when an insect touches one of them. The trapdoor springs open, and water rushes into the bladder. The stream of water propels the victim into the trap, and the door snaps shut behind it. The action takes place in less than one 100th of a second and makes a distinct popping sound, which can be heard with the aid of a sensitive underwater microphone. The trapped insect eventually dies, it is broken down by bacteria, and the breakdown products are absorbed by cells in the walls of the bladder.

Science fiction writers have contributed to superstitions and beliefs that deep in the tropical jungles there are plants capable of trapping humans and other large animals. No such plants have been proved to exist, however. The largest pitcher plants known hold possibly 1 liter (roughly 1 quart) of fluid in their pitchers, and small frogs have been known to decompose in them, but the trapping of anything larger than a mouse or possibly a small rabbit seems very unlikely.

AUTUMNAL CHANGES IN LEAF COLOR

When leaf cells break down and die, the leaves tend eventually to turn some shade of brown or tan due to a reaction between leaf proteins and tannins stored in the cell vacuoles. This is akin to the formation of leather when tannins react with animal hides. Before turning brown, however, the leaves may exhibit a variety of other colors.

The chloroplasts of mature leaves contain several groups of pigments, such as green *chlorophylls,* and *carotenoids*—which include yellow *carotenes* and pale yellow *xanthophylls*. Each of these groups plays a role in photosynthesis. The chlorophylls are usually present in much larger quantities than the other pigments, and their intense green color masks or hides the presence of the carotenes and xanthophylls. In the fall, however, the chlorophylls break down, and other colors are revealed. The exact cause of the chlorophyll breakdown is not known, but it does appear to involve, among other factors, a gradual reduction in day length. Water-soluble *anthocyanin* and *betacyanin* pigments may also accumulate in the vacuoles of the leaf cells in the fall.

Anthocyanins, the more common of the two groups, are red if the cell sap is slightly acid, blue if it is slightly alkaline, and of intermediate shades if it is neutral. Betacyanins are usually red; they apparently are restricted to several plant families, such as the cacti, the Goosefoot Family (to which beets belong), the Four-o'clock Family, and the Portulaca Family.

Some plants (e.g., birch trees) consistently exhibit a single shade of color in their fall leaves, but many (e.g., maple, ash, sumac) vary considerably from one locality to another or even from one leaf to another on the same tree, depending on the combinations of carotenes, xanthophylls, and other pigments present. Some of the most spectacular fall colors in North America occur in the Eastern Deciduous Forest, particularly in New England and the upper reaches of the Mississippi Valley. In parts of Wisconsin and Minnesota, one can observe the brilliant reds, oranges, and golds of maples; the deep maroons (and also yellows) of ashes; the bright yellows of aspen; and the seemingly glowing reds of sumacs and wahoos—all in a single locality. Some fall coloration is found almost anywhere in temperate zones where deciduous trees and shrubs exist.

ABSCISSION

Plants whose leaves drop seasonally are said to be **deciduous.** In temperate climates, new leaves are produced in the spring and are shed in the fall, but in the tropics, the cycles coincide with wet and dry seasons rather than with temperature changes. Even so-called evergreen trees shed their leaves; they do so a few at a time, however, so that they never have the bare look of deciduous trees in their winter condition. The process by which the leaves are shed is called **abscission.**

Abscission occurs as a result of changes that take place in an *abscission zone* near the base of the petiole of each leaf (Fig. 7.23). Sometimes, the abscission zone can be seen externally as a thin band of slightly different color on the petiole. Hormones that apparently inhibit the formation of the specialized layers of cells that facilitate abscission are produced in young leaves. As the leaf ages, hormonal changes take place, and at least two layers of cells become differentiated. Closest to the stem, the cells of the *protective layer,* which may be several cells deep, become coated and impregnated with fatty *suberin.* On the leaf side, a *separation layer* develops in which the cells swell, sometimes divide, and become gelatinous. In response to any of several environmental changes (such as lowering temperatures, decreasing day lengths or light intensities, lack of adequate water, or damage to the leaf), the pectins in the middle lamella of the cells of the separation layer are broken down by enzymes. All that holds

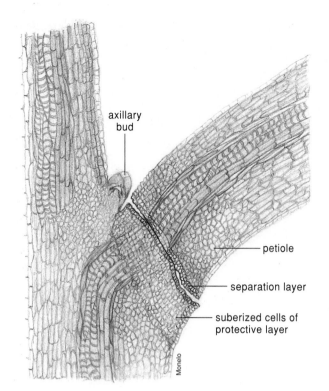

axillary bud

petiole

separation layer

suberized cells of protective layer

Monelo

FIGURE 7.23 The abscission zone of a leaf.

the leaf on to the stem at this point are some strands of xylem. Wind and rain then easily break the connecting strands, leaving tiny bundle scars within a leaf scar (see Fig. 6.1), and the leaf falls to the ground.

HUMAN AND ECOLOGICAL RELEVANCE OF LEAVES

Humans use shade trees and shrubs in landscaping for cooling as well as for aesthetic effects. The leaves of shade plants planted next to a dwelling can make a significant difference in energy costs to the home owner. Humans also use for food the leaves of cabbage, parsley, lettuce, spinach, and chard and the petioles of celery and rhubarb, to mention a few. A large number of spices and flavorings are derived from leaves, including thyme, marjoram, oregano, tarragon, peppermint, spearmint, wintergreen, basil, dill, sage, cilantro, and savory.

Many dyes (e.g., a yellow dye from bearberry, a reddish dye from henna, and a pale blue dye from blue ash) are extracted from leaves (see Appendix 3), although nearly all commercial dyes are now derived from coal tar. Many cordage fibers for ropes and twines come from leaves, with various species of *Agave* (century plants) accounting for about 80% of the world's production. Bowstring fibers are obtained from a relative of the common houseplant *Sansevieria,* and Manila hemp fibers, which are used both in fine-quality cordage and in textiles, are obtained from the leaves of a close relative of the banana. Panama hats are made from the leaves of the panama hat palm, and palms and grasses are used in the tropics as thatching material for huts and other buildings.

In the high mountains of Chile and Peru, the leaves of the yareta plant are used for fuel. They produce a resin that causes the leaves to burn with an unusually hot flame. Leaves of many plants produce oils. Petitgrain oil (from the leaves of a variety of orange tree) and lavender oil, for example, are used for scenting soaps and colognes. Patchouli and lemongrass oils are used in perfumes, as is citronella oil, which was once the leading mosquito repellent before synthetic repellents gained favor. Eucalyptus oil, camphor, cajeput, and pennyroyal (Fig. 7.24) are all used medicinally.

Leaves are an important source of drugs used in medicine and also of narcotics and poisons. Cocaine, obtained from plants native to South America, has been used medicinally and as a local anesthetic, but its use as a narcotic has, in recent years, become a major problem in western cultures. Andean natives chew coca leaves while working and are reported capable of performing exceptional feats of labor with little or no food while under the influence of cocaine. Apparently, the drug, which is highly addictive in forms such as "crack," anesthetizes the nerves that convey hunger pangs to the brain.

Belladonna is a drug complex obtained from leaves of the deadly nightshade, a native of Europe. The plant has been used in medicine for centuries, and several drugs are now isolated from belladonna. Included among the isolates are atropine, which is used in shock treatments, to dilate eyes, to relieve pain locally, and to check secretions. Scopolamine, also a belladonna derivative, is used in tranquilizers and sleeping aids. Another European plant, the foxglove, is the source of digitalis. This drug has been used for centuries in regulating blood circulation and heartbeat.

Tobacco is another widely used leaf. At present, more than 940 million kilograms (2 billion pounds) of tobacco for smoking, chewing, and use as snuff are produced annually around the world. Also at present, about 125,000 Americans die annually from lung cancer, and almost all of them have a history of cigarette smoking. Cigarette smoking is also evidently a principal contributor to cardiovascular diseases. During recent years, the increase in the use of chewing tobacco has seen a corresponding increase in the development of mouth and throat cancers. Federal law forbids concentrations of more than five parts per billion of nitrosamines (cancer-causing chemicals) in cured meats, but the levels of nitrosamines in the five most popular brands of chewing tobacco range from 9,600 to 289,000 parts per billion. Though humans have long used tobacco, it has only been in the last two decades that its health threat has become appreciated.

Similarly, marijuana, the controversial plant widely used as an intoxicant, has been utilized in various ways for thousands of years. The active principle, tetrahydrocannabinol (THC), although found in the leaves, is concentrated in hair secretions among the female flowers. In recent decades, marijuana has found increasing acceptance in the western

FIGURE 7.24   Flowers of a pennyroyal plant.

hemisphere; it appears, however, not to be the harmless intoxicant many thought it to be. Regular use of marijuana for a year has been shown to have the same effect on human lungs as smoking one and a half packs of cigarettes a day for 13 years. A decision to legalize its prescription in pill form for the alleviation of nausea caused by chemotherapy treatments has been criticized by some medical authorities.

The drug lobeline sulphate is obtained from the leaves of a close relative of garden lobelias. It is used in compounds taken by people who are trying to stop smoking. The leaves of several species of *Aloe*, especially *Aloe vera*, yield a juice that is used to treat various types of skin burns, including those accidentally received from X-ray equipment.

Several beverages are extracted by the brewing of leaves. Numerous teas have been obtained from a wide variety of plants, but most now in use come from a close relative of the garden camellia. Maté, the popular South American tea, is brewed from the leaves and twigs of a relative of holly. The alcoholic beverages pulque and tequila find their origin in the mashed leaves of *Agave* plants, and absinthe liqueur receives its unique flavor from the leaves of wormwood, a relative of western sagebrush, and other flavorings, such as anise.

Insecticides of various types are also derived from leaves. A type of rotenone and a substance related to nicotine are obtained from tropical plants; both are effective against a variety of insects. Mexico's cockroach plant has leaves that, when dried, are highly effective in killing cockroaches, flies, fleas, and lice. Water extracts of the leaves (as well as other parts) of India's neem tree are reported to control more than 100 species of insects, mites, and nematodes, with little effect on useful predator insects.

Carnauba and caussu waxes are obtained from the leaves of tropical palms. Leaves are used extensively by florists in floral arrangements and bouquets, and their uses for other aesthetic purposes are legion. Leaves may find more extensive use in the future as a direct source of food. It has been shown that a curd obtained by coagulating juice squeezed from alfalfa leaves contains more than 40% protein, and juices from other leaves have yielded better than 50% protein. Experiments are under way to make the leaf curd palatable for human consumption. Additional information on past and present uses of leaves is given in Chapter 24 and in Appendices 2, 3, and 4. The scientific names of the plants that are the sources of the various substances discussed are given in Appendix 1.

Summary

1. Leaves are similar to solar panels in that they are covered with a transparent epidermis that admits light to the interior, and many twist on their petioles so that their flat surfaces are inclined to the sun throughout the day.

2. The lower and often the upper surfaces of leaves have stomata that permit air circulation and facilitate photosynthesis. Leaves also respire, accumulate wastes, and eliminate excess moisture via transpiration. The shapes, sizes, and textures of leaves vary greatly.

3. All leaves originate as primordia. Most leaves consist of a blade and a petiole that may have paired stipules at the base. Leaves may be simple or compound, and all are associated with leaf gaps.

4. The venation and the compounding of leaves may be either pinnate or palmate. The main vein is the midrib in simple leaves and the rachis in compound leaves. Monocot leaves have parallel veins. Dichotomous venation occurs in a few leaves.

5. The epidermis is a surface layer of cells that is coated with a cuticle. Waxes, glands, hairs, and an occasional cellular crystal may also be present. The lower epidermis usually contains numerous stomata, each formed by a pair of guard cells that regulate both evaporation of water vapor from the leaf and gas exchange between the interior and the atmosphere.

6. The mesophyll between the upper epidermis and the lower epidermis is divided into an upper palisade layer, which consists of rows of parenchyma cells containing numerous chloroplasts, and a lower spongy layer in which the parenchyma cells are loosely arranged. Veins traverse the mesophyll.

7. Leaves may be specialized, usually in adaptation to specific combinations of environmental factors. Specializations include thick leaves, fewer stomata, water-storage leaves, tendrils, spines, food-storage leaves, flower pot leaves, window leaves, reproductive leaves, bracts, and insect-trapping leaves.

8. Environmental factors and hormonal changes that occur in autumn in the abscission zone at the petiolar base of each deciduous leaf cause leaves to drop. As green chloroplast pigments break down, revealing pigments of other colors, and as different pigments accumulate in cell vacuoles, leaves change color.

Review Questions

1. Leaves have no secondary xylem and phloem. Why not?

2. What are bracts?

3. Identify or define *hydathode, transpiration, guard cells, mesophyll, venation, glands,* and *compound leaf.*

4. How can one distinguish between the upper and lower epidermis in most leaves?

5. What is the function of bundle sheaths?

6. How do leaves in shaded areas differ from leaves in sunlit areas on the same plant?

7. What leaf modifications are associated with dry areas, wet areas (e.g., lakes), climbing, and reproduction?

8. Why do leaves turn different colors in the fall?

Discussion Questions

1. Can you think of any advantages or disadvantages to a plant in having very tiny or very large leaves?

2. In Chapter 3, it was noted that living cells are connected to one another by plasmodesmata that extend through tiny holes in the walls. If this is true, does a leaf really need veins for the acquisition and distribution of materials?

3. Is there any advantage to a leaf in having palisade mesophyll on the upper half and spongy mesophyll on the lower half?

4. Pines and other plants that grow in very cold climates have sunken stomata just like plants of the desert, yet the annual precipitation where they grow may be very abundant. Is there any advantage to them in having such a modification?

5. Since tendrils and spines can be either modified leaves or modified stems, how might you determine the origin of a given specimen?

Additional Reading

Addicott, F. T. 1982. *Abscission*. Berkeley, CA: University of California Press.

Dale, J. E. 1992. How do leaves grow? *Bioscience* 42: 323–32.

Feininger, A. 1984. *Leaves: One hundred seventy-seven photographs*. Mineola, NY: Dover Publications.

Juniper, B. E., et al. 1989. *The carnivorous plants*. San Diego, CA: Academic Press.

Martin, E. S., and M. E. Donkin. 1983. *Stomata*. Baltimore, MD: E. Arnold Publications.

Prance, G. T. 1985. *Leaves: The formation, characteristics and uses of hundreds of leaves found in all parts of the world*. New York: Crown Publishers, Inc.

Roth, I. 1984. *Stratification of tropical forests as seen in leaf structure*. Norwell, MA: Kluwer Academic Publications.

Steeves, T. A., and I. M. Sussex. 1989. *Patterns in plant development*, 2d ed. Englewood Cliffs, NJ: Prentice-Hall.

Zeiger, E., et al. 1987. *Stomatal function*. Palo Alto, CA: Stanford University Press.

A water lily flower (Nymphaea sp.).

8 Flowers, Fruits, and Seeds

Overview

In this chapter, the structure and parts of flowers are described, and some modifications are noted. A brief comparison between dicots and monocots is made. The nature and development of fruits are discussed next, and fruit structure and parthenocarpy are described. A listing and discussion of the various types of dry and fleshy fruits follows, and then the agents of fruit and seed dispersal are explored. The chapter concludes with an examination of seed structure and germination and some observations on dormancy and longevity of seeds, stratification, and vivipary.

Some Learning Goals

1. Know the parts of a typical flower and the function of each part.
2. Learn the features that distinguish monocots from dicots.
3. Understand the distinction between a fruit and a vegetable.
4. Know the regions of mature fruits.
5. Learn five types of fleshy and dry fruits, and know how simple, aggregate, and multiple fruits are derived from the flowers.
6. Learn the adaptations of fruits and seeds to the agents by which they are dispersed.
7. Diagram and label a mature dicot seed (e.g., bean) and a monocot seed (e.g., corn) in section to show the parts and regions.
8. Understand the changes that occur when a seed germinates and note the environmental conditions essential to germination.
9. Know the types of factors that control dormancy. Learn how dormancy may be broken both naturally and artificially.

Note to the Reader: Flowering plants are more abundant and more important to humanity than are all other kinds of plants combined; much of this text is devoted to various aspects of flowering plants. In early editions of this book, most basic features of flowers, fruits, and seeds were discussed in a single chapter. This followed a progressive examination of members of the Plant Kingdom, beginning with the simplest forms, so that the reader could appreciate how flowering plants developed and became the most advanced and complex of all plants. Some shorter courses, however, do not cover certain reproductive and evolutionary aspects of the Plant Kingdom, and instructors differ in their assessment of the relative importance of such matters in a short course.

In recognition of this dilemma, the original single chapter is represented in this edition by chapters 8 and 23. Chapter 8 emphasizes form and structure of flowers, fruits, and seeds, with brief reference to reproduction. Chapter 23 covers more details of the life cycle and evolution of flowering plants. Even if your course does not include Chapter 23 in the curriculum, it is recommended that you read it as an adjunct to Chapter 8.

A number of years ago, an Australian farmer glanced back at a field he was plowing and was startled to see what looked like flowers being tossed to the surface of the furrows. Closer inspection revealed that they were indeed flowers but both they and the plants to which they were attached were pale and devoid of chlorophyll. He reported his find to a university, where botanists determined that the farmer had stumbled upon the first known underground flowering plant. The plant proved to be an orchid that derived its nourishment from organic matter in the soil and was pollinated by tiny flies that gained access to the below-ground flowers via mud cracks that developed in the dry season (Fig. 8.1).

The underground-flowering orchid is but one of some 240,000 known species of flowering plants. Additional unknown numbers of undescribed species occur in remote areas, particularly in the tropics. The flowering plants are vital to humanity in that they provide countless useful products, with just 11 species—10 of them members of the Grass Family—furnishing 80% of the world's food (an amplification of this subject is given in chapters 24 and 25).

FIGURE 8.1 Flowers of an Australian orchid produced by plants that complete their entire life cycle below ground.
(Drawing courtesy Karen Hamilton)

FIGURE 8.2   A *Rafflesia* flower. These flowers may be as much as 1 meter (3 feet 3 inches) in diameter.
(Courtesy Charles H. Lamoureux)

The flowers themselves range in size from the minute flowers of the duckweed *Wolffia columbiana,* whose entire speck of a plant body is only 0.5 to 0.7 millimeter (0.02 to 0.03 inch) wide, with flowers little more than 0.1 millimeter long, to the enormous *Rafflesia* flowers (Fig. 8.2) of Indonesia that are up to 1 meter (3 feet 3 inches) in diameter and may weigh 9 kilograms (20 pounds) each.

Flowers may be any color or combination of colors of the rainbow, as well as black or white; they may have virtually any texture, from filmy and transparent to thick and leathery, spongy to sticky, hairy, prickly, or even dewy wet to the touch. Numerous flowers of trees, shrubs, and garden weeds are quite inconspicuous and lack odor (Fig. 8.3), but many others are strikingly beautiful, particularly when examined with a dissecting microscope. Their fragrances, which can be exhilarating to seductive or even putrid, are the basis of international perfume and pet repellent industries.

The habitats of flowering plants are as varied as their form. In addition to plants that live underground, as we have noted, there are plants that go through their life cycles dangling from wires or other plants (Fig. 8.4). Flowering plants occur in both fresh and salt water, in the cracks and crevices of rocks, in deserts and jungles, in frigid arctic regions, and in areas where the temperatures regularly soar to 45°C (113°F) in the shade. In fact, they occur almost anywhere they can receive their basic needs of light, moisture, and a minimal supply of minerals. A species of chickweed survives at an altitude of 6,135 meters (over 20,000 feet) in the Himalaya Mountains, and fumitory plants in the same area flower even when the night temperatures plummet to –18°C (0°F).

Flowering plants can go from the germination of a seed to a mature plant producing new seeds in less than a month, or the process may take as long as 150 years. Those that complete the cycle in a single season are called **annuals.** **Biennials** take two years to complete the cycle; **perennials,** however, may take several to many years to go from a germinated seed to a plant producing new seeds. Perennials may also produce flowers on new growth that dies back each winter, while other parts of the plant may persist indefinitely.

Flowering plants occur in two major classes, the *Dicotyledonae* and the *Monocotyledonae,* commonly referred to as **dicots** and **monocots,** respectively. Members of the two classes are distinguished from one another on the basis of features listed in Table 8.1.

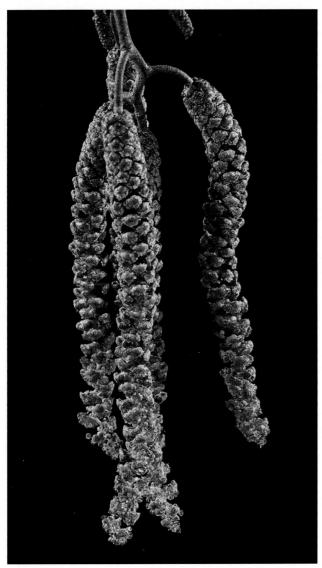

FIGURE 8.3   Male catkins of an alder. Each catkin consists of numerous tiny, inconspicuous, wind-pollinated flowers that have no petals.

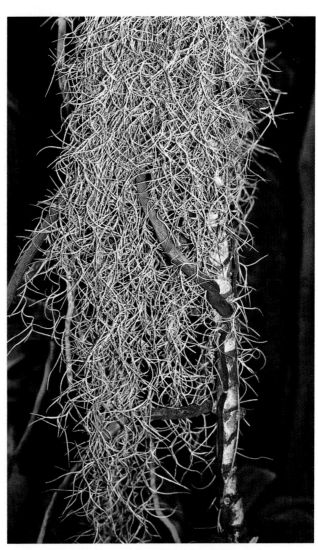

FIGURE 8.4   Spanish moss. Spanish moss is a nonparasitic flowering plant that goes through its life cycle suspended from other plants or objects such as wires. This plant should not be confused with *lichens* (see Chapter 18), which may also hang suspended from other objects.

Table 8.1

SOME DIFFERENCES BETWEEN DICOTS AND MONOCOTS

DICOTS	MONOCOTS
1. Seed with two cotyledons (seed leaves)	1. Seed with one cotyledon (seed leaf)
2. Flower parts mostly in fours or fives or multiples of four or five	2. Flower parts in threes or multiples of three
3. Leaf with a distinct network of veins	3. Leaf with more or less parallel veins
4. Vascular cambium and, frequently, cork cambium, present	4. Vascular cambium and cork cambium absent
5. Vascular bundles of stem in a ring	5. Vascular bundles of stem scattered
6. Pollen grains mostly with three *apertures* (thin areas in the wall — see figures 23.6 and 23.7)	6. Pollen grains mostly with one *aperture*

DIFFERENCES BETWEEN DICOTS AND MONOCOTS

Slightly less than three-fourths of all flowering plant species are dicots. Dicots include many annual plants and virtually all flowering trees and shrubs. Monocots, which are primarily herbaceous, include plants that produce bulbs (e.g., lilies), grasses and related plants, orchids, irises, and palms; they are believed to have developed from ancestors derived from primitive dicots (Table 8.1).

STRUCTURE OF FLOWERS

Regardless of form, all flowers share certain basic features. A typical flower develops several different parts, each with its own function (Fig. 8.5). Each flower, which begins as an embryonic *primordium* that develops into a *bud* (discussed in Chapter 6), occurs as a specialized branch at the tip of a stalk called a **peduncle.** The peduncle swells at its tip into a small pad known as the **receptacle.** The other parts of the flower, some of which are in *whorls,* are attached to this receptacle (a whorl consists of three or more plant parts, such as leaves, encircling another plant part, such as a twig, at the same point on an axis).

The outermost whorl consists, as a rule, of three to five small, usually green, somewhat leaflike **sepals.** The sepals of a flower are collectively referred to as the **calyx.** In many flowers, the calyx functions in protection, particularly while the flower is in the bud. The next whorl of flower parts consists of petals, which are known collectively as the **corolla.** When corollas are showy, they attract pollinators, such as bees, moths, or birds. Some may have special markings that are invisible to humans but can be seen by bees whose vision functions in the ultraviolet range of the light spectrum. The corolla is often missing in wind-pollinated plants. In peach flowers, the petals are distinct separate units, but in other flowers, such as petunias or jimson weeds, the petals are fused together so that the corolla consists of a single, flared, trumpetlike sheet of tissue (Fig. 8.6). Sepals also may be fused together.

Several to many **stamens** are attached to the receptacle around the base of the often greenish *pistil* located in the center of the flower. Each stamen consists of a semirigid but otherwise slender **filament** with a sac called an **anther** at the top. The development of **pollen grains** in anthers is described in Chapter 23. In most flowers, the release of pollen is facilitated by lengthwise slits that develop on the anthers, but pores that function in the same way develop on the anthers of members of the Heath Family and those of a few other groups.

The **pistil,** which often is shaped like a tiny vase that is closed at the top, consists of three regions that merge with one another. At the top is a slight swelling called the **stigma,** which is connected by a slender stalklike **style** to the swollen base called the **ovary;** the ovary later develops into a *fruit.*

If the calyx and corolla are attached to the receptacle at the base of the ovary, the ovary is said to be **superior,** but in other flowers, the receptacle grows up around the ovary so that the calyx and corolla appear to be attached at the top; such flowers are said to have **inferior** ovaries (ovary positions are shown in Fig. 23.11). Within the ovary is a cavity containing an egg-shaped **ovule,** which is attached to the wall of the cavity by means of a short stalk. The ovule (the development of which is described in Chapter 23) eventually becomes a **seed.**

While peach flowers are produced singly on their own peduncles, those of lilacs, grapes, bridal wreaths, and many others are produced in clusters called **inflorescences** (Fig. 8.7). In an inflorescence, the peduncle has numerous little stalks serving individual flowers. These additional little stalks, one to each flower, are called **pedicels.**

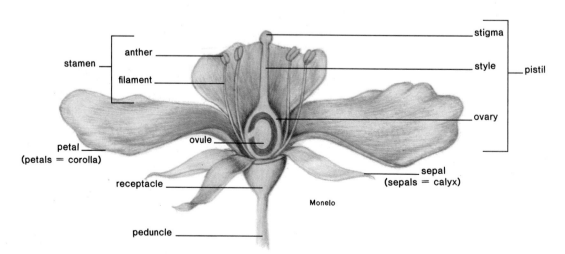

FIGURE 8.5 Parts of a typical flower. The interior structure of the ovule and the sexual processes involved are discussed in Chapter 23.

FIGURE 8.6 A jimson weed flower.

FRUITS

Introduction

In 1893, the U.S. Supreme Court, in the case of *Nix v. Hedden,* ruled that a tomato was legally a vegetable rather than a fruit. This was in keeping with the general conception in the public's mind that fruits tend to be relatively sweet and in the nature of a dessert food, while vegetables tend to be more savory as salad or main course foods. Be that as it may, botanically speaking, a **fruit** is any ovary that has developed and matured. It also usually contains seeds. By this definition, many so-called vegetables, including tomatoes, string beans, cucumbers, and squashes, are really fruits. Vegetables can consist of leaves (e.g., lettuce, cabbage), leaf petioles (e.g., celery), specialized leaves (e.g., onion), stems (e.g., white potato), roots (e.g., sweet potato), stems and roots (e.g., beets), flowers and their peduncles (e.g., broccoli), flower buds (e.g., globe artichoke), or other parts of the plant.

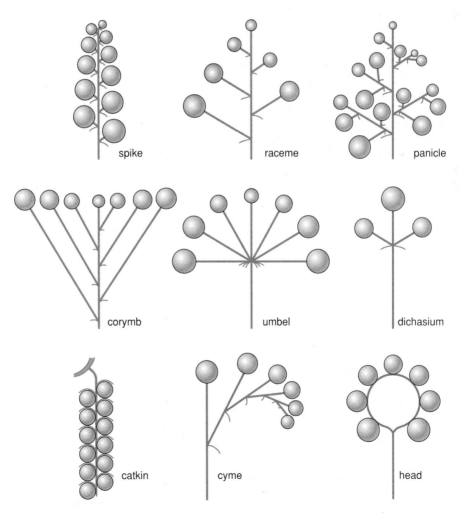

FIGURE 8.7 Inflorescence types. Each ball represents a flower.

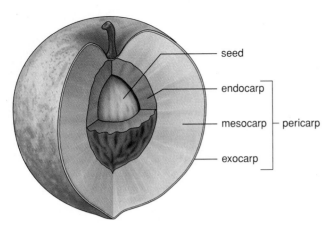

FIGURE 8.8   Regions of a mature fruit.

Fruit Regions

By the time an ovary has matured into a fruit, the bulk of it is divided into three regions, which are sometimes difficult to distinguish from one another (Fig. 8.8). The skin forms the **exocarp,** while the inner boundary around the seed(s) forms the **endocarp.** The endocarp may be hard and stony (as in a peach pit around the seed). It also may be papery (as in apples), or it may not be distinct from the **mesocarp,** which is the name given to the often fleshy tissue between the exocarp and the endocarp. The three regions are collectively called the **pericarp.** In dry fruits, the pericarp is usually quite thin.

Some fruits consist of only the ovary and its seeds. Others have adjacent flower parts, such as the receptacle or calyx, fused to the ovary, or different parts may be modified in various ways. Fruits may be either fleshy or dry at maturity, and they may split, exposing the seeds, or no split may occur. They may be derived from a single ovary or from more than one. Traditionally, all these features have been used in the classification of fruits, but not all fruits lend

All fruits arise from flowers and thus are found exclusively in the flowering plants. *Fertilization* (see Chapter 23) usually indirectly determines whether the ovary or ovaries (and sometimes additional tissues such as the receptacle) of a flower will develop into a fruit. If at least a few of the ovules are not fertilized, the flower normally withers and drops without developing further. Pollen grains contain specific stimulants called *hormones* (discussed in detail in Chapter 11) that may initiate fruit development, and sometimes a quantity of dead pollen is all that is needed to stimulate an ovary into becoming a fruit. It is the hormones produced by the developing seeds, however, that promote the greatest fruit growth. These hormones, stimulate the production of more fruit growth hormones by the ovary wall.

themselves to neat pigeonholing by such characteristics. Some of these problems are pointed out in the classification that follows.

Kinds of Fruits

Fleshy Fruits
Fruits whose mesocarp is at least partly fleshy at maturity are classified as fleshy fruits.

Simple Fleshy Fruits   Simple fleshy fruits are fruits that develop from a flower with a single pistil. The ovary may be superior or inferior, and it may be *simple* (derived from a single modified leaf called a **carpel**) or it may consist of one or more carpels and thus be *compound* (for a discussion of the derivation of carpels and compound ovaries, see Chapter 23). The ovary alone may develop into the fruit, or other parts of the flower may develop with it.

Drupe. A **drupe** is a simple fleshy fruit with a single seed enclosed by a hard, stony endocarp, or pit (Fig. 8.9). It usually develops from flowers with a superior ovary containing a single ovule. The mesocarp is not always obviously fleshy, however. In coconuts, for example, the *husk* (consisting of the mesocarp and the exocarp), which is usually removed before the rest of the fruit is sold in markets, is very fibrous (the fibers are used in making mats and brushes). The seed ("meat") of the coconut is hollow and contains a watery *endosperm* (see Chapter 23) commonly but incorrectly referred to as "milk." It is surrounded by the thick, hard endocarp typical of drupes. Other examples of drupes are the stone fruits (e.g., apricots, cherries, peaches, plums, olives, and almonds). In almonds, the husk, which dries somewhat and splits at maturity, is removed before marketing, and it is the endocarp that we crack to obtain the seed.

Berry. **Berries** usually develop from a compound ovary and commonly contain more than one seed. The entire pericarp is fleshy, and it is difficult to distinguish between the mesocarp and the endocarp (Fig. 8.10). Three types of berries may be recognized.

A *true berry* is a fruit with a thin skin and a pericarp that is relatively soft at maturity. Although most contain more than one seed, notable exceptions are dates and avocados, which have only one seed. Typical examples of true berries are tomatoes, grapes, persimmons, peppers, and eggplants. Some fruits that popularly include the word *berry* in their common name (e.g., strawberry, raspberry, blackberry) botanically are not berries at all.

Some berries are derived from flowers with inferior ovaries so other parts of the flower also contribute to the flesh. They can usually be distinguished by the remnants of flower parts or their scars that persist at the tip. Examples of such berries are gooseberries, blueberries, cranberries, pomegranates, and bananas. In the cultivated banana, fruit development is parthenocarpic (that is, the fruits develop without fertilization—see Chapter 23), so there are no seeds. Several other species of banana produce an abundance of seeds.

A.

B.

C.

FIGURE 8.9 Representative drupes. *A*. Peach. *B*. Almond. *C*. Olive.

The *pepo* has a relatively thick rind. Fruits of members of the Pumpkin Family, including pumpkins, cucumbers, watermelons, squashes, and cantaloupes, are pepos.

The *hesperidium* is a berry with a leathery skin containing oils. Numerous outgrowths from the inner lining of the ovary wall become saclike and swollen with juice as the fruit develops. All members of the Citrus Family produce this type of fruit. Examples are oranges, lemons, limes, grapefruit, tangerines, and kumquats.

Pome. **Pomes** are simple fleshy fruits, the bulk of whose flesh comes from the enlarged receptacle that grows up around the ovary. The endocarp around the seeds is papery or leathery. Examples are apples, pears, and quinces. In an apple, the ovary consists of the core and a little adjacent tissue. The remainder of the fruit has developed primarily from the receptacle (Fig. 8.11). Botany texts often refer to pomes, pepos, some berries, and other fruits derived from more than an ovary alone as *accessory fruits* or as fruits having *accessory tissue*.

Aggregate Fruits An **aggregate fruit** is one that is derived from a single flower with several to many pistils. The individual pistils develop into tiny drupes or other fruitlets, but they mature as a clustered unit on a single receptacle (Fig. 8.12). Examples are raspberries, blackberries, and strawberries. In a strawberry, the cone-shaped receptacle becomes fleshy and red, while each pistil becomes a little whitish *achene* (achenes are described in the section on dry fruits) on the strawberry surface. In other words, the strawberry, while being an aggregate fruit, is also partly composed of accessory tissue.

A.

B.

FIGURE 8.10 Representative berries. *A.* Grapes. *B.* Tomatoes.

Multiple Fruits **Multiple fruits** are derived from several to many individual flowers in a single inflorescence. Each flower has its own receptacle, but as the flowers mature separately into fruitlets, they develop together into a single larger fruit, as in aggregate fruits. Examples of multiple fruits are mulberries, osage oranges (Fig. 8.13), pineapples, and figs. Pineapples, like bananas, usually develop parthenocarpically (see Chapter 23), and there are no seeds. The individual flowers are fused together on a fleshy axis, and the fruitlets coalesce into a single fruit.

Figs mature from a unique "outside in" inflorescence. The individual flowers of the inflorescence are enclosed by the common receptacle, which has an opening to the outside at the tip (Fig. 8.14). Some varieties develop parthenocarpically, but others are pollinated by tiny wasps that crawl in and out through the opening. Some multiple fruits, such as those of the sweet gum, are dry at maturity.

Dry Fruits

Fruits whose mesocarp is definitely dry at maturity are classified as *dry fruits*.

Dry Fruits That Split at Maturity The fruits in this group are distinguished from one another by the manner in which they split.

Follicle. The **follicle** splits along one side or seam only, exposing the seeds within (Fig. 8.15). Examples are larkspur, columbine, milkweed, and peony.

Legume. The **legume** splits along two sides or seams (Fig. 8.16). Literally thousands of members of the Legume Family produce this type of fruit. Examples are peas, beans, garbanzo beans, lentils, carob, kudzu, and mesquite. Peanuts are legumes, but they are atypical in that the fruits develop and mature underground. The seeds are usually released in nature by bacterial breakdown of the pericarp instead of through an active splitting action.

Silique. **Siliques** also split along two sides or seams, but the seeds are borne on a central partition, which is exposed when the two halves of the fruit separate (Fig. 8.17 A). Such fruits, when they are less than three times as long as they are wide, are called *silicles* (Fig. 8.17 B). Siliques and silicles are typically produced by members of the Mustard Family, which includes broccoli, cabbage, radish, shepherd's purse, and watercress.

Capsule. **Capsules** are the most common of the dry fruits that split (Fig. 8.18). They consist of at least two carpels and split in a variety of ways. Some split along the partitions between the carpels, while others split through the cavities (*locules*) in the carpels. Still others form a cap toward one end that pops off and permits release of the seeds, or they form a row of pores through which the seeds are shaken out as the capsule rattles in the wind. Examples are irises, orchids, lilies, poppies, violets, and snapdragons.

FIGURE 8.11   Apples (representative of pomes). The bulk of the flesh is derived from the receptacle and the bases of the calyx, corolla, and stamen.

A.   B.

FIGURE 8.12   Blackberry. *A*. A flower. Note the numerous green pistils. *B*. Aggregate fruits.
(*B*. courtesy Robert A. Schlising)

FIGURE 8.13   An osage orange.

A.

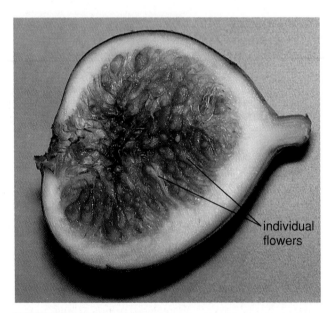

individual
flowers

FIGURE 8.14   Section through a developing fig.

Dry Fruits That Do Not Split at Maturity   In this type of dry fruit, the single seed is, to varying degrees, united with the pericarp.

Achene. The single seed of the **achene** is attached to its surrounding pericarp only at its base. The husk (pericarp) is relatively easily separated from the seed. Examples are sunflower "seeds" (the edible kernel plus the husk constitute the achene) (Fig. 8.19), buttercup, and buckwheat.

B.

FIGURE 8.15   Follicles. *A.* Milkweed. *B.* Magnolia. The fruit of the magnolia is actually an aggregate fruit consisting of approximately 40 to 80 individual one-seeded follicles on a common axis. The follicles and axis fall from the tree as a unit.

A.

FIGURE 8.16   Legumes of a coral tree.

Nut. **Nuts** are one-seeded fruits similar to achenes, but they are generally larger, and their pericarp is much harder and thicker. They develop with a cup, or cluster, of bracts at their base. Examples are acorns (Fig. 8.19), hazelnuts, hickory nuts, and chestnuts. Many nuts in the popular sense are not nuts, botanically speaking. We have already seen that peanuts are atypical legumes, and that coconuts and almonds are drupes. Walnuts and pecans are also drupes, whose "flesh" withers and dries after the seed matures. Brazil nuts are the seeds of a large capsule, and a cashew nut is the single seed of a peculiar drupe. It appears as a curved appendage at the end of a swollen pedicel, which is eaten raw in the tropics or made into preserves or wine. Pistachio nuts are also the seeds of drupes.

Grain (Caryopsis). The pericarp of the **grain** is tightly united with the seed and cannot be separated from it (Fig. 8.19). All members of the Grass Family, including corn, wheat, rice, oats, and barley, produce grains (also called **caryopses**).

B.

FIGURE 8.17   *A.* A silique after it has split open. The seeds are borne on a central, membranous partition. *B.* Silicles of *Lunaria* (dollar plant).

A.

C.

B.

D.

FIGURE 8.18   Capsules. *A.* Butterfly iris. *B. Bletilla* orchid. *C.* Autograph tree. *D.* Unicorn plant.

Samara. In **samaras,** the pericarp surrounding the seed extends out in the form of a wing or membrane, which aids in dispersal (Fig. 8.20). Samaras are produced in pairs in maples. In ashes, elms, and the tree of heaven, they are produced singly.

Schizocarp. The twin fruit called a *schizocarp* (Fig. 8.21) is unique to the Parsley Family. Members of this family include parsley, carrots, anise, caraway, and dill. Upon drying, the twin fruits break into two one-seeded segments.

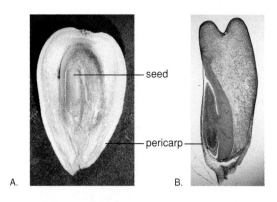

seed

pericarp

A.   B.

pericarp

seed

cup composed
of fused bracts

C.

FIGURE 8.19 Dry fruits that do not split at maturity. *A.* Achene of a sunflower. *B.* Grain (caryopsis) of corn. *C.* Nut (acorn) of an oak.

FIGURE 8.20 Samaras of a big-leaf maple.

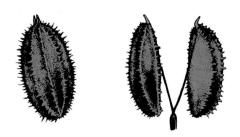

FIGURE 8.21 Schizocarps of carrots. A schizocarp separates at maturity into two one-seeded fruitlets.

FRUIT AND SEED DISPERSAL

Why are so many species of orchids rare, while dandelions, shepherd's purse, and other weeds occur all over the world? Why are some plants confined to single continents, mountain ranges, or small niches occupying less than a hectare (2.47 acres) of land? The answers to these questions involve many different factors, including climate, soil, the adaptability of the plant, and its means of seed dispersal. How fruits and seeds are transported from one place to another is the subject of the following sections. Other factors are discussed in chapters 13 and 25.

Dispersal by Wind

Fruits and seeds have a variety of adaptations for wind dispersal (Fig. 8.22). The samara of a maple has a curved wing that causes the fruit to spin as it is released from the tree. In a brisk wind, samaras may be carried up to 10 kilometers (6 miles) away from their source. In hop hornbeams, the seed is enclosed in an inflated sac that gives it some buoyancy in the wind. In some members of the Buttercup and Sunflower Families, the fruits have plumes, and in the Willow Family, the fruits are surrounded by cottony or woolly hairs that aid in wind dispersal. In button snakeroots and Jerusalem sage, the fruits are too large to be airborne, but they are spherical enough to be rolled along the ground by the wind.

Seeds themselves may be so tiny and light that they can be blown great distances by the wind. Orchids and heaths, for example, produce seeds that are as fine as dust and equally light in weight. In trees such as catalpas and jacarandas, the seeds themselves are winged rather than the fruits, which remain on the branches and split, releasing their contents. Dandelion fruitlets have plumes that radiate out at the ends like tiny parachutes; these catch even a slight breeze. In tumble mustard and other tumbleweeds, the whole aboveground portion of the plant may break off and be blown away by the wind, releasing seeds as it bumps along.

Dispersal by Animals

The adaptations of fruits and seeds for animal dispersal are legion. Birds, mammals, and ants all act as disseminating

FIGURE 8.22   Types of seeds and fruits dispersed by wind.

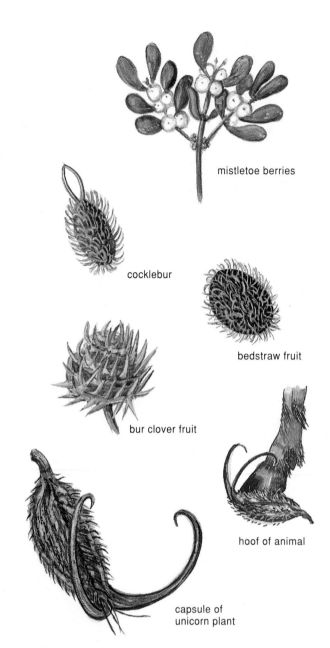

FIGURE 8.23   Types of seeds and fruits dispersed by animals and birds.

agents (Fig. 8.23). Shore birds may carry seeds great distances in mud that adheres to their feet. Other birds and mammals eat fruits whose seeds pass unharmed through their digestive tracts. In blackbirds, the seeds may remain in the tract for as little as 15 minutes, but in mammals, the period is more commonly about 24 hours. In the giant tortoises of the Galápagos Islands, seeds do not pass through the tract for two weeks or more, and the seeds usually will not germinate unless they have been subjected to such treatment (see Chapter 11). Some seeds and fruits are gathered and stored by rodents, such as squirrels and mice, and then are abandoned. Blue jays, woodpeckers, and other birds carry away nuts and other fruits, which they may drop in flight and abandon.

Many fruits and seeds catch in or adhere to the fur or feathers of animals and birds. Some, such as those of bed-

straw and bur clover, are covered with small hooks that catch in fur (or a hiker's socks). The large capsules of unicorn plants have two giant curved extensions about 15 centimeters (6 inches) long. These catch on the fetlock of a deer or other animal that happens to step on the fruit, and the seeds are scattered as the animal moves along. Twinflowers and flax have fruits with sticky appendages that adhere to fur on contact, and fruits of the puncture vine penetrate the skin and stick by means of hard little prickles (Fig. 8.24).

Bleeding hearts, trilliums, and several dozen other plants have on their seeds appendages that contain oils attractive to ants (Fig. 8.25). The Scandinavian scientist Sernander once estimated that more than 36,000 such seeds were carried by members of a single ant colony to their nest,

FIGURE 8.24   Fruits of a puncture vine clinging to a bicycle tire.

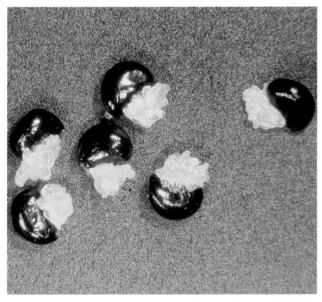

FIGURE 8.25   Seeds of the Pacific bleeding heart. The white appendages are *elaiosomes*, which are removed from the seeds by ants and used for food.

A.   B.

FIGURE 8.26   Sedge adaptation to water dispersal. *A*. A sedge plant. *B*. A sedge fruit. The seed is enclosed within an inflated covering that enables it to float on water.

where the ants stripped off the appendages for food but did not harm the seeds themselves.

Dispersal by Water

Some fruits are adapted to water dispersal by virtue of the fact that they contain trapped air. Many sedges, for example, have seeds surrounded by inflated sacs that enable the seeds to float (Fig. 8.26). Others have waxy material on the surface of the seeds, which temporarily prevents them from absorbing water while they are floating. Sometimes, a heavy downpour will create a torrent of water that dislodges masses of vegetation along a stream bank, carrying whole plants and their fruits to new locations. Large raindrops themselves may splash seeds out of their opened capsules.

Seeds and fruits of a few plants have thick spongy pericarps that absorb water very slowly. Such fruits are adapted to dispersal by ocean currents, even though saltwater eventually may penetrate enough to kill the delicate embryos. Enough fruits are beached before this occurs to ensure the survival of the species. The best known of ocean-dispersed plants is the coconut palm, whose large fruits apparently have been carried many hundreds of kilometers throughout the tropical seas of the world.

Other Dispersal Mechanisms

Fruits of many families mechanically eject seeds, sometimes with considerable force. For example, as the capsules of

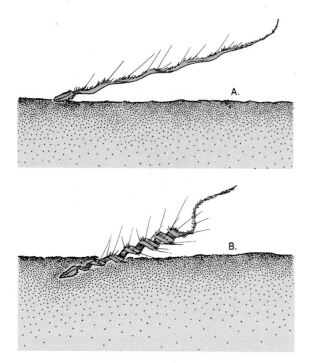

FIGURE 8.27   Filaree fruitlets. *A.* Under humid conditions. *B.* Under dry conditions. Alternate coiling and uncoiling causes the fruitlets to be "screwed" into the ground.

witch hazel dry, a splitting action occurs that may fling the seeds over 12 meters (40 feet) away. Seeds of some legumes and touch-me-nots also are explosively released, and those of dwarf mistletoes, when they are violently released in response to the heat of a warm-blooded animal coming close to the plants, can actually raise small welts on the skin. In manroots and a few other members of the Pumpkin Family, the seed release resembles a geyser eruption as a frothy substance containing the seeds squirts out of one end of the melonlike fruits.

In filarees, each carpel of the fruit splits away and curls back from a central axis. Each fruitlet consists of a single, pointed seed with a long, slender tail that is sensitive to changes in humidity (Fig. 8.27). At night when the humidity increases, the tail is relatively straight, but in the sun it coils up like a corkscrew, literally drilling the pointed seed into the ground as it does so and effectively planting it in the process.

Humans are by far the most efficient transporters of fruits and seeds. Many noxious weeds and plant diseases, as well as valuable food and medicinal plants, have been carried from one continent to another by explorers and travellers, over the past few hundred years in particular. Most countries now have strict regulations barring the importation of plant materials, except by special permit, and some plants are not allowed across borders under any circumstances. Even within a country, certain fruits and seeds that might carry diseases harmful to local agriculture may be barred from entry into a given region. In the United States, for ex-ample, Arizona, California, and Hawaii have border inspections in an attempt to prevent the importation of fruits, such as citrus or popcorn, across state lines, as such plants have in the past been carriers of diseases or pests that are presently under control.

SEEDS

Structure

If an ordinary kidney bean (a dicot) is examined closely, a small white scar called a *hilum* can be seen on the concave side. This marks the point at which the ovule was attached to the ovary wall. The *micropyle* may be visible as a small pore adjacent to the hilum. If this bean is placed in water for an hour or two, it may swell enough to split the seed coat. Once the seed coat is removed, the bean can be seen to consist of two halves, or **cotyledons,** with a tiny rudimentary bean plant between them toward one edge (Fig. 8.28). The cotyledons are food-storage organs that also function as the first "seed leaves" of the seedling plant. They and the tiny, rudimentary bean plant to which they are attached constitute the *embryo.* Some seeds (e.g., those of grasses and all other monocots) have only one cotyledon.

The tiny plant bears undeveloped leaves and a meristem at one end. This embryo shoot is called a **plumule.** The cotyledons are attached just below the plumule. The stem part of the axis above the cotyledons, which at this stage is almost nonexistent, is referred to as the **epicotyl,** while the part below the point of attachment is the **hypocotyl.** In an embryo, it is often difficult to tell where the stem ends and the root begins, but the tip that will develop into a root is called a **radicle.** When a kidney bean germinates, the hypocotyl lengthens below a crook so that the cotyledons are pulled above the ground, but in lima beans and peas, the hypocotyl remains short so that the cotyledons do not emerge above the surface (see Chapter 11).

In other seeds, a plumule-radicle axis may be in the center instead of to one side, and the cotyledon(s) may not play a significant role in food storage. In corn, for example, the bulk of the food-storage tissue is *endosperm* (see Chapter 23). Corn "seeds" (Fig. 8.29) also display features not seen in beans. The plumule and the radicle are enclosed in tubular, sheathing structures called the **coleoptile** and the **coleorhiza,** respectively. These protect the delicate tissues within as the seeds germinate. Development of the coleoptile and coleorhiza ceases after they have attained several millimeters in length, and the plumule and radicle burst through the tips.

Germination

Germination of a seed depends on the interplay of a number of factors, both internal and external. Many seeds require a period of **dormancy** before they will germinate. Dormancy

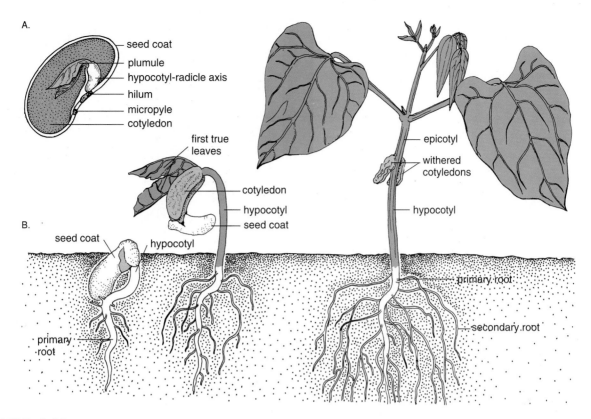

FIGURE 8.28 A common garden bean. *A*. Seed structure. *B*. Germination and development of the seedling.

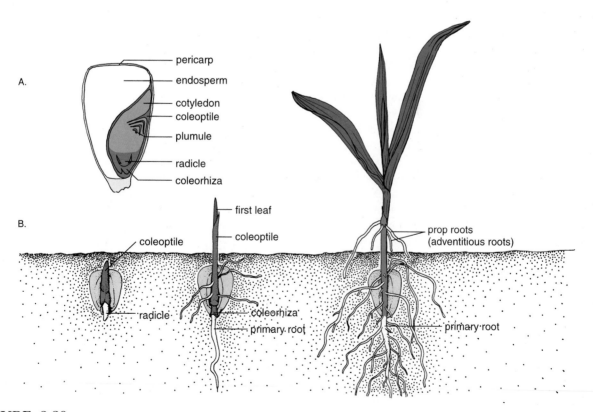

FIGURE 8.29 Corn *A*. Grain structure. *B*. Germination and development of the seedling.

is brought about by either mechanical or physiological circumstances or by both. In the Legume Family and others, the seeds may have seed coats so thick or tough that they prevent the absorption of water or oxygen. Some seeds even have a one-way valve that lets moisture out but prevents its uptake. Dormancy in such seeds may sometimes be broken artificially by *scarification,* which involves nicking or slightly cracking the seeds or dipping them in a concentrated acid for a few seconds to a few minutes. In nature, such seeds may remain dormant until cracks in the seed coat are brought about by the mechanical abrasion of rock particles in the soil, alternate thawing and freezing, or in some cases, bacterial action.

Dormancy may also be induced by growth-inhibiting substances present in the seed coat, the interior of the seed, or tissues of the fruit surrounding it. Many desert plants have inhibitors in the seed coat. These have to be washed away by soaking rains before germination will occur. The inhibitors function in survival of the species by preventing germination unless there has been sufficient rainfall for a seedling to become established.

Apples, pears, citrus fruits, tomatoes, and other fleshy fruits contain inhibitors that prevent germination of the seeds within the fruits. Once the seeds are removed and washed, they germinate readily. The embryos of some seeds, such as those of the American holly, consist of only a few unspecialized cells when the fruit ripens. The seeds will not germinate after the fruit has dropped until the embryo has developed fully with the aid of food materials stored in its endosperm. Such a process of development is called *after-ripening.*

In many woody plants of temperate areas, germination stimulators need to be present to initiate growth. These stimulators normally do not develop unless the seeds encounter a wet period accompanied by cold temperatures. Usually, this period needs to be a minimum of four to six weeks. The dormancy of such seeds can be broken artificially by placing them in a refrigerator, preferably in damp sand, for a few weeks. This technique is called *stratification.*

Even when mechanical and physiological barriers to germination are not present, a seed will not normally germinate unless environmental factors are favorable. Water and oxygen are essential to the completion of germination, and light or its absence also plays a role. Many seeds imbibe 10 times or more their total weight in water before the radicle emerges. Some seeds, such as those of castor beans (Fig. 8.30) and certain spurges, have appendages that function in water absorption and thereby speed up the germination process.

After water has been imbibed, enzymes begin to function in the protoplasm, which has now been rehydrated. Some enzymes convert stored proteins to amino acids, others convert fats and oils to soluble compounds, and still other enzymes aid in the conversion of starch to sugar. The soluble substances can then be conveyed to the embryo, and

FIGURE 8.30   Castor bean seeds. Note the small water-absorbing appendage (caruncle) at the end of each seed.

respiration, which in a dormant seed is almost imperceptible, can be greatly accelerated. Respiration releases the energy needed to initiate growth of the embryo, and a new plant begins to develop as mitosis and cell elongation take place.

Anaerobic respiration (discussed in Chapter 10) often initially furnishes the energy for the embryo, with aerobic respiration furnishing the energy as soon as the splitting of the seed coat admits oxygen. If seeds are kept waterlogged after planting, oxygen available to them is greatly reduced and germination then may fail to be completed. Most seeds require temperatures within certain ranges to germinate. These usually need to be above freezing but below 45°C (113°F). Germination percentages tend to be low approaching either extreme, however. Most crop plants have an optimum (ideal) germination temperature of between 20°C and 30°C (68°F to 86°F).

The role of light in germination varies with the kinds of plants concerned. Seeds of some varieties of lettuce will not germinate in the dark (see discussion of *phytochrome* in Chapter 11), while those of other seeds, such as the California poppy, germinate only in the dark. In lettuce seeds, the light apparently inactivates germination inhibitors, while in the California poppy, it stimulates inhibitor formation. (See the additional discussion of dormancy in Chapter 11.)

Longevity

From time to time, one reads or hears of seeds of wheat or other edible plants germinating after lying dormant in Egyptian pyramids or Native American tombs and caves for thousands of years, but none of these reports has been confirmed. In fact, there is evidence in a few instances that rats or rodents in recent times carried the seeds concerned to nests in the tombs and caves. However, reports of seeds of

The Seed That Slept for 1,200 Years

Rip Van Winkle would appreciate this. An Oriental Sacred Lotus (*Nelumbo nucifera*) seed collected from the sediment of a dry lake bottom near a small village in northeastern China has germinated after being dormant for over 1,200 years. It is one of the oldest living seeds ever found. The Beijing Institute of Botany donated ancient Sacred Lotus seeds to a team of UCLA scientists who dated the seeds with a non-destructive method called accelerator mass spectroscopy. Small amounts of tissue (less than 10 mg) were sampled for radiocarbon dating prior to germination studies. Before the use of this newer dating method, whole seeds had to be destroyed in the process of dating, thereby eliminating the possibility of testing the seeds for viability. After lying dormant in a bed of black clay at depths of 0.5–2.8 meters (1.5–9 feet), the germinated 1,200 year old seed (1,288 ± 271 yr.) was the oldest but not the only survivor from this subterranean tomb. Three other ancient lotus seeds found at various depths germinated, and the UCLA team determined that one was more than 600 years old and another was more than 300.

While reports have claimed seed germination of more ancient seeds recovered from dry archaeological sites in Egypt (such as King Tut's tomb in the pyramids of Giza and from the tombs of other pharaohs), experts now agree that these reports are unreliable. Apart from these lotus seeds, the oldest documented viable seeds are lupine seeds that were frozen in Arctic tundra and from Professor Beal's seed germination study (see text for discussion).

These Sacred Lotus seeds have managed to ward off the ravages of time. Existing in an impenetrable seed coat and mired in an oxygen deficient mud, the seeds have intact genetic and enzymatic systems that reactivated when they were split open and soaked in water. (Enzymes are proteins that speed up chemical reactions in the cell.) A key enzyme that repairs proteins was present during germination; this enzyme has been found in similar quantities in modern day Sacred Lotus seeds.[1] The repair enzyme functions in converting damaged amino acids back to their naturally occurring functional form. This is especially important in "repairing" the proteins of the cell membrane. Without intact membranes, cells are not able to function and such damage will lead to the death of the cell and eventually the organism.

The architecture of the fruit no doubt plays a key role in the longevity of these seeds. The fruits of the

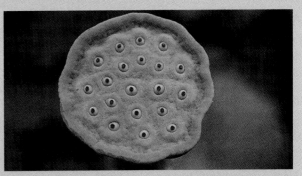

BOX FIGURE 8.1 A lotus pod with 20 seeds.

Sacred Lotus are round to oblong, 10–13 cm (4–5 inches) long, 8–10 cm (3–4 inches) in diameter, and each contains a single seed. The fruit wall or pericarp, which is impervious to water and is also airtight, is initially green and then turns purplish brown as it becomes dry and notably hard. Chinese botanists who first investigated similar ancient lotus seeds collected from the same deposits were unable to get them to germinate, even after 20 months of soaking in water. It wasn't until they scarified (filed open to permit water absorption) the seeds, as prescribed in a 1,400 year old Chinese manuscript, *Ch'i Min Yao Sr* (Important Technology for People in Harmony), that they were ultimately successful in germinating the seeds. The hard, airtight fruit walls are the most significant of the structural features that contribute to the exceptional longevity of the seeds.

The Sacred Lotus was introduced to China following the introduction of Buddhism from India in the first century B.C. It is regarded as a symbol of purity and strength, emerging from the mire of lake waters and opening its crimson flowers to the heavens. The earliest dated depiction of Buddha (in 240 A.D.) shows him surrounded with a halo and seated cross-legged on a lotus throne. From old Chinese manuscripts we learn that lotus has been cultivated as a crop plant (most parts are edible) for the past 4,000 years and traditional Chinese herbal medicine considered the Sacred Lotus a mainstay of their pharmaceutical collections. The large number of lotus seeds collected from the site suggests that the plant was under cultivation in this now dried-up lotus lake. It is likely that the 1,200 year old seed was derived from a plant cultivated by Buddhists at this site in northeastern China.

Unlike Rip Van Winkle who, following a drink of liquor, slept for only 20 years, these Sacred Lotus seeds are remarkable in their capacity to revive after more than 1,000 years. Lessons learned from these plants can shape our thinking about our own aging and the possibilities that may exist in the future for extending life at the margins.

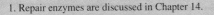

1. Repair enzymes are discussed in Chapter 14.

the aquatic lotus plant germinating after a little more than 1,000 years and a report documenting the germination of Arctic tundra lupine seeds that were frozen for an estimated 10,000 years have been confirmed.

Seeds remain viable (retain the capacity to germinate) for periods that vary greatly, depending on the species and the conditions of storage. Some seeds, such as those of certain willows, cottonwoods, orchids, and tea, remain viable for only a few days or weeks, regardless of how they are stored, but the period of viability of most seeds is extended by months or even years when they are stored at low temperatures and kept dry.

Packets of vegetable and flower seeds sold in stores are dated, giving the buyer a rough idea of how long a significant number of the seeds might be expected to remain viable. Generally, seeds of Pumpkin Family members (e.g., squash, cantaloupe, cucumber) retain a relatively high percentage of viability for several years, while those of members of the Lily Family (e.g., onion, leek, chives) retain a good percentage of viability for only two or three years. Properly stored wheat seeds have been reported to retain better than 30% viability for more than 30 years, and some weed seeds stored under conditions of low oxygen, high humidity, and cool temperatures have remained viable for even longer periods.

In 1879, William J. Beal, a botanist who pioneered the development of hybrid corn, buried 20 pint-sized bottles of weed seeds on the campus of what is now Michigan State University in East Lansing, Michigan. Each bottle contained 1,000 seeds of 20 different species of weeds. Every five years, a bottle of seeds was dug up and the seeds were planted, until the schedule was changed in 1920 to every 10 years. When the first bottle was dug up in 1884, seeds of most of the weeds germinated; in 1960, seeds of evening primrose, curly dock, and moth mullein still germinated; and in 1980, 29 moth mullein seeds, 1 mullein seed, and 1 mallow seed germinated—101 years after they were placed in the bottles. It is of interest to note that until 1980 no mallow seed had germinated since 1899. Only six of the original bottles now remain; they are not scheduled to be unearthed until the year 2040. Recent evidence indicates that the timing of the digging up of the seed bottles did not take into account the facts that certain temperature patterns are critical to the germination of many weed seeds and that if Beal's experiment had been conducted in a different way, the germination results would have been quite different.

A few species of both dicots and monocots produce seeds that have no period of dormancy at all. In some instances, the embryo, which develops from the zygote, continues to grow without pause in a phenomenon known as

FIGURE 8.31   Young seedlings of red mangrove whose seeds have no dormant period and germinate while the fruit is still on the tree. The seedlings grow to lengths of up to 25 centimeters (10 inches) before falling and becoming planted in the mud below. Ocean currents and tides also distribute mangrove seeds throughout tropical tidal zones. The dispersal has been so effective that there are 60,000 square kilometers (23,000 square miles) of mangroves in Southeast Asia alone. Mangrove wood is harvested for fuel, and in the past 40 years the groves, in some localities, have been reduced in area by more than 50%. Incidentally, the fruit is sweet and edible.

vivipary. In the red mangrove, a tropical tree associated with coastal waters and estuaries, each fruit contains a single seed in which the embryo continues to grow while the fruit is still on the tree, reaching a length of 25 centimeters (10 inches) or more before the seedling becomes detached and essentially plants itself in the mud below (Fig. 8.31).

Summary

1. Flowers occur in a wide variety of sizes, colors, textures, and habitats. Annuals complete their life cycles in one growing season, biennials take two seasons, and perennials may take several to many years to complete their cycles.

2. Dicots and monocots are distinguished from one another by the differences in numbers of flower parts and cotyledons, venation, presence or absence of cambia, vascular bundle arrangement, and pollen grain apertures.

3. A typical flower consists of a peduncle and a receptacle to which are attached sepals (calyx), petals (corolla), stamens, and pistil. Flowers may have no petals; or the petals may be fused together. Many flowers may be in an inflorescence.

4. Stamens consist of a pollen-bearing anther and a stalk or filament.

5. A pistil consists of a stigma, style, and ovary. The ovary may be superior or inferior and contains one or more ovules.

6. A fruit, which is unique to flowering plants, is a mature ovary. Hormones promote the greatest fruit growth. Parthenocarpic fruits are seedless and develop without fertilization, but not all seedless fruits are parthenocarpic.

7. A mature fruit has an outer exocarp, an inner endocarp around the seed(s), and a mesocarp between the exocarp and endocarp; the three regions may be fused together as a pericarp. Fruits may be fleshy or dry at maturity. A fruit may be derived from the ovary alone or from adjacent flower parts as well.

8. Fleshy fruits may develop from a flower with a single pistil (simple fruit); aggregate fruits develop from a single flower with more than one pistil; multiple fruits are derived from flowers in an inflorescence.

9. Simple fleshy fruits include drupes and berries; berries may be true berries, pepos, or hesperidiums. Pomes have flesh derived from both the receptacle and the ovary.

10. Accessory fruits consist of more than the ovary alone; some aggregate fruits (e.g., strawberries) are partly composed of accessory tissue. The individual fruitlets of a multiple fruit develop together into a single larger fruit.

11. Some dry fruits split as they mature; such fruits include follicles, legumes, siliques or silicles, and capsules.

12. Nonsplitting dry fruits include achenes, nuts, grains (caryopses), samaras, and schizocarps.

13. Fruits and seeds may have wings, plumes, and other adaptations for wind dispersal. Some fruits and seeds have adaptations for animal, bird, or water dispersal. Some fruits eject seeds with force, and some have modifications that drill seeds into the ground.

14. Humans disperse many seeds. Most countries and a few states have strict regulations governing the importation of plant materials, primarily to control the spread of pests and diseases.

15. A bean seed has a hilum, a micropyle, a seed coat, two cotyledons, and an embryonic bean plant consisting of a plumule and a radicle.

16. In grains, the plumule and the radicle are protected by a coleoptile and a coleorhiza, respectively.

17. Germination of a seed depends on the cessation of dormancy. Dormancy may be sustained by mechanical circumstances; scarification may break dormancy in such seeds.

18. Dormancy may also be induced by growth or germination inhibitors, or after-ripening may need to occur. Cold temperatures may be necessary for the germination of some seeds; stratification may break the dormancy of such seeds.

19. A seed will not germinate unless environmental factors including water, oxygen, light, and certain temperature ranges are favorable.

20. Seeds remain viable for a few days to more than 100 years. The viability of most seeds is extended by storage at low temperatures under dry conditions, but some weed seeds have their viability extended by storage under humid, cool conditions that include little oxygen.

21. Some plants produce seeds that undergo no dormancy at all. The growth of the embryo while the seed and fruit are still on the plant is termed *vivipary*.

Review Questions

1. Define *calyx, corolla, receptacle, peduncle, pedicel, pistil, filament, ovary,* and *carpel.*

2. Indicate the features by which dicots are distinguished from monocots.

3. What is the difference between a fruit and a vegetable?

4. What causes an ovary to develop into a fruit?

5. What are the various parts of a fruit?

6. How do fleshy fruits differ from dry fruits?

7. Distinguish among simple, aggregate, and multiple fruits.

8. Distinguish among achenes, grains, samaras, and nuts.

9. What adaptations do seeds and fruits have for dispersal by water and animals?

10. Define *plumule, radicle, coleoptile, coleorhiza, hypocotyl, after-ripening, stratification,* and *vivipary.*

Discussion Questions

1. Most wind-pollinated flowering plants have inconspicuous, nonfragrant flowers. How might nature be affected if all flowers were that way?

2. Do you believe the botanical distinction between fruits and vegetables is a good one? If you do, why? If not, how would you change it?

3. In the discussion of pomes, it was observed that the bulk of the flesh in an apple comes from the receptacle. What could you do to prove or disprove this?

4. Seed and fruit dispersal is achieved with the aid of wind, water, animals, mechanical means, and humans. If you were "designing" a new plant, can you think of any new way in which it might be dispersed?

5. When volcanic activity or coral polyps cause new islands to appear in the oceans, the islands eventually acquire some vegetation. Would you expect the types of dispersal mechanisms for the flowering plants on these islands to be the same as they were for ancient continents? Why?

Additional Reading

Beattie, A. J. 1990. Seed dispersal by ants. *Scientific American* 263(2): 76.

Bell, A. D. 1990. *Plant forms: An illustrated guide to flowering plant morphology.* New York: Oxford University Press.

Bewley, J.D., and M. Black (Eds.). 1985. *Seeds: Physiology of development and germination.* New York: Plenum Publishing Corp.

Bold, H.C., et al. 1987. *Morphology of plants and fungi,* 5th ed. New York: Harper & Row.

Duffus, C.M., and J.C. Slaughter. 1980. *Seeds and their uses.* New York: John Wiley & Sons, Inc.

Holm, E. 1979. *The biology of flowers.* New York: Penguin Books.

Murray, D. (Ed.). 1987. *Seed dispersal.* San Diego, CA: Academic Press.

Pijl, L. van der. 1982. *Principles of dispersal in higher plants,* 2d ed. New York: Springer-Verlag.

Raven, P., R.F. Evert, and S.E. Eichhorn. 1992. *Biology of plants,* 5th ed. New York: Worth Publishers.

Roth, I. 1978. *Fruits of angiosperms.* Forestburgh, NY: Lubrecht & Cramer.

Weberling, F. 1989. *Morphology of flowers and inflorescences.* (R. J. Pankhurst, Trans.) New York: Cambridge University Press.

Chapter Outline

Akaka Falls, near Hilo, Hawaii. Ti or ki plants (Cordyline fruticosa) *are in the left foreground. Ti leaves are used for ceremonial skirts, as food wrappers, and for sandals. The roots, which are rich in fructose, are used for food and to make an alcoholic beverage. The plants are popular as houseplants in areas outside the tropics.*

Water in Plants

9

This chapter begins by introducing molecular movement through a comparison between balls in motion in a room and molecular activity. This is followed by a discussion of diffusion, osmosis, turgor, plasmolysis, imbibition, and active transport.

The entry of water into the plant is then taken up; this is followed by a discussion of the movement of water through the plant, the evaporation of water into leaf air spaces, and transpiration. A discussion of mineral requirements for growth concludes the chapter.

Some Learning Goals

1. In simple terms, explain diffusion, osmosis, turgor, imbibition, and active transport.
2. Discuss the pressure-flow hypothesis and the cohesion-tension theory.
3. Know the pathway, movement, and utilization of water in plants.
4. Explain how a stoma opens and closes.
5. Know and understand mineral requirements for growth.

Nearly everyone has had the experience of driving along a highway or city street when someone in the car says, "What's that smell?" Soon a bakery, or a paper mill, or perhaps a dead skunk comes into view, and the smell gets stronger. Then it fades away as the source is left behind (Fig. 9.1). We take for granted the fact that there is a correlation between our proximity to an odor source and the intensity of the odor, but how and why does the odor reach us?

By way of an answer, let us imagine two adjacent rooms identical in height, width, and length, with no windows, doors, fixtures, or furniture. Now suppose we lift a small flap in the ceiling of one of them and drop in 100 tennis balls. These are ordinary tennis balls except for one extraordinary feature: They have perpetual motion motors in them that cause them to travel in any direction at 30 MPH. The tennis balls quickly make the room seem like a battlefield as they whiz around, bounce off the walls, floor, and ceiling, and also collide with one another, each time being deflected at a different angle. Shortly after they are introduced, the tennis balls will probably become randomly distributed throughout the room.

Now what would you expect to happen if we opened up a small hole in the wall between the two rooms? Eventually, a tennis ball should bounce or travel at just the right angle to go through the hole into the other room. Theoretically, it could then bounce straight back into the first room, but it seems unlikely that it would. Reason tells us that long before it might happen to strike the exact angle it needed to return, several more balls would come in from the first room. Given enough time, some balls might indeed bounce back into the first room, but in the long run, each of the two rooms

would end up with roughly 50 tennis balls, with an occasional ball going between the rooms in either direction. The situation just described is somewhat analogous to what takes place in nature on a molecular level.

MOLECULAR MOVEMENT

Molecules and ions (discussed in Chapter 2) are constantly in random motion. Visual evidence of this can be seen with an ordinary light microscope. If a drop of India ink is diluted with water and observed through a microscope under high power, the tiny carbon particles of the ink appear to be in constant motion. This motion, known as *Brownian movement*, is the result of the bombardment of the visible particles by invisible water molecules, which are in constant motion themselves.

Diffusion

The differing intensity of smells discussed earlier involves molecules behaving something like the tennis balls. Through their random motion, molecules tend to become distributed throughout the space available to them. Thus, if perfume molecules are kept in a bottle, they will become distributed throughout the bottle, but if the stopper is removed, they will eventually become dispersed throughout the room, even if there is no fan or other device to move the air.

This movement of molecules or ions from a region of higher concentration to a region of lower concentration is called **diffusion** (Fig. 9.2). Molecules that are moving from a region of higher concentration to a region of lower concentration are said to be moving *along a diffusion gradient,* while molecules going in the opposite direction are said to be going *against a diffusion gradient.* When the molecules, through their random movement in all directions, have become distributed throughout the space available, they are considered to be in a state of *equilibrium.* The rate of diffusion depends on several factors, including temperature and the density of the medium through which diffusion is taking place.

Except within the area immediately surrounding the source, unaided diffusion requires a great deal of time because molecules and ions are infinitesimal. Something that is less than a millionth of a millimeter in diameter is going to take a long time to move just 1 millimeter, even though the amount of movement may be great in proportion to the size of the particle concerned. In gases there is a great deal of space between the molecules and correspondingly less chance of the molecules bumping into each other and thus being slowed down. Accordingly, gas molecules occupy a space that becomes available to them relatively rapidly, while liquids do so more slowly and solids are slower yet.

Large molecules move much more slowly than small molecules. If you added a tiny drop of a dye (which has relatively large molecules) to one end of a bathtub of water without disturbing the water in any way, it would take years for the dye molecules to diffuse throughout the tub and reach

FIGURE 9.1

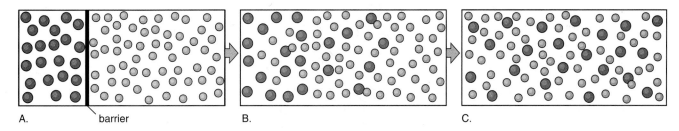

A.   barrier   B.   C.

FIGURE 9.2   Simple diffusion. *A*. A barrier separates two kinds of molecules. *B*. When the barrier is removed, random movement of individual molecules results in both kinds moving from a region of higher concentration to a region of lower concentration. *C*. Eventually, equilibrium (even distribution) is reached. The diffusion gradually slows down as equilibrium is approached.

a state of equilibrium. In nature, however, wind and water currents distribute molecules much faster than they ever could be distributed by diffusion alone. Diffusion rates are also affected by the density of the molecules concerned and by other factors, such as temperature.

Osmosis

Despite the fact that the protoplasts of living cells are bounded by membranes, it is now well-known that water (a **solvent**) moves freely from cell to cell. This has led scientists to believe that plasma, vacuolar, and other membranes have tiny holes or spaces in them, even though such holes or spaces are invisible to the instruments presently available, and it also has led to the construction of models of such membranes (see Fig. 3.7). Membranes through which different substances diffuse at different rates are described as **differentially permeable,** or **semipermeable.** All plant cell membranes appear to be differentially permeable.

In plant cells, **osmosis** is essentially the *diffusion of water through a differentially permeable (semipermeable) membrane from a region where the water is more concentrated to a region where it is less concentrated*. Osmosis ceases if the concentration of water on both sides of the membrane becomes equal.

A demonstration of osmosis can be made by tying a membrane over the mouth of a thistle tube that has been filled with a solution of 10% sugar in water (i.e., the solution consists of 10% sugar and 90% water). Fluid rises in the narrow part of the tube as osmosis occurs when the thistle tube is immersed in water (Fig. 9.3).

Although the simple definition just given of osmosis should suffice for our purposes, plant physiologists prefer to define and discuss osmosis more precisely in terms of *potentials*. It is possible to prevent osmosis from taking place by applying pressure. The minimum pressure required to prevent fluid from moving as a result of osmosis is referred to as the *osmotic potential* of the solution. In other words, osmotic

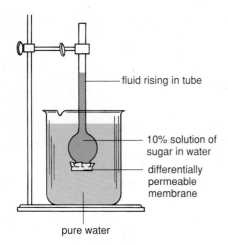

FIGURE 9.3   A simple osmometer made by tying a membrane over the mouth of a thistle tube.

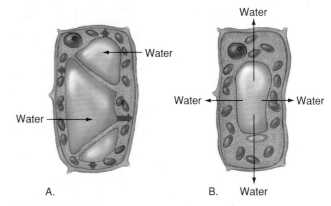

FIGURE 9.4   *A.* A turgid cell. Water has entered the cell by osmosis, and turgor pressure is pushing the cell contents against the cell walls. *B.* Water has left the cell and turgor pressure has dropped, leaving the cell limp.

potential is the pressure required to prevent osmosis from taking place.

Water enters a cell by osmosis until the osmotic potential is balanced by the resistance of the cell wall to expansion. Water gained by osmosis may keep a cell firm, or **turgid,** and the **turgor pressure** that develops against the walls as a result of water entering the vacuole of the cell is called *pressure potential.*

The release of turgor pressure can be heard each time you bite into a crisp celery stick or the leaf of a young head of lettuce. When we soak carrot sticks, celery, or lettuce in pure water to make them crisp, we are merely assisting the plant in bringing about an increase in the *turgor* of the cells (Fig. 9.4).

The *water potential* of a plant cell is essentially its osmotic potential and pressure potential combined. If we have two adjacent cells of different water potentials, water will move from the cell having the higher water potential to the cell having the lower water potential.

Osmosis is the primary means by which water enters plants from their surrounding environment. In land plants, water from the soil enters the cell walls and intercellular spaces of the epidermis and the root hairs and travels along the walls until it reaches the endodermis. Here it crosses the differentially permeable membranes and protoplasts of the endodermal cells on its way to the xylem, where it flows to the leaves, evaporates within the leaf air spaces, and diffuses out (*transpires*) through the stomata into the atmosphere. The movement of water occurs because there is a water potential gradient between relatively high soil water potential to successively lower water potentials in roots, stems, leaves, and the atmosphere.

Plasmolysis

If you place turgid carrot and celery sticks in a 10% solution of salt in water, they soon lose their rigidity and become limp enough to curl around your finger. The water potential inside the carrot cells is greater than the water potential out-

side, so diffusion of water out of the cells into the salt solution takes place. If you were to examine such cells with the aid of a microscope, you would see that the vacuoles, which are largely water, had disappeared and that the protoplasm had shrunk away from the walls and was clumped in the middle of the cell. Such cells are said to be *plasmolyzed.* This loss of water through osmosis, which is accompanied by the shrinkage of protoplasm away from the cell wall, is called **plasmolysis** (Fig. 9.5). If plasmolyzed cells are placed in fresh water before permanent damage is done, water reenters the cell by osmosis, and the cells become turgid once more.

Imbibition

Osmosis is not the only force involved in the absorption of water by plants. *Colloidal* materials (i.e., materials that contain a permanent suspension of fine particles) and large molecules, such as cellulose and starch, usually develop electrical charges when they are wet, and they attract water molecules, which adhere to the internal surfaces of the materials. Because water molecules are *polar,* they can become both highly adhesive to large organic molecules such as cellulose and cohesive with one another. (Polar molecules have slightly different electrical charges at each end due to their asymmetry—see the discussion in Chapter 2.) This process, known as **imbibition,** results in the swelling of tissues, whether they are alive or dead, often to several times their original volume. Imbibition is the initial step in the germination of seeds (Fig. 9.6).

The physical forces developed during germination can be tremendous, even to the point of causing a seed to split a rock weighing several tons (Fig. 9.7). It has been found, for example, that a pressure of 42.2 kilograms per square centimeter (600 pounds per square inch) is needed to break the seed coat of a fresh walnut from within, and that water being imbibed by a cocklebur seed develops a force of up to 1,000 times that of normal atmospheric pressure. Yet when water

A.

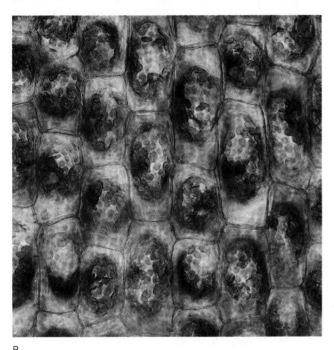

B.

FIGURE 9.5 A portion of a leaf of the water weed *Elodea*. *A*. Normal cells. *B*. Plasmolyzed cells.

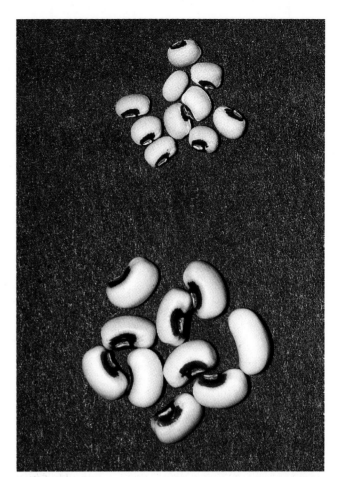

FIGURE 9.6 Black-eyed pea seeds before and after imbibition of water.

and oxygen reach walnut and cocklebur seeds, they germinate readily, as do seeds that fall into the crevices of rocks or have boulders roll over on them.

The huge stone blocks used in the construction of the pyramids of Egypt are believed to have been quarried by hammering rounded wooden stakes into holes made in the face of the stone and then soaking the stakes with water. As the stakes swelled, the force created by imbibition was sufficient to split the rock.

Active Transport

Return for a moment to our two rooms with the tennis balls. Suppose that, in addition to the 100 tennis balls, we drop in 50 slightly underinflated basketballs, also extraordinary in having perpetual motion motors that propel them in any direction at 12 MPH. They should also become randomly distributed throughout the room shortly after they are introduced. Assume, however, that the hole in the wall (which is large enough for the passage of a tennis ball) is not quite large enough to allow a basketball to pass through freely. The basketballs will then remain in the first room. But if we were to install a mechanical arm next to the hole in the second room, and if this arm could grab basketballs that came near the hole and squeeze them through in one direction, basketballs would be transported into the second room *through the expenditure of energy*. The basketballs obviously would gradually accumulate in the second room in greater numbers.

Plants expend energy, too. Plant cells generally contain a larger number of mineral molecules and ions than exist in the soil immediately adjacent to the root hairs. If it were not for the barriers imposed by the differentially permeable membranes, therefore, these molecules and ions would

FIGURE 9.7   A live oak that grew from an acorn lodged in a small crack in the rock. When it rained, the acorn imbibed water and the force of the swelling split the rock. A root is now slowly widening the split.

FIGURE 9.8   A mangrove tree. Mangroves flourish in tropical tidal zones where the salt content of the water is high enough to plasmolyze the cells of most plants. The mangroves still obtain water via osmosis, which takes place because the mangrove cells accumulate an unusually high concentration of organic solutes; some are also able to excrete excess salt.

move from a region of higher concentration in the cells to a region of lower concentration in the soil.

Most molecules needed by cells are polar, and those of solutes may set up an electrical gradient across a differentially permeable membrane of a living cell and require special transport proteins embedded in the membrane (see Fig. 3.7) to pass through. The transport proteins are believed to occur in two forms, one facilitating the transport of specific ions to the outside of the cell and the other facilitating the transport of specific ions into the cell.

The plants absorb and retain these substances against a diffusion (or electrical) gradient *through the expenditure of energy*. This process is called **active transport.** The precise mechanism of active transport is not fully understood. It apparently involves an enzyme and what has been referred to by some scientists as a proton "pump"—involving the plasma membrane of plant and fungal cells and sodium and potassium ions in animal cells. Both "pumps" are energized by special energy-storing ATP molecules (discussed in Chapter 10).

Some plants such as mangrove and saltbush and also certain algae thrive in areas where the water or soil contains enough salt to kill most vegetation. Such plants accumulate large amounts of organic solutes, including the carbohydrate *mannitol* and the amino acid *proline*. The organic solutes enable the plants to take up water via osmosis (Fig. 9.8), despite the otherwise adverse environment. Some mangroves also have salt glands in their leaves through which they excrete excess salt.

WATER AND ITS MOVEMENT THROUGH THE PLANT

If you were to cover the soil at the base of a plant with foil, place the pot where it would receive light, and then put the potted plant under a glass bell jar, you would notice moisture accumulating on the inside of the jar within an hour or two. Because of the foil barrier, the water could not have come directly from the soil; it had to have come through the plant. More than 90% of the water entering a plant passes through and evaporates—primarily into leaf air spaces and then through the stomata into the atmosphere (see Fig. 9.10), with usually less than 5% of the water escaping through the cuticle. This process of water vapor loss from the internal leaf atmosphere is called *transpiration* (Fig. 9.9).

The amount of water transpired by plants is greater than one might suspect. For example, mature corn plants

FIGURE 9.9   A potted plant sealed under a bell jar. The surface of the soil has been covered with foil. Note the accumulation of moisture on the inside of the glass. The moisture could only have come through the plant by the process of transpiration.

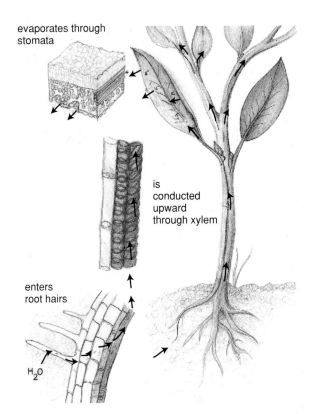

evaporates through stomata

is conducted upward through xylem

enters root hairs

H_2O

FIGURE 9.10   Pathway of water through a plant.

each transpire about 15 liters (4 gallons) of water per week, while four-tenths of a hectare (1 acre) of corn may transpire more than 1,325,000 liters (350,000 gallons) in a 100-day growing season. A hardwood tree utilizes about 450 liters (120 gallons) of water while producing 0.45 kilogram (1 pound) of wood (or 1,800 liters while producing 0.45 kilogram of dry weight substance), and the 200,000 leaves of an average-sized birch tree will transpire from 750 to more than 3,785 liters (200 to 1,000 gallons) per day during the growing season. Humans recycle much of their water via the circulatory system, but if they were to have requirements similar to those of plants, each adult would have to drink well over 38 liters (10 gallons) per day.

Why is so much water involved in the normal processes of living plants? Water constitutes about 90% of the weight of young cells. The thousands of enzyme actions and other chemical activities of cells take place in water, and additional although relatively negligible amounts are used in the process of photosynthesis. The exposed surfaces of the chlorenchyma cells within the leaf have to be moist at all

times, for it is through this film of water that the carbon dioxide molecules needed for the process of photosynthesis enter the cell from the air. Water is also needed for cell turgor, which gives rigidity to herbaceous plants.

Consider also what it must be like in the mesophyll of a flattened leaf that is fully exposed to the midsummer sun in areas where the air temperature soars to well over 38°C (100°F) in the shade. If it were not for the evaporation of water molecules from the moist surfaces, which brings about some cooling, and reradiation of energy by the leaf, the intense heat could damage the plant. Sometimes, the transpiration is so rapid that the loss of water begins to exceed the intake, and the stomata may close, thus preventing wilting (see also in Chapter 11 the discussion of the role of *abscisic acid* relative to excessive water loss).

How does water travel through the roots from 3 to 6 meters (10 to 20 feet) or more beneath the surface and then up the trunk to the topmost leaves of a tree that is more than 90 meters (300 feet) tall? We know that continuous tubular pathways of xylem run throughout a plant, extending from the young roots up through the stem and branches to the tiny veinlets of the leaves; we also know that the water gets to the start of this "plumbing system" by osmosis following a water potential gradient. But water is raised through the columns apparently by a combination of factors, and the process has been the subject of much debate for the past 200 years (Fig. 9.10).

One of the earliest explanations for the rise of water in a living plant was given in 1682 by the English scientist

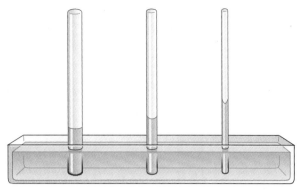

FIGURE 9.11 Capillarity in narrow tubes. The smaller the diameter of the tube, the greater the rise of the fluid.

Nehemiah Grew. He suggested that cells surrounding the xylem vessels and tracheids performed a pumping action that propelled the water along. This was questioned, however, when it was found that water will rise in lengths of dead stems also. Then, after Marcello Malpighi suggested it, the belief that capillary action moved the water became popular.

It is well-known that water will rise in a narrow tube and that the heights attained are inversely proportional to the diameter of the tube. It is also known that this rise occurs through the forces involved in the forming of a concave *meniscus* (curved surface) at the top of the water column (Fig. 9.11). But even though water can indeed rise 1 meter (3 feet) or more in a very narrow tube, air must be present above the column for the forces to work, and such is not the case in a plant. In fact, any air introduced into a water column in xylem interferes with the rise of water. Also, while capillarity might produce sufficient force to raise water a meter or two, the diameter of the tubes is not small enough to raise it more than that.

The pioneer plant physiologist Stephen Hales discovered and measured *root pressure* as one means whereby water is moved through plants. When some plants are pruned after growth has begun in the spring, water will exude from the cut ends. This is the result of root pressure, which involves osmosis and perhaps other forces. Some plants do not "bleed" when they are pruned, and the force exerted by root pressure has been shown generally to be less than 30 grams per square centimeter (a few pounds per square inch). This is considerably less than that needed to raise water to the tops of tall trees. Furthermore, root pressure seems to drop to negligible amounts in the summer, when the greatest amounts of water are moving through the plant.

The Cohesion-Tension Theory

Stephen Hales also identified a pulling force due to evaporation of water from leaves and stems. This has led to the most satisfactory explanation of the rise of water in plants thus far suggested, the **cohesion-tension theory.** Water molecules are electrically neutral, but they are asymmetrical in shape (see Fig. 2.3). This results in the molecules having very slight pos-

itive charges at one end and very slight negative charges at the other end. Such molecules are said to be *polar.* When the negatively charged end of one water molecule comes close to the positively charged end of another water molecule, weak hydrogen bonds hold the molecules together.

It is known that water molecules adhere to capillary walls (e.g., those of xylem tracheids and vessels) and cohere to each other, permitting a certain amount of tension. It is possible, for example, to fill a small glass with water, place a thin smooth sheet of cardboard over the mouth, and invert the glass without the water spilling, because the adhesion of the water molecules to the cardboard and the cohesiveness of the water molecules to one another hold the cardboard against the rim of the glass.

When water evaporates from the mesophyll cells in a leaf and diffuses out of the stomata (*transpires*), the cells involved develop a lower water potential than that of the adjacent cells. Because the adjacent cells then have a correspondingly higher water potential, replacement water moves into the first cells by osmosis. This continues across rows of mesophyll cells until a small vein is reached. As indicated in earlier chapters, each small vein is connected to a larger vein, the larger veins are connected to the main xylem in the stem, and that, in turn, is connected to the xylem in the roots that receive water, via osmosis, from the soil. As transpiration takes place, it creates a "pull," or tension, on water columns, drawing water from one molecule to another all the way through an entire span of xylem cells. The cohesion required to move water to the top of a tall tree is considerable, but the cohesive strength of the water columns is usually more than adequate. Any breaking of the tension through the introduction of a gas bubble results in a temporary or permanent blocking of water transport. This seldom is a problem, however, because small bubbles may be redissolved and larger gas bubbles rarely block more than a few of the numerous capillary tubes of xylem at any time the tissue is functioning.

Passage of water is partly through cell protoplasm and partly through spaces between living cells, between cellulose fibers in the cell walls, and through spaces in the centers of dead cells. Most water and solutes reaching the root xylem can travel across the epidermis and cortex via the cell walls until reaching the endodermis where the water and solutes are then forced by *Casparian strips* to cross the protoplasts of the endodermal cells on their way to the vessels or tracheids of the xylem (Fig. 9.12).

If rapid transpiration is occurring, the roots are likely to grow rapidly toward available water. In corn plants, for example, the main roots may grow at a rate of more than 6 centimeters (2.3 inches) a day. Solutes, as well as water, may move so rapidly during periods of rapid transpiration that there is little osmosis taking place across the endodermis. Scientists believe that at such times water may be pulled through the roots by *bulk flow,* which is the passive movement of a liquid from higher to lower water potential.

To recapitulate, "columns" of water molecules move through the plant from roots to leaves, and the abundant

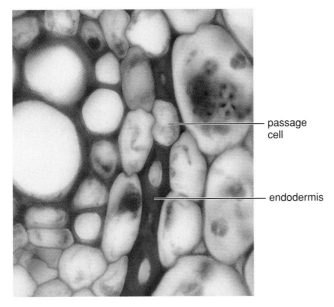

FIGURE 9.12   Part of the center of a buttercup root, showing endodermal cells with Casparian strips, ×600.

FIGURE 9.13   An aphid feeding on a young stem of linden (basswood). A droplet of "honeydew" is emerging from the rear of the aphid.

From Martin H. Zimmerman, "Movements of Organic Substances in Trees" *Science* 133: 73-79, 1961, American Association for the Advancement of Science.

water of a normally moist soil supplies these "columns" as the water continues to enter the root by osmosis (see Fig. 9.10); simply put, the difference between the water potentials (water "concentrations") of two areas (e.g., soil and the air around stomata) generates the force to transport water in a plant.

TRANSPORT OF FOOD SUBSTANCES (ORGANIC SOLUTES) IN SOLUTION

One of the most important functions of water in the plant involves the *translocation* (transportation) of food substances in solution by the phloem, a process that has recently come to be better understood. Many of the studies that led to our present knowledge of the subject utilized aphids (small, sucking insects) and organic compounds designed as radioactive tracers.

Most aphids feed on phloem by inserting their tiny tubelike mouthparts (*stylets*) through the leaf or stem tissues until they reach and puncture a sieve tube. The turgor pressure of the sieve tube then forces the fluid present in the tube through the aphid's digestive tract, and it emerges at the rear as a droplet of "honeydew." In some studies, research workers anaesthetized feeding aphids and cut their stylets so that much of the tiny tube remained where it had been inserted. Fluid exuded (sometimes for many hours) from the cut stylets and was then collected and analyzed (Fig. 9.13).

Carbon dioxide, a basic raw material of photosynthesis, can be synthesized with radioactive carbon. By exposing a photosynthesizing leaf to radioactive carbon dioxide, the pathway of manufactured food substances can be traced. The radioactive substances produce on photographic film an image corresponding to the food pathway. Data obtained from such studies reveals that food substances in solution are confined entirely to the sieve tubes while they are being transported. At one time, it was believed that ordinary diffusion and cyclosis (discussed in Chapter 3) were responsible for the movement of the substances from one sieve-tube member to the next, but it is now known that the substances move through the phloem at approximately 100 centimeters (almost 40 inches) per hour—far too rapid a movement to be accounted for by diffusion and cyclosis alone.

The Pressure-Flow Hypothesis

At present, the most widely accepted theory for movement of substances in the phloem is called the **pressure-flow** (or **mass flow**) **hypothesis.** According to this theory, food substances in solution (organic solutes) flow from a *source,* where water is taken up by osmosis (for example, sources include food-storage tissues, such as the cortex of roots or rhizomes, or food-producing tissues, such as the mesophyll tissues of leaves) and given up at a *sink,* which is a place where food is utilized, such as the growing tip of a stem or root. Food substances in solution (organic solutes) are moved along concentration gradients between sources and sinks (Fig. 9.14).

First, in a process called *phloem-loading,* sugar, by means of active transport, enters the sieve tubes of the smallest veinlets. This decreases the water potential in the sieve tubes, and water then enters these phloem cells by osmosis. Turgor pressure, which develops as this osmosis occurs, is responsible for driving the fluid through the sieve-tube network toward the sinks.

As the food substances (largely sucrose) in solution are actively removed at the sink, water also leaves the sink

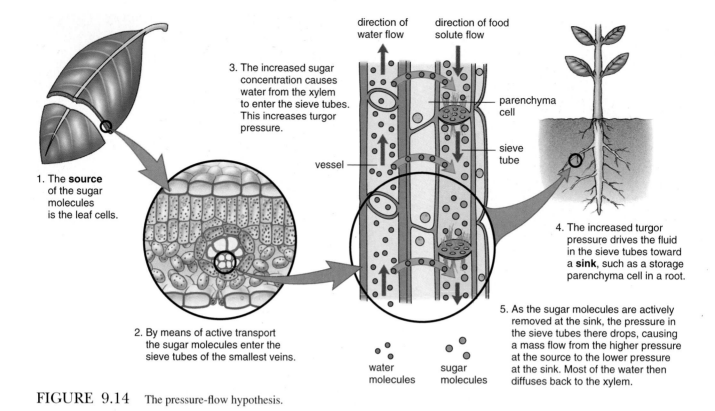

direction of water flow

direction of food solute flow

3. The increased sugar concentration causes water from the xylem to enter the sieve tubes. This increases turgor pressure.

parenchyma cell

sieve tube

vessel

1. The **source** of the sugar molecules is the leaf cells.

4. The increased turgor pressure drives the fluid in the sieve tubes toward a **sink**, such as a storage parenchyma cell in a root.

5. As the sugar molecules are actively removed at the sink, the pressure in the sieve tubes there drops, causing a mass flow from the higher pressure at the source to the lower pressure at the sink. Most of the water then diffuses back to the xylem.

2. By means of active transport the sugar molecules enter the sieve tubes of the smallest veins.

water molecules

sugar molecules

FIGURE 9.14   The pressure-flow hypothesis.

ends of sieve tubes, and the pressure in these sieve tubes is lowered, causing a mass flow from the higher pressure at the source to the lower pressure at the sink. Most of the water diffuses back to the xylem, where it then returns to the source and is transpired or recirculated. The pressure-flow hypothesis nicely explains how nontoxic dyes applied to leaves or substances entering the sieve tubes, such as viruses introduced by aphids, are carried through the phloem.

REGULATION OF TRANSPIRATION

The stomata, which often occupy 1% or more of the surface area of a leaf, *regulate transpiration and gas exchange,* although direct control of transpiration is exerted by the water vapor concentration of the atmosphere. The guard cells forming each stoma have relatively elastic walls with radially oriented microfibrils, making them analogous to pairs of sausage-shaped balloons joined at each end, each with a row of rubber bands around it. The part of the wall adjacent to the hole itself is considerably thicker than the remainder of the wall (Fig. 9.15). This thickness allows each stoma to be opened and closed by means of changes in the turgor of the guard cells. The stomatal pore is closed when turgor pressure is low and open when turgor pressure is high. Changes in turgor pressures in the guard cells, which contain chloroplasts, occur when they are exposed to

changes in light intensity, carbon dioxide concentration, or water concentration.

The turgor pressure changes take place when solute concentrations shift as osmosis and active transport between the epidermal cells and the guard cells occur. Examples include the taking up of potassium and chloride ions from other epidermal cells or from organic acids produced from starch. Energy is expended by guard cells to take up potassium ions from adjacent epidermal cells, leading to the stomata opening. Upon release of the potassium ions, the stomata close. With an increase in potassium ions, the water potential in the guard cells is lowered, and the osmosis that takes place as a result brings in water that makes the cells turgid. The departure of potassium ions also results in water leaving, making the cells less turgid and causing the stomata to close (see Fig. 9.15). Stomata will passively close whenever water stress occurs, but there is evidence that the hormone *abscisic acid* is produced in leaves subject to water stress and that this hormone causes membrane leakages, which induce a loss of potassium ions from the guard cells causing them to close (see also page 186).

The stomata of most plants are open during the day and closed at night, but water is conserved in a number of desert plants whose stomata open only at night when there is less water stress on the plants. Such plants have a specialized form of photosynthesis called *CAM photosynthesis,* since the carbon dioxide they require for typical photosynthesis is shut off during daylight hours by the closing of the stomata (see also page 169).

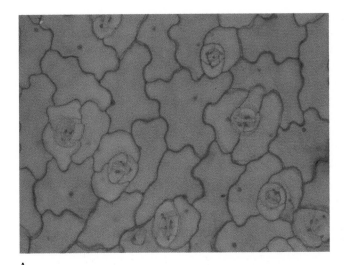

A.

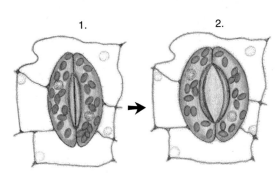

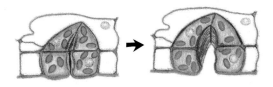

B.

FIGURE 9.15 *A*. Stomata in the epidermis of a stonecrop (*Sedum*) leaf, ca. ×500. *B*. How a stoma opens and closes. *1*. A closed stoma. *2*. As turgor pressure in the guard cells increases, the thinner outer walls stretch more than the thicker inner walls, causing the stoma to open.

FIGURE 9.16 A barrel cactus. The stems store carbon dioxide taken in at night and then release it for use in photosynthesis during the day.

Some desert plants are able to store carbon dioxide taken in at night in cell vacuoles in the form of organic acids, which can be converted back to carbon dioxide during the daytime when photosynthesis occurs (Fig. 9.16). Other desert plants have their stomata below the surface of the leaf or stem in little sunken chambers. These chambers often are partially filled with epidermal hairs, which further reduce water loss. Pine trees, which frequently occur in areas where the soil is frozen for part of the year limiting available water, also have sunken stomata (see Fig. 7.11). A few tropical plants that occur in damp, humid areas (e.g., ruellias; see also Fig. 4.12B) actually have stomata that are raised above the surface of the leaf, while plants that occur in water generally have no stomata on submerged surfaces.

Although transpiration is affected by light and carbon dioxide concentration, several other factors play at least an indirect role. For example, the stomata may be regulated by the water gradient being affected by air currents sweeping away the water molecules as they emerge from the stomata, which speeds up the rates of evaporation within.

Humidity plays an inverse but direct role in transpiration rates: High humidity reduces transpiration, and low humidity accelerates it. A direct correlation also exists between temperature and the movement of the water molecules out of the leaf. The transpiration rate of a leaf at 30°C (86°F), for example, is about twice as great as it is for the same leaf at 20°C (68°F). The various adaptive modifications of leaves and their surfaces and the availability of water to the roots also may play important roles in influencing the amount of water transpired. (See also the discussion of leaf modifications in Chapter 7.)

If a cool night follows a warm humid day, water droplets may be produced through structures called **hydathodes** at the tips of veins of the leaves of some plants. This loss of water in liquid form is called **guttation** (Fig. 9.17). Although the droplets resemble dew, the two should not be confused. Dew is water that is condensed from the air, while guttation water is literally forced out of the plant by root pressures. As the sun strikes the droplets in the morning, they dry up, leaving a residue of salts and organic substances, one of which is used in the manufacture of commercial flavor enhancers (e.g., the monosodium glutamate in products such as Accent). In the tropics, the amount of water produced by guttation can be considerable. In taro plants, used by the Polynesians to make poi, a single leaf may produce overnight as much as a cupful (about 240 milliliters) of water through guttation.

FIGURE 9.17   Droplets of guttation water at the tips of leaves of young barley plants.

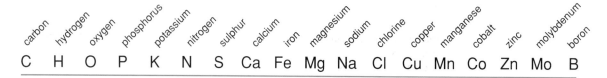

C. H O P K'N S CaFe; Mighty good (but) Not always Clean. CuM'n CoZ'n MoB(y)?

carbon	hydrogen	oxygen	phosphorus	potassium	nitrogen	sulphur	calcium	iron	magnesium	sodium	chlorine	copper	manganese	cobalt	zinc	molybdenum	boron
C	H	O	P	K	N	S	Ca	Fe	Mg	Na	Cl	Cu	Mn	Co	Zn	Mo	B

FIGURE 9.18   Elements essential as building blocks for compounds synthesized by plants.

MINERAL REQUIREMENTS FOR GROWTH

Growth phenomena are not controlled by internal means alone. Light, temperature, soil structure, minerals, and other external factors all play a role. In fact, growth depends on a complex combination of chemical and physical forces, both internal and external, which are in delicate balance with one another.

Plants may take up many elements from the soil, but besides the carbon and oxygen obtained from carbon dioxide and the hydrogen obtained from water, only 15 elements are essential as building blocks to most plants for the numerous compounds they synthesize. Sodium, a comparatively abundant element, is apparently required by few plants. The **essential elements** can be remembered by a sentence (Fig. 9.18) that includes the symbols of the elements involved.

Macronutrients and Micronutrients

The mineral elements are usually put into two categories: (1) *macronutrients*, which are used by plants in greater amounts and constitute from 0.5% to 3.0% of the dry weight of the plant, and (2) *micronutrients*, which are needed by the plant in very small amounts, often constituting only a few parts per million of the dry weight.

The macronutrients are nitrogen, potassium, calcium, phosphorus, magnesium, and sulphur, with the first four usually making up about 99% of the nutrient total. Those elements remaining, the micronutrients, are present in amounts ranging from bare traces—as in the case of sodium and cobalt, neither of which may actually be essential for some plants—to 1,500 parts per million, of iron and manganese, and up to 10,000 parts per million, of chlorine.

Specific organisms may require elements in addition to these widely required ones. For example, certain algae apparently require the elements vanadium, silicon, or iodine, while some ferns utilize aluminum. Several loco weeds absorb and accumulate selenium in amounts constituting up to 5 micrograms per gram of dry weight. Selenium, which is often fatally poisonous to livestock, appears to enhance the growth of these plants by reducing toxic effects of phosphates, but there is no direct evidence it is essential to them.

When any of these elements is deficient in the soil, the plants exhibit characteristic symptoms of the deficiency, which disappear after the problem has been corrected (Fig. 9.19). Table 9.1 shows some of the uses of essential elements in plants and describes the symptoms of deficiency for each element.

See the discussion of fertilizers and fertilizing in Appendix 4 for further information on ratios of nitrogen, phosphorus, and potassium (NPK) needed for plants and on organic versus inorganic fertilizers. See also page 273 for a discussion of compost and composting.

Table 9.1

USES OF ESSENTIAL ELEMENTS IN PLANTS

ELEMENT	SOME FUNCTIONS	DEFICIENCY SYMPTOMS
Nitrogen	Part of proteins, nucleic acids, chlorophyll	Relatively uniform loss of color in leaves, occurring first on the oldest ones
Potassium	Activates enzymes; concentrates in meristems	Yellowing of leaves, beginning at margins and continuing toward center; lower leaves mottled and often brown at tip
Calcium	Essential part of middle lamella; involved in movement of substances through cell membranes	Terminal bud often dead, young leaves often appearing hooked at tip; tips and margins of leaves withered; roots dead or dying
Phosphorus	Necessary for respiration and cell division; synthesis of high-energy cell compounds	Plants stunted; leaves darker green than normal; lower leaves often purplish between veins
Magnesium	Part of the chlorophyll molecule; activates enzymes	Veins of leaves green but yellow between them with dead spots appearing suddenly; leaf margins curling
Sulphur	Part of some amino acids	Leaves pale green with dead spots; veins lighter in color than the rest of the leaf area
Iron	Needed to make chlorophyll and in respiration	Larger veins remaining green while rest of leaf yellows; mainly in young leaves*
Manganese	Activates some enzymes	Dead spots scattered over leaf surface; all veins and veinlets remain green; effects confined to youngest leaves
Boron	Influences utilization of calcium ions, but functions unknown	Petioles and stems brittle; bases of young leaves break down

* The symptoms of iron deficiency may be caused by several factors, such as overwatering, cold temperatures, and nematodes (small roundworms) in the roots. The iron may be relatively abundant in the soil, but its uptake may be prevented or sharply reduced by these environmental conditions. Iron becomes more soluble under acid conditions—so much so that it can produce toxic conditions for most plants. Acid soil plants, on the other hand, have a much higher iron requirement than plants that require more nonacid conditions; thus, azaleas and other plants having high iron requirements can achieve normal growth in a nonacid soil if acidic fertilizers and/or materials such as conifer leaf compost are added to the soil.

It should be noted that all the micronutrients are harmful to plants when supplied in excessive quantities. Copper will kill algae in concentrations of one part per million, and boron has been used in weed killers. Even macronutrients are harmful if present in heavy amounts, although nonessential elements sometimes can counteract their toxicity.

FIGURE 9.19   Leaves of bean plants grown in media deficient in various elements to show deficiency symptoms. *A*. A normal plant that has been furnished with all the essential elements. The other plants were grown in media deficient in specific elements, as follows: *B*. Potassium. *C*. Phosphorus. *D*. Calcium. *E*. Nitrogen. *F*. Sulphur. *G*. Micronutrients. *H*. Magnesium. *I*. Iron.

Summary

1. Molecules and ions are in constant random motion and tend to distribute themselves evenly in the space available to them. They move from a region of higher concentration to a region of lower concentration by simple diffusion along a diffusion gradient; they may also move against a diffusion gradient. Evenly distributed molecules are in a state of equilibrium. Diffusion rates are affected by temperature, molecule size and density, and other factors.

2. Osmosis is the diffusion of water through a differentially permeable membrane. It takes place in response to concentration differences of dissolved substances.

3. Osmotic pressure or potential is the pressure required to prevent osmosis from taking place. The pressure that develops in a cell as a result of water entering it is called *turgor*. Water moves from a region of higher water potential (osmotic potential and pressure potential combined) to a region of lower water potential when osmosis is occurring. Osmosis is the primary means by which plants obtain water from their environment.

4. Plasmolysis is the shrinkage of the protoplasm away from the cell wall as a result of osmosis taking place when the water potential inside the cell is greater than it is outside.

5. Imbibition is the attraction and adhesion of water molecules to the internal surfaces of materials; it results in swelling and is the initial step in the germination of seeds.

6. Active transport is the expenditure of energy by a cell that results in molecules or ions entering or leaving the cell against a diffusion gradient.

7. Water that enters a plant passes through it and mostly transpires into the atmosphere via stomata. Water retained by the plant is used in photosynthesis and other metabolic activities.

8. The cohesion-tension theory postulates that water rises through plants because of the adhesion of water molecules to the walls of the capillary conducting elements of the xylem, cohesion of the water molecules, and tension on the water columns created by the pull developed by transpiration.

9. The translocation of food substances takes place in a water solution, and according to the pressure-flow hypothesis, such substances flow along concentration gradients between their sources and sinks.

10. Transpiration is regulated by humidity and the stomata, which open and close through changes in turgor pressure of the guard cells. These changes, which involve potassium ions, result from osmosis and active transport between the guard cells and the adjacent epidermal cells.

11. Aquatic, desert, tropical, and some cold zone plants have modifications of stomata or specialized forms of photosynthesis that adapt them to their particular environments.

12. Guttation is the loss of water in liquid form through hydathodes at the tips of leaf veins.

13. Growth phenomena are controlled by both internal and external means and by chemical and physical forces in balance with one another. Besides carbon, hydrogen, and oxygen, 15 other elements are essential to most plants. When any of the essential elements are deficient in the plant, characteristic deficiency symptoms appear.

Review Questions

1. Distinguish among diffusion, osmosis, active transport, plasmolysis, and imbibition.

2. Why do living plants need a great deal of water for their activities?

3. Explain how a tall tree gets water to its tips without the aid of mechanical pumps.

4. What is the difference between transpiration and guttation?

5. Explain the pressure-flow hypothesis.

6. What are macronutrients? List them.

7. When nutrients are deficient in the soil, how are the deficiencies manifested in plants?

Discussion Questions

1. Why is salted meat less likely to spoil than unsalted meat?

2. Why would dye molecules in a bathtub of water take a long time to diffuse completely throughout the tub, but perfume molecules released in a closed room take considerably less time to do the same thing?

3. Why does osmosis not cause submerged water plants to swell up and burst?

4. Some bodies of water, such as the Dead Sea, have considerably higher salt concentrations than those of the human body. If you were swimming in such water, would you expect your cells to become plasmolyzed? Why?

Additional Reading

Baker, D.A. 1989. *Transport of photoassimilates*. New York: Halsted Press.

Borghetti, M., et al. (Eds.). 1993. *Analysis of water transport in plants*. New York: Cambridge University Press.

Devlin, R.M., and F. Witham. 1983. *Plant physiology,* 4th ed. Belmont, CA: Wadsworth Publishing Co.

Galston, A.W. 1993. *Life processes of plants: Mechanisms for survival*. New York: W. H. Freeman.

Kluge, M., and I.P. Ting. 1978. Crassulacean acid metabolism. *Ecological studies* (Vol. 30). New York: Springer-Verlag.

Loughman, B.C., et al. 1990. *Structural and functional aspects of transport in roots*. Norwell, MA: Kluwer Academic Publications.

Salisbury, F.B., and C.W. Ross. 1992. *Plant physiology,* 4th ed. Belmont, CA: Wadsworth Publishing Co.

Chapter Outline

Flowers of Bougainvillea *hybrids, widely cultivated in the tropics and subtropics. Large colorful bracts surround the three cream-colored flowers.*

Plant Metabolism

10

Overview—————————

Photosynthesis and respiration are each presented at three different levels: The essence of the process is examined, the major steps are briefly introduced, and the processes are then examined in greater detail. One or two levels may be sufficient for some readers; others will want to explore all three. The chapter discusses the importance of the main features of each process and summarizes the light reactions, the carbon-fixing reactions, glycolysis, the Krebs cycle, and the electron transport system. It concludes with a tabular comparison between photosynthesis and respiration and makes a few brief observations on assimilation and digestion.

Some Learning Goals

1. Contrast the generalized equations of photosynthesis and respiration.
2. Understand what occurs in the light and carbon-fixing reactions of photosynthesis, and know the principal products of the reactions.
3. Explain what occurs in glycolysis, the Krebs cycle, and the electron transport chain of respiration.
4. Distinguish between aerobic respiration and fermentation.
5. Compare assimilation and digestion.

A number of years ago, I had a student who told me her father was an alchemist and that he was on the verge of discovering how to transform lead into gold. The belief that an element could be transformed into another element became widespread in medieval times, but with the development of modern science and technology, such romantic yet totally unscientific ideas have been discarded by educated people in general and relegated to the history books.

Some amazing molecular transformations of another kind, however, take place all around us every day in green organisms and have been taking place since long before humans appeared on this planet. By far, the most important of these transformations, *photosynthesis,* involves little more than fresh air, water, a green pigment, and light; yet parts of the water and air are converted in cells to a sugar without the aid of any cumbersome machinery. In addition, 24 hours a day, as long as any cell remains alive, stored energy is released from sugar by another process, *respiration.* These two processes take up the bulk of our discussion in this chapter.

PHOTOSYNTHESIS

Plants growing millions of years ago originally captured from the sun the energy within oil and coal, which was then transformed into fossil fuels by geological processes. These fossil fuels today provide about 90% of the energy needed to power trains, trucks, ships, aircraft, factories, and a myriad of electrically energized appliances, computers, and communications systems. The energy needs of transportation, in-

dustry, and the modern household seem insignificant, however, when compared with the combined energy requirements of all living organisms. Every living cell requires energy just to remain alive, and additional energy is needed for the cell to reproduce, grow, or do physical work as part of an organism. In addition, oxygen is vital to nearly all life in processes releasing that stored energy for use.

Photosynthesis, at least indirectly, is not only the principal means of keeping all forms of humanity functioning but also the sole means of sustaining life at any level—except for a few bacteria that derive their energy from sulphur salts and other inorganic compounds. This unique manufacturing process of green plants furnishes raw material, energy, and oxygen. In photosynthesis, energy from the sun is harnessed and packed into sugar molecules made from water and carbon dioxide with the aid of chlorophyll; oxygen is given off as a by-product of the process.

It has been estimated that all of the world's green organisms (including those in the oceans) together produce between 100 billion and 200 billion metric tons (110 billion and 220 billion tons) of sugar each year. To visualize that much sugar, consider that it is enough to make about 300 quadrillion sugar cubes with a total volume exceeding that of two million Empire State Buildings[1] (Fig. 10.1).

Much of the sugar produced by plants is converted to wood, fibers (such as cotton and linen), and other structural materials. The first products of photosynthesis may also be converted to disaccharides, polysaccharides such as starch, and other storage forms of carbohydrates that the digestive activities of living organisms break down to smaller molecules.

Photosynthetic sugars are also involved in the production of amino acids for proteins and a host of other cell constituents. In fact, photosynthesis produces more than 94% of the dry weight substance of green organisms, with the remainder coming from the soil or dissolved matter.

The capacity of plants to meet our energy needs probably will determine the ultimate size of human populations. In some heavily populated parts of the world, the food supply already is falling short of providing enough energy to sustain life, and starvation is widespread. Meanwhile, in the Western world, significant numbers of persons consume too much food and are spending large sums on weight reduction. We are, however, all approaching a point at which human populations in general must stabilize, or even those in the most affluent areas will exceed the capacity of the plants to sustain them.

The food energy produced by plants is not necessarily the ultimate limiting factor for human populations. Pollution eventually could interfere with photosynthetic oxygen production to the extent of diluting the oxygen content of the atmosphere.

A great deal of photosynthesis occurs in organisms living in the oceans. It is estimated that between 40% and 50% of

1. One quadrillion is 1,000,000,000,000,000. New York City's Empire State Building has 102 stories and is 381 meters (1,250 feet) tall.

FIGURE 10.1 Various forms of processed sugar.

the oxygen in the atmosphere originates in oceans and lakes. Laboratory tests have shown, however, that pollutants—such as the PCBs (polychlorinated biphenyls) used in electrical insulators—are capable, in concentrations as low as 20 parts per billion, of stopping many delicate algae from carrying on photosynthesis. The concentration of such substances in ocean waters at present is considerably less than one part per billion, but PCB concentrations of up to five or more parts per billion have already been reported in some estuaries.

While the use of PCBs and related chemicals has been curtailed in the United States, other countries are still using them, and residues are still washing off into rivers and on into the oceans. If the concentrations eventually were to build up to the point of stopping algal growth and reproduction, the oxygen consumption of an overpopulated world could eventually exceed the capacity of the remaining plant life to replenish it, with suffocation of nearly all animal life the grisly result. However, starvation would be far more likely to occur first.

Note to the Reader: Photosynthesis is undoubtedly the most important process on earth to life as we know it. It is also a complex process that can be summarized briefly or examined in detail. The following is a discussion of the subject at three different levels: (1) an examination of the essence of photosynthesis; (2) a brief introduction to the major steps of photosynthesis; and (3) a closer look at photosynthesis. The process of respiration, which is discussed after photosynthesis, is treated in similar fashion. The third level, in particular, contains detail that either may or may not be discussed in your course.

I.   The Essence of Photosynthesis

Photosynthesis is an energy-storing process that takes place in chloroplasts and other parts of green organisms in the presence of light. The light energy is stored in a simple sugar molecule, which is produced from carbon dioxide (CO_2) present in the air and water (H_2O) absorbed by the plant. When the carbon dioxide (CO_2) and the water (H_2O) are combined and form a sugar molecule ($C_6H_{12}O_6$, glucose) in a chloroplast, oxygen gas (O_2) is released as a by-product. The oxygen diffuses out into the atmosphere. The process is summed up in the following equation:

$$6\ CO_2 + 12\ H_2O + \text{light energy} \xrightarrow[\text{enzymes}]{\text{chlorophyll}}$$

$$C_6H_{12}O_6 + 6\ O_2 + 6\ H_2O$$

$$\text{glucose} \quad \text{oxygen} \quad \text{water}$$

Photosynthesis takes place in chloroplasts (see figs. 3.5 and 3.11) or in cells with chlorophyll diffused throughout the cytoplasm. The principals in the process are *carbon dioxide, water, light,* and *chlorophyll;* a brief examination of each follows.

Carbon Dioxide

Our atmosphere consists of approximately 79% nitrogen and about 20% oxygen. The remaining 1% is made up of a mixture of less common gases, including 0.039% (390 parts per million) carbon dioxide and a little hydrogen, helium, argon, and neon. The 0.039% carbon dioxide in the air surrounding the leaves of plants reaches the chloroplasts in the mesophyll cells by diffusing through the stomata, which, when open, allow the movement of carbon dioxide into the leaf interior. The carbon dioxide dissolves in the thin film of water on the outside walls of the cells. By diffusion, the carbon dioxide then passes through the walls into the cytoplasm and reaches the chloroplasts.

The amount of carbon dioxide constantly being taken from the atmosphere during daylight hours by all green plants is enormous. Just four-tenths of a hectare (1 acre) of corn (10,000 plants) accumulates more than 2,500 kilograms (5,512 pounds) of carbon from the atmosphere during a growing season. Over 10 metric tons (11 tons) of carbon dioxide are needed to furnish this much carbon. It also has been calculated that the total present atmospheric supply of carbon dioxide (more than 2.2 billion metric tons or about 50 metric tons over each hectare of the earth's surface) would be completely used up in about 22 years if it were not constantly being replenished.

A large reservoir of carbon dioxide in the oceans maintains the atmosphere's percentage of the gas at an ambient 0.039, while plant and animal respiration, the burning of material containing carbon, volcanoes, and similar sources replace it at roughly the same rate at which it is removed during photosynthesis.

There is evidence that the concentration of carbon dioxide in the atmosphere is increasing. A 1970 study by scientists at the Massachusetts Institute of Technology on critical environmental problems predicted a 20% increase by the year 2000. Since the 1980s, global warming as a result of this increase has become a major political and scientific issue. (See the section on the greenhouse effect in Chapter 25 for a discussion of the problem.)

A small increase in carbon dioxide levels could benefit plants, providing the level does not climb beyond 0.2% (which is extremely unlikely). Increases in yields of between 100% and 200% have been obtained by fertilizing plants with carbon dioxide, and some large commercial greenhouses have run pipes over plant beds to supplement their natural supply. The changes in climate that can result from an increase in atmospheric carbon dioxide, however, could have seriously adverse effects that would more than offset positive benefits to plants.

Water

Less than 1% of all the water absorbed by plants is used in photosynthesis; most of the remainder is transpired or incorporated into protoplasm, vacuoles, and other materials. The water used is the source of the oxygen released as a by-product of photosynthesis, even though carbon dioxide also contains oxygen. This has been demonstrated by using isotopes of oxygen to make both the carbon dioxide and the water used in photosynthesis. When an oxygen isotope is used only in the water, it appears in the oxygen gas released. If, however, it is used only in the carbon dioxide, it is confined to the sugar and water produced and never appears in the oxygen gas, demonstrating clearly that the water is the sole source of the oxygen released.

If water is in short supply, it may indirectly become a limiting factor in photosynthesis; under such circumstances, the stomata usually close and sharply reduce the carbon dioxide supply.

Light

Light exhibits properties of both waves and particles. Energy reaches the earth from the sun in waves of different lengths, the longest waves being radio waves and the shortest being gamma rays. About 40% of the radiant energy we receive is in the form of visible light. If this visible light is passed through a glass prism, it splits up into its component colors. Reds are on the longer wavelength end and violets are on the shorter wave end, with yellows, greens, and blues between (Fig. 10.2). Although nearly all of the visible light colors can be used in photosynthesis, those in the violet to blue and red-orange to red wavelengths are used most extensively. Many in the green range are reflected. Leaves commonly absorb about 80% of the visible light that reaches them.

The intensity of light varies with the time of day, season of the year, altitude, latitude, and atmospheric conditions. On a clear summer day at noon in a temperate zone,

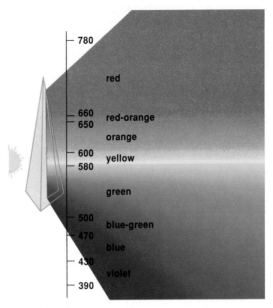

FIGURE 10.2   Visible light that is passed through a prism is broken up into individual colors with wavelengths ranging from 390 nanometers (violet) to 780 nanometers (red).

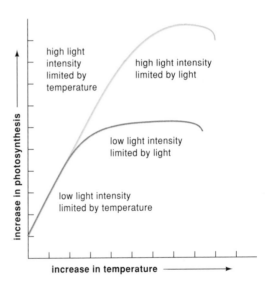

FIGURE 10.3   How temperature and light interact to affect photosynthesis. When temperature is a limiting factor, the rate of photosynthesis is independent of light intensity. When light is a limiting factor, the rate of photosynthesis is independent of temperature.

2. A foot-candle is the light cast by a standard candle at a distance of 1 foot. Foot-candles have been used as units of measurement of light intensity for many years. Note, however, that such units are subject to variables in light perception by human eyes and other variables in candles. They also present problems in conversion to metric measurements. Accordingly, modern scientists have turned to other measurements of light (e.g., radiant energy per square meter). Because these other forms of measurement are not essential to our introductory discussion, the term *foot-candle* is retained here.

sunlight attains an intensity in the vicinity of 10,000 foot-candles.[2] In contrast, consider that a good reading lamp produces only about 25 foot-candles.

Plants vary considerably in the light intensities they need for photosynthesis to occur at optimal (ideal) rates, and other factors, such as the temperature and amount of carbon dioxide available, can be limiting (Fig. 10.3). For example, an increase in photosynthesis will not occur in some plants receiving more than 3,000 foot-candles of light unless supplemental carbon dioxide is provided. With supplements, however, rates of photosynthesis will continue to increase up to about 5,000 foot-candles. Herbaceous plants on a forest floor can survive with less than 2% of full daylight, and some mosses are reported to thrive on intensities as low as 0.05% to 0.01%. Most land plants that naturally grow in the open need at least 30% of full daylight to thrive. The optimal amount for some species of trees approaches full daylight, while shade plants often do well in 10% of full daylight.

Light that is too intense may change the way in which some of a cell's metabolism takes place. For example, in high light intensities, *photorespiration* may occur. Photorespiration is a special type of respiration that uses oxygen and releases carbon dioxide but differs from common aerobic respiration in its chemical pathways (see page 169).

Photooxidation, which involves the destruction ("bleaching") of chlorophyll by light, may also occur. Although carbon dioxide is released in this process, most of the chemical steps involved are quite different from those of normal respiration. In the fall, photooxidation plays a significant role in the breakdown of chlorophyll in leaves, resulting in the autumn colors discussed in Chapter 7. High light intensities may also cause an increase in transpiration, resulting in the closing of stomata. A sharp reduction in the available carbon dioxide supply inevitably follows.

Chlorophyll

There are several different types of chlorophyll molecules, all of which contain one atom of magnesium. They are very similar in structure to the heme of hemoglobin, the iron-containing red pigment that transports oxygen in blood. Each molecule has a long lipid tail, which anchors the chlorophyll molecule in the lipid layers of the thylakoid membranes (Fig. 10.4).

Chloroplasts of most plants contain two kinds of chlorophyll associated with the thylakoid membranes.

Chlorophyll *a* is bluish green and has the formula $C_{55}H_{72}MgN_4O_5$. Chlorophyll *b* is yellowish green in color and has the formula $C_{55}H_{70}MgN_4O_6$. Usually, a chloroplast has about three times more chlorophyll *a* than *b*. When a molecule of chlorophyll *b* absorbs light, it transfers the energy to a molecule of chlorophyll *a*. Chlorophyll *b*, then, makes it possible for photosynthesis to take place over a broader spectrum of light than would be possible with chlorophyll *a* alone.

Other such pigments include carotenoids (yellowish to orange pigments found in all plants), phycobilins (blue or red pigments found in blue-green bacteria and red algae), and several other types of chlorophyll. Chlorophylls *c*, *d*, and possibly *e* take the place of chlorophyll *b* in certain algae, and several other photosynthetic pigments are found in bacteria. The various chlorophylls are all closely related and differ from one another only slightly in the structure of their molecules.

In chloroplasts, about 250 to 400 pigment molecules are grouped as a light-harvesting complex called a **photosynthetic unit,** with countless numbers of these units in each granum. Two types of these photosynthetic units function together in the chloroplasts of green plants, bringing about the first phase of photosynthesis, the *light reactions* (discussed in the next section).

II. Introduction to the Major Steps of Photosynthesis

The process of photosynthesis takes place in two successive series of steps called the **light reactions** and the **carbon-fixing reactions.**

The Light Reactions

In the 1930s, Robin Hill, a biochemist in England, discovered that a solution of fragmented and whole chloroplasts, isolated from leaves that had been ground up and centrifuged, could

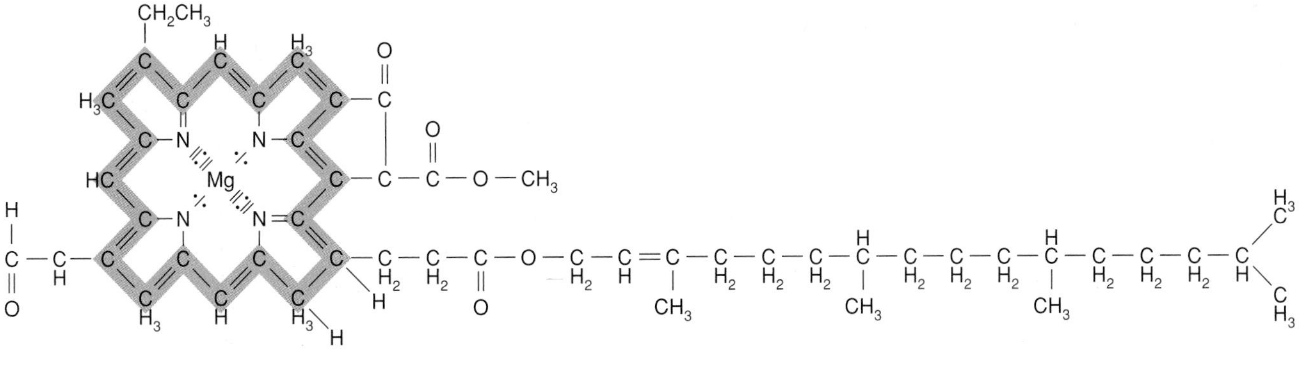

FIGURE 10.4 The structure of a molecule of chlorophyll *a,* the most important of the pigments involved in photosynthesis. The boxlike ring structure on the left, with magnesium and nitrogen inside, functions in capturing light energy. The tail, which extends into the interior of a thylakoid membrane, is incapable of absorbing water; all chlorophyll molecules are, however, fat soluble.

briefly produce oxygen if an electron acceptor was present to receive electrons from water. In 1951, it was shown that *nicotinamide adenine dinucleotide phosphate (NADP⁺)*, a substance derived from the B vitamin *niacin*, was a natural electron acceptor in this reaction, which, in honor of its discoverer, has become known as the *Hill reaction.*

The Hill reaction and other research gave insight into a major phase of photosynthesis, called the **light reactions,** which involve light striking chlorophyll molecules that are embedded in the thylakoids of the grana in the chloroplasts (see Fig. 3.11). The light initiates reactions that result in the conversion of some of the light energy to chemical energy. In the process, (1) water molecules are split apart, producing hydrogen ions and electrons, and oxygen gas is released; (2) energy-storing molecules commonly known as **ATP (adenosine triphosphate)**[3] are created; and (3) the hydrogen from the split water molecules is involved in the creation of NADPH, which carries hydrogen and which is used in the second major phase of photosynthesis, the *carbon-fixing reactions.*

The Carbon-Fixing Reactions

The **carbon-fixing reactions** were previously called *dark reactions* because light is not directly involved in their func-

3. There are millions of ATP molecules in living cells. When an ATP molecule releases its energy, it gives up one of its three phosphate groups and becomes ADP, or adenosine diphosphate. An ADP molecule becomes an ATP molecule again when it regains a third phosphate group and stores energy. ATP is an important participant in many reactions involving the transfer of energy.

tioning. They are a series of reactions that take place outside of the grana in the stroma of the chloroplast (see page 36), if the products of the light reactions are available.

The carbon-fixing reactions may develop in different ways, depending on the particular kind of plant involved. In 1961, Dr. Melvin Calvin of the University of California received a Nobel Prize for discovering how the most widespread type of carbon-fixing reactions occurs, and this process is now often referred to as the **Calvin cycle.**

In this cycle, carbon dioxide (CO_2) from the air is combined with a 5-carbon sugar (RuBP, ribulose bisphosphate), and then the combined molecules are converted, through several steps, to 6-carbon sugars, such as glucose ($C_6H_{12}O_6$). Energy and other items involved in these steps are furnished by the ATP molecules and NADPH produced during the light reactions. Some of the 6-carbon sugars that are produced during the carbon-fixing reactions are recycled, while others are stored as starch or other polysaccharides (simple sugars strung together in chains). A simplified summary of the photosynthetic reactions is portrayed in Figure 10.5. More detailed diagrams of the light and carbon-fixing reactions are shown in figures 10.8 through 10.13.

Two molecules of a 3-carbon sugar compound (3 PGA—an abbreviation of 3-phosphoglyceric acid) are shown as the first stable substance produced when carbon dioxide from the air and RuBP are combined and then converted during the carbon-fixing reactions. Many tropical plants produce a 4-carbon compound at this point instead. This 4-carbon pathway is discussed, along with another variation found mostly in desert plants, in the next section.

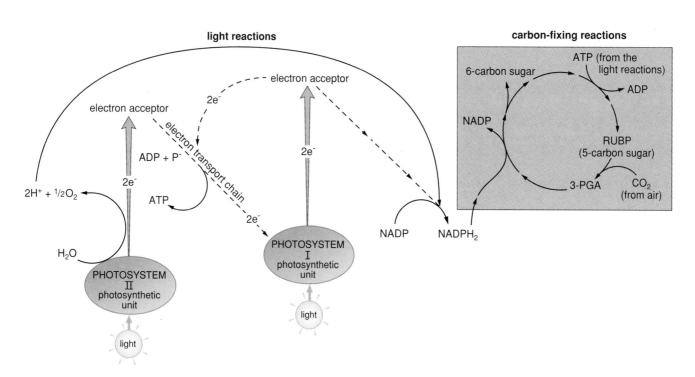

FIGURE 10.5   A simple summary of photosynthetic reactions.

III. A Closer Look at Photosynthesis

A great deal has been learned about photosynthesis since 1772 when Joseph Priestly (1733–1804), a naturalist in England, received a medal for demonstrating that vegetation "restored" oxygen so that a mouse could live in air that had been used up by a burning candle. Seven years later, Jan Ingen-Housz (1730–1799) of Holland, who visited England to treat the royal family for smallpox, carefully repeated Priestly's demonstrations. He showed that the air was "restored" only when green parts of plants were in the presence of sunlight.

In 1782, Jean Senebier (1742–1809), a Swiss pastor, discovered that the photosynthetic process required carbon dioxide, and in 1796, Ingen-Housz showed that carbon went into the nutrition of the plant. The final component of the overall photosynthetic reaction was explained in 1804 by another Swiss, Nicholas Theodore de Saussure (1767–1845), who showed that water was involved in the process.

A little current information about the details of photosynthesis is given in the following modest amplification of the preceding "second level" outline. Those who wish more information are referred to the reading list at the end of the chapter.

The Light Reactions Reexamined

As noted earlier, light has characteristics of both waves and particles. Over 300 years ago, Sir Isaac Newton, while experimenting with a prism, produced a spectrum of colors from visible white light and postulated that light consisted of a series of discrete particles he called "corpuscles."

Newton's theory only partially explained light phenomena, however, and by the middle of the 19th century, James Maxwell and others showed that light and all other parts of the extensive electromagnetic spectrum travel in waves.

By the late 1800s, with the discovery that a *photoelectric effect* can be produced in all metals, the wave theory also became inadequate to explain certain attributes of light. When a metal is exposed to radiation of a critical wavelength, it becomes positively charged because the radiation forces electrons out of the metal atoms. The ability of light to force electrons from metal atoms depends on its particular wavelength—its energy content—and not on its intensity or brightness. The shorter the wavelength, the greater the energy and vice versa.

In 1905, Albert Einstein proposed that the photoelectric effect results from discrete particles of light energy he called **photons.** In 1921, he received the Nobel Prize in physics for this work. Both waves and particles (photons) are today almost universally recognized as aspects of light. The energy (quantum) of a photon is not the same for all kinds of light; light of longer wavelengths has lower energy, and light of shorter wavelengths has proportionately higher energy.

Chlorophylls, the principal pigments of photosynthetic systems, absorb light primarily in the violet to blue and also in the red wavelengths; they reflect green wavelengths, which is why leaves appear green. This was first ingeniously demonstrated in 1882 by T. W. Engelmann. He focused a tiny spectrum of light on a filament (single row of cells) of *Spirogyra,* a freshwater alga. The alga had been mounted in a drop of water on a microscope slide containing bacteria that move toward an oxygen source. As shown in Figure 10.6, the bacteria assembled in greatest numbers along the algal filament in the blue and red portions of the spectrum, demonstrating that oxygen production is directly related to the light the chlorophyll absorbs. An analysis using living material is called a *bioassay.* Information gained from bioassays often is significant.

Each pigment has its own distinctive pattern of light absorption, which is referred to as the pigment's *absorption spectrum* (Fig. 10.7). When a pigment absorbs light, the energy levels of some of the pigment's electrons are raised. When this occurs, the energy may be emitted immediately as light (a phenomenon called *fluorescence*). In chlorophyll, this is characteristically in the red part of the visible light spectrum, so that an extract of chlorophyll placed in light (especially ultraviolet or blue light) will appear red. The absorbed energy may also be emitted as light after a delay (a phenomenon called *phosphorescence*); it may otherwise be converted to heat, or it may be stored in a chemical bond as it is in photosynthesis.

Oxidation-Reduction Reactions The complex process of photosynthesis involves several types of chemical reactions. These include *oxidation-reduction reactions.* **Oxidation** is *a loss of an electron or electrons;* it involves the removal of electrons from a compound. **Reduction** is *a gain of an electron or electrons;* it involves the addition of electrons from a compound.

When an electron is removed, a proton may follow, with the result that hydrogen is often removed during oxidation and added during reduction. Oxygen is usually the oxidizing agent (the final acceptor of the electron), but oxidations can occur without oxygen being involved.

Oxidation-reduction reactions are also vital to the process of respiration, which is discussed later in this chapter.

Photosystems The two types of photosynthetic units present in most chloroplasts make up **photosystems** known as *photosystem I* and *photosystem II,* respectively (Fig. 10.8). Each photosynthetic unit of photosystem I consists of 200 or more molecules of chlorophyll *a,* small amounts of chlorophyll *b,* carotenoid pigment with protein attached, and a pair of special *reaction-center* molecules of *chlorophyll* called P_{700}. Although all pigments in a photosystem can absorb photons, the reaction-center molecules are the only ones that can actually use the light energy. The remaining photosystem pigments are called *antenna pigments* because together they function somewhat like an antenna in gathering and passing light to the reaction-center molecule (Fig. 10.9).

There are also iron-sulphur proteins that are the primary electron acceptors for photosystem I (i.e., iron-sulphur

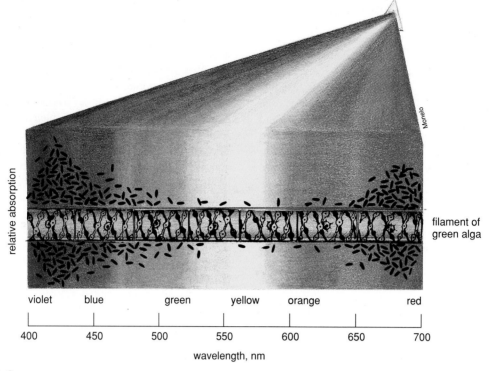

relative absorption

violet   blue   green   yellow   orange   red

| | | | | | |
400   450   500   550   600   650   700

wavelength, nm

filament of
green alga

FIGURE 10.6   Engelmann's experiment. A tiny spectrum of light was focused on a microscope slide with a row of algal cells suspended in water containing bacteria that move toward an oxygen source. The bacteria assembled mostly in the areas exposed to the red and blue portions of the spectrum.

proteins first receive electrons from P_{700}). A photosynthetic unit of photosystem II consists of chlorophyll a, β-carotene (the precursor of Vitamin A) attached to protein, a little chlorophyll b, and a pair of reaction-center molecules of chlorophyll a; these special molecules are called **P_{680}**.

The letter P stands for pigment and the numbers *700* and *680* of the reaction-center molecules of chlorophyll a refer to peaks in the absorption spectra of light with wavelengths of 700 and 680 nanometers, respectively. These peaks differ slightly from those of the otherwise identical chlorophyll a molecules of the photosynthetic units. A primary electron acceptor called *pheophytin* (or *Pheo*) is also present in photosystem II. One reaction-center molecule was found by Johann Deisenhofer and Hartmot Michel of Germany to have over 10,000 atoms. They received a Nobel Prize in 1988 for their work in determining the atomic structure.

Photolysis   When a photon of light strikes a P_{680} molecule of a photosystem II reaction center (located near the inner surface of a thylakoid membrane), the light energy boosts an electron to a higher energy level. This is referred to as *exciting* an electron. Excited electrons are unstable and lose much of their energy in the form of heat. Four photons strike the P_{680} molecule at the same time, but only one electron at a time can be accepted by the P_{680} molecule.

The excited electron is picked up by the primary acceptor molecule pheophytin; the electron then crosses the

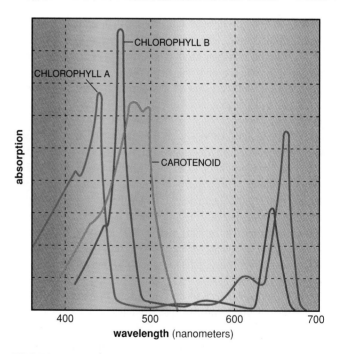

absorption

CHLOROPHYLL B

CHLOROPHYLL A

CAROTENOID

| | | |
400   500   600   700

wavelength (nanometers)

FIGURE 10.7   The absorption spectra of chlorophyll *a*, chlorophyll *b*, and a carotenoid. The maximum absorption of the chlorophylls is in the blue and red wavelengths. The maximum absorption of carotenoids is in the blue-green to green parts of the visible spectrum.

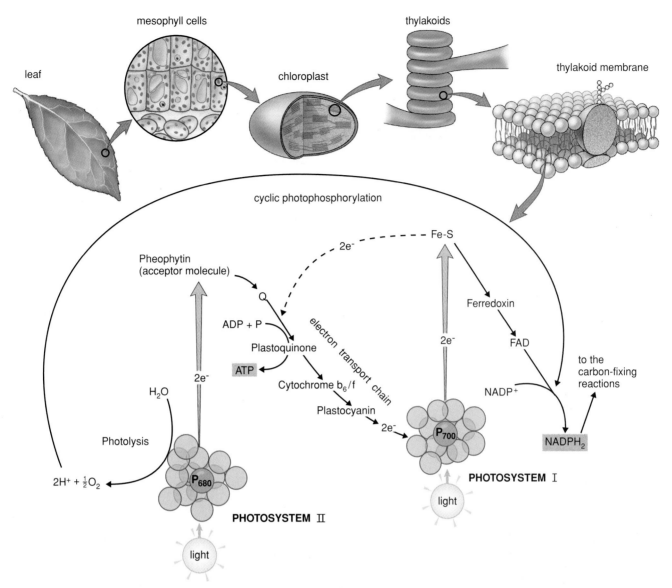

FIGURE 10.8 The light reactions of photosynthesis, which occur in more than one way. In widespread noncyclic photophosphorylation, which involves two photosystems that convert light energy to chemical energy in the form of ATP and NADPH$_2$, water molecules are split, releasing oxygen and the H$^+$ used to make NADPH$_2$. The ATP and NADPH$_2$ are used in the carbon-fixing reactions that convert CO$_2$ to sugars (see Fig. 10.10). Only photosystem I is involved in cyclic photophosphorylation, which is found in prokaryotic organisms. In this relatively simple system electrons boosted from a photosystem I reaction-center molecule are shunted back into the reaction center via the electron transport chain. ATP is produced from ADP but no NADPH$_2$ or oxygen are produced.

thylakoid membrane and is passed to another acceptor, *Pq (plastoquinone)* near the outer surface of the thylakoid membrane.[4] The electrons lost by the P$_{680}$ molecule are replaced by electrons extracted from water by *protein Z,* which contains manganese required to split water molecules. As two water molecules are split, a molecule of oxygen and four protons are simultaneously produced. The splitting of the water molecules is mediated by an enzyme on the inside of the thylakoid membrane and is called *photolysis.*

4. Except for the high speed (trillionths of a second) at which it usually carries out its function, an acceptor molecule operates something like a pickup order telephoned to a clerk in a store in the sense that an item is picked up from a source, temporarily held, and then transferred elsewhere.

Photophosphorylation The acceptor molecule Pq releases the excited electron to an electron transport system, which functions something like a downhill bucket brigade, that moves electrons from H$_2$O to a temporary, high-energy storage molecule, *nicotinamide adenine dinucleotide phosphate (NADP$^+$),* an electron acceptor for photosystem I.

The electron transport chain consists of iron-containing pigments called **cytochromes** and other electron transfer molecules plus a copper-containing protein, *plastocyanin.* While electrons pass along the electron transport system and protons are being moved through the coupling factor, ATP molecules are being formed from ADP and phosphate in the process of *photophosphorylation.*

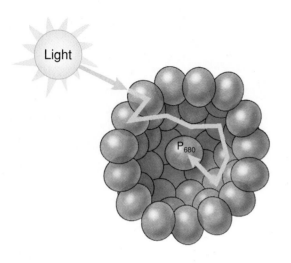

FIGURE 10.9 A photosystem II model depicting antenna pigments gathering and passing light to a special reaction-center molecule (P680).

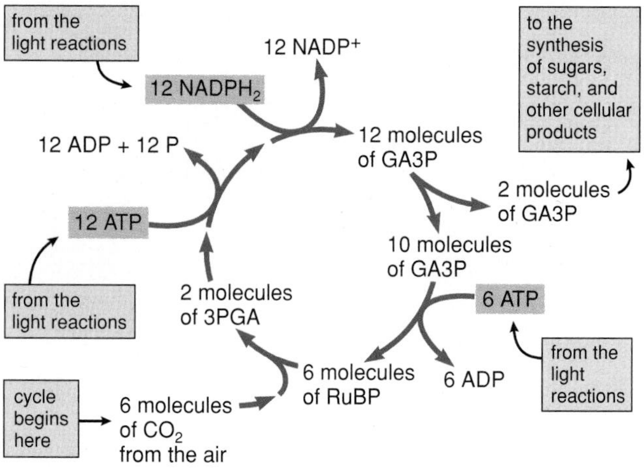

FIGURE 10.10 The Calvin Cycle (carbon-fixing reactions) of photosynthesis. The cycle takes place in the stroma of chloroplasts where each step is mediated by an enzyme. Carbon dioxide molecules from the air enter the cycle one at a time, making six turns of the cycle necessary to produce one molecule of a 6-carbon sugar such as glucose ($C_6H_{12}O_6$).

A somewhat similar series of events occurs in photosystem I. When a photon of light strikes a P_{700} molecule in a photosynthetic unit, the energy excites an electron, which is transferred to an iron-sulphur acceptor molecule designated Fe-S. This then passes the electron to another iron-sulphur acceptor molecule, *ferredoxin* (Fd), which in turn releases it to a carrier molecule called *flavin adenine dinucleotide* (*FAD*) and finally to $NADP^+$. The $NADP^+$ is reduced to NADPH.

The electrons removed from the P_{700} molecule are replaced by electrons from photosystem II via the electron transport chain just outlined. This overall movement of electrons from water to photosystem II to photosystem I to $NADP^+$ is called *noncyclic* electron flow, because it goes in one direction only. The production of ATP that correspondingly occurs is designated *noncyclic photophosphorylation*.

Photosystem I can also work independently of photosystem II. When it does, the electrons boosted from P_{700} reaction-center molecules (of photosystem I) are passed from an acceptor molecule called $\mathbf{P_{430}}$ to the electron transport chain between the two photosystems (instead of to ferredoxin and $NADP^+$) and back into the reaction center of photosystem I. This is *cyclic* electron flow. ATP generated by cyclic electron flow is called *cyclic photophosphorylation,* but no NADPH or oxygen are produced nor are water molecules split. A summary of the light reactions is shown in Figure 10.8.

Chemiosmosis The location of the enzyme that mediates the splitting of water molecules (on the inside of the thylakoid membrane) results in a proton gradient being formed across the membrane. The movement of protons across the membrane is thought to be a source of energy for the synthesis of ATP. The action has been described as similar to the movement of molecules during osmosis and has been called *chemiosmosis* or the *Mitchell Theory,* after its author, Peter Mitchell. In this concept, protons move across a thylakoid membrane through protein channels called *ATPase,* or *coupling factor.* Coupled to the proton movement, ADP and phosphate (P) combine, producing ATP.

The Carbon-Fixing Reactions Reexamined

We have seen how ATP and NADPH are synthesized during the light reactions. Both of these substances play key roles in the synthesis of carbohydrate from carbon dioxide of the atmosphere, which reaches the interior of chlorenchyma tissues via stomata, during the *carbon-fixing reactions.* As indicated earlier, the carbon-fixing reactions are really a whole series of reactions, each mediated by an enzyme in this major phase of photosynthesis. The carbon-fixing reactions take place in the stroma of the chloroplasts, and, as long as the products of the light reactions are present, they need no light to occur although they normally take place during daylight hours. (Some of the enzymes involved in the carbon-fixing reactions may actually need light for their activation or conversion to a form in which they can actively catalyze reactions of the carbon-fixing reactions.)

The Calvin Cycle There are three known mechanisms through which carbon dioxide is converted to carbohydrate during the carbon-fixing reactions. The most widespread is the *3-carbon pathway* or *Calvin cycle* (Fig. 10.10).

The main steps of the Calvin cycle (carbon-fixing reactions) are as follows:

1. Six molecules of carbon dioxide (CO_2) from the air combine with six molecules of ribulose 1, 5-bisphosphate (RuBP, the 5-carbon sugar continually being formed while photosynthesis is occurring), with the aid of the enzyme rubisco (RuBP carboxylase).[5]

2. The resulting six 6-carbon unstable complexes are immediately split into two 3-carbon molecules known as 3-phosphoglyceric acid (3PGA), the first stable compound formed in photosynthesis.

3. The NADPH (which has been temporarily holding the hydrogen and electrons released during the light reactions) and ATP (also from the light reactions) supply energy to convert the 3PGA to 12 molecules of glyceraldehyde 3-phosphate (GA3P, 3-carbon sugar phosphate).

4. Ten of the 12 glyceraldehyde 3-phosphate molecules are restructured and become six 5-carbon molecules of RuBP, the sugar with which the Calvin cycle was initiated.

The net gain of two glyceraldehyde 3-phosphate molecules (GA3P) either can contribute to an increase in the carbohydrate content of the plant (glucose, starch, cellulose, or related substances) or can be used in pathways that lead to the net gain of lipids and amino acids.

The 4-Carbon Pathway   Plants of at least 100 genera that occur primarily in the tropics produce a 4-carbon compound, *oxaloacetic acid* instead of 3-carbon 3-phosphoglyceric acid during the initial steps of the carbon-fixing reactions. Oxaloacetic acid is produced when a 3-carbon compound, phosphoenolpyruvate (PEP), and carbon dioxide are combined with the aid of an enzyme, phosphoenolpyruvate carboxylase, in mesophyll cells. Depending on the species, the oxaloacetic acid may then be converted to aspartic, malic, or other acids.

These 4-carbon organic acids, which are not substitutes for 3PGA, migrate to bundle sheath cells surrounding the vascular bundles (veins) of leaves, where they are converted to pyruvic acid and carbon dioxide. The pyruvic acid returns to the mesophyll cells and interacts with ATP, forming additional PEP molecules. The carbon dioxide combines with RuBP in the bundle sheath cells and is converted to 3PGA and related molecules in the Calvin cycle. (Fig 10.11)

Because the PEP system produces 4-carbon compounds, plants having this system are called 4-carbon, or C_4, plants, to distinguish them from plants that have only the 3-carbon, or C_3, system. C_4 plants have several characteristic features:

5. Rubisco is a huge, complex enzyme that may constitute up to 30% of the protein present in a living leaf. It is the most abundant protein known.

1. Unlike C_3 plants, which have one form of chloroplast and can run only the Calvin cycle, C_4 plants possess two forms of chloroplasts and have an alternate pathway for carbon dioxide utilization. There are large chloroplasts, often with few to no grana, in the bundle sheath cells surrounding the veins. These large chloroplasts contain numerous starch grains, and the grana, when present, are poorly developed. The much smaller, very numerous chloroplasts of the mesophyll cells usually lack starch grains and have well-developed grana.

2. High concentrations of PEP carboxylase are found in the mesophyll cells. This is significant because PEP carboxylase permits the conversion of carbon dioxide to carbohydrate at much lower concentrations than does rubisco (found only in the bundle sheath cells), the corresponding enzyme in the Calvin cycle.

3. The optimum temperatures for C_4 photosynthesis are much higher than those for C_3 photosynthesis, thus allowing C_4 plants to thrive under conditions that would adversely affect, or even kill, C_3 plants (Fig. 10.12).

Obviously, the C_4 pathway furnishes carbon dioxide to the Calvin cycle in a more roundabout way than does the C_3 pathway, but it should be noted that the advantage of this extra pathway is a major reduction of *photorespiration* in C_4 plants (see page 163). During photorespiration, which is in competition with the Calvin cycle and takes place in light while the Calvin cycle is functioning, RuBP reacts with oxygen and eventually releases carbon dioxide, while in photosynthesis, RuBP and carbon dioxide are involved in the synthesis of carbohydrate.

In C_4 plants, the C_4 pathway in the mesophyll cells results in carbon dioxide being picked up even at low concentrations (via the enzyme PEP carboxylase) and in carbon dioxide being concentrated in the bundle sheath, where the Calvin cycle takes place almost exclusively. Thus, the Calvin cycle enzyme, rubisco, will catalyze the reaction in which RuBP will react with carbon dioxide rather than oxygen. Consequently, C_4 plants have photosynthetic rates that are two to three times higher than those of C_3 plants.

CAM Photosynthesis   *Crassulacean acid metabolism* (*CAM*) photosynthesis is found in plants of more than 20 families, including cacti, stonecrops, orchids, bromeliads, and many succulents growing in regions of high light intensity. A few succulents do not have CAM photosynthesis, however, and several nonsucculent plants do. Plants with CAM photosynthesis typically do not have well-defined palisade mesophyll in the leaves, and, in contrast to the chloroplasts of the bundle sheath cells of C_4 plants, those of CAM photosynthesis plants resemble the mesophyll cell chloroplasts of C_3 plants.

CAM photosynthesis is similar to C_4 photosynthesis in that 4-carbon compounds are produced during the

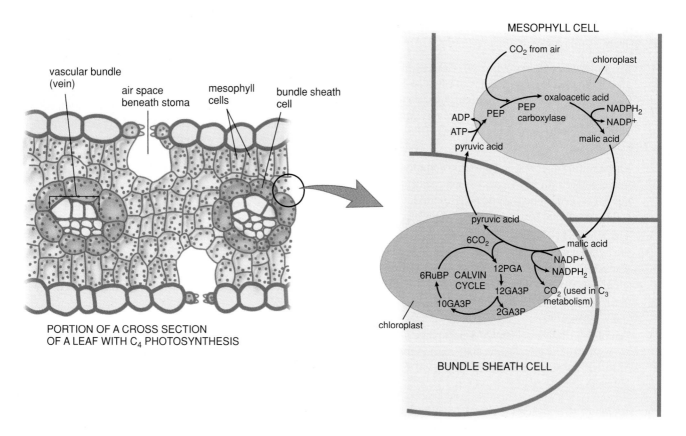

FIGURE 10.11   An illustration of the C$_4$ photosynthesis pathway. Carbon dioxide is converted to organic acids in mesophyll cells. After the acids move into bundle sheath cells, some carbon dioxide is released and enters the Calvin cycle where it becomes a 3-carbon compound that moves back to a mesophyll cell; there it is converted to PEP, which accepts carbon dioxide from the air.

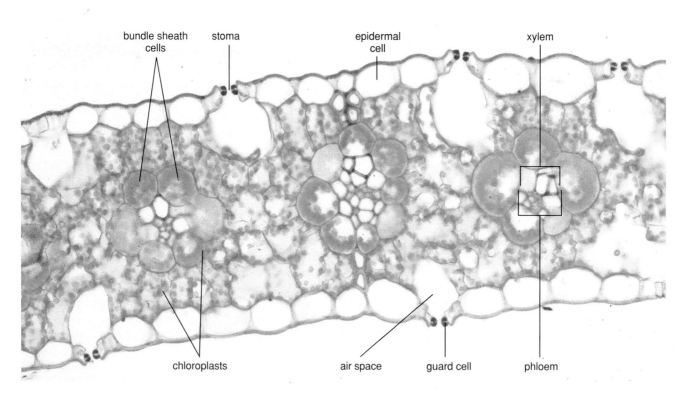

FIGURE 10.12   A portion of a cross section of a leaf of corn (*Zea mays*), a C$_4$ plant. Compare this with leaves of typical C$_3$ plants, as illustrated in figures 7.5 and 7.10A.

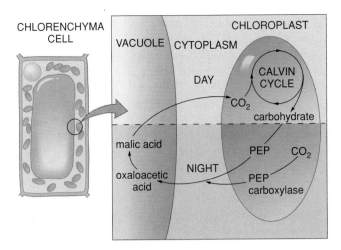

FIGURE 10.13 Crassulacean acid metabolism (CAM) photosynthesis found in orchids, pineapples, and many desert plants. CAM is similar to C_4 photosynthesis but the plants have their stomata closed during daylight heat, thus conserving water. The organic acids accumulate at night and break down during the day, releasing carbon dioxide, which then enters the Calvin cycle and C_3 metabolism while the stomata are closed.

carbon-fixing reactions. In these plants, however, malic acid accumulates in the chlorenchyma tissues at night and is converted back to carbon dioxide during the day. The enzyme PEP carboxylase is responsible for converting the carbon dioxide plus PEP to malic acid at night. During daylight, the malic acid diffuses out of the cell vacuoles in which it was stored and is converted back to carbon dioxide for use in the Calvin cycle. A much larger amount of carbon dioxide can be converted to carbohydrate each day than would otherwise be possible, since the stomata of CAM photosynthesis plants generally admit very little carbon dioxide during the day because they are closed (closed stomata greatly reduce water loss). This arrangement permits the plants to function well under conditions of both limited water supply and high light intensity (Fig. 10.13).

RESPIRATION

The solar energy that is converted to chemical energy by the process of photosynthesis is stored in various organic compounds—for example, wood and coal. If the organic compounds are burned, the energy is released very rapidly in the form of heat and light, and much of the usable energy is lost. Living organisms, however, "burn" their energy-containing compounds in numerous, small, enzyme-controlled steps that release tiny amounts of immediately usable energy, usually storing the released energy in ATP molecules, which permits the available energy to be used more efficiently and the process to be controlled more precisely.

I. The Essence of Respiration

Respiration is an energy-releasing process that takes place in all active cells 24 hours a day, regardless of whether photosynthesis happens to be occurring simultaneously in the same cells. It is initiated in the cytoplasm and completed in the mitochondria. The energy is released from simple sugar molecules that are broken down during a series of steps controlled by enzymes. No oxygen is needed to initiate the process, but in **aerobic respiration** (the most widespread form of respiration), the process cannot be completed without oxygen gas (O_2). The controlled release of energy is the significant event; carbon dioxide (CO_2) and water (H_2O) are the by-products. Aerobic respiration is summed up in the following equation:

$$C_6H_{12}O_6 + O_2 \xrightarrow{\text{enzymes}} CO_2 + H_2O + \text{energy}$$
glucose oxygen carbon water
dioxide

Anaerobic respiration and **fermentation** are two forms of respiration carried on by certain bacteria and other organisms in the absence of oxygen gas. These forms of respiration release much less energy than aerobic respiration does. The two forms differ from one another in the manner in which hydrogen released from the glucose is combined with other substances (see the discussion in a later section). Fermentation is very important industrially, particularly in the brewing industry. Two well-known forms of fermentation are illustrated by the following equations:

$$C_6H_{12}O_6 \xrightarrow{\text{enzymes}} 2\,C_2H_5OH + 2\,CO_2 + \text{energy (ATP)}$$
glucose ethyl carbon
alcohol dioxide

$$C_6H_{12}O_6 \xrightarrow{\text{enzymes}} 2\,C_3H_6O_3 + \text{energy (ATP)}$$
glucose lactic
acid

The relatively small amount of energy released during these forms of respiration is partly stored in two ATP molecules. The actual amount of energy stored is only roughly 29% of the approximately 48 kcal of energy released in anaerobic respiration.

II. Introduction to the Major Steps of Respiration

Glycolysis

In most forms of carbohydrate respiration, the first major phase takes place in the cytoplasm and requires no oxygen gas (O_2). This phase, called **glycolysis,** involves three main steps and several smaller ones, each controlled by an enzyme. During the process, a small amount of energy is released, and some hydrogen atoms are removed from compounds derived from a glucose molecule. The essence of this complex series of steps is as follows:

1. In a series of reactions, the glucose molecule becomes a fructose molecule carrying two phosphates (P).
2. This sugar (fructose) molecule is split into two 3-carbon fragments (GA3P).
3. Some hydrogen, energy, and water are removed from these 3-carbon fragments, leaving **pyruvic acid** (Fig. 10.14).

The energy needed to start the process of glycolysis is supplied by two ATP molecules. By the time pyruvic acid has been formed, however, four ATP molecules have been produced from the energy released along the way, for a net gain of two ATP molecules. A great deal of the energy originally in the glucose molecule remains in the pyruvic acid. The hydrogen ions and electrons released during the process are picked up and temporarily held by a hydrogen acceptor molecule, *nicotinamide adenine dinucleotide* (**NAD**). What happens to them next depends on the kind of respiration involved: *aerobic respiration, true anaerobic respiration,* or *fermentation.*

Aerobic Respiration

In **aerobic respiration** (the most common type), glycolysis is followed by two major stages: the *Krebs cycle* and the *electron transport chain*. Both stages occur in the mitochondria and involve many smaller steps, each of which is controlled by enzymes (see Fig. 10.14).

The Krebs Cycle

The **Krebs cycle** was named after Hans Krebs, a British biochemist who received a Nobel Prize in 1953 for his unraveling of many of the complex reactions that take place in respiration. This cycle is sometimes referred to as the *organic acid* (or *citric acid*) *cycle* because of the important role played by several organic acids during the process.

Before entering the Krebs cycle, which takes place in the fluid *matrix* located within the compartments formed by the cristae of mitochondria (see Fig. 3.10), carbon dioxide is released from the pyruvic acid produced by glycolysis. What remains is restructured to a 2-carbon acetyl group. This acetyl group combines with an acceptor molecule called *coenzyme A (CoA)*. This combination (called *acetyl CoA*) then enters the Krebs cycle, which is a series of chemical reactions that are catalyzed by enzymes. Little of the energy originally trapped in the glucose molecule is released during glycolysis. As the Krebs cycle proceeds, however, small amounts of energy and hydrogen are successively removed. The removal takes place from a series of organic acids until the remaining energy has been transferred to compounds such as NADH (*nicotinamide adenine dinucleotide*), ATP, and FADH (*flavin adenine dinucleotide*). Carbon dioxide is produced as a by-product while the cycle is proceeding.

The Electron Transport Chain

Much of the energy originally in the glucose molecule now has been transferred to the hydrogen and electron acceptors NAD and FAD, which became NADH$_2$ and FADH$_2$, respectively. Hydrogen's electrons next are passed along an *electron transport chain* consisting of special acceptor molecules arranged in a precise sequence on the inner membrane of mitochondria.

The acceptor molecules include iron-containing proteins called *cytochromes*. Energy is released in small increments at each step along the chain, with water being formed as the hydrogen ions and electrons finally combine with oxygen from the air. ATP is produced from ADP and P as the energy is released at several points along the chain. Again, as the final step in aerobic respiration (see Fig. 10.14), water is produced by oxygen acting as the ultimate electron acceptor and combining with hydrogen.

By the time the process is complete, the recoverable energy locked in a molecule of glucose has been released and is stored in ATP molecules. This stored energy is then available for use in the synthesis of other molecules and for growth, active transport, and a host of other metabolic processes. Aerobic respiration produces a net gain of 36 ATP molecules from one glucose molecule, using up six molecules of oxygen and producing six molecules of carbon dioxide and a net total of six molecules of water. For each mole (180 grams) of glucose aerobically respired, 686 kcal of energy is released, with about 39% of it being stored in ATP molecules and the remainder being released as heat.

A. Anaerobic Respiration and Fermentation

Many bacteria carry on both fermentation and true anaerobic respiration simultaneously, making it difficult to distinguish between the two processes. Some texts use the terms *anaerobic respiration* and *fermentation* interchangeably to designate respiration occurring in the presence of little or no oxygen gas.

In many instances in the biological realm, glucose molecules may undergo glycolysis without enough oxygen being available to permit the completion of aerobic respiration. In such cases, the hydrogen released during glycolysis is simply transferred from the hydrogen acceptor molecules back to the pyruvic acid, after it has been formed, creating ethyl alcohol, lactic acid, or similar substances. A little energy is released during either fermentation or true anaerobic respiration, but most of it remains locked up in the alcohol, lactic acid, or other compounds produced.

In *true anaerobic respiration,* the hydrogen removed from the glucose molecule during glycolysis is combined with an inorganic ion, as is the case when sulphur bacteria (discussed in Chapter 17) convert sulphate (SO$_4$) to sulphur (S) or another sulphur compound or when certain cellulose bacteria produce methane gas (CH$_4$) by combining the hydrogen with carbon dioxide.

Oxygen gas is not required to make these compounds, but few organisms can live long without oxygen, and many that carry on fermentation can also respire aerobically. If oxygen becomes available, the remaining energy can be

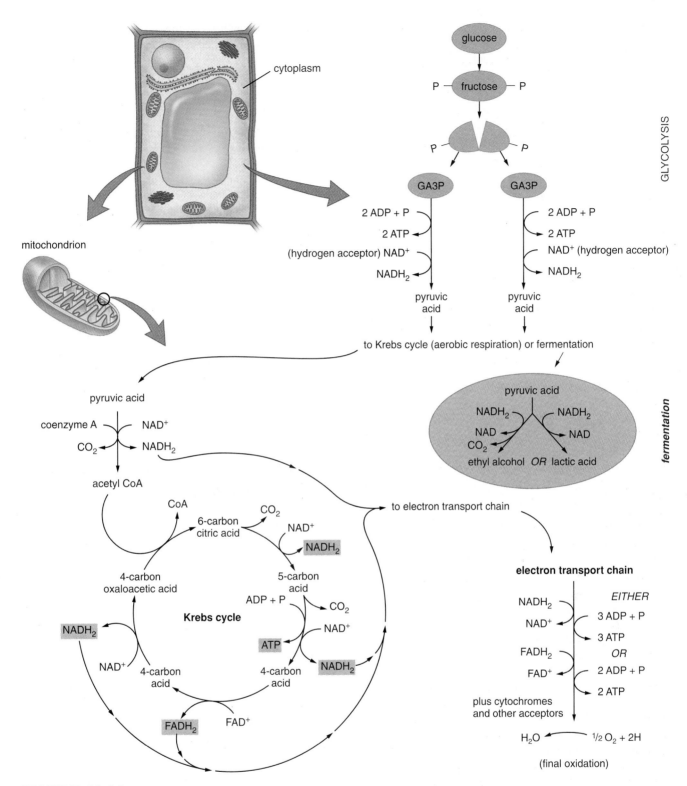

fermentation

FIGURE 10.14 A summary of respiration. In the first phase, glycolysis, which takes place in the cytoplasm, a sugar molecule is converted to two pyruvic acid molecules. The subsequent phases take place within the matrix of a mitochondrion. In aerobic respiration, the pyruvic acid is broken down in the Krebs cycle and energy is transferred to compounds such as NADH, ATP, and FADH. Carbon dioxide and hydrogen are also released. The released hydrogen is carried by an electron transport chain and combined with oxygen, forming water. In anaerobic respiration and fermentation pyruvic acid is converted in the absence of oxygen gas to ethyl alcohol or lactic acid with little release of energy.

released by breaking these compounds down further. During anaerobic respiration or fermentation, only about 7% of the total energy in the glucose molecule is removed, and so much of that energy goes into the making of the alcohol or the lactic acid or is dissipated as heat that there is a net gain of only two ATP molecules (compared with 36 ATP molecules produced in aerobic respiration).

Living cells can tolerate only certain concentrations of alcohol. In media in which yeasts are fermenting sugars, for example, once the alcohol concentration builds up beyond 12%, the cells die and fermentation ceases. This is why most wines have an alcohol concentration of about 12%.

B. Factors Affecting the Rate of Respiration

Temperature Temperature plays a major role in the rate at which the various respiratory reactions occur. For example, when air temperatures rise from 20°C (68°F) to 30°C (86°F), the respiration rates of plants double and sometimes even triple. The faster respiration occurs, the faster the energy is released from sugar molecules, with an accompanying decrease in weight. In growing plants, this weight loss is more than offset by the production of new sugar by photosynthesis. In harvested fruits, seeds, and vegetables, however, respiration continues without sugar replacement; some water loss also occurs. Respiring cells convert energy stored as starch or sugar primarily to ATP, but much of the energy is lost in the form of heat, with only 39% being stored as ATP. Most fresh foods are kept under refrigeration, not only to lower the respiration rate and retard water loss, but also to dissipate the heat. Keeping the temperatures down is also important to prevent the growth and reproduction of food-spoiling molds and bacteria, which may thrive at warmer temperatures.

Heat inactivates most enzymes at temperatures above 35°C (95°F), but a few organisms, such as various blue-green bacteria or algae in the hot springs of Yellowstone National Park and similar places, have adapted in such a way that they are able to thrive at temperatures of up to 60°C (140°F)—heat that would kill other organisms of comparable size almost instantly.

Water Water inside the cells and their organelles acts as a medium in which the enzymatic reactions can take place. Living cells often have a water content in excess of 90%, but the cells of mature seeds may have a water content of less than 10%. When water content becomes this low, respiration does not cease completely, but it continues at a drastically reduced rate, resulting in only very tiny amounts of heat being released and of carbon dioxide being given off. Seeds may remain viable (capable of germinating) for many years if stored under dry conditions. If they come in contact with water, however, they swell by imbibition. Respiration rates then increase rapidly. If the wet seeds happen to be in an unrefrigerated storage bin, the temperature may increase to the point of killing the seeds. In fact, if fungi and bacteria begin to grow on the seeds,

temperatures from their respiration can become so high that spontaneous combustion can sometimes occur.

Oxygen If, due to flooding, the roots of trees and houseplants have their oxygen supply sharply reduced, their respiration rates may be decreased and their growth retarded. They may even die if the condition persists too long. In the case of stored foods, however, reducing the oxygen in the storage areas to bring about lower rates of respiration may be desirable, and consequently, it has become a common commercial practice to reduce the oxygen present in warehouses where crops are stored to as little as 1% to 3% by pumping in nitrogen gas, while maintaining low temperatures and humidity. Oxygen concentration is not reduced below 1% because that can result in an undesirable increase in fermentation.

III. A Closer Look at Respiration

Respiration, like photosynthesis, is a very complex process, and, as with photosynthesis, it is beyond the scope of this book to explore the subject in great detail. The following amplification of information already discussed is modest, and those who wish further information are referred to the reading list at the end of the chapter.

Glycolysis Reexamined

As previously discussed, this initial phase of all forms of respiration brings about the conversion of each 6-carbon glucose molecule to two 3-carbon pyruvic acid molecules via three main steps, each mediated by enzymes:

1. *phosphorylation,* whereby the 6-carbon sugars receive phosphates;

2. *sugar cleavage,* which involves the splitting of 6-carbon fructose into two 3-carbon sugar fragments; and

3. *pyruvic acid formation,* which involves the oxidation of the sugar fragments.

Energy needed to initiate the process is furnished by an ATP molecule, which also furnishes the phosphate group for the phosphorylation of the sugar glucose to yield glucose 6-phosphate. Another ATP, with the aid of the enzyme fructokinase, yields fructose 1,6-bisphosphate. As a result of the cleavage of the fructose 1,6-bisphosphate, two different 3-carbon sugars are produced, but ultimately only two glyceraldehyde 3-phosphate (GA3P) molecules remain. These two 3-carbon sugars are oxidized to two 3-carbon acids, and, in the successive production of several of these acids, phosphate groups are removed from the acids. The phosphate groups combine with ADP, producing a net direct gain of two ATP molecules during glycolysis. In addition, hydrogen is removed as GA3P is oxidized. This hydrogen is picked up by a hydrogen acceptor molecule, **nicotinamide adenine dinucleotide (NAD⁺),** which becomes NADH. Glycolysis, which requires no oxygen gas, is summarized in Figure 10.14.

The Krebs Cycle Reexamined

Before a pyruvic acid molecule enters the Krebs cycle, which takes place in the mitochondria, a molecule of carbon dioxide is removed and a molecule of NADH is produced, leaving an acetyl fragment. The 2-carbon fragment is then bonded to a large molecule called *coenzyme A*. Coenzyme A consists of a combination of the B vitamin pantothenic acid and a nucleotide. Pantothenic acid is one of several B vitamins essential to respiration in both plants and animals; others include thiamine (vitamin B_1), niacin, and riboflavin. The bonded acetyl fragment and coenzyme A molecule is referred to as *acetyl CoA*. The following equation summarizes the fate of the two pyruvic acid molecules following glycolysis and leading to the Krebs cycle:

$$2 \text{ pyruvic acid} + 2 \text{ CoA} + 2 \text{ NAD}^+ \longrightarrow$$
$$2 \text{ acetyl CoA} + 2 \text{ NADH} + 2 \text{ CO}_2$$

In addition to pyruvic acid, fats and amino acids can also be converted to acetyl CoA and enter the process at this point. The NADH molecules donate their hydrogen to an electron transport chain (discussed in the next section), and the acetyl CoA enters the Krebs cycle (see Fig. 10.14).

In the Krebs cycle, acetyl CoA is first combined with oxaloacetic acid, a 4-carbon compound, producing citric acid, a 6-carbon compound. Oxaloacetic acid is neither a starting substance nor an end product of the Krebs cycle but rather an intermediate product found in small amounts that keep the cycle going. The cycle then progresses in circular fashion. A carbon dioxide is soon removed, producing a 5-carbon compound. Then another carbon dioxide is removed, producing a 4-carbon compound. This 4-carbon compound, through additional steps, is converted back to oxaloacetic acid, the substance with which the cycle began, and the cycle is repeated.

Each full cycle uses up a 2-carbon acetyl group and releases two carbon dioxide molecules while regenerating an oxaloacetic acid molecule for the next turn of the cycle. Some hydrogen is removed during the process and is picked up by a flavoprotein hydrogen acceptor molecule, *flavin adenine dinucleotide* (FAD), and also by NAD. One molecule of ATP, three molecules of NADH, and one molecule of $FADH_2$ are produced for each turn of the cycle. The Krebs cycle may be summarized as follows:

oxaloacetic + acetyl CoA + ADP + P + 3 NAD^+ + FAD $\longrightarrow$
acid

oxaloacetic + CoA + ATP + 3 $NADH_2$ + $FADH_2$ + 2 CO_2
acid

The hydrogen carried by NAD and FAD can be traced mostly to the acetyl groups and to water molecules added to some compounds in the Krebs cycle. The FAD and $FADH_2$ are now known to be intermediate compounds, with *ubiquinol* a final product of the Krebs cycle.

The Electron Transport Chain and Oxidative Phosphorylation

After completion of the Krebs cycle, the glucose molecule has been totally dismantled and some of its energy has been transferred to ATP molecules. A considerable portion of the energy was transferred to NAD^+ and FAD when they were used to pick up hydrogen and electrons from the molecules derived from glucose as they were broken down during glycolysis and the Krebs cycle. This energy is released as the hydrogen and electrons are passed along an *electron transport chain*. This chain, like the electron transport chain of photosynthesis, functions something like a high-speed bucket brigade in passing along electrons from their source to their destination. Several of the electron carriers in the transport chain are cytochromes. They are very specific and, as electrons flow along the chain, they can transfer their electrons only to other specific acceptors. When the electrons reach the end of the chain, they are picked up by oxygen and combine with hydrogen ions, forming water.

Part of the energy that is released during the movement of electrons along the electron transport chain can be used to make ATP in a process called *oxidative phosphorylation*. If hydrogen ion and electron transport begins with NADH, which was produced inside the mitochondria (i.e., during the conversion of pyruvic acid to acetyl CoA and during the Krebs cycle), sufficient energy is produced to yield three ATP molecules from each NADH molecule. Similarly, if hydrogen ion and electron transport begins either with $FADH_2$ or with NADH, produced outside the mitochondria (i.e., during glycolysis), two ATP molecules are produced.

The manner in which ATP is produced during the operation of the respiratory electron transport chain involves essentially the same *chemiosmotic* concept that was applied earlier to proton movement across thylakoid membranes. Photophosphorylation (see page 167) needs to be mentioned at this point.

In recent years, there has been some evidence to support the theory, proposed in the 1960s by Peter Mitchell, a British biochemist, concerning the electron transport chain. This theory holds that oxidative phosphorylation is energized by a gradient of protons (H^+) that flow by **chemiosmosis** across the inner membrane of a mitochondrion. Mitchell, who received a Nobel Prize for his work in 1978, surmised that protons are "pumped" from the matrix of the mitochondria to the region between the two membranes (see Fig. 3.7) as electrons flow from their source in $NADH_2$ molecules along the electron transport chain, which is located in the inner membrane. The protons are believed to "diffuse" back into the matrix via channels provided by an enzyme complex, releasing energy that is used to synthesize ATP.

If we retrace our steps through the entire process of aerobic respiration, we find that glycolysis yields four molecules of ATP and two molecules of NADH (from which four

Greenhouse Gases and Plant Growth

Like a greenhouse, the earth has an atmosphere that traps the radiant energy from the sun and warms the planet, making it suitable for life. Without this protective atmosphere the planet would lose heat into space, with the result being a drastic cooling of the planet's surface. The gases that trap the heat, like the glass of a greenhouse, are called "greenhouse gases." Consisting of carbon dioxide, methane, nitrous oxides, and chlorofluorocarbons, these gases account for less than 1% of the atmosphere, yet are crucial to the sustenance of life on earth.

It is difficult to conceive that human activity could alter the atmosphere of the earth. But ever since the beginning of the industrial revolution some 200 years ago, greenhouse gas levels, especially carbon dioxide, have risen significantly and are projected to increase even more in the next century. This transformation of earth's atmosphere is taking place by the burning of fossil fuels and by destroying the vegetational cover of the earth through massive deforestation. How do scientists know that greenhouse gases are increasing? By drilling into the ice caps at the poles and taking ice core samples that are then dated, trapped air bubbles in ice layers hundreds of years old can be analyzed and compared to recent atmospheric levels of greenhouse gases. What these studies show is that the level of carbon dioxide has dramatically increased from 280 ppm (parts per million) in 1800 to roughly 350 ppm in 1990, with a projected doubling of CO_2 to 700 ppm by the middle to the end of the next century. This represents a yearly rate of change of approximately 1.8 ppm.

Why should we be concerned with increases in carbon dioxide and other greenhouse gases? There are several reasons. First, climate scientists tell us that with a doubling of CO_2 levels, the earth's surface temperature will increase an average of between 1.5 and 4.5 °C. With warming occurring on a global scale, snow fields and glaciers will begin to melt and contribute to a rise in sea level that will inundate thousands of miles of coastal areas and force vast numbers of people to seek shelter inland. Sea level rise will also occur by ocean water expanding as it warms. Coastal areas that now support extensive agriculture will be unfarmable because of the flooding. Secondly, an increased CO_2 level will have effects on plant life that could be detrimental to world agriculture. Plant biologists studying the effects of an enriched CO_2 atmosphere tell us that some plants will grow more robust because of a "CO_2 fertilizing" effect. However, to sustain this increased plant productivity will require sufficient nutrients, light and water—commodities that will be available only at a high price. Because farmers in most of the world cannot afford to increase fertilizer applications and buy irrigation water, plant growth in a CO_2-enriched environment will be limited to current production levels. A warmer climate may cause more frequent droughts and therefore, reduce crop yields in today's major grain-producing regions of the world, notably the Great Plains of the United States. Extreme climate fluctuations would limit the yields of certain crops, thus negating the potential for greater productivity through the "CO_2 fertilization" effect.

How is carbon dioxide released into the atmosphere? Although all living organisms give off CO_2 during respiration, this does not affect global CO_2 levels significantly. The primary culprit is the burning of fossil fuels—oil, natural gas, and coal. Deforestation and burning of timber are other sources of CO_2 emissions. On a per capita basis, the United States is the leading nation in CO_2 emissions (19.7 tons per person), followed by Canada (17.4), and Saudi Arabia (13.1). China (1.9), Egypt (1.4), and India (0.7) have significantly lower per capita CO_2 emissions. These figures clearly reveal the United States' and Canada's dependence on fossil fuels, while less industrialized countries such as China, Egypt, and India consume much less energy per capita.

Plants created our present atmosphere by absorbing CO_2 and releasing O_2 during photosynthesis. Because of the oxygen-generating abilities of plants, the gradual buildup of an oxygen atmosphere occurred some 600 million years ago and facilitated the development of large animals on earth. Plants have maintained the delicate balance of our atmosphere for millions of years. Yet this balance is now threatened by human activity. Plant response to these perturbations of the atmosphere is one of the most important reasons for studying plants today.

more molecules of ATP are formed), for a total of eight ATP from the conversion of glucose to two pyruvic acid molecules. Two ATP are used in the process, however, leaving a net gain of six ATP.

When two pyruvic acid molecules are converted to two acetyl CoA in the mitochondria, two more NADH molecules (which will generate six molecules of ATP) are produced. The two acetyl CoA molecules metabolized in the

Table 10.1

SUMMARY COMPARISON OF PHOTOSYNTHESIS AND RESPIRATION

PHOTOSYNTHESIS	RESPIRATION
1. Stores energy in sugar molecules	1. Releases energy from sugar molecules
2. Uses carbon dioxide and water	2. Releases carbon dioxide and water
3. Increases weight	3. Decreases weight
4. Occurs only in light	4. Occurs in either light or darkness
5. Occurs only in cells containing chlorophyll	5. Occurs in all living cells
6. Produces oxygen in green organisms	6. Utilizes oxygen (aerobic respiration)
7. Produces ATP with light energy	7. Produces ATP with energy released from sugar

Krebs cycle yield two molecules of ATP, two molecules of FADH$_2$ (from which four ATP are formed), and six molecules of NADH (which cause the formation of 18 molecules of ATP), making a Krebs cycle total of 24 ATP. Thus, a grand total of 36 ATP is produced for the aerobic respiration of one glucose molecule. The 36 ATP molecules represent about 39% of the energy originally present in the glucose molecule. The remaining energy is lost as heat or is unavailable. Aerobic respiration is still about six times more efficient than anaerobic respiration.

A condensed comparison between photosynthesis and respiration is shown in Table 10.1.

ASSIMILATION AND DIGESTION

Sugars produced through photosynthesis may undergo many transformations. Some sugars are used directly in respiration, but others not needed for that purpose may be transformed into lipids, proteins, or other carbohydrates. Among the most important carbohydrates produced from simple sugars are sucrose, starch, and cellulose. Much of the organic matter produced through photosynthesis is eventually used in the building of protoplasm and cell walls. This conversion process is called *assimilation*. Energy released through respiration is used in each building step, not only in the formation of fats, proteins, and complex carbohydrates but also in the manufacture of such products as resins, gums and oils, drugs, pigments, acids, and tannins. The capacity to assimilate is a fundamental attribute of living organisms, but much is still not known about the details of the process.

When photosynthesis is occurring, sugar may be produced faster than it can be used or transported away to other parts of the plant. When this happens, the excess sugar may be converted to large, insoluble molecules, such as starch or oils temporarily stored in the chloroplasts, and then later changed back to a soluble form that is transported to other cells. The conversion of starch and other insoluble carbohydrates to soluble forms is called **digestion** (Fig. 10.15). The

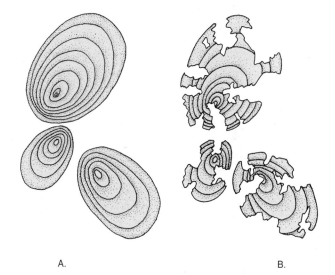

FIGURE 10.15 Starch grains. *A.* Before digestion. *B.* After partial digestion.

process is nearly always one of **hydrolysis,** in which water is taken up and, with the aid of enzymes, the links of the chains of simple sugars that comprise the molecules of starch and similar carbohydrates are broken. The disaccharide malt sugar (maltose), for example, is transformed to two molecules of glucose with the aid of an enzyme (maltase), by the addition of one molecule of water, as follows:

$$\underset{\text{maltose}}{C_{12}H_{22}O_{11}} + \underset{\text{water}}{H_2O} \xrightarrow[\text{(enzyme)}]{\text{maltase}} \underset{\text{glucose}}{2\ C_6H_{12}O_6}$$

Fats are broken down to their component fatty acids and glycerol and proteins to their amino acid building blocks in similar fashion. Digestion is carried on in any cell where there may be stored food, with very little energy being released in the process. In animals, special digestive organs also play a role in digestion, but plants have no such additional "help" in the process. In both plants and animals, digestion within cells is similar and is a normal part of metabolism.

Digestion in plants takes place within the cells where the carbohydrates, fats, or proteins are stored (except in insect-trapping plants), while in animals digestion usually occurs outside of the cells in the digestive tract. Apart from the location, the process is essentially similar in plants and animals.

SUMMARY

1. Photosynthesis is an energy-storing process that combines carbon dioxide and water in the presence of light with the aid of chlorophyll; oxygen is a by-product. All life depends on photosynthesis, which takes place in chloroplasts.

2. Carbon dioxide is a constant 0.039% of the atmosphere. Photosynthesis, which takes place primarily in blue-violet and red-orange light wavelengths, uses less than 1% of plant-absorbed water.

3. All types of chlorophyll contain a magnesium atom. Most chlorophyll in leaves is chlorophyll *a*. Other pigments widen the spectrum of light in which it is possible for photosynthesis to occur.

4. In chloroplasts, various pigment molecules are grouped into photosynthetic units, which function in the light reactions, the first major phase of photosynthesis.

5. During the light reactions of photosynthesis, water molecules are split and oxygen gas is released. Hydrogen ions and electrons are released from water and transferred to create NADPH and energy-storing ATP.

6. In the carbon-fixing reactions, carbon dioxide is combined with RuBP, and this combination is converted to 6-carbon sugars. The ATP and NADPH, produced during the light reactions, furnish the energy and materials needed for the carbon-fixing reactions. The first substance produced in the carbon-fixing reactions is usually a 3-carbon compound; some tropical and desert plants produce a 4-carbon compound instead.

7. The two types of photosynthetic units present in most chloroplasts are photosystems I and II. Each photosystem has a reaction-center molecule of chlorophyll *a* that boosts an electron to a higher energy level as it reacts to light energy.

8. In photosystem II, the excited electron is picked up by an electron acceptor molecule and released to an electron transport system consisting of cytochromes and other molecules; these pass it along to a special transfer molecule. An electron donor replaces the electron lost from the chlorophyll *a* reaction-center molecule. The donor's losses are replaced with electrons from water molecules, which are split, releasing oxygen.

9. Electron transport while the photosystems are operating and proton movement across thylakoid membranes are both involved in ATP production. Events occurring in photosystem I are similar; the excited electron is either returned to the special reactive molecule (cyclic electron flow) or passed on to $NADP^+$, which forms NADPH with electrons from the original splitting of water molecules in photosystem II (noncyclic electron flow).

10. In the carbon-fixing reactions taking place in the stroma of chloroplasts, the combination of RuBP and carbon dioxide is initially split into two 3PGA molecules; then GA3P is formed with hydrogen ions and electrons from NADPH and energy from ATP. Ten of the 12 GA3P molecules are converted back to RuBP. The remaining two GA3P molecules are the basis for an increase in the carbohydrate content or net gain in lipids or amino acids of a plant.

11. In the carbon-fixing reactions of C_4 plants, 4-carbon oxaloacetic acid is initially produced instead of 3-carbon PGA. In the leaf mesophyll of C_4 plants, there are large chloroplasts, which contain rubisco in the bundle sheaths, and small chloroplasts, which contain higher concentrations of PEP carboxylase that facilitate the conversion of carbon dioxide to carbohydrate at much lower concentrations than is possible in C_3 plants.

12. CAM photosynthesis occurs in succulent plants whose stomata are closed and admit little CO_2 during the day. Regular photosynthesis occurs as the 4-carbon compounds that accumulate at night are converted back to carbon dioxide during the day.

13. Respiration is an energy-releasing process that takes place in the mitochondria and cytoplasm of cells. The energy is released, with the aid of enzymes, from simple sugar and organic acid molecules.

14. In aerobic respiration, stored energy release requires oxygen; CO_2 and water are by-products of the process.

15. Anaerobic respiration and fermentation do not require oxygen gas, and much less energy is released. The remaining energy is in the ethyl alcohol, lactic acid, or other such substances produced. Some released energy is stored in ATP molecules. Respiration rates are affected by temperatures, available water, and environmental oxygen.

16. Glycolysis requires no molecular oxygen; two phosphates are added to a 6-carbon sugar molecule and the prepared molecule is split into two 3-carbon sugars (GA3P). Some hydrogen, energy, and water are removed from the GA3P, producing pyruvic acid. There is a net gain of two ATP molecules. Hydrogen ions and electrons released during glycolysis are picked up by NAD, which becomes NADH.

17. In aerobic respiration, the pyruvic acid loses some CO_2, is restructured, and becomes acetyl CoA. Energy,

CO₂, and hydrogen are removed from the acetyl CoA in the Krebs cycle, which involves enzyme-catalyzed reactions of a series of organic acids.

18. NADH passes the hydrogen gained during glycolysis and the Krebs cycle along an electron transport chain; small increments of energy are released and partially stored in ATP molecules, and the hydrogen is combined with oxygen gas, forming water in the final step of aerobic respiration.

19. Hydrogen removed from glucose during glycolysis is combined with an inorganic ion in anaerobic respiration. The hydrogen is combined with the pyruvic acid or one of its derivatives in fermentation. Both processes occur in the absence of oxygen gas, with only about 7% of the total energy in the glucose molecule being released, for a net gain of two ATP molecules.

20. Two molecules of NADH and two ATP molecules are gained during glycolysis when two 3-carbon pyruvic acid molecules are produced from a single glucose molecule. Another molecule of NADH₂ is produced when the pyruvic acid molecule is restructured and becomes acetyl CoA prior to entry into the Krebs cycle.

21. In the Krebs cycle, acetyl CoA combines with 4-carbon oxaloacetic acid, producing first a 6-carbon compound, next a 5-carbon compound, and then several 4-carbon compounds. The last 4-carbon compound is oxaloacetic acid. Two CO₂ molecules are also released during this process.

22. Some hydrogen removed during the Krebs cycle is picked up by FAD and NAD⁺; one molecule of ATP, three molecules of NADH, and one molecule of FADH₂ are produced during one complete cycle. Energy associated with electrons and/or with hydrogen picked up by NAD⁺ and FAD is gradually released as the electrons are passed along the electron transport chain; some of this energy is transferred to ATP molecules during oxidative phosphorylation.

23. Energy used in ATP synthesis during oxidative phosphorylation is believed to be derived from a gradient of protons formed across the inner membrane of a mitochondrion while electrons are moving in the electron transport chain by chemiosmosis.

24. Altogether, 38 ATP molecules are produced during the complete aerobic respiration of one glucose molecule; two are used to prime the process so that there is a net gain of 36 ATP molecules.

25. The conversion of the sugar produced by photosynthesis to fats, proteins, complex carbohydrates, and other substances is termed *assimilation*. Digestion takes place within plant cells with the aid of enzymes. During digestion, large insoluble molecules are broken down by hydrolysis to smaller soluble forms that can be transported to other parts of the plant.

Review Questions

1. What happens in the light reactions of photosynthesis?
2. What roles do water, light, carbon dioxide, and chlorophyll play in photosynthesis?
3. What is glycolysis?
4. What are the differences among aerobic respiration, anaerobic respiration, and fermentation?
5. How do temperature, water, and oxygen affect respiration?
6. Explain digestion and assimilation.

Discussion Questions

1. Since plants apparently benefit from small increases in the carbon dioxide available to them, would you be doing your favorite houseplant a favor by breathing repeatedly on it?
2. Why were the carbon-fixing reactions of photosynthesis formerly called the *dark reactions*?

Additional Reading

Danks, S. M., et al. 1985. *Photosynthetic systems: Structure, function, and assembly*. New York: Wiley-InterScience.

Douce, R., and D. A. Day (Eds.). 1985. *Higher plant cell respiration*. New York: Springer-Verlag.

Fong, F. K. (Ed.). 1982. *Light reaction path of photosynthesis*. New York: Springer-Verlag.

Galston, A. W. 1993. *Life processes of plants: Mechanisms for survival*. New York: W. H. Freeman.

Goodwin, T. W. 1988. *Plant pigments*. San Diego, CA: Academic Press.

Gregory, R. P. 1989. *Photosynthesis*. New York: Routledge, Chapman and Hall, Inc.

Gutteridge, S. 1990. Limitations of the primary events of CO₂ fixation in photosynthetic organisms: The structure and mechanism of rubisco. *Biochimica et Biophysica Acta* 1015: 1–14.

Jones, H. G. 1992. *Plants and microclimate: A quantitative approach to plant physiology*. San Diego, CA: Academic Press.

Lehninger, A. L., D. L. Nelson, and M. M. Cox. 1992. *Principles of biochemistry*, 2d ed. New York: Worth.

Salisbury, F. B., and C. W. Ross. 1992. *Plant physiology*, 4th ed. Belmont, CA: Wadsworth Publishing Co.

Schulze, E. D., and M. M. Caldwell (Eds.). 1994. *Ecophysiology of photosynthesis*.

Ting, I. P. 1985. Crassulacean acid metabolism. *Annual Review of Plant Physiology* 36: 595–622.

Walker, D. 1979. *Energy, plants and man: An introduction to photosynthesis in C₃, C₄ and CAM plants*. Philadelphia: International Ideas, Inc.

Chapter Outline

A single fiddleneck of a tropical tree fern (Sadleria cyatheoides). As the fern frond (leaf) matures, the fiddleneck uncoils and expands into a beautiful, feathery structure.

11 Growth

O v e r v i e w—————————————————————

This chapter introduces growth phenomena with a discussion of the distinctions among growth, differentiation, and development. This is followed by a discussion of plant hormones (auxins, gibberellins, cytokinins, abscisic acid, ethylene) and their roles in plant growth and development. The chapter explores plant movements, including spiraling, twining, contraction, nastic, tropic, turgor, taxic, and miscellaneous other movements. The discovery and functions of photoperiodism and phytochrome are briefly surveyed. The chapter concludes with a discussion of the relationship of temperature to growth and dormancy.

Some Learning Goals

1. Know distinctions among growth, differentiation and development and among plant hormones and vitamins.
2. Identify the types of plant hormones, and describe the major functions of each; discuss commercial applications for each.
3. Distinguish among the various types of plant movements, and know the forces behind them.
4. Understand photoperiodism, and make distinctions among short-day, long-day, intermediate-day, and day-neutral plants.
5. Explain what phytochrome is and how it functions.
6. Discuss the role of temperature in plant growth.
7. Learn dormancy and stratification, and give examples.

From time to time, I grow a few vegetables and berries in my backyard. When Italian squash plants are producing fruit, I have occasionally seen a young squash just beginning to develop, and I have made a mental note to harvest it within a day or two. Then, other matters have distracted me and I have forgotten to follow through, only to discover a few days later that my squash has grown into an enormous "monster" the size of a watermelon. I then have wondered, "How did it grow that big that fast?"

The words **grow, growing,** and **growth** are used in several ways. If, for example, you see a rubber balloon being inflated with gas, you may refer to its gradual increase in diameter as its growth in size. Or if there is a leak in the roof, you might say that the puddle in your living room is growing bigger. In the biological realm, however, growth is always associated with cells. It may be defined simply as an "increase in mass due to the division and enlargement of cells" and may be applied to an organism as a whole or to any of its parts.

Many plants, such as radishes and pumpkins, go through a sequence of growth stages. At first, they grow rapidly, then for a while they show little if any increase in volume, and eventually they stop growing completely. Finally, tissues break down, and the plant dies. Such growth is said to be *determinate*. Parts of other plants, such as the stems of redwood trees, may exhibit *indeterminate* growth and continue to be active for thousands of years.

All living organisms begin as a single cell, which usually divides and keeps dividing until something consisting of possibly billions of cells is formed. As the various cells mature, they usually become larger and then *differentiate*—that is, they develop different forms adapted to specific functions, such as conduction, support, or secretion of special substances. If **differentiation** did not occur, the result would be a shapeless blob with no distinct tissues and little, if any, coordination.

The term **development** is applied to the process of growth and differentiation of cells into tissues, organs, and organisms. What brings about growth and differentiation, and how do they occur? We already know they are controlled by the genes of the chromosomes, which dictate not only the kinds of cells that will develop but also their proportionate numbers and the ultimate size of the organism.

The sizes of the cells themselves are limited by several factors. They are dependent on molecular diffusion within for the transport of oxygen and food and for the removal of wastes, since they have no circulatory organs of their own. As a spherical cell enlarges, the total area of its surface, through which substances enter and leave, does not increase at the same rate as its volume. When a cell's surface area has increased fourfold, for example, its volume has increased roughly six times, obviously putting a great strain on its capacity to admit and exchange materials through its surface.

Most plant cells have become adapted to this problem through various changes in shape that increase the surface area (e.g., development of a convoluted outer surface; flattening of the cell) and also through the development of large vacuoles, with the cytoplasm becoming a thin layer adjacent to the cell wall. These adaptations alone, however, do not increase the strength of the bounding membranes, and most cells divide before their enlargement becomes a significant problem.

PLANT GROWTH REGULATORS AND ENZYMES

The basic units of heredity known as *genes* in the nucleus of each living cell program the synthesis and development of *enzymes,* which control every metabolic step within cells. As previously noted, enzymes are organic catalysts that speed up chemical reactions, usually in the cells where they are produced. They are nearly all proteins and are produced in minute quantities. They generally are not transported from one part of an organism to another.

Genes also dictate the production of **hormones,** which influence many growth phenomena. Hormones, which in plants are produced mostly in growing shoot tips, are chemical substances that differ from enzymes in structure. Like enzymes, they occur in minute quantities, but, unlike enzymes, they ordinarily are transported from their point of origin to another part of the plant where they have specific effects, such as causing stems to bend, initiating flowering, or even inhibiting growth.

Vitamins, which function in activating enzymes, are organic molecules of varied structure that are synthesized in the membranes and cytoplasm of cells. They are essential in relatively small amounts for the normal growth and development of all organisms. Most vitamins essential to all living organisms are produced by plants, although a few are also synthesized by animals. Vitamin A, for example, does not occur in plants. Carotene pigments found in chloroplasts act as *precursors* (simple molecules that produce new molecules after reacting with other molecules) for vitamin A in animals.

Because some of the effects of vitamins are similar to those of hormones, they are sometimes difficult to distinguish from one another. The term *plant growth regulators* is now used for both vitamins and hormones when one wishes to avoid distinguishing between the two.

Plant Hormones

The major known types of plant hormones are *auxins, gibberellins, cytokinins, abscisic acid,* and *ethylene.*

Auxins

In 1881, Charles Darwin and his son Francis were fascinated by the fact that *coleoptiles* (more or less tubular, closed sheaths protecting the emerging stems of germinating seeds of the Grass Family) bend toward a light source. They noticed that the bending did not occur if they placed little metal foil caps over the coleoptile tips, but the bending resumed when the caps were removed. They concluded that the coleoptile tips must be sensitive to light. Charles Darwin

passed away in 1882 but others following up on his work demonstrated that a water-soluble "influence" was produced in the coleoptile tips and that electrical triggering of the bending, which had been suspected, did not exist.

In 1926, Frits Went, a young plant physiologist in Holland, cut off the tips of oat coleoptiles and placed the tips, cut surface down, on flat portions of a growth medium called **agar** (a substance obtained from marine algae). After a few hours, he removed the tips and cut the gelatinlike agar into little square blocks, which he then placed on decapitated coleoptiles. He found that if he placed the block squarely over the cut top of a coleoptile, it grew straight up, but if he placed the block off center, the tip bent away from the side on which the block had been placed (Fig. 11.1). He also found that the more coleoptile tips he placed on the agar to begin with, the more pronounced was the response. His experiments demonstrated conclusively that something produced in the coleoptile tips moved out into the agar and that this substance was responsible for the bending observed. Went named the substance **auxin** (from the Greek word *auxein,* meaning "to increase"), a term now in widespread use.

Auxins were the first plant hormones to be discovered, with several others coming to light later. At least three major groups apparently promote (but do not control) the growth of plants. Depending on the concentration, some may also have an inhibitory effect.

Auxin molecules have a structure similar to those of the amino acid tryptophan, which is found in both plant and animal cells. For many years, auxins were thought to be synthesized from tryptophan by living cells. In 1991, however, it was demonstrated that mutant plants incapable of producing tryptophan can develop auxin without it, and the precise nature of the synthesis remains to be determined.

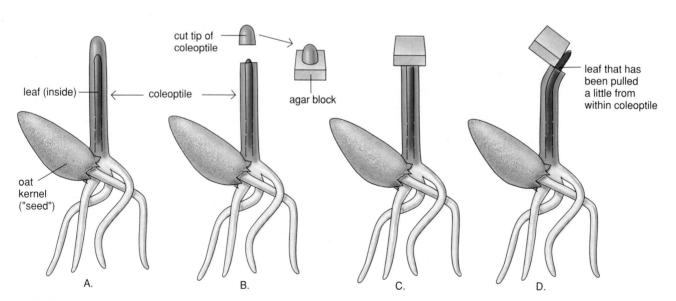

FIGURE 11.1   Went's experiment with oat coleoptiles. *A.* A germinated oat "seed" with an intact coleoptile. *B.* The tip of a coleoptile was cut off, placed on a small agar block, and left for an hour or two. *C.* When this agar block was placed squarely on a decapitated coleoptile, growth was vertical. *D.* When the agar block was placed off center so that only half of the decapitated coleoptile was in contact with it, the tip bent away from it. This experiment demonstrated that something affecting growth diffused from the coleoptile tip into the agar and from the agar to the decapitated coleoptile.

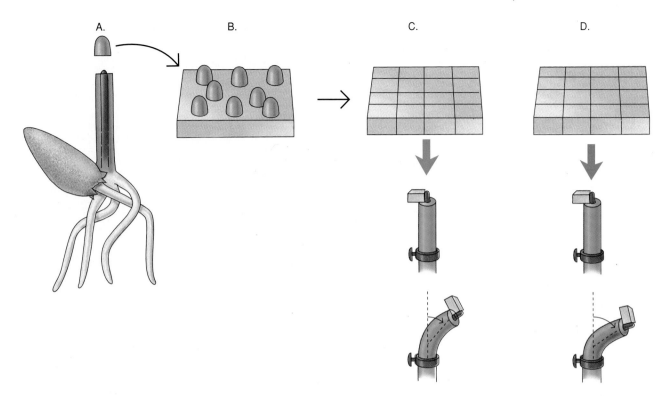

FIGURE 11.2   How a bioassay of auxin is made. *A.* A specific number of coleoptile tips are cut and placed on a measured portion of agar (*B*) for a set period of time. *C.* The coleoptile tips are then removed and the agar is cut into blocks of specific size. An agar block is placed off center on a decapitated coleoptile held by a clamp; the leaf within the coleoptile is pulled up slightly to support the agar block. After another set period the angle of curvature is measured. *D.* The angle of curvature is compared to that produced when a similar agar block containing a known amount of auxin is placed off center on another coleoptile for the same period of time.

Production of auxins occurs mainly in apical meristems, buds, young leaves, and other active young parts of plants. It is sometimes difficult to predict how cells will respond to auxins because responses vary according to the concentration, location, and other factors. For example, if an auxin of specific concentration promotes shoot growth in a certain plant, the same auxin of identical concentration will usually inhibit its root growth. At appropriate concentrations, auxins normally stimulate the enlargement of cells by increasing the plasticity (irreversible stretching) of cell walls.

Hormone concentrations were measured at first by *bioassays,* which relate the response of a sensitive plant part to the amount of hormone applied to the part. For example, a fixed number of coleoptile tips can be placed on a specific sample of agar for a certain time period, and a measured cut block of the agar can then be placed off center on a decapitated coleoptile. The concentration can be determined from the angle of bending of the coleoptile that occurs within a certain time frame when it is compared with the bending brought about by a similar block containing a known amount of auxin (Fig. 11.2).

Hormone concentrations today are more frequently determined by vaporizing a sample and moving it through a tube of liquid or powdered material by a technique known as *gas chromatography.* As the sample moves through the column, the hormone and other components of the sample separate, and the amount of hormone present can then be determined relatively precisely.

Auxins also may have many other effects, including triggering the production of different growth regulators (especially ethylene), causing the Golgi bodies (dictyosomes) to increase rates of secretion, playing a role in controlling some phases of respiration, and influencing numerous developmental aspects of growth. These aspects include flower initiation, sex determination, growth rate, fruit growth and ripening, abscission, rooting, aging, dormancy, and apical dominance, which is discussed on page 188. Sensitivity to auxins is less in many monocots than it is in dicots, and shoots are less sensitive than roots. Higher concentrations, however, will kill almost any living plant tissues. The effects of auxins combined with those of other regulators produce many of the growth phenomena.

Movement of auxins from the cells where they originate requires the expenditure of energy stored in ATP molecules. The migration is relatively slow (about 1 centimeter per hour) although the active transport mechanism involved carries the hormone up to 10 times faster than it would move by simple diffusion alone. The movement is *polar,* in that the auxins flow away from their source, usually downward from a stem tip, with movement in roots proceeding toward the tip. This polar movement occurs even when a stem is inverted. The auxins apparently are ordinarily not carried through the sieve tubes of the phloem "plumbing" but proceed from cell to cell, particularly through parenchyma cells surrounding vascular bundles.

In the past, it was believed that several natural auxins occurred in nature, but until recently, most scientists thought that *indoleacetic acid* (IAA) was the only active auxin. Furthermore, it was believed that other organic acids that act like auxins were actually converted to IAA in the plants. Current research, however, indicates that plants do actually produce three other growth regulators that are not converted to IAA but bring about many of the same responses as IAA. One, *phenylacetic acid* (PAA), is often more abundant but less active than IAA. Another, *4-chloroindoleacetic acid* (4-chloroIAA), is found in germinating seeds of legumes. A third, *indolebutyric acid* (IBA), occurs in the leaves of corn and various dicots. It previously was known only in synthetic form. At least three auxin precursors also have activities similar to those of auxins.

IAA and a number of organic acids that are not plant hormones regulate growth and have also been produced synthetically. They are in widespread use in agriculture and horticulture. At the proper concentrations, IAA and other growth regulators stimulate the formation of roots on almost any plant organ. Nurseries apply these substances in paste or powdered form to the bases of cut segments of stems to stimulate a more vigorous production of roots than would occur naturally (Fig. 11.3). Synthetic growth regulators presently in use include NAA (naphthalene acetic acid), 2,4-D (2,4-dichlorophenoxyacetic acid), and MCPA (2-methyl-4-chlorophenoxyacetic acid).

Orchardists spray fruit trees with auxins to promote uniform flowering and fruit set, and they later spray the fruit to prevent the formation of abscission layers and subsequent premature fruit drop. A substantial saving in labor expenses results from being able to go through an orchard or a pineapple plantation and pick all of the fruit at one time. If auxins are applied to flowers before pollination occurs, seedless fruit can be formed and developed. Some orchardists use auxins for controlling the number of fruits that will mature in order not to have too many small ones, and some even control the shapes of plants for sales appeal.

Weeds were controlled by hand labor and by the use of caustic or otherwise poisonous substances until it was discovered that synthetic auxins, including 2,4-D, 2,4,5-T, 2,4,5-TP, NAA, and MCPA, when sprayed at low concentrations, kill some weeds. Broad-leaved plants such as dandelions and plantains (for reasons that are not clearly understood) are more susceptible to low concentrations of these plant growth regulators than are narrow-leaved plants such as grasses. They have been used on lawns to kill weeds without noticeably affecting the grass. The precise mechanisms of action of these herbicides (plant killers) are still largely unknown.

Thus far, there is little evidence that 2,4-D has directly adverse effects on humans and other animal life, but such is

FIGURE 11.3   Effect of auxin applied to the base of a *Gardenia* cutting four weeks after application. *Left.* A treated cutting. *Right.* An untreated cutting.

not the case with 2,4,5-T, which was banned in 1979 for most uses in the United States by the Environmental Protection Agency. The controversy over the use of 2,4,5-T began in the 1960s during the Vietnam War when Agent Orange, a 1-to-1 mixture of 2,4-D and 2,4,5-T, was used to defoliate jungles in Vietnam. Offspring of Vietnam Vets who had been exposed to Agent Orange were born with birth defects and subsequent tests and experiments with 2,4,5-T have produced leukemia, miscarriages, birth defects, and liver and lung diseases in laboratory animals. The diseases and defects are apparently caused by TCDD (2,3,7,8-tetrachlorodibenzoparadioxin), a dioxin contaminant that evidently is unavoidably produced in minute amounts during the manufacture of 2,4,5-T. TCDD harms or even kills laboratory animals in doses as low as a few parts per billion.

Gibberellins

In 1926, Eiichi Kurosawa of Japan reported the discovery of a new substance that was causing what was referred to as *bakanae* or "foolish seedling" effect on rice. The stems of rice seedlings infected with a certain fungus grew twice as long as those of uninfected plants, but the stems were weakened so that they eventually collapsed and died. Kuro-sawa found that extracts of the fungus applied to uninfected plants brought about the same growth stimulations as the fungal disease. It was not until nine years later that other Japanese scientists were able to purify the substance, which Kurosawa named **gibberellin,** after the scientific name of the fungus that produced it (*Gibberella fujikuroi*).

Today, more than 80 different gibberellins (usually ab-breviated to GA) have been isolated from immature seeds (especially those of dicots), root and shoot tips, young leaves, and fungi. The majority of gibberellins produced by plants are inactive, apparently functioning as precursors to active forms. No single species has thus far been found to have more than 15 gibberellins, which probably also occur in algae, mosses, and ferns. None are known in bacteria.

Acetyl coenzyme A, which is vital to the process of respiration, functions as a precursor in the synthesis of gib-berellins. Interest in their hormonal properties is high, for their ability to increase growth in plants is far more impres-sive than that of auxin alone, although traces of auxin appar-ently need to be present for gibberellins to produce their maximum effects.

Most dicots and a few monocots grow faster with an application of gibberellins, but coniferous trees such as pines and firs generally show little, if any, response. If a gib-berellin, at the appropriate concentration, is applied to a cab-bage, the plant may become 2 meters (6 feet) tall, and bush bean plants can become pole beans with a single application. There is not usually, however, a corresponding stimulation of root growth. Many genetically dwarf plants grow as tall as their normal counterparts, after an application of the appro-priate gibberellin. Gibberellins not only dramatically in-crease stem growth, but they are also involved in nearly all of the same regulatory processes in plant development as auxins (Fig. 11.4). In some kinds of plants, flowering can be brought about by applications of gibberellins, and the dor-mancy of buds and seeds can be broken. Some gibberellins appear to lower the threshold of growth; that is, plants may start growing at lower temperatures than usual after an ap-plication of gibberellin. For example, an application to a lawn could cause it to turn green two or three weeks earlier in the spring.

In some varieties of lettuce and cereals, the seeds nor-mally germinate only after being subjected to a period of cold temperatures. An application of gibberellins can elimi-nate the cold requirement for germination. It is common for plants after germination of seed to produce juvenile foliage at first and then different adult foliage as the plants mature. In such plants, gibberellin treatment of buds that would nor-mally develop into adult branches causes the foliage to de-velop in juvenile form.

Several *growth retardants* available commercially in-hibit or block the synthesis of gibberellins. Applications of growth retardants result in stunted plants because stem elon-gation is inhibited. When applied to certain commercially grown crops such as chrysanthemums, flowers with thicker, stronger stalks are produced.

FIGURE 11.4 Effect of gibberellins on growth. *Left.* Untreated cabbage plants. *Right.* Similar cabbage plants that have been treated with gibberellins.

(Photo © Sylvan H. Wittwer/Visuals Unlimited)

Gibberellins have been used experimentally to increase yields of sugar cane and hops and have revolutionized the production of seedless grapes through increasing the size of the fruit and lengthening fruit internodes, which results in slightly wider spaces between grapes in the bunches. Better air circulation between grapes reduces their susceptibility to fungal diseases, and the need for hand thinning is eliminated. They have been used in navel orange orchards to delay the aging of the fruit's skin, and have been found to increase the length and crispness of celery petioles (the parts that are most in demand as a raw food). Gibberellins are now used to in-crease seed production in conifers and to enhance starch di-gestion in breweries by increasing the rate of malting.

Field experiments have shown that if gibberellins are applied to plants at the same time selected herbicides are ap-plied, the gibberellins may reverse the effects of the herbi-cides. The relatively high cost of gibberellins has limited the extent of their use in horticulture and agriculture.

Cytokinins

By the close of the 19th century, botanists suspected that there must be something that regulates cell division in plants. Their conjecture led to the discovery in 1913, by Gottlieb Haberlandt of Austria, of an unidentified chemical in the phloem of various plants. This chemical stimulated cell division and initiated the production of cork cambium. In the 1950s, several substances that promote cell division were found in coconut milk (a liquid *endosperm;* endosperm is discussed in Chapter 23), but it was not until 1964 that the identity of such substances was determined in kernels of corn. These various stimulants to cell division came to be known as **cytokinins.**

The several cytokinins now known differ somewhat in their molecular structure and possibly also in origin, but they are similar in composition to adenine. You will recall that adenine is a building block of one of the four nucleotides found in DNA, although none of the cytokinins appears to be derived from DNA. Some cytokinins do, however, occur in certain forms of RNA. Cytokinins are found most abundantly in meristems and other developing tissues, particularly those of young fruit.

If auxin is present during the cell cycle, cytokinins promote cell division by speeding up the progression from the G_2 phase to the mitosis phase, but no such effect takes place in the absence of auxin. Cytokinins also play a role in the enlarging of cells, the differentiation of tissues, the development of chloroplasts, the stimulation of cotyledon growth, and the delay of aging in leaves and in many of the growth phenomena also brought about by auxins and gibberellins. Despite their role in cell division and enlargement, however, there is a total absence of evidence that cytokinins initiate or promote animal cancers or have any other effect on animal cells.

Experiments have shown that cytokinins prolong the life of vegetables in storage. Related synthetic compounds have been used extensively to regulate the height of ornamental shrubs and to keep harvested lettuce and mushrooms fresh. They have also been used to shorten the straw length in wheat, so as to minimize the chances of the plants blowing over in the wind, and to lengthen the life of cut flowers. Many have not yet been approved for general agricultural use.

Abscisic Acid

Although the promotion of growth by auxins and gibberellins had been amply demonstrated by the 1940s, plant physiologists began to suspect that something else produced by plants could have opposite (inhibitory) effects. In 1949, Torsten Hemberg, a Swedish botanist, showed that substances produced in dormant buds blocked the effects of auxins. He called these growth inhibitors *dormins.*

In 1963, three groups of investigators working independently in the United States, Great Britain, and New Zealand discovered a growth-inhibiting hormone, which was in 1967 officially called **abscisic acid (ABA).** Later, it was shown that ABA and dormins were one and the same.

ABA is synthesized in plastids, apparently from carotenoid pigments. It is found in many plant materials but is particularly common in fleshy fruits, where it evidently prevents seeds from germinating while they are still on the plant. When ABA is applied to seeds outside of the fruit, germination usually is delayed. Because the stimulatory effects of other hormones are inhibited by ABA, cell growth is usually also inhibited.

ABA was originally believed to promote the formation of abscission layers in leaves and fruits. However, the evidence suggesting that ethylene (discussed next) is far more important than ABA in abscission now is overwhelming, and, despite its name, ABA has little, if any, influence on the process.

When ABA is applied to active plant buds, the leaf primordia become bud scales and the bud goes into a dormant winter condition. Since nursery plants being shipped stand to suffer less damage in a dormant condition than in an active one, there is practical value to the application of ABA and similar growth inhibitors to such nursery stock. The inhibiting effects of ABA can be reversed by the application of gibberellins.

ABA apparently helps leaves respond to excessive water loss. When the leaves wilt, ABA is produced in amounts several times greater than usual. This interferes with the transport or retention of potassium ions in the guard cells, causing the stomata to close. When the uptake of water again becomes sufficient for the leaf's needs, the ABA breaks down and the stomata reopen. Despite the growth-inhibiting effects of ABA, there is no evidence that it is in any way toxic to plants.

Ethylene

In 1934, R. Gane discovered that the simple gas **ethylene** was produced naturally by fruits; it had been known for some time before that the ripening of green fruits could be accelerated artificially by placing them in ethylene. It is now known that ethylene is produced not only by fruits but also by flowers, seeds, leaves, and even roots. Several fungi and a few bacteria are also known to produce it, and its regulatory effects make it a hormone in the broad sense of the word. Ethylene is produced from the amino acid methionine; oxygen is required for its formation.

The production of ethylene by plant tissues varies considerably under different conditions. A surge of ethylene lasting for several hours becomes evident after various tissues, including those of fruits, are bruised or cut, and applications of auxin can cause an increase in ethylene production of two to 10 times. As pea seeds germinate, the seedlings produce a surge of ethylene when they meet interference with their growth through the soil. This apparently causes the stem tip to form a tighter crook, which may aid the seedling in pushing to the surface.

When a storm stirs up a green field or animals pass across it, plants respond to these and other mechanical stresses with an increase in the production of fibers and

FIGURE 11.5   Effect of ethylene on holly twigs. Two similar twigs were placed under glass jars for a week; at the same time, a ripe apple was placed under the jar on the right. Ethylene produced by the apple caused abscission of the leaves.

collenchyma tissue. Cell elongation is also inhibited, resulting in shorter, sturdier plants. The responses, called *thigmomorphogenesis,* are under the control of genes that are activated by touch. Several substances, including enzymes, ethylene, and a protein called *calmodulin,* are involved. Calmodulin, which requires calcium for its actions, plays a role in several kinds of plant reactions to stimuli.

Ethylene apparently can trigger its own production. If minute amounts are introduced to the tissues that produce the gas, a tremendous response by the tissues often results. These tissues may then produce so much ethylene that the part concerned can be adversely affected. Flowers, for example may fade in the presence of excessive amounts, and leaves may abscise (Fig. 11.5).

In ancient China, growers used to ripen fruits in rooms where incense was being burned, and citrus growers used to ripen their fruits in rooms equipped with a kerosene stove. In houses where gas heating is used, occupants often experience great difficulty in growing houseplants, and greenhouse heaters using such fuel create a dilemma for owners attempting to promote plant growth.

In the days before electric street lights became commonplace, gas lights were used, and in some cities in Germany, leaves fell from the shade trees if they were located near a gas line that leaked. In all of these instances, minute amounts of ethylene gas resulting from the fuel combustion or leakage brought about the results, both good and bad. In fact, as little as one part of ethylene per 10 million parts of air may be sufficient to trigger responses.

Today, commercial use of ethylene is extensive. It has been used for many years to ripen harvested green fruits, such as bananas, mangoes, and honeydew melons, and to cause citrus fruits to color up before marketing. Since ethylene production almost ceases in the absence of oxygen, apple and other fruit growers have found that if they place unripe fruit in sealed warehouses after harvest, pump out the air, reduce temperatures to just above freezing, and replace the air with inert nitrogen gas or carbon dioxide, the fruit will remain metabolically inactive for long periods. The growers can then remove the fruit in batches throughout the year, add as little as one part per million of ethylene, and have ripe fruit at any time there is a demand. This is why apples are always available in supermarkets, even though harvesting is usually confined to a few weeks in the fall.

Fruits that respond to ethylene usually have a major increase in respiration just before ripening occurs. The increase in ethylene production at that time is often up to 100 times greater than it was a day or two earlier. The accompanying major increase in respiration is called a *climateric,* and fruits that exhibit such phenomena are called *climateric fruits.* Some fruits such as grapes are *nonclimateric* and do not respond in this way to ethylene.

A few growers still use natural ethylene to ripen pears and peaches by wrapping each fruit individually in tissue paper. The paper retards the escape of ethylene from the fruit and hastens ripening. "Resting" potato tubers will sprout following brief applications of ethylene, and seeds may be stimulated to germinate if given a short exposure to the gas just before sprouting, although treatment after sprouting inhibits growth.

Ethylene is used in Hawaii to promote flowering in pineapples, and it causes members of the Pumpkin Family to produce more female flowers and thus more fruit. Because trees grown in containers in nurseries are usually crowded, they are often tall and spindly, but applications of small amounts of ethylene to the trunks while they are enclosed in plastic tubes causes a marked thickening, making the trees sturdier and less likely to break.

Other Hormones or Related Compounds

A number of compounds called *oligosaccharins,* which are released from cell walls by enzymes, influence cell differentiation, reproduction, and growth in plants and therefore must be considered hormones. However, oligosaccharins produce their effects at concentrations of up to 1,000 times less than those of auxins, and the effects (e.g., growth promotion, inhibition of flowering) are not only highly specific but the responses to them are essentially identical in all species. *Brassosteroids,* which have a gibberellinlike effect on plant stem elongation, are known from legumes and a few other plants. Yams, which incidentally are a source of DHEA—a hormone whose production by humans tends to decrease with age—are also the source of *batasins.* Batasins promote dormancy in *bulbils,* which are produced from axillary buds in lilies and a few other plants.

HORMONAL INTERACTIONS

Apical Dominance

For centuries, gardeners and nursery workers have often deliberately removed terminal buds of plants to promote bushier growth, knowing that the buds are involved in *apical dominance.* Apical dominance is the suppression of the growth of the lateral or axillary buds, believed to be brought about by an auxinlike inhibitor in a terminal bud. It is strong in trees with conical shapes and little branching toward the top (e.g., many pines, spruces, firs) (Fig. 11.6) and weak in trees with stronger branching (e.g., elms, ashes, willows).

There is evidence that, as the tissues produced by the apical meristem of the terminal bud increase in length, the source of the inhibitor moves farther away and the concentration of the inhibitor gradually drops in the vicinity of the lateral buds. We presume that the inhibitory effect eventually falls below a threshold, allowing the lateral buds to grow. Each lateral bud then itself becomes a terminal bud that produces its own inhibitor, which in turn suppresses the new lateral buds developed beneath it. These presumptions, however, have not yet been confirmed and some studies suggest that a cytokinin deficiency in the lateral buds actually plays a greater role in apical dominance than auxinlike in-

FIGURE 11.6 A Jeffrey pine tree. The trunk of this tree, which would normally be single, is forked because of the earlier removal of the terminal bud.

hibitors. Further studies are needed before the precise causes of apical dominance are fully understood.

Senescence

The breakdown of cell components and membranes that eventually leads to the death of the cell is called **senescence.** As mentioned in Chapter 8, the leaves of deciduous trees and shrubs senesce and drop through a process of abscission every year. Even evergreen species often retain their leaves for only two or three years (with the notable exception of bristlecone pines, which hold their leaves for up to 30 years), and the aboveground parts of many herbaceous perennials senesce and die at the close of each growing season.

Why do plant parts senesce? Some studies have suggested that certain plants produce a senescence "factor" that behaves like, or is actually, a hormone, but we are not yet certain of the precise mechanisms involved. We do know, however, that both ABA and ethylene promote senescence, while auxins, gibberellins, and cytokinins delay senescence in a number of plants that have been studied. Other factors, such as nitrogen deficiency and drought, also hasten senescence.

OTHER INTERACTIONS

The development of roots and shoots in both tissue culture and in girdled stems is also regulated by a combination of auxins and cytokinins. Experiments with living pith cells of tobacco plants have shown that such cells will enlarge when supplied with auxin and nutrients, but they will not divide unless small amounts of cytokinin are added. By varying the amounts of cytokinin, it is possible to stimulate the pith cells to differentiate into roots or into buds from which stems will develop. This last regulatory effect can be used to offset apical dominance. The extent of the suppression depends on how close the axillary buds are to the terminal bud and on the amount of inhibitor involved. Experiments with pea plants have shown that the axillary buds will begin to grow as little as four hours after a terminal bud has been removed. If cytokinins are applied in appropriate concentration to axillary buds, however, they will begin to grow, even in the presence of a terminal bud.

Gibberellins, cytokinins, and enzymes all play a role in the germination of seeds. After water has been imbibed by cereal seeds, gibberellins are released by the embryo and stimulate the secretion of enzymes that digest endosperm. In many dicot seeds, cytokinins stimulate the production of starch-digesting enzymes.

PLANT MOVEMENTS

We noted in Chapter 2 that all living organisms exhibit movement but that most plant movements are relatively slow and imperceptible unless seen in time-lapse photography or demonstrated experimentally. Now we need to examine how and why plant movements occur.

Growth Movements

Growth movements result from varying growth rates in different parts of an organ. They are mainly related to young parts of a plant and, as a rule, are quite slow, usually taking at least two hours to become apparent although the plant may have begun microscopic changes within minutes of receiving a stimulus. The stimulus may be either internal or external.

Movements Resulting Primarily from Internal Stimuli

Helical (Spiraling) Movements Charles Darwin once attached a tiny sliver of glass to the tip of a plant growing in a pot. Then he suspended a piece of paper blackened with carbon over the tip, and, as the plant grew, he raised

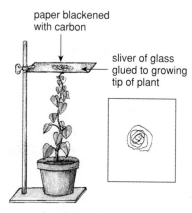

paper blackened with carbon

sliver of glass glued to growing tip of plant

tip spirals as it grows, tracing its pattern on the paper

FIGURE 11.7 Charles Darwin's demonstration of spiraling growth.

the paper just enough to allow the tip to touch the paper without hurting the plant. He found that the growing point traced a spiral pattern in the blackened paper. We know now that such helical movements of growth are common to many plants (Fig. 11.7).

Nodding Movements Members of the Legume Family, such as garden beans, whose ethylene production upon germinating causes the formation of a thickened crook in the hypocotyl, exhibit a slow oscillating movement (i.e., the bent hypocotyl "nods" from side to side like an upside-down pendulum) as the seedling pushes up through the soil. This nodding movement apparently facilitates the progress of the growing plant tip through the soil.

Twining Movements Although twining movements are mostly stimulated internally, external forces, such as gravity and contact, may also play a role in these movements. They occur when cells in the stems of climbing plants, such as morning glory, elongate to differing extents, causing visible spiraling in growth (in contrast with the spiraling movements previously mentioned, which are not visible to the eye). Tendril twining, which is initiated by contact, results from an elongation of cells on one side of the stem and a shrinkage of cells on the opposite side, followed by differences in growth rates (Fig. 11.8). Some tendrils are stimulated to coil by auxin, while others are stimulated by ethylene.

Contraction Movements In Chapter 6, we noted that the bulbs of a number of dicots and monocots have contractile roots, which pull them deeper into the ground. In lilies, for example, seeds germinating at the surface ultimately produce bulbs that end up 10 to 15 centimeters (4 to 6 inches) below ground level because of the activities of contractile roots. There is some evidence that temperature fluctuations at the surface determine how long the contracting will continue. When the bulb gets deep enough that the differences between daytime and nighttime

FIGURE 11.8 Typical twining of a tendril produced by a manroot plant. Note that the direction of coiling reverses near the midpoint.

temperatures are slight, the contractions cease. The aerial roots of some banyan trees straighten out by contraction after the roots have made contact with the ground. The "shrinking" of roots has been shown to take place at the rate of 2.2 millimeters (0.1 inch) a day in sorrel.

Nastic Movements When flattened plant organs, such as leaves or flower petals, first expand from buds, they characteristically alternate in bending down and then up as the cells in the upper and lower parts of the leaf alternate in enlarging faster than those in the opposite parts. Such nondirectional movements (i.e., movements that do not result in an organ being oriented toward or away from the

direction of a stimulus) are called **nastic.** In prayer plants (see Fig. 11.15) and others where the leaves or flowers fold up and reopen daily, turgor movements are also involved, as are the external stimuli of light and temperature. These movements are controlled by a biological "clock" on cycles of approximately 24 hours and are referred to as *circadian rhythms* (discussed under *sleep movements* on page 192). In a few instances, nastic movements apparently are initiated by external stimuli alone.

Movements Resulting from External Stimuli

Permanent, directed movements resulting from external stimuli are commonly referred to as **tropisms.**

Phototropism The main shoots of most plants growing in the open tend to develop vertically, although the branches often grow horizontally. If a box is placed over a plant growing vertically and a hole is cut to admit light from one side, the tip of the plant will begin to bend toward the light within a few hours. If the box is later removed, a compensating bend develops, causing the tip to grow vertically again. Such a growth movement toward light is called a *positive phototropism* (Fig. 11.9). A similar bending away from light is called a *negative phototropism.* The shoot tips of most plants are positively phototropic, while roots are either insensitive to light or negatively phototropic.

Leaves often twist on their petioles and, in response to illumination, become perpendicularly oriented to a light source. In fact, many plants have solar-tracking leaves with the blades oriented at right angles to the sun throughout the day. Some botanists have referred to solar-tracking move-ments as *heliotropisms* but, unlike in the phototropic re-sponses of stems and roots, growth is not involved.

It has been widely reported that sunflowers exhibit heliotropic movements and face the sun throughout the day. This is not true, except, perhaps, when the plants are very young; sunflowers face east as they develop and remain facing in that direction until the seeds are mature and the plant dies.

Strictly speaking, the twisting of petioles that facili-tates heliotropic movements should be called *phototorsion,* because *motor cells* (in *pulvini,* which are special swellings at the bases of leaves or leaflets) at the junction of the blade and the petiole control the movement (see *turgor movements* on page 192). When viewing a tree from directly overhead or observing a vine growing on a fence, it is possible to note how surprisingly little overlap of the leaves occurs. Each leaf is oriented so that it receives the maximum amount of light available (Fig. 11.10).

We have already noted that if the tip of a coleoptile is covered or removed, the structure will not bend toward light and that auxin is produced in the tip (see Fig. 11.1). We have also noted that auxin promotes the elongation of cells, at least in certain concentrations. For some time, it was be-lieved that stem tips bent toward light because auxin was

FIGURE 11.9   A cyclamen plant that received light from one direction for several weeks. Note how all visible plant parts are oriented on the side that received light.

FIGURE 11.10   Wild grape vines in the fall. Note how little overlap of leaves has occurred.

destroyed or inactivated on the exposed side, leaving more growth-promoting hormone on the side away from the light, causing the cells there to elongate more and produce a bend. Careful experiments have shown, however, that stem tips growing in the open have the same total amount of auxin present as do stem tips from the same species receiving light from one side. Other experiments indicate that the auxin migrates away from the light, accumulating in greater amounts on the opposite side, promoting greater elongation of cells on the "dark" side. Apparently, an active transport system enables the auxin to migrate against a diffusion gradient.

Different intensities of light may bring about different phototropic responses. In Bermuda grass, for example, the stems tend to grow upright in the shade and parallel with the ground in the sun. In other plants, such as the European rock rose, which grows among rocks or on walls, the flowers are positively phototropic, but, once they are fertilized, they become negatively phototropic. As the pedicels (stalks) elongate, the developing fruits are buried in cracks and crevices, where the seeds then may germinate.

Phototropic responses are not confined to flowering plants. A number of mushrooms, for example, show marked positive responses to light, and certain fungi that grow on horse dung have striking phototropic movements, which are discussed in Chapter 19.

Gravitropism   Growth responses to the stimulus of gravity are called **gravitropisms.** The primary roots of plants are positively gravitropic, while shoots forming the main axis of plants are negatively gravitropic (Fig. 11.11). There is evidence that plant organs perceive gravity through the movement of large starch grains in special cells of the root cap (discussed in Chapter 6). The starch grains, called *statoliths*, are also found in coleoptile tips and in the endodermis. When a potted plant is placed on its side, the statoliths will, within a few minutes, begin to float or tumble down until they come to rest on the side of the cells closest to the gravity stimulus. In roots, the cells on the side opposite the stimulus begin elongating within 10 seconds to an hour or two, bringing about a downward bend, while the opposite occurs in stems (figs. 11.12 and 11.13).

Some cell biologists have suggested that mitochondria and Golgi bodies (dictyosomes) also respond to gravity, but precisely how they or the starch grains bring about the response is the subject of conjecture. It may be that auxin and ABA, along with calcium ions, are agents

FIGURE 11.11 This *Coleus* plant was placed on its side the day before the photograph was taken. The stems bent upward within 24 hours of the pot being tipped over.

involved in modifying cell elongation so as to produce these gravitropic bendings.

The stimulus of gravity can be negated by rotating a plant placed in a horizontal position. A simple device called a *clinostat* uses a motor and a wheel to rotate a potted plant slowly about a horizontal axis (Fig. 11.14). As the plant rotates, both the stem and roots continue to grow horizontally instead of exhibiting characteristic gravitropic responses. Obviously, neither the starch grains in the statoliths nor other organelles can settle while the plant is moving, and this apparently prevents transport of auxin, which would bring about cell elongations that produce curvatures of root and stem.

Other Tropisms A plant or plant part response to contact with a solid object is called a *thigmotropism*. One of the most common thigmotropic responses is seen in the coiling of tendrils and in the twining of climbing plant stems (see figs. 7.13 and 11.8). Such responses can be relatively rapid, with some tendrils wrapping around a support two or more times within an hour. The coiling results from cells in contact becoming slightly shorter while those on the opposite side elongate.

Roots often enter cracked water pipes and sewers. In fact, roots have been known to grow upward for considerable distances in response to water leaks. Some researchers have called such growth movements *hydrotropisms,* but most plant physiologists today doubt that responses to water and several other "stimuli" are true tropisms. Other external stimuli that produce responses designated tropic by some scientists are chemicals (*chemotropism*), temperature (*thermotropism*), wounding (*traumotropism*), electricity (*electrotropism*), dark (*skototropism*), and oxygen (*aerotropism*).

It has been noted that greater concentrations of roots occur on the north and south sides of wheat seedlings, and it has been suggested that magnetic forces may be involved; the term *geomagnetotropism* has been proposed for this phenomenon. Some of these tropic or tropiclike responses have been artificially induced, but others take place regularly in nature. For example, germinating pollen grains produce a long tube that follows a diffusion gradient of a chemical released within a flower; this is considered a chemotropism. Thermotropic responses to cold temperatures may be seen in the shoots of many common weeds, which grow horizontally when cold temperatures prevail and return to erect growth when temperatures become warmer.

Turgor Movements

Turgor movements result from changes in internal water pressures. The cells concerned may be in normal parenchyma tissue of the cortex, or they may be in special swellings called *pulvini* located at the bases of leaves or leaflets. Some turgor movements may be quite dramatic, taking place in a fraction of a second. Others may require up to 45 minutes to become visible.

"Sleep" Movements (Circadian Rhythms)
Members of several flowering plant families, particularly the Legume Family (which includes peas and beans), exhibit movements in which either leaves or petals fold as though "going to sleep" (Fig. 11.15). The folding and unfolding usually takes place in regular daily cycles, with folding most frequently taking place at dusk and unfolding occurring in

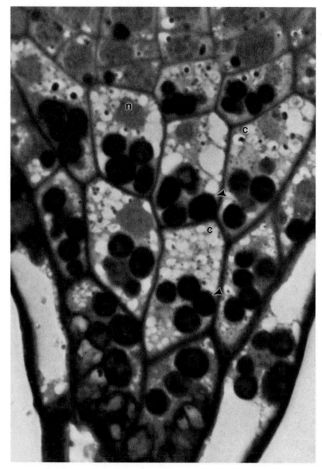

FIGURE 11.12   A root cap of a tobacco plant. The force of gravity is at the bottom of the picture. Note that the amyloplasts (more or less spherical dark objects) are toward the bottom of each cell. The amyloplasts are believed to function as statoliths in the perception of gravity by roots, ×2,000.

(Light micrograph courtesy John Z. Kiss)

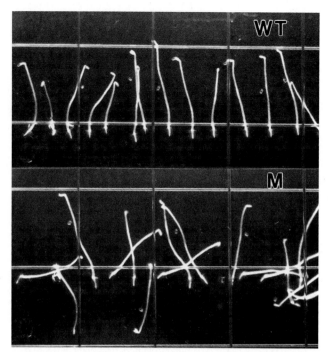

FIGURE 11.13   Tobacco seedlings grown in the dark. The source of gravity is at the bottom of the pictures. Normal "wild type" seedlings in the top row are all more or less perpendicular to the gravity. The seedlings in the bottom row are mutants with much less starch than normal plants. The mutant seedlings are disoriented, suggesting that any amyloplasts of typical mass function as statoliths in the perception of gravity, ×0.5.

(Courtesy John Z. Kiss)

the morning. Such cycles, which have been more extensively documented in the Animal Kingdom, have come to be known as **circadian rhythms.**

Circadian rhythms appear to be controlled internally by the plants, although they are also geared to changing day lengths and seasons. They do not generally accelerate when temperatures increase. The actual timing of circadian rhythms varies with the species, although most plants exhibiting sleep movements do so at dusk and at dawn. The flowers of several species, however, open their flowers at different hours of the night.

About 200 years ago, the famous Swedish botanist Linnaeus planted wedge-shaped segments of a circular garden with plants that exhibited sleep movements. The plants were arranged in successive order of their sleep movements throughout a 24-hour day. One could tell the approximate time simply by noting which part of the garden was "asleep" and which was "awake." A few others copied the garden clock idea, but the expense and difficulty of obtaining all the plants

FIGURE 11.14   A clinostat, which is a tool used by plant biologists to negate the effects of gravity. Growing plants or seedlings are slowly rotated so that the statoliths in cells that perceive gravity do not settle to the bottom, and typical growth or bending of stems or roots away from gravity does not occur.

A.

B.

FIGURE 11.15   A prayer plant. *A*. The plant at noon. *B*. The same plant at 10 P.M., after "sleep" movements of its leaves have occurred.

from different parts of the world and replanting them each year proved to be too great for the practice to be continued.

The movements are produced by turgor changes caused by the passage of water in and out of cells at the bases of the leaves or leaflets. The function of these movements is not clear, and the rhythms are also not confined to sleep movements. Species of certain warmer marine water algae called *dinoflagellates* (discussed in Chapter 18) glow in the dark through *bioluminescence*, a process by which chemical energy is reconverted to light energy. One species always glows brightly within two or three minutes of midnight, even if it is maintained in culture in continuous dim light. This particular dinoflagellate also glows when culture containers in which it is suspended are jarred, and it displays two other types of circadian rhythms. One rhythm involves cell division, with peak mitotic activity occurring just before dawn, and the other rhythm pertains to photosynthesis, which reaches a maximum around noon.

Another type of rhythm is seen in the giant bamboos of Asia, which send up huge flowering stalks every 33 or 66 years, even if the plants have been transplanted to other continents and are growing under different conditions. These flowering stalks use up all the energy reserves of the bamboo, and they die shortly after appearing. This has especially been a problem in cities where nearly all of the bamboo plantings were propagated from a single source and then all died simultaneously after flowering.

Water Conservation Movements

Leaves of many grasses have special thin-walled cells (*bulliform cells*) below parallel, lengthwise grooves in their sur-faces. During periods when sufficient water is not available, these cells lose their turgor and the leaf rolls up or folds (Fig. 11.16). Experiments with certain prairie grasses have shown that the rolling effectively reduces transpiration to as low as 5% to 10% of normal.

Contact Movements

The sudden movements of bladderworts (discussed in Chapter 7) involve turgor changes apparently triggered by electric charges released upon contact or as a result of variations in light or temperature. The springing of the trap of the Venus flytrap (also discussed in Chapter 7) was thought to be brought about in similar fashion, but recent research has shown that the trap closes when its outer epidermal cells expand rapidly, and it reopens when the inner epidermal cells expand in the same way. About one-third of the ATP available in the cells is used in each movement, so repeated stimulation of the trap by touching the trigger hairs readily fatigues the trap if sufficient time for ATP replenishment is not given between stimulations.

The sensitive plant (*Mimosa pudica*) has well-developed swellings (*pulvini*) at the bases of its many leaflets and a large pulvinus at the base of each leaf petiole. When the leaf is stimulated by touch, heat, or wind, there is a type of chain reaction in which potassium ions migrate from one-half of each pulvinus to the other half. This is followed by a rapid shuttling of water from the pulvinar parenchyma cells of one-half to those of the other half. The loss of turgor results in the folding of both the leaflets and the leaf as a whole (Fig. 11.17).

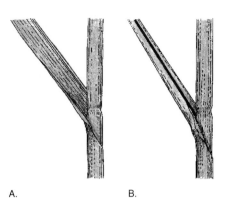

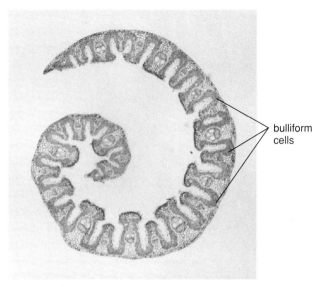

bulliform
cells

A.　　　　　B.　　　　　　　　　　　　C.

FIGURE 11.16　Water conservation movement in a grass leaf when insufficient water to maintain normal turgor is available. *A*. The leaf when adequate water is available. *B*. The leaf after it has rolled up. *C*. Enlargement of a cross section of a rolled leaf showing the large, thin-walled *bulliform* cells (arrows), which partially collapse under dry conditions and thus bring about the rolling of the leaf blade.

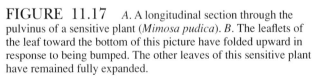

parenchyma
cells

phloem

xylem

A.

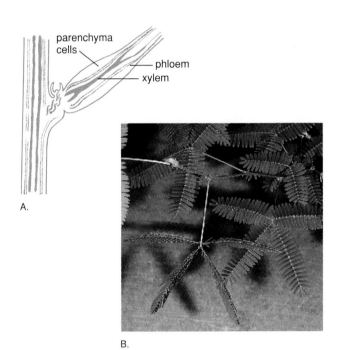

B.

FIGURE 11.17　*A*. A longitudinal section through the pulvinus of a sensitive plant (*Mimosa pudica*). *B*. The leaflets of the leaf toward the bottom of this picture have folded upward in response to being bumped. The other leaves of this sensitive plant have remained fully expanded.

A.

B.

FIGURE 11.18　A bush monkey flower. The white, two-lobed structure in the center is the stigma. *A*. The stigma as it appears before pollination. *B*. The stigma as it appears before pollination. *B*. The stigma 2 seconds after being touched by a pollinator; the lobes have rapidly folded together.

If a part of the stem of a sensitive plant is cut off and then immediately reattached with a water-filled piece of rubber tubing, it can be shown that something is transmitted through the water from above the cut. Within a few minutes after a leaf above the cut is stimulated to fold, a leaf or two below the cut will also fold. Although potassium ions have

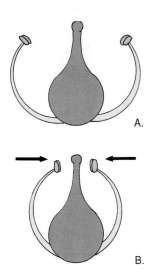

FIGURE 11.19 Contact movements of the stamens of a barberry flower. *A.* Position of the stamens before contact. *B.* Position of the stamens after they have jerked inward in response to contact.

been shown to leave cells in pulvini that are losing turgor, it is not known if the reaction is transmitted across water in the rubber tubing by these ions, by electrical charges, or by something else. A normal response to a stimulus by a sensitive plant (*Mimosa pudica*) takes from a few seconds to less than two minutes.

The redwood sorrel, whose leaflets also have pulvini, flourishes on the floor of redwood forests where the light may be only 1/200th that of full sunlight. Occasionally, light of sufficient intensity to damage delicate leaves may temporarily penetrate the overhead canopy. When this occurs, the leaflets begin to fold downward within 10 seconds and are completely folded in about six minutes, unfolding once again when the brighter light is gone.

Contact movements are not confined to leaves. Many flowers exhibit movements of stamens and other parts, most such movements facilitating pollination. The pollen-receptive stigmas found in flowers of bush monkey flower shrubs have two lobes. These lobes fold inward, enclosing pollen grains that have landed on them (Fig. 11.18). The stigmas of African sausage tree flowers and Asian cone flowers exhibit similar turgor movements upon contact, while the pollen-bearing stamens of barberry and moss rose flowers snap inward suddenly upon contact, dusting visiting insects with pollen (Fig. 11.19).

Orchid flowers exhibit some of the most spectacular of all contact movements, including those by which little sacs of pollen are forcibly attached to the bodies of visiting insects. In a few instances, the benefactors even receive dunkings in small buckets of fluid from which they escape via a trap door. (Brief discussions and illustrations of a few examples are given in Fig. 23.6.)

Taxes (Taxic Movements)

The *taxis*, a type of movement that involves either the entire plant or its reproductive cells (see chapters 18 through 23), occurs in several groups of plants and fungi but not among flowering plants. In response to a stimulus, the cell or organism, either propelled or pushed by **flagella** (whiplike appendages) or **cilia** (short whiplike appendages), moves toward or away from the source of the stimulus.

Stimuli for taxic movements include chemicals, light, oxygen, and gravitational fields. In ferns, for example, the female reproductive structures produce a chemical that prompts a *chemotaxic* response in the male reproductive cells (sperms)—that is, the sperms swim toward the source of the chemical. Certain one-celled algae exhibit *phototactic* responses, swimming either toward or away from a light source. Other similar organisms exhibit *aerotactic* movements in response to changes in oxygen concentrations.

Miscellaneous Movements

Slime molds, discussed in Chapter 19, inhabit damp logs and debris. During their active stages, their protoplasm, which has no rigid cell walls, migrates over dead leaves and other substrates in a crawling-flowing motion, somewhat like thin, slowly moving gelatin. Certain blue-green bacteria (e.g., *Oscillatoria*) wave slowly back and forth or slide up and down against each other in *gliding movements,* and diatoms (one-celled algae with thin glass shells) give the appearance of being jet-propelled. It is not certain just how these movements occur, but there is evidence that submicroscopic fibrils produce rhythmic waves that bring about the motion. In the case of diatoms, some materials are apparently forced out of the cells, and the friction set up by the process may propel the organisms through the water in which they are found.

Dehydration movements do not involve living cells or hormones, the forces being purely physical. They are caused by imbibition or by the drying out of tissues or membranes. The individual fruitlets of filarees and other members of the Geranium Family have long, stiff, pointed extensions, which are sensitive to changes in humidity. As humidity decreases during the day, they coil up, and then at night, when humidity increases, the coils relax. This alternating coiling and uncoiling results in the pointed fruitlets planting themselves into the ground in a corkscrewlike fashion (see Fig. 8.27).

A number of fruits that are podlike and dry at maturity split in various ways with explosive force, flinging seeds as far as 12 meters (39 feet) from the plant. Examples of such fruits are those of the garbanzo bean, witch hazel, vetches, and Mexican poppies. In ice plants and stonecrops, rain or dew causes the mature fruits to open due to the swelling of membranes. Pressures inside the squirting cucumber build up to the point where, upon abscission of the fruit, the seeds are expelled from the stalk distances of up to 10 or more meters (33 feet). Dwarf mistletoes, which are parasitic on coniferous trees, produce tiny sticky fruits that

FIGURE 11.20   Photoperiodism. The poinsettia with red bracts received less than 8.5 hours of light per day. The green poinsettias received more than 10 hours of light per day and did not flower.

(Photo © Sylvan Wittwer/Visuals Unlimited)

are explosively released and adhere to the trunks and branches of trees in the vicinity.

PHOTOPERIODISM

In the early 1900s, two American plant physiologists, Wightman W. Garner and Henry A. Allard, became curious about a tobacco plant growing at a research center in Beltsville, Maryland. The plant grew 3 to 4 meters (10 to 13 feet) tall during the summer, but it failed to flower in August when the normal-sized tobacco plants flowered. They brought the giant plant into a greenhouse for protection during the winter and were surprised when it flowered in December.

Garner and Allard decided to conduct some experiments with the tobacco plants. If the plants were started in pots in the fall in a greenhouse, they grew only about 1 meter (3 feet) tall before flowering. They wondered what would happen if they kept the tobacco plants, as well as some soybean seedlings that would not bloom in Beltsville before September, in complete darkness from 4 p.m. until 9 a.m., thus allowing them only seven hours of daylight. They found that all the plants flowered, and further investigations showed that the length of day (actually the length of the night) was directly related to the onset of flowering in many plants. They published the results of their investigations in 1920 and later called the phenomenon they had discovered **photoperiodism** (Fig. 11.20).

The critical length of day (i.e., the maximum or minimum length of day) for the initiation of flowering is often about 12 to 14 hours, although it can vary considerably. Plants that will not flower unless the day length is shorter than the critical length (e.g., in the fall or spring) are called **short-day plants.** They include asters, chrysanthemums, dahlias, goldenrods, poinsettias, ragweeds, sorghums, salvias, strawberries, and violets. Plants that will not flower unless periods of light are longer than the critical length are called **long-day plants.** These include garden beets, larkspur, lettuce, potatoes, spinach, and wheat. Such plants usually flower in the summer but will also flower when left under continuous artificial illumination. Accordingly, leafy vegetables, such as lettuce and spinach, need to be harvested in the spring and grown again in the fall in temperate latitudes if *bolting* (producing flowering stalks) is to be avoided. Potato breeders in the United States grow their plants in northern states where the long summer days initiate flowering, but since the potatoes themselves are produced independently of flowering, the plants may be grown for crop purposes at any latitude where other conditions are favorable.

Indian grass and several other grasses have two critical photoperiods; they will not flower if the days are too short and they also will not flower if the days are too long. Such species are referred to as **intermediate-day plants.** Other plants, particularly those of tropical origin, will flower under any length of day, providing, of course, they have received the minimum amount of light necessary for normal growth. Such plants are called **day-neutral plants** and include garden beans, calendulas, carnations, cyclamens, cotton, nasturtiums, roses, snapdragons, sunflowers, and tomatoes, as well as many common weeds, such as dandelions. With some plants, small differences in day length may be critically important. Some varieties of soybeans, for example, will not flower when days are 14 hours long but will flower if the day length is increased to 14 1/2 hours. This difference could amount to less than 320 kilometers (200 miles) of latitude, so certain varieties grown in the southern states might not produce fruit in the northern states and vice versa.

Commercial florists and nursery owners have made extensive use of photoperiods, manipulating with artificial light the flowering times of poinsettias, some lilies, and other plants so as to have them flower at times of the biggest demand, such as Christmas or Mother's Day. The light intensities used to lengthen the days artificially can be very low—often less than 10 foot-candles (i.e., one-1,000th the intensity of full sunlight; note the discussion of foot-candles in Chapter 10)—and can come from incandescent bulbs, which have more red wavelengths than fluorescent lamps and therefore have more effect on phytochromes, which are involved in photoperiodism and are discussed in the next section.

A number of vegetative activities of plants are also affected by photoperiods. In the shortening days of the fall, for example, many woody plants will begin undergoing the changes that lead to the dormancy of buds, regardless of whether "Indian summer" temperatures may be prevailing. In the spring, certain seeds respond to photoperiods in their germination, both long-day and short-day species having been discovered. Plants that produce tubers (e.g., Jerusalem artichoke) may develop them only under short-day conditions, even though they are long-day plants with respect to flowering. Usually, these photoperiodically controlled responses are valuable to the plants in that they prepare them for changes in the seasons and thus ensure their survival and perpetuation.

PHYTOCHROME

Experiments that were conducted after Garner and Allard's discoveries prompted researchers to look for a pigment that controlled photoperiodism. They had already found that the initiation of flowering could be inhibited if plants were exposed to even a brief period of red light during the night, which suggested the existence of a light-sensitive pigment. Within a few years, such a pigment was discovered. It was not visible, however, and a special pigment-analysis instrument had to be constructed to detect it. In 1959, after the pigment had been isolated, it was named **phytochrome.**

Phytochrome is an extraordinary pale blue proteinaceous pigment that apparently occurs in all higher plants and is associated with the absorption of light. Only minute amounts are produced, mostly in meristematic tissues. It occurs in two stable forms, either of which can be converted to the other: P_{red}, or P_r, is a form that absorbs red light; $P_{far-red}$, or P_{fr}, is an active related form that absorbs the far-red light found at the edge of the visible light spectrum. When either form absorbs light, it is converted to the other form, so P_r becomes P_{fr} when it absorbs red light, and P_{fr} becomes P_r when it absorbs far-red light. P_r is stable indefinitely in the dark. The normal effect of light in nature is to cause more P_r to become P_{fr} than vice versa. P_{fr} converts back to P_r in the dark over a period of several hours, or else it becomes inactivated, but its conversion in the presence of appropriate light is instantaneous (Fig. 11.21).

Phytochrome pigments play a role in a great variety of plant responses: In addition to being involved in photoperiodism, they have roles in several aspects of plant development, changes in plastids, the production of anthocyanin pigments, and the detection of shading by other plants.

One of the most studied phenomena associated with phytochrome pigments involves the germination of seeds. Some seeds, for example, do not germinate in darkness, but the red part of the spectrum in sunlight converts P_r to P_{fr}, which in turn somehow unblocks the germinating mechanism. If such seeds (e.g., those of "Grand Rapids" lettuce) are given suitable moisture and oxygen conditions and then exposed only to red light, they will germinate readily, but if they are given only far-red light, they will not. Furthermore, if they are given alternating flashes of red and far-red light, the type of light in the final flash determines whether they will germinate. If the last flash, which may last only a few seconds, is red light, the seeds will germinate, but if it is far-red light, they will not.

When a seedling first emerges from the soil, light changes the P_r to P_{fr}, triggering a reduction in the production

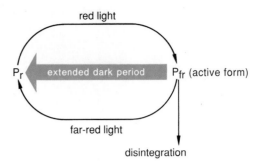

FIGURE 11.21   Phytochrome interconversions.

of ethylene by the cells. This, in turn, permits the crooks in the hypocotyls to relax, and the plant straightens up. Elongation of stems appears to be inhibited by P_{fr}. The mechanism is so sensitive that as little as three seconds of clear moonlight may produce a shortening of internodes of **etiolated** oat seedlings (those that are spindly and pale from having been grown in the dark). When stems of a tree are shaded, however, more far-red light reaches them, causing the P_{fr} to be converted to P_r, thereby lowering the inhibition so that the stems grow longer and out from under the shade.

A FLOWERING HORMONE?

After photoperiodism was discovered, many scientists, including M. H. Chailakhyan, a Russian botanist, conducted experiments to determine which part of a plant was sensitive to day length. They soon found it was mainly the leaf, although they also noted that buds of long-day plants exhibited similar responses. If they completely enclosed the leaf of a short-day plant in black paper for all but eight hours of a 24-hour day, while exposing the rest of the plant to long days, flowering was initiated, but if they treated a long-day plant the same way, flowering did not occur (Fig. 11.22). This suggested that something in the leaf was being carried to an area where flowers were initiated. In the 1930s, Chailakhyan gave support for this theory by showing that, when part of a plant was exposed to the appropriate day length to initiate flowering and then grafted to a plant that had not been so exposed, something would cross the graft so that both plants would flower. Chailakhyan suggested the name *florigen* for the "something."

In the 1950s, further evidence was obtained through experiments in which the leaves of some plants were removed immediately after exposure to critical photoperiods, while the leaves of similar plants were removed several hours after exposure. The plants whose leaves were removed immediately after exposure did not flower, but those whose leaves were removed later produced almost as many flowers as others whose leaves were not removed. This indicated that something initiating flowering moved out of the leaves, but its departure was prevented if the leaves were removed before it had a chance to do so.

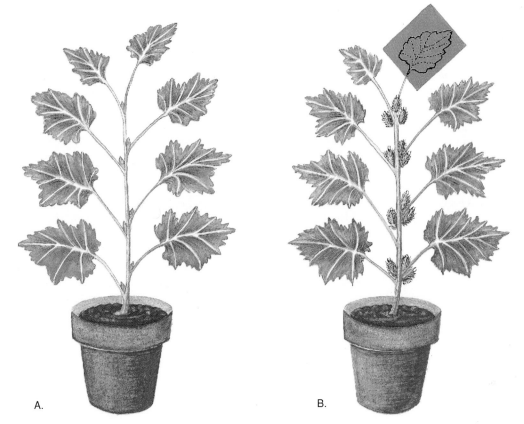

FIGURE 11.22 An experiment illustrating the effect of subjecting one leaf of a short-day plant (cocklebur) to short days while the rest of the plant is exposed to long days. *A.* The short-day plant exposed to long days. No flowers were produced. *B.* The same plant exposed to long days while one leaf was covered with black paper 16 hours a day for a few weeks. The plant flowered, presumably because some substance that initiates flowering was produced in the shaded leaf and then diffused or was carried to the stem tip where flower buds are produced.

After the existence of phytochrome had been demonstrated, it was theorized that it might be involved in photoperiodism, with P_{fr} inhibiting flowering in short-day plants and promoting flowering in long-day plants. Presumably, the P_{fr} would accumulate in the light in short-day plants and revert back to P_r or be broken down during dark periods. After the length of the night increased sufficiently, all or most of the P_{fr} would disappear, and flowering then would cease to be inhibited, with the converse taking place for long-day plants. P_{fr} has been demonstrated to disappear in many plants in as little as three or four hours of darkness, however, indicating that phytochrome interconversions alone cannot explain photoperiodism.

On the basis of the evidence, it was theorized that plants produce one or more flowering hormones ("florigen"), which may then be transported to the apical meristems where flower buds are initiated. Despite many years of research and all the circumstantial evidence, however, a flowering hormone still has not yet been isolated from a plant, nor has it otherwise been proved to exist. This has led to speculation that photoperiods may bring about a shunting of nutrients that initiate flowering or that flowering is triggered by changes in relative proportions of other hormones.

The existence of a flowering hormone remains strictly hypothetical, however, and a simple explanation for the phenomenon of photoperiodism may never be proved to exist.

TEMPERATURE AND GROWTH

Each species of plant has an optimum temperature for growth, although the optimum may vary throughout the life of the plant, and a minimum temperature below which growth will not occur. Each species also has a maximum temperature above which injury may result (or, at least, growth may cease).

Dr. Frits Went, the discoverer of auxin, experimented with the growth of tomato plants under conditions in which night temperatures were lowered to 17°C (63°F) while day temperatures were maintained at 25°C (77°F). These plants were found to grow better than plants kept at a steady temperature around the clock. Went applied the term **thermoperiodism** to this phenomenon. It has since been shown that the optimal thermoperiod may change with the growth stage of the plant. Young pepper plants have been shown to develop best when night temperatures are approximately 25°C (77°F), but as they grow older the optimum night temperature for their development drops more than 11°C (20°F). Lower night temperatures often result in a higher sugar content in plants and may also produce greater root growth, although some plants, such as peas, seem to be unaffected in this regard.

The growth of many field crops is roughly proportional to prevailing temperatures—making it possible to predict harvest times—although the number of days until maturity varies considerably with the locale. In 1855, the Swiss botanist A. P. de Candolle established a basis for the harvest dates of crops. His method, based on summing the temperature means for each day, makes it possible to follow a crop's progress with some precision. This method has been refined by multiplying the temperature means by the number of hours of daylight.

In Hawaii, a high degree of precision has been attained in predicting the date of the pineapple harvest by measuring the rate of growth within the temperature range at which the plants grow and multiplying the number of hours per day at each temperature. These methods, which work best for crops that are not closely regulated by photoperiods, do not, however, take into account the requirements of some plants for different optimum temperatures at different stages of their development.

DORMANCY AND QUIESCENCE

As the days grow shorter in the fall, cells produce sugars and amino acids that lower the point at which cold damage will occur. Protective bud scales develop on buds, and leaves become senescent. As cell metabolism slows down, many plants that are now prepared for winter become **dormant.** Dormancy may be defined as a period of growth inactivity in seeds, buds, bulbs, and other plant organs even when environmental requirements of temperature, water, or day length are met.

Quiescence is a state in which a seed cannot germinate unless environmental conditions normally required for growth are present. We have already discussed the formation of abscission layers in the fall as plants undergo change from an active condition to either a dormant or a quiescent one. In temperate and cold climates, most plants and their seeds go through a dormant period lasting from a few days to several months, and in each case, the dormancy nearly always has some survival value for the species. Many of the stone fruits (e.g., cherries, peaches, and plums) need at least several weeks of rest at temperatures below 7°C (45°F) before the buds will develop into flowers or new growth, and in some instances, the buds will not develop before the tree has encountered a period of freezing temperatures. This dormancy adapts the plants to the conditions prevailing in the cooler temperate zones of the world. In response to prevailing conditions, seeds of desert plants germinate only after appreciable rain has fallen, and seeds of plants native to vernal pools

and other wet areas may germinate only if they are immersed in water.

The change from dormancy to a state in which germination will occur in seeds is sometimes controlled by a variety of factors referred to as *after-ripening,* depending on the species concerned. Some plants become sensitive to photoperiods only after they have been subjected to a period of chilling, either while they are still seeds or shortly after germination. Growers often place moistened seeds of such plants in a refrigerator for several weeks. This process is known as *stratification* and is one of several practices employed to break dormancy. In some biennial plants, the cold treatment can be replaced by an application of gibberellins. Natural stratification occurs when winter grains are sown in the fall, and it has been shown that the extent of flowering and yield in such grains is profoundly reduced in the absence of a period of natural chilling.

Some plants have their dormancies broken only after they have received a series of similar stimulations, so a single stimulation, such as an unseasonal rainstorm, will not trigger growth. Other plants, such as poor man's pepper, require both certain temperatures and photoperiods for germination, and some weeds require specific thermoperiods plus some mechanical abrasion of the seed coat to break dormancy. Still other seeds may require specific enzyme action or a combination of conditions acquired only by passing through the digestive tracts of birds, bats, or other animals before they will germinate.

When a mechanical tomato picker was being developed in California during the early 1960s, tomato breeders wanted to have fruits that would stand up to the bumping they would receive in the machinery, and so they tried to produce suitable varieties. Their search for new genes to use in their breeding programs led them to the wild tomatoes of the Galápagos Islands. When they brought the seeds back to California, they experienced difficulty in getting them to germinate, even after subjecting them to a wide variety of stratification treatments. Then they remembered that tomatoes are eaten by the giant tortoises of the Galápagos Islands. They tried feeding the tomatoes to giant tortoises in a zoo. After passing through the animals, the seeds germinated normally. Apparently an enzyme in the tortoises' digestive tracts broke the dormancy of the seeds.

Germination is prevented by a thick or restrictive seed coat in the seeds of some plants, such as cocklebur, and the seeds of some legumes are almost stonelike because of the exceptional hardness of the outer covering or seed coat. Growers often scarify (nick or file) seed coats to allow water and gases to enter. In nature, various decay organisms corrode the seed coats, or the seeds themselves may secrete enzymes that aid in this process. The seeds of certain lupines have remarkable valves that restrict the uptake of water, preventing germination before the seed coat is eroded by other means. Many plants have seeds that will not germinate until they have been exposed to fire, and one such Australian legume with a thick seed coat will not readily germinate until heat from a fire causes a plug in the seed to pop out.

Dormancy may be restricted to certain parts of the plant rather than occurring in the plant as a whole. In coffee, for example, only the flower buds may show dormancy. In other woody plants, the onset of dormancy progresses from axillary or lateral buds toward the base of a branch on to the terminal bud at the tip. Some seeds go through two cycles of dormancy before full growth occurs. In wake-robins, the radicle (embryo root) of the seed emerges and becomes established after being subjected to a period of cold, but the epicotyl does not emerge until a second season of cold has been encountered.

In many seeds, growth inhibitors, such as ABA, are present in the seed coat, often in abundance. Before germination can occur, these inhibitors may need to be leached out by rain, broken down by enzymes, or have their effects overcome by other hormones. Studies have shown that gibberellin, which has the capacity to break dormancy when applied artificially, is produced by the germinating seeds of both cone-bearing and flowering plants. Ethylene, which is also effective in breaking dormancy, is produced by various germinating seeds, including those of oats, peanuts, and clover. Cytokinins do not generally appear to be involved in breaking dormancy, but a few instances are known of their doing so artificially. Although auxins are formed during the early stages of germination, they probably are not involved in the control of dormancy.

Many aspects of growth discussed in this chapter are applied to, or are involved in, vegetative propagation, which is discussed in Chapter 14 and Appendix 4.

Summary

1. Growth is defined as an "increase in volume due to the division and enlargement of cells." As cells mature, they differentiate into forms adapted to specific functions.

2. Development is a change in form as a result of growth and differentiation combined. Cells themselves assume different shapes and forms, which adapt them to the problem of their total volume increasing at a greater rate than their total surface area.

3. Many growth phenomena are influenced by hormones, which are produced in minute amounts in one part of an organism and usually transported to another part, where they have specific effects on growth, flowering, and other plant activities. Vitamins are organic molecules that function in activating enzymes; they are sometimes difficult to distinguish from hormones.

4. Darwin and his son noted in 1881 that coleoptiles bend toward a light source. Frits Went, in 1926, followed up on the Darwinian observations and demonstrated that something in coleoptile tips moved out into agar when decapitated tips were placed on it. He called the substance *auxin* and showed that it could cause coleoptiles to bend.

5. Three groups of plant hormones that promote the growth of plants have been found; others are known only for their inhibitory effects. Auxins stimulate the enlargement of cells and are involved in many other growth phenomena. Synthetic auxinlike compounds have been used as weed killers and defoliants. Dioxin, which has caused disease and defects in laboratory animals, is a toxic contaminant of 2,4,5-T, which is now banned in the United States.

6. Gibberellins promote stem growth without corresponding root growth; cytokinins promote cell division and can be used to stimulate the growth of axillary buds. Abscisic acid causes buds to become dormant and apparently helps leaves respond to excessive loss of water. Ethylene gas hastens ripening of fruits and is used commercially to ripen green fruits.

7. Senescence is the breakdown of cell parts that leads to the death of the cell.

8. Plant movements, such as spiraling, nodding, twining, contraction, and nastic, are growth movements that result primarily from internal stimuli.

9. Tropisms are permanent, directed growth movements that result from external stimuli, such as light (phototropism), gravity (gravitropism), contact (thigmotropism), and chemicals (chemotropism).

10. Turgor movements result from changes in internal water pressures; they may be very rapid or take up to 45 minutes to become visible. Turgor movements include sleep movements in which leaves or flowers fold daily in what is known as a circadian rhythm. Other types of plant movements include water conservation movements, contact movements, taxes, and dehydration movements.

11. Photoperiodism is a response of plants to the duration of day or night. Short-day plants will not flower unless the day length is shorter than a critical day length, and long-day plants will not flower unless the day length is longer. Intermediate-day plants have two critical photoperiods; the flowering of day-neutral plants is independent of day length.

12. Phytochrome is a light-sensitive pigment that occurs in all higher plants and plays a role in many different plant responses. It occurs in two forms, each of which can be converted to the other by the absorption of light. Daylight generally results in more P_r being converted to P_{fr} (the active form) than vice versa. P_r absorbs red light and P_{fr} absorbs far-red light. P_{fr} will convert back to P_r over a period of several hours in the dark; P_r is stable indefinitely in the dark.

13. The precise mechanism of photoperiodism is not yet fully understood. Phytochrome interconversions may

play a role in photoperiodism but cannot be used alone to explain it. A hormone initiating flowering was believed to be present in leaves, but it has never been isolated from plants.

14. Each species of plant has an optimum temperature for growth.

15. Dormancy is a period of growth inactivity in seeds, buds, bulbs, and other plant organs even when appropriate environmental conditions are met. Quiescence is a state in which a seed is unable to germinate until appropriate environmental conditions prevail.

16. The change from dormancy to a state in which germination will occur in seeds is controlled by a variety of factors, including temperature, moisture, photoperiod, thickness of seed coat, enzymes, and growth inhibitors. Stratification is the breaking down of dormancy in seeds by keeping them refrigerated and moist for several weeks.

Review Questions

1. What are the differences among hormones, enzymes, and vitamins?

2. Auxins, gibberellins, and cytokinins all promote growth. How does one distinguish among them?

3. How do hormonal herbicides function?

4. List several commercial applications of ethylene gas.

5. What are statoliths, and what role do they play in plant growth?

6. Distinguish among internally and externally induced growth movements, turgor movements, and dehydration movements.

7. How do day-neutral and intermediate-day photoperiod plants differ in their requirements?

8. What is the difference between dormancy and quiescence? How may dormancy be artificially broken?

9. What are the requirements for a seed to germinate?

10. How does phytochrome differ from other plant pigments?

Discussion Questions

1. If green plants require light in order to produce food needed for growth, why are seedlings that are germinated in the dark taller than those of the same age grown in the light?

2. If you remove a terminal bud from a stem, will growth in length stop altogether? Explain.

3. Would it be technically correct to say that some plants go to sleep at night? Explain.

4. Many plants produce seeds that require a period of dormancy before they will germinate. Of what value is the dormancy to the plant? Where might you expect to find plants with seeds that do not undergo dormancy?

5. The rapid turgor movements of the sensitive plant (*Mimosa pudica*) are rare in the Plant Kingdom. If all plants evolved such movements, would it be because the phenomenon has possible survival value or other value to the plants? Explain.

Additional Reading

Briggs, W. R., R. L. Jones, and V. Walbot (Eds.). 1992. *Annual review of plant physiology and plant molecular biology.* Vol. 43. Palo Alto, CA: Annual Reviews, Inc.

Davies, P. J. (Ed.). 1987. *Plant hormones and their role in growth and development.* Norwell, MA: Kluwer Academic Publications.

Foskett, D. E. 1994. *Plant growth and development: A molecular approach.* San Diego, CA: Academic Press.

Hart, J. W. 1990. *Plant tropisms and other growth movements.* New York: Routledge, Chapman, and Hall.

Moore, T. C. 1989. *Biochemistry and physiology of plant hormones,* 2d ed. New York: Springer-Verlag.

Nickell, L. G. 1981. *Plant growth regulators—agricultural uses.* New York: Springer-Verlag.

Salisbury, F. B., and C. W. Ross. 1992. *Plant physiology,* 4th ed. Belmont, CA: Wadsworth Publishing Co., Inc.

Steeves, T. A., and I. M. Sussex. 1989. *Patterns in plant development,* 2d ed. New York: Cambridge University Press.

Thimann, K. V. 1977. *Hormone action in the whole life of plants.* Amherst, MA: University of Massachusetts Press.

Thimann, K. V. 1980. *Senescence in plants.* Elkins Park, PA: Franklin Book Co., Inc.

Pollen grains of a western tarweed (Calycadenia *sp.*), *×3,000. (Scanning electron micrograph courtesy Robert L. Carr.)*

Meiosis and Alternation of Generations

12

A s a young child, I used to augment my allowance by growing a few vegetables and other plants from seed in my yard; the produce was then sold—mostly to my mother. Occasionally, an unharvested plant would go to seed and I would save the seeds for the next growing season. I noticed that, in at least some instances, the second generation of plants did not always quite resemble the parents, either in appearance or taste. Although I was mildly curious, I did not understand until many years later that a primary basis for the variation was a process known as *meiosis.*

As discussed in Chapter 3, plants grow through an increase in the number of cells brought about by mitosis and cell division. Mitosis ensures, in very precise fashion, that the number of chromosomes and the nature of the DNA in the daughter cells will be identical to those of the parent cell. But living organisms do not grow indefinitely, although a few may remain alive for a long time. In order to perpetuate the species, they must reproduce or they will become extinct.

This reproduction may take place through natural vegetative propagation, which is discussed in Chapter 14, or by means of special cells called *vegetative spores,* which are produced by mitosis. Such processes, in which the cells involved are identical in their chromosomes to the cells from which they arise, are forms of **asexual reproduction** (the prefix *a* of *asexual* means "without," so *asexual reproduction* simply means "reproduction without sex").

Nearly all plants, however, also undergo **sexual reproduction,** which in flowering and cone-bearing plants ultimately results in the formation of **seeds.** In sexual reproduction, sex cells called **gametes** are produced. Two gametes, called **egg** and **sperm** in higher plants and animals, form a single cell called a **zygote** when they fuse together. The zygote is the first cell of a new individual (Fig. 12.1).

If gametes had the same number of chromosomes as all the other cells of the organism, then the zygote, which results from the union of two gametes, would have double the number of its parents' chromosomes. If such zygotes were then to develop into mature organisms, each of the new organism's cells would have double the original number of chromosomes, and when they produced gametes, the zygotes resulting from the union of this next generation of gametes would have four times the number of chromosomes of the grandparents. If that sort of thing were to continue for 20 generations, a species having 16 chromosomes in each cell to begin with would end up with no fewer than 8,388,608 chromosomes per nucleus! The problem does not actually arise, however, because at a certain point in the life cycle of all sexually reproducing organisms, the unique process of **meiosis** occurs.

The process of meiosis brings about the development of gametes that have only half the number of chromosomes of any body developing from the zygote. When the gametes form zygotes as they unite in pairs, the original chromosome number is restored.

Since some aspects of meiosis are similar to those of mitosis (the two processes are compared in Fig. 12.2), it is strongly recommended that you review mitosis in Chapter 3 before going through the phases of meiosis outlined in the following section.

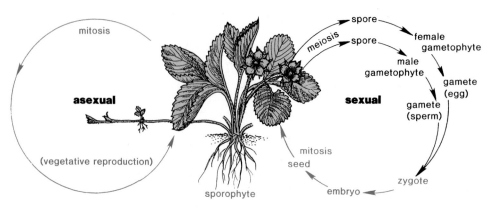

FIGURE 12.1   Asexual and sexual reproduction in a strawberry plant. More detailed life cycles are shown in Figure 12.6.

Keep in mind that all living cells, before undergoing meiosis, have two sets of chromosomes, one from the male parent, the other from the female parent. Generally, the members of each pair of chromosomes are identical to each other in length, in the amount of DNA present, and in having a centromere at the same precise location. However, as discussed in the next chapter, their genes may control contrasting characteristics. Such chromosomes are called *homologues,* or **homologous chromosomes.** When a cell undergoes meiosis, four cells are produced from two successive divisions, which take place without pausing. Because of the remarkable events that occur during the process, the four cells, depending on the organism involved, are rarely, if ever, identical to the original cell or to each other.

THE PHASES OF MEIOSIS

In meiosis, as in mitosis, a doubling of the strands of DNA of each chromosome takes place before the process of meiosis actually begins, so all the chromosomes each initially have two identical parallel strands held together by a centromere. Like mitosis, meiosis is a continuous process that has been divided up into arbitrary phases for convenience. Again, bear in mind that the lines between one phase and another are indistinct.

Four main phases are recognized in each of the two divisions, and the first phase is further subdivided. Since the main phases of the first division bear the same names as those of the second division, it has become customary to designate first division phases with a Roman numeral I and those of the second division with a Roman numeral II. The number of the chromosomes is reduced to half during the first division, but no further reduction in number takes place throughout the second division. Accordingly, some people have referred to division I as a *reduction division* and to division II as an *equational division.*

Meiosis generally takes much longer than mitosis to be completed. In lilies, for example, mitosis runs its full course in about 24 hours, while meiosis takes about two weeks. In other organisms, meiosis may take months or even years to be completed. After the process has begun in human females, for example, it sometimes takes as long as 50 years to reach its conclusion since it remains in a state of arrest for most of that time.

Division I (Meiosis I or Reduction Division)

Prophase I
The main features of prophase I are as follows: (1) The chromosomes coil, becoming shorter and thicker as they do so, and their two-stranded nature becomes apparent; the chromosomes also become aligned in pairs. (2) The nuclear envelope and the nucleolus disassociate. (3) Parts of each closely associated pair of chromosomes may be exchanged between the pair, and then the chromosomes separate (Fig. 12.2; also see Fig. 12.4 A, B).

As in the prophase of mitosis, the beginning of this phase is marked by the appearance of chromosomes as faint threads in the nucleus. These threads gradually coil like a spring so that they become as much as 100 times shorter in length, and correspondingly the coil becomes obviously thicker than the thread alone. As the chromosomes become shorter and thicker, they become aligned in pairs, and eventually two parallel threads, the **chromatids,** can be distinguished for each chromosome. Each pair of chromosomes, therefore, has four chromatids. The four chromatids of a homologous pair each have their own centromere (see figs. 12.2 and 12.4 B).

As prophase I progresses, parts of the chromatids of the homologous chromosomes may break and be exchanged with each other. Depending on the length of the chromosomes, this exchange may occur at one to several points throughout the length of the paired homologous chromosomes, or if the chromosomes are short, it may not occur at all. The evidence for this exchange of parts appears a little later in prophase I, when the chromosomes of each homologous pair appear to push each other apart. It can then be seen that adjacent chromatids have *crossed over* and are sticking together at one to several points.

Immediately following **crossing-over,** there is an exchange of some of the DNA contributed by the two parents, which is the basis for some of the variability seen in the offspring. An X-shaped figure called a **chiasma** (plural: **chiasmata**) results from each crossover; the relative positions of the chiasmata provide a basis for the study of where various genes are located on the chromosomes (Fig. 12.3).

After the chiasmata have appeared, the chromatids slowly separate, each remaining the same length as it was originally and retaining the same amount of DNA but now possessing "traded" material. The process is something like exchanging three fenders and the chrome grille of a red car with those of an otherwise identical blue car that has a black grille. When you are through with the exchange, you still will have structurally identical cars, but each will have a different combination of fender colors and grilles.

By the end of prophase I, the nuclear envelope and the nucleolus have disassociated and disappeared, and spindle fibers are beginning to form. As in mitosis, some spindle fibers are attached to the centromeres of the chromosomes, while others extend from pole to pole.

Metaphase I
The main features of metaphase I are as follows: (1) The chromosomes become aligned in pairs at the equator of the cell. (2) The now complete spindle becomes conspicuous (figs. 12.2 and 12.4 C).

Metaphase I resembles metaphase of mitosis, except that when the chromosomes move to the invisible, circular, essentially platelike equator, the centromeres of the

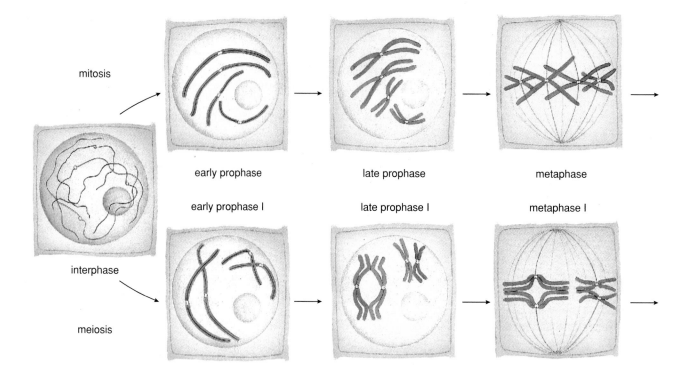

mitosis

early prophase   late prophase   metaphase

interphase

meiosis

early prophase I   late prophase I   metaphase I

FIGURE 12.2   A comparison of mitosis and meiosis.

chromosomes are paired, with the homologous chromosomes lined up directly opposite one another on each side of the equator (Fig. 12.5). Also, at this stage, as in mitosis, each chromosome's two chromatids are held together at their centromeres and function as a single unit.

Anaphase I

The main feature of anaphase I is as follows: One whole chromosome (consisting of two chromatids) from each pair migrates to a pole (see figs. 12.2 and 12.4 D).

In anaphase of mitosis, the two chromatids of each chromosome, which are being held together at their centromeres, separate as the centromeres become detached from each other, and a chromatid of each pair migrates to an opposite pole. Anaphase I of meiosis is fundamentally different from anaphase of mitosis in that the chromatids of each chromosome remain cohered at their centromeres and do *not* separate from one another. Instead, a whole, two-stranded chromosome from each pair migrates to an opposite pole. Accordingly, when the chromosomes arrive at the poles, they still consist of two chromatids, but only half the total

number of chromosomes is at each pole. If a particular chromosome was involved in crossing-over in prophase I, one of the chromatids will consist of a mixture of original DNA and DNA from a homologous chromosome.

Telophase I

Depending upon the species, the chromosomes may now either partially revert to interphase, becoming longer and thinner as they do so, or they may proceed directly to division II (see figs. 12.2 and 12.4 E). Normally, no new nuclear envelopes are organized around the chromosomes, but the nucleoli usually do reappear. This phase has been completed when the original cell has become two cells or two nuclei.

Division II (Meiosis II or Equational Division)

In the second division of meiosis, the events of the various phases are similar to those of mitosis except that there is no duplication of DNA during any part of the interphase that may occur between the two divisions. Also, the number of

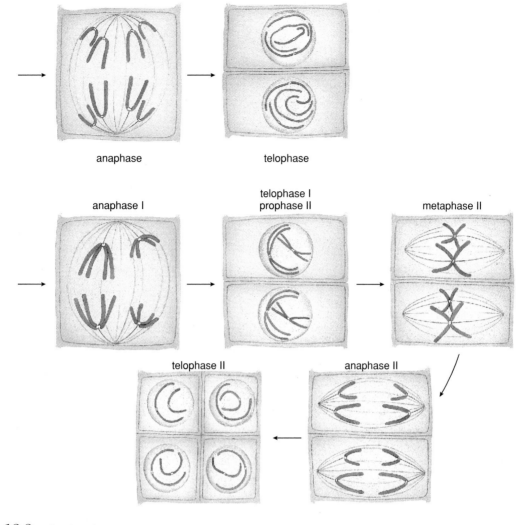

anaphase   telophase

anaphase I   telophase I / prophase II   metaphase II

telophase II   anaphase II

FIGURE 12.2 Continued

chromosomes in each cell at the end of meiosis is half the number that existed at the beginning of meiosis. In some organisms where there is no interphase, the process proceeds almost directly from telophase I to metaphase II.

Prophase II
The main feature of prophase II is as follows: The chromosomes of both groups become shorter and thicker simultaneously, and their two-stranded nature once more becomes apparent (see figs. 12.2 and 12.4 F).

Metaphase II
The main features of metaphase II are as follows: (1) The centromeres of the chromosomes become aligned along the equator. (2) New spindles become conspicuous and complete.

The two spindles may be formed either perpendicular to the ones that had formed in metaphase I or along the same plane. A spindle fiber is attached to each centromere, and other spindle fibers extend from pole to pole (see figs. 12.2 and 12.4 G).

Anaphase II
The main feature of anaphase II is as follows: The centromeres and chromatids of each chromosome separate and migrate to opposite poles (see figs. 12.2 and 12.4 H).

Telophase II
The main features of telophase II are as follows: (1) The coils of the chromatids (now called *chromosomes* again) relax so the chromosomes become longer and thinner.

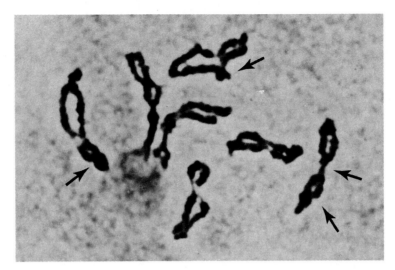

FIGURE 12.3 Chiasmata, which result from crossing-over, in chromosomes of crested wheat grass at prophase I of meiosis. (Photomicrograph by William Tai)

A. **Early Prophase I.** Chromosomes begin to coil and contract and appear as threads.
B. **Prophase I.** Homologous chromosomes become aligned in pairs, and some crossing over is visible.
C. **Metaphase I.** Bivalents become aligned along the equator.
D. **Anaphase I.** Homologous chromosomes separate and migrate to opposite poles.
E. **Late Telophase I.** Chromosomes are at the poles, and the original cell becomes two cells.

A.

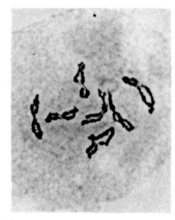

B.

C.

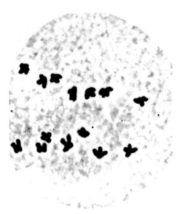

D.

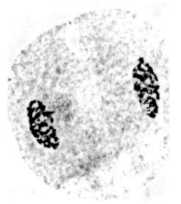

E.

FIGURE 12.4 Various phases of meiosis in crested wheat grass.

(2) New nuclear envelopes and nucleoli reappear for each group of chromosomes (see figs. 12.2 and 12.4 I).

This phase is accompanied by the formation of new cell walls between each of the four groups of chromosomes. The set of chromosomes present in each of the four cells formed by the end of telophase II constitutes half the original number, and if crossing-over and exchange of material between chromatids has occurred, none of the four cells will have exactly the same combination of DNA. In many organisms, these four cells are called *meiospores;* they develop into various structures or bodies from which gametes are produced by mitosis.

ALTERNATION OF GENERATIONS

As we have noted, meiosis occurs at some point in the life cycle of all organisms that reproduce sexually. The chromosomes that result from the process constitute a complete set in each cell, since one member of every original pair ends up in each cell. The original chromosomal complement, consisting of two complete sets of chromosomes, is restored when gametes unite, forming a zygote.

Any cell having one set of chromosomes is said to be **haploid,** and any cell with two sets of chromosomes is said to be **diploid.** By the time meiosis is complete, four *haploid* cells have been produced from one *diploid* cell. The gametes of any organism are haploid, while a zygote of the same organism is diploid. This holds true regardless of the number of chromosomes peculiar to a given organism. We can state that an organism having *n* (a specified quantity) chromosomes in its haploid cells will have twice as many, or *2n*, chromosomes in its diploid cells.

In most animals, the only haploid cells (i.e., cells with *n* or a single set of chromosomes) are the gametes (egg and sperm) and the cells that become the gametes. In plants and other green organisms, however, this is generally not so. In a complete life cycle involving sexual reproduction, there is an alternation between a diploid (*2n*) **sporophyte** phase and a haploid (*n*) **gametophyte** phase. This alternation is commonly referred to as **Alternation of Generations.**

F. **Prophase II.** The chromosomes coil and contract again; because of crossing over, the chromatids are no longer identical with each other.

G. **Metaphase II.** The chromosomes of each cell become aligned along their respective equators.

H. **Anaphase II.** The chromatids separate completely and migrate to the poles.

I. **Telophase II.** The four groups of chromatids (now called chromosomes again) are at the poles; new cell walls begin to form.

J. **Formation of cell walls.** Cell walls form; the four cells will become pollen grains.

(Photomicrographs by William Tai)

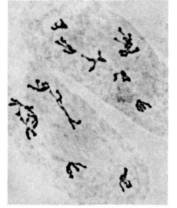

F.

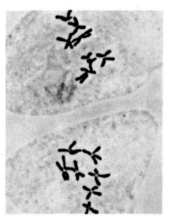

G.

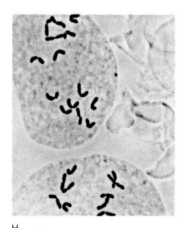

H.

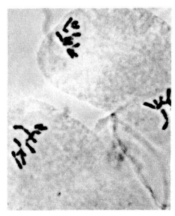

I.

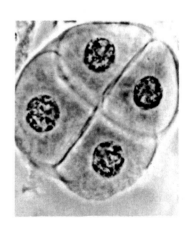

J.

FIGURE 12.4   Continued

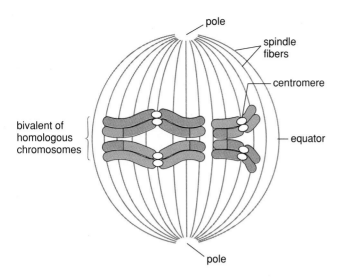

FIGURE 12.5   A diagram of homologous chromosomes at metaphase I of meiosis.

The diploid body itself is called a **sporophyte.** It develops from a zygote and eventually produces **spore mother cells (meiocytes),** each of which undergoes meiosis, producing four **spores.** The haploid bodies that develop from these spores are called **gametophytes.** These eventually form sex structures, or cells, in which gametes are produced by mitosis.

As becomes evident in chapters to follow, the gametophytes of many primitive forms constitute a large part of the visible organism, but as we progress up through the Plant Kingdom to more complex plants, they become proportionately reduced in size until they may be only microscopic. The switch from one generation to the next takes place as spores are produced when spore mother cells undergo *meiosis* and again when a zygote is produced through fusion of gametes, or **fertilization** (also called **syngamy**).

Although the sporophyte generation of some primitive organisms may consist of a single cell (the zygote), the basic plan of Alternation of Generations can be seen in the Protoctistan, Fungal, and Plant Kingdoms. It becomes most conspicuous, however, in the Plant Kingdom, and it differs from one organism to the next in the forms of the various bodies and cells.

The accompanying diagram (Fig. 12.6) and the following six rules pertaining to Alternation of Generations in the majority of plants and other green organisms, apply to the life cycle of any sexually reproducing organism discussed in this book.

1. The first cell of any *gametophyte generation* is normally a *spore* (*sexual spore* or *meiospore*), and the last cell is normally a *gamete.*

2. Any cell of a gametophyte generation is usually *haploid* (*n*).

3. The first cell of any *sporophyte generation* is normally a *zygote,* and the last cell is normally a *spore mother cell* (*meiocyte*).

4. Any cell of a sporophyte generation is usually *diploid* (2*n*).

5. The change from a sporophyte to a gametophyte generation usually occurs as a result of *meiosis.*

6. The change from a gametophyte to a sporophyte generation usually occurs as a result of *fertilization* (fusion of gametes), which is also called *syngamy.*

The word *generation* as used in *Alternation of Generations* simply means *phase of a life cycle* and should not be confused with the more widespread use of the word pertaining to time or offspring.

Summary

1. Reproduction may take place through natural vegetative propagation or by spores (asexual reproduction) or by sexual processes (sexual reproduction). In sexual reproduction, two gametes unite, forming a zygote, which is the first cell of a new individual.

2. The process of meiosis ensures that gametes will have half the chromosome number of the zygotes and also usually ensures that offspring will not be identical to the parents in every respect.

3. Meiosis takes place by means of two successive divisions, each of which, like mitosis, is divided into arbitrary phases even though the process is continuous. In prophase I, the chromosomes become paired, often exchange parts, and then separate. The similar chromosomes of each pair are referred to as being homologous. Exchange of parts of chromatids may be affected by the parts initially crossing over to and from one another, forming chiasmata, and then tearing apart.

4. In metaphase I, the chromosomes become aligned at the equator in pairs, and in anaphase I, whole chromosomes from each pair migrate to opposite poles. In telophase I, the chromosomes either partially revert to their interphase state or initiate the second division, which is essentially similar to mitosis.

5. In prophase II, the chromosomes of each of the two groups become shorter and thicker again, with both groups becoming aligned at their respective equators in metaphase II. In anaphase II, the chromatids of each chromosome separate and migrate to opposite poles, and in telophase II, the chromatids (now called *chromosomes* again) lengthen, and new nuclear envelopes and nucleoli appear for all four groups. New cell walls are produced between each of the four groups.

6. If the chromatids have exchanged parts earlier, none of the four groups may have identical combinations of DNA, and each group has half the original number of chromosomes. Each of the four cells constitutes a

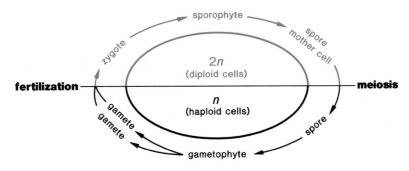

FIGURE 12.6   A typical life cycle of plants or related organisms that undergo sexual reproduction.

spore (sexual spore or meiospore), which may develop into a body or structure within which gametes may be produced by mitosis.

7. Any cell having one set of chromosomes is said to be haploid or to have *n* chromosomes; any cell having two sets of chromosomes is said to be diploid or to have 2*n* chromosomes.

8. In the life cycle of an organism that undergoes sexual reproduction, there is an alternation between a haploid phase and a diploid phase. The haploid body is called a *gametophyte* and the diploid body is called a *sporophyte*.

9. The change from the haploid phase to the diploid phase occurs when two gametes (each *n*) unite, forming a zygote (2*n*) in the process of fertilization (syngamy). The change from the diploid to the haploid phase occurs as a result of a spore mother cell (meiocyte) undergoing meiosis, when a 2*n* cell becomes four *n* cells. This switching of phases is referred to as Alternation of Generations.

Review Questions

1. Indicate when during meiosis each of the following events occurs: (a) crossing-over, (b) chromatids separating at their centromeres and migrating to opposite poles, and (c) chromosomes aligning themselves in pairs at the equator.

2. Is there any difference between mitosis and the second division phases of meiosis?

3. What is the significance of meiosis with respect to sexual reproduction?

4. What is meant by saying that the cells of a sporophyte phase are diploid or 2*n*?

5. At what stage of a life cycle does the chromosome number of cells switch from *n* to 2*n*?

Discussion Questions

1. If an organism reproduces very freely by asexual means, is there any advantage to its also reproducing sexually?

2. Would it make any difference if the events of the second division of meiosis occurred before the events of the first division?

3. Would it be correct to say that a bivalent has four times the amount of DNA present in the chromosomes of a cell in anaphase II?

4. Would you assume that the length of a chromosome might have something to do with the number of crossovers it might form with its homologue?

5. Two mitotic divisions may take place in as little as a few hours, while meiosis may take much longer. What reasons can you suggest for this?

Additional Reading

Becker, W. M., and D. W. Deamer. 1991. *The world of the cell.* Redwood City, CA: Benjamin/Cummings Publishing Co., Inc.

Darnell, L. 1991. *Cell biology and composition.* New York: W. H. Freeman.

DeRobertis, E. D. P., and E. M. DeRobertis Jr. 1987. *Cell and molecular biology,* 8th ed. Philadelphia: Lea and Febiger.

Herrmann, H. 1990. *Cell biology.* New York: HarperCollins Pubs., Inc.

John, B. 1990. *Meiosis.* New York: Cambridge University Press.

Raven, P. H., R. F. Evert, and S. E. Eichorn. 1992. *Biology of plants,* 5th ed. New York: Worth Publishers.

Wolfe, S. L. 1993. *Cell and molecular biology.* Belmont, CA: Wadsworth Publishing Co.

Narrow-tubed monkey flowers (Mimulus angustatus), *a vernal pool species of northern California. (Courtesy Robert A. Schlising)*

13 Genetics

When a plant is being grown as a crop, the more parts of the plant that are edible or otherwise useful, the more efficient the use of the land on which the crop is grown. Some parts of many crop plants are not used, however, except possibly as ingredients in compost. The roots of lettuce, for example, are not useful for either food or fuel. In tomatoes and grapes, only the fruits are edible; in potatoes, only the tubers are used, and in cotton, the fibers and the seeds constitute the useful parts.

In 1928, the Russian biologist G. D. Karpechenko reported on an experiment in which he tried to cross two food plants in the Cabbage Family: the radish, with an edible root, and the cabbage, with edible leaves. One might assume that such a cross, if successful, would produce an ideal crop plant. Karpechenko found that these two plants, although related, did not cross readily, but he did succeed in obtaining a few seeds from such crosses. When these seeds were planted, the results were dramatic. Huge plants grew and produced fertile seeds of their own. From a crop point of view, however, the cross was a dismal failure because these large plants had leaves mostly like those of a radish and roots like those of a cabbage—they were virtually useless!

Such crossing experiments are a part of **genetics,** the science that deals with heredity, or natural inheritance. Genetics is one of the youngest of the biological sciences, having developed after the details of meiosis and mitosis became understood at the turn of the century. It has since become a vast and very important field of study, having the potential not only for improving food production in a hungry world but also for solving various problems related to the inheritable human diseases, the effects of radiation, the control of population growth, and many other matters of direct human interest and concern.

MENDEL AND HIS WORK

Genetics as a science originated about a generation before its significance became appreciated in the scientific community as a whole. An Austrian monk, Gregor Mendel (Fig. 13.1), taught between 1853 and 1868 in what became the Czechoslovakian city of Brno. In the monastery there, he carried on a wide range of studies and experiments in physics, mathematics, and natural history. He also became an authority on bees and meteorology and kept notes on experiments involving two dozen different kinds of plants. Today, he is best known for the studies he conducted with a number of varieties of pea plants.

Mendel published the results of his studies on pea plants in a biological journal in 1866, and he sent copies of his paper to leading European and American libraries. He also sent copies to at least two eminent botanists of the time.

FIGURE 13.1   Gregor Mendel.

(Photo ca. 1870s, Brünn, Moravia. Courtesy, Brooklyn Botanic Gardens, New York. Print at Hunt Institute, Pittsburgh, PA)

His work, nevertheless, was completely ignored or overlooked until 1900, when three botanists (Eric von Tschermak of Austria, Carl Correns of Germany, and Hugo de Vries of Holland), working independently in their own countries, reached the same conclusions as Mendel. Each, as a result of library research, came across Mendel's original paper.

Peas, unlike most flowering plants, are self-pollinating and self-fertile. In other words, a pollen grain of a pea flower is capable of germinating on its own stigma, fertilizing one of its own ovules, and developing a fertile pea seed. Although Mendel did not know about mitosis and meiosis, he had observed that if a pea plant grew to a certain height, its offspring, if grown under the same conditions, would always reach approximately the same height, generation after generation. He wondered what would happen if he crossed a tall plant with a short one. Would the offspring be intermediate in height?

To make such a cross, Mendel would need to prevent self-pollination by reaching into a flower of one plant and removing the stamens before the pollen had matured. Then he could take pollen from another plant and apply it by hand to the stigma of the first flower. He also needed to do something to prevent insects from accidentally bringing pollen from other flowers after the cross had been made, and so he covered the experimental flowers with a cloth or bag.

When Mendel made such crosses, the results were astonishing. All of the offspring were tall. There were no short or intermediate plants. He found, however, that if he allowed the offspring plants to pollinate themselves, their offspring, in turn, occurred in a ratio of approximately three tall plants to one short plant (Fig. 13.2).

Mendel then tried crosses between peas with smooth seeds and those with wrinkled seeds. He also crossed green-seeded plants with yellow-seeded ones and ultimately used seven different pairs of contrasting characteristics in all.

The original plants involved in making the crosses were referred to as the *parents,* or the *parental generation,* and their offspring were referred to as the *first filial generation* (*filial* means "of or relating to a son or daughter"), usually abbreviated to **F₁**. The offspring of F_1 plants were called the *second filial generation,* or **F₂.**

Mendel counted the number of offspring in each cross, something none of his predecessors had done. In addition, he studied the inheritance of single pairs of contrasting characteristics, succeeding where others had failed because he did not try to interpret inheritance in terms of whole organisms.

Besides being astute and meticulous in his work, Mendel was also fortunate in choosing the plants and the characteristics he did, because we now know that many other characteristics are controlled by complex factors that do not give the simple, clear-cut results he obtained. Mendel was also exceptionally fortunate in choosing peas for his experiments because, although he did not know it, peas have just seven chromosomes in each set, and each of the pairs of characteristics he studied is carried on different homologous chromosomes. A summary of the data accumulated by

Mendel in his work with peas is provided in Table 13.1. As Mendel began to collect data in his experiments, he realized that there must be something inside the pea plants that made them have yellow or green seeds and red or white flowers. He referred to this unknown agent as a *factor*. He also deduced that each plant must have two such factors for each characteristic, since even though all the offspring of an F_1 generation appeared more or less identical, the F_2 generation revealed some plants with one characteristic and others with the contrasting characteristic. These discoveries and deductions came to be a principle or law known as the *law of unit characters*. Stated simply, this law says that "factors, which always occur in pairs, control the inheritance of various characteristics." Today, it is known that the paired factors, which later became known as **alleles,** are **genes** (discussed on page 24). The alleles are always at exactly the same position (*locus*) on homologous chromosomes. We also know now that there may be hundreds or even thousands of alleles on each pair of homologous chromosomes.

From his data and observations, Mendel also deduced that one factor, or allele, of a pair may overcome or conceal the expression of the other. For example, in the F_1 generation of a cross involving green-podded and yellow-podded parents, all the plants had green pods, yet both kinds of parents obviously had to contribute something to the cross since some of the F_2 plants had green pods while others had yellow pods. This deduction led to another principle known as the *law of dominance*. This principle says that "in any given pair of factors or genes, one may suppress or mask the expression of the other." The expressed factor (allele) is referred to as the **dominant,** while the one not expressed is referred to as the **recessive.** In the green-podded × yellow-podded cross, the green pod allele is dominant and the yellow pod allele is recessive.

Mendel's crosses also made it clear that a plant's having green pods did not indicate whether both members of the pair of factors (alleles) controlling pod color were present, because the dominant could suppress or mask the presence of the recessive one. To distinguish between the appearance of an organism and the genetic information it contains, we use the terms **phenotype** and **genotype.** *Phenotype* refers to the physical appearance of the organism, and *genotype* refers to the genetic information controlling the phenotype. We customarily use words to describe phenotypes and letters to designate genotypes.

In Mendel's crosses, for example, seven pairs of phenotypes are shown as parents—these phenotypes include yellow seeds, green seeds, smooth seeds, wrinkled seeds, green pods, yellow pods, and so on. In designating the genotypes of each of these plants, a capital letter is used to indicate the dominant member of a pair of alleles; the same letter in lowercase is used to indicate the corresponding recessive. In the green-podded × yellow-podded pea cross, for example, green is dominant. If a plant has a green phenotype, its genotype could be either *GG* or *Gg.* The yellow genotype would be *gg.* The plant is said to be **homozygous** if both alleles of a pair are identical (e.g., *GG* or *gg*) and

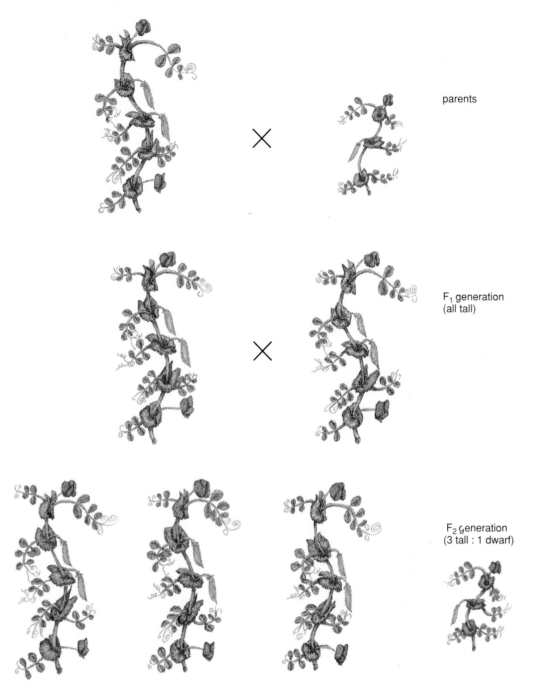

parents

F₁ generation
(all tall)

F₂ generation
(3 tall : 1 dwarf)

FIGURE 13.2 A cross between a tall variety and a dwarf variety of peas.

heterozygous if the pair is composed of contrasting alleles (e.g., *Gg*).

Although the principle of dominance was amply demonstrated by Mendel's crosses, we know now that in many other instances neither member of a pair of alleles completely dominates the other. In other words, in some species, there is an absence of dominance. In snapdragons, for example, the F_1 plants of a cross between a red-flowered parent and a white-flowered parent are all pink-flowered because the characteristics of both alleles blend, yielding an intermediate condition. When an F_2 generation is produced

from such a cross, the flowers of the offspring appear in a ratio of 1 red: 2 pink: 1 white. This lack of dominance is also referred to as *incomplete dominance* (Fig. 13.3).

Recall that the cells contributed by the parents in sexual reproduction (egg and sperm in higher organisms) are called *gametes* and that the offspring develop from *zygotes* formed when the gametes unite. Also recall that gametes have only one set (half the number) of chromosomes present in the body cells of the organism. When meiosis occurs, the members of pairs of alleles become separated and are not matched up again until a zygote is formed. Mendel called

Table 13.1

SUMMARY OF MENDEL'S DATA

PARENTS	F₁	F₂
Yellow × green seeds	All yellow	6,022 yellow: 2,001 green
Smooth × wrinkled seeds	All smooth	5,474 smooth: 1,850 wrinkled
Green pod × yellow pod	All green	428 green: 152 yellow
Long stem × short stem	All long	787 long: 277 short
Axial flowers × terminal flowers	All axial	651 axial: 207 terminal
Inflated pods × constricted pods	All inflated	882 inflated: 299 constricted
Red flowers × white flowers	All red	705 red: 224 white

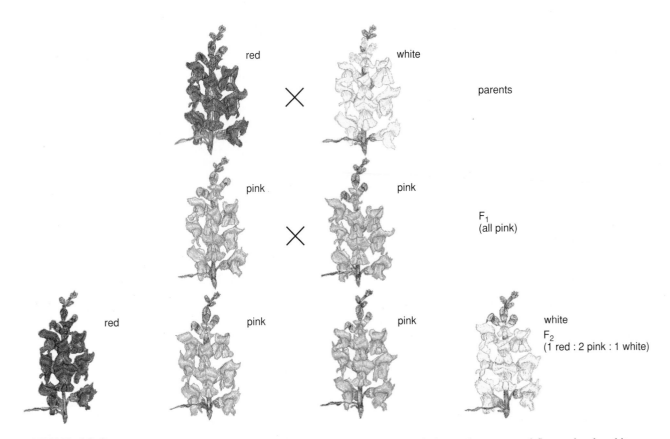

FIGURE 13.3   Absence of dominance (also referred to as incomplete dominance). A cross between a red-flowered and a white-flowered variety of snapdragons.

this separation of paired factors (alleles) into separate cells the *law of segregation.*

The Monohybrid Cross

With these first three laws in mind, we'll examine a cross between a homozygous green-podded pea plant and a homozygous yellow-podded pea plant (Fig. 13.4). The genotype of the homozygous dominant parent (green-podded) is ***GG;*** the

genotype of the homozygous recessive parent (yellow-podded) is ***gg.*** Recall that gametes have only one member of a pair of *homologous chromosomes.* The gametes of the green-podded parent (***GG***), therefore, will be ***G;*** the gametes of the yellow-podded parent (***gg***) will be ***g.*** No matter which egg unites with any sperm of the other parent, all the zygotes of this cross will be heterozygous, with the genotype ***Gg.*** This means that all the individuals of the F₁ generation will be green-podded phenotypes, and all will be, like the

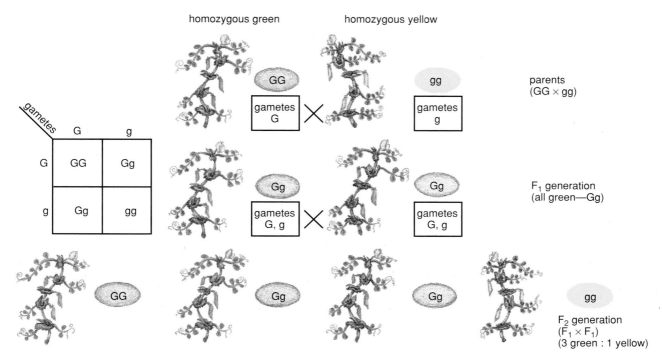

parents
(GG × gg)

F₁ generation
(all green—Gg)

F₂ generation
(F₁ × F₁)
(3 green : 1 yellow)

FIGURE 13.4 A monohybrid cross between a green-podded pea plant and a yellow-podded pea plant. Green (*G*) is dominant; yellow (*g*) is recessive.

zygotes, heterozygous. A cross of this type, involving two parents differing by only a single genetic trait, is called a **monohybrid** cross.

When making monohybrid crosses, one usually tries to cross members of the F_1 generation with one another to produce an F_2 generation. Each F_1 parent produces two kinds of gametes in roughly equal numbers. All of the F_1 plants involved are green-podded phenotypes, and their genotypes are all *Gg*. When gametes are produced, half will be *G* and half will be *g*.

A *G* egg may unite with either a *G* sperm nucleus, producing a *GG* zygote, or, purely at random, the same egg may unite with a *g* sperm nucleus, producing a *Gg* zygote. The same type of random unions occurs with *g* eggs so that either *Gg* or *gg* zygotes are produced. When all the offspring of a large number of such crosses are counted, a genotypic ratio of 1 *GG*: 2 *Gg*: 1 *gg* can be seen. The phenotypic ratio will be approximately three green-podded plants to one yellow-podded plant. These four equally possible F_2 offspring are shown at the bottom of Figure 13.4.

The Dihybrid Cross

Thus far in our discussion, the crosses have involved single pairs of alleles. But, homologous chromosomes may contain hundreds or even thousands of alleles. If crosses are made involving parents that are homozygous for two or more pairs of genes, the nature of meiosis dictates that the genes will not necessarily be inherited together. In fact, the two or more sets of genes are separated when meiosis occurs, and, if they are on separate chromosomes, they are combined again in the formation of zygotes. Another law of Mendel's, the *law of independent assortment,* recognizes this and states that the "factors [alleles] controlling two or more contrasting pairs of characteristics segregate independently and the gametes combine at random." This law is illustrated in **dihybrid crosses,** which involve two pairs of alleles located on separate pairs of chromosomes.

Let's assume that the alleles for green pods and yellow pods are located on one pair of chromosomes while those for tall plants and dwarf plants are located on another pair. Recall that tallness in peas is dominant and dwarfness is recessive. The phenotypes of the homozygous dominant parent will be green-podded and tall, while those of the recessive parent will be yellow-podded and dwarf; their corresponding genotypes will be *GGTT* and *ggtt*. All the gametes from the dominant parent will be *GT;* all those from the recessive parent will be *gt*. The dihybrid genotypes of the zygotes will all be *GgTt* and the phenotypes of the F_1 generation will all be green-podded and tall.

If these dihybrid members of the F_1 generation are then crossed with one another to produce an F_2 generation, four kinds of gametes will be produced in equal numbers by each parent because of the independent segregation that occurs during meiosis. These gamete genotypes are *GT, Gt, gT,* and *gt.* Since any one of the four kinds of gametes can unite randomly with any of the four kinds of gametes of the other parent, 16 combinations are possible.

In order to avoid the confusion of having to remember all possible combinations of gametes, a diagram called a *Punnett square* is used to determine the genotypes of the zygotes. The Punnett square diagram looks somewhat like a checkerboard, with one set of gametes across the top and the other set down one side (Fig. 13.5).

Nine different zygote genotypes are possible, produced in a ratio of 1 *GGTT:* 2 *GGTt:* 2 *GgTT:* 4 *GgTt:* 1 *GGtt:* 2 *Ggtt:* 1 *ggTT:* 2 *ggTt:* 1 *ggtt.* Four kinds of phenotypes are possible in a ratio of 9 green-podded, tall: 3 green-podded, dwarf: 3 yellow-podded, tall: 1 yellow-podded, dwarf. These ratios are expected when large numbers of individuals are involved. With small numbers, chance alone may not produce the expected ratios. Human populations, for example, are divided relatively equally into males and females, but in smaller groups such as families, the general population ratio of one male to one female may not be evident.

The Backcross

If a scientific theory is valid, one should be able to test it experimentally. Mendel himself tested his predictions by means of **backcrosses.** He had found in pea flowers that red color is dominant and that white is recessive (as contrasted with snapdragons in which color dominance is lacking), so all the F_1 offspring of a cross between a homozygous red-flowered parent and a homozygous white-flowered parent would be red. The genotypes of such a cross would be *RR* (red) × *rr* (white), and the F_1 generation would be all *Rr.* The gametes of such F_1 offspring would be either *R* or *r*, with each type being produced in equal numbers. If pollen from an F_1 hybrid is placed on the stigma of a white-flowered plant (which, since it shows the recessive trait, has to be homozygous), there is one chance in two that a dominant (*R*) gamete will form a zygote with a recessive (*r*) one and an equal chance that two recessive (*r*) gametes will form a zygote, with the offspring being produced in equal numbers of heterozygous red (*Rr*) and white (*rr*).

Backcrosses involving the homozygous recessive parent and the F_1 offspring are routinely made today, and invariably they yield the same results, namely, a 1:1 ratio of phenotypes. The same 1:1 ratios are produced even when there is no dominance. In snapdragons, for example, when the pink-flowered F_1 offspring are backcrossed with a red-flowered parent, a ratio of 1 red: 1 pink results. If the white parent is used, the backcross produces a ratio of 1 white: 1 pink.

Linkage

There may be thousands of genes on each chromosome, and the closer the genes are to one another on any given chromosome, the more likely they are to be inherited together. The tendency of genes on the same chromosome to be inherited together is called **linkage.**

In 1906, just a few years after the basic details of meiosis became known, W. Bateson and R. C. Punnett (after whom the Punnett square is named) of Cambridge University in England became the first to report on linkage in sweet peas. They knew from earlier work with sweet peas that purple flower color was dominant and red was recessive. They also knew that pollen grains with a shape that was oblong in outline ("long pollen") were dominant and that spherical pollen grains ("round pollen") were recessive.

When Bateson and Punnett crossed a *homozygous purple, long* plant with a *homozygous red, round* plant, all the F_1 offspring were purple and long, as expected. When they crossed the F_1 plants with one another to obtain an F_2 generation, however, the F_2 offspring phenotypes were produced in numbers that differed markedly from the expected 9:3:3:1 ratio. Puzzled by these results, they tried a backcross, crossing F_1 plants (heterozygous purple, long) with the recessive parent (homozygous red, round). Again, the expected 1:1:1:1 ratio was not produced. Instead, they obtained a ratio of 7 purple, long: 1 purple, round: 1 red, long: 7 red, round. They could not explain adequately what they had observed. In 1910, however, T. H. Morgan, who had observed similarly puzzling ratios in his experiments with fruitflies, theorized that linkage and crossing-over were responsible, an explanation that is still considered valid today.

If the genes for purple and red and those for long and round were on separate pairs of chromosomes, the genotypes of the F_1 would all be *PpLl* and the homozygous recessive would be *ppll.* This should have produced 1 *PpLl:* 1 *Ppll:* 1 *ppLl:* 1 *ppll* instead of the 7:1:1:7 ratio actually obtained. Apparently, the F_1 parent produced many more *PL* and *pl* gametes than *Pl* and *pL* gametes. If we designate the parents in the backcross $\frac{PL}{pl}$ and $\frac{pl}{pl}$ to indicate that *P* and *L* are on one chromosome and *p* and *l* are on another, we would expect the F_1 parent $\frac{PL}{pl}$ to have produced just two kinds of gametes, namely *PL* and *pl,* and the backcross to have produced only two kinds of phenotypes—purple, long and red, round—in equal numbers. But some purple, round and red, long phenotypes also were produced. The only plausible explanation for this seems to be that some *crossing-over* (see prophase I of meiosis in Chapter 12) occurs, with *P* and *L* genes occasionally being switched with *p* and *l* genes.

Chromosomal Mapping

As a result of thousands of experiments with many organisms, particularly fruitflies and corn, it is now known that crossing-over between particular pairs of linked genes occurs regularly. The frequency of crossing-over varies widely, depending on the genes involved. We must conclude, therefore, that each gene has a specific location (*locus*) on a chromosome. Indeed, by calculating percentages of crossing-over through counting large numbers of offspring phenotypes, we can effectively map the relative locations of genes on chromosomes. If, for example, we assume that the space within which 1% crossing-over occurs is one unit of map

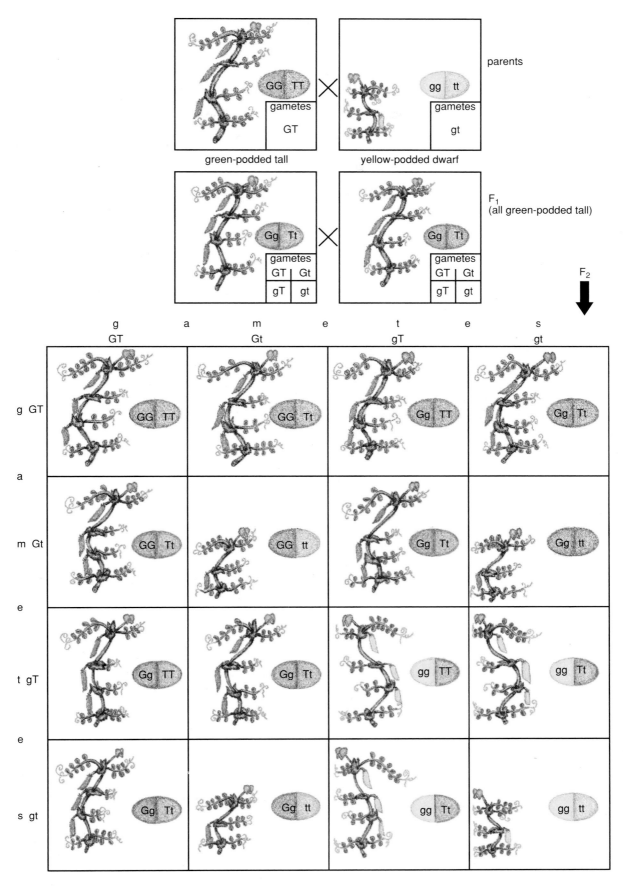

FIGURE 13.5 A dihybrid cross between a green-podded, tall pea plant and a yellow-podded, dwarf pea plant. Green (**G**) and tall (**T**) are dominant; yellow (**g**) and dwarf (**t**) are recessive.

distance between linked genes, then 7% crossing-over between genes A and B would mean that they are 7 units of map distance apart on their chromosome.

A ◄——— 7 units ———► B

If we find there is 10% crossing-over between A and a third linked gene C and that B and C are shown by crossover percentages to be 17 units apart, we can deduce that A lies between B and C on its chromosome.

C ◄——— 10 units ———► A ◄— 7 units —► B
17 units

However, if there is only 3% crossing-over between B and C, we have to conclude that B lies between A and C.

C ◄——— 10 units ———► A
◄ 3 units ► B ◄— 7 units —►

In Bateson and Punnett's backcross, two of every 16 F_2 offspring (12.5%) were produced through recombinations due to crossing-over. We can deduce from this that the genes controlling flower color and pollen shape in sweet peas are 12.5 map units apart on the chromosomes. The locations of many hundreds of pairs of genes have been determined and mapped in corn and in fruitflies (Fig. 13.6), and the mapping of chromosomes in other organisms is currently being undertaken at institutions around the world.

The Hardy-Weinberg Law

Before Mendel's work became known, most biologists believed that inherited characteristics were a blending of those furnished by the parents, and it was difficult to understand why unusual characteristics did not eventually become so diluted that they essentially disappeared. After Mendel, biologists asked why dominant genes did not eventually completely eliminate recessive ones in breeding populations. G. H. Hardy, a mathematician, and W. Weinberg, a physician, pointed out the reason in 1908, and their observation became known as the *Hardy-Weinberg law*.

The Hardy-Weinberg law, which essentially specifies the criteria for genetic equilibrium in large populations, states that "the proportions of dominant genes to recessive genes in a normally interbreeding population will be retained from generation to generation unless there are forces that change those proportions."

In small populations, there can be a random loss of individual genotypes if, for example, breeding does not take place or if breeding is not totally random with respect to genotype. This can lead to loss of a gene.

The proportions can also be upset by mutation. In mutation, a particular gene mutates to another form and then mutates back to its original form at a slower rate.

Selection, however, is by far the most significant cause of changes in the proportions of dominant and recessive genes in a population. The manner in which selec-

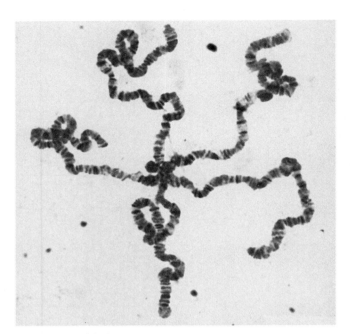

FIGURE 13.6 Salivary gland chromosomes of a common fruitfly. The staining bands seen on the chromosomes have been correlated, in some instances, with specific genes through chromosome mapping.

(Photomicrograph courtesy Wesley Dempsey)

tion operates is discussed under the topic of evolution in Chapter 15.

Interactions Between Genes

Most of our discussion of genetic inheritance to this point has centered around pairs of alleles that govern single sets of contrasting characteristics. All gene action, however, involves the production of proteins and enzymes that appear to affect the expression of characteristics in addition to those directly controlled by given pairs of alleles, and the expression of only a small minority of characteristics is actually controlled by a single pair of alleles.

Instead, unpaired genes evidently influence each other in various, sometimes subtle, ways and so the phenotype is seldom the direct expression of all its alleles. One allele, for example, may influence the expression of several different characteristics, or a single allele may suppress the expression of another allele in a different pair. Also, many genes, in proportion to the number present, may influence a single characteristic. For example, the more genes there are involved in the expression of fruit size and color, the larger or more intensely colored the fruit may become and vice versa. Trying to understand how such gene interactions take place is part of the ongoing research conducted by today's geneticists.

MOLECULAR GENETICS

As indicated in earlier chapters, genetic information is carried and passed from one cell to another in precise fashion by large complex nucleic acids found primarily in the nuclei of living cells. These nucleic acids (DNA and RNA) are discussed further in the pages that follow.

Structure of DNA

The thousands of atoms in a DNA molecule are found in chains of building blocks called *nucleotides*. Each nucleotide consists of three parts: (1) a nitrogenous base; (2) a 5-carbon sugar, deoxyribose; and (3) a phosphate group. Both the nitrogenous base and the phosphate group are bonded to the sugar (Fig. 13.7). Four kinds of nucleotides occur in DNA. Each kind has a unique nitrogenous base, but all have the same types of phosphate groups and sugar. Two of the nucleotide bases—*adenine* and *guanine*—are called *purines;* they have a molecular structure that resembles two linked rings. The other two nucleotide bases—*cytosine* and *thymine*—are called *pyrimidines;* they have a molecular structure consisting of a single ring.

Each nucleotide in a DNA molecule is bonded to the next one in such a way that the nucleotides form a chain, with the sugar of one attached to the phosphate group of the next; the nitrogenous bases are oriented as side groups on the chain. Each species has its own unique DNA, with the nucleotides occurring in a different sequence for each kind of DNA molecule. The possibilities for variety in the sequences seem virtually unlimited (Fig. 13.8).

In 1951, two cell biologists, James D. Watson and Francis Crick, initiated a series of brilliant investigations that eventually led to the construction of a model of a DNA molecule. When they began their work, they had certain information on which to start building their conclusions. They knew, for example, that Linus Pauling had postulated that the structure of DNA might be similar to the structure of protein and that he had shown part of the structure of some proteins to be helical and maintained by hydrogen bonds between the amino acids. They also knew, from studies by Erwin Chargraff and others at Columbia University the previous year, that there was a ratio of 1:1 between nucleotides with adenine and nucleotides with thymine in DNA molecules and that there was also a ratio of 1:1 between nucleotides with guanine and those with cytosine. Furthermore, they knew, from studies the same year by Rosalind Franklin and Maurice H. F. Wilkins, in which specialized X-ray equipment was utilized, that a DNA molecule was composed of regularly repeating units that appeared to be helically arranged.

When they had pieced together all the facts, Watson and Crick concluded that a DNA molecule consists of a huge, entwined double helix exactly 20 angstrom units wide, the strands of which appear to be wrapped around an invisible pole—one strand spiraling in one direction and the other strand spiraling at the same angle in the opposite direction. They also concluded that the nucleotides are linked in ladderlike fashion between the two strands.

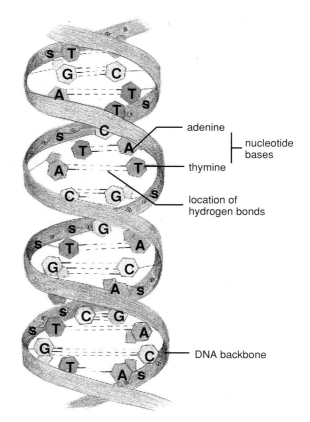

FIGURE 13.7 Structure of a DNA molecule. In this enlargement of a small portion of a DNA molecule, the rungs of the twisted ladder formed by the two entwined spiraling strands consist of nitrogen-containing bases supported by alternating units of sugar (S) and phosphate (P) molecules. The purines adenine or guanine (A or G) occur opposite the pyrimidines thymine or cytosine (T or C). The purines and the pyrimidines opposite each other are held together by hydrogen bonds linking the nitrogenous bases of the paired molecules. The double helix is 2 nanometers wide.

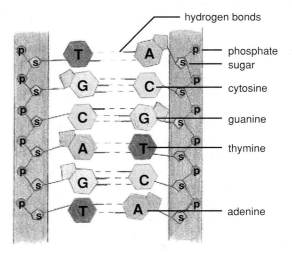

FIGURE 13.8 The pairing of nucleotides in a tiny portion of a strand of DNA. The variations in sequences of pairs are virtually unlimited.

Watson and Crick knew, however, that if purines were linked across the strands of the helix, the size of the combined molecules would exceed the 20 angstrom width of the DNA molecule and that two linked pyrimidines would not be wide enough to reach all the way across. If a purine and a pyrimidine were linked to each other, however, the width would fit the model perfectly. Accordingly, they concluded that the ladder rungs have to consist exclusively of purine-pyrimidine pairs. This meant that adenine-cytosine and guanine-thymine pairings could not occur, but adenine-thymine (or thymine-adenine) and guanine-cytosine (or cytosine-guanine) could occur. This nicely explained the 1:1 ratios shown by Chargraff and his colleagues, and such a model was proposed in 1953. Watson, Crick, and Wilkins shared a Nobel Prize for their work, Franklin having died before she, too, could be a part of the honor. Subsequent research has continued to support their conclusions to the extent that their model of a DNA molecule is now accepted as an authentic representation.

Replication (Duplication) of DNA

If DNA is indeed the determinant of inheritable characteristics, it must be able to duplicate itself precisely in order to pass along information from generation to generation. Since all DNA molecules have only four different nucleotides, the differences between the DNA of one organism and that of another must lie in the sequence of the four possible types of ladder rungs and their total number, as well as in the ratios of adenine and thymine to guanine and cytosine. If these four nucleotides are synthesized in living cells, what tells the cell exactly how to put them together to form the appropriate DNA molecules?

Watson and Crick observed that if the two strands of DNA are "unzipped" by an enzyme breaking the hydrogen bonds between the nucleotides down the middle of the ladder, each separated chain provides all the information needed for putting together a new ladder (Fig. 13.9). In other words, since guanine can pair only with cytosine and vice versa, a guanine nucleotide synthesized by the cell can lock on, or bond, only to a cytosine nucleotide, and an adenine nucleotide can bond only to a thymine nucleotide. If a cell "unzips" the two strands in its DNA molecules, separate nucleotides can line up next to each of the single chains in precise sequence and be bonded together to form a new chain, with the separate chains functioning as molds, or templates, for self-duplication. This theory of Watson and Crick, also proposed in 1953, has since been supported by evidence obtained by the use of nucleotides containing isotopes of carbon and nitrogen atoms. The evidence also reveals that a number of different enzymes are involved in catalyzing the replication of DNA.

RNA and Synthesis of Proteins

Once the chemical nature of DNA and the way in which it duplicates itself became understood, attention was focused on the manner in which DNA determines a cell's or an organism's function and structure. It was known that proteins are large complex molecules composed of 20 different kinds of amino acids linked together. If each of the four nucleotides in DNA coded only one amino acid, there could be no more than four amino acids. If all possible combinations of nucleotide pairs (16) were used to code amino acids, the number would still not be enough to code the 20 known amino acids. If a minimum of three nucleotides were involved for each amino acid, however, up to 64 combinations would be possible—more than enough to code 20 different amino acids. Such an arrangement was postulated and later confirmed; each trio of nucleotides along a DNA strand does indeed code a specific amino acid. It was also confirmed that most proteins are synthesized in the cytoplasm, while DNA is largely confined to the nucleus. Obviously, then, the DNA cannot function as a direct template for protein synthesis. Some intermediary agent or agents have to be involved.

In the early 1940s, it was suspected that ribonucleic acid (RNA) is involved in protein synthesis because it had been found that cells producing large amounts of protein contain large amounts of RNA and vice versa. RNA is similar to DNA, but it differs in its sugar component having an additional oxygen atom; it also has a different pyrimidine, *uracil*, instead of thymine. RNA usually occurs as a single strand and only very rarely as a completely double strand.

Specific Types of RNA and Their Functions

Messenger RNA (mRNA)

Messenger RNA is a large molecule composed of several hundred to several thousand nucleotides. A strand of mRNA forms along a portion of a DNA molecule like a row of jigsaw pieces fitting together with another row, with uracil taking the place of thymine. A sequence of three nucleotide bases, known specifically in mRNA as a *codon*, codes one amino acid. Codons can be read from one direction only, thus preventing the possibility of specifying a wrong amino acid. Once the mRNA molecule has been formed within the nucleus, enzymes split off or modify portions of the mRNA molecule. Specific enzymes revise or remove any parts that do not specify the correct protein sequence or are imprecise in any way. The mRNAs are then carried out into the cytoplasm through a nuclear pore.

In the cytoplasm, each mRNA unit becomes attached by one end to a ribosome, which is itself composed of another form of RNA (ribosomal RNA or rRNA) and protein. Usually, several ribosomes become associated with the mRNA molecule and move along in sequence, "reading" the coded information and participating in the bonding together of amino acids, which have been activated in the cytoplasm at roughly the same time. One mRNA molecule apparently is "read" by two or three such groups of ribosomes and then destroyed. It has been estimated that the average life of an

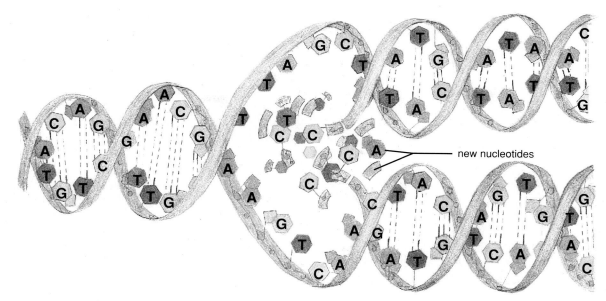

FIGURE 13.9   Replication (duplication) of DNA. A molecule replicates by "unzipping" along the hydrogen bonds, which link the pairs of nucleotides in the middle, and then each half serves as a template for nucleotides available in the cell. As the new nucleotides fit into the appropriate places and are linked together, a double helix is re-formed for each original half.

mRNA molecule in the much-studied, digestive-tract bacterium *Escherichia coli* is about two minutes, but it is generally much longer (up to several hours) in plant cells.

Ribosomal RNA (rRNA)

Ribosomes are composed of two somewhat spherical subunits, each having its own specific kinds of proteins and RNA. The larger of the two subunits contains a "slot" into which a transfer RNA (tRNA) molecule can fit. How one tRNA molecule replaces another in this slot has not yet been determined. Ribosomal RNA is formed along one strand of the DNA helix in the same manner as mRNA and tRNA molecules. Its role in protein synthesis is not certain, but it is presumed that the ribosome precisely positions amino acids, mRNA, tRNA, and the protein being formed as synthesis is taking place. Ribosomal RNA has a long life.

Transfer RNA (tRNA)

A molecule of tRNA consists of roughly 80 nucleotides bonded together in a single strand that doubles back on itself. There is at least one form of tRNA for each of the 20 amino acids. Each specific form of tRNA has an *anticodon loop,* where the strand doubles back on itself. This anticodon is a sequence of three nucleotides that will "recognize" a codon and pair with it on a strand of mRNA. An amino acid is bound on the open, or acceptor, end of the tRNA strand with the aid of an enzyme that specifically identifies both a single tRNA strand and its proper amino acid (Fig. 13.10).

Once the tRNA molecule has "captured" an amino acid, the tRNA becomes bound to a *ribosome* at the same point to which an mRNA molecule has become attached. Then, as the ribosome moves along the mRNA strand, the tRNA releases its amino acid and becomes detached, with another tRNA amino acid complex taking its place. In this way, the amino acids are lined up in sequence according to the mRNA code, and each is linked to the amino acid of the next tRNA molecule by a polypeptide bond until a complete protein molecule has been assembled. The tRNA thus relays amino acids in response to a specific "blueprint" coded by the DNA and carried by the mRNA. The amino acids of the various proteins of the cell are attached to each other precisely according to the information delivered from the nucleus. Specific enzymes ensure that the appropriate amino acids are incorporated within each protein.

After a tRNA strand becomes detached, it is then available for reuse in "capturing" another amino acid molecule. Apparently, tRNA molecules can go through this cycle repeatedly, although enzymes eventually may break them down. Transfer RNA, like rRNA, has a long life.

How Do Genes Produce Phenotypes?

We know how a single gene can control a specific enzyme for facilitating the production of color from a colorless precursor in the petals of a flower. But despite the vast amount of information gathered so far about specific genes and their actions, we still have little idea of how all the myriad of genes interact to produce something as "simple" as a bacterium (which has 600 or more proteins), let alone a complex plant such as a dandelion. It will be interesting to see the extent to which scientists are able to use modern technology and the Information Superhighway to unravel these mysteries during the 21st century.

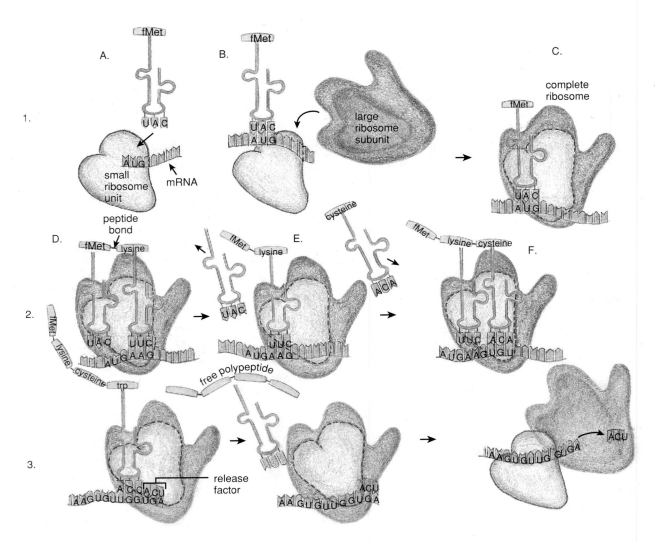

FIGURE 13.10 How a protein is synthesized. A protein is assembled with the aid of RNA molecules and ribosomes. Each ribosome consists of two subunits, one larger than the other. The process can be divided into three phases: 1. *Initiation. A.* An mRNA molecule from chromosomal DNA becomes attached to a small ribosomal subunit. *B.* A tRNA molecule bearing formylated methionine (Met) hydrogen bonds to the initiation codon (AUG) on the mRNA. *C.* The large ribosomal subunit locks into place. 2. *Elongation.* *D.* A second tRNA molecule, with an amino acid attached, moves into the ribosome, and its anticodon bonds to the mRNA. *E.* A peptide bond forms between the two amino acids that are now side by side. At the same time, the bond between the first amino acid and its tRNA molecule is broken, and the tRNA is released. *F.* As the first tRNA molecule is released, the second tRNA molecule moves into its place, and a third tRNA moves in behind it while another peptide bond is formed. A polypeptide chain builds as the process is repeated. 3. *Termination.* When a *termination codon* is reached, the polypeptide chain is separated from the last tRNA molecule, which is then released. Finally, a *release factor* triggers the separation of the two ribosomal subunits.

Summary

1. Genetics, the science of heredity and natural inheritance, includes crossing experiments, such as Karpechenko's cross of a cabbage and a radish.

2. The science of genetics originated with Gregor Mendel, an Austrian monk, who performed experiments with peas. Mendel's crosses involved pea varieties that had pairs of contrasting characteristics carried on a separate pair of homologous chromosomes. The offspring of the parental plants crossed were designated the F_1 generation, and the offspring of crosses between members of the F_1 generation were designated the F_2 generation. F_1 plants were also called hybrids.

3. Mendel, who was exceptionally fortunate in his choice of characteristics to study, meticulously counted the number of all offspring. He found that the F_1 hybrids

of his pea crosses all resembled their parents in appearance, but the F_2 offspring were produced in a ratio of three plants resembling one parent to one plant resembling the other parent.

4. Mendel called the agents for the characteristics inside the plants *factors* and deduced that each plant must have two factors for each characteristic. His law of unit characters states that "factors, which always occur in pairs, control the inheritance of various characteristics." The factors later became known as *alleles,* which are pairs of genes.

5. Mendel called the suppressing member of a pair of factors a *dominant* and its counterpart a *recessive.*

6. Phenotypes, which are described with words, denote appearance, and genotypes, for which letters are used, designate genetic makeup. Dominants are shown with capital letters and recessives with lowercase letters.

7. A homozygous plant has identical members of a pair of alleles. A heterozygous plant has contrasting members of a pair of alleles. The F_1 generation phenotypes may be intermediate between the parents as a result of absence of dominance (incomplete dominance).

8. The offspring of a monohybrid cross are produced in a ratio of 3:1, when dominants and recessives are involved. The offspring of a dihybrid cross involving dominance and two pairs of genes carried on separate pairs of chromosomes usually are produced in a phenotypic ratio of 9:3:3:1.

9. Backcrosses, which involve $F_1 \times$ the recessive parent, are used to check the results of genetic experiments. Backcrosses normally produce phenotypic ratios of 1:1 (monohybrid) or 1:1:1:1 (dihybrid).

10. Linked genes are inherited together. Bateson and Punnett discovered that expected 9:3:3:1 ratios in the F_2 phenotypic generation of certain sweet pea crosses did not occur. The 7:2:2:7 ratio they obtained was postulated to be the result of linkage and occasional crossing-over.

11. Chromosomal mapping involves calculating crossover percentages to determine the relative position of genes on chromosomes. The closer the genes are to each other on a chromosome, the less likely they are to be involved in a crossover and vice versa.

12. The Hardy-Weinberg law explains why recessive characteristics in a population do not eventually disappear. Selection is the most significant cause of changes in the proportions of dominant and recessive genes in a population.

13. Genes interact with one another, and so a phenotype is seldom the direct expression of its alleles.

14. Watson, Crick, and Wilkins were awarded a Nobel Prize in 1953 for developing a model of DNA that is now accepted as authentic. DNA replicates (duplicates itself) by "unzipping" down its nucleotide "ladder rungs" and having new nucleotides replace the old ones until two new chains of DNA have been formed.

15. New proteins are formed precisely according to coded information contained in the DNA of the nucleus by means of RNA, which occurs in three forms. Messenger RNA forms along part of a DNA molecule and is carried out of the nucleus into the cytoplasm, where it becomes attached to a ribosome. Ribosomes "read" the coded information and participate in assembling amino acids together to form a protein, with the aid of transfer RNA molecules. Ribosomes themselves are largely composed of ribosomal RNA.

Review Questions

1. When did genetics originate as a science?

2. Why did Mendel succeed where others had failed?

3. Who were Tschermak, Correns, and de Vries?

4. Define *hybrid, F_1 generation, phenotype, genotype, homozygous, heterozygous, dominant,* and *recessive.*

5. What are Mendel's laws?

6. In a variety of flowering plants, blue flowers and dwarf habit are controlled by dominant genes; white flowers and tall habit are the recessives. The gene pairs are carried on separate pairs of chromosomes. If a homozygous blue, dwarf plant is crossed with a homozygous white, tall plant,

   a. what are the phenotypes and genotypes of the F_1 generation?

   b. how many different *kinds* of genotypes could occur in the F_2 generation?

   c. what phenotypic ratio would you expect to occur in the F_2 generation?

   d. what genotypes could result from the cross *BBDD $\times$ BbDd?*

   e. what phenotypes could result from the cross *bbDd $\times$ BBDd?*

7. In summer squashes, disk-shaped fruit is dominant over spherical fruit and white fruit color is dominant over yellow fruit color. If a homozygous white, disk-shaped variety is crossed with a homozygous yellow, spherical variety, how many different homozygous genotypes are possible in the F_2 generation?

8. A cross is made between a white-flowered plant and a red-flowered plant of the same variety. The F_2 generation consists of 32 white-flowered plants, 64 pink-flowered plants, and 29 red-flowered plants. Explain.

9. In tomatoes, smooth skin is dominant and fuzzy skin is recessive; also, tallness is dominant and dwarfism is recessive. A homozygous smooth, tall variety is

crossed with a homozygous fuzzy, dwarf variety, and the F_1 is backcrossed to a homozygous fuzzy, dwarf variety. The offspring of the backcross are 95 smooth, tall; 3 fuzzy, tall; 4 smooth, dwarf; 96 fuzzy, dwarf. Is linkage involved? If so, what is the percentage of crossing-over?

10. In squashes, white color, which is controlled by a single gene, is dominant, and yellow is recessive. Give the phenotypes and the genotypes for each of the following crosses: *Ww × Ww; Ww × ww; WW × ww.*

11. If red flowers are dominant and white flowers are recessive in an *annual* variety of *self-pollinating* beans, assume that you plant one heterozygous bean seed on an island, and it germinates and thrives. What will the colors and ratios of the phenotypes be in the F_4 generation?

12. In snapdragons, red flower color shows incomplete (lack of) dominance over white flower color. If you wanted to produce seeds that would produce only pink-flowered plants when sown, how would you go about it?

13. If one plant is homozygous for three dominant characteristics and another plant is correspondingly recessive for the same characteristics, what proportion of the F_2 generation will resemble each parent if the two plants are crossed?

14. What would happen in the production of gametes if a species had an odd number of chromosomes in each of its cells?

15. What is meant when genes are said to be linked?

16. What is chromosomal mapping and how is it done?

17. What is the Hardy-Weinberg law?

18. Describe the structure and duplication of DNA.

19. How are proteins synthesized?

20. What is the nature and function of each form of RNA?

Discussion Questions

1. Mendel is said to have been exceptionally lucky in his discoveries. Do you think most scientists who make significant discoveries are lucky? Explain.

2. The peas Mendel worked with were largely self-fertile. Of what advantage or disadvantage to a species is such a phenomenon?

3. Does DNA really control the synthesis of proteins? Explain.

Additional Reading

Brown, T. A. 1990. *Genetics: A molecular approach.* New York: Chapman and Hall.

Gardner, E. J., et al. 1991. *Principles of genetics,* 8th ed. New York: Wiley.

Grant, V. 1991. *The evolutionary process,* rev. ed. New York: Columbia University Press.

Klug, W. S. 1994. *Concepts of genetics,* 4th ed. New York: Macmillan.

Maynard-Smith, J. 1989. *Evolutionary genetics.* New York: Oxford University Press.

Olby, R. C. 1985. *The origins of Mendelism,* 2d ed. Chicago: University of Chicago Press.

Otte, D., and J. A. Endler (Eds.). 1989. *Speciation and its consequences.* Sunderland, MA: Sinauer Associates, Inc.

Russell, P. J. 1992. *Genetics,* 3d ed. New York: HarperCollins College Pubs., Inc.

Tamarin, R. 1991. *Principles of genetics,* 3d ed. Dubuque, IA: Wm. C. Brown Publishers.

Chapter Outline

A flower of butterfly pea (Clitoria ternata), *a tropical vine whose flower construction ensures that both the anthers and pistil touch the backs of visiting insects. The seeds are believed to be toxic to livestock.*

Plant Biotechnology and Propagation

14

After a brief introduction to genetic engineering, the techniques and applications of recombinant DNA technology are discussed. Then some uses of hybridization, polyploidy, and mutations in plant breeding are explored, and the basics of tissue culture and mericloning are reviewed. The chapter closes with an overview of the principal types of vegetative propagation and grafting.

Some Learning Goals

1. Understand the functions of enzymes involved in the development of recombinant DNA.
2. Be able to delineate the basic steps involved in gene splicing.
3. Know several applications of genetic engineering.
4. Learn the roles of hybridization, polyploidy, and mutations in traditional plant breeding.
5. Understand the positive and negative aspects of the "Green Revolution."
6. Be able to explain tissue culture, mericloning, and related techniques to others.
7. Learn the basic types of vegetative propagation.

Some institutions with important around-the-clock services, such as hospitals and hotels, often have backup generators, which function during power failures. Backup generators probably will be needed for many types of power equipment throughout the foreseeable future, but a scientist at the University of California at Davis has predicted that, within a few years, backup generators for some types of lighting may become obsolete. He believes shrubs that glow in the dark will line airport runways, freeways, and sidewalks. The glowing shrubs are expected to be developed, through the use of *genetic engineering* techniques, by introducing into the shrubs genes from luminescent bacteria (Fig. 14.1). The initial glow is expected to be intensified, also through genetic engineering.

The genetic engineering techniques involved in the development of new crops, medicines, disease control or elimination, waste management, oil spill cleanups, and shrubs that glow, as well as a varied assortment of manipulations or uses of plants or plant cells and tissues, are all part of the field of **biotechnology.**

Prior to the 1970s, most plant biotechnology involved traditional forms of plant breeding, such as hybridization (crossing of different varieties) and selection for desirable traits. Such breeding involved transfer of pollen from anthers to stigmas, which, in turn, normally brings about a sexual blending of characteristics. Modern biotechnological techniques, however, are largely asexual, providing much more precise control over the features expressed in the offspring. A brief look at both traditional and modern biotechnology forms the body of this chapter.

GENETIC ENGINEERING OR RECOMBINANT DNA TECHNOLOGY

If genes for desirable traits in plants (e.g., superior nutritional quality, metabolic efficiency, capacity for nitrogen fixation) are isolated from selected plants and can be incorporated into bacteria, the bacteria may produce large amounts of recombinant DNA for introduction into other plants. If the desirable trait has been successfully incorporated, the plants, in turn, can be propagated asexually.

The process of genetic engineering, or gene "splicing," involves removing a gene from its normal location and inserting it into a circular strand of DNA; the restructured DNA is then transferred into cells of another species.

The procedure begins with the isolation of pure bacterial plasmid DNA. A **plasmid** is a small, circular DNA fragment. As indicated in Chapter 17, bacterial cells differ from other cells in having no nucleus or membrane-bound organelles, such as plastids or mitochondria. Each bacterial cell does, however, have a long strand of DNA and also usually up to 40 or more plasmids.

Isolation of Plasmid DNA

Isolation of DNA is accomplished by breaking up the bacterial cells and extracting the DNA material with solvents such as phenol. The chromosomes can be separated from the plasmids by placing the extracted material in salt solutions that precipitate the chromosomes but not the plasmids.

Restriction Enzymes

Next, the linkages between adjoining nucleotides of plasmids are broken by special bacterial enzymes known as *restriction enzymes,* which break a circular plasmid at a specific nucleotide sequence, leaving a linear strand with "sticky" ends where the nucleotides are unpaired. The ends are called *sticky* because they attract complementary nucleotide sequences.

Repair Enzymes

Recombinant plasmids are made by mixing large numbers of plasmid DNA segments with fragments of the desired DNA, which has been broken by the same restriction enzyme so that the sticky ends are complementary to each other. When so mixed, a fragment from one source may associate with a fragment from the other source, and with the aid of a *repair enzyme* (DNA ligase), the two fragments may link, yielding a circular recombinant DNA molecule with parts of both original molecules. The recombinant plasmids are then inserted into bacterial cells where they replicate, thus introducing new traits into a particular strain of bacteria. The new inserted gene can make a new protein (Fig. 14.2).

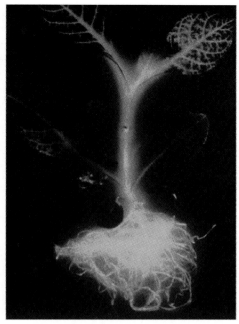

A.

B.

FIGURE 14.1 A tobacco plant that has been genetically engineered to contain genes from a firefly. The plant glows when it is bathed in ATP, which provides the energy that activates the enzyme luciferase. The luciferase "turns on the glow" of luciferin.

(*A.* Courtesy Keith V. Wood; *B.* From D.W. Ow et al. 1986. "Transient and stable expression of the firefly luciferase gene in plant cells and transgenic plants." *Science* 234 (November 14, 1986): 856–59. Copyright © 1986 by the AAAS.)

Protein Sequencers and Gene Synthesizers

Bacteria are easily propagated, and it is relatively simple to obtain large amounts of DNA with enormous potential for scientific studies. The simplest bacterium, however, has at least 600 different proteins, and in the past it has taken several months and considerable expense to isolate a desirable protein from all the others.

The development during the 1980s of two machines, *protein sequencers* and *gene synthesizers,* has dramatically reduced the costs in time and funds for *cloning,* or producing genetically identical genes. Protein sequencers can reveal the precise sequence of amino acids in a protein. With that information, one can produce synthetic gene fragments that code for the protein. The fragments can then be introduced into the DNA of bacteria. Modified bacteria can be cultured to produce large quantities of the desired protein.

Cell Bombardment and Electroporation

Cell bombardment, an increasingly used method of gene transfer, relies on shooting DNA-coated particles into plant cells. The new DNA becomes incorporated in about 2% of the cells. Monsanto Company has used cell bombardment to introduce into corn a gene from BT (*Bacillus thuringiensis*) bacteria that makes the corn resistant to the European corn borer—a major pest that annually damages about 20% of the U.S. corn crop at a cost of nearly $1 billion.

In 1995, approval was also given for the release of a potato that was genetically altered to produce in its leaves the BT gene that kills Colorado potato beetles. BT genes are effective against a wide variety of caterpillars but harmless to almost all other organisms. It is anticipated that attempts will also be made to control in this fashion tomato hornworms, cabbage loopers, and similar pest larvae of many other food plants.

Electroporation involves the incubation of cells that are placed in an electrical field. Short electrical pulses cause the cell walls to become sufficiently permeable to admit foreign DNA, which is introduced without force into the cells.

These methods of gene transfer have the advantages of reliability and broad application; the disadvantages are that complex DNA insertions can be time-consuming and expensive; copies of DNA also can become rearranged in an undesirable manner.

Other Applications of Genetic Engineering

A common soil bacterium, *Agrobacterium tumefaciens,* which causes slow-growing tumors known as *crown gall* in woody dicots, has the capacity to use tDNA to transfer DNA into plant cells. The tumor-inducing plasmid can be altered by the deletion of the disease-causing parts and their

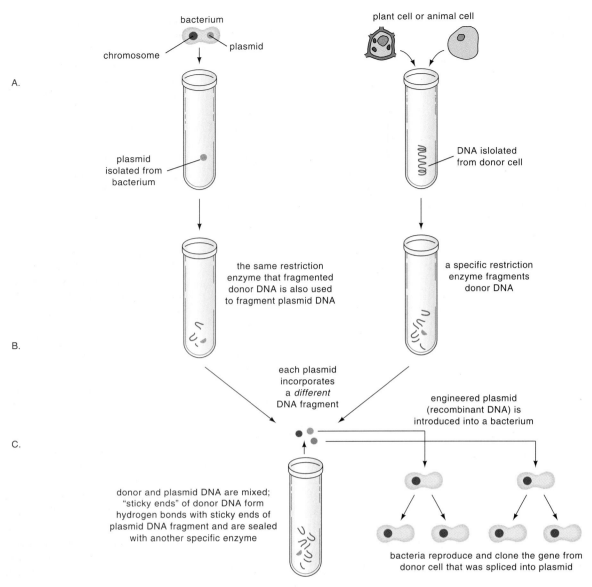

FIGURE 14.2 How recombinant bacteria are produced. *A.* DNA isolated from a plant of animal cell, and a plasmid isolated from a bacterial cell, are fragmented with the same restriction enzyme and then mixed. *B.* Some "sticky ends" of DNA fragments from both cells bond together, forming recombinant DNA. *C.* The recombinant DNA is inserted into a bacterium, which multiplies the recombinant DNA as it divides.

(After Lewis, Ricki, *Life*. 1992. Wm. C. Brown Publishers, Dubuque, IA.)

replacement with desirable genes. Use of crown gall bacteria for gene transfer has the advantage, in many instances, of incorporating only one precisely defined segment of DNA into a plant cell. An important goal of agricultural genetic engineers is the adaptation of the crown gall bacterium gene transfer system to the improvement of cereal crops.

Bacteria have been used for many years in the production of industrial chemicals, but genetic engineering may greatly increase such use. More complex organisms have more complex genetic complements, however, and much additional research is needed before the elimination of diseases and defects in higher plants and animals becomes commonplace as a result of genetic engineering.

Viruses, some of which naturally infect plants, are also easily propagated and are subject to similar engineering techniques. It is possible that viruses, discussed in Chapter 17, will play a major role in the genetic manipulation of higher plants in the future.

Early Experiments

Pioneering experiments with higher animals and plants were already under way by the late 1970s and early 1980s. In 1980, for example, J. Gordon of Yale University injected thousands of recombinant plasmids into newly fertilized mouse eggs that were in a dish and then implanted the eggs in the oviducts of about 10 female mice. The mice produced 78 offspring, two of which proved to have traces of the new DNA in their cells.

In 1981, scientists of the University of Wisconsin and the U.S. Department of Agriculture reported transferring a gene from a French bean seed into a cell of a sunflower plant and called the tissue that developed after the transfer "sunbean." The gene, which directs major protein production, is stable in the sunflower cell, and high levels of bean protein production are anticipated when technology is developed to regenerate a sunflower plant from the "sunbean" cells.

Among other genetically engineered organisms developed during the 1980s are the so-called ice-minus bacteria, which produce a protein coat that deters the formation of ice crystals and reduces frost damage to crops. Although this organism has enormous potential for extending growing seasons and ranges for a variety of crops, its developers experienced strong opposition to field testing, even though they received permission to do so. Much of the opposition was based on a fear of the unknown and the belief that the release of genetically engineered organisms outside the laboratory could be potentially disastrous if unanticipated effects, such as uncontrollable growth or virulent mutations, were to occur.

Later Developments

The United States in 1986 withdrew and then reinstated the first license for a genetically engineered vaccine that contained a herpes virus that could not duplicate itself. To date, there is no cure for herpes, and a vaccine to prevent it would be a major step in the anticipated eventual suppression or elimination of it and numerous other diseases.

In 1995, a researcher at Texas A&M University reported transferring into tobacco and potato plants a gene that in mice creates antibodies that protect against hepatis B. The hepatitis B vaccine series currently costs $120, but with genetically engineered plants doing all the work, scientists believe the cost of the vaccine can be reduced to about $1. Such cost reduction is particularly significant for countries such as China, where 150 million mostly poor people carry the disease.

In 1986, the Environmental Protection Agency granted the Mycogen Corporation of San Diego, California, permission to conduct small-scale field tests of a genetically engineered pesticide. In 1989, a Cornell University entomologist sprayed cabbage plants with a genetically disabled virus specific for cabbage loopers. The purpose of the Cornell test was to see if the genetically disabled virus would disappear quickly after attacking insects. If this proves to be the case, scientists later hope to splice into the cabbage looper virus a new gene that will be more deadly to insect pests but will disappear after spreading quickly throughout the insects. If these experiments are successful, dependence on the roughly three million pounds of dangerous pesticides used annually in the United States could be significantly reduced.

In 1995, the Canadian Department of Agriculture and Agri-Food approved for spring planting in Saskatchewan two genetically altered varieties of canola seed that are capable of surviving otherwise fatal herbicides. As a result, it is believed one application of a biodegradable herbicide may effectively replace several herbicide sprays per season.

Scientists from the United States and Australia have produced insect-resistant seeds by introducing into common peas a gene that prevents weevils from digesting starch. The gene is active only while the seed is dormant and becomes inactive after germination. In 1995, second generation plants that are seed weevil-resistant were produced. In these plants, the weevils are not killed but essentially starve to death. It is hoped that the gene can now be introduced into the seeds of many other crops.

Doctors at the National Institutes of Health in Bethesda, Maryland, received federal approval in 1989 to begin injecting terminal cancer patient volunteers with billions of their own white blood cells that had been engineered with bacterial genes that function as a tracking device. Although the "tracking" genes are not therapeutic, they do serve to aid doctors in monitoring how well the cells are fighting cancer.

Earlier experiments of a slightly different nature with 20 victims of a rapidly growing skin cancer used white blood cells, known as tumor-infiltrating lymphocytes, that had been cultured outside the body. A remarkable shrinkage of the tumors was demonstrated in 12 of the patients.

In California, a genetically engineered sun lotion that promises to protect humans from three types of cancer-causing radiation is being developed.

A list of some of the medicinal drugs already produced with the aid of genetic engineering is shown in Table 14.1.

Calgene has also successfully engineered a tobacco plant that is highly resistant to Roundup, a common commercial herbicide. If most crop plants could be engineered to become that resistant, herbicides selective for only undesirable weed plants could be sprayed on crop fields without affecting the desirable plants. In 1994, a Roundup-resistant soybean was developed that first was released to farmers in 1996.

Plant pathologists at the U.S. Department of Agriculture have isolated genes that can be bred into popular varieties of beans to make them resistant to 33 kinds of rust fungus. This is especially important because rust fungi can mutate and become resistant to fungicides in a single season. It takes numerous breeding steps to introduce the full complement of rust resistance genes into the beans, but once the process has been completed, spraying for fungi may be eliminated or at least significantly reduced.

Certain fish that inhabit Arctic and Antarctic regions produce an organic form of antifreeze that prevents water molecules in their bodies from freezing. Genetic engineers have successfully cloned the gene responsible for the antifreeze, which they expect soon to be able to produce and introduce into ice cream to prevent it from forming ice crystals after it has slightly thawed and then been refrozen.

Table 14.1

DRUGS PRODUCED USING RECOMBINANT DNA TECHNOLOGY

DRUG	USE
Atrial natriuretic factor	Dilates blood vessels, promotes urination
Epidermal growth factor	Accelerates healing of wounds and burns; promotes healing of gastric ulcers
Erythropoietin	Stimulates production of red blood cells in treatment of anemia
Factor VIII	Promotes blood clotting in treatment of hemophilia
Fertility hormones (follicle stimulating hormone, luteinizing hormone, human chorionic gonadotropin)	Treats infertility
Human growth hormone	Promotes growth of muscle and bone in pituitary dwarfs
Insulin	Allows cells to take up glucose in treatment of diabetes
Interferon	Destroys some cancer cells and some viruses
Lung surfactant protein	Helps alveoli in lungs to inflate in infants with respiratory distress syndrome
Pegaspargase	Treats lymphoblastic leukemia
Renin inhibitor	Lowers blood pressure
Somatostatin	Decreases growth in muscle and bone in pituitary giants
Super oxide dismutase	Prevents further damage to heart muscle after heart attack
Tissue plasminogen activator	Dissolves blood clots in treatment of heart attacks and arterial blockages

Primary source: R. Lewis. 1992. *Life*. Dubuque, IA: Wm. C. Brown Publishers.

Some other ongoing research projects that involve genetically engineered organisms are the production of an enzyme that greatly reduces the cost and improves the quality of certain expensive antibiotics; the development of an anti-inflammatory agent previously produced in only trace amounts; the production of human insulin (for use by diabetics) by a bacterium containing human genes for it; the production of a human growth hormone; and the production of a yeast that would enhance fermentation processes and improve wine flavors. A genetically altered yeast that can shorten by several weeks brewing time for beer has already been developed in Japan and Germany.

One bacterium discovered in a Mexican swamp produces a powerful digestive enzyme that can remove protein stains more rapidly than any known detergent and is not affected by temperatures and pH changes that break down other cleaning enzymes. Genetic engineers are now trying to develop a strain of the bacterium that has multiple genes for production of the "super" enzyme.

A Few Pros and Cons of Genetic Engineering

The successful artificial addition of DNA to certain bacteria since the 1970s and subsequent progress in genetic engineering technology have produced considerable controversy. Some people fear that such tinkering by scientists could accidentally create uncontrollable monsters. Such fear is largely unfounded and is not based on scientific principles. Other people are enthusiastic and hail genetic engineering as the forerunner of eventual suppression or elimination of human defects and diseases.

In 1994, the U.S.D.A. deregulated a herbicide-resistant cotton variety produced by Calgene. The cotton, known as BXN, contains an introduced gene that is resistant to the herbicide bromoxynil. Some scientists view with alarm deregulation that potentially could increase pesticide usage. They worry about the possible effects on the environment and human health, and they believe that the cultivation of herbicide-tolerant cotton will lead to increased and prolonged use of a dangerous chemical. U.S.D.A. scientists and those of the Environmental Protection Agency counter that BXN cotton will allow the lowering of herbicide dosages by 40%, and they predict a fourfold decline in herbicide use by the year 2012.

The promise of genetic engineering for improving human quality of life and longevity is unquestionably enormous. The potentials for some restoration of ecological imbalances or, conversely, further human ecological disruption also exist. Some critics point, for example, to incidents such as the accidental release in South America of killer bees that have now moved north into the United States and fear equivalent inadvertent accidents with genetic engineering. Others fear deliberate genetic manipulation by unscrupulous governments, which could ultimately lead to catastrophes.

Given the nature of gene-splicing technology it seems unlikely, if all the present checks and balances are preserved, that accidents resulting in significantly harmful engineered organisms will occur. As long as avarice, politics, and physical errors or errors of human judgment are potentially involved, however, both constant vigilance and public knowledge of the issues will be essential for the substantial positive promises of genetic engineering to arrive at full fruition.

TRADITIONAL PLANT BREEDING

Most of the varieties of crop, ornamental, and other economically important plants available today have been developed from wild ancestral parents through controlled plant breeding. Such breeding appears to possess significant potential for reducing world hunger and improving renewable natural resources in the future. This potential, in turn, has led a number of field botanists to team up with anthropologists, medical doctors, and interpreters to explore wild plants for new sources of drugs.

One activity of such teams involves the interviewing of native medical practitioners and members of indigenous tribes in the tropics of the Americas and Africa in attempts to glean information about their utilization of local plants. Tasks of this kind have taken on a measure of urgency as undescribed species of tropical plants and other organisms are being lost daily to large-scale clearing of forests and savannahs and as the destruction eliminates unknown numbers of plants with possible potential for use as food, medicines, and fiber and for other uses. When the teams have sifted through all the gathered information, they hope to save significant material of new *gene pools* for use in breeding yields and crop plants with improved disease resistance. (A gene pool consists of all the genes of all the individuals in a population.) In the case of food plants, the teams also hope to improve the quality and quantity of protein produced by the plants. Several ways of improving existing varieties and of developing new ones came into use long before recombinant DNA technology was developed. The basic methods include *hybridization, polyploidy, mutation,* and *tissue culture.*

The Green Revolution

The Green Revolution of the 1960s saw a great increase in production of food grains due to new high-yielding varieties, new pesticides, and better management techniques.

Norman Borlaug, who is known as the "Father of the Green Revolution," was awarded a Nobel Prize in 1970 for developing new strains of wheat in Mexico. The new high-yielding strains, which were produced from crosses with a dwarf variety from Japan, were widely planted. Prior to the beginning of the project in 1944, Mexico had imported wheat for many years, but the new varieties were so successful that, within 20 years, Mexico had quadrupled its wheat production and had sufficient amounts to export to other countries. India and Pakistan experienced similar gains.

Although the Green Revolution has been a dramatic success, the success has not been unqualified. The high-yielding crops generally perform well only in areas where irrigation water is available, and they require extensive applications of fertilizer, most of which is produced from fossil fuels. In 1973, when OPEC brought about a major increase in the cost of fossil fuels, the agricultural areas of the world discovered that the increased costs suddenly changed what had been profits into deficits. In addition, the agricultural methods employed had not taken into account the longer-term effects of farming without appropriate ecological considerations, and the growing areas had become degraded with pesticides, herbicides, salts, and topsoil erosion. Since 1973, many small farmers have been forced out of business, and there is much debate over how to proceed with food and energy production in the coming decades.

Hybridization

When different varieties (or in some cases, species) are crossed, the F_1 generation often displays *hybrid vigor;* that is, the plants are bigger or better yielding than either parent. The effect usually disappears in the next generation, however, so F_1 seeds need to be planted each season to produce predictable results. Most corn grown in North America today comes from F_1 hybrid seed. When a new variety is being developed through hybridization, the experimental work is carried through at least two and, more often, through up to seven or more generations before development is complete.

During the experiments, backcrossing is used to reinforce in the offspring desirable characteristics of one parent. *Inbreeding* (brought about by repeated self-pollination) and managed *outcrossing* (cross-pollination between individuals of the same variety or species) normally play significant roles. Repeated inbreeding develops homozygous purebred strains, and as undesirable recessive genes come together, yields and vigor may decline. Through careful selection, however, desirable dominant and recessive genes are retained. Then pure strains are crossed, often resulting in vigorous F_1 hybrids. These hybrids, in turn, may be crossed with one another, often producing even more vigorous double hybrids (Fig. 14.3).

Since cultivation of wheat began thousands of years ago in the Mediterranean region, hundreds of different varieties have been developed. One widely used group of bread wheat varieties apparently originated from a hybrid formed between a species that had 28 chromosomes ($n = 14$) and a grass that had 14 chromosomes ($n = 7$). The hybrid was sterile because the cells contained nonhomologous chromosomes that could not pair properly during meiosis, but a doubling of the chromosomes followed, presumably when chromosomes failed to separate during mitosis or meiosis, resulting in a fertile hybrid with 42 ($n = 21$) chromosomes. Many varieties were bred to counteract the susceptibility of wheat to black stem rust, a fungus that has cost farmers millions of dollars in losses. New resistant varieties constantly must be developed as older varieties lose their resistance to mutant rust strains that seem to evolve almost as fast as resistant wheats can be produced.

Polyploidy

We noted in Chapter 12 that any cell that has one set of chromosomes (e.g., a gamete) is *haploid* and that any cell that has two sets of chromosomes (e.g., a zygote) is *diploid.* But

FIGURE 14.3   Hybrid corn plants.

solution at just the right strength (usually 0.1% to 0.3%) to kill most of them and then checking for the small percentage (usually about 5%) of surviving seedlings whose cells have twice the original number of chromosomes.

Some tetraploids are artificially induced without the use of colchicine. In tomatoes, for example, wounding of the plant causes it to produce *callus* tissue whose cells may have double the basic complement of chromosomes. In corn, tetraploids may be produced by applying heat to seeds. Polyploidy is further discussed under the "Role of Hybridization in Evolution" in Chapter 15.

Mutations

Mutations, which involve a change in a gene or chromosome and occur naturally in all living organisms, can be artificially induced by radiation and chemicals. Like those occurring naturally, artificially induced mutations are largely harmful, and the expense of screening large populations for a rare desirable mutant form generally is not practical. A few useful modifications have been obtained, however, through induced mutation. These include higher yielding strains of *Penicillium* mold, blight resistant varieties of oats, and an improved grape from a variety whose individual fruits previously were bunched too compactly. Additional discussion of the role of mutations in evolution is given in Chapter 15.

TISSUE CULTURE AND MERICLONING

Tissue Culture

A *tissue culture* is essentially a mass of **callus** tissue (an undifferentiated group of cells) growing on an artificial medium. It can be started from almost any part of the plant, although tissues taken from the vicinity of meristems usually produce the best results. With the proper media, the callus tissue eventually differentiates into shoots and roots or sometimes into plant embryos. The technique was originally used in the culture of cells from carrot, tobacco, and citrus, but it is now used with a widespread variety of plants.

Using aseptic practices (i.e., practices that seek to eliminate bacteria and other disease organisms), the technique is designed to isolate unique tissues with the potential to become new disease-free plants; it also permits rapid multiplication of the materials. The best results are obtained when the procedures are conducted in a room or chamber that has been sterilized with ultraviolet light or has been steam-heated before work begins. All instruments, glassware, and media need to be sterilized in advance. An *autoclave,* which is something like a large pressure cooker, is generally used for this purpose (Fig. 14.4). The ingredients for the media are measured very precisely and usually include distilled water, various inorganic salts, and organic substances, such as yeast extracts, vitamins, sugar, and hormones.

many plants, including about half of all flowering plants, have more than two sets of chromosomes in their cells and are therefore **polyploid.** Polyploids that have three, four, six, or eight sets of chromosomes are said to be *triploid, tetraploid, hexaploid,* or *octoploid,* respectively. Polyploidy can arise in various ways. Hybridization followed by a doubling of the chromosome number, as exemplified by the bread wheat just mentioned, results in a polyploid called an *alloploid.* The chromosome sets can also be doubled when pairs of chromosomes fail to separate during mitosis or meiosis. The resulting polyploid is called an *autoploid;* a diploid plant could become a tetraploid by either alloploidy or autoploidy.

Since tetraploids are often bigger or more vigorous than their diploid counterparts, plant breeders sometimes try to develop tetraploids artificially by treating seeds with *colchicine,* an alkaloidal drug derived from the corms of the autumn crocus. The colchicine interferes with spindle formation during mitosis, so pairs of chromosomes do not separate during anaphase, and if the cell with the double set of chromosomes survives and becomes functional, a tetraploid plant may be the ultimate result. Developing tetraploids with colchicine usually involves treating seeds with a colchicine

FIGURE 14.4 An autoclave, in which pressurized steam is used to sterilize media, glassware, and instruments.

The considerable potential for crop improvement offered by tissue culture lies in the relative ease with which selection for desirable traits can be accomplished in large populations of cells cultured in a few glass containers instead of in populations of plants on large acreages of land and in varied environmental conditions. Manipulation of the cells often involves subjecting them to a variety of stresses, such as herbicide and other poisons, heat, cold, disease bacteria and fungi, and radiation, and then isolating cells that survive from among those that are susceptible.

Shoot Meristem Culture (Mericloning)

Orchids are among the most prized of all cultivated plants. They also produce the smallest seeds known, and it often takes seven or more years from seed until the first flower appears. Normally, the seeds have to become associated with a specific fungus before they will germinate, and such fungi may not occur outside of the orchid's natural habitat. In ad-

dition, because of variables inherited from the parents, orchid hybridizers do not know if they have been successful in developing plants with superior characteristics until the orchids flower several years later. The growers have partially overcome both problems, however, with the use of the shoot meristem, or *mericloning,* technique.

In mericloning, certain hormones and other substances are substituted for the fungus in artificial growing media. The technique usually involves the removal of the apical meristem and the youngest one or two leaf primordia, with the use of a dissecting microscope under aseptic conditions (Figs. 14.5 and 14.6). The meristem is then placed on a sterile medium and kept at room temperature under lights (Fig. 14.7). Adventitious roots usually develop within a month. The tissue that develops prior to the appearance of shoots and roots is subdivided as it grows, and each subdivision is then capable of becoming a new plant. By this process, commercial growers produce in a relatively short time thousands of plants, all genetically identical to the plant from which the meristem originated.

Artificial Seeds

Scientists working with tissue cultures are aware that, with the addition of certain hormones and nutrients to the culture medium, embryos will develop from the callus of some (but apparently not all) species of seed plants. Genetically identical embryos mass-produced through tissue culture can then be packaged with a food supply, hormones, and a biodegradable protective coat. The need for fertilizers and pesticides can also be reduced by adding nitrogen-fixing bacteria, fertilizers, pesticides, and other enhancements to the package. At present, the production of such artificial seeds is more costly than conventional seed production, but several companies are exploring more efficient production techniques. Already, however, the costs are partially offset by the fact that the artificial seeds produce a crop of identical plants, which significantly lessens the expense of harvesting.

Protoplast Fusion

Tissue culture also lends itself to various genetic engineering techniques, including the transfer of organelles from one cell to another and the hybridization of cells through *protoplast fusion.* This latter technique involves a culture of two different kinds of plant cells; the walls of the cells are digested with enzymes, leaving naked protoplasts. A small percentage of the protoplasts may, on their own, undergo fusion in the culture, resulting in hybrid cells with genes from both parent cells.

Other protoplasts can be stimulated by lasers or electrical charges to unite or be induced to do so with chemicals, such as diluted polyethylene glycol (antifreeze). The hybrid cells are isolated from those with identical genotypes by culturing them on a medium on which only the hybrid cells can grow. Regeneration of plants from fused protoplasts has

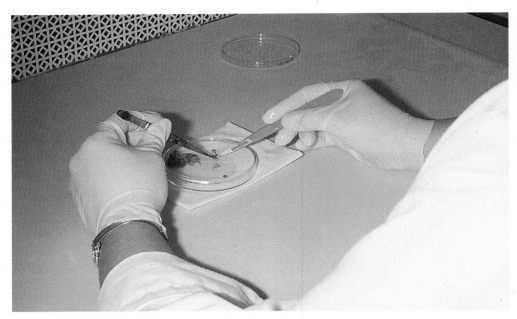

FIGURE 14.5   A skilled technician removing the apical meristem from an orchid plant.

proved to be difficult for monocots such as corn, wheat, and rice but seems promising for dicots.

If the organelles (e.g., chloroplasts) of one cell are more desirable than those of the other cell, the nucleus of the cell with less desirable organelles can be destroyed by radiation. A plant resulting from the fusion of two different protoplasts is called a *somatic hybrid;* those with the cytoplasm of two cells but only one nucleus are referred to as *cybrids.* Somatic hybrids frequently are less desirable than the parents, but several exceptions produced by protoplast fusion have resulted in potentially valuable grain crop plants.

Clonal Variants

Since tissue culture techniques came into widespread use, scientists have occasionally been surprised to find in their cell cultures cells that differ from others in appearance. For reasons that are not clear, such variants appear much more frequently than mutants do in field-grown crops. Some of the variants have superior color or other desirable vegetative characteristics, and scientists are now developing new varieties of vegetables and other food plants by searching for such variants in cell cultures.

Some cell cultures are derived from pollen grains or other haploid tissues or cells. Each cell in such cultures has only one set of chromosomes and therefore cannot develop spore mother cells that could undergo meiosis and develop gametes. However, the drug colchicine can be used to interfere with spindle formation during mitosis of such cells. This can result in two sets of chromosomes remaining in a single cell, and if this diploid cell divides and the tissue differentiates, a homozygous plant is the ultimate result. Producing a homozygous plant in this manner can be accomplished

within a year or two, which is about four years less time than it would take to produce a similar plant by ordinary breeding and selection in the field.

TRADITIONAL VEGETATIVE PROPAGATION

The value of being able to take a piece of stem of a desirable plant from one locality to another and induce it to grow into a new plant identical to its parent or of being able to divide rhizomes, tubers, and other parts of plants and have them each develop into new plants has been recognized since ancient times. *Grafting,* which involves the permanent union of parts of two plants, also has its origins in antiquity. Plants with desirable features, such as superior fruits or flowers or attractive leaves, may be grafted onto related plants whose chief attraction may lie in a vigorous root system or one that is resistant to disease.

The Chinese were apparently practicing grafting by the year 1000 B.C., and Theophrastus discussed grafting and other forms of vegetative propagation in his book *Causes of Plants,* written in the third century B.C. During the 14th and 15th centuries A.D., large numbers of plants were imported into European gardens and maintained by grafting. Even though the internal nature of the graft union was not fully understood, many forms of grafting were developed; no fewer than 119 different methods of grafting were described in 1821 by A. Thouin. Today, the practice of various forms of vegetative propagation is almost universal and is an important part of the economies of the world. Some of the more widespread techniques of vegetative propagation are

FIGURE 14.6 Flasks of sterile growing medium to which meristematic orchid tissue has been added. The flasks are rotated under lights. Roots and shoots appear within a few weeks.

FIGURE 14.7 The plants that have developed from the cultured meristematic tissue are separated and further cultured to maturity.

discussed in the following sections; details of grafting are discussed in Appendix 4.

Stem Cuttings

Occasionally, a basic change may occur in the chromosomes of a cell, and the result is a bud or a whole plant that develops differently from the rest of the tree or variety. Such a *mutation* is likely to be undesirable or to result in the death of the cell or tissue. In a few instances, however, the mutation may produce superior characteristics, such as better flowers or fruit, dwarf forms, or disease resistance. If you were to plant seeds from such a mutant form, the plant that would develop could differ as much from the mutant parent as you do from either your mother or your father. Since pollen from one parent has to be transferred to a flower of another parent for fruits and seeds to form, the mutant plant carries characteristics from both parents and thus may pass on to future generations a variety of combinations of characteristics.

When a mutation is desirable, the element of chance inheritance can be eliminated by cultivating tissue directly from the mutant plant, so that each of the offspring has characteristics (phenotypes) identical to those of the single parent; the desirable feature can thus be perpetuated. Wash-

ington navel oranges, for example, have come from a mutation that appeared on a sweet orange tree growing near Bahia, Brazil, around 1820. The grower, whose identity is not known, apparently recognized the superior nature of the fruit on a branch of one of the trees and, either by bud grafting or by making cuttings, multiplied the new form. Fifty years later, a missionary sent a dozen budded navel orange trees from Brazil to the U.S. Commissioner of Agriculture in Washington, D.C., and in 1873, two trees were shipped from Washington to Riverside, California. From these California trees, one of which is still living, almost the entire navel orange industry of the world has been developed through vegetative propagation.

In propagation by *cuttings,* which may have been used by the Brazilian orange grower, pieces of a plant are induced to produce roots and are then planted to grow on their own.

A large number of plants can be started with cuttings from a few trees or shrubs in a relatively small amount of space and time, and little skill is required. Cuttings also do not have the problem of incompatibility, which is found with grafting. It is not always desirable to propagate by cuttings, however, particularly if the variety concerned does not have a sturdy, disease-resistant root system. Some plants are also very difficult to start from cuttings because they do not readily form roots. Next to planting seeds, however, cuttings have become the most widely used means of multiplying plants in the world today.

Most cuttings are made from stems that are 0.6 to 2.5 centimeters (0.25 to 1.00 inch) thick and contain at least two nodes. They may vary in length from 7.5 centimeters to

The New Harvest: Bioengineered Foods

While George Washington probably never cut down a cherry tree, he certainly grew tomatoes in his Mt. Vernon garden. Tomatoes were a favorite gardening plant in colonial America and remain popular today as Americans spend about $3.5 billion a year to buy fresh tomatoes. But the fresh summer tomatoes grown by so many backyard gardeners have a new cousin—a genetically modified variety called the Flavr Savr™. Introduced to markets in 1994, the Flavr Savr™ tomato was the first genetically engineered food to reach supermarket shelves. Calgene, a biotechnology company in Davis, California, took 12 years to develop and receive FDA approval for the Flavr Savr™ tomato.

Why would Calgene spend this amount of time and money to create the Flavr Savr™? It will come as no surprise that of the most common items bought fresh in supermarkets, tomatoes rank last in customer satisfaction according to a survey conducted by the U.S. Department of Agriculture. So if a company could produce a better tasting tomato, it stands to make a good profit. The problem with commercially grown tomatoes is this— most fresh market tomatoes are harvested green, prior to the development of the tomatoes' full flavor potential. They are then transported in refrigerated trucks to storage houses where ethylene gas is used to induce the red coloration natural to the ripening process. Only then are

Vine ripened Flavr Savr™ tomatoes.
(Photo courtesy Calgene, Inc., Davis, CA)

tomatoes shipped to market. The typical result is a tomato that "looks red but eats green," with a taste far removed from home-grown, vine-ripened tomatoes.

Picking "green" is done because growers are in a race with nature. They need to get the tomatoes to market before spoilage occurs. If tomatoes are allowed to ripen on the vine, they begin to get soft because tomatoes slowly self-destruct for dispersal of their seeds.

Calgene reasoned that they could give growers an additional three to five days of vine-ripening time before harvesting—and flavor could be built on the vine resulting in a more flavorful tomato—if this softening process

---

60.0 centimeters (3 inches to 2 feet). With deciduous plants, the cuttings are made—while the buds are dormant—from healthy wood of the previous season that has been growing in the sun. Pieces may be cut transversely, or a small part of wood from an older branch may be left attached at the bottom end. If a number of cuttings are being made at one time, they may be tied together with rubber bands, placed bottom ends down in cool, damp sand or sawdust from which water will drain easily, covered, and stored until spring. They should be checked from time to time to see if buds are developing. If they are, the cuttings should either be planted or refrigerated, as development of buds without corresponding root growth will kill the material.

In the spring, the bottom ends of the cuttings can be inserted 4 to 6 centimeters (3.0 to 4.5 inches) in damp sand (to minimize problems involving soil fungi) or directly into the soil. Here adventitious roots develop, and normal growth

should occur. Evergreen plants tend to produce adventitious roots much more slowly, taking from several months to a year to produce sufficient growth to become established. Since cuttings may dry out, they need to be misted with water frequently, and they also need to be kept out of the sun and handled rapidly while they are being prepared. Leaves from the lower part of the cutting should be removed. If bottom heating is available, the additional warmth usually accelerates the development of the roots as long as high humidity can be maintained.

Leaf Cuttings

Houseplants, such as African violets, peperomias, begonias, sansevierias, and others with somewhat succulent leaves can be propagated in various ways from their leaves (Fig. 14.8). One commonly used method is to cover a jar or pan

could be retarded. Their solution was to slow down softening by inserting a new gene into the tomato plant.

Calgene scientists knew that the cells of tomatoes are held together by a natural chemical called pectin that acts like a cement. During ripening, an enzyme called polygalacturonase (PG) begins to dissolve the pectin, creating the softening. Their idea was to take the PG gene that specifies the enzyme polygalacturonase, cut it out of the chromosome with molecular scissors called restriction enzymes, and reinsert it **backwards** into the chromosome. This backward gene, called an **antisense gene,** produces a product (messenger RNA) that interferes with the naturally occurring PG messenger RNA. As a result, the amount of the PG enzyme produced is reduced significantly in the ripening tomato—and softening is delayed.

The process of inserting new genes into plants is these days relatively simple. Once Calgene researchers identified and cloned the PG gene, they reversed the gene and spliced this reversed, antisense PG gene into the DNA of a normal soil-inhabiting bacterium called *Agrobacterium*. It so happens that *Agrobacterium* can transfer genetic material into plants, a process that occurs when genes are inserted that cause the plant to produce a tumorous growth called Crown Gall disease. Making use of this natural infection (and DNA transfer) process, the Calgene team mixed *Agrobacterium* carrying the antisense PG gene with tomato leaf pieces in a petri dish. Because the *Agrobacterium* infection process is not 100% efficient (only approximately 30% of tomato cells actually become infected), a screening test must be performed to determine which tomato leaf cells contain the new gene. This is impossible to determine just by looking at the plants, so a "marker" gene is piggybacked along with the antisense PG gene. The marker gene used is a bacterial gene that codes for a protein that gives the bacterium resistance to the antibiotic *kanamycin*. The gene is called the kanamycin resistance gene (K_r). By growing the infected leaf pieces in a solution containing kanamycin, only those cells that have incorporated the K_r gene into their chromosome structure will survive this screening test. The uninfected cells will be killed because they do not carry the kanamycin resistance gene. The infected cells carrying both the antisense PG gene and the K_r gene can now be grown into mature plants and the seeds harvested.

These fundamental genetic engineering techniques can be applied to any number of genes that produce in a plant desirable characteristics such as improved flavor and nutritional value, resistance to insects and viruses, and so forth. Although the Flavr Savr™ tomato was the pioneer in the agricultural biotechnology revolution, other products like Monsanto's Roundup Ready soybeans (which withstand herbicides that kill a wide variety of weeds) have followed and many more are in the pipeline awaiting regulatory approval.

FIGURE 14.8   Leaf cuttings. *Left.* Disks of *Peperomia* leaves that have been placed on moist filter paper. *Right.* A *Peperomia* leaf with its petiole in water.

of water with foil or plastic film, punch small holes in the covering, and insert the petioles so that they are immersed at the base. The water should be changed occasionally. After a few weeks, adventitious roots develop at the base of the petiole. The leaves may then be planted, leaving the blade exposed; new plants gradually appear from beneath the soil.

Rex begonias are sometimes propagated by punching out disks from the leaf blades with an instrument that resembles a cookie cutter; each disk should be about 1.9 centimeters (0.75 inch) in diameter. The disks are placed on damp filter paper in petri dishes, where they develop both roots and shoots. Rex begonia leaves may also be induced to form new plants by making cuts about 3 millimeters (0.16 inch) deep across the main veins on the lower surface. Leaves are then placed lower side down on moist potting medium, and the edges of the leaf are pinned down with toothpicks. Eventually, new plants develop at the vein cuts, while the original leaf withers.

Other houseplants, the raspberry-blackberry group (excluding red raspberries), citrus plants, rhododendrons, and camellias may be propagated by cutting a short piece of stem containing a single node with a leaf and its axillary bud and inserting it into a mixture of sand and peat moss, with the bud 1.3 centimeters (0.5 inch) or more below the surface. As with all evergreen or herbaceous plants, it is essential that the humidity be kept high during propagation. If a misting device is not available, the area should be covered with glass or polyethylene sheeting—allowing, however, for some air circulation—and the leaf surfaces should occasionally be sprayed with water.

Root Cuttings

A variety of plants may be propagated from root cuttings. The roots should be obtained after a growing season has been completed and the plants are dormant. The roots are cut into pieces from 2.5 to 7.5 centimeters (1 to 3 inches) long and placed in a potting medium or damp sand, either in a horizontal position or with the bottom end down. Since it is often difficult to tell the top end from the bottom once a root has been divided, some growers make transverse cuts at the top end and slanting cuts at the bottom end to avoid confusion.

The production of adventitious roots by cuttings is apparently brought about through the stimulation of auxin production. It is common practice to add to the natural auxin of the plant by the application of either additional IAA or synthetic growth-promoting substances, particularly indolebutyric acid (IBA) or naphthalene acetic acid (NAA), which often seem to be more effective than natural auxin in stimulating root growth. These substances are available at nurseries and florists in powder or paste form, and their use not only significantly increases the number of successful cuttings but also initiates more vigorous root growth. Cytokinins have also been used in the propagation of leaf disks. Since fungi are frequently a problem in vegetative propagation, fungicides have been found useful for inhibiting their growth.

Layering

Tip Layering

Tip layering is used with blackberries, boysenberries, and other plants with somewhat flexible stems. The technique involves bending the tips of the canes over so that they touch the ground and covering them with a small mound of soil. Roots form on the portion of buried stem, and eventually shoots appear also. These new plants can then be separated from the parent stems. Variations of tip layering include forcing a stem to lie horizontally and covering it with small mounds of soil at intervals and heaping soil around the base of a plant so that the individual stems produce roots there. Once roots have been established, the individual plantlets or

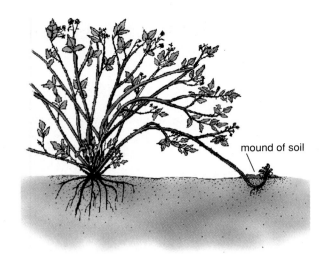

FIGURE 14.9   Tip layering. The tips of canes are bent to the ground and covered with a small mound of soil. When a new plant has developed at the tip, it can be cut from the parent plant and grown independently.

pieces of stem can be cut off from the original parent and grown independently (Fig. 14.9).

Air Layering

Tropical trees and shrubs are sometimes propagated by air layering. With this method, sphagnum moss is wrapped around a branch or main stem that has been lightly gashed with a sterilized knife, the moss is dampened with water, and then the area is covered with polyethylene film. Adventitious roots eventually form on the portion of the stem covered by the moss. Sometimes, the stem is girdled by removing about 2.5 centimeters (1 inch) of the bark all around the stem before the moss is applied. The moss should not be kept too wet or disease organisms may rot the stem (Fig. 14.10).

Propagation from Specialized Stems and Roots

Modified plant organs, such as rhizomes, tubers, and corms, were discussed in chapters 5 and 6. Virtually all of these may be divided, and each piece is capable of developing into a new plant (Fig. 14.11). If any piece of a rhizome or tuber containing at least one node is planted, it should develop into an independent plant. Strawberry and saxifrage runners and the runners of mints root at intervals and develop shoots above the roots. As soon as the new plants appear, they may be separated and grown independently. The scales of lily bulbs are quite fleshy and may be used in vegetative propagation. In later summer, while the weather is still warm, the scales can be separated and planted in moist soil about 2.5 to 5.0 centimeters (1 to 2 inches) deep. Three to five bulblets usually form at the base of each scale in a few weeks; these can be transplanted the following spring.

FIGURE 14.10 Steps in air layering. *A.* Vertical cuts are made on the branch. *B.* Damp sphagnum moss is wrapped around the cut area. *C.* Polyethylene film is wrapped around the moss. *D.* The ends of the film are taped shut. Adventitious roots should develop in the cut area and become visible through the moss in the ensuing weeks.

(USDA Photos)

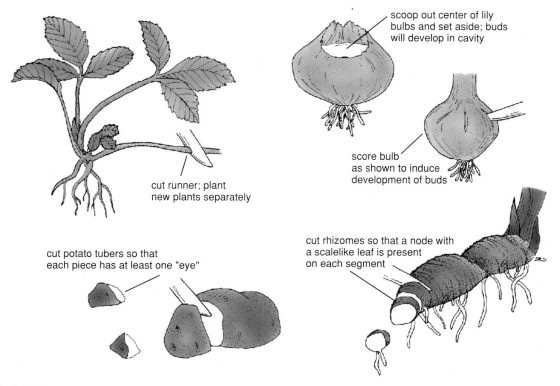

FIGURE 14.11 Propagation of specialized stems.

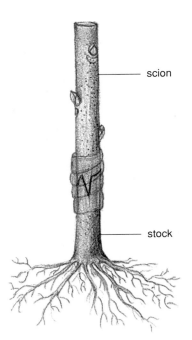

FIGURE 14.12 A simple graft. The rooted portion (*stock*) and portion to be grafted on to the stock (*scion*) are cut so that the two parts will fit together with the cambium of both portions in close contact.

Bulbs of hyacinth, grape hyacinth, squills, and similar plants can be induced to form bulblets by inverting them and cutting down about a third of the way with a sterilized knife, as though sectioning an apple, or by scooping out the center with a sterilized melon baller. After they are cut, the bulbs should be placed in dry sand in a warm place for about two weeks. During this time, callus tissue forms along the cuts. After the callus has formed, the bulbs should be removed from the sand and stored in trays at room temperature under humid but well-aerated conditions. Small bulblets form along the cuts during the ensuing two or three months. These are left attached, and the whole original bulb should be planted. After a year's growth, the bulblets can be separated and replanted independently. In addition to using sterile instruments, it may be necessary to use fungicides to prevent fungal diseases from damaging the materials.

Some bulbs, such as those of daffodils, tulips, and amaryllis, have dry scales on the outside and fleshy scales on the inside. Unlike those of the lilies, these bulbs do not respond to being separated. These plants and plants with corms (e.g., gladioli) naturally produce bulblets or cormlets, however, which may be separated and grown independently. The number of bulblets produced appears to be related to soil depth. Most develop when the original bulb is about 7.5 centimeters (3 inches) deep. Few, if any, are produced when the bulb is more than 17 centimeters (8 inches) deep.

Plants with fleshy storage roots, such as sweet potatoes or dahlias, can be propagated by dividing the root so that each section has a shoot bud. This is done in early spring, just before planting, when shoot buds have appeared.

Sweet potatoes can also be cut in pieces and induced to form new plants by suspending each segment with toothpicks from the top of a jar partially filled with water. The segments form adventitious shoots and additional roots in a relatively short time.

GRAFTING

Simple grafting basically involves the insertion of a short portion of stem, called a scion, into another stem with a root system, the **stock** (Fig. 14.12). A segment of stem, called an *interstock,* is sometimes grafted between the stock and the scion, or a single bud with a little surrounding tissue may be grafted onto a stock. In all grafts, *the cambia of the stock and scion must be placed tightly in contact with each other, and all parts being grafted must be related to one another.* For example, different varieties of apple may be grafted together, but it is not possible to graft elm or maple wood to an apple tree.

After the graft is made, the fit may be sufficiently tight without additional support, but it is customary to tie flat rubber strips, raffia, plastic tape, or waxed string around the union and cover it with grafting wax. The grafting wax is needed to prevent the tissues from drying and also to reduce the possibility of disease organisms gaining access to the interior of the stem. If the grafts are buried beneath the surface of the ground, they should be checked after growth begins to see if the wrapping is breaking down. If they are not below ground, the wrapping should be removed when it threatens to hinder the increase in girth that occurs when the cambium adds new tissue.

Discussions and illustrations of the major types of grafting are given in Appendix 4.

Summary

1. Genetic engineering involves the alteration of the DNA of a cell by artificially mixing large numbers of enzymatically broken DNA strands from two sources. Fragments from one source may associate with those from another source, creating recombinant DNA with parts of both original strands. The recombinant plasmids are inserted into live bacterial cells where they replicate. Since bacteria are easy to propagate, large amounts of recombinant DNA may be obtained in this fashion. Such recombinant DNA may be used in the future for improving crops and possibly eliminating diseases of both plants and animals.

2. Most economically important plants have been developed from wild ancestors through controlled plant breeding from available gene pools. Hybridization (which often produces hybrid vigor),

inbreeding, outcrossing, polyploidy, and mutations have all been employed in the development of new varieties. Norman Borlaug, the "Father of the Green Revolution," successfully increased productivity of crops through hybridization, but his methods are controversial because of growth requirements and adverse long-range effects on the ecology of agricultural lands.

3. Tissue culture, which involves the stimulation of isolated cells and tissues to grow on artificial media under sterile conditions, offers potential for crop improvement with considerably less space than is needed with field techniques. Tissue culture also lends itself to genetic engineering, protoplast fusion, artificial seed production, and clonal variant techniques.

4. Shoot meristem culture (mericloning) is employed in the mass production of certain plants, such as orchids; normal plant development from seed to flower is significantly reduced, and variations inherent in sexual reproduction are eliminated.

5. Vegetative propagation, which involves the growing of new plants from portions of other plants, began in antiquity. Cuttings of stems, leaves, and roots can be induced to grow into plants that are identical to the parent material. Such propagation is particularly useful in the perpetuation of favorable mutations, such as those of the navel orange. Layering may be used to propagate plants from the tips of flexible stems or from aerial stems. Almost any specialized plant stem or root can be propagated from portions of the parent material.

6. Grafting involves the union of an isolated stem portion (scion) with a rooted stem (stock). The cambium of both the stock and scion must be in close contact for the graft to be successful.

Review Questions

1. What is genetic engineering, and how is it accomplished?

2. How do restriction and repair enzymes perform their functions?

3. Give some current uses of genetic engineering.

4. Are there any potentially negative aspects to genetic engineering? Explain.

5. Explain the roles of hybridization, polyploidy, and mutations in plant breeding.

6. What is tissue culture, and how is it used commercially?

7. What is meant by the term *mericloning*?

8. Explain the difference between artificial and true seeds.

9. Give the major ways in which vegetative propagation is accomplished. What advantages are there to vegetative propagation as opposed to using seeds to propagate?

10. What is the difference between a stock and a scion?

Discussion Questions

1. Genetic engineering has already greatly improved the production of certain medicines, but research has been heavily restricted, despite the availability of funds for the work. Do you think restrictions should be lifted? Why?

2. Tissue culture and mericloning techniques have dramatically facilitated the production of desirable plants. What might be any long-range disadvantages to this and other forms of vegetative propagation?

3. What advantages might there be to having several varieties of apple or plum grafted on the same tree?

Additional Reading

Bajaj, Y. P. 1992. *High tech and micropropagation,* Nos. 1 & 2. New York: Springer-Verlag.

Bajaj, Y. P. (Ed.). 1989. *Plant protoplasts and genetic engineering.* New York: Springer-Verlag.

Bhojwani, S. S. (Ed.). 1991. *Plant tissue culture: Applications and limitations.* New York: Elsevier Science, Inc.

Blick, B. R., and J. J. Pasternak. 1994. *Molecular biotechnology: Fundamentals and applications of recombinant DNA.* Washington, DC: American Society of Microbiologists.

Endress, R. 1994. *Biotechnology of secondary products of plant cells.* New York: Springer-Verlag.

Hartmann, H. T., and D. E. Kester. 1990. *Plant propagation: Principles and practices,* 5th ed. Englewood Cliffs, NJ: Prentice-Hall.

Kyte, L. 1988. *Plants from test tubes: An introduction to micropropagation.* Portland, OR: Timber Press.

Lewis, R. 1992. *Life.* Dubuque, IA: Wm. C. Brown Publishers.

Stoner, C. H. 1992. *Biotechnology for hazardous waste.* Chelsea, MI: Lewis Publications, Inc.

Torrey, J. G. 1985. The development of plant biotechnology. *American Scientist* 73: 354–63.

Kahili ginger (Hedychium gardnerianum) *and Hawaiian tree ferns* (Cibotium *sp.*).

15 Evolution

Overview————————

The discussion of evolution begins with an abbreviated history of Charles Darwin and the principles or tenets involved in the theory of evolution through natural selection. Mechanisms of evolution, including mutations, hybridization and introgression, polyploidy, apomixis, and reproductive isolation are discussed. The chapter concludes with an examination of the role of isolation and mutation in modifying Darwinism and with an examination of the evidence for evolution.

Some Learning Goals

1. Understand how introgressive hybridization may lead to the development of new species.
2. Know the contributions of Charles Darwin to theories of organic evolution and the tenets of natural selection as he understood them.
3. Explain the significance of mutation and reproductive isolation to evolution.
4. Give reasons, past and present, for the controversy over evolutionary theory.

FIGURE 15.1   Charles Darwin. (Courtesy National Library of Medicine)

Few historical events since 1859 have had a greater impact on society in general and the biological sciences in particular than Charles Darwin's book, *On the Origin of Species by Means of Natural Selection, or the Preservation of Favoured Races in the Struggle for Life.* Darwin's theory of evolution through natural selection has stimulated an enormous amount of thinking and research and has provided an explanation based on natural laws for the diversity of life around us. It also initially caused a great deal of controversy because, even though Darwin believed to his death in a divine creator, he also believed that the creator had used natural laws to bring all living things into being gradually over long periods of time, which ran against a literal interpretation of the Bible. Most of his contemporaries, guided by this literal interpretation of the biblical account of creation were convinced that all living things had been created in six days and had existed unchanged since the beginning.

Although the controversy has subsided, some disagreement still exists today. When the popular *Scofield Reference Bible* was first published in 1909, it included in the margin opposite the account of creation a date, "4004 B.C.," arrived at by the 17th century Irish archbishop James Ussher, who based his calculation on faulty interpretation of biblical genealogies. Ussher's date has been deleted from the margins of editions of the *Scofield Reference Bible* published since 1967, and the editors have observed that little evidence exists for fixing dates of biblical events prior to 2100 B.C.

Many *scientific creationists,* who since the 1970s increasingly have sought to have a nonevolutionary interpretation of the living world included in public school biology textbooks, do not necessarily believe the earth was created in 4004 B.C. or later, nor do they refuse to recognize the existence of minor variations in living organisms. The majority, however, believe the earth is not more than 30,000 years old and reject the foundations of evolution as incompatible with a literal interpretation of the biblical account of creation. In doing so, scientific creationists reject most of the evidence for evolution, including that which has accumulated since Darwin.

Part of the remaining disagreement also stems from a failure of people in diverse fields to define terms and to distinguish between fact and theory in evolutionary matters. Evolution itself, for example, has been broadly defined by some as simply being synonymous with change. We are told that anything from cars to computers to cultures is evolving and that even our thought processes evolve. To distinguish, therefore, between change in inanimate or intangible entities and progressive change in living organisms over time, we need to refer to the latter as *organic evolution.* Even this way we do not pin down the subject entirely, for there are variations in organic evolutionary theory. That understanding that pervades and unifies most biological thought today, however, finds its origin in the observations of Charles Darwin (Fig. 15.1) and of a contemporary of his, Alfred Wallace, who independently arrived at the same conclusions as Darwin.

CHARLES DARWIN

Charles Darwin was born in Shrewsbury, England, in 1809. His father, Dr. Robert Darwin, was a successful country physician and relatively wealthy. In 1825, at the age of 16, young Darwin was sent to the University of Edinburgh Medical School to follow in his father's footsteps. He did not do well in his studies, however, and dropped out after two years. The following year, he went to Cambridge University to study for the ministry but did not excel academically in this field either. Part of the reason for his poor scholastic showing apparently was due to his spending much time in the countryside collecting beetles, rocks, and other specimens or talking at length with his Cambridge biology professors, who held him in high regard. He barely attained what today would be called a C average, but he managed to graduate with a degree in theology in 1831. He was then 22 years old but was still not really sure what he wanted to do with his life.

While Darwin was pondering such matters, King William IV of England commissioned a sailing vessel, the HMS *Beagle,* to undertake a voyage around the world for the purpose of charting coastlines, particularly those of South America. Young Darwin's Cambridge biology professors recommended him for the (unpaid) post of assistant naturalist and captain's companion on the voyage. His father was opposed to the idea, but his uncle sided with him, and Darwin accepted the position.

The voyage, which began December 27, 1831, and took five years, gave Darwin an opportunity to collect specimens on both sides of South America, as well as in the Galápagos Islands and along the coasts of Australia and New Zealand. He kept a daily journal and spent countless hours alone on horseback collecting and observing the living world around him. This gave him ample opportunity to think about the forms and distribution of the myriad new organisms he encountered. His thoughts slowly led to the development of ideas that later blossomed into his theory of evolution through natural selection.

Upon his return to England in 1836, Darwin obtained the financial support of his father and his cousin Emma, whom he married. Although still young, he retired to the country and began working on his collections and journal. He also carried on a voluminous correspondence with other biologists and made extensive investigations into pollinating mechanisms, earthworm ecology, geographical distributions of plants and animals, and several other areas of natural history.

Throughout all of his activities, he was guided by a concept that he had adapted from an essay written by Malthus in 1798 on human populations and food supplies. Darwin realized that, although human beings might artificially improve or increase their food supply through selective breeding and cultivation, plants and animals could not do so and were therefore vulnerable to a process of selection in nature, which would explain changes in natural populations. He was reluctant to publish his ideas, however, and did not begin putting together his book on the origin of species until 1856.

Meanwhile, an English naturalist by the name of Alfred R. Wallace (1823–1913), who made major contributions to our knowledge of animal geography, independently concluded that natural selection contributed to the origin of new species and sent Darwin a brief essay on the topic in 1858. At the urging of friends, Darwin published a statement of his views jointly with Wallace, and, in 1859, Darwin's classic book *On the Origin of Species by Means of Natural Selection* was published.

Our understanding of the term *species* has been modified and refined since Darwin's time, but, as discussed in the next chapter, most biologists today think of a species in terms of a "population of individuals capable of interbreeding freely with one another but not freely with those of another species."

TENETS OF NATURAL SELECTION

In essence, Darwin's theory of evolution through natural selection is based on four principles or tenets:

1. *Overproduction*. Many living organisms produce enormous quantities of reproductive cells or offspring. For example, a single maple tree produces thousands of seeds each year, most being capable of becoming a new tree. Some fungi produce trillions of spores, each with the potential to become a new fungus.

2. *Struggle for Existence*. All the germinating seeds, spores, and other reproductive cells of living organisms compete for available moisture, light, nutrients, and space. The amounts of these elements available in nature are not sufficient to support all of these organisms, and many die as a result.

3. *Inheritance and Accumulation of Favorable Variations*. All living organisms vary. Those hereditary variations with survival value are inherited from generation to generation and accumulate with time, while other variations not important to the survival of the species are gradually eliminated.

4. *Survival and Reproduction of the Fittest*. Those forms of organisms best adapted to the environment have the best chance to survive and reproduce, while others less well adapted may succumb. A tree with thicker bark, for example, has a variation favorable to surviving cold temperatures; hence it may survive to reproductive age and leave more offspring, which will bear the inherited features of the thicker-barked tree.

One of the criticisms of Darwin's theory, published in 1859, was that it did not explain how hereditary variations originated and developed. It should be remembered, however, that Mendel's findings were not published until 1866, and the details of mitosis and meiosis did not become known until 1900 to 1906. Today, with our far greater knowledge of how variations occur and are inherited, we have come to understand the mechanisms of evolution in populations much better than was possible in Darwin's time.

All organisms, from aardvarks to zinnias, occur in populations that are composed of a few to billions of individuals.

Yet even within very large populations, it is virtually impossible to find two that are identical down to the last molecule. As humans, we are well aware of this within our own species, but variation exists in all living organisms, even though in simpler one-celled forms the differences may be much less obvious. Whether the differences are obvious or subtle, however, the general characteristics of a population will eventually change, if the environment or other factors favor certain hereditary variations and bring about the elimination of others.

In some instances, the environment may alter the phenotype slightly without affecting the genetic constitution of an organism. Plants that grow relatively tall at sea level may become dwarfed when they are transplanted to cooler areas in the mountains or drier areas near deserts, yet they are capable of breeding freely with plants at the original location if they are returned to that area.

If we apply fertilizers to plants, stimulate their growth with hormones, or prune them, the changes are not passed on to the offspring, for no permanent change occurs in a given population unless there is *inheritable variation*. The changes in transplanted, fertilized, or pruned plants are not transmitted to the offspring because the gametes of those plants will carry the same genetic information they would have carried if the transplanting, fertilizing, or pruning had not occurred.

Before the details of genetics became known, many prominent biologists believed that hereditary changes in populations over long periods of time (evolution) occurred as a result of the inheritance of acquired characteristics. One of the more prominent supporters of this widespread idea was Jean Baptiste Lamarck (1744–1829). He believed that giraffes acquired their long necks over many generations as a result of the gradual increase in neck length as shorter-necked animals stretched to reach leaves on the branches of trees. Slight stretching of the neck was supposed to have been passed on to the offspring as it occurred, and eventually numerous tiny increases due to individual stretching added up to the present great neck length of giraffes (Fig. 15.2).

If acquired characteristics could be inherited, we should be able to demonstrate it experimentally, and, indeed, many workers have attempted to do so, but all have failed. For example, one biologist surgically removed the tails of mice for many successive generations, but the average length of the tails of the last generation was exactly the same as that of the first generation. The experiment demonstrated that repeatedly removing tails in no way affects the hereditary characteristics carried in the genes within the cells. This is also the reason fruit trees that are pruned annually never produce seeds that grow into dwarfed trees, even after many generations.

Despite the experiments just discussed, short-tailed mice and dwarf fruit trees do occasionally occur but for reasons quite different from those proposed by Lamarck and his contemporaries. They come about as a result of a sudden change in a gene or chromosome. Such a change is called a **mutation,** a term introduced in 1901 by the Dutch botanist Hugo de Vries.

FIGURE 15.2   Lamarck and his contemporaries believed that the long neck of a giraffe developed over time as the animals stretched to reach higher leaves, and that the little length increases were inherited and became cumulative. This theory of inheritance of acquired characteristics was experimentally disproved and discarded when the true mechanisms of gene inheritance became known.

Changes within chromosomes may occur in several ways. A part of a chromosome may break off and be lost (*deletion*), or a piece of a chromosome may become attached to another (*translocation*). In some instances, a part of a chromosome may break off and then become reattached in an inverted position (*inversion*) (Fig. 15.3). A mutation of a gene itself may involve a change in one or more nucleotide pairs (nucleotides are discussed in Chapter 13).

Mutation rates vary considerably from gene to gene, but mutations occur constantly in all living organisms at an average estimated to be roughly one mutant gene for every 200,000 produced. *Mutator genes* that increase the rate of mutation in other genes have been discovered, but generally the mutation rate for a specific gene remains relatively constant unless changes in the environment (e.g., increases in cosmic radiation) occur. Most mutations are harmful, many times to the point of killing the cell. About 1% or fewer are either harmless or produce a characteristic that may help the organism survive changes in its environment.

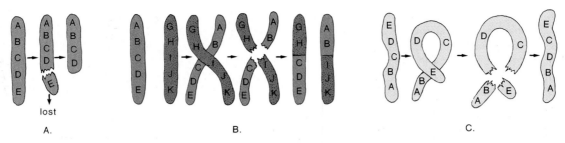

FIGURE 15.3   Some types of chromosomal changes that can occur. *A*. **Deletion.** Part of a chromosome breaks off and is lost. *B*. **Translocation.** Part of one chromosome becomes attached to another. *C*. **Inversion.** Part of a chromosome breaks off, becomes inverted from its original position, and then is reattached.

MODERN APPROACHES TO EVOLUTION

Modern theories of evolution combine Darwin's ideas with those of Mendel and include the Hardy-Weinberg law, discussed on page 220. They recognize today that evolution is based primarily on changes in gene frequencies resulting from mutation, *reproductive isolation*, and recombination of genes through hybridization.

THE ROLE OF HYBRIDIZATION IN EVOLUTION

Hybridization (the production of offspring by two parents that differ in one or more characteristics) seldom occurs naturally, but when it does take place, the hybrids may be significant or important in evolutionary change, depending on how the characteristics of the parents were combined. If, for example, the environment changes (e.g., average temperatures drop or annual precipitation increases), hybrids may have gene combinations that are better suited to the new environment than those of either parent.

If two species hybridize occasionally, *introgression* (backcrossing between the hybrids and the parents) may occur. If the backcrossing occurs repeatedly, some characteristics of the parents may eventually disappear from the population, if the new combinations of genes in the offspring happen to be better suited to the environment than those of the parents and as natural selection favors the offspring. Both parents and hybrids may, however, also evolve in other ways.

Polyploidy occurs occasionally in nature when, during mitosis, a new cell wall fails to develop between two daughter cells, even though the chromosomes have divided. As observed during our discussion of plant breeding in Chapter 14, this situation can result in a cell with twice the original number of chromosomes. If mitosis were to occur normally after such a cell was formed and the cell divided repeatedly until a complete organism resulted, that organism would have double the original number of chromosomes in all its cells.

The hybrids resulting from a cross between two species are often sterile because the chromosomes do not pair up properly in meiosis. If polyploidy does occur in such a hybrid, however, the extra set of chromosomes present from each parent provides an opportunity for any one chromosome to pair with its homologue in meiosis, and thus possibly the problem of sterility may be overcome. This type of polyploidy apparently occurred frequently in the past (in terms of geological time), and it is believed that more than 100,000 species of flowering plants that exist today originated in this fashion (Fig. 15.4).

Sterile hybrids also may reproduce asexually. Asexual propagation, called **apomixis,** may include the development and production of seeds without fertilization, as well as other forms of vegetative reproduction discussed in chapters 4 through 7 and 14, and in Appendix 4.

When species reproduce mainly by apomixis but sometimes also hybridize so that new combinations of genes are occasionally produced, they can be highly successful in nature. Dandelions and wild blackberries, for example, reproduce apomictically, as well as by other means, and are among the most successful plants known (Fig. 15.5).

REPRODUCTIVE ISOLATION

If new genes are produced in a freely interbreeding population, they will gradually be spread throughout the population, and the nature of the whole population will change in time. If some barrier divides the population, however, the two new populations eventually may become distinct from each other, sometimes in a relatively short period of time.

The log of a Portuguese sailing vessel of more than 600 years ago indicates that, for unknown reasons, some rabbits native to Portugal were released on one of the Canary Islands during a visit. When 20th century biologists examined the island descendants of those rabbits, which in Portugal forage during the day, they found them to be smaller than their continental ancestors, to be nocturnal in foraging habits, and to have larger eyes. In addition, attempts at breeding them back to their European ancestors failed because, in the short space of 600 years, a new species of rabbit had evolved.

FIGURE 15.4 Fireweed (*Epilobium angustifolium*)—a polyploid found primarily in mountainous areas of North America below 3,050 meters (10,000 feet). One subspecies has four sets of chromosomes, with a diploid chromosome number of $2n = 36$. A second subspecies has 8 sets of chromosomes, with a diploid chromosome number of $2n = 72$.

FIGURE 15.5 A common dandelion. In addition to reproducing by ordinary vegetative and sexual means, dandelions reproduce apomictically. Apomixis is a form of sexual reproduction through which seeds are produced without fertilization.

How do two populations of organisms that initially have the same gene pool come to have gene pools different enough to prevent their interbreeding? Geographic or other isolation of the two populations from each other prevents the flow of genes between the two populations, and random mutations, which are rarely identical, then spread only throughout the population in which they arise. In time, the genetic changes may become so great that, even when the isolation is removed, gene flow between the two populations no longer can occur.

In the United States, there are two closely related species of small trees or shrubs called *redbuds* that look very much alike. The eastern redbud occurs on the borders of streams, mostly east of the Mississippi River between the Canadian border and Florida; the western redbud is native to stream areas in California, Utah, Nevada, and Arizona. The two species can be artificially hybridized, but each is so

adapted to its own wild habitat and associated climate that specimens of either species succumb when transplanted to the other's wild habitat (Fig. 15.6).

Several other factors contribute to the development of new species from geographically isolated populations with a common ancestry. When separation first occurs it is most unlikely that both populations will have genes that are identical in all respects, and a small population will have only a small percentage of the genetic variation present throughout the original population. In addition, geographically isolated populations normally will be subjected to selection pressures from numerous subtle to conspicuous differences in environment. The Canary Island rabbits, for example, initially probably found it difficult to compete with other animals for food during the day, and as mutations for improved night vision occurred, those acquiring the genes were able to survive while those without them perished.

Isolation leading to the development of new species is not limited to physical barriers, such as mountains or oceans, or to climate. Soils may play a role, or a mutant form within

A.

B.

FIGURE 15.6 Two species of redbud, both native to North America. *A.* Eastern redbud (*Cercis canadensis*). *B.* Western redbud (*Cercis occidentalis*).

A.

B.

FIGURE 15.7 *A.* Dutchman's breeches (*Dicentra cucullaria*). *B.* Squirrel corn (*Dicentra canadensis*).

a population may flower at a different time, preventing exchange of genes between it and nonmutant forms.

In the temperate deciduous woods of eastern North America and in the Columbia River basin in the Pacific Northwest, there are many populations of early spring-flowering herbs called *Dutchman's breeches,* which are discussed further in Chapter 16. These have delicate, highly dissected leaves, which are nearly indistinguishable from those of related but slightly later-flowering squirrel corn plants often found growing right among the Dutchman's breeches in eastern North America (Fig. 15.7). So closely do the plants resemble each other vegetatively that early botanists and many lay persons assumed they were the same species.

It is believed, however, that at some point in geological history a mutation or mutations occurred that caused some plants to begin flowering after other plants had already set seed. As a result, one group became reproductively isolated from the other group. In due course, other mutations af-

fecting the form and fragrance of the flowers and the shape and pigmentation of food-storage bulblets beneath the surface also occurred, but the plants continued to occupy the same habitats. In short, we now have two closely related but distinct species growing together without interbreeding, simply because it is no longer possible for them to do so.

Other isolating mechanisms may be mechanical. In orchids, for example, the pollen is usually produced in little sacs called *pollinia* that stick to the heads or bodies of visiting insects. If pollination is to occur, the pollinia must be inserted within a concave stigma. Each species of orchid is constructed so that it is highly unlikely that a pollinium of one species will be inserted within the stigma of another species. As a result, many species of related orchids can be *sympatric* (occupy the same territorial range) without genes being exchanged.

Four closely related sympatric species of Peruvian *Catasetum* orchids, which can easily be artificially hybridized, have no known natural hybrids, despite their being pollinated by a single species of bee. Microscopic examination of the pollinators has shown that the pollinium of one species is attached to the insect's head, that of another is attached to the insect's back, that of a third to the abdomen, and that of the fourth only to the left front leg. Even after visits to hundreds of flowers, none of the pollinia is misplaced (Fig. 15.8).

Even if pollen from one species is placed on or within the stigma of another species, however, fertilization frequently does not follow because the sperm is chemically or mechanically prevented from reaching the egg. Other isolating mechanisms include the failure of hybrids to survive or breed and the failure of embryos to develop.

RATES OF EVOLUTION

Darwin believed that evolution by natural selection was a slow and gradual process. A number of contemporary biologists, however, favor variations of *punctuated equilibrium* theories that hold that major changes have taken place in spurts of maybe 100,000 years or less, followed by periods of millions of years during which changes, if any, have been minor. They base their theories on fossil records, which have large "chains" of missing organisms. Although missing-link fossils are occasionally discovered, the record does little to support Darwin's concept of gradual, long-term change, even though it is widely acknowledged that possibly as few as one organism in a million ever became a fossil. Others opposed to theories of evolution through sudden change argue that because such a tiny percentage of organisms becomes fossilized and usually only the harder parts of organisms (e.g., bones, teeth) are preserved, drawing definite conclusions from fossil evidence about evolution through either gradual or sudden change is not warranted.

As indicated in Chapter 21, the conditions necessary for an organism to become a fossil are very specialized and limited in occurrence and probably also were in the past. This makes it quite improbable that large numbers of missing-link fossils will ever be found. Proponents of evolution through periods of rapid change argue that, under conditions of changing climates or other situations exerting strong selection pressures on forms with adaptive mutations, new species of organisms could arise in less than 100 generations, making 100,000 years ample time for considerable evolution to occur. Since it is not possible to prove or disprove the various theories experimentally, the debate on evolution rates will undoubtedly continue indefinitely, until or unless new evidence convincingly supports one theory more than another.

FIGURE 15.8 *Catasetum* orchids.

DISCUSSION

If we now know and understand some of the mechanisms of evolution, why does there yet remain any controversy over the broad subject itself? Obviously, ignorance is a factor, but it is not the sole reason. Science deals with tangible facts and evidence that can be experimentally tested; beliefs stemming from metaphysics and religion are outside the realm of science to prove or disprove.

Evidence in support of evolution is drawn from several additional areas, including the form and ecology of living organisms and the way they are related to each other today. Other evidence comes from the structure and relationships of proteins and other molecules and from fossils, which are remnants of previously living forms (Fig. 15.9). The simplest fossils are generally found in the oldest geological strata, while more complex forms tend to be found in younger strata.

Still further evidence is drawn from the geographical distribution of organisms. Many groups are confined to a single continent or island. In some instances, where similar

FIGURE 15.9   A fossil fern.

organisms occur on more than one land mass, there is evidence that the land masses concerned were once linked together, which would have permitted terrestrial migration. Still other conclusions are drawn from the physiology and chemistry of the organisms and, in animals, particularly, from the presence of vestigial organs (structures that are apparently now functionless but are presumed to represent previously functional organs in predecessor species). Among the best known of such structures in humans are the appendix and wisdom teeth.

Most scientists who have studied the evidence feel that some form of evolutionary process is the only plausible explanation for the unity of life at the molecular and cellular level and the extraordinary diversity of the organisms now around us. There is little unanimity of thought, however, as to precisely how evolution proceeded in the past. One authority is convinced that a certain group evolved from another, while other equally eminent authorities maintain that the exact reverse occurred. Part of the reason for such paradoxes is that the historical record is quite incomplete.

Although the fossil evidence for the evolution of a few organisms, such as certain mollusks and the horse, is fairly substantial, other fossil evidence is very fragmentary. As mentioned in the previous section, possibly fewer than one in each million organisms that once existed ever became a fossil, and there are probably no fossils at all of many herbaceous and soft-bodied organisms. With such evidence, scientists can deal only in probabilities or possibilities, and it is inevitable that various, sometimes conflicting, interpretations result.

Objective scientists freely acknowledge that numerous problems concerning the interpretation of the geological past and the pathways of organic evolution exist, but they ask their detractors to suggest more plausible alternatives. At this point, some people apply the tenets of religious faith, which like history, are not subject to scientific experimentation. Some see no conflict between science and religion, and others are convinced that the two are mutually exclusive. Depending on individual points of view, an impasse may result, with persons of different persuasions—including, unfortunately,

some scientists—becoming dogmatic on the topic. Virtually no objective thinkers will deny, however, the extraordinary impact theories of evolution have had on modern peoples and on their concepts of the living world around them.

Summary

1. The theory of organic evolution received its greatest impetus from Charles Darwin's *On the Origin of Species.*

2. Darwin's theory is based on four principles or tenets: (1) overproduction; (2) struggle for existence; (3) variation and inheritance; and (4) survival and reproduction of the fittest.

3. Before 1900, many biologists, notably Lamarck, believed that evolution occurred as a result of the inheritance of acquired characteristics. This theory was discredited experimentally.

4. Mechanisms of evolution include mutations in chromosomes or genes, hybridization (particularly introgression), polyploidy, apomixis, and reproductive isolation.

5. If new genes are produced in a freely interbreeding population, they will gradually be spread throughout the population, and the nature of the whole population will change in time. If a population is divided by a barrier, genes occurring in the one population will not spread throughout the isolated population as before. In time, because of the isolation, each new population may develop into separate species incapable of breeding with one another.

6. Darwin believed that evolution through natural selection was a gradual process over great periods of time. Some contemporary biologists believe that evolution has taken place in spurts between long periods of little change, based on evidence from the fossil record. Interpretation of the fossil record is, however, controversial and will be debated indefinitely.

7. Additional evidence in support of evolution is drawn from the form, the ecology, and relationships of living organisms; molecular structure; geographical distribution; fossils; and vestigial organs.

8. Most scientists feel some form of evolution is the only plausible explanation for the unity of life at the molecular and cellular level and the great diversity of life, but there is little agreement among them as to precisely how evolution proceeded in the past.

9. There is a variety of opinions and convictions on origins, but few advocates can deny the major impact that theories of evolution have had on modern peoples and on their concepts of life.

Review Questions

1. What is the difference between organic evolution and other forms of evolution?

2. How did Darwin's theory of evolution differ from Lamarck's?

3. What basic modifications have been made in evolutionary theory since Darwin's time?

4. Why is a hybrid often sterile?

5. What is meant by a *mechanism of evolution*?

6. Why is reproductive isolation necessary for evolution to occur?

7. What evidence is there to support modern concepts of evolution? Are there any problems with the evidence?

Discussion Questions

1. One of Darwin's tenets was that there is a struggle for existence among living organisms. Do plants struggle with one another to survive? If so, how do they do it?

2. Some populations change noticeably in form within a hundred years. If only one gene in every 200,000 per cell mutates and if most mutations are harmful, how is such change possible?

3. Do you think there might be viable alternatives to organic evolutionary theory to account for the diversity of life around us? Explain.

Additional Reading

Avers, C. J. 1989. *Process and pattern in evolution*. New York: Oxford University Press.

Ayala, F. J. 1991. *Origin of species*, 2d ed. In J. J. Head (Ed.), *Carolina Biology Readers*. Burlington, NC: Carolina Biological Supply Company.

Babb, N. 1994. *Evidence for evolution*. New York: Vantage Press, Inc.

Baltscheffsky, H., et al. (Eds.). 1987. *Molecular evolution of life*. New York: Cambridge University Press.

Darwin, C. 1975. *On the origin of species*, facsimile 1st ed. of 1859. New York: Cambridge University Press.

Grant, V. 1991. *The evolutionary process: A critical review of evolutionary theory*, rev. ed. New York: Columbia University Press.

Kerkut, G. A. 1960. *Implications of evolution*. Oxford and New York: Pergamon Press.

Stebbins, G. L. 1977. *Processes of organic evolution*, 3d ed. Englewood Cliffs, NJ: Prentice-Hall.

Stone, E. M., and R. J. Schwartz. 1990. *Intervening sequences in evolution and development*. New York: Oxford University Press.

A flower of Chinese hibiscus (Hibiscus rosa-sinensis), the most widely cultivated of some 200 hibiscus species. In India the flowers are used for shining shoes. Other species are grown commercially as a source of fiber for canvas, rope, and sails.

16 Plant Names and Classification

Overview

This chapter begins with a discussion of the problems involved in the use of common names for plants, as illustrated by a survey of such names for two related species of American spring-flowering perennials. It continues with a brief historical account of the events that led to the development and acceptance of Linnaeus's Binomial System of Nomenclature. The history of the development of a five-kingdom classification is presented, along with a list of the divisions and classes included in the kingdoms covered in this text. A dichotomous key to the kingdoms and divisions of organisms is provided. The chapter concludes with a brief discussion of cladistics.

Some Learning Goals

1. Understand several problems associated with the application of common names to organisms.
2. Know what the Binomial System of Nomenclature is, how it developed, and how it is currently used.
3. Learn several reasons for recognizing more than two kingdoms of living organisms.
4. Understand the bases for Whittaker's five-kingdom system.
5. Construct a dichotomous key to 10 different objects (e.g., pencils, golf balls, crayons, balloons, clocks).

The general public thinks biologists are slightly pompous or weird when they refer to peanuts as *Arachis hypogaea* or an African hoopoe bird as *Upupa epops*. The biologists aren't, however, merely showing off or being difficult. Rather, they're identifying organisms by their *scientific names* through a system that has evolved over the last few centuries. It has now become vital to us, regardless of the location or language, to be able to distinguish among the 10 million existing types of organisms, as well as those that have become extinct, and to do so in a way that will identify them anywhere.

At present, all living organisms are given a single two-part Latin scientific name, and most also have *common names*. Only one correct scientific name applies to all individuals of a specific kind of organism, no matter where they're found, but many common names may be given to the same organism, and one common name may be applied to a number of different organisms.

The scientific name *Dicentra cucullaria,* for example, has been given to a pretty spring-flowering plant native to eastern North America and the Columbia River basin in the West. Its unique flower shape, which recalls the baggy pants of a traditional Dutch costume, has resulted in its being given the common name *Dutchman's breeches.* It also, however, has the common names *little-boys' breeches, kitten's breeches, breeches-flower, boys-and-girls, Indian boys-and-girls, monkshood, white eardrops, soldier's cap, colicweed, little blue stagger, white hearts, butterfly banners, rice roots,* and *meadow bells.* In addition to these English common names, the plant has Indian names, and in the Canadian province of Quebec it has French names.

FIGURE 16.1 Plantain (*Plantago major*). These plants have at least 300 different common names.

Often growing with *Dicentra cucullaria* is a related plant, with similar leaves but with slightly differently shaped flowers, having the scientific name *Dicentra canadensis.* Because of the similarities between the two plants and their close association in the woods where they occur, some people in the past assumed that they were merely two different forms of the same plant. This is reflected in the fact that some of the common names of *Dicentra cucullaria* (e.g., colicweed, Indian boys-and-girls, little blue stagger) have also been applied to *Dicentra canadensis,* in addition to the names *squirrel corn, turkey corn, turkey pea, wild hyacinth, fumitory, staggerweed,* and *trembling stagger.* But the problem of the diversity and the overlapping of common names for these two plants doesn't stop there, for the names *monkshood, soldier's cap, rice roots, meadow bells,* and *turkey corn* have also been applied to completely unrelated plants. Both species of *Dicentra* are shown in Figure 15.7.

In Europe, with its many languages, common names can become very numerous indeed. The widespread weed with the scientific name *Plantago major,* for example, is often called *broad-leaved plantain* in English, but it also has no fewer than 45 other English names, 11 French names, 75 Dutch names, 106 German names, and possibly as many as several hundred more names in other languages, with literally dozens of these common names also applying to quite different plants (Fig. 16.1). If it weren't for the early recognition by biologists and others of the urgent need for worldwide uniformity in naming and classifying all organisms, utter chaos eventually might have prevailed in communications concerning them.

DEVELOPMENT OF THE BINOMIAL SYSTEM OF NOMENCLATURE

The first person known to attempt to organize and classify plants was Theophrastus, the brilliant student of Aristotle and Plato. His third century B.C. classification of nearly 500 plants into trees, shrubs, and herbs, along with his distinctions between plants on the basis of leaf characteristics, was used for hundreds of years. It was not until the 13th century that a distinction was made between monocots and dicots on the basis of stem structure.

After this, herbalists added many plants to Theophrastus's list, and by the beginning of the 18th century, details of fruit and flower structure, in addition to form and habit, were used in classification schemes. European scholarly institutions were beginning to bulge with thousands of plants that explorers had brought back from around the world, and confusion over scientific and common names was increasing.

The use of Latin in schools and universities had become widespread, and it was then customary to use descriptive Latin phrase names for both plants and animals. All organisms were grouped into **genera** (singular: **genus**), with the first word of the Latin phrase indicating the particular genus to which the organisms belonged. For example, all known mints were given phrase names beginning with the word *Mentha,* the name of the genus. Likewise, the phrases for lupines began with *Lupinus,* and those for poplars began with *Populus.* A complete phrase name for spearmint was *Mentha floribus spicatis, foliis oblongis serratis.* Roughly translated, it means "Mentha with flowers in a spike (an elongated but compact flower cluster); leaves oblong, saw-toothed." A phrase name for the closely related peppermint was *Mentha floribus capitatus, foliis lanceolatis serratis subpetiolatis,* meaning "Mentha with flowers in a head; leaves lance-shaped, saw-toothed, with very short petioles."

Linnaeus

At this point, the Swedish botanist Carolus Linnaeus (1707–1778; Fig. 16.2) began improving the way organisms were named and classified. The system he established worked so well that it has persisted to the present. In fact, Linnaeus's system is now used throughout the entire world.

Linnaeus, who was nicknamed the "Little Botanist" at school, inherited his passion for plants from his father, who was a minister and an amateur gardener. He is said to have been much impressed at the age of 4 by his father's remarks about the uses of neighborhood plants in his home community of Råshult, located in southern Sweden. After a brief

FIGURE 16.2 Linnaeus.
(Courtesy National Library of Medicine)

tenure as a student at the University of Lund, he spent most of his time making excursions to Lapland, Holland, France, and Germany. Eventually, he became the professor of botany and medicine at Uppsala, where he inspired large numbers of students, 23 of whom became professors themselves. He frequently led large field trips into the countryside, accompanied by a musical band.

When Linnaeus began his work, he set out to classify all known plants and animals according to their genera. In 1753, he published a two-volume work entitled *Species Plantarum,* which was to become the most important of all works on plant names and classification. In this work, he not only included a referenced list of all the Latin phrase names previously given to the plants, but, when necessary, he changed some of the phrases to reflect relationships, placing one to many specific kinds of organisms called **species**[1] in each genus. He limited each Latin phrase to a maximum of 12 words, and in the margin next to the phrase he listed a single word, which, when combined with the generic name, formed a convenient abbreviated designation for the species.

---

1. Note that *species* is like the word *sheep* in that it is spelled and pronounced the same in either singular or plural usage. There is no such thing as a plant or animal "specie." Since Linnaeus's time, a species has been defined as "a population of individuals capable of freely interbreeding in nature but not generally interbreeding with members of another species."

The word in the margin for spearmint was *spicata,* and the word for peppermint was *piperita*. Accordingly, the abbreviated name for spearmint was *Mentha spicata* and for peppermint *Mentha piperita*.

Although Linnaeus originally considered the phrase names the official names of the plants, he and those who followed him eventually replaced all the phrase names with abbreviated ones.

Because of their two parts, these abbreviated names became known as *binomials* and the method of naming became known as the *Binomial System of Nomenclature*. Today, all organisms are named according to this system, which also includes the authority for the name, either in abbreviated form or in full, after the Latin name. For example, the full scientific name for spearmint is written *Mentha spicata* L., the L. standing for Linnaeus; the full scientific name for the common dandelion is written *Taraxacum officinale* Wiggers, after Fredericus Henricus Wiggers, who was the first (in his *Flora of Holstein,* published in 1780) to describe the species.

Besides establishing the Binomial System of Nomenclature, Linnaeus tried to do more than publish just a long list of plant names. After all, such lists aren't very useful if they can't be used to identify the plants concerned. Linnaeus organized all known plants into 24 **classes,** which were based mainly on the number of stamens (pollen-bearing structures) in flowers. All plants with five stamens per flower, for example, were placed in one class, while those with six stamens were placed in another class.

Plants and other organisms that don't produce flowers (e.g., mosses, fungi) were put in a class of their own. His arrangement was for convenience and was artificial because it did not necessarily reflect natural relationships. For the first time, however, it became possible for workers to identify plants previously unknown to them. This arrangement was used in *Species Plantarum* and other works by Linnaeus (Fig. 16.3).

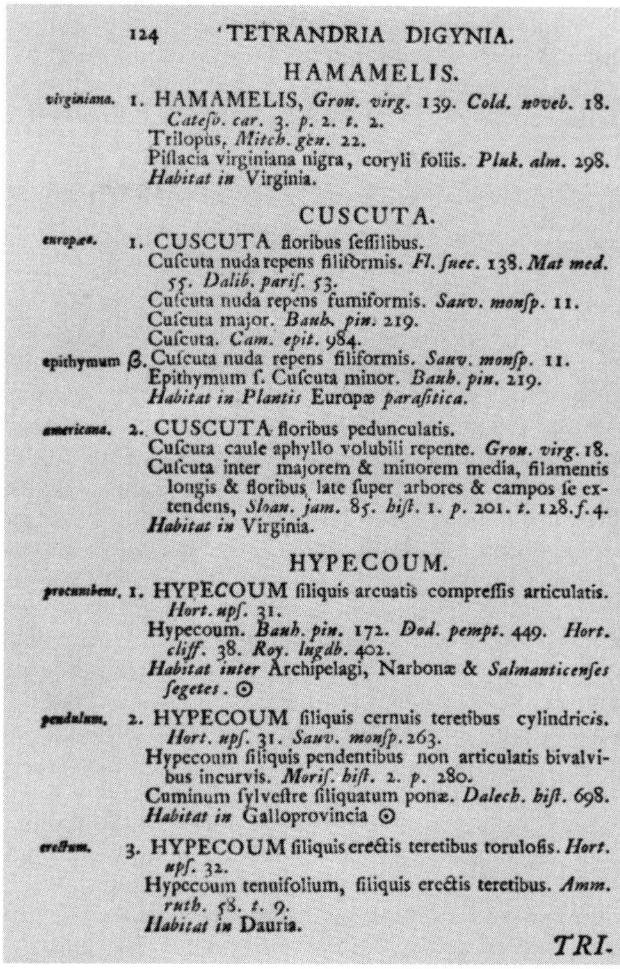
124 TETRANDRIA DIGYNIA.

HAMAMELIS.

virginiana. 1. HAMAMELIS, *Gron. virg.* 139. *Cold. noveb.* 18. *Catesb. car.* 3. *p.* 2. *t.* 2.
Trilopus, *Mitch. gen.* 22.
Pistacia virginiana nigra, coryli foliis. *Pluk. alm.* 298.
Habitat in Virginia.

CUSCUTA.

europæa. 1. CUSCUTA floribus sessilibus.
Cuscuta nuda repens filiformis. *Fl. suec.* 138. *Mat med.* 55. *Dalib. paris.* 53.
Cuscuta nuda repens fumiformis. *Sauv. monsp.* 11.
Cuscuta major. *Bauh. pin.* 219.
Cuscuta. *Cam. epit.* 984.
epithymum β. Cuscuta nuda repens filiformis. *Sauv. monsp.* 11.
Epithymum s. Cuscuta minor. *Bauh. pin.* 219.
Habitat in Plantis Europæ *parasitica.*

americana. 2. CUSCUTA floribus pedunculatis.
Cuscuta caule aphyllo volubili repente. *Gron. virg.* 18.
Cuscuta inter majorem & minorem media, filamentis longis & floribus late super arbores & campos se extendens, *Sloan. jam.* 85. *hist.* 1. *p.* 201. *t.* 128. *f.* 4.
Habitat in Virginia.

HYPECOUM.

procumbens. 1. HYPECOUM siliquis arcuatis compressis articulatis. *Hort. ups.* 31.
Hypecoum. *Bauh. pin.* 172. *Dod. pempt.* 449. *Hort. cliff.* 38. *Roy. lugdb.* 402.
Habitat inter Archipelagi, Narbonæ & *Salmanticenses segetes.* ⊙

pendulum. 2. HYPECOUM siliquis cernuis teretibus cylindricis. *Hort. ups.* 31. *Sauv. monsp.* 263.
Hypecoum siliquis pendentibus non articulatis bivalvibus incurvis. *Moris. hist.* 2. *p.* 280.
Cuminum sylvestre siliquatum ponæ. *Dalech. hist.* 698.
Habitat in Galloprovincia ⊙

erectum. 3. HYPECOUM siliquis erectis teretibus torulosis. *Hort. ups.* 32.
Hypecoum tenuifolium, siliquis erectis teretibus. *Amm. ruth.* 58. *t.* 9.
Habitat in Dauria.

TRI-

FIGURE 16.3 A page from *Species Plantarum* by Linnaeus.

The International Code of Botanical Nomenclature

In 1867, more than 100 years after *Species Plantarum* was published, about 150 European and American botanists met in Paris to try to standardize rules governing the naming and the classifying of plants. They agreed to use the works of Linnaeus as the starting point for all scientific names of plants and decided that his binomials, or the earliest ones published after him, would have priority over all others.

International congresses of botanists have met at varying intervals since 1867 and have revised and expanded these rules. Today, the modified rules comprise what is known as the *International Code of Botanical Nomenclature,* which is a single book with a common index to its English, French, and German translations of the various rules and recommendations. It now specifically recognizes Linnaeus's *Species Plantarum* as the starting point for scientific names of plants and spells out details of naming and classifying, which are adhered to by botanists of all nationalities. A similar code for animals has been developed and established by zoologists. It, too, uses a work of Linnaeus as the starting point for scientific names.

Development of the Kingdom Concept

If you were to ask the average person on the street the differences between plants and animals, you would probably be told that plants are green, don't eat each other, and don't move. Animals, on the other hand, are not green, do eat plants or each other, and do move. A distinction between plants and animals in general has probably always existed in the minds of intelligent beings, and it was natural when classification schemes were first developed that all living organisms would be placed, according to the highest category of kingdom, in either the *Plant Kingdom* or the *Animal Kingdom.*

While this distinction still works well for the more complex plants and animals, it breaks down for some of the so-called simpler organisms. For example, there are more than 300 species of single-celled organisms called *euglenoids* (see Fig. 18.7) inhabiting a variety of freshwater habitats. These microscopic creatures have small whiplike tails called *flagella* that pull them through the water, and they also can ingest food particles through a groove called a *gullet*. Both of these features would be considered animal-like, and so euglenoids in the past have been treated as animals in some textbooks. Many of these organisms also, however, have chloroplasts, and if light is present, they can carry on photosynthesis efficiently enough to eliminate the need for ingesting food. Accordingly, they have often been treated as plants in botany books.

A similar problem has existed with slime molds, which resemble masses of protoplasm slowly flowing or creeping across dead leaves and debris, engulfing and feeding on bacteria and other substances as they go, and looking somewhat like amoebae, which botanists and zoologists alike have traditionally called animals. When slime molds reproduce, however, dramatic changes take place. They become stationary and develop reproductive bodies that are distinctly funguslike, causing some to consider them fungi.

In an attempt to overcome this problem, the biologists J. Hogg and Ernst Haeckel proposed a third kingdom in the 1860s. All organisms that did not develop complex tissues (e.g., algae, fungi, and sponges) were placed in a third kingdom called *Protoctista*. This third kingdom included such a variety of organisms, however, that in 1938 another biologist by the name of Herbert F. Copeland proposed it, too, be divided. He assigned the name *Monera* to all single-celled organisms with prokaryotic cells (the differences between prokaryotic and eukaryotic cells are discussed in chapters 3 and 17), leaving the algae, fungi, and single-celled organisms with eukaryotic cells in the Kingdom Protoctista.

Although many biologists thought Copeland's four-kingdom system of classification was a definite improvement, it too had problems—particularly the fact that basic differences in the mode of nutrition existed among organisms within the Kingdom Protoctista. As a result, many biologists today favor variations of a five-kingdom system proposed by R. H. Whittaker in 1969 (Table 16.1).

In Whittaker's system, three kingdoms of more complex organisms based on three basic forms of nutrition (photosynthesis, ingestion of food, and absorption of food in solution) are recognized, along with two kingdoms of *protists*, which are distinguished on the basis of differences in cellular structure.

Even the five-kingdom arrangement is not without its critics and problems. Carl Woese, a microbiologist and leading authority on bacteria, argued cogently in the early 1980s that Kingdom Monera should be divided into two kingdoms, based on some previously unrecognized fundamental differences between two major groups of bacteria (discussed in Chapter 17). More recently, James Lake and his colleagues have proposed dividing the archaebacteria into two kingdoms, making three kingdoms of prokaryotes in all. The slime molds, which have no cell walls during their active state but do develop walls when they reproduce, still do not fit well into the five-kingdom system, and other organisms traditionally regarded as fungi may be more closely related to protists. Much information still needs to be accumulated to understand natural relationships, and until this occurs, taxonomic categories will remain subject to different interpretations.

CLASSIFICATION OF MAJOR GROUPS

Since Linnaeus's time, a number of classification categories have been added between the levels of kingdom and genus. Genera are now grouped into **families**, families into **orders**, orders into **classes**, classes into **divisions**,[2] and divisions into **kingdoms**.

Depending on which system of classification is used, there may be between 12 and nearly 30 divisions of "plants" recognized.

Table 16.2 presents a classification of major groups of living organisms utilizing a modification of Whittaker's five-kingdom system. It is the classification followed throughout this book. Hypothetical derivations and relationships are indicated in Figure 16.4.

Viruses, which don't have cellular structure or most of the other attributes of living organisms, are not included in this classification; they are discussed in Chapter 17 after the bacteria. Lichens, each of which is a combination of an alga and a fungus, are discussed with the fungi in Chapter 19. If we were to give a complete classification of an onion according to this particular arrangement, it would look like this:

> Kingdom: Plantae
>> Division: Magnoliophyta
>>> Class: Liliopsida
>>>> Order: Liliales
>>>>> Family: Liliaceae
>>>>>> Genus: *Allium*
>>>>>>> Species: *Allium cepa* L.

The second part of the scientific name of a species, called the *specific epithet,* is followed by the name of the author, usually in abbreviated form. The author is the person (or persons) who originally named the plant or placed the species in a particular genus. Binomials are printed in italics (or the words are underlined). In the case of *Allium cepa,* the L. following the name stands for Linnaeus, since this plant was one of the 7,300 species he included in his *Species Plantarum.*

---

2. *Division* is equivalent to the term *phylum,* which is used in animal classification. Although use of the word *phylum* for all living organisms has been proposed, the International Code of Botanical Nomenclature presently permits only the use of the word *division* for plants.

Table 16.1

FOUR CLASSIFICATIONS OF ORGANISMS INTO KINGDOMS

TWO KINGDOMS (TRADITIONAL)	THREE KINGDOMS (HOGG AND HAECKEL)	FOUR KINGDOMS (COPELAND)	FIVE KINGDOMS (WHITTAKER)	FEATURES
		Monera Bacteria	**Monera** Bacteria	Cells prokaryotic
	Protoctista Bacteria Algae Slime molds Flagellate fungi True fungi Protozoa Sponges	**Protoctista** Algae Slime molds Flagellate fungi True fungi Protozoa Sponges	**Protoctista** Algae Slime molds Flagellate fungi Protozoa Sponges	Cells eukaryotic
			Fungi True fungi	Absorb food in solution
Plantae Bacteria Algae Slime molds Flagellate fungi True fungi Bryophytes Vascular plants	**Plantae** Bryophytes Vascular plants	**Plantae** Bryophytes Vascular plants	**Plantae** Bryophytes Vascular plants	Produce food via photosynthesis
Animalia Protozoa Sponges Multicellular animals	**Animalia** Multicellular animals	**Animalia** Multicellular animals	**Animalia** Multicellular animals	Ingest food

In addition to these categories of classification, various "in-between" categories, such as *subdivision, subclass,* and *suborder,* have been used, and species themselves are sometimes further divided into *subspecies, varieties,* and *forms.* A few authors also recognize "super" categories, such as *superkingdom* and *superorder.* One classification widely used in the past recognizes only two divisions in Kingdom Plantae (Division Bryophyta for nonvascular plants and Division Tracheophyta, with a number of subdivisions, for all vascular plants). These subdivisions are treated as full divisions in this and a number of other texts.

Since human judgment has to enter into the compilation of classifications, there undoubtedly will never be complete agreement on them, but they are nevertheless useful, indeed necessary, for making some order of the diversity of life around us.

Those who specialize in identifying, naming, and classifying organisms and trying to sort out natural relationships are called *taxonomists* (or *systematists*). One of the activities of taxonomists is the construction of keys to help others identify organisms with which they may not be familiar. Most such keys are *dichotomous;* that is, they give the reader pairs of statements based on features of the organisms. By carefully examining an organism and choosing from each pair the statements that most closely apply to the organism, we can arrive at an identification. If, for example, we agree with the first of the two statements given for each number in the key, then we either use the identification at the end of the line or proceed to the next indented statement beneath. If we disagree with the first of the paired statements, then we look for the matching alternate statement and proceed from there.

The key on pages 262–263, which deals with major groups of organisms, illustrates natural relationships. It covers only general features and does not indicate various details or occasional exceptions that would appear in keys for lesser groups. All organisms in any given group are presumed to be more closely related to each other than to organisms in another group.

CLASSIFICATION OF ORGANISMS IN FIVE KINGDOMS

Kingdom Monera
Subkingdom Archaebacteriobionta (archaebacteria)
   Division Archaebacteria (methane, salt, and sulfolobus bacteria)
Subkingdom Eubacteriobionta (true bacteria)
   Division Eubacteria
       Class Eubacteriae (unpigmented, purple, and green sulphur bacteria)
       Class Cyanobacteriae (blue-green bacteria)
       Class Chloroxybacteriae (chloroxybacteria)

Kingdom Protoctista
Subkingdom Phycobionta
   Division Chrysophyta
       Class Xanthophyceae (yellow-green algae)
       Class Chrysophyceae (golden-brown algae)
       Class Bacillariophyceae (diatoms)
       Class Cryptophyceae (cryptophytes)
   Division Pyrrophyta (dinoflagellates)
   Division Euglenophyta (euglenoids)
   Division Chlorophyta
       Class Chlorophyceae (green algae)
       Class Charophyceae (stoneworts)
   Division Phaeophyta (brown algae)
   Division Rhodophyta (red algae)
Subkingdom Mastigobionta
   Division Chytridiomycota (chytrids)
   Division Oomycota (water molds)
Subkingdom Myxobionta
   Division Acrasiomycota (cellular slime molds)
   Division Myxomycota (true slime molds)
   [Phylum Protozoa—protozoans]
   [Phylum Porifera—sponges]

Kingdom Fungi
   Division Zygomycota (coenocytic fungi)
   Division Eumycota (septate fungi)
       Class Ascomycetes (cup fungi)
       Class Basidiomycetes (club fungi)
       Class Deuteromycetes (imperfect fungi)
       [Class Lichenes (lichens)]

Kingdom Plantae
   Division Hepaticophyta (liverworts)
   Division Anthocerotophyta (hornworts)
   Division Bryophyta (mosses)
   Division Psilotophyta (whisk ferns)
   Division Lycophyta (club mosses)
   Division Sphenophyta (horsetails)
   Division Pterophyta (ferns)
   Division Pinophyta (gymnosperms)
    Subdivision Cycadicae (cycads)
    Subdivision Pinicae
      Class Ginkgoatae (*Ginkgo*)
      Class Pinatae (conifers)
    Subdivision Gneticae (*Gnetum, Ephedra, Welwitschia*)
   Division Magnoliophyta (flowering plants)
      Class Magnoliopsida (dicots)
      Class Liliopsida (monocots)

                                   Division Trachaeophyta

Kingdom Animalia (multicellular animals)

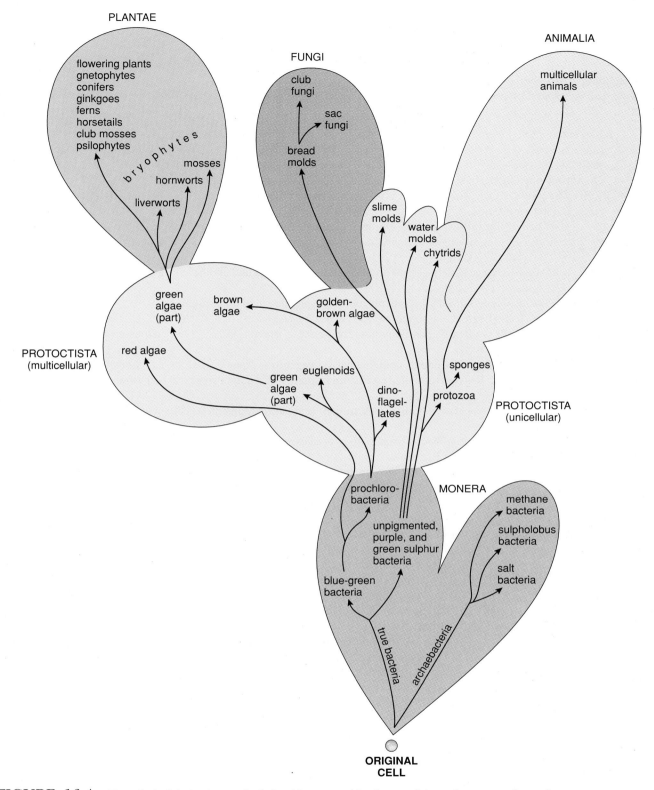

FIGURE 16.4 Hypothetical derivations and relationships among kingdoms and the major groups of organisms.

KEY TO MAJOR GROUPS OF ORGANISMS (EXCLUSIVE OF KINGDOM ANIMALIA)

1. Unicellular, prokaryotic organisms with cell walls . . . .**Kingdom Monera**
   2. Cell walls without muramic acid . . . . *Subkingdom Archaebacteriobionta*
   2. Cell walls with muramic acid . . . . *Subkingdom Eubacteriobionta*

1. Unicellular, colonial, filamentous, or multicellular eukaryotic organisms, with or without cell walls.
   3. Organisms whose female (and usually male) reproductive structures consist of a single cell or of sterile cells surrounding the one-celled reproductive structures; zygotes not developing into embryos.
      4. Organisms unicellular, filamentous, or plasmodial (i.e., with naked protoplasm).
         5. Reproductive cells with flagella (or food reserve floridean starch) . . . . **Kingdom Protoctista**
            6. Cells with plastids . . . . *Subkingdom Phycobionta*
               7. Plastids with yellow, brown, or orange pigments more conspicuous than the chlorophyll pigments.
                  8. Food reserves oils or carbohydrates other than starch; two flagella both located at one end of the cell . . . . .Division Chrysophyta
                  8. Food reserve starch; cells with a flagellum at one end and another at right angles to it in a central groove . . . . .Division Pyrrophyta
               7. Plastids with chlorophyll pigments more conspicuous than other pigments.
                  9. Cells flexible; carbohydrate food reserve *paramylon* . . . . .Division Euglenophyta[3]
                  9. Cells not flexible; carbohydrate food reserve starch . . . . .Division Chlorophyta (in part)
            6. Cells without plastids.
               10. Vegetative bodies not amoebalike; organisms aquatic . . . .*Subkingdom Mastigobionta*
                  11. Vegetative bodies consisting primarily of a single spherical cell, often with rhizoids . . . . .Division Chytridiomycota
                  11. Vegetative bodies consisting of branching, filamentous cells with numerous nuclei . . . . .Division Oomycota
               10. Vegetative bodies amoebalike; organisms terrestrial . . . .*Subkingdom Myxobionta*
         5. Reproductive cells without flagella . . . .**Kingdom Fungi**
            12. Filaments of the vegetative bodies branched and with numerous nuclei, but not partitioned into individual cells . . . . .Division Zygomycota
            12. Filaments of the vegetative bodies partitioned into individual cells, each with one to several nuclei . . . . .Division Eumycota
      4. Organisms multicellular, not filamentous or plasmodial.
            13. Organisms with accessory pigments essentially similar to those of higher plants; carbohydrate food reserve starch . . . . .Division Chlorophyta (in part)
            13. Organisms with some accessory pigments differing from those of higher plants; food reserves carbohydrates other than ordinary starch.
               14. Organisms brownish in color due to presence of brown pigments; carbohydrate food reserve *laminarin* . . . . .Division Phaeophyta
               14. Organisms reddish in color due to presence of red pigments; carbohydrate food reserve *floridean starch* . . . . .Division Rhodophyta
   3. Organisms with multicellular reproductive structures; zygotes developing into embryos . . . .**Kingdom Plantae**
         15. Plants without true xylem or phloem . . . . .Division Bryophyta
         15. Plants with true xylem and phloem.
            16. Plants with true leaves absent; *enations*[4] present; stems branching dichotomously . . . . .Division Psilotophyta
            16. Plants with true leaves present; enations absent; stems branching in various ways or unbranched
               17. Plants with leaves having a single vein (*microphylls*).

---

3. Only about a third of the euglenoids develop chloroplasts.
4. See page 360 for a discussion of *enations*.

---

18. Stems not ribbed and not containing silica; leaves photosynthetic . . . . .Division Lycophyta
18. Stems ribbed and containing silica; leaves reduced to scales and nonphotosynthetic . . . . .Division Sphenophyta
17. Plants with leaves usually having more than one vein (*megaphylls*).
    19. Plants reproducing by means of spores produced on the leaves . . . . .Division Pterophyta
    19. Plants reproducing by means of seeds developed from ovules.
        20. Plants without flowers; seeds produced in cones or conelike structures . . . . .Division Pinophyta
            21. Leaves pinnate and large, resembling those of palms . . . . .Subdivision Cycadicae
            21. Leaves not pinnate or palmlike . . . . . . .Subdivision Pinicae
                22. Leaves fan-shaped, with numerous dichotomously forking veins . . . . .Class Ginkgoatae
                22. Leaves not fan-shaped.
                    23. Wood containing no vessels; resin canals present . . . . .Class Pinatae
                    23. Wood containing vessels; resin canals absent . . . . .Subdivision Gneticae
        20. Plants with flowers; seeds produced in enclosed ovaries . . . . .Division Magnoliophyta
            24. Flowers with parts mostly in fours and fives; cotyledons two . . . . .Class Magnoliopsida
            24. Flowers with parts mostly in threes; cotyledon one . . . . .Class Liliopsida

Once again, it must be emphasized that a key of this type, while calling attention to basic differences between major groups of organisms, may also oversimplify the differences, partly because it does not list all the characteristics of each member of a given group. Moreover, it does not necessarily identify exceptions to the rule or some of the intermediate forms that frequently are transferred back and forth as research uncovers new evidence or as new lines of speculation develop.

Keys at this level of classification also are not completely practical because, to be able to arrive at an identification, one sometimes has to have available specific stages in the life cycles of the plants. Furthermore, chemical tests that are not easily performed or details that may be difficult to see with a light microscope may be required. Nevertheless, such a key is generally preferable to a simple listing of groups and their characteristics for purposes of distinguishing among them.

CLADISTICS

Since the 1970s, taxonomists have increasingly used **cladistics** in their attempts to determine natural relationships. Cladistics is a system of examining natural relationships among organisms, based on features shared by the organisms. The relationships are portrayed in straight line diagrams (evolutionary trees) called *cladograms* (Fig. 16.5).

In cladistics, the value or form of a feature is referred to as a *character state*. In a feature such as flower color, for example, the character could be flower color, and the state could be purple or red. After the specific chemical or other nature of a character is determined, hypotheses (assumptions) are made about which states are primitive (ancestral) and which are derived (have evolved from something else). Obviously, the hypotheses need to be tested, since at least two different assumptions can be made for any one state. To test the hypotheses, we often try to find additional characters that may support one cladogram more than they support others. Those cladograms that are not supported may be eliminated and those that are supported may be subjected to further testing.

In trying to choose the best of several to many cladograms, taxonomists use the principle of *parsimony*. Parsimony, as it applies to cladistics, is based on a principle of logic called *Occam's razor,* which states that "one should not make more assumptions than the minimum needed to explain anything." In other words, to choose the best cladogram, we need to make the fewest reasonable assumptions possible about evolutionary changes.

Cladistics has been used to infer evolutionary relationships among many different groups of organisms in all the kingdom survey chapters that follow in this book.

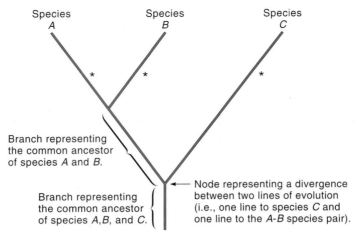

Species
A

Species
B

Species
C

*

*

*

Branch representing
the common ancestor
of species A and B.

Branch representing
the common ancestor
of species A,B, and C.

Node representing a divergence
between two lines of evolution
(i.e., one line to species C and
one line to the A-B species pair).

* Terminal branches representing the evolution of individual species
after divergence from ancestors shared with other species.

FIGURE 16.5   A simple cladogram showing relationships of three species derived from a common ancestor. The cladogram indicates that Species A and Species B are more closely related to each other than either is to Species C.

Summary

1. All organisms are identified by a two-part Latin name (binomial); most also have one to many common names. Biologists recognized the need for worldwide uniformity in naming and classifying organisms.

2. Theophrastus (third century B.C.) was the first person of note to attempt to classify plants. A distinction was made between monocots and dicots in the 13th century; herbalists later added many plants to Theophrastus's original list. By the 18th century, descriptive Latin phrase names were used for all organisms.

3. Linnaeus compiled a comprehensive list of all known Latin phrase names for plants according to their genera in *Species Plantarum,* published in 1753. Linnaeus placed specific kinds of related organisms (species) in each genus; he also listed a single word next to the Latin phrase, which, when combined with the name of the genus, formed an abbreviation for each species.

4. The abbreviated binomial, plus the authority for it, eventually replaced the phrase name; this method of naming plants became known as the Binomial System of Nomenclature. Linnaeus also organized all known plants into 24 classes, based primarily on the number of stamens in flowers.

5. In 1867, European and American botanists met in Paris and agreed to use Linnaeus's 1753 publication and binomials as the starting point for all scientific names of plants. The rules for naming and classifying plants drawn up at that meeting and periodically revised today constitute the International Code of Botanical Nomenclature, now followed by botanists of all nationalities.

6. All living organisms were at first placed in either the Plant or Animal Kingdom. In the 1860s, Hogg and Haeckel proposed a third kingdom (Protoctista) for all organisms that did not develop complex tissues. In 1938, H. F. Copeland, on the basis of cellular differences, divided the Protoctista into Kingdoms Monera and Protoctista.

7. Because of differences in the mode of nutrition of organisms in Kingdom Protoctista, most biologists now favor five kingdoms based both on forms of nutrition and cellular structure, as proposed by R. H. Whittaker in 1969. Slime molds still do not fit well into the five-kingdom system, but this system appears to be the most satisfactory arrangement so far proposed.

8. Since Linnaeus's time, several classifications have been added between the levels of kingdom and genus.

9. Bacteria are placed in Kingdom Monera; all algae, euglenoids, dinoflagellates, flagellate fungi, and slime molds are placed in Kingdom Protoctista; true fungi are placed in Kingdom Fungi; liverworts, hornworts, mosses, and all vascular plants are placed in Kingdom Plantae.

10. Viruses, which are not classified within the five kingdoms, are discussed in Chapter 17; lichens, which are a combination of an alga and a fungus, are discussed in Chapter 19.

11. The second part of a binomial, the specific epithet, is followed by the name of the author, usually in abbreviated form.

12. Subcategories (e.g., subspecies, suborder) are sometimes used in classification.

13. Taxonomists may construct keys to aid in the identification of organisms. Most keys are dichotomous. Keys to major groups of organisms deal only in generalities and do not indicate occasional exceptions or allow for intermediate forms that may be transferred back and forth between groups as new knowledge about them is gained.

14. Cladistics is a system of classifying and inferring evolutionary relationships based on an examination of shared features.

Review Questions

1. What are the advantages and drawbacks of using scientific names as compared with common names?

2. What is the International Code of Botanical Nomenclature?

3. What is meant by the term *binomial*?

4. Which divisions include organisms found mainly in water?

5. List some characteristics by which the following may be distinguished from one another, and then construct a simple dichotomous key that could aid someone from another planet in identifying them: blondes, oranges, hammers, bicycles, rosebushes, automobiles, brunettes, cats, rats, raspberry bushes, dogs, wrenches, peaches.

Discussion Questions

1. Since phrase names are generally more descriptive of organisms than binomials are, do you think they should be revived?

2. After Linnaeus organized all the genera known to him into classes, a number of other categories of classification between the levels of kingdom and genus were added. Are these other categories really useful? Explain.

3. Do you think biologists are justified in dividing the traditional Plant and Animal Kingdoms into five kingdoms, and would it make any significant difference to the world at large if they had not? Explain.

Additional Reading

Bisby, F. A., and G. F. Russell (Eds.). 1994. *Designs for a global plant species information system.* Fair Lawn, NY: Oxford University Press.

Cronquist, A. 1988. *The evolution and classification of flowering plants,* 2d ed. Boston: Houghton Mifflin Co.

Jeffrey, C. 1990. *Biological nomenclature,* 3d ed. New York: Routledge, Chapman and Hall, Inc.

Soltis, P. S., et al. (Eds.). 1991. *Plant molecular systematics.* New York: Chapman and Hall.

Stace, C. A. 1989. *Plant taxonomy and biosystematics,* 2d ed. New York: Routledge, Chapman and Hall, Inc.

Stearn, W. T. 1957. *An introduction to the Species Plantarum and cognate botanical works of Carl Linnaeus.* London: Printed for the Ray Society and sold by Bernard Quaritch.

Stuessy, T. 1990. *Plant taxonomy: The systematic evaluation of comparative data.* New York: Columbia University Press.

Stuessy, T. 1994. *Case studies in plant taxonomy. Exercises in applied pattern recognition.* New York: Columbia University Press.

Woodland, D. W. 1991. *Contemporary plant systematics.* Englewood Cliffs, NJ: Prentice-Hall.

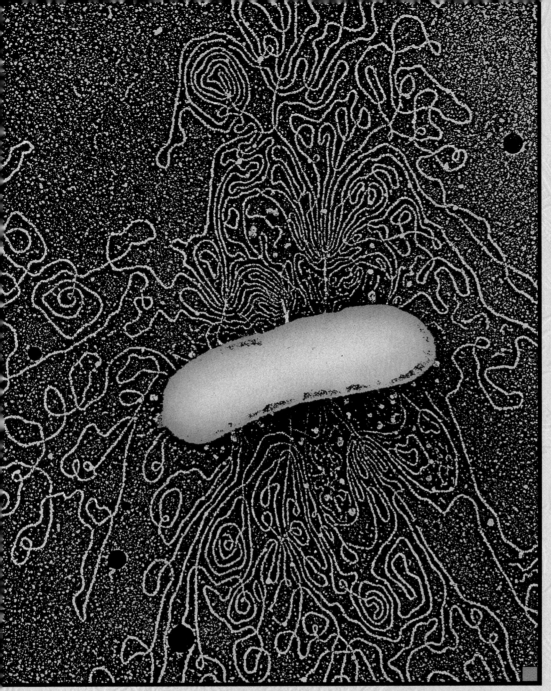

A sausage-shaped bacterial cell that has ruptured. The long DNA strand, which is normally tightly coiled inside, has unraveled and is spilling out. (Photomicrograph © Gopal Murti, Science Photo Library/Photo Researchers, Inc.)

Chapter Outline

17 Kingdom Monera and Viruses

Overview————————————————

Following an introduction involving luminescent bacteria associated with flashlight fish, the chapter gives an overview of the features of Kingdom Monera. A brief investigation of bacteria and their reproduction is followed by a discussion of archaebacteria and true bacteria, with a survey of representatives and human relevance of heterotrophic and autotrophic true bacteria. Sections on composting, disease transmission, bacterial diseases, and Koch's Postulates are included. Blue-green bacteria and prochlorobacteria are explored next, and human relevance of blue-green bacteria is examined. The chapter concludes with an overview of the nature, reproduction, and human relevance of viruses.

Some Learning Goals

1. Know the basic forms of bacteria. Explain how a prokaryotic cell differs from a eukaryotic cell and why prokaryotic organisms are difficult to classify.
2. Learn the forms of nutrition in bacteria.
3. Know at least 10 bacteria useful to humans and understand how they are useful.
4. Understand the various ways in which disease bacteria are transmitted; describe how each type of disease bacterium functions in causing disease.
5. Learn Koch's Postulates.
6. Know how the major groups of bacteria differ from one another in form, pigmentation, and reproduction. Also know how viruses differ from bacteria in form and reproduction.

The astonishing variety and beauty of undersea life, with representatives from all five kingdoms, has attracted the attention of snorkelers and scuba divers in ever-increasing numbers, particularly since Jacques Cousteau and other marine biologists have added to our knowledge of this vast domain with their sophisticated investigations and color cinematography. The shallow tropical waters of the Great Barrier Reef of Australia, the Virgin Islands National Park, Hanauma Bay of Hawaii, and countless other reefs teem with brilliantly colored marine life. Scuba divers who explore the oceans at night in areas such as Indonesia's Banda Sea, the region just north of Darwin, Australia, or Israel's Gulf of Elat are frequently rewarded with a spectacular sight not seen by those who explore by day. They find themselves in the midst of a shimmering, constantly changing light display.

Countless numbers of lights, bright enough to read by, flicker, dart, glide slowly, or even congregate in glimmering groups. The lights come from large pouches beneath the eyes of small fish, which can turn them off at will by covering them with special folds of tissue. Unlike their deep-sea relatives, these *flashlight fish*, as they are called, do not produce the light within their own bodies. Instead, it comes from thousands of microscopic, luminescent bacteria that enjoy a *symbiotic,* or specifically a *mutualistic,* relationship with the fish.

In this instance, the fish furnish some food and oxygen to the bacteria from numerous blood vessels in the pouches, while the bacteria produce the light used by the fish not only to search for food at night but also to attract live food and to confuse predators by flashing the lights on and off while swimming in erratic patterns, both organisms deriving benefit from the relationship.

Luminescent bacteria are but minor representatives of the most numerous and geologically ancient of the earth's living organisms. Fossils of bacteria have been found in geological strata that are estimated to be 3.5 billion years old. Since fossils of the first organisms with eukaryotic cells date back "only" 1.3 billion years, bacteria are almost three times as old as any other organism known to have existed on this planet.

The approximately 4,000 species of bacteria recognized today occur in astronomical numbers in almost every conceivable natural habitat. Each gram of garden soil, for example, is estimated to contain as many as 2 billion bacteria. They are found on, and in, all plants and animals, in all types of soils, throughout both fresh and salt waters, in polar ice caps and bubbling hot springs, in coal and petroleum, throughout the atmosphere, in bottles of India ink, and almost anywhere else one may care to look for them.

Just how many species of bacteria there are is uncertain, since it is difficult to classify simple one-celled organisms on the basis of visible features alone. For example, a disease-causing bacterium with the scientific name of *Streptococcus pneumoniae* occurs in 84 known strains, all of which look alike but each of which causes a different form of pneumonia or related disease. Should each strain then be called a species? *Microbiologists,* who study microscopic organisms of all kinds, are not sure, but they tend to recognize clusters of strains based on what they *do* rather than how they *look.*

Although bacteria that cause disease and spoilage receive most of the publicity, more than 90% of bacterial species are either harmless or useful to humans. They are used in industry for a wide variety of purposes and are so important ecologically that life as we know it today would be vastly different without the activities of these tiny organisms. Both industrial uses and the role of bacteria in ecology are discussed in later sections.

FEATURES OF KINGDOM MONERA

The members of Kingdom Monera all have prokaryotic cells. Such cells have no nuclear envelopes, but each has a long strand of DNA. In addition, there may be small DNA fragments called *plasmids,* ribosomes, and membranes present in bacterial cells. Membrane-bound organelles, such as plastids, mitochondria, Golgi bodies, and endoplasmic reticulum, are lacking (Table 17.1; see Fig. 17.14).

Table 17.1

SOME DIFFERENCES BETWEEN EUKARYOTIC AND PROKARYOTIC CELLS

EUKARYOTIC CELLS	PROKARYOTIC CELLS
1. Have a nuclear envelope	1. Lack a nuclear envelope
2. Have two to hundreds of chromosomes per cell; DNA is double-stranded	2. Have a single long strand of DNA plus (usually) several to 40 plasmids
3. Have membrane-bound organelles (e.g., plastids, mitochondria)	3. Lack membrane-bound organelles
4. Have 80S ribosomes[1]	4. Have 70S ribosomes[1]
5. Have asexual reproduction by mitosis	5. Have asexual reproduction by fission
6. Have sexual reproduction by fusion	6. Sexual reproduction unknown

1. S is a Svedberg unit, which is used to measure the rate at which a particle suspended in a fluid settles to the bottom when centrifuged. An 80S ribosome is larger than a 70S ribosome.

In a number of members of Kingdom Monera, cells may occur in a common matrix of gelatinous material as variously shaped colonies or in the form of chains or **filaments** (threads), but each cell is completely independent, with protoplasmic connections between them being absent. Some species are **motile** (capable of independent movement), usually by means of simple **flagella** (whiplike tails, which often occur in pairs) that propel or pull the cell through the water. Several filamentous species exhibit a gliding motion in which the filaments glide back and forth independently or against each other, but most species are nonmotile.

Nutrition in Kingdom Monera is primarily by the absorption of food in solution through the cell wall, but some bacteria are *chemosynthetic* (capable of obtaining their energy through chemical reactions involving various compounds or elements) while a few true bacteria, including blue-green bacteria and chloroxybacteria, exhibit forms of photosynthesis.

Reproduction is predominantly asexual, by means of **fission,** a form of cell division in which one cell essentially pinches in two. Fission does not involve mitosis since there are no nuclei or other organelles. The strand of DNA duplicates itself, however, and the DNA is distributed between the two new cells originating from the parent cell. Sexual reproduction is unknown, but genetic recombination occurs in several groups. The genetic recombination is facilitated by means of *pili* (minute tubes) that form connections between cells.

Cellular Detail and Reproduction of Bacteria

Bacterial cells are *prokaryotic*. Although such cells have no nuclear envelopes or organelles, folds of the plasma and other membranes apparently perform some of the functions of the organelles of eukaryotic cells. Ribosomes that are about half the size of those in the cytoplasm of eukaryotic cells are also present. Bacterial cells have a *nucleoid,* which is a single, long, very condensed DNA molecule in the form of a ring. The nucleoid is usually attached at one point to the plasma membrane. In addition, up to 30 or 40 small, circular DNA molecules called *plasmids* may be present. The plasmids replicate independently of the large DNA molecule or chromosome, and the entire complement of plasmids often consists of copies of one or at most very few different plasmids. The chromosome and sometimes all or part of the plasmids replicate before the cell divides (Fig. 17.1).

Mitosis as seen in eukaryotic cells does not occur. Instead, there is an internal reorganization of material during which the two DNA molecules migrate to opposite ends of the cell. Then, at approximately the middle of the cell, the plasma membrane and cell wall, as a result of forming a transverse wall, divide the cell in two. Finally, the two new cells separate and enlarge to their original size, or in some bacteria, the cells remain attached to each other in chains. Fission is found so universally in bacteria that they have been called *schizomycetes,* which means "fission fungi" (Fig. 17.2). However, bacteria are not currently believed to be closely related to the fungi, as was the case when the name was originally applied.

Under ideal conditions of moisture, food supply, and temperature, a bacterium may undergo fission every 10 to 20 minutes. If a single bacterium of average size were to duplicate itself every 20 minutes for as little as 36 hours, a mass of bacteria numbering 322,981,536,679,200,000,000,000, 000,000,000 and weighing 126,464,618,590 metric tons (137,438,953,472 tons) would be produced. If this mass were not compacted, its volume would exceed that of the earth. Of course, bacteria do not continue to reproduce at their maximum rate for very long because several factors prevent it. These factors include the exhaustion of food supplies and the accumulation of toxic wastes.

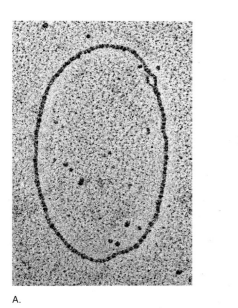 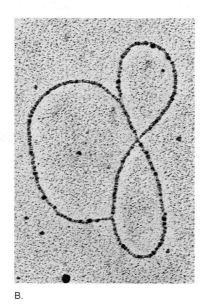 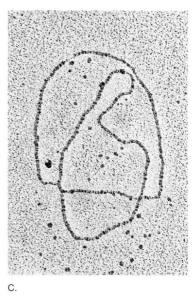

A.   B.   C.

FIGURE 17.1   Duplication of a nucleoid in a bacterial cell. The duplication begins (*A*) at a single point or bubble (*upper right*) that expands (*B*) until the two new rings of DNA separate (*C*), ×150,000.

(Electron micrographs courtesy Tsuyoski Kakefuda)

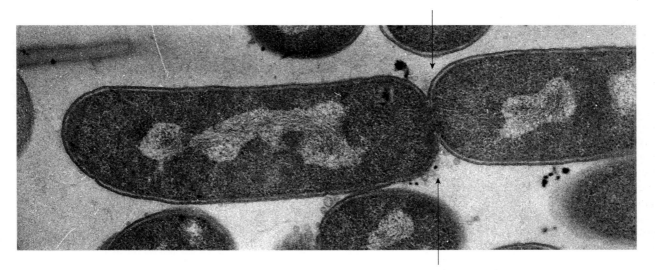

FIGURE 17.2   A dividing bacterial cell. The new wall is growing inward, ×48,750.

(Electron micrograph courtesy R.G.E. Murray)

Bacteria do not produce gametes or zygotes, nor do they undergo meiosis. However, at least three forms of genetic recombination occur:

1. In *conjugation*, part of the DNA strand or chromosome is transferred from a donor cell to a recipient cell through a tiny, hollow, tubelike **pilus** while the two cells are in contact (Fig. 17.3). Once in the recipient cell, the DNA fragment becomes part of the new cell. Bacterial cells that are capable of con-

jugating have different sexual forms of DNA, and only cells of different mating types may undergo conjugation.

2. In *transformation,* a living cell acquires fragments of DNA released by dead cells into the medium in which it occurs and incorporates them into its own cell.

3. In *transduction,* fragments of DNA are carried from one cell to another by viruses (discussed toward the end of this chapter).

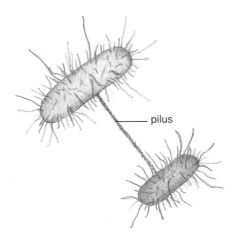

pilus

FIGURE 17.3   Conjugation in bacteria. Part of the DNA strand of the donor cell migrates through the hollow, tubelike *pilus* to the recipient cell, where it is incorporated into the DNA of its new cell. Shorter pili cover the surfaces of both cells.

Size, Form, and Classification of Bacteria

Except for the blue-green bacteria, which are discussed separately, and a few other giant bacteria, which attain lengths of up to 60 micrometers, most are less than 2 or 3 micrometers in diameter, and a few of the smaller species approach 0.15 micrometer. The latter are so small that 6,500 of them arranged in a row would not quite extend across the head of a pin. Because of their small size, bacteria are not visible individually to the unaided eye and are studied in the laboratory with electron microscopes or with the highest magnifications of light microscopes.

The thousands of different kinds of bacteria occur primarily in three forms: *cocci* (singular: *coccus*), which are spherical or sometimes elliptical; *bacilli* (singular: *bacillus*), which are rod-shaped or cylindrical; and *spirilli* (singular: *spirillum*), which are in the form of a helix, or spiral (Fig. 17.4). Further classification is based on numerous visible features including, for example, presence of pigments, development of slimy or gummy capsulelike sheaths around cells, presence of hairlike or budlike appendages, development of internal thicker-walled *endospores,* and gliding movements by which threadlike groups of bacteria appear to slide back and forth.

Some bacteria have slender flagella, usually about 5 to 10 micrometers in length, which propel them through fluid media. Others have somewhat shorter tubelike pili, which resemble flagella but do not function in locomotion. The pili apparently enable bacteria to attach themselves to surfaces or to each other.

Bacteria are also grouped into two large categories based on their reaction to a dye. After a heat-fixed smear of cells has been stained blue-black with a violet dye a dilute iodine solution, alcohol, or acetone is added. When so

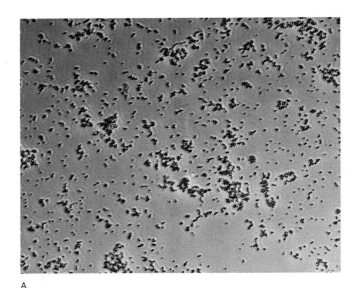

A.

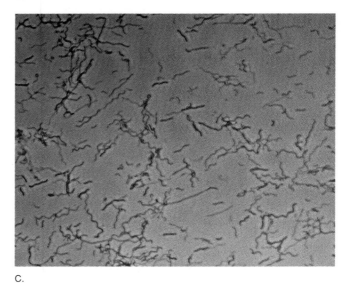

B.

C.

FIGURE 17.4   Three basic forms of bacteria. *A.* Cocci. *B.* Bacilli. *C.* Spirilli, ×1,500.

treated, some species rapidly lose their color (but absorb pink safranin dye), and are called *gram-negative*. Others, called *gram-positive,* retain most of the blue-black color. Gram's stain, named after Christian Gram who discovered it in 1884, has many variations, but all the variations produce similar results. Other characteristics used in classifying bacteria are indicated in the discussions that follow.

SUBKINGDOM ARCHAEBACTERIOBIONTA— THE ARCHAEBACTERIA

The *archaebacteria* represent one of two quite distinct lines of the most primitive known living organisms. They are fundamentally different in their metabolism from the other line of bacteria, the *eubacteria,* and also differ in the unique sequence of bases in their RNA molecules, the lack of muramic acid in their walls, and in the production of distinctive lipids.

In the early 1980s, these basic differences led University of Illinois microbiologist Carl Woese and his colleagues, who have conducted considerable research on the archaebacteria, to suggest that the organisms belong in a kingdom of their own. At present, three distinct groups of bacteria are included in the archaebacteria.

The Methane Bacteria

The *methane bacteria,* which comprise the lion's share of the archaebacteria, are killed by oxygen and so are active only under anaerobic conditions found in swamps, ocean and lake sediments, hot springs, animal intestinal tracts, sewage treatment plants, and other areas not open to the air. Their energy is derived from the generation of methane gas from carbon dioxide and hydrogen.

Methane, or "marsh gas," is a principal component of natural gas and may have been a major part of the earth's atmosphere in early geologic times. It still is present in the atmospheres of the planets Jupiter, Saturn, Uranus, and Neptune and is the main ingredient of firedamp, which causes serious explosions in mines. Methane will burn when it constitutes only 5% to 6% of the air, and a flitting, dancing light called *ignis fatuus,* or "will-o'-the-wisp," which is occasionally seen at night over swamps and marshy places, is said to be due to the spontaneous combustion of the gas.

The Salt Bacteria

Commercial salt evaporation ponds and other shallow areas in bodies of water with high salt content often have a unique appearance from above. They can be strikingly red due to the presence of a distinctive group of archaebacteria. These are the *salt bacteria,* whose metabolism enables them to thrive under conditions of extreme salinity that instantly kills other living cells. The bacteria carry on a simple form of photosynthesis with the aid of a membrane-bound red pigment

FIGURE 17.5   The north end of Utah's Lake Bonneville, which has a very high salt content. The reddish areas are due to a red pigment produced by salt bacteria. The pigment, called *bacterial rhodopsin,* is involved in a form of photosynthesis.

called *bacterial rhodopsin.* The concentration of salt inside the cells is much lower than in their surroundings, but their metabolism is so closely tied to their environment that the bacteria die if placed in waters with lower salt concentrations (Fig. 17.5).

The Sulpholobus Bacteria

The *sulpholobus bacteria* constitute a third group of archaebacteria whose members occur in sulphur hot springs. The extraordinary metabolism of these bacteria permits them to thrive at very high temperatures—mostly in the vicinity of 80°C (176°F), with some doing very well at only 10°C below the boiling point of water. Their environment is also exceptionally acidic, the hot springs often having a pH of less than 2 (the neutral point on the 14-point pH scale is 7). One genus of bacteria (*Thermoplasma*) in this group is bounded by only a plasma membrane and has no cell wall. It is found only in the embers of coal tailings. Another genus (*Thermoproteus*) appears to be confined to the geothermal areas of Iceland.

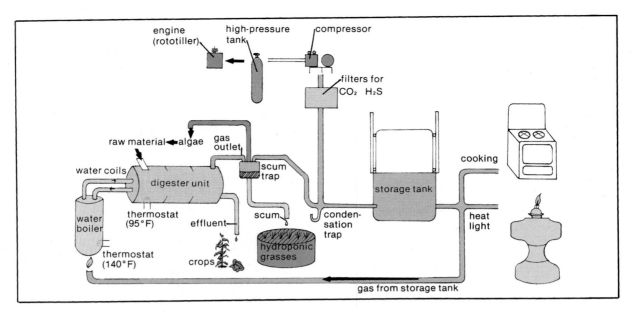

FIGURE 17.6 Integrated organic digester operation for the production of methane gas.
(Reprinted from *Producing Your Own Power* © 1974 by Rodale Press, Inc. Permission granted by Rodale Press, Inc., Emmaus, PA, 18049.)

Recently, James Lake and his colleagues discovered that the shape of the ribosomes of sulpholobus bacteria is significantly different from those of other archaebacteria, eubacteria, and eukaryotes and that the chemistry of sulphur-dependent bacteria also distinguishes them from other archaebacteria. Lake and his colleagues have proposed a third kingdom of prokaryotes named *Eocytes* (dawn cells). Other microbiologists are hesitant to base kingdom status on ribosome shape, but the controversy over the significance of these discoveries will undoubtedly continue as further research into the matter is pursued.

Human Relevance of the Archaebacteria

Archaebacteria are significant to humans in several ways. In the future, methane bacteria may be used on a large scale to furnish energy for engine fuels and for heating, cooking, and light, since the methane gas they produce can be substituted for natural gas. Methane has an octane number of 130 and has been used as a motor fuel in Italy for over 40 years. It is clean, nonpolluting, safe, and nontoxic, and it prolongs the life of automobile engines, also making them easier to start. The methane is given off by the bacteria as they "digest" organic wastes in the absence of oxygen. Nine kilograms (20 pounds) of horse manure or 4.5 kilograms (10 pounds) of pig manure fed daily into a methane digester will produce all the gas needed for the average American adult's cooking needs. Considerably less green plant material is required to produce the same amount.

The digester basically consists of an airtight drum connected by a pipe to a storage tank with a means of drawing off the sludge left after the gas has been produced (Fig. 17.6). The sludge itself makes an excellent fertilizer, although many sludges that originate from municipal and large-scale agricultural plants carry with them toxic levels of metals.

France had over 1,000 methane plants in operation by the mid-1950s, and in India, where cows produce over 812 million metric tons (800 million tons) of manure per year, many villages satisfy their fuel needs with manure-fed methane gas digesters. They are now being employed on a small scale in rural areas in the United States. Methane is the primary source of hydrogen in the commercial production of ammonia.

SUBKINGDOM EUBACTERIOBIONTA— THE TRUE BACTERIA

The Unpigmented, Purple, and Green Sulphur Bacteria

The *eubacteria* (true bacteria) have *muramic acid* in their cell walls and are also fundamentally different from archaebacteria in their RNA bases, metabolism, and lipids. They constitute quite a heterogeneous assemblage, with most being **heterotrophic** (organisms that cannot synthesize their own food and therefore depend on other organisms for it).

The majority of heterotrophic bacteria are **saprobes** (living organisms that obtain their food from nonliving organic matter). Saprobic bacteria, along with fungi, are primarily responsible for decay and for recycling all types of organic matter in the soil. Some of their recycling activities are discussed in the section on the nitrogen cycle in Chapter 25. Other heterotrophic bacteria are *parasites* (living organisms that depend on other living organisms for their food).

Several parasites that cause important human diseases are discussed later in the chapter.

Autotrophic Bacteria

A few groups of true bacteria are similar to green plants in being **autotrophic;** that is, they are capable of synthesizing organic compounds from simple inorganic substances. Some carry on photosynthesis without, however, producing oxygen as a by-product. Included in the photosynthetic bacteria are the *purple sulphur bacteria,* the *purple nonsulphur bacteria,* and the *green sulphur bacteria.* The *blue-green bacteria* and the *chloroxybacteria,* which do produce oxygen through photosynthesis, are discussed separately later in this chapter. Cells of the first two groups appear purplish, or occasionally red to brown, because of the presence of a mixture of greenish, yellow, and red pigments. Their greenish *bacteriochlorophyll* pigments (there are several closely related ones) are very similar to the chlorophyll *a* of higher plants.

No plastids are present in bacteria and their pigments are, instead, located in thylakoids or small spherical bodies. In their photosynthesis, the purple bacteria substitute hydrogen sulphide, the bad-smelling gas given off by rotten eggs, for the water used by higher plants. The generalized equation for the process is as follows:

$$CO_2 + 2\,H_2S \xrightarrow[\text{light}]{\text{bacterio-}\atop\text{chlorophyll}} (CH_2O)n + H_2O + 2S$$

carbon hydrogen   carbo-   water sulphur
dioxide sulphide   hydrate

The equation for the purple nonsulphur bacteria is similar, but hydrogen from organic molecules is used instead of hydrogen sulphide. Green sulphur bacteria do use hydrogen sulphide, but their chlorophyll, called *chlorobium chlorophyll,* differs significantly in its chemistry from the chlorophylls of higher plants.

Other groups of bacteria are *chemoautotrophic;* that is, they are capable of obtaining the energy they require from various compounds or elements through chemical reactions involving the oxidation of reduced inorganic groups, such as NH_3, H_2S, and Fe^{++}, or the oxidation of hydrogen gas (as discussed in the section on oxidation-reduction reactions in Chapter 10).

Chemoautotrophic bacteria include *iron bacteria,* which transform soluble compounds of iron into insoluble substances that accumulate as deposits (e.g., in water pipes); *sulphur bacteria,* which can convert hydrogen sulphide gas to elemental sulphur and sulphur to sulphate; and *hydrogen bacteria,* which flourish in soils where they use molecular hydrogen derived from the activities of anaerobic and nitrogen-fixing bacteria. Nitrifying and nitrogen-fixing bacteria are a group of ecologically important bacteria that are discussed, along with their roles in the nitrogen cycle, in Chapter 25.

HUMAN RELEVANCE OF THE UNPIGMENTED, PURPLE, AND GREEN SULPHUR BACTERIA

Compost and Composting

Long before the existence of bacteria was known or even suspected, primitive agriculturists made rough piles of weeds, garbage, manure, and other wastes. They watched as the piles became reduced to a fraction of their original volume and changed into **compost,** a dark, fluffy material that conditions and enriches soil when mixed with it. Today, the compost pile is at the heart of the activities of numerous organic gardeners and farmers. With many communities instituting ordinances forbidding the burning of leaves and refuse and with space for waste disposal becoming scarce, a significant number of cities and towns have turned to the composting of street leaves and other materials. They have found that they not only save space but also produce a useful, ecologically compatible product (Fig. 17.7).

Because bacteria are ubiquitous, a single leaf or nonliving tissues in a pile of any size will eventually be decomposed. Ideally, however, a compost pile is 2 meters wide, 1.5 meters deep, and at a minimum, 2 meters long (6 feet wide, 4.5 feet deep, and at least 6 feet long). If the pile is of lesser length or breadth or greater depth, the conditions that favor the breakdown of materials (created by the bacteria themselves) develop at a slower pace.

Any accumulation of household garbage, leaves, weeds, grass clippings, and/or manure may be heaped together, and it has been demonstrated that no so-called starter culture of microorganisms is needed to initiate decomposition. Once the everpresent decay organisms have material to decompose, their numbers increase rapidly, and much heat is generated. In fact, the temperature in the center of the pile often rises to 70°C (158°F). As it rises, populations of organisms adapted to higher temperatures replace those not as well adapted. The remains of the latter are then added to the

FIGURE 17.7 Compost. *Left.* Garbage and leaves before composting. *Right.* After composting.

Table 17.2

SOME DISEASES OF HUMANS CAUSED BY BACTERIA OR VIRUSES

BACTERIAL DISEASES	VIRAL DISEASES
Anthrax	AIDS
Botulism	Chicken pox/shingles
Brucellosis	Common colds
Bubonic plague	Diabetes (some)
Cholera	Encephalitis
Diphtheria	Genital herpes
Dysentery (some)	Infectious hepatitis
Gonorrhea	Influenza
Leprosy	Measles
Mastitis	Mononucleosis
Meningitis	Mumps
Pneumonia (some)	Pneumonia (some)
Syphilis	Poliomyelitis
Tetanus	Rabies
Tuberculosis	Rubella
Typhoid fever	Smallpox
Whooping cough	Yellow fever

compost. The high temperatures also kill many weed seeds and most disease organisms.

The proportion of carbon to nitrogen present in a compost pile largely determines the pace at which the microbial activities proceed, a ratio of 30 carbon to 1 nitrogen being optimal. Microbial growth stops if the proportion of carbon gets much greater, and nitrogen is lost in the form of ammonia if the ratio drops lower.

If the materials are kept moist and turned occasionally to aerate the pile, composting may be completed in as little as two weeks. Shredding the materials exposes much more surface area for the microorganisms to work on and speeds up the process. Composters generally avoid or keep to a minimum a few materials, such as *Eucalyptus* and walnut leaves, which contain substances that inhibit the growth of other plants, when the compost is intended for use as a soil conditioner or crop fertilizer. Bamboo and some types of fern fronds are particularly resistant to decay bacteria and will decompose much more slowly than other materials. Domestic cat manure may contain stages of a parasite that can infect humans, although the organism should be killed if proper composting techniques are employed. Generally, however, almost any organic materials are suitable.

While the value of compost to the soil is indisputable, its value as a fertilizer has certain limitations. The nitrogen content of good compost is about 2% to 3%, phosphorus 0.5% to 1.0%, and potash 1% to 2%, as compared with five to ten times those amounts in chemical fertilizers, which also require less labor to produce. Nevertheless, with the spiraling costs of chemical fertilizers, increasing awareness of the problems accompanying their use, and greater public en-

lightenment concerning the accumulation and disposal of solid wastes, compost is undoubtedly destined to play a major role in pertinent agricultural practices of the future.

True Bacteria and Disease

It has been calculated that plant diseases alone cause American farmers losses of more than $4 billion per year. Although many plant diseases are caused by fungi, a number involve bacteria. Bacteria are involved in diseases of pears, potatoes, tomatoes, squash, melons, carrots, citrus, cabbage, and cotton. Bacteria also cause huge losses of foodstuffs after they have been harvested or processed. They are even better known, however, as the culprits in many serious diseases of animals and humans (Table 17.2).

Since they are so tiny and often incapable of independent movement, how are they transmitted, and what is it about their activities that causes them to fell creatures billions of times their size? We have learned much about their activities in the past century and know now that they can gain access to human tissues in a variety of ways.

Modes of Access of Disease Bacteria

Access from the Air Every time a person coughs, sneezes, or just speaks loudly, an invisible cloud of tiny saliva droplets is produced (Fig. 17.8). Each droplet contains bacteria or other microorganisms and a tiny amount of protein. The moisture usually evaporates almost immediately, leaving minute protein flakes to which live bacteria adhere. As normal breathing takes place, these may soon find their way into the respiratory tracts of other humans or animals, particularly

FIGURE 17.8   Stop-action photograph of a sneeze. (Reprinted with the permission of Marshall W. Jennison, Syracuse University)

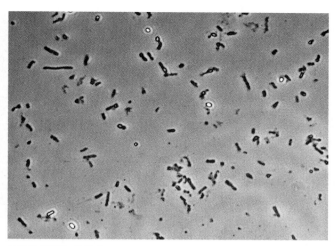

FIGURE 17.9   Botulism bacteria, ×1,500. (Courtesy Robert McNulty)

if they have been confined within the air of a room. Fortunately, the natural resistance of most of those who acquire the bacteria in this way prevents the bacteria from multiplying to the point of causing a disease. When the resistance is not there, however, a number of diseases, including diphtheria, whooping cough, some forms of meningitis and pneumonia, and strep throat can develop.

Psittacosis, a disease carried by birds, is caused by the inhalation of *chlamydias,* which are exceptionally minute organisms unable to manufacture their own ATP molecules. They are apparently energy parasites, depending on their host cells for the energy needed to carry on their own functions. Some chlamydias are transmitted sexually and in recent years have become a widespread human problem.

Access Through Contamination of Food and Drink

Food Poisoning and Diseases Associated with Natural Disasters   In the past, open sewers and unsanitary conditions for food preparation caused a number of bacterial diseases, including cholera, dysentery, *Staphylococcus* and *Salmonella* food poisoning, and typhoid and paratyphoid fevers to reach epidemic proportions. Although *Staphylococcus* food poisoning, which is seldom fatal, is still fairly common today, civilized countries now rarely see epidemics caused by other bacteria unless a natural disaster, such as a flood or a typhoon, disrupts normal sewage disposal. The diseases are more often spread by carriers who handle food, or by houseflies.

The United States and other countries with ocean shorelines ban the harvesting of shellfish at certain times because *Salmonella* bacteria may multiply enough in clams and mussels to cause illness in humans who eat them—particularly if they are not well cooked. The *Salmonella* bacteria multiply in the intestinal tract or spread from there to other parts of the body.

Legionnaire's Disease   Some bacteria are very common on algae in freshwater streams, lakes, and reservoirs. One such bacterium, *Legionella pneumophila,* causes Legionnaire's disease, which killed 34 members of the American Legion attending a convention in a Philadelphia hotel in 1976. An estimated 50,000 Americans are infected annually by Legionnaire's disease bacteria, which nearly always pass through the human digestive tract harmlessly without multiplying. On rare occasions, however, something unknown triggers their reproduction, mostly in older males who are heavy smokers or alcoholics. The results are often fatal.

Botulism   The most deadly of all known biological toxins is produced by a bacterium with the scientific name of *Clostridium botulinum* (Fig. 17.9). The name comes from *botulus,* the Latin word for *sausage,* since the bacterium was first discovered after people had died from eating some contaminated sausage at a picnic.

Unlike *Salmonella* food poisoning, **botulism** is not an infection but is poisoning from a substance produced by bacteria that can grow and multiply anaerobically (in the absence of oxygen) in improperly processed or stored foods. Home-canned beans, beets, corn, and asparagus in particular have been known to permit the development of botulism bacteria. Just 1 gram (0.035 ounce) of the toxin is enough to kill 14 million adults, and a little more than half a kilogram (1.1 pound) could eliminate the entire human race.

The bacteria, which are present in most soils, produce unusually heat-resistant spores and are likely to be present on any soil-contaminated foods. They may not be destroyed during canning unless the food is heated for 30 minutes at 80°C (176°F) or boiled for 10 to 15 minutes. They ordinarily will not grow in foods that have been preserved in brines containing at least 10% salt (sodium chloride) or in fairly acid media, such as the juices produced by most stone fruits. The toxin is absorbed directly from the stomach and the intestines, affecting nerves and muscles and causing paralysis. While some antidotes are available, they are ineffective after symptoms have become advanced.

Deaths reported from botulism in the United States reached a peak of about 25 per year in the 1930s but have declined to five or six per year since then. There is evidence that some deaths of infants due to unknown causes are actually due to botulism. For reasons that are not clear, botulism bacteria, which pass harmlessly through the digestive tracts of humans over the age of 1 year, may germinate in the intestines of infants less than a year old. Between 1976 and 1980, 170 cases of sudden infant death reported worldwide appear to have been due to botulism, and the evidence now suggests that botulism is responsible for about 5% of the 8,000 cases of sudden infant death that occur each year in the United States alone.

Access Through Direct Contact

Syphilis and Gonorrhea   Bacteria responsible for diseases, such as syphilis, gonorrhea, anthrax, and brucellosis, enter the body through the skin or mucous membranes (i.e., those membranes lining tracts with openings to the exterior of the body). Both syphilis and gonorrhea are transmitted through sexual intercourse or other forms of direct contact and rarely through the use of public washroom towels or toilets. In 1976, gonorrhea accounted for the largest number of reported cases of any communicable disease in the United States, and currently, despite increasing precautions taken by the public since the escalation of AIDS, more than 1 million Americans become infected every year. The symptoms of both syphilis and gonorrhea, which include persistent sores or discharges from the genitalia, sometimes disappear after a few weeks, leading victims to believe the body has healed itself. More often than not, however, symptoms reappear in different parts of the body at a later date. Both diseases are curable when treated promptly, but it is very important that such treatment be sought, since failure to do so can lead to sterility, blindness, and even death.

Anthrax   Anthrax, which is primarily a disease of cattle and other farm animals in addition to wild animals, is sometimes transmitted to humans, particularly workers in the wool and hide industries. Like syphilis and gonorrhea, it can be effectively eliminated if treated early enough but may be fatal if allowed to progress. Brucellosis, another disease of farm animals, is occasionally transmitted to humans through direct contact or through the consumption of contaminated milk. It is sometimes called *undulant fever* because of a daily rise and fall of temperature apparently associated with the release of toxins by the bacteria.

Access Through Wounds

Tetanus and Gas Gangrene   When one steps on a dirty nail or is wounded in such a way that dirt is forced into body tissues, tetanus (lockjaw) bacteria, which are common soil organisms, may gain access to dead or damaged cells. There they can multiply and produce a deadly toxin so powerful that 0.00025 gram (0.00000088175 ounce) is enough to kill an adult. In contrast, about 150 times that amount of strychnine is needed to achieve the same result. Control of tetanus through immunization is very effective and has become widespread. Only 21 cases of tetanus were reported in the United States in 1989. Several related bacteria that gain access to the body in the same way are responsible for poten-

tially fatal gas gangrene, which used to be feared on the battlefields in the past but is now controlled through the use of antibiotics and aseptic techniques.

Access Through Bites of Insects and Other Organisms

Bubonic Plague   Bubonic plague (Black Death) and tularemia are two bacterial diseases transmitted by fleas, deerflies, ticks, or lice that have been parasitizing infected animals, particularly rodents.

Rat fleas, found on infected rats that inhabit dumps, sewers, barnyards, and ships (Fig. 17.10), acquire the bacteria for plague and then pass them on to humans through their bites. The disease has been found in ground squirrels and other rodents in the United States, particularly in the West, since 1900.

In the past, bubonic plague has spread with great speed and reached devastating epidemic proportions. In 1665 in London, hundreds of thousands of humans perished from the disease, and between 1347 and 1349, it is believed to have killed one-fourth of the entire population of Europe (about 25 million persons). Today, plague is rare in North America, but it still occasionally manifests itself in port cities of Asia, Europe, and South America. Control depends on control of rats and fleas, which are virtually impossible to eradicate entirely, and the use of vaccines, which produce immunity for about 6 to 12 months.

Tularemia, Rickettsias, and PPLOs   Tularemia is primarily a disease of animals, but infected ticks or deerflies may transmit the disease to humans through bites. It is an occupational disease of meat handlers and is fatal in 5% to 8% of the cases. Ticks, lice, and fleas may also transmit rickettsias, which cause typhus and spotted fevers. Rickettsias are extremely tiny bacteria that live within eukaryotic cells.

Pleuropneumonialike organisms (PPLOs) are also minute bacteria that may be transmitted by various means. These have no cell walls and therefore are quite plastic. They're found in many plants, in hot springs, and in the moist surfaces of the respiratory and intestinal tracts of animals and humans. They're responsible for a form of human pneumonia and for numerous plant diseases. They're the only group of prokaryotes known to be resistant to penicillin.

Lyme Disease   Lyme disease, which was known in Europe before 1900, has spread rapidly throughout the United States since 1975 when an outbreak occurred at Lyme, Connecticut. Its arthritislike effects are caused by a bacterium (*Borrelia burgdorferi*) hosted by deer and field mice. It is injected into the human bloodstream by deer ticks.

Koch's Postulates

Since there are so many bacteria present everywhere, how can one be certain that a given bacterium obtained from an infected person is actually the organism responsible for the observed disease? Robert Koch, a German physician, became known during the latter half of the 19th century for his investigations of anthrax and tuberculosis. As a result of his work, Koch formulated rules for proving that a particular microorganism is the cause of a particular disease. His rules, with minor modifications, are still followed today. They

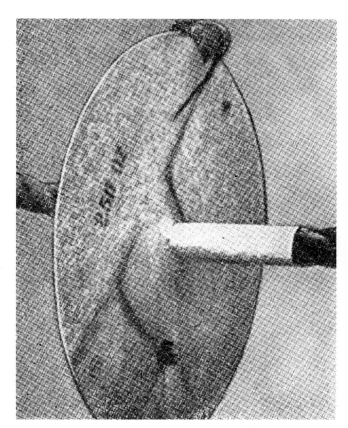

FIGURE 17.10 A rat climbing over a barrier on a ship's mooring line.
(Courtesy U.S. Public Health Service)

A.

B.

FIGURE 17.11 A tomato hornworm (*A*) before, and (*B*) 3 days after spraying with *Bacillus thuringiensis.*

have come to be known as *Koch's Postulates,* and their essence is as follows:

1. The microorganism must be present in all cases of the disease.
2. The microorganism must be isolated from the victim in pure culture (i.e., in a culture containing only that single kind of organism).
3. When the microorganism from the pure culture is injected into a susceptible host organism, it must produce the disease in the host.
4. The microorganism must be isolated from the experimentally infected host and grown in pure culture for comparison with that of the original culture.

True Bacteria Useful to Humans

For many years, we've controlled insect pests of food plants mostly through the use of toxic sprays. Residues of the sprays remaining on the fruits and vegetables have accumulated in human tissues, often with adverse effects, while at the same time many organisms have become immune or resistant to the toxins. In addition, the sprays have killed useful organisms, and precipitation runoff has washed the toxins into streams, lakes, and oceans, harming or killing aquatic organisms.

As we've become aware of the undesirable effects of the use of toxic sprays, we've looked for alternative means of controlling crop pests. Today, many harmful pests and even weeds can be significantly limited through the use of "biological controls," which are discussed in Appendix 2.

Bacillus Thuringiensis *and* B. Popillae

Three biological control bacteria have been registered for use by the U.S. Department of Agriculture. One, *Bacillus thuringiensis* (often referred to as *BT*), has been remarkably effective against a wide range of caterpillars and worms, including peach tree borers, European corn borers, bollworms, cabbage worms and loopers, tomato and fruit hornworms (Fig. 17.11), tent caterpillars, fall webworms, leaf miners, alfalfa caterpillars, leaf rollers, gypsy moth larvae, and cankerworms.

The bacteria, which are easily mass-produced by commercial companies, are sold in the form of a stable wettable dust containing millions of spores. When the spores are sprayed on food plants, they are harmless to humans, birds, animals, earthworms, or any living creatures other than moth or butterfly larvae. When a caterpillar ingests any tissue with BT spores on it, the bacteria quickly become active and multiply within the digestive tract, soon paralyzing the gut. The caterpillar stops feeding within two or three hours and slowly turns black, dropping off the plant in two to four days.

The toxin-producing gene from *Bacillus thuringiensis* was recently introduced into another bacterium, *Pseudomonas fluorescens,* which is now used on corn to control black cutworms.

In 1983, a variety of *Bacillus thuringiensis* (var. *israelensis*) called *BTI* was introduced into the commercial market for the control of mosquitoes. The bacterium attacks only mosquito larvae ("wigglers") and one or two other pests. Six years of experiments performed on more than 70 species of fish, snails, shrimp, and insects demonstrated no adverse effects on either plants or animals other than mosquitoes (and a couple of lesser pests) even at dosages 100 times more powerful than those needed to kill mosquito larvae. Use of this bacterium to control mosquitoes in the future may constitute a significant step in lessening the ecological damage and disruption that so frequently accompanies the use of toxic chemicals for pest control.

Another bacterium, *Bacillus papilliae,* is also marketed in a powder form. When it is applied to soil where grubs of the highly destructive Japanese beetle are present, it causes what is known as *milky spore disease* in the grubs, which die in a few days. The spores are carried throughout the topsoil by rainwater, by foraging grubs, and by organisms that feed on the grubs.

Bioremediation

Naturally occurring bacteria have shown much potential in the developing science of *bioremediation,* which involves the study of the use of living organisms in the cleanup of toxic wastes and pollution. One bacterium produces enzymes that break down nitroglycerin and trinitrotoluene, which are contaminants in the soil around some munitions factories and explosive sites.

Preliminary tests indicate that the bacteria can decompose such waste and contaminant materials into harmless residues within six months. Other bacteria, such as *Pseudomonas capacia,* can perform similar feats in oil spills and chemical dumps containing degreasers, such as trichloroethylene (TCE), creosote, and even 2,4,5-T, the Agent Orange defoliant of Vietnam infamy. The pollution-fighting bacteria may, however, need a few nutrients added to the contaminants to stimulate their activities. Some scientists believe bacteria may also eventually be used to break down nuclear wastes. Bacteria already are being developed through *genetic engineering* (discussed in Chapter 14) to deal with other specific pollution problems, and their use in managing various forms of pollution may in the future become widespread.

Other Useful Bacteria

Human eyes contain *rhodopsin,* a protein that is so sensitive it reacts to light in less than 1 millionth of a second. Certain bacteria that contain a form of rhodopsin have proved to be invaluable in research on the chemistry of vision and have led to our understanding of how eyes convert light energy into vision.

Bacteria play a major role in the dairy industry. Milk, which is composed of proteins, carbohydrates, fats, minerals, vitamins, and about 87% water, has exceptional nutritive value for animals and is also an excellent medium for the growth of many kinds of bacteria. Milk is sterile when secreted within the udder of the cow, but it picks up bacteria as it leaves the cow's body. It spoils rapidly if it is not obtained and stored under strictly sanitary conditions. Even after it has been pasteurized and refrigerated, the numbers of bacteria in it will increase the longer it stands. If milk is left in open containers in household refrigerators, for example, bacterial growth will not be kept in check for much longer than 24 hours.

Except for milk itself, either alone or in mixtures (e.g., ice cream), all dairy products are manufactured from raw or pasteurized milk by the controlled introduction of various bacteria. Such products include buttermilk, acidophilus milk, yogurt, sour cream, kefir, and cheese. Whey, the watery part of the milk separated from curd during cheese making, is utilized, along with starches and molasses, for the commercial production of lactic acid. Lactic acid is used extensively in the manufacture of textile and laundry products, in the leather tanning industry, as a solvent in lacquers, and in the treatment of calcium and iron deficiencies in humans.

Beneficial bacteria, such as *Lactobacillus acidophilus* (a common organism in healthy digestive tracts), aid in digestion, reduce the risk of cancer, and may even reduce cholesterol levels. Acidophilus bacteria also have been widely used, along with antibiotics, to control or eliminate human female yeast infections. Researchers at the University of Minnesota are working on developing strains of intestinal bacteria that have an external layer of sticky polysaccharides that make them resistant to being eliminated by less desirable bacteria in the digestive tract.

In their metabolism of various sugars, proteins, and other organic substances, bacteria also produce waste products that have important industrial uses. Such products are often produced in large quantities by culturing bacteria in huge vats. These products include acetone, used in the manufacture of photographic film; explosives; solvents (e.g., nail polish remover); butyl alcohol, used in the manufacture of synthetic lacquers; dextran, used as a food stabilizer and as a blood plasma substitute; sorbose, used in the manufacture of ascorbic acid (vitamin C); and citric acid, used as a lemon flavoring for foods. Some vitamin and medicinal preparations also involve bacterial synthesis.

Bacteria are used in the curing of vanilla pods, cocoa beans, coffee, and black tea and in the production of vinegar, sauerkraut, and dill pickles. Fibers for linen cloth are separated from flax stems by bacteria, and green plant materials are fermented in silos to produce ensilage for cattle feed. In recent years, the production of several important amino acids by bacteria has been exploited commercially. More than 6,800 metric tons (7,500 tons) of one amino acid, glutamic acid, are produced in the United States each year. This is in demand as a flavor-enhancing agent in the form of monosodium glutamate.

In 1989, Patricia Mertz of the University of Miami discovered that *Brevibacterium,* the genus of bacteria responsible for the odor of Limburger and Brie cheeses, also is the source of foot odor in certain people. Bactericides are being tested to improve foot pads that merely absorb odor rather than kill the bacteria.

CLASS CYANOBACTERIAE— THE BLUE-GREEN BACTERIA

Introduction

In the past, algae as a group have been distinguished from other organisms in being photosynthetic, in having relatively simple structures, and, for those reproducing sexually, in having sex structures consisting of a single cell. As more has become known about cellular details, however, it is apparent that the differences between the *blue-green bacteria* (formerly known as the blue-green algae) and true algae are quite basic. Blue-green bacteria have prokaryotic cells, like all other members of Kingdom Monera, while all algae that are assigned to Kingdom Protoctista have eukaryotic cells. In fact, blue-green bacteria are so much like other true bacteria (Fig. 17.12) that most biologists now have stopped referring to them as algae and consider them true bacteria.

The main distinctions between organisms traditionally regarded as bacteria and blue-green bacteria are (1) blue-green bacteria have chlorophyll *a,* which is found in higher plants, and oxygen is produced when they undergo photosynthesis; (2) blue-green bacteria also have blue *phycocyanin* and red *phycoerythrin* phycobilin pigments; (3) blue-green bacteria are the only organisms that can both fix nitrogen and produce oxygen—a paradox, since nitrogen-fixation is essentially an anaerobic process. Except for the chloroxybacteria, other bacteria capable of carrying on photosynthesis do not produce oxygen, and they do not have chlorophyll *a.*

Distribution

Blue-green bacteria are found in almost as diverse habitats as are other true bacteria. They are common in temporary pools or ditches, particularly if the water is polluted. They are not found in acidic waters, but they are abundant in other fresh and marine waters around the globe, from the frozen lakes of Antarctica to warm tropical seas. In the open oceans, blue-green bacteria are the principal photosynthetic organisms in plankton, the tiny cells of *Synechococcus* commonly occurring in concentrations of 10,000 cells per milliliter. *Trichodesmium* is a marine, filamentous, nitrogen-fixing, blue-green bacterium that forms extensive mucilage-producing colonies in some tropical waters. Other bacteria multiply in the mucilage and become food for protozoa.

A different species of blue-green bacterium is found in each temperature range of the hot springs of Yellowstone National Park, where water temperatures approach 85°C (185°F). There the bacteria precipitate chalky, insoluble carbonate deposits, which become a rocklike substance called *travertine.* The deposits accumulate at the rate of up to 2 to 4 millimeters per week, with other blue-green bacteria often forming brilliantly colored streaks in the travertine.

Blue-green bacteria are often the first photosynthetic organisms to appear on bare lava after a volcanic eruption, and they also thrive in the tiny fissures of desert rocks. Some are found in jungle soils or on the shells of turtles and snails, while others live symbiotically in various types of organisms, including amoebae, protozoans, diatoms, certain sea anemones and their relatives, and some fungi and in the roots of tropical palmlike plants called *cycads.* They also flourish in tiny pools of water formed at the bases of the leaves of tropical grasses and other plants, and they are well-known components of "compound" organisms called *lichens,* which consist of a fungus and a photosynthetic partner.

Form, Metabolism, and Reproduction

The cells of blue-green bacteria often occur in chains or in hairlike filaments, which are sometimes branched. Several species occur in irregular, spherical, or platelike colonies, the individual cells being held together by the gelatinous sheaths they secrete (Fig. 17.13). The sheaths may be colorless or pigmented with various shades of yellow, red, brown, green, blue, violet, or blue-black, which makes some colonies quite striking in appearance.

The cells themselves appear blue-green in color in about half of the approximately 1,500 known species. This color, which is due to the presence of green chlorophyll *a* and blue *phycocyanin,* is often masked by the presence of other pigments. Several yellow or orange carotenoid pigments similar to those of higher plants are usually present, and varying amounts of *phycoerythrin* (a red phycobilin pigment in a form unique to the blue-green bacteria) may give the cells a distinct red color. The periodic appearances of large numbers of blue-green bacteria with considerable amounts of phycoerythrin are believed to have given the Red Sea its name. The organisms produce a nitrogenous food reserve called *cyanophycin.* The production of such food reserves is atypical for prokaryotic organisms. Blue-green bacteria also produce and store carbohydrates and lipids. Flagella are unknown in the blue-green bacteria, but some of these organisms are nevertheless capable of movement. *Oscillatoria* filaments (see Fig. 17.13), for example, seem to rotate on axes and move in a gliding fashion, apparently by the twisting of minute fibrils inside the cell walls while secreting mucilage that reduces friction. New cells are formed through fission, while new colonies or filaments may arise through *fragmentation* (breaking up) of older ones.

In the common genera *Nostoc* and *Anabaena* (see Fig. 17.13), which form chains of cells, fragmentation often occurs at special, larger, colorless, nitrogen-fixing cells called **heterocysts,** which are produced at intervals in the chains. Members of these two genera also may produce thick-walled cells called *akinetes,* which can resist freezing and other adverse conditions. When favorable conditions return, this survival feature enables the cells to germinate and become new

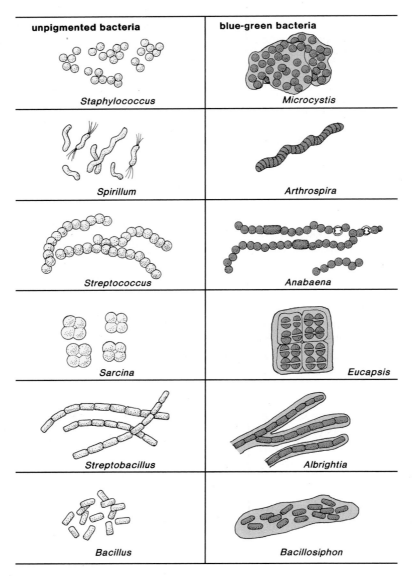

FIGURE 17.12   Similarity of form between various unpigmented bacteria and blue-green bacteria, ×2,000. (After Pelczar, M.J., Jr., and R.D. Reid. 1972. *Microbiology*. 3d ed. Redrawn by permission of the McGraw-Hill Book Co.)

chains or filaments. Some akinetes have been known to germinate after lying dormant for more than 80 years.

Blue-green bacteria do not produce gametes or zygotes and do not undergo meiosis. Genetic recombination has, however, been reported—apparently taking place in similar fashion to that reported for other bacteria—but its occurrence evidently is rare.

Blue-Green Bacteria, Chloroplasts, and Oxygen

Blue-green bacteria that occur symbiotically in other organisms commonly lack a cell wall and appear to function essentially as chloroplasts. When a eukaryotic cell containing chloroplasts divides, the chloroplasts divide at the same time. The cells of blue-green bacteria occurring within the cells of other organisms divide in similar fashion, leading to speculation that chloroplasts originated as blue-green bacteria, or *chloroxybacteria* (discussed after the blue-green bacteria), living within other cells.

Fossils of photosynthetic organisms believed to be 3.5 billion years old and closely resembling present-day blue-green bacteria, have been found in Australia. It was not until about 3 billion years ago that these organisms began producing oxygen as a by-product of photosynthesis. The oxygen slowly began to accumulate, becoming substantial about 1 billion years ago. At the same time the oxygen was accumulating, other photosynthetic organisms appeared, and forms of aerobic respiration developed.

Within the last half billion years, enough *ozone,* which is a by-product of oxygen, accumulated to become an effective shield against most of the harmful ultraviolet radiation coming from the sun. Photosynthetic organisms, which had been protected from the radiation by their watery

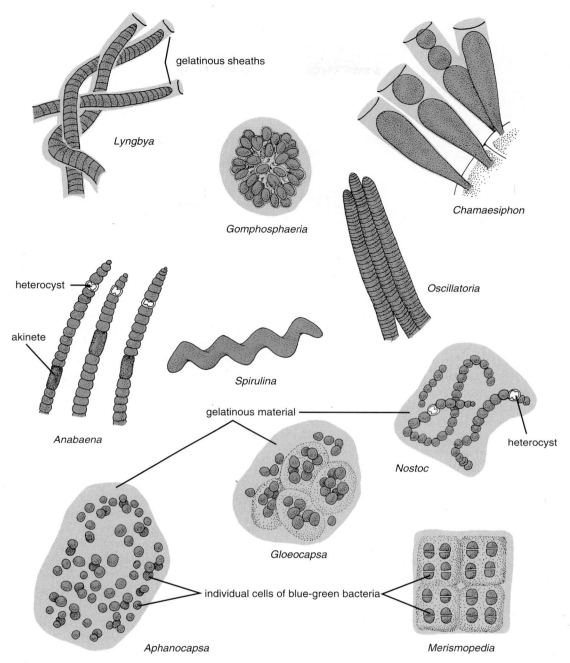

FIGURE 17.13 Representative blue-green bacteria, ×2,500.

environments, were then able to survive on land.[1] Blue-green bacteria thus appear to have played a fundamental role in almost the entire history of living organisms.

Human Relevance of the Blue-Green Bacteria

Blue-green bacteria are included among the many aquatic and photosynthetic organisms at the bottom of various food

1. See also the discussion in Chapter 18 of the role of certain green algae in the transition of aquatic organisms to land and the discussion in Chapter 25 of the effects of ozone depletion on our environment.

chains. They store energy in the sugar they produce through photosynthesis. Small fish and crustaceans feed on them, only to be eaten by larger fish, which then become food for other aquatic organisms or humans.

During warmer months, various algae and blue-green bacteria may become abundant in bodies of fresh water, especially if the water is polluted. A floating scum or mat called a *bloom* may cover 2,000 or more square kilometers (800 square miles) of the surfaces of larger lakes in late summer (Fig. 17.14). Blue-green bacteria, in particular, tend to become abundant in blooms when agricultural runoffs are high in nitrogen and phosphorus. The water in the vicinity of a bloom often acquires a fishy or otherwise objectionable odor or taste.

FIGURE 17.14 A bloom of cyanobacteria on a lake in upstate New York.

(© John D. Cunningham/Visuals Unlimited)

Anabaena, Microcystis, and other genera of blue-green bacteria produce toxic substances that can kill both domestic and wild animals. Cattle, hogs, horses, sheep, rabbits, dogs, and even poultry have been poisoned through drinking water that had an algal-bacterial bloom. Many fish that are immune to the toxins of certain blue-green bacteria become poisonous to their predators after feeding on the bacteria.

While the blue-green bacteria are alive, they produce oxygen, and the oxygen content of the water is temporarily increased. Later, however, decay bacteria decomposing the bodies of organisms that have died may deplete the available oxygen so much that fish and other organisms are killed.

Blue-green bacteria can be attacked by specific viruses called *cyanophages,* and there has been considerable interest in using the viruses to control blooms in lakes and ponds. Many of the blue-green bacteria responsible for the blooms, however, have thick, mucilaginous sheaths that may prevent viruses and fungi from penetrating the cells.

Blooms that develop in calm ocean waters in the tropics can be even more extensive than those of fresh waters. One marine bloom reported in 1965 between the shore and the Great Barrier Reef of Australia was 1,600 kilometers (almost 1,000 miles) long and covered 52,000 square kilometers (20,000 square miles) of ocean.

Most blue-green bacteria are not palatable to humans, but there are exceptions. Species of *Spirulina* have been used for food by the natives of the Lake Chad region of central Africa and areas around Mexico City for centuries. *Spirulina,* which has a significant vitamin content, is now cultured commercially and sold in health food stores. The Japanese use several blue-green bacteria, including two species of *Nostoc,* as side dishes, and the Chinese treat *Nostoc commune* as a delicacy. A few colorless forms are mild parasites in humans and animals.

Some strains of certain *Nostoc* species produce antibiotics that kill related strains of the same species. *Scytonema hofmannii* and other blue-green bacteria produce antibiotics called *cyanobacterins* that kill many different forms of both blue-green bacteria and eukaryotic algae. These particular blue-green bacteria undoubtedly play a role in their own survival by inhibiting growth of competing organisms.

Swimmers in Hawaii occasionally suffer from "swimmers' itch," a severe skin inflammation that is caused by a species of *Lyngbya* that sometimes becomes abundant. Ironically, the toxin produced by these organisms has been demonstrated to suppress leukemia and several other types of cancer.

In human water supplies, blue-green bacteria frequently clog filters, corrode steel and concrete, cause natural softening of water, and produce undesirable odors or coloration in the water. Many communities control blue-green bacteria in reservoirs through the addition of very dilute amounts of copper sulphate.

More than 40 species of blue-green bacteria are known to fix nitrogen from the air at roughly the same rates as the nitrogen-fixing bacteria of the leguminous plants discussed in Chapter 25. They may be more important than originally thought in this regard. In Southeast Asia, so much usable nitrogen is produced in the rice fields by naturally occurring blue-green bacteria that rice is often grown for many years on the same land without the addition of fertilizer.

CLASS CHLOROXYBACTERIAE— THE CHLOROXYBACTERIA

In 1976, Ralph A. Lewin of the Scripps Institute of Oceanography announced the discovery of unicellular, prokaryotic organisms with bright green cells that were living on marine animals called *sea squirts* found in shallow marine waters of Baja California. These organisms, which were given the name *Prochloron,* have the chlorophylls *a* and *b* of higher plants but no trace of the phycobilin accessory pigments associated with blue-green bacteria. Instead, their accessory pigments are confined to the carotenoid pigments found in higher plants. Also, unlike the single membrane thylakoids of blue-green bacteria, the thylakoids of the new organisms have double membranes.

Lewin considered the pigment differences between blue-green bacteria and the bright green cells of *Prochloron* to be basic enough to warrant recognition at the division level, and he proposed a new division, the *Prochlorophyta.* Many microbiologists are reluctant to recognize these organisms as belonging to a separate Moneran division because the prokaryotic cell structure and chemistry are similar to those of blue-green and other true bacteria (Fig. 17.15), but others agree with Lewin's assessment of the significance of the pigment system. While the pigment system is, indeed, significant, the chloroxybacteria are treated here as a class of true bacteria because their remaining structure and features are essentially indistinguishable from those of other true bacteria.

In 1984, Dutch biologists discovered a similar organism, which they named *Prochlorothrix,* in shallow lakes

FIGURE 17.15   A section through a cell of the prochlorobacterium *Prochloron,* which lives on the surface of sea squirts (marine animals). Note the absence of a nucleus and other organelles, and the concentric layers of membranes that perform some of the functions of organelles, ×15,750.

(Electron micrograph by Jean Whatley)

FIGURE 17.16   Papavoviruses in a human wart, ×52,000.

(Electron micrograph courtesy Richard S. Demaree Jr.)

in the Netherlands. It differs from *Prochloron* in being free-living and filamentous. In the late 1980s, Sally Chisholm of the Massachusetts Institute of Technology identified yet another marine chloroxybacterium that flourishes in dim light at a depth of about 100 meters (328 feet). This organism now appears to be one of the two most numerous bacteria living in ocean waters.

The discovery of chloroxybacteria adds weight to the theory that chloroplasts may have originated from such cells living within the cells of other organisms, especially since the pigments involved are identical to those of higher plants.

VIRUSES

Introduction

Smallpox is a communicable disease that apparently was widespread for thousands of years, periodically killing countless numbers of individuals. As it developed in its victims, it appeared as blisterlike lesions on the skin, which later often became permanent pits or depressions. In 1901, an outbreak of smallpox in New York caused 720 deaths. Since that time,

however, it has been eliminated in the United States through vaccination, and the World Health Organization believes it has now been eradicated throughout the world.

Vaccination involves the introduction of a weakened form of disease agent to the body. The body's natural defenses, in fighting the agent, build up an immunity to the disease. The first known vaccinations were performed in England by Benjamin Jesty, a farmer, and Edward Jenner, a country physician. They both had noticed that farmhands working with cows having cowpox, a milder disease than smallpox, did not contract smallpox during the devastating epidemics of the disease that occurred in Europe from time to time.

In 1796, Jenner scratched a boy's skin with fluid he had obtained from a cowpox blister on the hand of a milkmaid. Six weeks later, he deliberately inoculated the boy with fluid from a blister of a smallpox victim, but the boy developed no symptoms of the disease. Jenner thus had performed a successful vaccination against a dread disease more than 50 years before Louis Pasteur developed the germ theory of disease.

Size and Structure

We know now that smallpox was caused by something considerably smaller in size than bacteria. During Pasteur's time, virtually all infectious agents, including bacteria, protozoans, and yeast, were called **viruses.** One of Pasteur's associates, Charles Chamberland, discovered that porcelain filters would block out bacteria but would not keep an unseen agent from passing through. The agent caused rabies,

Plant Viruses

The book *Hot Zone* and the movie *Outbreak* have created an awareness of emerging viruses and their dangers to the human population. The Ebola, Hanta, and HIV viruses are now everyday words that have become synonymous with death. There is another group of viruses that also has a significant impact—plant viruses—which cause an estimated $15 billion worth of crop loss per year worldwide. They infect plants and cause hundreds of diseases such as tomato spotted wilt disease, tobacco mosaic disease, maize stripe disease, and apple chlorotic leaf spot disease.

In total, nearly 400 plant viruses have been identified and classified by the International Committee on Taxonomy of Viruses (ICTV). A further 320 have been identified, but are awaiting final classification.

Surprisingly, the first viruses ever identified were in plants. In 1898, a Dutch professor of microbiology, Dr. Martinus Beijerinck, was working to identify the disease that caused tobacco leaves to become mottled with light green and yellow spots. He demonstrated that the condition was not caused by a bacterium as was commonly thought at the time, but rather by some other unknown pathogen in the sap of the tobacco plant. He proved this by collecting sap from a diseased plant that was then passed through a filter capable of straining out any bacteria. When the filtered solution was reinjected into the leaf veins of healthy plants and the disease was transmitted, he had made his point. He called this filtered sap a *contagium vivum fluidium* (a contagious living fluid), and introduced the term **virus** to describe its property of being able to reproduce itself within living plants. Dr. Beijerinck's virus was later named tobacco mosaic virus (TMV), consistent with the now-established practice of naming plant viruses both by the plant it infects and by describing the major disease symptom (e.g., mosaic, wilting, spotted, etc.).

Not only were the first viruses discovered in plants, but the understanding of their biochemical nature was first recognized through research on tobacco mosaic virus. Today we know that viruses are submicroscopic, infectious particles that are composed of a protein coat and a nucleic acid center. They can be seen only with an electron microscope. As obligate parasites, they can reproduce themselves only within a living cell. The biochemical nature of viruses remained unknown until 1935 when Dr. Wendell Stanley, an organic chemist in the United States, succeeded in crystallizing the protein coat of tobacco mosaic virus (TMV). Stanley, however, did not recognize the nucleic acid content of the virus that was later shown to be RNA. The fact that RNA could exist **separately** from DNA was a discovery that has had great influence on the development of molecular biology thought.

Today we know that tobacco mosaic virus is a rigid rod, 300 nm × 15 nm, composed of a protein coat of approximately 2,100 helically arranged protein subunits surrounding an axial canal that contains a single-stranded RNA molecule consisting of 6,400 nucleotides. It, like all plant viruses, is classified according to the type of nucleic acid that it contains, either DNA or RNA but never both, whether the nucleic acid is single- or double-stranded, and the shape of the virus particle (spheres, stiff rods, flexible rods).

---

another serious disease of both animals and humans. Agents of disease that could pass through filters became known as *filterable viruses,* although the word *filterable* is no longer used. Today, we know that not only smallpox and rabies are caused by these viruses but also measles, mumps, chicken pox, polio, yellow fever, influenza, fever blisters, warts, and the common cold.

Only those organisms that have certain unique features are now called *viruses*. These features, which include a complete lack of cellular structure, make viruses quite different from anything else in the five kingdoms of living organisms now recognized. In fact, some question whether viruses are even living organisms. In 1946, Wendell Stanley, an American chemist, received a Nobel Prize for demonstrating that a virus causing tobacco mosaic, a common plant disease, could be isolated, purified, and crystalized and that the crystals could be stored indefinitely and would always produce the disease in healthy plants any time they were placed in contact with them. We also know that viruses do not grow by increasing in size or dividing, nor do they respond to external stimuli. They cannot move on their own, and they cannot carry on independent metabolism.

Viruses are incredibly numerous. In 1989, for example, marine biologists at the University of Bergen in Norway discovered that a teaspoon of sea water typically contains more than 1 billion viruses. They are about the size of large molecules, varying in diameter from about 15 to 300 nanometers (Fig. 17.16). Thousands of the smallest ones could fit inside a single bacterium of average size.

Viruses consist of a nucleic acid core surrounded by a protein coat. The architecture of the protein coats varies considerably, but many have 20 sides and resemble tiny geodesic domes, while others have distinguishable head and tail regions. The nucleic acid core consists of either DNA or

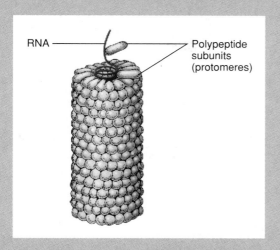

RNA

Polypeptide subunits (protomeres)

TMV is highly contagious, so much so that it can be transmitted to healthy plants merely from the fingers of smokers of cigarettes that were made from infected tobacco. This is unlike most other plant viruses that can survive no more than a few hours outside their living host. Plant viruses can gain entry into a plant only through an open wound or puncture, and are typically transmitted by insect vectors such as aphids, leafhoppers, whiteflies, and mites. Aphids are the most important vectors, infecting healthy plants when they insert their mouthparts, called stylets, into phloem tubes for feeding. During feeding, they inject salivary secretions containing the virus particles into the plant's sieve-tubes.

Once injected, viruses are transmitted within the phloem and move throughout the plant. However, viruses cannot move directly through cell walls. Rather, cell-to-cell movement of viral particles occurs via the plasmodesmata (sing., plasmodesma), which are cylindrical pores through cell walls that are membrane-lined. Plasmodesmata create cytoplasmic bridges that cross cell walls to connect adjacent cells and this transport route explains why many viral infections are systemic, affecting the entire organism.

Viruses seldom kill the plant outright, but rather weaken it by causing abnormalities in leaves (such as mottling or changes in leaf color, shape, or vein patterns), changes in flower color, or irregularities in fruit size, shape, or color. Viruses can also cause fruits to ripen prematurely and have an unpleasant taste or reduced sugar content. Crop yields of fruits and vegetables, as well as quality, can be reduced.

There are not many options for controlling plant viral diseases. The most effective control is achieved by sanitation—removing and burning diseased plants and killing virus-carrying insects. Additionally, naturally resistant varieties of some plants have been developed. Chemicals remain an ineffective treatment for plant viruses because of the cost and environmental concerns.

Viral diseases affect many important agricultural crops in addition to tobacco. Crop losses worldwide are enormous each year. With the world population increasing at about 1.6% yearly, any disease that threatens agricultural productivity and the ability of the human population to feed itself must be taken seriously. Although not as spectacular or newsworthy as Ebola or HIV, plant viruses are silent killers because they rob humanity by directly affecting the food supply.

RNA—never both. Viruses have been classified in several ways. Originally, they were grouped according to their hosts and the types of tissues or organs they affected. Now they are separated first according to the DNA or RNA in their cores. Then they are grouped according to size and shape, the nature of their protein coats, and the number of identical structural units in their cores.

Bacteriophages

Viruses that attack bacteria have been studied extensively. These are called **bacteriophages,** or simply **phages** (Fig. 17.17 A, B). Some resemble the space exploration vehicles portrayed in the literature and films of science fiction. They consist of a head on top of a thin cylindrical core, which is surrounded by a sheathing coat. Toward the base of the core are six spiderlike fibril "legs," which anchor the virus in place.

Viral Reproduction

Viruses can replicate (reproduce themselves) only at the expense of their host cells. In doing so, they first become attached to a susceptible cell by their "legs." Then they penetrate to the interior, some types leaving their coats on the outside. Inside the cell, their DNA or RNA directs the synthesis of new virus molecules, which are then assembled into complete viruses. These are released from the host cell, usually as it dies (Fig. 17.18).

Some viruses (e.g., those causing influenza) can mutate (see Chapter 13) very rapidly, allowing them to attack organisms that previously had been immune to them. As a result, new vaccines constantly have to be developed to combat new strains of viruses. Some viruses greatly affect the metabolism of their host cells. For example, in botulism bacteria, the botulism toxins are produced only if specific

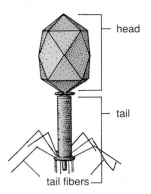

A.

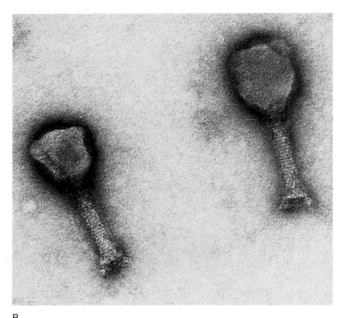

B.

FIGURE 17.17 Phage viruses. Insert drawing shows some of the detail.

(Electron micrograph courtesy D. Kay)

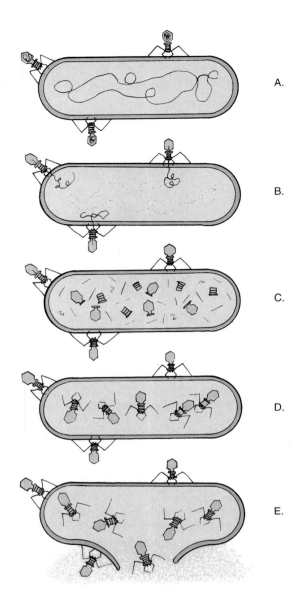

A.

B.

C.

D.

E.

FIGURE 17.18 Stages in the development of a phage virus within a bacillus bacterium. *A*. The virus becomes attached to the bacterium. *B*. The DNA of the virus enters the cell. *C*. Various components of the virus are synthesized from the DNA of the bacterium. *D*. The viral components are assembled into units. *E*. The assembled viruses are released as the bacterial wall breaks down.

(After Pelczar, M.J., Jr., and R.D. Reid. 1972. *Microbiology,* 3d ed. Redrawn by permission of the McGraw-Hill Book Co.)

phages are present and active. Evidence is mounting that many forms of cancer, which usually involves abnormal cell growth, are caused by viruses. Scientists also suspect that all living organisms carry viruses in an inactive form in their cells, and they are trying to discover what causes the inactive viruses suddenly to become active.

Cells of higher animals that are invaded by viruses produce a protein called *interferon,* which is released into the fluid around the cells or into the bloodstream. Minute amounts of interferon in contact with cells cause the cells to produce a protective protein that prevents or inhibits the propagation of numerous types of viruses within the protected cells and also inhibits viruses from causing tumors that transform normal cells into tumor cells.

Because of these properties of interferon, research is currently being conducted to discover how to produce it in much greater quantities for use in controlling cancers and

many other viral diseases. Especially promising is the use of bacteria as hosts for donor DNA. This process, in effect, turns the bacteria into interferon synthesis centers (see Fig. 14.2). In 1981, scientists at the University of Washington and the Genentech Corporation of San Francisco announced that they had succeeded in producing a form of interferon by splicing interferon genes into yeast cells. Because yeast cells are larger than bacteria, the process can potentially produce much larger quantities of interferon at considerably lower cost than is possible using bacteria.

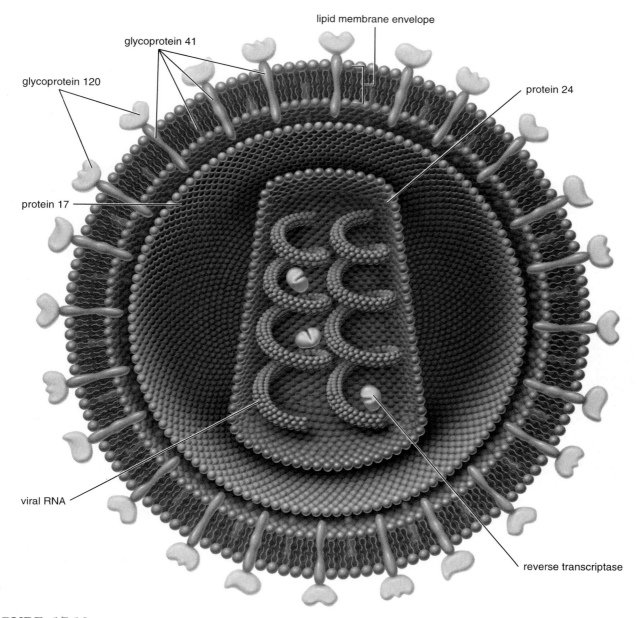

FIGURE 17.19 A virus is a nucleic acid coated with protein. The human immunodeficiency virus (HIV), which causes AIDS, consists of RNA surrounded by several layers of proteins. Once inside a human cell (usually a T cell, part of the immune system), the virus uses an enzyme to convert its RNA to DNA, which then inserts into the host DNA. HIV damages the human body's protection against disease by killing T cells and by using these cells to make more of itself.

Human Relevance of Viruses

The economic impact of viruses in both developing and industrialized countries is enormous. The annual loss in work time due to common cold and influenza viruses alone amounts to millions of hours. While discomfort, adverse effects on employment, and even deaths due to viral diseases, such as chicken pox, measles, German measles, mumps, and yellow fever, have declined dramatically since immunizations against the diseases became widespread, they still take their toll. Another viral disease, infectious hepatitis, periodically still claims victims. Guillain-Barré Syndrome and Epstein-Barr infections are debilitating diseases caused by

viruses that are apparently carried by nearly everybody, but what triggers them into action is as yet still unknown.

AIDS (acquired immune deficiency syndrome), a fatal disease caused by one or more viruses called *retroviruses* that are related to those that cause cancer, was unknown until its discovery in 1959 in a blood sample in what is now Zaire. Research has shown that two forms of the AIDS virus, HIV-1 and HIV-2, diverged from a common ancestor in the early 1950s (Fig. 17.19). Thirty years later, the AIDS virus had spread all over the world. Retroviruses mutate so rapidly that they are capable of evolving about a million times faster than cellular organisms. Since the initial appearance of AIDS, increasing research is being applied to the development of a vaccine with the use of a genetically engineered virus. If successful, the vaccine will cause the human body to develop a defense against AIDS viruses without the disease itself developing.

The production of vaccines for a number of other diseases is undertaken by a flourishing worldwide multimillion dollar industry. Other mass-produced viruses are used to infect ticks, insects, and other disease organisms of both animals and plants. Some viruses cause great losses when they infest creamery vats during the manufacture of dairy products or culture vats during the production of antibiotics. One phage attacks nitrogen-fixing bacteria in the roots of leguminous plants, while other phages attack diphtheria and tuberculosis bacteria. One natural virus called *Abby* (an abbreviation of Abington, which is one of 19 strains of nuclear polyhedrosis viruses) is found only in the caterpillars of gypsy moths. Gypsy moth caterpillars have been particularly destructive in forests of both North America and Europe, and it is hoped that the destruction may be greatly reduced by the dissemination of this virus in the forests under attack.

Summary

1. Kingdom Monera consists of prokaryotic organisms, which are grouped into archaebacteria and eubacteria. The bacteria occur as single cells, in colonies, or in the form of chains or filaments. Some cells may be motile or they may exhibit a gliding motion; most are nonmotile.

2. Bacterial nutrition occurs primarily by absorption of food in solution through the cell wall, but some bacteria are photosynthetic or chemosynthetic.

3. Reproduction is asexual by means of fission; some genetic recombination occurs by means of pili between cells.

4. Bacteria are mostly less than 2 or 3 micrometers in diameter. They occur as spheres (cocci), rods (bacilli), and spirals (spirilli) and are further classified on the basis of sheaths, appendages, and motion. They are also classified as gram-positive or gram-negative.

5. Prokaryotic cells have no nuclear envelopes or organelles. Each cell has one long DNA molecule and sometimes up to 30 or 40 small, circular DNA molecules called *plasmids,* which replicate independently of the large DNA molecule or chromosome.

6. Neither meiosis nor mitosis occurs, but fission takes place with the development of a transverse wall near the middle of the cell; gametes and zygotes are not produced. Conjugation facilitates genetic recombination. Transformation involves the incorporation of fragments of DNA released by dead cells; transduction involves the viral transfer of fragments of DNA from one cell to another.

7. Heterotrophic bacteria are saprobes or parasites. Autotrophic bacteria are photosynthetic but do not produce oxygen; chemoautotrophic bacteria obtain their energy through oxidation of reduced inorganic groups.

8. Any nonliving tissues will eventually be decomposed to compost by bacteria and fungi. Compost is definitely good for the soil but has limited value as a fertilizer.

9. Bacteria cause huge losses through plant diseases and food spoilage and many serious diseases in animals and humans. They gain access to their hosts by various means.

10. Koch formulated postulates (rules) for proving that a particular microorganism is the cause of a particular disease.

11. Bacteria useful to humans include *Bacillus thuringiensis, Bacillus thuringiensis* var. *israelensis,* and *Bacillus popilliae.* Bacteria also play a major role in the manufacture of dairy products, such as yogurt, sour cream, kefir, and cheese.

12. Bacteria are used in the manufacture of industrial chemicals, vitamins, flavorings, food stabilizers, and a blood plasma substitute; they play a role in the curing of vanilla, cocoa beans, coffee, and tea and in the production of vinegar and sauerkraut; and they aid in the extraction of linen fibers from flax stems, in the production of ensilage for cattle feed, and in the production of several important amino acids.

13. Blue-green bacteria are virtually ubiquitous in their occurrence.

14. The cells of blue-green bacteria occur in a variety of forms. They are distinguished from other bacteria in having chlorophyll *a,* in producing oxygen, and in having blue and red phycobilin pigments. They produce cyanophycin, a nitrogenous food reserve.

15. Blue-green bacteria have no flagella, but some species have gliding movements. Fragmentation may occur at heterocysts. Akinetes may also be produced.

16. Blue-green bacteria cells may have been the origin of chloroplasts, since they divide as chloroplasts do when their host cells divide.

17. Blue-green bacteria may become very abundant in bodies of polluted fresh water. Toxic substances are produced when the bacteria die and are decomposed. At least 40 species of blue-green bacteria are known to fix nitrogen.

18. Chloroxybacteria are similar in form to blue-green bacteria but have pigmentation similar to that of higher plants and lack phycobilins.

19. Vaccination was first performed by Jesty and Jenner in connection with smallpox. Smallpox is now believed to have been eradicated.

20. Viruses, which have no cellular structure, are about the size of large molecules. Some can be isolated, purified, and crystallized yet remain virulent. They cannot grow or increase in size and cannot be reproduced outside

of a living cell, upon whose DNA they depend for duplication.

21. Viruses consist of a core of nucleic acid surrounded by a protein coat. They are classified on the basis of the DNA or RNA in their core, their size and shape, the number of identical structural units in their cores, and the nature of their protein coats.

22. Bacteriophages are viruses that attack bacteria. In replicating, they become attached to a susceptible cell, which they penetrate, with their DNA or RNA directing the synthesis of new virus molecules from host material; the assembled new viruses are released when the host cell dies.

23. Cells of higher animals being invaded by viruses produce interferon, a protein that causes cells to produce a protective substance that inhibits duplication of viruses and also inhibits the capacity of viruses to transform normal cells into tumor cells.

24. Viral diseases, such as chicken pox, measles, mumps, and yellow fever, have declined since immunizations against the diseases have become widespread. Mass-produced viruses are used to infect insects and other pests of both plants and animals. Some viruses cause losses in creamery vats.

Review Questions

1. What is symbiosis? Give examples other than those mentioned in the text.
2. Why are bacteria not classified on the basis of visible features alone?
3. How do bacteria exchange DNA?
4. How does fission differ from mitosis?
5. Is photosynthesis the same in bacteria as it is in higher plants? Explain.
6. How do chemoautotrophic bacteria differ from photosynthetic bacteria?
7. What is the difference between nitrification and nitrogen fixation?
8. If decay bacteria use nitrogen, how does composting accumulate any nitrogen?
9. How are disease bacteria transmitted?
10. What are Koch's Postulates?
11. Why are many bacteria considered useful?
12. What do blue-green bacteria and other bacteria have in common? How do they differ?
13. How do blue-green bacteria survive freezing and desiccation?
14. What is an algal-bacterial bloom?
15. How do viruses differ from bacteria?
16. What is a vaccination?
17. What is a phage?
18. How do viruses multiply?

Discussion Questions

1. If a virulent phage were to eliminate all the bacteria in North America for one year, how would our lives be affected?
2. What would be the feasibility and the advantages or disadvantages of using only blue-green bacteria and other nitrogen-fixing bacteria for our agricultural nitrogen needs?
3. Methane gas produced by bacteria is proving to be sufficient to meet all the fuel needs of villages in India. Do you think we could produce and use methane in a similar fashion in the United States? Explain.
4. If blue-green bacteria are capable only of asexual reproduction, does this mean that species of these organisms can never change in form?
5. As long as viruses can multiply, why should there be any question as to whether they are living?

Additional Reading

Bos, L. 1992. *Introduction to plant virology.* New York: State Mutual Book and Periodical Service.

Brock, T. D., and M. Madigan. 1990. *Biology of microorganisms,* 6th ed. Englewood Cliffs, NJ: Prentice-Hall.

Cann, A. 1993. *Molecular virology.* San Diego, CA: Academic Press.

Carr, N. G., and B. A. Whitton (Eds.). 1982. *The biology of Cyanobacteria.* Berkeley, CA: University of California Press.

Dimmock, N. J., and S. B. Primrose. 1987. *Introduction to modern virology.* London: Blackwell Scientific Publications.

Dixon, B. 1994. *Power unseen: How microbes rule the world.* San Francisco: W. H. Freeman.

Gunter-Schlegel, H. 1994. *General microbiology,* 7th ed. New York: Cambridge University Press.

Holt, J. G., and N. R. Krieg (Eds.). 1984. *Bergey's manual of determinative bacteriology,* vol. 1. Baltimore: Williams and Wilkins.

Kim, C. W. 1991. *Microbiology,* 10th ed. East Norwalk, CT: Appleton and Lange.

Pelczar, M. J., Jr. 1993. *Microbiology: Concepts and applications,* 5th ed. New York: McGraw-Hill.

Prescott, L., et al. 1993. *Microbiology,* 2d ed. Madison, WI: Brown and Benchmark.

Stolp, H. 1988. *Microbial ecology.* New York: Cambridge University Press.

Stoner, C. H. 1993. *Biotechnology for hazardous waste treatment.* Boca Raton, FL: Lewis Publications.

Tortora, G. J. 1991. *Microbiology: An introduction,* 4th ed. Menlo Park, CA: Benjamin/Cummings Publishing Co., Inc.

Woese, C. R. 1981. Archaebacteria. *Scientific American* 244(6): 98–122.

Giant kelp (Macrocystis pyrifera), *which grows to lengths of 30 meters (100 feet). One specimen from the Pacific Coast was 45.7 meters (148 feet) long. Giant kelps are harvested for their algin, a useful substance discussed in this chapter.*

(Photo by Digital Stock.)

18 Kingdom Protoctista

Overview

After giving an introduction and summary of the features of members of Kingdom Protoctista, this chapter discusses the divisions of algae, slime molds, chytrids, and water molds. A brief discussion of the life cycle of a representative of each division is included.

First, the Chrysophyta (yellow-green algae, golden-brown algae, diatoms, and cryptophytes) are discussed. Next, the Division Pyrrophyta (dinoflagellates) is briefly examined and the role of members of the division in red tides and bio-luminescence is explored. Asexual and sexual reproduction in the Chlorophyta (green algae) is shown in Chlamydomonas, Ulothrix, Spirogyra, *and* Oedogonium; *mention is made of* Chlorella, *desmids,* Acetabularia, Volvox, *and* Ulva. *After giving a brief overview of the Euglenophyta (euglenoids), the chapter takes up the Phaeophyta (brown algae) and Rhodophyta (red algae). Differences between the divisions of algae are shown in a table that lists food reserves, special pigments, and flagella. This is followed with a digest of the human and ecological relevance of the algae. The chapter concludes with a brief examination of the slime molds, chytrids, and water molds.*

Some Learning Goals

1. Know features that the members of Kingdom Protoctista share with one another, and note the basic ways in which they differ.
2. Understand how a diatom differs in structure and form from other members of the Division Chrysophyta.
3. Diagram the life cycles of *Chlamydomonas, Ulothrix, Spirogyra,* and *Oedogonium;* indicate where meiosis and fertilization occur in each.
4. Learn at least two features that distinguish the Chrysophyta, Pyrrophyta, Euglenophyta, Chlorophyta, Phaeophyta, and Rhodophyta from one another.
5. Know what *holdfasts, stipes, blades, bladders,* and *thalli* are.
6. Learn the function of each of the three thallus forms of a red alga, such as *Polysiphonia,* and know at least 20 economically important uses of algae, in addition to the numerous uses of algin that are given.
7. Understand why the slime molds, chytrids, and water molds are not true fungi, and what they have in common with other members of Kingdom Protoctista.

I was once the sole passenger on a ship carrying cargo from an Indian Ocean port to Boston. I had turned in early one night as we were nearing the equator, but I awoke around 2:00 A.M. and went to the bathroom to get a glass of water. I had not turned on the light and was startled to notice that the water in the toilet bowl was "winking" at me. Numerous little lights in the water were flashing on and off! I remembered that ships out of the harbor take in water from their surroundings for their sewage lines, and I hurried out on deck to see where we were headed. I was at once treated to a beautiful display of *bioluminescence* (discussed in Chapter 11) caused by millions of

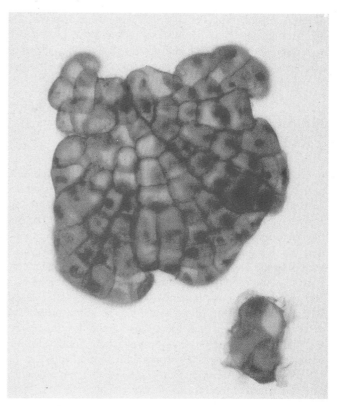

FIGURE 18.1 *Coleochaete,* a green alga that has several features of higher plants and is believed to be an indirect ancestor of land plants, ×500.

microscopic algae called *dinoflagellates.* As the waves broke, both at the bow and in the wake, there was a sparkling, shimmering glow as the tiny organisms, through their respiration, produced light. The dinoflagellates constitute but one of nine divisions of algae and other relatively simple eukaryotic organisms discussed in this chapter.

The fossil record indicates that less than 1 billion years ago all living organisms were confined to the oceans, where they were protected from drying out, ultraviolet radiation, and large fluctuations in temperatures. They also absorbed the nutrients they needed directly from the water in which they were immersed. The fossil record also suggests many times that, beginning about 400 million years ago, green algae, which are important members of Kingdom Protoctista, began making the transition from water to land, eventually giving rise to green land plants.

Coleochaete (Fig. 18.1), a tiny, freshwater green alga that grows as an **epiphyte** (an alga or plant that attaches itself in a nonparasitic manner to another living organism) on the stems and leaves of submerged plants, shares several features with higher plants and probably was an indirect ancestor of today's land plants. The features include somewhat parenchymalike cells, the development of a cell plate and phragmoplast during mitosis, a protective covering for the zygote, and the production of a ligninlike compound. Lignin adds mechanical strength to cell walls, and the discoverers of the substance suggest that it originated as a protection against microbes. The ligninlike compound was present 100 million years earlier than the presumed evolution of plant life from water to land, and its presence in algal cells could explain how such organisms adapted to land habitats.

FEATURES OF KINGDOM PROTOCTISTA

In contrast with members of Kingdom Monera, which have prokaryotic cells, all members of Kingdom Protoctista have *eukaryotic* cells. The organisms comprising this kingdom are very diverse and heterogeneous, but none have the distinctive combinations of characteristics possessed by members of Kingdoms Plantae, Fungi, or Animalia. Many, including the euglenoids, protozoans, and some algae, consist of a single cell, while other algae are multicellular or occur as colonies or filaments. Nutrition is equally varied, with the algae being photosynthetic, the slime molds and protozoans ingesting their food, the euglenoids either carrying on photosynthesis or ingesting their food, and the oomycetes and chytrids absorbing their food in solution.

Individual life cycles vary considerably, but reproduction is generally by cell division and sexual processes. Many *protists* (mostly single-celled members of Kingdom Protoctista) are motile, usually by means of flagella, but other members of the kingdom, especially those that are multicellular, are nonmotile, although most of the multicellular members produce some motile cells.

SUBKINGDOM PHYCOBIONTA—ALGAE

Children fortunate enough to have lived near ocean beaches where seaweeds are cast ashore by the surf have often enjoyed stamping on the bladders ("floats") of kelps to hear the distinct popping sound as they break. Some have collected and pressed beautiful, feathery red seaweeds, and others who have waded around the shores of freshwater lakes or in slow-moving streams have encountered slimy feeling pond scums. Anyone who has kept tropical fish in glass tanks have sooner or later had to scrape a brownish or greenish film from the inner surfaces of the tank, and those who have lived in homes with their own swimming pools have learned that the plaster of the pool soon acquires colored patches on its surface if chemicals are not regularly added to prevent them from appearing.

Although some of the seaweeds have flattened, leaflike blades, algae have no true leaves or flowers. They are involved in our everyday lives in more ways than most people realize (see the discussion on the human and ecological relevance of the algae, which begins on page 308). Seaweeds and some pond scums, fish tank films, and colored patches in swimming pools include but a few of the numerous kinds of algae all assigned to Kingdom Protoctista. The algae are grouped into several major divisions based on the form of their reproductive cells and combinations of pigments and food reserves.

DIVISION CHRYSOPHYTA—THE GOLDEN-BROWN ALGAE

If the roughly 6,000 members of this division were not primarily microscopic, many surely would become collectors' items in the art and antique shops of the world because of their exquisite form and ornamentation (Fig. 18.2; see also figs. 18.3 and 18.4).

The algae in this division can be grouped into four classes: *yellow-green algae* (Xanthophyceae), *golden-brown algae* (Chrysophyceae), *diatoms* (Bacillariophyceae), and *cryptophytes* (Cryptophyceae). Superficially, the organisms of each class may appear unrelated to each other, but they do share several features, including food reserves, specialized pigments, and other cell characteristics. Some members of each class produce a unique "resting" cell called a *statospore* (Fig. 18.3). These cells resemble miniature glass apothecary bottles, complete with plugs that dissolve or "uncork," releasing the protoplast inside. Many statospores are striking in form, with finely sculptured ornamentations on the surface.

Diatoms

Diatoms (Fig. 18.4) are the best-known and the most economically important members of the division. These mostly unicellular algae occur in astronomical numbers in both fresh and salt water but are particularly abundant in colder marine habitats. In fact, as a rule, the colder the water, the greater the number of diatoms present, and huge populations of diatoms are found on and within ice in both Antarctica and the Arctic. A major constituent of the foam that accumulates at the wave line on beaches is an oil produced by diatoms.

Diatoms also usually dominate the algal flora on damp cliffs, the bark of trees, bare soil, and the sides of buildings. More than 5,600 living species are recognized, with almost as many more known only as fossils. Some can withstand extreme drought, and one species is known to have become active after lying dormant in dry soil for 48 years.

Diatoms look like ornate, little, glass pillboxes, with half of the rigid, crystal-clear wall fitting inside the other overlapping half. Many marine diatoms are circular in outline when viewed from the top, while freshwater species viewed the same way tend to resemble the outline of a kayak. As much as 95% of the wall content is silica, an ingredient of glass, deposited in an organic framework of pectin or other substances.

The diatom walls usually have exquisitely fine grooves and pores that are exceptionally minute passageways connecting the protoplasm with the watery environment outside the shell. These fine grooves and pores are so uniformly spaced they have been used to test the resolution of

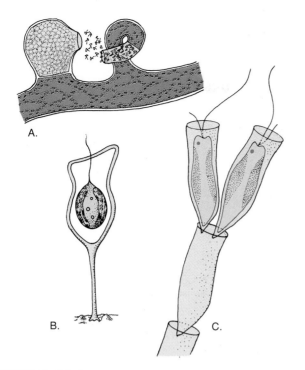

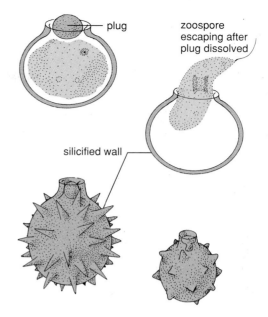

FIGURE 18.2   Some representatives of the Division Chrysophyta. *A. Vaucheria. B. Stipitococcus. C. Dinobryon.*

FIGURE 18.3   Statospores, which often resemble apothecary bottles, are formed by many of the golden-brown algae.

(After G.M. Smith. 1950. *The Freshwater Algae of the United States,* 2d ed. Redrawn by permission of the McGraw-Hill Book Co.)

microscope lenses. When *phycologists* (scientists who specialize in the study of algae) want to examine the markings on diatom walls, they often obtain a better view by dissolving the protoplasm with hot acid.

Each diatom may have one, two, or many chloroplasts per cell. In addition to chlorophyll *a*, the accessory pigments chlorophyll c_1 and chlorophyll c_2 are typically present. The chloroplasts usually are golden-brown in color because of the dominance of **fucoxanthin,** a brownish pigment also found in the brown algae. Food reserves are oils, fats, or the carbohydrate *chrysolaminarin.*

Freshwater diatoms that have a lengthwise groove called a *raphe* glide backward and forward with somewhat jerky motions, at a rate of up to three times their length in 5 seconds. It is believed that the movements occur in response to external stimuli such as light. Extremely tiny fibrils apparently take up water as they are discharged into the raphe or pores through which moving cytoplasm protrudes. When they come in contact with a surface, they stick and contract, moving the cell as the caterpillar treads on a tractor do and leaving a trail something like that of a snail.

Reproduction in diatoms is unique. Before any cell division can occur, an adequate source of silicon must be present in the surrounding medium. In culture, the number of cells produced is directly proportional to the amount of silicon added to an otherwise nutrient-balanced medium. Division in culture is also rhythmic, with all the cells dividing at the same time.

After a protoplast, which is diploid, has undergone mitosis and division, the two halves of the cell separate, with a

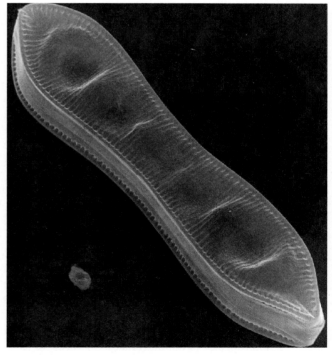

FIGURE 18.4   A diatom.

(Scanning electron micrograph courtesy J.D. Pickett-Heaps)

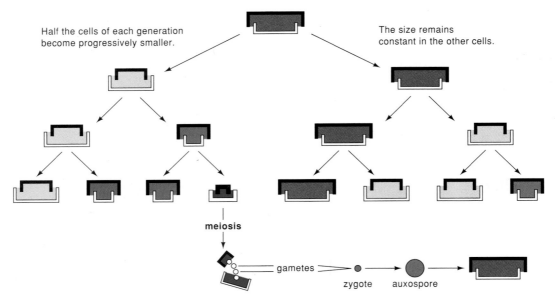

FIGURE 18.5 Reproduction in diatoms. The two rigid halves of each pillboxlike cell separate and a new rigid half shell forms *inside* each original half. This results in half the new cells becoming smaller with each generation. At one point, however, the diploid nucleus of a reduced-size cell undergoes meiosis, and four gametes are formed. The zygote produced when the two gametes unite becomes considerably enlarged. The enlarged cell, called an *auxospore,* develops into a diatom of the same size as the original diatom.

In the figure: "Half the cells of each generation become progressively smaller." ... "The size remains constant in the other cells." ... meiosis ... gametes ... zygote ... auxospore

daughter protoplast remaining in each half. Then a new half wall fitting *inside* the old half is formed. This occurs for a number of generations, with the result that half of the cells become progressively smaller. Eventually, however, a protoplast undergoes meiosis, producing four gametes, which then escape. These fuse with other gametes, becoming zygotes called *auxospores*. Auxospores are like any other zygotes except that they restore the original size of the diatom by rapidly ballooning before forming rigid pillboxlike walls (Fig. 18.5).

DIVISION PYRROPHYTA— THE DINOFLAGELLATES

Occasionally, visitors to an ocean beach in midsummer may notice a distinctly reddish tint to the water, usually as a result of a phenomenon known as a *red tide*. Red tides are caused by the sudden and not fully understood multiplication of unicellular organisms called *dinoflagellates* (Fig. 18.6). There are over 3,000 species of dinoflagellates, 300 of them capable of producing red tides. When a red tide appears, some biologists collect cups of seawater for examination with a microscope. (The material can be preserved indefinitely with the addition of a few drops of formaldehyde, vinegar, or other weak acid.)

Each cup of seawater collected usually contains a large number of dinoflagellate "shells," which are the best

known representatives of the Division Pyrrophyta. Some resemble armor-plated spaceships, while others may be smooth or have fine lengthwise ribs. The "armor" plates are located just inside the plasma membrane and are composed mostly of cellulose of varying thickness.

Dinoflagellates have two flagella that are distinctively arranged, usually attached near each other in two adjacent and often intersecting grooves. One flagellum, which acts as a rudder, trails behind the cell. The other, which encircles the cell at right angles to the first groove, gives the cell a spinning motion as it undulates in its groove like a tiny snake.

Most dinoflagellates have two or more disc-shaped chloroplasts, which contain distinctive brown pigments in addition to various other pigments, including chlorophylls a and c_2. About 45% of the species are, however, nonphotosynthetic, and some ingest food particles, whether or not chlorophyll is present. Some have an *eyespot* (a pigmented organelle that is sensitive to light), and all have a unique nucleus in which the chromosomes remain condensed and clearly visible throughout the life of the cell. The chromosomes contain a disproportionately large amount of DNA— as much as 40 times that found in human cells. The food reserve is starch, which in dinoflagellates is stored outside the chloroplasts.

Dinoflagellates occur in most types of fresh and salt water, but those that cause red tides have received the most publicity because about 40 of the species also produce powerful neurotoxins that accumulate in shellfish such as oysters,

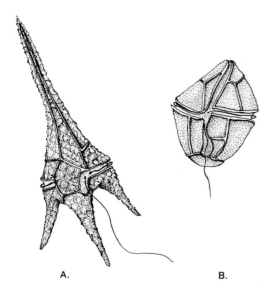

A.   B.

FIGURE 18.6   Dinoflagellates. *A. Ceratium.*
B. Gonyaulax.

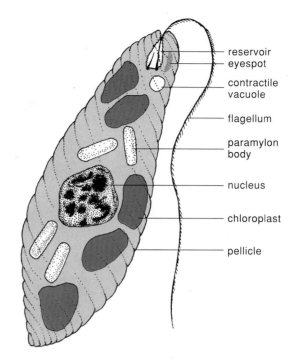

- reservoir
- eyespot
- contractile vacuole
- flagellum
- paramylon body
- nucleus
- chloroplast
- pellicle

FIGURE 18.7   A single *Euglena.*

mussels, scallops, and clams. The poisons apparently don't harm the shellfish, but about 2,000 persons a year become ill—15% fatally—from eating contaminated shellfish. Although fish in open waters can swim away from affected waters, large numbers still die and wash ashore after a red tide, and the devastation to caged fish can be enormous. Even pelicans, dolphins, whales, and manatees have been poisoned by dinoflagellate toxins in their food chains.

The toxins, which are of three major types, have been studied for possible use in chemical warfare. They are so potent that as little as half a gram (enough to barely cover two of the periods printed on this page) can be fatal. In humans, they can cause nausea, abdominal cramps, muscular paralysis, amnesia, hallucinations, diarrhea, and respiratory failure resulting in death.

The havoc to the fishing industry caused by major red tides is so great that several laboratories have been conducting research with dinoflagellates and their marine habitats to try to determine the causes of their blooms and to find a way of preventing such destruction in the future.

Until 1970, blooms of toxin-producing dinoflagellates were known only from temperate waters of the Northern Hemisphere. By 1990, however, the blooms were also occurring with increasing frequency throughout both temperate and tropical waters of the Southern Hemisphere.

Reproduction is by cell division. Sexual reproduction appears to be rare.

DIVISION EUGLENOPHYTA— THE EUGLENOIDS

Barnyard pools and sewage treatment ponds often develop a rich green bloom of algae. A superficial examination of water from such a pool with the aid of a microscope usually reveals large numbers of active green cells, and a closer inspection probably will reveal one to several of the more than 750 species of *euglenoids,* of which *Euglena* is a common example (Fig. 18.7).

A *Euglena* cell, which is spindle-shaped and has no rigid wall, can be seen to change shape even as the organism moves along. Just beneath the plasma membrane are fine strips that spiral around the cell parallel to one another. The strips and the plasma membrane are devoid of cellulose and together are called a *pellicle.* A single functional flagellum, which has numerous tiny hairs along one side, pulls the cell through the water. A second very short flagellum is present within a *reservoir* at the base of the functional flagellum.

Other features of *Euglena* include the presence of a *gullet,* or groove, through which food can be ingested, and in about a third of the 500 species there are several to many mostly disc-shaped chloroplasts present. A red **eyespot** is located in the cytoplasm near the base of the flagella, and a carbohydrate food reserve called *paramylon* normally is present in the form of small whitish bodies of various shapes.

Reproduction is by cell division. The cell starts to divide at the flagellar end and eventually splits lengthwise, forming two complete cells. Sexual reproduction has been suspected but has never been confirmed.

Some species of *Euglena* can live in the dark if appropriate food and vitamins are present. Others are known to reproduce faster than their chloroplasts under certain circumstances, so some chloroplast-free cells are formed. As long as a suitable environment is provided, these cells also can survive indefinitely. In the past, when only two

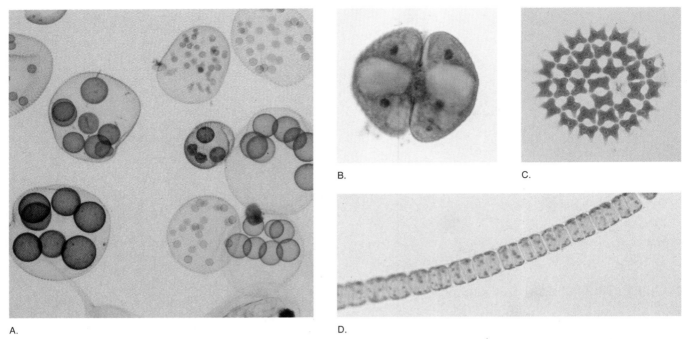

FIGURE 18.8   Representative green algae. *A. Volvox.* The cells form hollow, spherical colonies that spin on their axes as the flagella of each cell beat in such a way that the motion is coordinated. New colonies are formed within the old ones, ×1,000. *B. Cosmarium,* a desmid. These algae consist of single cells that often have a constriction in the center, ×1,200. *C. Pediastrum.* A colonial alga that forms flat plates, ×1,200. *D.* A filament of *Ulothrix,* ×800.

kingdoms were recognized, *Euglena's* capacity to satisfy its energy needs through either photosynthesis or ingestion of food resulted in its being treated as a plant in botany texts and as an animal in zoology texts.

DIVISION CHLOROPHYTA— THE GREEN ALGAE

Division Chlorophyta includes about 7,500 species of organisms commonly known as the *green algae.* They occur in a rich variety of forms and are very widespread, with some of the most beautiful chloroplasts of all photosynthetic organisms. Some are unicellular and microscopic. Others form threadlike filaments, platelike colonies, netlike tubes, or hollow balls (Fig. 18.8).

Some green algae are seaweeds, resembling lettuce leaves or long green ropes. Several unicellular species grow in greenish patches or streaks on the bark of trees, while others grow in large numbers on the fur of sloths and other jungle animals, providing them with a form of camouflage. Still others thrive in snowbanks, live in flatworms and sponges, or are found on the backs of turtles. They are the most common member in lichen "partnerships" (discussed in Chapter 19). The greatest variety, however, is found in freshwater ponds, lakes, and streams. Ocean forms are also varied; there they are an important part of the **plankton**

(free-floating, mostly microscopic organisms) and thus of food chains.

The cells of green algae resemble those of higher plants in having the same kinds of chlorophylls (*a* and *b*) and other pigments in their chloroplasts. The green algae also are believed to have been ancestral to the higher plants, and like the higher plants, they store their food within the chloroplasts in the form of starch. Although most green algae have a single nucleus in their cells, one group, the *bryopsids,* has multinucleate cells. Most green algae can undergo both asexual and sexual reproduction. The manner in which they do so is illustrative of the forms of reproduction found in most of the organisms discussed in the chapters to follow, and so several different representative green algae will be examined in some detail here.

Chlamydomonas

A lively little alga, *Chlamydomonas* (Fig. 18.9) is a common inhabitant of quiet freshwater pools. It has an ancient history among eukaryotic organisms, with fossil relatives occurring in rock formations reported to be nearly 1 billion years old. *Chlamydomonas* is unicellular, with a somewhat oval cell surrounded by a cellulose wall. A pair of whiplike flagella at one end propels the cell very rapidly. The flagella are, however, difficult to see with ordinary light microscopes, and the cell itself is usually less than 25 micrometers (one 10,000th of an inch) long. Near the base of the flagella are two or

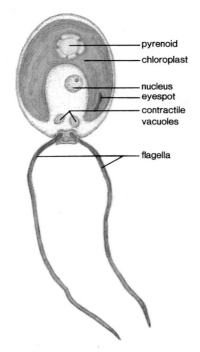

pyrenoid

chloroplast

nucleus
eyespot
contractile
vacuoles

flagella

FIGURE 18.9   *Chlamydomonas.*

more vacuoles, which can contract and expand. They apparently regulate the water content of the cell.

A dominant feature of each *Chlamydomonas* is a single, usually cup-shaped chloroplast, which at least partially hides the centrally located nucleus. One or two roundish glistening **pyrenoids** are located in each chloroplast. Pyrenoids are proteinaceous structures thought to contain enzymes associated with the synthesis of starch. Most species also have a red eyespot on the chloroplast near the base of the flagella. The eyespot is sensitive to light, but it is merely part of an organelle within a single cell and nothing like an eye in structure.

Asexual Reproduction

Before a *Chlamydomonas* reproduces asexually, the cell's flagella degenerate and drop off or are reabsorbed. Then the nucleus divides by mitosis, and the entire protoplasm becomes two cells within the cellulose wall. The two daughter cells, which develop flagella, escape and swim away as the parent cell wall breaks down. Once they have grown to their full size, they may repeat the process. Sometimes mitosis occurs more than once, producing 4, 8, or up to 32 little cells with flagella inside the parent cell. Occasionally, flagella do not develop, and the cells remain together in a colony. When growth conditions change, however, each cell of the colony may develop flagella and swim away. This type of reproduction brings about no changes in the number of chromosomes present in the nucleus, and all the cells remain *haploid.*

Sexual Reproduction

Under certain combinations of light, temperature, and additional unknown environmental forces, many cells in a population of *Chlamydomonas* may congregate together. Careful study of such events has revealed that pairs of cells appear to be attracted to each other by their flagella and function as gametes, which sometimes are of two types. The cell walls break down as the protoplasts slowly emerge and mate, fusing together and forming zygotes. A new wall, often relatively thick and ornamented with little bumps, forms around each zygote. This may remain dormant for several days, weeks, or even months, but under favorable conditions a dramatic change occurs. The protoplast, which is now *diploid,* undergoes meiosis, producing four haploid **zoospores** (motile cells that do not unite with other cells; many different kinds of algae produce zoospores). When the

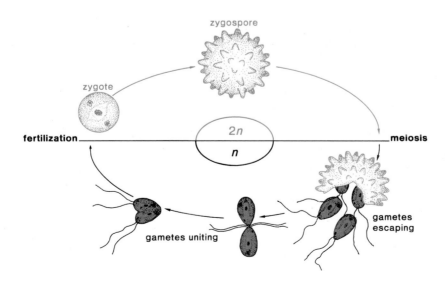

FIGURE 18.10   Sexual life cycle of *Chlamydomonas.*

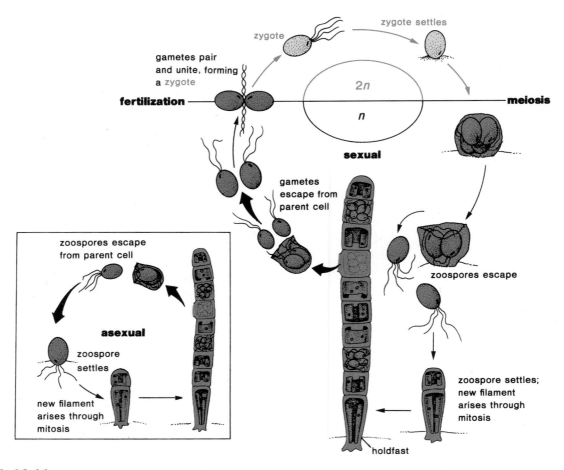

FIGURE 18.11 Life cycle of *Ulothrix*.

old zygote wall breaks down, the zoospores swim away and grow to full-sized *Chlamydomonas* cells (Fig. 18.10).

Ulothrix

An examination of dead twigs, rocks, and other debris in cold freshwater ponds, lakes, and streams (and a few marine habitats) often reveals a threadlike alga called *Ulothrix* (Fig. 18.11), whose name is derived from the Greek words *oulos* (woolly) and *thrix* (hair). Each alga consists of a single row of cylindrical cells attached end to end and forming a thread or *filament*. The basal cell of each filament is slightly longer than the other cells and functions as an attachment cell, or **holdfast.** The nucleus of each cell is surrounded by a wide chloroplast that is shaped like a wide, partial to nearly complete bracelet. Each chloroplast contains one to several pyrenoids. Any of the cells, except the holdfast, may divide, and as they do so the filaments grow longer.

Asexual Reproduction

The protoplast of any cell except the holdfast appears to clump and condense inside the rigid cell wall, divide by mi-

tosis, and become *zoospores*. The zoospores of *Ulothrix* are quite similar to *Chlamydomonas* cells in that they have contractile vacuoles and an eyespot, but they have four flagella instead of two.

Frequently, a protoplast divides one to several times before becoming zoospores, but after zoospores are formed they usually escape from the parent cell through a pore in the wall. After swimming about for a few hours to several days, they settle on submerged objects, shed their flagella, and divide. One of the first two daughter cells becomes a holdfast, while the other continues to divide and forms a new filament. In some instances, the protoplasts do not produce flagella after they have condensed and divided like developing zoospores, but they are otherwise capable of germinating and producing new filaments. Such cells, which are released when the parent cell wall breaks down, are called *aplanospores*.

Sexual Reproduction

Both asexual and sexual reproduction start out in the same way. The protoplast of any cell except the holdfast appears to condense and then divides. Up to 64 zoosporelike cells,

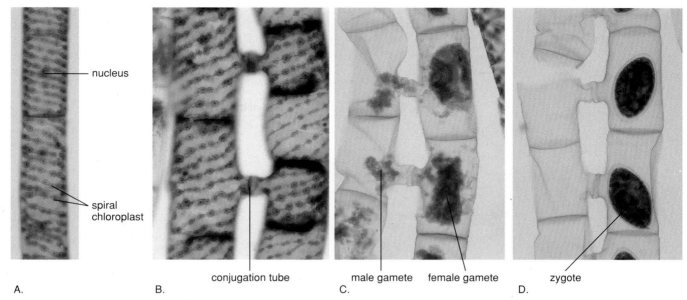

A.   nucleus   spiral chloroplast   B.   conjugation tube   C.   male gamete   female gamete   D.   zygote

FIGURE 18.12 *Spirogyra* (watersilk). *A.* A portion of a vegetative filament showing the ribbonlike chloroplasts spirally arranged in each cell. The centrally located darker object in each cell is a nucleus. *B.* Papillae have grown out from opposite cells of two closely adjacent filaments and formed conjugation tubes. *C.* The condensed protoplasts in the cells on the *left* are functioning as male gametes that are migrating through the conjugation tubes to the stationary female gametes in the cells on the *right*. *D.* Zygotes have been produced in the cells on the *right* as a result of fusion of gametes.

each, however, with two flagella, may be produced. When these cells escape from the parent cell walls, they function as gametes, uniting in pairs with gametes from other filaments and forming zygotes. The zygotes form thick walls and become dormant. Eventually, their protoplasts undergo meiosis, giving rise to zoospores, which then can become new filaments.

Although the gametes of *Ulothrix* come from different filaments, they are identical in size and appearance. Sexual reproduction involving such gametes is called **isogamy.** It is found only in simpler organisms. As in *Chlamydomonas,* the zygotes of *Ulothrix* are the only diploid (2*n*) cells in the life cycle. All the other cells are haploid (*n*).

Spirogyra

One's initial reaction to a first encounter with *watersilk,* as *Spirogyra* is called, may be less than ecstatic because of the slimy feel of the watery sheaths surrounding its filaments. The microscope, however, reveals a beautiful alga with some of the most striking chloroplasts known.

These common freshwater algae, which form unbranched filaments of cylindrical cells, are found most frequently floating in masses at the surface of quiet waters. Each cell contains one or more long, slightly frilly, ribbon-shaped chloroplasts that look as though they had been spirally wrapped around an invisible pole occupying most of the cell's interior. Some species of *Spirogyra* have as many

as 16 of these chloroplasts in each cell. Every one of these elegant green ribbons has pyrenoids at regular intervals throughout its length.

Asexual Reproduction

Unlike the two algae just discussed, *Spirogyra* does not form any zoospores or other cells with flagella. Any cell is capable of dividing, but the only asexual reproduction resulting in new filaments is brought about through the breakup or *fragmentation* of existing filaments. This frequently occurs as a result of a storm or other disturbance.

Sexual Reproduction

In colonies of *Spirogyra,* the filaments usually are produced so close to each other that they may actually be touching. When sexual reproduction begins, the individual cells of adjacent filaments form little dome-shaped bumps, or **papillae** (singular: **papilla**), opposite each other. As these papillae grow, they force the filaments apart slightly, and then they fuse at their tips, forming small cylindrical **conjugation tubes** between each pair of cells. The condensed protoplasts then function as gametes. Usually, those of one filament will seem to flow or crawl like amoebae through the conjugation tubes to the adjacent cells, where each protoplast fuses with the stationary gamete, forming a zygote. Each moving protoplast is considered a male gamete, while the stationary ones function as female gametes (Fig. 18.12).

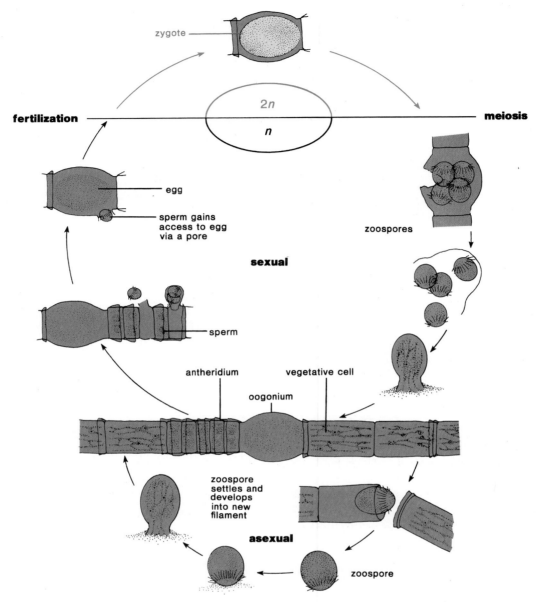

zygote

2n

n

fertilization

meiosis

egg

sperm gains
access to egg
via a pore

zoospores

sexual

sperm

antheridium

vegetative cell

oogonium

zoospore
settles and
develops
into new
filament

asexual

zoospore

FIGURE 18.13   Life cycle of *Oedogonium*.

The zygotes usually develop thick walls and remain dormant for some time, often over the winter. Thick-walled zygotes are characteristic of most freshwater green algae. Eventually, their protoplasts undergo meiosis, producing four haploid cells. Three of these disintegrate, and a single new *Spirogyra* filament grows from the interior of the old zygote shell. The type of sexual reproduction shown by *Spirogyra* is called **conjugation.**

Oedogonium

Aquatic flowering plants and other algae often provide surfaces to which *Oedogonium* (pronounced ee-doh-goh-nee-um), a filamentous green alga, may attach itself. It is,

however, strictly epiphytic and in no way parasitic. The basal cells of the unbranched filaments form holdfasts, and the terminal cell of each filament is rounded. The remaining cells are cylindrical and attached end to end. The name *Oedogonium* comes from words meaning "swollen reproductive cell" and is quite apt, as the female reproductive cells do indeed bulge noticeably. Each cell contains a large netlike chloroplast, which rolls and forms a tube something like a loosely woven basket around and toward the periphery of each protoplast. There are pyrenoids at a number of the intersections of the net.

Asexual Reproduction

Akinetes (thick-walled overwintering cells) may occasionally be formed, but more commonly zoospores are produced singly in cells at the tips of the filaments (Fig. 18.13). Unlike the zoospores of most other algae, those of *Oedogonium* have about 120 small flagella forming a fringe around the cell toward one end. The zoospores look like tiny, balding, faceless heads. Their rapid rotating movements can be quite entertaining when viewed through a microscope. After they escape from their filaments, they eventually settle and form new filaments in the same manner as *Ulothrix*.

Sexual Reproduction

Oedogonium shows more sexual specialization than any of the other three green algae previously discussed. Short box-like cells called **antheridia** (singular: **antheridium**) are formed in the filaments alongside the ordinary vegetative cells. A pair of male gametes, or **sperms,** is produced in each antheridium. The sperms resemble the zoospores but are smaller. Certain cells become swollen and round to elliptical in outline. These cells, called **oogonia** (singular: **oogonium**), each contain a single female gamete or **egg,** which occupies nearly all of the cell. As the egg matures, a pore develops on the side of the oogonium. When sperms escape from the antheridia, they are attracted to the oogonia by a substance released by the eggs. One sperm eventually enters the oogonium through the pore and unites with the egg, forming a zygote.

In some species of *Oedogonium,* oogonia are produced only on female filaments and antheridia only on male filaments. Sometimes, the male filaments are dwarf. They attach themselves to the female filaments, and then both produce hormones that influence the other's development.

Zygotes may remain dormant for a year or more, but they eventually undergo meiosis, producing four zoospores, each of which is capable of developing into a new filament. The sexual reproduction exhibited by *Oedogonium,* in which one gamete is *motile* (capable of spontaneous movement) while the other gamete is larger and stationary, is called **oogamy.**

Other Green Algae

As indicated earlier, the green algae constitute a very diverse group, with an extraordinary variety of forms and chloroplast shapes. Each species obviously has to reproduce in order to perpetuate itself. Most undergo both sexual and asexual reproduction, but a few do not. For example, the worldwide algae that make parts of some tree trunks appear as though they had received a light brushing or spattering of green paint usually are unicellular or colonial forms that reproduce only asexually.

These and *Chlorella,* another widespread green alga composed of tiny spherical cells, reproduce by forming either daughter cells or aplanospores through mitosis. The daughter cells often remain together in packets, while the aplanospores of *Chlorella,* which are formed inside the parent cells, grow to full size as the parent cell wall breaks down.

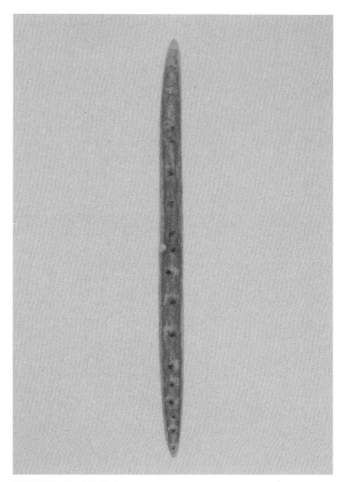

FIGURE 18.14 A single cell of the common desmid, *Closterium.*

Chlorella is very easy to culture and is a favorite organism of research scientists. It has been used in many major investigations of photosynthesis and respiration, and in the future it may become important in human nutrition. (See the section on human and ecological relevance of the algae, which begins on page 308.)

Chlorella could also play a key role in long-range space exploration. Present exploration is severely limited by the weight of oxygen tanks and food supplies needed on a spacecraft, and so scientists have turned to *Chlorella* and similar algae as portable oxygen generators and food sources. Future spacecraft may be equipped with tanks of such algae. These would carry on photosynthesis, using available light and carbon dioxide given off by the astronauts, furnishing the astronauts with oxygen. As the algae multiplied, the excess could either be eaten or fed to freshwater shrimp, which could, in turn, become food for the astronauts. Still other algae and bacteria could recycle other human wastes. Such a self-perpetuating closed system, as it is called, has already been successfully tested with mice and

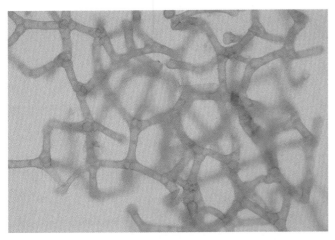

FIGURE 18.15 A portion of a water net (*Hydrodictyon*), a green alga whose cells form a network in the shape of a tube.

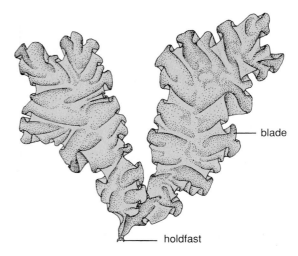

FIGURE 18.17 Sea lettuce (*Ulva*).

FIGURE 18.16 Mermaid's wineglass (*Acetabularia*).

other animals. Many algae, however, are known to produce traces of deadly carbon monoxide gas, and until this problem is resolved, humans will not be subjected to such research.

Desmids (Fig. 18.14), whose 2,500 species of crescent-shaped, elliptical, and star-shaped cells are mostly free-floating and unicellular, reproduce by conjugation. In the beautiful *water nets* (*Hydrodictyon*) (Fig. 18.15), sexual reproduction is isogamous, with up to 100,000 flagellated gametes being produced in a single cell.

Sexual reproduction is also isogamous in the *mermaid's wineglass* (*Acetabularia*), a marine alga consisting of a single huge cell shaped like a delicate mushroom (Fig. 18.16). Each cell is up to 5 centimeters (2 inches) long. This alga has been used in classic experiments demonstrating the influence of the nucleus on the form of the cell. If the cap of an alga is removed and the nucleus is replaced with a nu-

cleus taken from another species, the base regenerates a cap identical to the previous one. If this new cap is also removed, however, the next cap that develops shows form characteristics of both species. If the intermediate cap is then removed, the next cap that develops is identical to that of the species from which the nucleus originally came. Clearly, the original nucleus directs development of cytoplasmic substances regulating cap form, and when these are gone, the replacement nucleus exerts its own influence.

Volvox (see Fig. 18.8 A) is representative of a line of green algae that forms colonies, apparently by means of single cells similar to those of *Chlamydomonas,* held together in a secretion of gelatinous material. In some colonies, the cells are actually connected to one another by cytoplasmic strands. The flagella of individual cells beat separately but pull the whole colony along. A *Volvox* colony may consist of several hundred to many thousands of cells arranged so they resemble a hollow ball, which appears to spin on its axis as it moves. Reproduction may be either asexual or sexual, with smaller daughter colonies being formed inside the parent colony. The daughter colonies are released when the parent colony breaks apart.

Sea lettuce (*Ulva*) is a multicellular seaweed with flattened, crinkly edged green blades that may attain lengths of 1 meter (3 feet) or more (Fig. 18.17). The blades, which may be either haploid or diploid, are anchored to rocks by means of a holdfast at the base. Diploid blades produce spores that develop into haploid blades bearing gametangia. The gametes from the haploid blades fuse in pairs, forming zygotes, each of which can potentially grow into a new diploid blade. Except for the reproductive structures, the haploid and diploid blades of sea lettuce are indistinguishable from one another.

Cladophora is a branched, filamentous green alga whose species are represented in both fresh and marine

FIGURE 18.18 *Chara,* a stonewort.
(Courtesy John Z. Kiss)

waters. Unlike other green algae, the cells of *Cladophora* and its relatives are mostly multinucleate.

Stoneworts (Fig. 18.18), which are aquatic and often calcium-encrusted, loosely resemble small horsetail plants (discussed in Chapter 21). Most botanists now place them in their own division (Charophyta), but some include them with the green algae because of their pigmentation and a few other green algal features. Their sexual reproduction is oogamous, and the antheridia in which the sperms are produced are multicellular. Other features of both their vegetative growth and reproduction are more complex than those of any of the other green algae. Some botanists have considered them more closely related to mosses than to algae.

DIVISION PHAEOPHYTA— THE BROWN ALGAE

Most seaweeds that are brown to olive-green in color are assigned to this division, which includes between 1,500 and 2,000 species of *brown algae* of various sizes. Many are relatively large, and none are unicellular or colonial. Only four of the 260 known genera occur in fresh water, the vast majority being found in colder ocean waters, usually in shallower areas, although the giant kelp (Fig. 18.19) may be found in water 30 meters (100 feet) or more deep. One giant kelp measured 274 meters (710 feet) in length, which is believed to be a record for any single living organism.

Many have a body that is differentiated into a **holdfast,** a **stipe,** and flattened leaflike **blades** (Fig. 18.20). The holdfast is a tough, sinewy structure resembling a mass of intertwined roots. It holds the seaweed to rocks so tenaciously that even the heaviest pounding of surf will not readily dislodge it.

The stalk that constitutes the stipe is often hollow, with a meristem either at its base or at the blade junctions. Since the meristem produces new tissue at the base, the oldest parts of the blades are at the tips.

The blades, which like most of the rest of the body are photosynthetic, may have gas-filled floats called *bladders* toward their bases. The bladder gases may include as much as 10% by volume of deadly carbon monoxide. The function of the carbon monoxide is not known.

A.

B.

FIGURE 18.19 Kelps. *A.* The tip of a giant kelp. *B.* Sea palms (*Postelsia*). The heaviest pounding surf seldom dislodges these brown algae from the rocks.

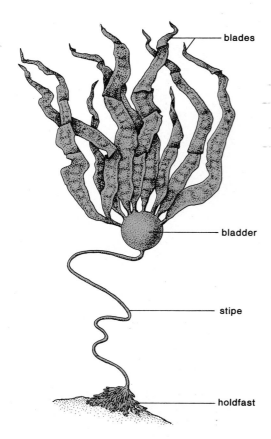

FIGURE 18.20 Parts of the brown alga *Nereocystis*, a kelp.

blades

bladder

stipe

holdfast

FIGURE 18.21 *Sargassum,* a floating brown alga from which the Sargasso Sea got its name. It is also found in other marine waters.

The color of the brown algae, which can vary from light yellow-brown to almost black, reflects the presence of varying amounts of the brown pigment *fucoxanthin,* in addition to chlorophylls *a* and *c* and several other pigments in the chloroplasts.

The main food reserve is a carbohydrate called *laminarin.* **Algin** or *alginic acid* (see Table 18.2 and the discussion of human and ecological relevance of the algae, page 308) occurs on or in the cell walls and can represent as much as 40% of the dry weight of some kelps. Reproductive cells are unusual in that the two flagella are inserted laterally (i.e., on the side) instead of at the ends. The only motile cells in the brown algae are the reproductive cells.

In some localities off the coast of British Columbia and Washington, herring deposit spawn (eggs) in layers up to 2.5 centimeters (1 inch) thick on both surfaces of giant kelp blades in late spring. In the past, and to a limited extent at present, native North Americans have harvested these spawn-covered blades, sun dried them, and used them for winter or feast food. Even today, some schoolchildren are given small pieces of the dried or preserved material for lunchbox snacks.

The Sargasso Sea gets its name from a brown rockweed, *Sargassum,* which is washed up in large quantities on the shores along the Gulf of Mexico after tropical storms (Fig. 18.21). A species occurring in the Pacific Ocean has been used, in chopped form, as a poultice on cuts received from coral by native Hawaiians. This and several other brown algae reproduce asexually by fragmentation, while some produce aplanospores. Sexual reproduction takes several different forms, depending on the species. The conspicuous phases of the life cycles are usually diploid.

In the common rockweed, *Fucus* (Fig. 18.22), separate male and female **thalli** (singular: **thallus;** the term for a body that is usually flattened, multicellular, and not organized into leaves, stems, and roots) are produced. Somewhat puffy fertile areas called *receptacles* develop at the tips of the branches of the thallus. The surface of each receptacle is dotted with pores (visible to the naked eye) that open into special, spherical hollow chambers called *conceptacles.*

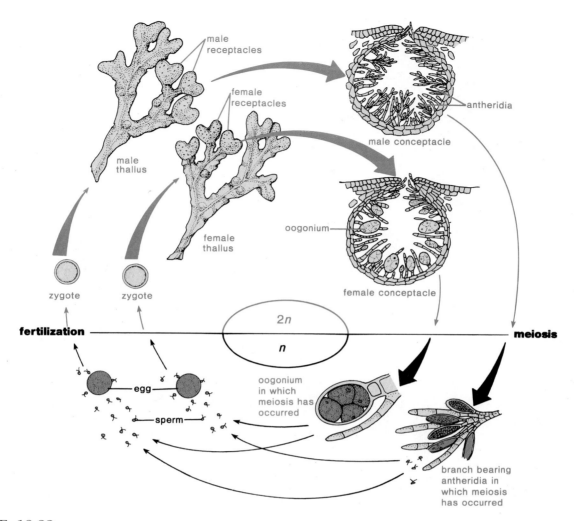

FIGURE 18.22 Life cycle of the common rockweed, *Fucus*.

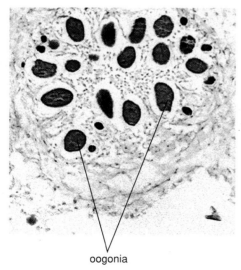

FIGURE 18.23 A sectioned female conceptacle of *Fucus*, ×100.

Within the conceptacles, **gametangia** (cells or structures in which gametes are produced) are formed. Eight eggs are produced in each *oogonium* (female gametangium) as a result of a single diploid nucleus undergoing meiosis followed by mitosis (Fig. 18.23). Meiosis also occurs in each *antheridium* (male gametangium), but three mitotic divisions follow meiosis, so 64 sperms are produced. Eventually, both eggs and sperms are released into the water, where fertilization takes place, and the zygotes develop into mature thalli, completing the life cycle.

DIVISION RHODOPHYTA— THE RED ALGAE

Like the brown algae, most of the more than 5,000 species of *red algae* are seaweeds (Fig. 18.24). However, they tend to occur in warmer and deeper waters than their brown

FIGURE 18.24 Representative red algae. *A. Botryocladia. B. Stenogramme. C. Gigartina. D. Gelidium.*

A.

B.

C.

D.

counterparts. Some grow attached to rocks in intertidal zones, where they may be exposed at low tide. Others grow at depths of up to 200 meters (656 feet) where light barely reaches them, and in 1984 a new species of red algae was discovered at a depth of 269 meters (884 feet), where the light is only 0.0005% of peak surface sunlight. A few are unicellular, but most are filamentous. The filaments frequently are so tightly packed that the plants appear to have flattened blades or to form branching segments. Some develop as beautiful feathery structures that have the appearance

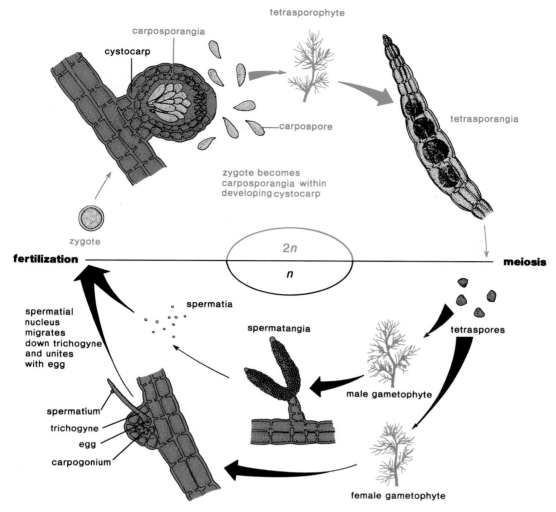

FIGURE 18.25 Life cycle of the red alga *Polysiphonia*.

of delicate works of art.[1] None match the large kelps in size, the largest species seldom exceeding 1 meter (3 feet) in height.

The red algae have relatively complex life cycles, often involving three different types of body structures.

1. Most seaweeds produce their own "glue" and are easy to mount on paper for display. Fresh specimens can be laid directly on clean, high rag-content paper, covered with a layer or two of cheesecloth, and pressed between sheets of blotting paper for a day or two until they are dry. Feathery types will make better specimens if they are placed in a shallow pan of water so that their delicate structures float out, as they do in the ocean. Paper should be slid under them and then carefully lifted so that the seaweed spreads out naturally on the surface. Once the specimens are dry, the cheesecloth (which keeps them from sticking to the blotting paper) is removed and they remain glued to the paper. They can then be displayed or stored indefinitely. Green and brown seaweeds can be treated in similar fashion.

Certain marine algae, which form crusty growths or jointed-appearing upright structures on rocks, accumulate calcium salts as they grow and often contribute to the development of coral reefs. These coralline algae need no special treatment to be displayed, although some may lose their natural pinkish or purplish color when they die.

Meiosis usually occurs on a special body called a *tetrasporophyte,* while gametes are produced on separate male and female bodies. All of the reproductive cells are nonmotile and are carried passively by water currents. Zygotes may migrate from one cell to another through special tubes, which form bizarre loops in some species.

In *Polysiphonia* (Fig. 18.25), a feathery red alga that is widespread in marine waters, the three types of thalli (male gametophyte, female gametophyte, and tetrasporophyte) all outwardly resemble one another. They are about 2 to 15 centimeters (1 to 6 inches) tall and are branched into many fine, threadlike segments. The male sex structures, called *spermatangia,* slightly resemble dense clusters of tiny grapes on slender branches of the male gametophyte thallus. Each spermatangium, as it is released from its branch, functions as a nonmotile male gamete, or *spermatium.*

The female sex structures, called *carpogonia,* are produced on the female gametophyte thallus. Each carpogonium consists of a single cell that looks something like a microscopic bottle with a long neck called the *trichogyne.*

Table 18.1

COMPARISON OF THE DIVISIONS OF ALGAE

DIVISION	FOOD RESERVES	SPECIAL PIGMENTS[1]	FLAGELLA
Chrysophyta 6,000 species of yellow-green algae, golden-brown algae, diatoms, and cryptophytes	oils, fats, chryso-laminarin	chlorophyll c_1 chlorophyll c_2	2, 1, or none, usually unequal, at or near apex
Pyrrophyta 3,000 species of dinoflagellates	starch	chlorophyll c_2	2; 1 trailing, 1 girdling; few with none
Euglenophyta 750 species of euglenoids	paramylon	chlorophyll b	2, 1, or 3, at or near apex
Chlorophyta 7,500 species of green algae	starch	chlorophyll b	2, 4, or more at apex, or none
Phaeophyta 1,500 to 2,000 species of brown algae	laminarin, mannitol	chlorophyll c	2, lateral and unequal
Rhodophyta More than 5,000 species of red algae	floridean starch, mannitol	chlorophyll d (some species), phycobilins	none

1. All divisions of algae have chlorophyll a and yellow to orange pigments called carotenes and xanthophylls, although the combinations and specific types of xanthophylls vary with the divisions. The fucoxanthin of the golden-brown algae and the brown algae is a xanthophyll.

A single nucleus at the base of the carpogonium functions as the female gamete, or *egg*. Since the spermatia have no flagella, they cannot move of their own accord, but currents may carry them considerable distances. If a spermatium should brush against a trichogyne, it may become attached. The walls between the spermatium and the trichogyne then break down, the nucleus of the spermatium migrates to the egg nucleus, and the two nuclei unite, forming a zygote.

Next, the zygote develops into a cluster of clublike *carposporangia* toward the base of a cystocarp (an urn-shaped body that the female gametophyte thallus forms around the carposporangia). Diploid asexual spores called *carpospores* are produced in the carposporangia and released, to be carried away by ocean currents.

When a carpospore lodges in a suitable location (e.g., a rock crevice or the hull of a ship), it usually germinates and grows into a *tetrasporophyte,* which closely resembles a gametophyte thallus. Tetrasporangia are formed along the branches of the tetrasporophytes. Each tetrasporangium undergoes meiosis, giving rise to four haploid *tetraspores.* When tetraspores germinate, they develop into male or female gametophytes, thereby completing the life cycle.

The red to purplish colors of most red algae are due to the presence of varying amounts of red and blue accessory pigments called *phycobilins,* which are closely related to those found in the blue-green bacteria. The similarity has led to the belief that the red algae may have been derived from the blue-green bacteria. Several other pigments, including chlorophyll a and sometimes chlorophyll d, are also present in the chloroplasts. The principal reserve food is a carbohydrate called *floridean starch.* A number of red algae also produce **agar** and other important gelatinous substances discussed in the next section.

A comparative summary of the food reserves, special pigments, and flagella of the divisions of eukaryotic algae is given in Table 18.1.

HUMAN AND ECOLOGICAL RELEVANCE OF THE ALGAE

Diatoms

We have already noted that numbers of blue-green bacteria are at the bottom of aquatic food chains, but members of all the protoctistan algal divisions play similar roles. Diatoms, for example, are consumed by fish that feed on plankton. Up to 40% of a diatom's mass consists of oils that are converted into cod and other liver oils, which are rich sources of vitamins for humans. The oils also may in the past have contributed to petroleum oil deposits. The National Renewable Energy Laboratory in Golden, Colorado, is presently working on a project to convert oil from cultured diatoms into a "clean" diesel fuel substitute.

Diatoms also have other extensive and more direct industrial uses. In the past, they have apparently been at least as numerous as they are now. As billions upon billions of them have reproduced and died, their microscopic, glassy shells have accumulated on the ocean floor, forming deposits of *diatomaceous earth*. These deposits have accumulated to depths of hundreds of meters in some parts of the world and are quarried in several areas where past geological activity has raised them above sea level. At Lompoc, California, beds of diatomaceous earth are more than 200 meters (650 feet) deep, while in the Santa Maria oil fields of California, deposits reach a depth of 1,000 meters (3,280 feet).

Diatomaceous earth is a light, porous, and powdery looking material that contains about 6 billion diatom shells per liter (1.057 quarts), yet a liter weighs only eight 10ths of a kilogram (1.76 pounds). It also has an exceptionally high melting point of 1,750°C (3,182°F) and is insoluble in most acids and other liquids. These properties make it ideal for a variety of industrial and domestic uses, including many types of filtration. The sugar industry uses diatomaceous earth in sugar refining, and its use for swimming pool filters is widespread. It is also used in silver and other metal polishes, in toothpaste, and in the manufacture of paint that reflects light, which is used on highway markers and signs and on the automobile license plates of some states. It is packed as insulation around blast furnaces and boilers. It is also manufactured into the construction panels used in prefabricated housing.

Pacific Coast fish farmers have experienced losses of salmon and cod when dense concentrations of *Chaetoceros* diatoms have developed in the aquaculture pens. The diatoms have long hollow spines that break off and penetrate the fish gills, disrupting gas exchange and causing bleeding. This damage, in turn, may permit secondary infections and excessive mucus production to occur.

Uses of Green Algae

A few green algae have occasionally been used for food, but members of this division have generally been used less by humans than have those of other divisions. With dwindling world food supplies, this could change. Sea lettuce has been used for food on a limited scale in Asian countries for some time, and several countries are experimenting with the suitability of plankton for human consumption. Except for vitamin C, *Chlorella* contains most of the vitamins needed in human nutrition, and since it is so easy to culture, it may become an important protein source in many parts of the globe. *Chlorella* has also been investigated as a potential oxygen source for atomic submarines, in addition to its possible use in space exploration.

Algin

Commercially produced ice cream, salad dressing, beer, jelly beans, latex paint, penicillin suspensions, paper, textiles, toothpaste, ceramics, and floor polish today all share a common ingredient, *algin,* produced by the giant kelps and other brown algae. It is now used in so many products (Table 18.2) that one might wonder how the world used to get along without it.

Algin has the unique ability to regulate water "behavior" in a wide variety of products. It can, for example, control the development of ice crystals in frozen foods, regulate the penetration of water into a porous surface, and generally stabilize any kind of suspension, such as an ordinary milkshake or other thick fluid containing water. It is produced by several kinds of seaweeds, but a major source is the giant kelp found in the cooler ocean waters of the world, usually just offshore where there are strong currents (see Fig. 18.19 A). This large seaweed, which has been known to attain a length of 92 meters (300 feet), sometimes grows at the rate of 3 to 6 decimeters (1 to 2 feet) per day. Specially equipped oceangoing vessels (Fig. 18.26) harvest the kelp by mowing off the top meter (3 feet) of growth, taking the chopped material aboard and then transferring it to processing centers onshore where it is extracted and refined.

Minerals and Food

Brown algae also produce a number of other useful substances. Many seaweeds, but particularly kelps, build up concentrations of iodine to as much as 20,000 times that of the surrounding seawater. Although it is cheaper to obtain iodine from other sources in the United States, dried kelp has been used in other parts of the world in the treatment of goiter, which results from iodine deficiency. Kelps are relatively high in nitrogen and potassium and have been used as fertilizer for many years. Before such use, the seaweeds need to be rinsed to rid them of salt. They also have been used as livestock feed in northern Europe and elsewhere. In the Orient, many marine algae are used for food—in soups, confections, meat dishes, vegetable dishes, and beverages. In Japan, acetic acid is produced through fermentation of seaweeds.

During the Irish famine of 1845 to 1846, *dulse,* a red seaweed, became an important substitute for the potato crop that had been destroyed by blight. Dulse also occurs along both the Atlantic and Pacific coasts of North America. In Maine and eastern Canada, where it is a popular snack food, dulse is referred to as "Nova Scotia Popcorn." Another red seaweed, *purple laver* (*nori*), which occurs in both American and Asian waters, is used extensively for food, particularly in the Orient. In Japan, it is cultured on nets or bamboo stakes set out in shallow marine bays (Fig. 18.27). It is harvested when the thin, crinkly, gelatinous blades are several centimeters (2 to 3 inches) in diameter and is used in meat and macaroni dishes, soups, and dry, spiced delicacies.

Irish moss is another important edible red alga. It is also used in bulking laxatives, cosmetics, and pharmaceutical

Table 18.2

SOME USES OF ALGIN

Food

1. As a thickening agent in toppings, pastry fillings, meringues, potato salad, canned foods, gravies, dry mixes, bakery jellies, icings, dietetic foods, flavored syrups, candies, puddings
2. As an emulsifier and suspension agent in soft drinks and concentrates, salad dressings, barbecue sauces, frozen food batters
3. As a stabilizer in chocolate drinks, eggnog, ice cream, sherbets, sour cream, coffee creamers, party dips, buttermilk, dairy toppings, milkshakes, marshmallows

Paper

1. Provides better ink and varnish holdout on paper surfaces; provides uniformity of ink acceptance, reduction in coating weight and improved holdout of oil, wax, and solvents in paperboard products; makes improved coating for frozen food cartons
2. Used for coating greaseproof papers

Textiles

1. Thickens print paste and improves dye dispersal. Reduces weaving time and eliminates damage to printing rolls or screens

Pharmaceuticals and Cosmetics

1. As a thickening agent in weight-control products, cough syrups, suppositories, ointments, toothpastes, shampoos, eye makeup
2. As a smoothing agent in lotions, creams, lubricating jellies; as a binder in manufacture of pills; as a blood anticoagulant
3. As a suspension agent for liquid vitamins, mineral oil emulsions, antibiotics; as a gelling agent for facial beauty masks, dental impression compounds

Industrial Uses

1. Used in manufacture of acidic cleaners, films, seed coverings, welding rod flux, ceramic glazes, boiler compounds (prevents minerals from precipitating on tubes), leather finishes, sizing, various rubber compounds (e.g., automobile tires, electric insulation, foam cushions, baby pants), cleaners, polishes, latex paints, adhesives, tapes, patching plaster, crack fillers, wall joint cement, fiberglass battery plates, insecticides, resins, tungsten filaments for light bulbs, digestible surgical gut (disappears by time incision is healed), oil well-drilling mud
2. Used in clarification of beet sugar; mixed with alfalfa and grain meals in dairy and poultry feeds

Brewing

1. Helps create creamier beer foam with smaller, longer-lasting bubbles

preparations. Blancmange is a dessert made from Irish moss and milk. *Carrageenan* is a mucilaginous substance extracted from Irish moss and used as a thickening agent in chocolate milk and other dairy products. *Funori,* obtained from yet another red alga, is used as a laundry starch, as an adhesive in hair dressings, and in some water-based paints.

Agar

One of the most important of all algal substances is **agar,** produced most abundantly by the red alga *Gelidium.* This substance, which has the consistency of gelatin, is used (with nutrients added) around the world in laboratories and medical institutions as a culture medium for the growth of bacteria. When various nutrients are added to it, it can also be used as a culture medium for the growth of both plant and animal cells. Full-sized plants have been induced to develop from pollen grains sown on nutrient agar. Orchid tissues are cultured commercially on it and induced to grow into full-sized plants (as discussed in the section on mericloning in Chapter 14), and its use in making the capsules containing drugs and vitamins is now worldwide. It is also used as an agent in bakery products to retain moistness, as a base for cosmetics, and as an agent in gelatin desserts to promote rapid setting.

Current research involving red algae and other seaweeds indicates they contain a number of substances of

FIGURE 18.26   A specially equipped vessel harvesting kelp off the California coast. The ship moves backward through the kelp beds as the machinery at the stern mows the kelp and conveys it into the hold.

(Courtesy Kelco Company)

potential medicinal value. More than 20 seaweeds have been used in preparations designed for the expulsion of digestive tract worms, control of diarrhea, and the treatment of cancer. Some have shown considerable potential as antibiotics and insecticides. Chemical relatives of DDT have been found to be produced by certain red algae. The sea hare and other marine animals feed on such algae, and it is possible that such animals may degrade (break down) the DDT-like compounds to simpler substances—a feat unknown among land animals.

OTHER MEMBERS OF KINGDOM PROTOCTISTA

Protozoans (Phylum Protozoa) and *sponges* (Phylum Porifera) are included within Kingdom Protoctista; they have, however, traditionally been regarded as animals and are not covered in this book. The *slime molds,* which appear to be related to the protozoa, are discussed next.

SUBKINGDOM MYXOBIONTA— THE SLIME MOLDS

In the Ripley's Believe It or Not pavilion at the Chicago World's Fair in 1933 there was an exhibit of "hair growing on wood." The "hairs," while indeed superficially resembling short human hair, were actually the reproductive

FIGURE 18.27   Beds of purple laver ("nori") ready to harvest in shallow marine water at Sendai, Japan.

(Courtesy I.A. Abbott)

FIGURE 18.28 Reproductive bodies (sporangia) of the slime mold *Stemonitis.*

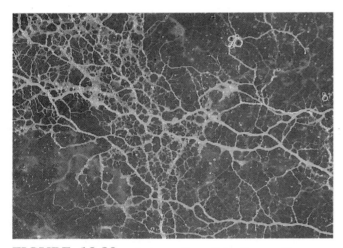

FIGURE 18.29 A plasmodium of the slime mold *Physarum.*

structures of a species of **slime mold** (Fig. 18.28). These curious organisms are totally without chlorophyll and thus are incapable of producing their own food. They are a bit of a puzzle to biologists because they are distinctly animal-like during much of their life cycle but also distinctly funguslike when they reproduce.

The tiny roundish *spores* of the more than 500 species of slime mold average only 10 to 12 micrometers in diameter and thus are individually invisible to the naked eye. Nevertheless, they are present nearly everywhere and are especially abundant in airborne dusts.

If you place almost any dead leaf or piece of bark in a dish and add a food source, such as a few dry oatmeal flakes to which a drop or two of water has been added, the odds are that slime mold spores will be present and will germinate. If the dish is covered, the spores will sometimes germinate in as little as 15 minutes, but germination usually takes several hours or longer. Within a few days, a curious glistening mass of active slime mold material looking something like the netted veins of a leaf may appear. This material, whose "veins" tend to merge into the shape of a fan at its leading edges, is a slime mold **plasmodium** (Fig. 18.29).

If the plasmodium is examined with a microscope, it becomes apparent that there are no cell walls present. The protoplasm in the veins, particularly toward the center, flows very rapidly and rhythmically. The protoplasmic movement may stop momentarily at regular intervals and then resume its flow, sometimes in the opposite direction.

Plasmodia are often white, but they also may be brilliantly colored in shades of yellow, orange, blue, violet, or black. A few are colorless and more or less transparent. They are found on damp forest debris, under logs, on old shelf or bracket fungi, sometimes on older mushrooms, and in other moist places where there is dead organic matter. They tend to flow forward at a rate of up to 2.5 centimeters (1 inch) or more per hour, often against slow moisture seepage, feeding on bacteria and other organic particles as they go. They contain many diploid nuclei, all of which divide often and simultaneously as growth occurs. With an adequate food supply, a plasmodium may increase to 25 times its original size in just one week.

FIGURE 18.30 Common slime mold sporangia. *A. Lamproderma. B. Lycogala.*
(*B.* courtesy L.L. Steimley)

Dramatic events take place when significant changes in food supplies, available moisture, light, and other environmental features occur. The plasmodium usually is converted, often quite rapidly, into many separate small **sporangia** (Fig. 18.30), each containing thousands of minute, one-celled **spores**. The sporangia often are globe-shaped, but in some species they develop as long or wide stationary bodies that may resemble a jumbled network of tubes, or they may resemble erect hairs or end up as a shapeless blob. The sporangia may have slender stalks, depending on the species. Others exhibit combinations of body forms. The spores are often distributed throughout a jumbled mass of threads called a *capillitium.*

When a spore is formed, a single nucleus and a little cytoplasm become surrounded by a wall. Meiosis takes place in the spore, and three of the four resulting nuclei degenerate. When the spore germinates, one or more amoeba-like cells called *myxamoebae* emerge. Sometimes, these have flagella, in which case they are called *swarm cells.* Either form may become like the other through the development or loss of flagella. At first, myxamoebae or swarm cells feed on bacteria and other food particles. Sooner or later, however, they function as gametes, fusing in pairs and forming zygotes. A new plasmodium usually develops from the zygote, although occasionally zygotes or small plasmodia may fuse and form larger plasmodia (Fig. 18.31).

Cellular Slime Molds

Two or more divisions of slime mold are recognized. Most of the species have typical plasmodia and follow the patterns just discussed. About two dozen species of *cellular slime molds* are, however, evidently not closely related to the other slime molds. Individual amoebalike cells of cellular slime molds feed independently, dividing and producing separate new cells from time to time. When the population reaches a certain size, they stop feeding and clump together, forming a mass called a *pseudoplasmodium.* The pseudoplasmodium resembles and crawls like a garden slug. It eventually becomes stationary and is transformed into a sporangiumlike mass of spores.

Human and Ecological Relevance of the Slime Molds

The slime molds, like the bacteria, contribute to ecological balance in forests and woodlands by breaking down organic particles to simpler substances; they also reduce bacterial populations in their habitats. One atypical species occasionally attacks cabbages and another causes powdery scab of potatoes. Yet another species causes a disease of watercress, but slime molds otherwise are of little economic significance.

SUBKINGDOM MASTIGOBIONTA, DIVISION CHYTRIDIOMYCOTA— THE CHYTRIDS

If you were to immerse dead leaves or flowers, old onion bulb scales, dead beetle wing covers, or other organic material in water that had been mixed with a little soil, within a day or two thousands of microscopic **chytrids** (pronounced kítt-ridds) probably would appear on the surfaces

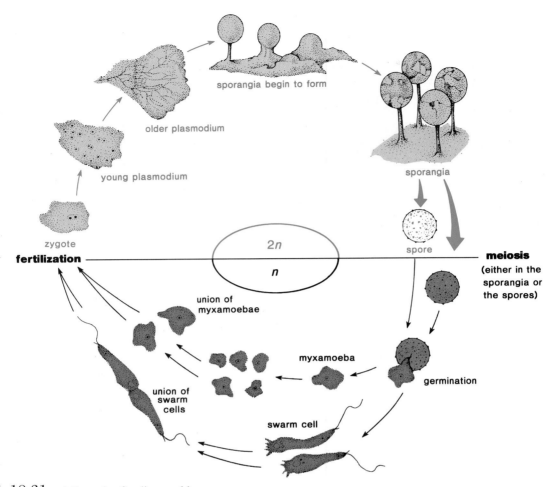

FIGURE 18.31 Life cycle of a slime mold.

(After C.J. Alexopoulos, and C.W. Mims, 1979. *Introductory Mycology,* 3d ed. © John Wiley & Sons, Inc. New York. Redrawn by permission of John Wiley & Sons, Inc.)

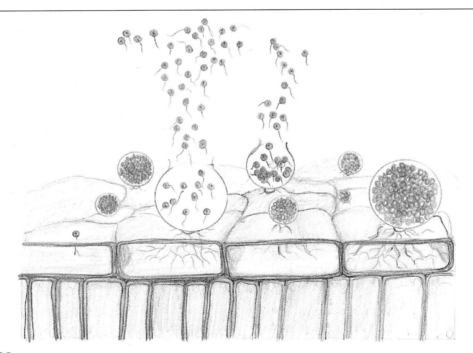

FIGURE 18.32 Chytrids on a dead leaf submerged in water.

of the immersed objects (Fig. 18.32). These simple, fungus-like organisms include many parasites of other protists, aquatic fungi, aquatic flowering plants, and algae. Some parasitize pollen grains, and many other species are saprobic. Some of the most common chytrids consist of a spherical cell with colorless branching threads called **rhizoids** at one end. The rhizoids anchor the organism to its food source.

Other chytrids may develop short hyphae or even a complete mycelium whose hyphae contain many nuclei. Such multinucleate mycelia without crosswalls are said to be **coenocytic** (pronounced seé-no-sitt-ik). The cell walls of some chytrids have been reported to contain cellulose in addition to chitin.

Many chytrid species reproduce only asexually through the production of zoospores within a spherical cell. The zoospores, which each have a single flagellum, settle upon release and grow into new chytrids. Some species undergo sexual reproduction by means of the fusion of two motile gametes with haploid nuclei or by the union of two nonmotile cells whose diploid zygote nuclei undergo meiosis. The zygote commonly is converted into a resting spore. The origin of chytrids is unknown, but the presence of a flagellum on the motile cells has led some authorities to suggest they may have originated from the protozoa.

Division Oomycota—
THE WATER MOLDS

Those who have kept tropical fish aquaria or have seen salmon at the end of a spawning run often have noticed cottony growths on cuts and bruises on a fish's body or have seen similar growths on the eyes of sick fish. These aquatic organisms, called **water molds** or *oomycetes,* are also often found on dead insects such as houseflies, and they soon develop on marijuana seeds placed in water. They range in form from single spherical cells to branching, threadlike, *coenocytic* hyphae. Coenocytic hyphae, which are not divided into individual cells, may form large *mycelia.* Motile cells with two flagella are produced at various stages of their life cycles (Fig. 18.33).

During asexual reproduction in water molds, crosswalls form just below the tips of certain hyphae. Numerous zoospores, each with two flagella, are produced in these special tip chambers. The zoospores, after emerging through a terminal pore, eventually give rise to new water mold mycelia. Sexual reproduction involves **oogonia** and **antheridia,** which arise on side branches under the influence of hormones produced within the organism. The mycelium is diploid, and meiosis takes place in the oogonia and antheridia. Zygotes formed in the oogonia eventually give rise to new mycelia (Fig. 18.34).

Water molds share several features with brown algae from which they appear to have evolved. Features common to water molds and brown algae include eggs (oogamy), cel-

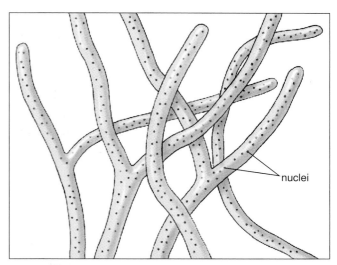

FIGURE 18.33   Coenocytic hyphae of a water mold.

lulose in the cell walls, a predominantly diploid life cycle, and zoospores with two flagella.

Human and Ecological Relevance of the Water Molds

Two important members of this division cause serious diseases of higher plants. Although dew or rainwater is needed for their reproduction, neither grows under water.

One oomycete, called *downy mildew of grapes,* undergoes its life cycle on grape leaves, usually killing the leaves and, if not controlled, the vine. This disease seriously threatened the French wine industry toward the end of the 19th century, after it reached the vineyards on imported American cuttings. Within a few years, it was controlled when it was discovered that Bordeaux mixture, a combination of copper sulphate and lime (calcium oxide), inhibited the growth of downy mildew. This unappetizing-looking mixture, which is the first substance known to have been used as a fungicide, was originally sprayed on grape vines to discourage grape thieves.

The other important oomycete is called *late blight of potato.* In the past, this virulent organism plagued potato farmers so much that at times it altered the course of history. In the summer of 1846, the entire potato crop of Ireland was wiped out in one week, causing a famine during which over a million people starved to death and many more emigrated to the United States.

The organism produces its mycelium inside the leaves, and so once its spores land on the surface and send hyphae into the interior tissues, it is relatively immune to spraying. However, spraying leaves with Bordeaux mixture or other fungicides before the spores have germinated checks development of the oomycete.

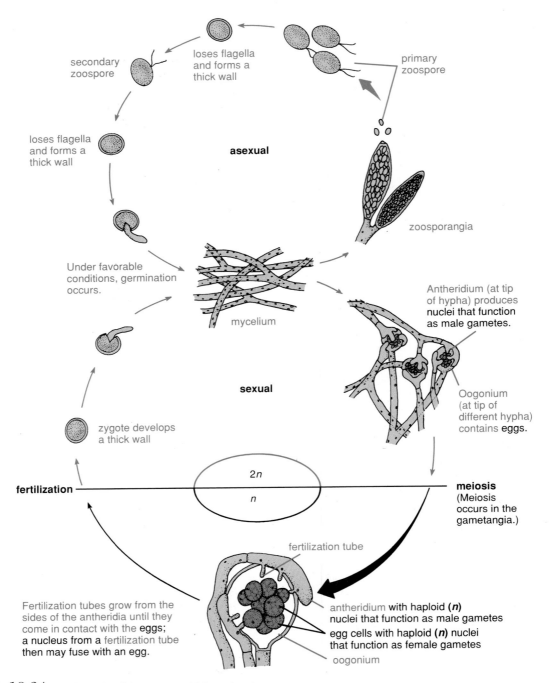

FIGURE 18.34 Life cycle of the water mold *Saprolegnia*.

Image labels:
- primary zoospore
- secondary zoospore
- loses flagella and forms a thick wall
- asexual
- loses flagella and forms a thick wall
- zoosporangia
- Under favorable conditions, germination occurs.
- mycelium
- Antheridium (at tip of hypha) produces nuclei that function as male gametes.
- sexual
- Oogonium (at tip of different hypha) contains eggs.
- zygote develops a thick wall
- 2n
- n
- fertilization
- meiosis (Meiosis occurs in the gametangia.)
- fertilization tube
- Fertilization tubes grow from the sides of the antheridia until they come in contact with the eggs; a nucleus from a fertilization tube then may fuse with an egg.
- antheridium with haploid (n) nuclei that function as male gametes
- egg cells with haploid (n) nuclei that function as female gametes
- oogonium

Summary

1. All living organisms were confined to the ocean less than 1 billion years ago. About 400 million years ago, green algae began making the transition from water to land and gave rise to land plants.

2. Kingdom Protoctista includes many diverse organisms that all have eukaryotic cells. Members may be unicellular or multicellular and occur as either colonies or filaments. Modes of nutrition include photosynthesis, ingestion of food, a combination of both, or absorption of food in solution. Some members are nonmotile but most are motile.

3. The four classes of Chrysophyta are the yellow-green algae, the true golden-brown algae, the diatoms, and

the cryptophytes. Some members of each class produce statospores.

4. Diatoms are very abundant; they have a glassy shell that consists of two halves that fit together like a pillbox.

5. Fucoxanthin gives a golden-brown color to many diatoms. Diatoms move in caterpillar fashion by contact of the cytoplasm with a surface as it protrudes through the pores.

6. Diatoms reproduce asexually by mitosis and sexually through the fusion of gametes that form an auxospore (zygote).

7. Dinoflagellates (Pyrrophyta) are unicellular organisms with two flagella. Some cause red tides that can kill fish and poison humans. Dinoflagellates in tropical waters exhibit bioluminescence (emission of light) when disturbed.

8. Euglenoids (Euglenophyta) have no rigid cell wall; only one functional flagellum, a gullet, and paramylon as a carbohydrate food reserve. Reproduction is by cell division. Sexual reproduction has not been confirmed.

9. Green algae (Chlorophyta) have cells with the same pigments and reserve food (starch) as those of higher plants. The chloroplasts of *Chlamydomonas* have pyrenoids; the cells have two or more vacuoles and often a red eyespot. Asexual reproduction is by mitosis; sexual reproduction is isogamous.

10. *Ulothrix* is a filamentous green alga that is attached to objects by means of a holdfast; each cell contains a bracelet-shaped chloroplast. Asexual reproduction is by zoospores; sexual reproduction is isogamous.

11. *Spirogyra* is a floating, filamentous green alga with spiral, ribbonlike chloroplasts. Asexual reproduction is by fragmentation. Sexual reproduction is by conjugation.

12. *Oedogonium* is an epiphytic, filamentous green alga; it has cylindrical, netted chloroplasts. In asexual reproduction, zoospores are produced. Sexual reproduction is oogamous.

13. Other green algae include *Chlorella,* desmids, *Acetabularia, Volvox, Ulva,* and *Cladophora.*

14. Stoneworts are either included with the green algae or placed in Division Charophyta.

15. Many brown algae (Phaeophyta) are large seaweeds; their thalli often may be differentiated into a stipe, flattened blades, and a holdfast.

16. The color of brown algae is largely due to fucoxanthin; the main carbohydrate food reserve is laminarin. Some brown algae produce algin, a gelatinous substance useful to humans. The reproductive cells have lateral flagella. The common rockweed, *Fucus,* produces eggs and sperms that form zygotes in the water.

17. Most red algae (Rhodophyta), represented by *Polysiphonia,* are seaweeds with life cycles that involve three different types of thalli and nonmotile gametes.

18. Red and blue phycobilins are partially responsible for the colors of red algae; chlorophyll *d* may be present in the chloroplasts. The main carbohydrate food reserve is floridean starch. Some red algae produce agar, an economically important gelatinous substance.

19. Algae are ecologically and economically important. Diatomaceous earth is used for filtering, polishes, insulation, and reflectorized paint. *Chlorella* is a potential food and oxygen source. Algin is used as a stabilizer and thickening agent in hundreds of products.

20. Some brown algae are a source of fertilizer, iodine, and food. Red algae are a source of agar and food and have potential medicinal value.

21. Subkingdom Myxobionta includes the slime molds, which are animal-like in their vegetative state; they consist of a multinucleate mass of protoplasm called a *plasmodium* that flows over damp surfaces, ingesting food particles.

22. Slime molds form stationary sporangia containing spores, from which myxamoebae or swarm spores emerge upon germination. Myxamoebae or swarm spores function as gametes, with new plasmodia developing from the zygotes.

23. Cellular slime molds produce a pseudoplasmodium, which crawls like a slug that converts to a stationary, sporangiumlike mass of spores.

24. Subkingdom Mastigobionta includes aquatic chytrids and water molds. Chytrids consist mostly of spherical cells with rhizoids at one end; they reproduce only asexually, with zoospores having a single flagellum.

25. Water molds, which have coenocytic mycelia, include organisms that cause diseases of fish and other aquatic organisms. Asexual reproduction involves zoospores; gametes are produced in oogonia and antheridia.

Review Questions

1. How do cells of diatoms differ from those of other organisms?

2. What forms of sexual and asexual reproduction occur in the green algae?

3. Which groups of algae produce the following important products: (a) agar, (b) algin, (c) nerve poisons, (d) abrasives for polishes?

4. How would you distinguish *Chlamydomonas* from *Euglena*?

5. *Spirogyra, Ulothrix,* and *Oedogonium* all form filaments. How can you tell them apart?

6. In the green algae studied, where in the life cycles does the chromosome number change from *n* to *2n* and vice versa?

7. Where and how is algin obtained?

8. Is there any difference in structure between the holdfasts of microscopic green algae and those of brown algae?

9. Why are some green algae red and some red algae green?

10. Which divisions of algae have only unicellular representatives?

Discussion Questions

1. Some algae are attached to solid objects or other organisms, while others are free-floating. What are the advantages and disadvantages of each type of growth?

2. The variety and sizes of algae found in the oceans are considerably greater than those of freshwater forms. What reasons can you suggest?

3. Seaweeds that grow in intertidal zones, where they may be exposed between tides, are often more gelatinous than their continually submerged counterparts. Explain.

4. Why do some algae grow on one side of a tree and not all around the trunk?

5. Should the bladders (floats) of some of the kelps give these algae any advantage over those algae that do not possess such structures?

Additional Reading

Abbott, I. A. 1984. *Limu: An ethnobotanical study of some edible Hawaiian seaweeds,* 3d ed. Lawai, HI: Pacific Tropical Botanical Garden.

Abbott, I. A., and E. Y. Dawson. 1978. *How to know the seaweeds,* 2d ed. Pictured Key Nature Series. Dubuque, IA: Wm. C. Brown Publishers.

Abbott, I. A., and G. J. Hollenberg. 1976. *Marine algae of California.* Palo Alto, CA: Stanford University Press.

Becker, E.W. (Ed.). 1985. *Production and use of microalgae.* Forestburgh, NY: Lubrecht & Cramer.

Bold, H. C., and M. J. Wynne. 1985. *Introduction to the algae,* 2d ed. Englewood Cliffs, NJ: Prentice-Hall.

Buetow, D. E. (Ed.). 1968–1989. *The biology of Euglena: General biology and ultrastructure,* 4 vols. San Diego, CA: Academic Press.

Chapman, V. J., and D. J. Chapman. 1980. *Seaweeds and their uses,* 3d ed. New York: Chapman and Hall.

Fritsch, F. E. 1935, 1945. *Structure and reproduction of the algae,* 2 vols. Ann Arbor, MI: Books Demand.

Hallegraeff, G. M. 1993. A review of harmful algal blooms and their apparent global increase. *Phycologia* 32(2): 79–99.

Irvine, D., and D. M. John. (Eds.). 1985. *Systematics of the green algae.* San Diego, CA: Academic Press.

Lee, R. E. 1989. *Phycology,* 2d ed. New York: Cambridge University Press.

Palmer, C. M. 1962. *Algae in water supplies.* Washington, DC: Government Printing Office, U.S. Department of Health, Education and Welfare, Public Health Service.

Prescott, G. W. 1978. *How to know the freshwater algae,* 3d ed. Pictured Key Nature Series. Dubuque, IA: Wm. C. Brown Publishers.

Round, F. E. 1984. *The ecology of algae.* New York: Cambridge University Press.

Scagel, R. F., et al. 1984. *Plants: An evolutionary survey.* Belmont, CA: Wadsworth Publishing Co.

Smith, G. M. 1950. *Freshwater algae of the United States,* 2d ed. New York: McGraw-Hill.

Sze, P. 1992. *A biology of the algae,* 2d ed. Dubuque, IA: Wm. C. Brown Publishers.

Turner, N. J. 1974. *Food plants of British Columbia Indians. Part 1: Coastal peoples.* Victoria, Canada: British Columbia Provincial Museum.

Chapter Outline

*Fly agaric mushrooms (*Amanita muscaria*).*

Kingdom Fungi and Lichens

19

Overview────────────────────

After a short introduction, the distinctions between Kingdoms Protoctista and Fungi are discussed. A summary of the features of Kingdom Fungi and a review of how the kingdom came to be recognized are given.

Kingdom Fungi includes two divisions and several classes of true fungi. Life cycles and discussions of the economic importance of representative members of each division are presented. Nematode-trapping fungi, Pilobolus, *truffles, morels, ergot, yeasts, stinkhorns, puffballs, bracket fungi, bird's-nest fungi, smuts, rusts, poisonous and hallucinogenic fungi, shiitake mushrooms, mushroom culture, antibiotics, industrial products obtained from fungi, and fungi in nature are discussed. The chapter concludes with an examination of various forms of lichens. Natural dyeing is briefly explored, and the economic importance of lichens is reviewed.*

Some Learning Goals

1. Know the general features that distinguish Kingdom Fungi from the other kingdoms.
2. Distinguish the divisions and classes of fungi from one another on the basis of their cells or hyphae and their reproduction.
3. Identify and describe sporangium, conidium, coenocytic, mycelium, dikaryotic, zygospore, ascus, and basidium.
4. Assign each of the following to its class of true fungi: athlete's foot, Dutch elm disease, *Pilobolus, Penicillium* mold, stinkhorn, yeast, ergot, chestnut blight fungus, puffball, smut.
5. Know five economically important fungi in each of four different classes of true fungi.
6. Understand how lichens are classified and identified.
7. Learn the basic structure of a lichen.

FIGURE 19.1 Typical fungal hyphae, as seen with the aid of a dissecting microscope.

When it comes to breaking down organic materials of all kinds and making them available for recycling, fungi and bacteria are the most important organisms known. The major oil spill that occurred in Alaska in 1989 intensified efforts to find better and faster ways of cleaning up such disasters than was possible in the past. Although there are bacteria that break down oil molecules, very few such organisms survive the saltwater of the ocean for any length of time. Attention now, however, is being given to a fungus that breaks down many types of oils to harmless simpler molecules, such as carbon dioxide.

This fungus, known as *Corollospora maritima,* occurs naturally on almost all the beaches of the world and flourishes in ocean surf, which, as mentioned in the previous chapter, contains oils produced by diatoms. Ways of culturing the fungus inexpensively and in large quantities are being explored, and also strains that produce the oil-degrading enzymes in significantly larger amounts are being sought. This chapter examines the nature, reproduction, and other features of this and the thousands of other known true fungi found all over the world.

The word *fungus* may evoke images of mushrooms or some sort of powdery or spongy, creeping growth. Although mushrooms are indeed fungi, and while many fungi do appear to be creeping along the ground, the forms of fungi seem almost infinite. There are about 100,000 known species of mushrooms, rusts, smuts, mildews, molds, stinkhorns, puffballs, truffles, and other organisms in Kingdom Fungi, and over 1,000 new species are described each year. There may be as many as 200,000 species yet undiscovered or undescribed, and our daily ravaging of rain forests and other fungal habitats undoubtedly is dooming many others to extinction before they can be found.

As they grow, most fungi produce an intertwined mass of delicate threads, which tend to branch freely and also often fuse together. The individual threads, which usually are more or less tubular, are called **hyphae** (singular: **hypha;** hyphae may be partitioned into cylindrical cells or they may become extensive, branched, multinucleate cells) (Fig. 19.1).

A mass of hyphae is collectively referred to as a **mycelium.** When appropriate food is available, fungi grow very rapidly. In fact, despite the microscopic size of individual hyphae, all those produced by a single fungus in one day when laid end to end could extend more than 1 kilometer (0.6 mile). Some fungi thrive in freezers if the temperature is not lower than −5°C (23°F), while others reproduce freely under temperature conditions of 55°C (131°F) or higher.

Besides being vital to the natural recycling of dead organic material, fungi, along with bacteria, also cause huge economic losses through food spoilage and disease; these topics are explored in the sections on human and ecological relevance of fungi later in the chapter. Scientists who study fungi are known as *mycologists,* and consumers of fungi are called *mycophagists* (from the Greek word *myketos* for "a fungus").

DISTINCTIONS BETWEEN KINGDOMS PROTOCTISTA AND FUNGI

In the past, the true fungi, slime molds, and bacteria were all placed in a single division of the Plant Kingdom. Once the fundamental differences between prokaryotic and eukaryotic cells became known, however, the bacteria were placed in the prokaryotic Kingdom Monera. Then it became increasingly apparent that the metabolism, reproduction, and general lines of diversity of fungi were different from those of members of the Plant Kingdom, with the fungi evidently having been independently derived from ancestral unicellular organisms. Accordingly, fungi were placed in their own kingdom.

All true fungi are filamentous or unicellular *saprobes* (decomposer organisms that live on dead organic matter) or are parasites. Both saprobes and parasites absorb their food in solution through their cell walls. The slime molds, which engulf their food and were discussed in Chapter 18, appear to be related to protozoa, which are single-celled protists included in Kingdom Protoctista. Chytrids and water molds are also funguslike, but like most protists their reproductive cells have flagella, and they, too, are presently included in Kingdom Protoctista.

The members of Kingdom Fungi are placed in two divisions and several classes. They are all filamentous and do not have motile cells. All true fungi produce hyphae, which grow at their tips. Structures such as mushrooms are formed from hyphae tightly interwoven and packed together. The cell walls of true fungi consist primarily of chitin, a material also found in the shells of arthropods (e.g., insects, crabs). Fungi exhibit a variety of forms of sexual reproduction. The food substances, which most fungi absorb through their cell walls, are often broken down with the aid of enzymes secreted to the outside by the cells. Because of the great variety of form and reproduction throughout Kingdom Fungi, a neat pigeonholing of all the members into distinct groups is difficult. Broad groups can, however, be recognized.

KINGDOM FUNGI— THE TRUE FUNGI

Division Zygomycota— The Coenocytic True Fungi

Although black bread molds are probably the best known members of this division, they are not the only fungi that grow on bread. So many organisms can, in fact, contribute to bread spoilage that nearly all commercially baked goods have, in the past, had chemicals, such as calcium propionate, added to the dough to prevent or retard the growth of such organisms. There is now a trend toward eliminating preservatives from bread, pies, and other bakery items since the chemicals that have made the goods a less suitable medium for the growth of fungi apparently are unhealthy for humans after prolonged use. Alternative ways of retarding fungal growth and spoilage are being sought.

Rhizopus (Fig. 19.2), a well-known representative black bread mold, has spores that are exceedingly common everywhere. They have been found in the air above the North Pole, over jungles, on the inside and outside of buildings, in soils, in clothing, in automobiles, and hundreds of kilometers out to sea, easily carried there by prevailing winds and breezes.

When a spore lands in a suitable growing area, it germinates and soon produces an extensive mycelium, which, like that of the water molds, is *coenocytic* (not partitioned into individual cells) and contains numerous haploid nuclei. After the mycelium has developed, certain hyphae called **sporangiophores** grow upright and produce globe-shaped **sporangia** at their tips (Fig. 19.3). Numerous black **spores** are formed within each sporangium. When these spores are released through the breakdown of the sporangium wall, they may blow away and repeat the cycle.

Such reproduction, involving no union of gametes, is asexual, but black bread molds also reproduce sexually by conjugation. Although there is no visible difference in form, black bread mold mycelia occur in two different mating strains. When a hypha of one strain encounters a hypha of the other, swellings called *progametangia* develop opposite each other on the hyphae. These protuberances grow toward each other until they touch. A crosswall is formed a short distance behind each tip, and the two *gametangia* merge, becoming a single large multinucleate cell in which the nuclei of the two strains fuse in pairs. A thick ornamented wall then develops around this cell with its numerous diploid nuclei. This structure, called a *zygospore,* is the characteristic sexual spore of members of this division. A zygospore may lie dormant for months, but eventually it may crack open, and one or more sporangiophores with sporangia at their tips grow out. Meiosis apparently takes place just before this occurs and thousands of black spores are produced in the sporangia. In some species, the spores are produced externally on hyphae instead of being formed in sporangia. Zoospores are unknown in this division.

One dung-inhabiting genus of fungi in this class has the scientific name of *Pilobolus,* derived from two Greek words meaning "cap thrower." The name is quite appropriate, as the mature sporangia are catapulted a distance of up to 8 meters (26 feet), where they adhere to grass or other vegetation (Fig. 19.4). When the vegetation is ingested by animals, the spores germinate in the digestive tract and are already growing in the dung when it is released.

Pilobolus fungi release their sporangia with force precisely in the direction of light, to which they are very

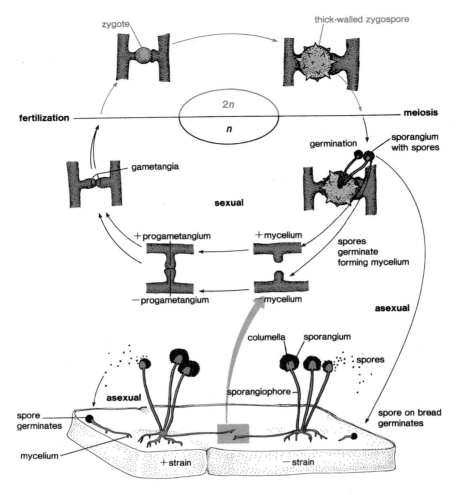

FIGURE 19.2 Life cycles of the black bread mold *Rhizopus.*

sensitive. This action can be demonstrated by placing some horse dung in a glass dish that has a lid and then covering the dish with black paper. If an opening is cut in the paper (the pattern of the opening can be elaborately frilly or angular if desired) and the dish is set where it will receive sufficient light for a few days, any *Pilobolus* sporangia that have been produced and forcibly discharged will form a black pattern on the glass corresponding closely to the cutout area.

Human and Ecological Relevance of the Coenocytic True Fungi

One species of bread mold is used in Indonesia and adjacent areas to produce a food called *tempeh.* Basically, tempeh consists of boiled, skinless soybeans that have been inoculated with a bread mold and allowed to sit for 24 hours. The mycelium that develops holds the soybeans together and produces enzymes that increase the content level of several of the B vitamins. The tempeh is fried, roasted, or diced for soup and is prepared fresh daily. Other bread mold species are used with soybeans to make a Chinese cheese called *sufu.*

At least two species of coenocytic fungi are now used commercially to carry out important steps in the manufacture of birth control pills and anesthetics, while others are utilized in the production of industrial alcohols and as meat tenderizers. Another species produces a yellow pigment used for coloring margarine.

Division Eumycota— The Noncoenocytic True Fungi Class Ascomycetes—The Sac Fungi

Should you happen one summer to be touring or visiting in the south of France, you might be startled to see men and women with coils of rope slung around their shoulders pushing pigs in wheelbarrows. They are not on their way to a hog-hanging festival but are instead wheeling the pigs into the woods to help them find *truffles.*

Truffles are gourmet "mushrooms," which grow mostly between 2.5 and 15.0 centimeters (1 and 6 inches) beneath the surface of the ground, usually near oak trees.

FIGURE 19.3 A zygomycete sporangium. Spores are visible in the split. Scanning electron micrograph, ×2,000.
(Courtesy Kenneth D. Whitney)

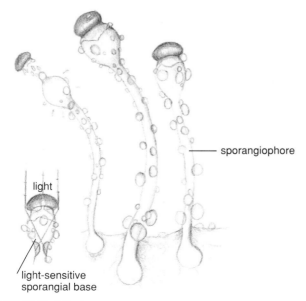

sporangiophore

light

light-sensitive
sporangial base

FIGURE 19.4 The dung fungus *Pilobolus*. The spores are released with force toward a light source.

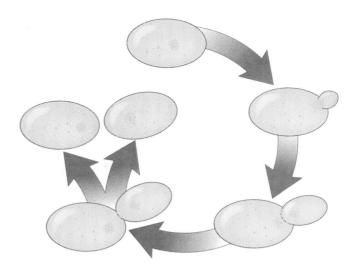

FIGURE 19.5 How a yeast cell reproduces asexually by *budding*.

They are somewhat prunelike in appearance and may be more than 10 centimeters (4 inches) in diameter, although most are smaller. They give off a tantalizing aroma that has been shown to contain pig sex *pheromones* (chemicals that produce specific responses). Pigs can detect truffles a meter (3 feet) below the surface and more than 15 meters (50 feet) away. The animals, in an attempt to get to them, strain at the ropes that are tied around their necks. The owners dig up the truffles, which were selling in the United States in 1989 for about $200 a pound, and reward the pigs with acorns or other less interesting food. Trained dogs are now often substituted for pigs in truffle hunting.

Truffles are the reproductive bodies of representatives of a large and varied class of true fungi called *sac fungi*. More than 20,000 species, including yeasts, powdery mildews, brown fruit rots, ergot, morels, microscopic parasites of insects, and canned fruit fungi, have thus far been described. Most produce mycelia superficially similar to those of the water molds and coenocytic true fungi, but the hyphae are partitioned into individual cylindrical cells. Each crosswall has one or more pores through which the single to many tiny nuclei can pass. Little bodies that serve as plugs are located near the pores. They seal off individual cells if they become damaged.

Asexual reproduction is by means of **conidia** (singular: **conidium**). Conidia are spores that are produced externally—outside of a sporangium—either singly or in chains at the tips of hyphae called *conidiophores* (see Fig. 19.31). In the case of yeasts, asexual reproduction is by **budding** (Fig. 19.5). When a yeast cell buds, a small protuberance appears to balloon out slowly from the cell and become pinched off as it grows to full size.

Sexual reproduction always involves the formation of tiny fingerlike sacs called **asci** (singular: **ascus**). When hyphae of two different "sexes" become closely associated in the more complex sac fungi, male *antheridia* may be formed

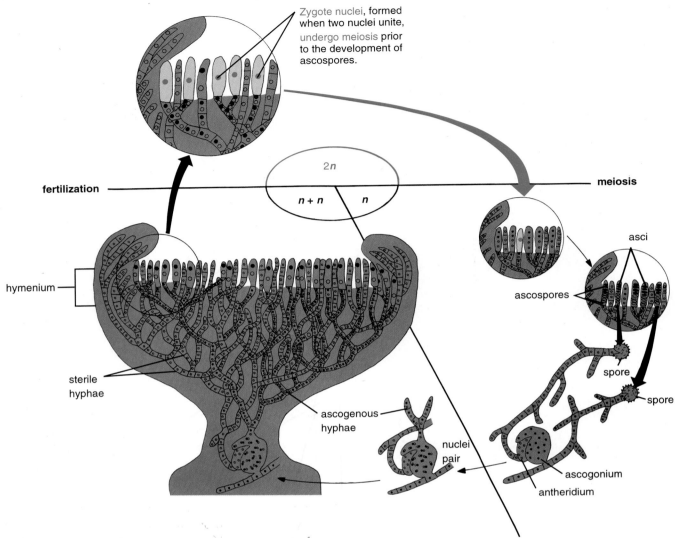

FIGURE 19.6   Life cycle of a sac fungus. When hyphae of two different strains of a sac fungus become closely associated, male antheridia may be formed on one and female ascogonia on the other. Male nuclei migrate into an ascogonium and pair but do not fuse with the female nuclei present. Then new hyphae (ascogenous hyphae), whose cells each contain a pair of nuclei, grow from the ascogonium. In a process involving fusion of the pairs of nuclei (followed by meiosis), fingerlike asci, each containing four or eight haploid nuclei, are formed in a layer called a *hymenium,* which lines an ascocarp. The haploid nuclei become ascospores, which are discharged into the air. They are potentially capable of initiating new mycelia and repeating the process.

on one and female *ascogonia* on the other. Male nuclei migrate into an ascogonium and pair but do not fuse with the female nuclei present. Then new hyphae (*ascogenous hyphae*), whose cells each contain one male and one female nucleus, grow from the ascogonium. The cells divide in a unique way, producing new cells with one of each kind of nucleus. At maturity, the pairs of nuclei in the cells at the tips of each ascogenous hypha unite, and tubular asci develop in a layer (referred to as the *hymenium*) at the surface of a structure called an *ascocarp*. The diploid zygote nucleus undergoes meiosis, and the resulting four haploid nuclei usually divide once more by mitosis so there is a row of eight nuclei in each ascus. These nuclei become *ascospores* as they are walled off from one another with a little cytoplasm (Fig. 19.6).

Thousands of asci may be packed together in an ascocarp, which often is cup-shaped (Fig. 19.7) but also may be completely enclosed (Fig. 19.8) or flask-shaped with a little opening at the top. Truffles are really enclosed ascocarps. Cup-shaped ascocarps may be several centimeters (2 to 3 inches) in diameter and may be brilliantly colored on the inside. When ascospores are mature, they are often released with force from the asci. If an open ascocarp is jarred at maturity, it may appear to belch fine puffs of smoke consisting of thousands of ascospores.

When an ascospore lands in a suitable area, it may germinate, producing a new mycelium, and then repeat the process. In many instances, however, a number of asexual generations involving conidia are produced between the sexual cycles.

FIGURE 19.7   An ascocarp of a sac fungus.

FIGURE 19.9   A morel.
(© Doug Sherman/Geofile)

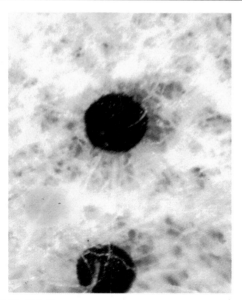

FIGURE 19.8   Closed ascocarps (*cleistothecia*) of a powdery mildew on an oak leaf, ×200.

Human and Ecological Relevance of the Sac Fungi

Morels, which have been called by some people "the world's most delicious mushrooms," and *truffles* have been prized as food for centuries. Neither is, however, a true mushroom. True mushrooms are included in the *club fungi* discussed in the next section. Wealthy Romans and Greeks used to insist on preparing morels personally according to various recipes

they had concocted, and they are still prized as gourmet food today. Prior to 1982, numerous unsuccessful attempts were made to cultivate them under controlled conditions, but morels now can be mass-produced commercially and are beginning to be marketed. Morels (Fig. 19.9) are the size of a medium to smallish mushroom, are tan in color, and have a wrinkled, somewhat conelike top on a stalk that resembles a miniature tree trunk. The numerous depressions between the wrinkles each contain thousands of asci. Although morels are perfectly edible by themselves, there are unconfirmed reports that some persons have become ill after consuming them with alcohol. Caution in this regard is advised.

A related "mushroom" called a *false morel* or *beefsteak morel* is considered a delicacy by many but has caused death in others. This points to the peculiar unpredictability of the interaction between the human metabolism and certain fungi. For some people, the fungi are quite safe to eat; for others they are not—a literal example of "one man's meat being another man's poison."

When rye and, to a lesser extent, other grains come into flower in a field, they may become infected with *ergot* fungus (Fig. 19.10). This fungus seldom causes serious damage to the crop, but as it develops in the maturing grain it produces several powerful drugs. If the infected grain is harvested and milled, a disease called **ergotism** may occur in those who eat the contaminated bread. The disease can affect the central nervous system, often causing hysteria, convulsions, and sometimes death. Another form of ergotism causes gangrene of the limbs, which may result in loss of the affected limb. It can also be a serious problem for cattle grazing in infected fields. They frequently abort their fetal calves and may succumb themselves.

Ergotism was common in Europe in the Middle Ages. Known then as *St. Anthony's Fire,* it killed 40,000 people in an epidemic in A.D. 994. In 1722, the cavalry of Czar Peter the Great was felled by ergotism just as the czar was about to conquer Turkey, and the conquest never took place. Some historians have implicated ergotism in violent social upheavals such as the French Revolution and the Salem Witchcraft Trials. As recently as 1951, an outbreak hospitalized 150 victims in a French village. Five persons died and 30 became temporarily insane, imagining they were being chased by snakes and demons.

In small controlled doses, ergot drugs have proved medically useful. They stimulate contraction of the uterus to initiate childbirth and have been used in abortions and in the treatment of migraine headaches. Ergot is also an initial source for the manufacture of the hallucinogenic drug lysergic acid diethylamide, popularly known as *LSD.*

Sac fungi play a basic role in the preparation of baked goods and alcohol. Enzymes produced by yeast aid in fermentation, producing ethyl alcohol and carbon dioxide in the process. The carbon dioxide produced in bread dough forms bubbles of gas, which are trapped, causing the dough to rise and giving bread its porous texture. Part of the flavor of individual wines, beers, ciders, sake, and other alcoholic beverages is imparted by the species, or strain, of yeast used to ferment the fruits or grains.

Yeasts are at least indirectly involved in the manufacture of a number of other important products. Ephedrine, a drug also produced by mormon tea, a leafless western desert shrub, is obtained commercially from certain yeasts. The drug is used in nose drops and in the treatment of asthma. Yeasts are also a rich source of B vitamins and are used in the production of glycerol for explosives. Ethyl alcohol is used in industry as a solvent and in the manufacture of synthetic rubber, acetic acid, and vitamin D. Yeast contains about 50% protein and makes a nourishing feed for livestock. Its enzyme, invertase, catalyzes the conversion of sucrose into glucose and fructose and is used to soften the centers of chocolate candies after the chocolate coating has been applied.

Several very important plant diseases are found in this class of fungi. *Dutch elm disease,* a disease originally described in Holland, has devastated the once numerous and stately elm trees in many towns and cities in the midwestern and eastern United States and has now spread to the West

FIGURE 19.10   Ergotized rye. The prominent darker objects scattered throughout the ears of rye are kernels infected with ergot fungus.

(Copyright © Loran Anderson)

and South. When Dutch elm disease was first discovered, limited control of the disease was achieved by spraying the trees with DDT to kill the elm bark beetles that spread the disease from tree to tree. DDT killed many useful organisms as well, however, and its use for such purposes was eventually banned.

Other sprays have since been used, again with limited success, and biological controls are now being sought. One such control was reported in 1980 by Gary Strobel of Montana State University. He injected *Pseudomonas* bacteria into 20 diseased trees and saved seven of them when the bacteria multiplied and killed the Dutch elm disease fungus. In

FIGURE 19.11   A peach leaf infected with peach leaf curl.

FIGURE 19.12   A common stinkhorn. Note the slimy mass of spores on the cap.

(Courtesy Leland Shanor)

1987, Dr. Stroebel injected a genetically altered strain of *Pseudomonas* bacteria with more powerful toxins against Dutch elm disease into 15 elm trees but then had to cut the trees down because he had not previously obtained permission from the Environmental Protection Agency for his experiment. Whether this control technique will meet with widespread success remains to be seen, but it is the type of control that is much to be preferred to poisonous sprays that upset delicate ecological balances (see Appendix 2, which deals with biological controls).

Chestnut blight has virtually eliminated the once numerous American chestnut trees from the eastern deciduous forests. Attempts to control the disease have met with very limited success thus far. Much better success has been obtained in controlling *peach leaf curl,* a disease that attacks the leaves of some stone fruits, especially peach trees (Fig. 19.11). Sprays that contain copper or zinc salts apparently inhibit the germination of spores of peach leaf curl and have

been effective in preventing serious damage when applied to trees before the dormant buds swell and open in the spring.

Class Basidiomycetes— The Club Fungi

At the end of my first year in college, I did odd jobs during the summer, including various types of yard work. On one occasion while cleaning dead leaves and debris from a shaded garden area, I came across two or three growths, about the width of a pencil and the length of a finger, rising above the surface of the ground. On closer inspection, I observed that these growths had the consistency and appearance of a sponge and the tips were partially covered with a slimy and evil-smelling substance. The growths turned out to be fungi called *stinkhorns* (Fig. 19.12), whose odor attracts flies, which disseminate the sticky spores that stick to their bodies.

A.

B.

C.

FIGURE 19.13 *A.* Common woods mushrooms. *B.* Earth stars. *C.* A bracket fungus.

(*B.* courtesy Perry J. Reynolds)

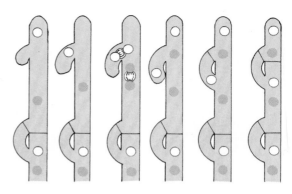

FIGURE 19.14 Development of a clamp connection. *Left to right.* Protrusion appears in wall of cell; one nucleus migrates into loop; both nuclei undergo mitosis; loop carrying daughter nucleus turns to cell wall; clamp connection established; daughter nuclei pair in two cells.

toadstools (the only distinctions between mushrooms and toadstools are based on folklore or tradition—botanically there is no difference), puffballs, earth stars (Fig. 19.13 B), shelf or bracket fungi (Fig. 19.13 C), rusts, smuts, jelly fungi, and bird's-nest fungi. They are called *club fungi* because in sexual reproduction, they produce their spores at the tips of swollen hyphae that often resemble baseball bats or clubs. These swollen hyphal tips are called **basidia** (singular: **basidium**). The hyphae, like those of the cup or sac fungi, are divided up into individual cells. These cells, however, have either a single nucleus or, in some stages, two nuclei. The crosswalls have a central pore that is surrounded by a swelling, and both the pore and swelling are covered by a cap. This cap, with some exceptions, blocks passage of nuclei between cells but allows cytoplasm and small organelles to pass through.

Asexual reproduction is much less frequent in club fungi than in the other classes of fungi. When it does occur, it is mainly by means of conidia, although a few species produce buds similar to those of yeasts, and others have hyphae that fragment into individual cells, each functioning like a spore and forming a new mycelium after germination.

Sexual reproduction in many mushrooms begins in the same way as it does in members of the two classes previously discussed. When a spore lands in a suitable place—often an area with good organic material and humus in the soil—it germinates and produces a mycelium just beneath the surface. The hyphae of the mycelium are divided into cells that each contain a single haploid nucleus. Such a mycelium is said to be **monokaryotic.** Monokaryotic mycelia of club fungi often occur in four mating types, usually designated simply as types 1, 2, 3, and 4. Only types 1 and 3 or types 2 and 4 can mate with each other.

If the growth of the hyphae of compatible mating types happens to bring them close together, cells of each mycelium may unite, initiating a new mycelium in which each cell has two nuclei. Such a mycelium is said to be

Stinkhorns are interesting but relatively unimportant representatives of another large class of true fungi, the *club fungi*. Included in this class are mushrooms (Fig. 19.13 A) or

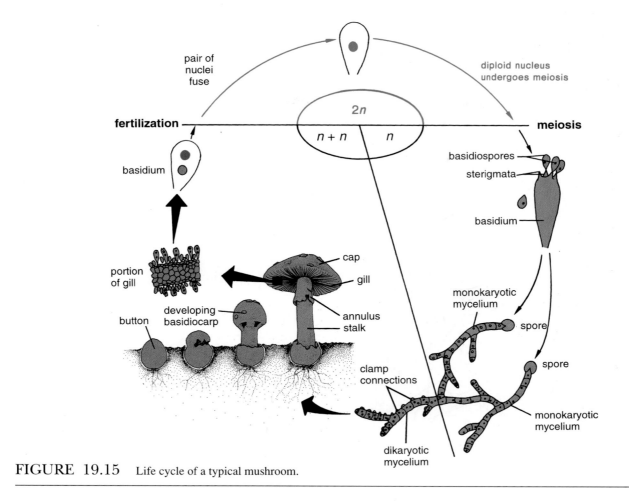

FIGURE 19.15   Life cycle of a typical mushroom.

Labels in figure: pair of nuclei fuse; diploid nucleus undergoes meiosis; 2n; n + n; n; fertilization; meiosis; basidium; basidiospores; sterigmata; basidium; portion of gill; cap; gill; annulus; stalk; monokaryotic mycelium; spore; developing basidiocarp; button; spore; clamp connections; monokaryotic mycelium; dikaryotic mycelium

dikaryotic. Dikaryotic mycelia usually have little walled-off bypass loops called *clamp connections* between cells on the surface of the hyphae (Fig. 19.14). The clamp connections develop as a result of a unique type of mitosis that ensures that each cell will have within it one nucleus of each original mating type.

After developing for a while, the dikaryotic mycelium may become very dense and form a compact, solid-looking mass called a *button.* This pushes above the surface and expands into a *basidiocarp,* commonly called a **mushroom** (Fig. 19.15). Most mushrooms have an expanded umbrella-like *cap* and a *stalk* (**stipe**). Some have a ring called an *annulus* on the stalk. It is the remnant of a membrane that extended from the cap to the stalk and tore as the cap expanded. Some mushrooms, such as the notorious *death angel mushroom,* also have a cup called a *volva* at the base (Fig. 19.16). Thin, fleshy-looking plates called **gills** radiate out from the stalk on the underside of the cap. Microscopic examination of a gill reveals the compacted hyphae of which it is composed and innumerable **basidia** oriented at right angles to the flat surfaces of the gill.

As each basidium matures, the two nuclei unite and then undergo meiosis. The four nuclei that result from meiosis become **basidiospores** when they migrate through four (in a few species, two) tiny pegs at the tip of the basidium, walls forming around the nuclei in the process. The tiny pegs, called *sterigmata* (singular: *sterigma*), serve as stalks for the basidiospores. One large mushroom may produce several billion basidiospores within a few days. These are forcibly discharged into the air between the gills, and they blow away with the slightest breeze.

If you remove a mushroom stalk and place the cap gill-side down on a piece of paper, covering it with a dish to eliminate air currents, the spores will fall and adhere to the paper in a pattern perfectly reflecting the arrangement of the gills. The dish and cap can be removed a day later, and the *spore print* (Fig. 19.17) can be made more or less permanent with the application of a little clear varnish or shellac. Professional mycologists use such spore prints as an aid to identification, employing white paper for dark-colored spores and black paper for white or light-colored spores.

FIGURE 19.16 A death angel mushroom (*Amanita*).
Note the egglike volva at the base.
(Courtesy Dr. T. Duffy)

FIGURE 19.17 A spore print.

FIGURE 19.18 A bolete. These mushrooms have pores
instead of gills beneath their caps.

In nature, some of the basidiospores eventually repeat the reproductive cycle. Often, a dikaryotic mycelium radiates out from its starting point, periodically producing basidiocarps in so-called *fairy rings*. If conditions are favorable, the mycelium continues to grow at the edges for many years while dying in the center as food resources are depleted. Some mycelia have been known to grow in this fashion for over 500 years.

Some mushrooms produce their spores on the surfaces of thousands of tiny pores instead of on gills (Fig. 19.18).

Shelf or *bracket fungi* (Fig. 19.19) grow out horizontally from the bark or dead wood to which they have become attached, some adding a new layer of growth each year. Perennial species can become large enough and so securely attached that they can support the weight of an adult.

Other members of this class produce spores within parchmentlike membranes, forming somewhat ball-like basidiocarps called *puffballs*. Puffballs, which generally are edible, range in diameter from a few millimeters to 1.2 meters (0.125 inch to 4 feet) (Fig. 19.20). They have no stalks and rest in contact with the ground. Literally trillions of spores may be produced by a large puffball. These are released through a pore at the top or from random locations when the outer membrane breaks down. *Earth stars* (see Fig. 19.13 B), which are similar to puffballs, differ from them in having a ring of appendages at the base that look like a set of woody petals around a flower.

FIGURE 19.19   A shelf, or bracket, fungus (*Phacolus*).
(Courtesy Richard Critchfield)

FIGURE 19.21   A bird's-nest fungus.

FIGURE 19.20   Giant puffballs. The largest weighed
6.12 kilograms (13.5 pounds).
(Courtesy Louise White, Redding, CA, *Record-Searchlight*)

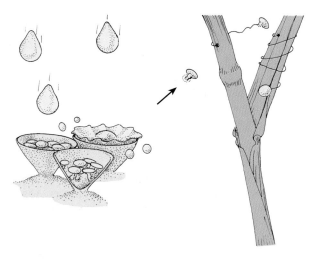

FIGURE 19.22   How the "eggs" in a bird's-nest fungus
are dispersed.
(After Harold J. Brodie. 1951. "The splash-cup dispersal mechanism in
plants." *Canadian Journal of Botany* 29: 224–34. Reproduced by
permission of the National Research Council of Canada from the
Canadian Journal of Botany, 29: 224–34)

Bird's-nest fungi (Fig. 19.21), which grow on wood or
manure, form nestlike cavities in which "eggs" containing
basidiospores are produced. In some species, each "egg" has
a sticky thread attached to it. When raindrops fall in the
nests, the eggs may be splashed out, and as they fly through
the air, the sticky threads catch on nearby vegetation, whip-
ping the eggs around it (Fig. 19.22). When animals graze on
the vegetation, the spores pass unharmed through the intestinal tract.

Smuts are parasitic club fungi that do considerable
damage to corn, wheat, and other grain crops. In corn smut
(Fig. 19.23), the mycelium grows between the cells of the
host. The hyphae absorb nourishment from these cells and
also secrete substances that stimulate them to divide and enlarge, forming tumors on the surfaces of the corn kernels.
These eventually break open, revealing millions of sooty
black spores, which are blown away by the wind. Some
smuts affect only the flowering heads or grains, while others
infect the whole plant.

FIGURE 19.23   Corn smut fungus on an ear of corn.

Rusts, which also are parasites, attack a wide variety of plants. Some rusts grow and reproduce on only one species of flowering or cone-bearing plant. Others, however, require two or more different hosts to complete their life cycles. *Black stem rust of wheat,* which has reduced wheat yields by millions of bushels in a single year in the United States alone, has plagued farmers ever since wheat was first cultivated thousands of years ago. More than 300 races of black stem rust are now known. This rust requires both common barberry plants and wheat to complete its life cycle (Fig. 19.24).

Since two hosts are necessary for black stem rust of wheat to complete its life cycle, control of the disease could be accomplished through eradication of common barberry bushes. An estimated 600 million such plants have been destroyed in the United States since 1918, in an attempt to eradicate the disease, but it has proved impossible to eliminate the species altogether. Producing rust-resistant strains of wheat has also helped, but even as new strains are developed, the rusts themselves hybridize or mutate, producing new races capable of attacking previously immune varieties of cereals—a striking example of evolution in action.

Another serious rust with two hosts is the *white pine blister rust,* which has caused huge losses of valuable timber trees in both the eastern and western United States. Basidiospores infect the pine trees, and when the basidiospores germinate, other types of spores are produced. These different spores in turn infect currant and gooseberry bushes, and the spores formed on the currants and gooseberries eventually give rise to new basidiospores, completing the cycle. The U.S. Forest Service had a program of gooseberry bush eradication in operation for many years in an attempt to alleviate the problem, but the program was only partially successful. Spraying programs have more recently been implemented with some success, and rust-resistant trees also are being selected and bred as alternatives.

Other rusts with two hosts include *apple rust* (alternate host: cedar trees), *poplar leaf spot* (alternate host: larch or tamarack trees), and *corn rust* (alternate host: sorrel plants).

A relative of black stem rust was recently discovered in the Rocky Mountains of Colorado. This rust causes its rock cress host plants to produce fake flowers that look and smell so real many insects are fooled by them. When bees and butterflies visit the fake flowers, they find a sugary, nectarlike secretion. While gathering the sticky fluid, they inadvertently also pick up fungal sex cells, which are spread to other rock cress plants.

Human and Ecological Relevance of the Club Fungi

Of the approximately 25,000 described species of club fungi, fewer than 75 are known to be poisonous. Many of the latter are, however, common and not readily distinguishable by amateurs from edible species. Also, some edible forms, such as the inky cap and shaggy mane mushrooms, which cause no problems by themselves, may make one very ill if consumed with alcohol. Few of the poisonous forms normally are fatal, but unfortunately some—such as the death angel, which causes 90% of the fatalities attributed to mushroom poisoning—are relatively common.

Poisoning from death angels and similar species is due to alpha-amatin (an alkaloid). Symptoms of the poison, which completely blocks RNA synthesis, usually take from 6 to 24 hours to appear. Until a few years ago, successful treatment was impossible by the time the intense stomachache, blurred vision, violent vomiting, and other symptoms occurred.

At present, hope for survival of the victims of mushroom poisoning lies with the administration of a drug known as *thioctic acid.* Even thioctic acid does not always prevent a fatality, however, and either the drug or the mushroom poison usually leave the patient with hypoglycemia (a blood sugar deficiency). Some wild mushroom lovers have fed parts of their collections to dogs or cats, and when nothing happened to the animals after an hour or two, they have eaten the mushrooms themselves. Both they and the animals later succumbed. Others have mistakenly believed that the toxic substances are destroyed by cooking. Records show, however, that before the discovery of thioctic acid as an antidote, death ensued in 50% to 90% of those who had eaten just one or two of the deadly mushrooms, cooked or raw, and

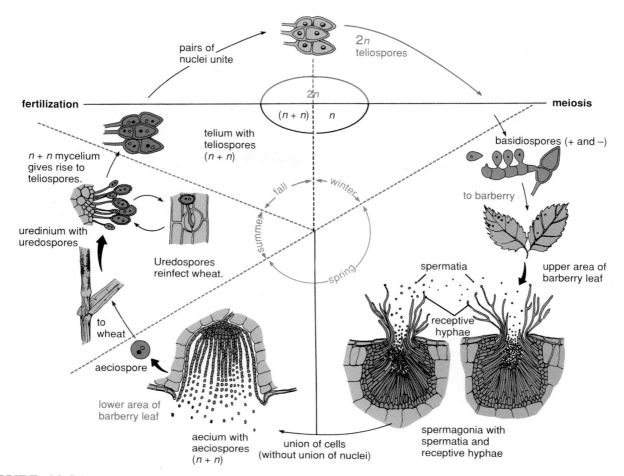

FIGURE 19.24 Life cycle of black stem rust of wheat.

that even as little as 1 cubic centimeter (less than 0.5 cubic inch) can be fatal.

A number of widespread but very unreliable beliefs exist concerning distinctions between edible and poisonous mushrooms. One holds that a silver coin placed in the cooking pan will turn black while the mushrooms are cooking if any poison is present. Some members of both edible and poisonous species will turn such a coin black, and some will not. Another superstition holds that edible species can be peeled, while poisonous ones cannot. Again this is a fallacy, for death angels peel quite easily. Still other erroneous beliefs are that poisonous mushrooms appear only in the fall or the early spring, that all mushrooms eaten by snails and beetles are edible, that all purplish-colored mushrooms are poisonous, and that all mushrooms growing in grassy areas are edible. Again, these notions simply are not supported by the facts. It is foolhardy for anyone to eat mushrooms that have not been correctly identified by a knowledgeable authority.

Some poisonous mushrooms cause hallucinations in those who eat them. During the Mayan civilization in Central America, *teonanacatl* ("God's flesh") sacred mushrooms (Fig. 19.25) were used in religious ceremonies. The consumption of these mushrooms, which has continued among native groups in Mexico and Central America to the present, results in sharply focused, vividly colored visions.

FIGURE 19.25 Teonanacatl mushrooms.
(Courtesy Drug Enforcement Administration)

Similar use of the striking, fly agaric mushroom (Fig. 19.26) in Russia and for a while in the Indus valley of India dates back to over 2,000 years ago. Users appear to go into a state of intoxication. It is believed that the ancient Norwegian beserkers, who occasionally exhibited fits of exceptionally savage behavior, did so after consuming fly agaric mushrooms. Related species that occur in the United States have not produced the same effects but have, instead, caused users to become nauseated and to vomit. In Siberia, users have noted that the intoxicating principle is passed out in the urine, and some persons have adopted the practice of drinking the urine of persons who have consumed fly agaric mushrooms. Reindeer, incidentally, are reported to be obsessed with both fly agaric and human urine.

More than 90% of a mushroom is water, and mushrooms generally contain smaller quantities of nutritionally valuable substances than do most foods. An apparent exception is the legendary *Shiitake* mushroom grown for centuries in China and Japan on oak logs and now cultured in the United States. It has more than double the protein of ordinary, commercially grown mushrooms and is very rich in calcium, phosphorus, and iron. It also contains significant amounts of B vitamins, vitamin D_2 (ergosterol), and vitamin C and has excellent flavor.

Ancient Chinese royalty believed that eating Shiitake mushrooms would promote healthful vigor and retard the aging process. It appears there could be some substance to this belief, as recent research suggests that mushrooms in general may be one of the richer sources of RNA, which has been shown to retard the aging process in cells through its metabolic conversion to ATP.

Lentinacin, an agent capable of lowering human cholesterol levels, has been obtained from Shiitake mushrooms, and purified extracts from spores of the mushrooms have demonstrated antiviral activity against influenza and polio viruses in laboratory animals. The extracts evidently induce the formation of *interferon,* a virus-combating substance produced by animal cells.

Many types of mushrooms have been cultivated for food since ancient times. In the second century B.C., a Greek doctor by the name of Nicandros taught people how to grow mushrooms underneath fig trees on soil fertilized with manure. Andrea Cesalpino, a noted Italian botanist and physician of the 16th century, cultivated mushrooms by scattering pulverized poplar bark on very rich soil near poplar trees with which the mushrooms were associated. In more recent times, Italians have cultivated mushrooms on waste material from olives, on coffee grounds, on remnants of oak leaves after tannins have been extracted for leather tanning, and on laurel berries. Today, jelly fungi and various mushrooms are cultivated in the Orient on media composed of wood and manure.

In Geneva, Switzerland, there is a special market for wild mushrooms where more than 50 species are sold under the supervision of a state mycologist. Although wild mushrooms are consumed by many people in the United States, few such species are sold in markets and only one species

FIGURE 19.26 Fly agaric mushrooms.
(Courtesy Perry J. Reynolds)

FIGURE 19.27 Mushroom beds in a commercial mushroom farm.

of mushroom (*Agaricus bisporus*) is cultivated to any extent. It has been grown in basements, caves, and abandoned mines, but contrary to popular belief, light does not affect its growth.

Large-scale mushroom-growing operations generally use windowless warehouses with stacked rows of shallow planting beds because temperatures, humidity, and other climatic factors are easier to control in such buildings (Fig. 19.27). The mushrooms are grown on compost made from straw and horse or chicken manure. Prior to use, the compost is pasteurized for a week to destroy microorganisms, unwanted fungi, and insects and their eggs. Then it is inoculated with spawn, which is compact mycelium grown from germinated basidiospores sown on bran of wheat or other grains. After inoculation, the spawn is covered with a thin layer of soil. The moisture content of the compost is controlled with regular, light waterings, and a humidity of about 75% is maintained.

The mycelium grows in temperatures ranging from just above freezing to 33°C (91.4°F), but commercial growers try to keep the temperatures between 9°C and 13°C (48°F to 55°F) because the mushrooms are less subject to disease or insect attacks and are also firmer when grown under cool conditions. The mushroom buttons appear one to three weeks after spawn is planted and continue appearing for six weeks or more. About 1 kilogram (2.2 pounds) of mushrooms is obtained from each 10 square decimeters (slightly over 1 square foot) of growing area. An estimated 59,000 metric tons (65,000 tons) of mushrooms are produced annually in the United States.

Fungi are constantly breaking down dead wood and debris and returning the components to the soil where they can be recycled. Sometimes, as we have seen, they can be very destructive from a human viewpoint, attacking everything from living plants and harvested or processed food to shoe leather, paper, cloth, construction timbers, paint, petroleum products, upholstery, and even glass, particularly in warmer climates.

Class Deuteromycetes— The Imperfect Fungi

Any fungus for which a sexual stage has not been observed is classified as an *imperfect fungus.* Many otherwise unrelated fungi are grouped together in this artificial class, which includes several well-known disease organisms as well as fungi important in disease control and food processing. If a member of this group is studied further and a sexual stage is discovered, the fungus is reassigned to its appropriate class.

All imperfect fungi reproduce by means of conidia. In one such group, which is found growing on dead leaves and debris at the bottom of streams with rapidly moving water, the spores often have four long extensions on them. These arms are arranged in such a way that three may catch like a tripod on a flat surface, keeping the fungus from being washed downstream.

One interesting group of imperfect fungi parasitizes protozoans and other small animals in various ways. Some develop an unbranched body inside their victim and slowly absorb nutrients until the host dies. They then produce a chain of spores that may stick to or be eaten by another victim. Others capture their prey on sticky hyphae to which passing amoebae adhere. One soil-dwelling group captures *nematodes* (eelworms) in hyphal rings or loops. In some, the rings are a little smaller than the circumference of a nematode, which tapers at both ends. When a nematode randomly happens to stick its head through such a loop, it frequently tries to struggle forward instead of backing out and becomes trapped (Fig. 19.28). The fungus then produces small rhizoidlike outgrowths called *haustoria,* which grow into the worm's body and digest it.

More specialized species of nematode-trapping fungi have loops that constrict around the worm less than a 10th of a second after contact. The loops are spaced so that sometimes the tail of the worm gets caught in a second loop while the worm is thrashing around after being caught by the head. A number of these fungi can easily be grown on agar media, but they do not form loops unless a nematode is placed in the dish. The nematodes evidently produce a substance that promotes the development of the loops.

Several species of imperfect fungi are found in ant and termite nests. The insects cultivate their fungus gardens by bringing in bits of leaves, other plant debris, caterpillar droppings, and their own feces, and in so doing, they form a rich growing medium. They also lick the hyphae and constantly probe them, using them for food as they grow.

Some wood-boring beetles have pouches in their bodies that function as fungus spore containers. The spores germinate in the wood tunnels, their mycelia forming a lining that produces yeastlike cells on which the beetles and their larvae feed. Many higher plants have *mycorrhizal fungi* associated with their roots. These fungi greatly increase the absorptive surface area around the roots and may be far more important than root hairs in this regard, particularly in mature roots.

Human and Ecological Relevance of the Imperfect Fungi

Among the best known of the imperfect fungi are the *Penicillium* molds (Fig. 19.29), which secrete *penicillin,* the well-known and widely used **antibiotic** (a substance produced by a living organism that interferes with the normal metabolism of another living organism). Sir Alexander Fleming of England noticed in 1929 that certain bacteria would not grow in the vicinity of the mycelium of a *Penicillium* mold, and he gave the name *penicillin* to the element in the mold that prevented the bacterial growth.

Fleming apparently did not grasp the significance of his findings and the findings did not particularly excite the medical profession until the outbreak of World War II some 10 years later. At that time, a team of British and American scientists at Northern Regional Laboratories in Peoria, Illinois, set out to see if they could coax *Penicillium* molds into producing more of this antibacterial substance, primarily because war casualties created a need for greater quantities of more effective medicines that could keep wounds from becoming infected.

The scientists began with cultures from the original mold observed by Fleming, but the amounts of penicillin it produced were so small that it was very expensive to obtain significant quantities for human use. They appealed to the general public, asking them to send in any material they found with a greenish or bluish mold on it. They received whole trainloads of moldy trash from all over the United States.

The breakthrough in the research came, however, when a different species of *Penicillium* mold—one that yielded 25 times the penicillin produced by the original culture—was found on a moldy cantaloupe from a local market. The scientists set to work germinating individual spores of this new mold on culture media, and by careful selection

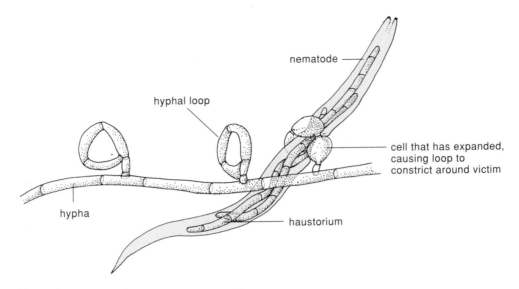

FIGURE 19.28   A nematode-trapping fungus with a victim.
(After Drechsler, C. 1937. "Some hyphomycetes that prey on free-living terricolous nematodes." *Mycologia* 29: 447–552. Redrawn by permission.)

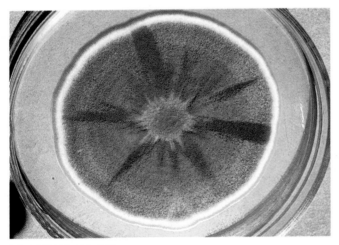

FIGURE 19.29   A *Penicillium* colony.

FIGURE 19.30   Blue cheese. The dark areas are parts of the mycelium of a species of *Penicillium* that gives the cheese its unique flavor.

they eventually were able to isolate a strain that produced more than 80 times the original quantity of penicillin. Later, when this strain was subjected to X-radiation, still other forms were produced that upped the penicillin output to 225 times that of Fleming's mold. Today, most of the penicillin produced around the world comes from descendants of that cantaloupe mold. Literally hundreds of other antibiotics effective in combating human and animal diseases have been discovered since the close of World War II, and the production of these drugs is a vast worldwide industry.

Penicillium molds are also used in other ways. Some are introduced into the milk of cows, sheep, and goats at stages in the production of "smelly" cheeses, such as blue (Fig. 19.30), Camembert, Roquefort, Gorgonzola, and Stilton. The molds produce enzymes that break down proteins and fats in the milk, giving the cheeses their characteristic flavors.

Since the early 1980s, organ transplants have been aided by the discovery and production of a "wonder drug" from an imperfect fungus found in soil. Called *cyclosporine*, the drug suppresses immune reactions that cause rejection of transplanted organs, without risking the development of leukemia and other undesirable side effects associated with other drugs.

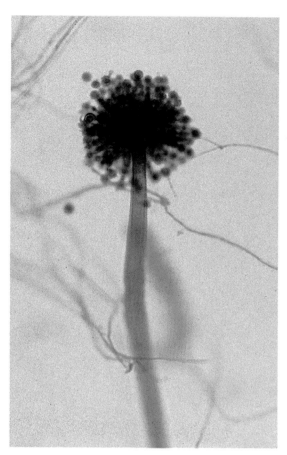

FIGURE 19.31 A conidiophore of *Aspergillus* seen with the aid of a microscope. Rows of conidiospores (also called *conidia*) are being produced at the tips of branches of the conidiophore.

Aspergillus is a genus of imperfect fungi whose species produce dark brown to blackish or yellow spores (Fig. 19.31). It is closely related to the *Penicillium* molds and is extensively used in industry. One or more species are used commercially for the production from sugar of citric acid, a substance for flavoring foods and for the manufacture of effervescent salts that were originally obtained from oranges. Citric acid is also used in the manufacture of inks and in medicines, and it is even used as a chicle substitute in some chewing gums.

Aspergillus fungi also produce gallic acid used in photographic developers, dyes, and indelible black ink. Other species are used in the production of artificial flavoring and perfume substances, chlorine, alcohols, and several acids. Further uses are in the manufacture of plastics, toothpaste, and soap and in the silvering of mirrors.

One species of *Aspergillus* is used in the Orient and elsewhere to ferment soybeans to make soy sauce, or *shoyu*. A Japanese food called *miso* is made by fermenting soybeans, salt, and rice with the same fungus. More than 0.5 million tons of miso are consumed annually.

A number of diseases of both humans and animals are caused by *Aspergillus* species. The diseases, called *aspergilloses,* attack the respiratory tract after the spores have been inhaled. One type thrives on and in human ears. Other diseases caused by different genera of imperfect fungi include those responsible for the widespread problems of athlete's foot and ringworm, for white piedra (a mild disease of beards and mustaches), and for tropical diseases of the hands and feet that cause the limbs to swell in grotesque fashion. One serious disease called *valley fever,* found primarily in the drier regions of the southwestern United States, usually starts with the inhalation of dust-borne spores of an imperfect fungus that produces lesions in the upper respiratory tract and lungs. The disease may spread elsewhere in the body, with sometimes fatal results.

Aspergillus flavus, which grows on moist seeds, secretes *aflatoxin,* the most potent natural carcinogen known. The toxin causes liver cancer, and no more than 50 parts per billion is allowed in human food. In humid climates, such as those of the southeastern United States and adjacent Mexico, improperly stored grain can become moist enough to support the fungus. Carcinogenic foods such as peanuts, peanut butter, and peanut-based dairy feeds may result. Dairy cattle feed is even more strictly controlled because concentrations of aflatoxin can accumulate in milk.

Two imperfect fungi show promise as biological controls of pest organisms. One has already been used with some success in controlling scale insects in Florida and other warm, humid regions. Another may be used to combat water hyacinths, which have caused serious clogging of waterways in areas of the world with mild to tropical climates.

LICHENS

A student who was interested in natural dyeing came to me a few years ago and asked if she could experiment with the dye potential of local plants as a special project. In the course of her experimentation, she obtained beautiful shades of yellow, brown, and green from two dozen common local plants, using simple recipes.[1] During the following summer, she extended the project to include lichens growing on the

1. Most natural dye recipes call for simmering at least 1 liter (approximately 1 quart) or 2 of loosely packed fresh or dry material covered with water in a large enamel kettle until most of the coloring appears to be in the water. This usually takes from one to several hours. The solid waste is then strained out and discarded. A *mordant* (substance that helps fibers take up dye permanently) is then often added to the liquid in amounts varying from 1 teaspoon to 0.5 cup. Commonly used mordants include alum, detergent ammonia, copper sulphate, tin (stannous chloride), and white vinegar. White wool or other fibers are washed in warm water and detergent, rinsed in warm water, and then left to soak in hot water for up to an hour. The fibers are then quickly transferred to the hot dyeing liquid and simmered for an hour or left overnight. The dyed material is then rinsed in warm water until no more dye diffuses out, washed again with detergent, and dried. Reaction of the dyed material to light can be tested by placing it in direct sunlight for several days. (See Appendix 3 for a number of sources of natural dyes.)

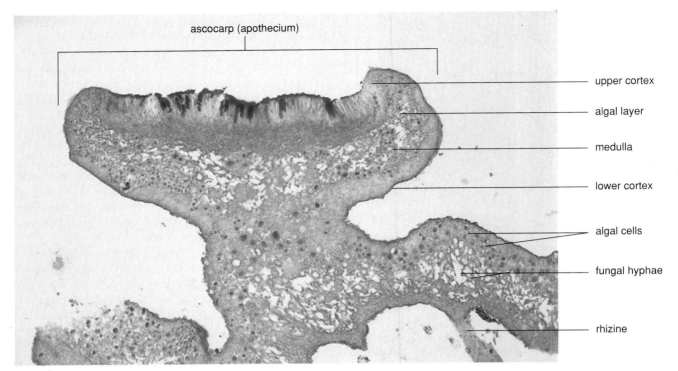

ascocarp (apothecium)

upper cortex

algal layer

medulla

lower cortex

algal cells

fungal hyphae

rhizine

FIGURE 19.32   A section through a foliose lichen.
(Photomicrograph by G.S. Ellmore)

trees and rocks at a camp where she served as a counselor. The rich colors she obtained from the lichens were even more spectacular, which is perhaps not surprising since these organisms were in the past a major source of dyes and still are used in a minor way in commercial dyeing.

Lichens have traditionally been referred to as prime examples of symbiotic relationships. Each consists of a fungus and an alga intimately associated in a spongy body called a **thallus.** The thallus can range in diameter from less than 1 millimeter to more than 2 meters (0.04 inch to 6.5 feet). The alga supplies the food for both organisms, while the fungus protects the alga from harmful light intensities, produces a substance that accelerates photosynthesis in the alga, and absorbs and retains water and minerals for both organisms. The evidence suggests, however, that it would probably be more correct to say that the fungus parasitizes the alga in a controlled fashion, actually destroying algal cells in some instances.

There are about 25,000 known species of lichens. The algal component is either a green alga or a blue-green bacterium, and a few lichens have two species of algae present. Three genera of green algae and one genus of blue-green bacterium are involved in 90% of all lichen species, and one species of alga may be found in many different lichens. Each lichen, however, has its own unique species of fungus. With the exception of about 20 tropical species of lichens that have a club fungus and one species (associated with bald cypress trees) that has a bacterial component, all lichens have members of the sac fungi for their fungal components. It is possible to isolate and culture the components separately. When this is done, however, the fungus takes on a very different compact but indefinite shape, and the algae grow faster than they do when they are part of a lichen. The fungal component is very rarely found growing independently in nature, while the algal component may do so. Lichen species, therefore, are identified according to the fungus present.

Lichens grow very slowly, at a maximum rate of 1 centimeter (0.4 inch) and a minimum rate of 0.1 millimeter (0.004 inch) per year. They are capable of living to an age of 4,500 or more years and are tolerant of environmental conditions that kill most other forms of life. They are found on bare rocks in the blazing sun or bitter cold in deserts, in both arctic and antarctic regions, on trees, and just below the permanent snow line of high mountains where nothing else will grow. One species grows completely submerged on ocean rocks. They even attach themselves to manufactured substances, such as glass, concrete, and asbestos.

Part of the reason for the wide range of tolerance of lichens is the presence in the lichens' thalli of gelatinous substances that enable them to withstand periods of rapid drying alternating with wet spells. While they are dry, their water content may drop to as low as 2% of their dry weight, and the upper part of the thallus becomes opaque enough to exclude much of the light that falls on them. In this state, most environmental extremes do not affect them at all as they temporarily become dormant and do not carry on photosynthesis.

A.

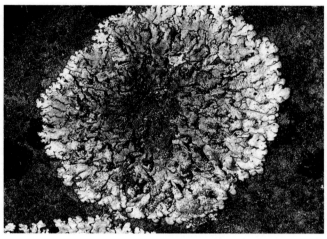

B.

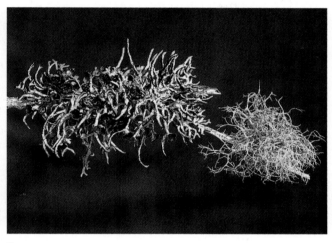

C.

FIGURE 19.33 Three types of lichen thalli. *A.* Crustose lichens on the surface of a rock. *B.* A foliose lichen. *C.* Fruticose lichens.

The lichen thallus usually consists of three or four layers of cells or hyphae (Fig. 19.32). At the surface is a protective layer constituting the *upper cortex.* The hyphae are so compressed they resemble parenchyma cells, and it is here that the gelatinous substances aid in the layer's protective function. Below the upper cortex is an *algal layer* in which algal cells are scattered among strands of hyphae. Next is a *medulla* consisting of loosely packed hyphae, which occupies at least half the volume of the thallus. A number of substances produced by the lichen are stored here. A fourth layer, called the *lower cortex,* is frequently but not always present. It resembles the upper cortex but is usually thinner and is often covered with anchoring strands of hyphae called *rhizines.*

Lichens have been loosely grouped into three major growth forms, which have no basis in their natural relationships but are convenient as categories in the first step in their identification (Fig. 19.33). *Crustose lichens* are attached to or embedded in their substrate over their entire lower surface. They often form brightly colored crusty patches on bare rocks and tree bark. The hyphae of some lichens that grow on sedimentary rocks penetrate as much as 1 centimeter (0.4 inch) into the rock. Others grow just beneath the cuticle of the leaves of tropical hardwood trees, with no apparent harm to the leaves. *Foliose lichens* have somewhat leaflike thalli, which often overlap one another. They are weakly attached to the substrate. The edges are frequently crinkly or divided into lobes. *Fruticose lichens* may resemble miniature upright shrubs, or they may hang down in festoons from branches. Their thalli, which are usually branched, are basically cylindrical in form and are attached at one point.

It should be stressed that while lichens may be attached to trees or other plants, the majority in no way parasitize them. A few, however, have been shown to partially parasitize the cortex of their hosts.

Although the fungal component of a lichen reproduces sexually, lichens are naturally dispersed in nature primarily by asexual means. In about a third of the species, small powdery clusters of hyphae and algae called *soredia* (singular: *soredium*) are formed and cut off from the thallus in a set pattern as it grows. Rain, wind, running water, and animals act as agents of dispersal. In other lichens, specialized parts of the thallus may break off or be separated by decay.

Sexual reproduction in lichens is similar to that in the sac fungi, except that in lichens the ascocarps produce spores continuously for many years. No one has yet observed the initiation of a new thallus in nature, but it is believed that they arise after ascospores carried by the wind come in contact with independently living algae, germinate, and parasitize them. Lichen algae reproduce by mitosis and simple cell division.

Human and Ecological Relevance of Lichens

One environmental factor from modern civilization has a marked effect on lichens. They are exceptionally sensitive to pollution, particularly that of sulphur dioxide. In fact, their sensitivity to this air pollutant is so great that it has been possible to calculate the amount of sulphur dioxide present in the air solely by mapping the occurrence or disappearance of certain lichens in a given area. Such studies have shown that some species of lichens have disappeared entirely from industrial areas and are now facing extinction due to air pollution. They are also very sensitive to nuclear radiation and have been used on at least one occasion in monitoring radioactive contamination when a satellite fell to earth in a remote area.

Lichens contribute to the degradation of historic ruins and other exposed rock or stone materials. Archaeological sites such as those of Troy (in modern Turkey) and Machu Picchu in Peru have suffered damage from the acids produced by lichens, and older gravestones sometimes slowly become covered by various lichen species.

Lichens provide food for many lower animals as well as large mammals. Reindeer and caribou can survive exclusively on fruticose lichens in Lapland when other food is unavailable, while North African sheep graze on a crustose lichen. By themselves, lichens do not make good food for human consumption, but they have been used as a food supplement (e.g., in soups) in parts of Europe. Most have acids that make them unpalatable, and some (e.g., rock tripe) have had harsh laxative effects on those who have tried them.

More than half of the lichens investigated have antibiotic properties. One lichen substance has been used in Europe in combination with another antibiotic in the treatment of tuberculosis. Europeans have also used lichen antibiotics to produce salves that have been effective in treating cuts and skin diseases.

Lichens were used for dyes by the Greeks and Romans, and a lichen dye industry persisted for many centuries in Europe. Native Americans and others also used lichens for dyes. Coal tar dyes have now largely replaced those of lichens, but lichens are still used in the manufacture of Scottish tweeds and East Indian cotton fabrics. Lichens are used in the preparation of the litmus paper that is used in elementary chemistry laboratories as an acid-alkaline indicator. The paper turns red under acid conditions and blue under alkaline ones.

Soaps are scented with extracts from lichens, and such extracts are still used in the manufacture of some European perfumes. Because of their resemblance to miniature trees and shrubs, some fruticose lichens are used by toy makers for the scenery of model railroads and car tracks. The importance of lichens in initiating soil formation is discussed in Chapter 25.

Summary

1. Most fungi produce threadlike hyphae. A mass of such threads is called a *mycelium.*

2. In the past, members of Kingdom Fungi were grouped with the slime molds and bacteria in a single division of the Plant Kingdom. Bacteria are now in Kingdom Monera, and the chytrids, water molds, and slime molds are in Kingdom Protoctista. Each group appears to have been independently derived.

3. Kingdom Fungi, whose members differ from those of other kingdoms in reproduction, metabolism, and lines of diversity, consists of two divisions and several classes of organisms.

4. True fungi are all filamentous or unicellular with chitinous walls; they are nonmotile and most produce spores. Sexual reproduction is varied. Fungi may thrive at temperatures below freezing or well above human comfort levels. They are major natural decomposers and cause huge economic losses through food spoilage and diseases.

5. The Division Zygomycota (coenocytic fungi)—represented by *Rhizopus*—has coenocytic mycelia that produce sporangiophores with sporangia containing numerous spores at their tips. Sexual reproduction involves gametangia produced on hyphae of opposite strains that merge, creating a single large cell, which becomes a zygospore and in which nuclei fuse in pairs. Meiosis occurs prior to the zygospore's cracking open.

6. Conidia may be produced externally on hyphae in the Zygomycota, but conidia production is more common in the other fungal divisions. *Pilobolus* releases its sporangia with force toward a light source. Bread molds are used to make tempeh and sufu and are used for various industrial purposes.

7. The hyphae of true fungi are partitioned into cells with pores in their crosswalls.

8. The Class Ascomycetes (sac fungi) includes truffles, yeasts, powdery mildews, brown fruit rots, ergot, morels, and insect parasites. Asexual reproduction is by conidia or budding. Sexual reproduction involves the formation of asci following union of male and female structures. Ascogenous hyphae may develop after the union, and these, in turn, develop into asci contained in cup-shaped or enclosed ascocarps. Ascospores are produced, following meiosis, in the asci. Upon release, each ascospore may germinate, forming a new mycelium.

9. Ergotism is a disease produced when rye infected with ergot is ingested. Yeasts produce carbon dioxide and alcohol used in baking and brewing, the drug ephedrine, and proteins and vitamins used for human and livestock consumption. Dutch elm disease, chestnut blight, and peach leaf curl are caused by ascomycetes.

10. The Class Basidiomycetes (club fungi) includes mushrooms, bracket fungi, rusts, smuts, puffballs, stinkhorns, earth stars, jelly fungi, and bird's-nest fungi. The hyphae are partitioned into cells with either one nucleus (monokaryotic) or two nuclei (dikaryotic).

11. Asexual reproduction in Basidiomycetes is usually by conidia. Sexual reproduction involves the fusion of cells of two mating types of monokaryotic hyphae. The union initiates the development of a dikaryotic mycelium that usually has clamp connections between each cell.

12. A dense dikaryotic mycelium may push above the surface, forming a button that expands into a basidiocarp or mushroom. The cap of the mushroom is usually on a stalk, which may have an annulus around it and sometimes a volva at the base.

13. Gills that radiate out from the stalk under the cap consist of hyphae that produce club-shaped basidia, in each of which fusion of two nuclei occurs, followed by meiosis. Four basidiospores usually develop on the outside of each basidium. The basidiospores may germinate and repeat the cycle.

14. Puffballs and earth stars are stalkless club fungi that may each produce trillions of spores. Shelf fungi may add a new layer of growth each year.

15. Bird's-nest fungi produce their basidiospores in "eggs" that may become attached to vegetation ingested by animals. The eggs pass through the intestinal tract of the animals unharmed.

16. Smuts infest cereals. Rusts, which may have more than one host, infest cereals and other important plants.

17. Most club fungi are not poisonous. The death angel mushroom causes about 90% of the fatalities attributed to mushroom poisoning. Thioctic acid is a partial antidote to mushroom poisoning.

18. For safety, poisonous and edible mushrooms should be distinguished from each other only by authorities.

19. Some mushrooms have been used for intoxication and hallucinatory purposes. Most mushrooms are not high in nutritional value. Only one species of mushroom is commercially cultivated in the United States.

20. Fungi and bacteria play a major role in natural recycling processes. They are also destructive of many economically important substances.

21. Members of the Class Deuteromycetes (imperfect fungi) all lack known sexual reproductive phases but otherwise resemble members of the other classes of true fungi.

22. Some imperfect fungi occurring in soil trap and digest nematodes. The class Ascomycetes includes the *Penicillium* molds, which produce penicillin, and *Aspergillus,* some species of which are used in the production of shoyu, miso, and other commercial products. *Aspergillus* and other imperfect fungi also cause diseases, such as athlete's foot and ringworm.

23. Some imperfect fungi are cultivated by ants and termites in their nests and form mycorrhizal associations with the roots of higher plants.

24. The bodies (thalli) of lichens consist of an alga and a fungus in intimate symbiotic relationship. The algal component is usually a green alga or a blue-green bacterium, which may occur in many different species of lichens. Each lichen species has its own unique sac fungus component by which it is identified.

25. Lichens grow very slowly. They can stand great temperature ranges and may live to be thousands of years old. They are exceptionally sensitive to industrial pollution.

26. The lichen thallus usually consists of several layers of hyphae, including an upper cortex, an algal layer containing algal cells, a medulla of loosely packed hyphae, and a lower cortex, which is often covered with anchoring rhizines, but which may be absent.

27. Lichens occur in three growth forms: crustose, foliose, and fruticose.

28. Lichens reproduce sexually but are naturally dispersed primarily by asexual means. Some produce soredia, which become detached and are dispersed by rain, wind, water, and animals.

29. Lichens provide food for animals, especially in cold areas. Most lichens are unpalatable to humans. Many lichens have antibiotic properties. Some lichens are used for perfume, litmus paper, and dyeing Scottish tweeds and East Indian fabrics, and as scenery for model railroads.

Review Questions

1. Why are fungi placed in a kingdom separate from protists and bacteria?
2. How do the cells of fungi differ from those of other organisms?
3. What does *coenocytic* mean? To which groups of organisms does it apply?
4. What means of asexual reproduction do the fungi exhibit?
5. If you were looking at hyphae of a bread mold, a cup fungus, and a club fungus through a microscope, could you distinguish among the three? How?
6. Which fungi produce zoospores?
7. Define *conidia, haustoria, zygospore, sporangiophore, ascus, mycelium, ascocarp, monokaryotic, dikaryotic,* and *clamp connection.*
8. Is there any one best way to tell a poisonous mushroom from an edible one?
9. What is ergotism?
10. Discuss the human relevance of the sac fungi.
11. What is a spore print?
12. To which division and/or class of fungi do each of the following belong: truffles, puffballs, fungi involved with LSD, stinkhorns, peach leaf curl, late blight of potato, bird's-nest fungi, jelly fungi, downy mildew of grape, death angel, and shelf or bracket fungi?
13. What is a fairy ring, and why does it form?
14. How do rusts and smuts attack their hosts?
15. Under what conditions do commercially produced mushrooms grow best?
16. Why are some fungi referred to as "imperfect"?
17. Discuss the economic importance of imperfect fungi.
18. What is the relationship between the alga and the fungus in a lichen?
19. What makes up the basic structure of a typical lichen?
20. How is it possible for lichens to live to such great ages?

Discussion Questions

1. Since fungi produce so many trillions of spores, why is it we are not overrun by them?
2. Why are fungal diseases much more common now in the United States than they were at the time of the Declaration of Independence?
3. If all rusts had only one host, would they (theoretically, at least) be easier to control?
4. Since most lichens are not good for food and are not very important economically, should we be as concerned about some of them becoming extinct as we are about whales becoming extinct?

Additional Reading

Ahmadjian, V. 1993. *The lichen symbiosis.* New York: John Wiley and Sons, Inc.

Alexopoulos, C. J., and C. W. Mims. 1979. *Introductory mycology,* 3d ed. New York: John Wiley and Sons, Inc.

Arora, D. K., and L. Ajello (Eds.). 1991. *Handbook of applied mycology,* vol. 2. New York: Marcel Dekker, Inc.

Batra, L.R. (Ed.). 1967. Insect-fungus symbiosis: Nutrition, mutualism and commensalism. *Scientific American* 217(5): 112–20.

Bliss, A. 1994. *North American dye plants,* rev. ed. Loveland, CO: Interweave Press.

Christensen, C. M. 1975. *Molds, mushrooms, and mycotoxins.* Minneapolis: University of Minnesota Press.

Cole, G. T., and H. C. Hoch (Eds.). 1991. *The fungal spore and disease initiation in plants and animals.* New York: Plenum Publishing Corp.

Crowder, W. 1926. Marvels of the mycetozoa. *National Geographic Magazine* 49(4): 421–44.

Hale, M. E., Jr. 1979. *How to know the lichens,* 2d ed. Dubuque, IA: Wm. C. Brown Publishers.

Ingold, C. T., and H. J. Hudson. 1993. *The biology of fungi,* 6th ed. New York: Chapman and Hall.

Lawrey, J. D., and M. E. Hale Jr. 1984. *Biology of lichenized fungi.* Westport, CT: Praeger Publications.

Littlefield, L. J. 1981. *Biology of the plant rusts: An introduction.* Ames, IA: Iowa State University Press.

Smith, A. H., and N. Weber. 1980. *The mushroom field hunter's guide,* enlarged ed. Ann Arbor: University of Michigan Press.

Chapter Outline

Moss-covered rocks in a small stream.

Introduction
to the Plant Kingdom:
Bryophytes

20

FEATURES OF THE PLANT KINGDOM

Most older botany texts and a few newer ones apply the term *plants* to many of the simpler organisms mentioned in previous chapters. Today, however, the majority of botanists confine this term to mosses, ferns, cone-bearing plants, flowering plants, and various relatives of each of these groups. These groups are the subjects of the next several chapters.

Plants and green algae, which were discussed in Chapter 18, share several major pigments (e.g., chlorophyll *a*, chlorophyll *b*, carotenoids). Plants and green algae also have in common starch as the primary food reserve and cellulose in their cell walls. When their cells divide, they develop phragmoplasts and cell plates, both of which are unknown in other organisms. These shared features suggest plants and green algae were derived from a common ancestor.

Fossils reveal that plants first appeared about 400 million years ago, and any hypothetical ancestor probably progressed from an aquatic to a land habitat even earlier. By the time plants became established on land, they had developed several features that kept them from drying out: (1) plant surfaces have a fatty cuticle that retards water loss; (2) the gametangia and spore-producing structures (sporangia) of plants are multicellular and are surrounded by a sterile cell jacket; and (3) a plant zygote develops into an **embryo** within tissues that originally surround the egg.

Plant Kingdom members are more complex and varied in form than those of other kingdoms, and their tissues are correspondingly more specialized for photosynthesis, conduction, support, anchorage, and protection. The sporophyte phases of the life cycles are predominant in the more advanced members, and reproduction is primarily sexual.

The multicellular embryos of plants are unknown in the other kingdoms. Because of this, all organisms used to be placed in either the Plant Kingdom or the Animal Kingdom. The Plant Kingdom was then divided into two subkingdoms. The monerans, protists, and fungi were put in the Subkingdom Thallophyta ("thallus plants"), and all other organisms except animals were put in the Subkingdom Embryophyta ("embryo plants").

In this text, three divisions of essentially nonvascular plants, the *bryophytes,* and several divisions of **vascular plants** (plants with xylem and phloem) are included in the Plant Kingdom. The bryophytes are discussed in this chapter, and the vascular plants are examined throughout the remaining chapters of the book.

INTRODUCTION TO THE BRYOPHYTES

In the midst of heavy battles in France during World War I, nurses at what is now referred to as a M.A.S.H. unit ran out of bandages for the wounded soldiers. In desperation, they substituted some soft green plant material they found growing in the water at the edge of a nearby lake. To their surprise, the material turned out to be a great substitute for the bandages: Fewer infections developed in the wounds with the plant bandages than in those with the cotton bandages.

The material the nurses used was a species of *Sphagnum* moss (bog or peat moss), which has since been experimentally demonstrated to have antiseptic properties. This moss, which has specialized water-absorbing "leaves" (see Fig. 20.10), has been used as a packing material in the past and is still widely used as a soil conditioner. The "bandage" *Sphagnum* is but one of about 23,000 species of **bryophytes,** which include *mosses, liverworts,* and *hornworts,* many of which frequently make a soft and cool-looking green covering on damp banks, trees, and logs that are shaded for at least part of the day (Fig. 20.1).

Some bryophytes can withstand long periods of desiccation and may be found on bare rocks in the scorching sun (Fig. 20.2), while others occur on frozen alpine slopes. The

FIGURE 20.1   Mosses and ferns growing on a tree trunk.

habitats of bryophytes range in elevation from sea level near ocean beaches to 5,486 or more meters (18,000 or more feet) in mountains.

Some are restricted to very specific habitats. For example, a few species are found only on the antlers and bones of dead reindeer. Others are confined to the dung of herbivorous animals, while still others grow only on the dung of carnivores. A few tropical bryophytes thrive only on large insect wing covers. The pygmy mosses, which appear annually on bare soil after rains, are only 1 to 2 millimeters (0.04 to 0.08 inch) tall and can complete their whole life cycle in a few weeks.

Bryophytes of all classes often have mycorrhizal fungi associated with their rhizoids. In some instances, the fungi apparently are at least partially parasitic. One species of completely colorless liverwort that lives underneath mosses is totally dependent nutritionally on its fungal associate.

The widespread peat mosses, which are ecologically very important in bogs and in the transformation of bogs to dry land, sometimes form floating mats over water and keep

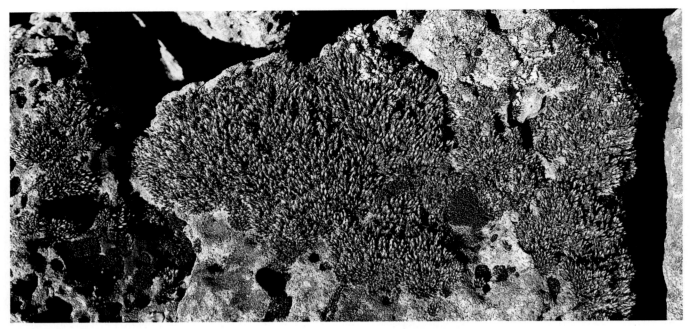

FIGURE 20.2   *Grimmia,* a rock moss that survives on bare rocks, often in scorching sun.

conditions acid enough to inhibit the growth of bacteria and fungi. Organisms that die in such waters or bogs are often preserved for hundreds or even thousands of years.

The luminous moss, which glows an eerie golden-green in reflected light, is found in caves near the entrances and in other dark, damp places. The upper surfaces of its cells are slightly curved, and each cell functions as a tiny magnifying glass, concentrating the dim light on the chloroplasts at the base. This allows photosynthesis to take place in light otherwise too faint for it to occur (Fig. 20.3).

None of the bryophytes have true xylem or phloem, and to be able to reproduce all bryophytes must have external water, usually in the form of dew or rain. Many mosses do have special water-conducting cells called *hydroids* in the centers of their stems, and a few have food-conducting cells called *leptoids* surrounding the hydroids. Neither type of cell, however, conducts as efficiently as vessel members of xylem and sieve-tube members of phloem, and most water is absorbed directly through the surface. The absence of xylem and phloem makes most bryophytes soft and pliable, and it is not surprising, therefore, that birds often use them to line their nests. In one study in the Appalachian Mountains of Virginia in 1975, for example, David Breil and Susan Moyle found birds native to the area used at least 65 species of bryophytes in the construction of their nests.

Alternation of Generations is more conspicuous in bryophytes and ferns than in most other organisms. In mosses, the "leafy" plant is a major part of the *gametophyte* generation, which produces the *gametes.* The *sporophyte* generation, which produces the *spores,* grows from the tip of

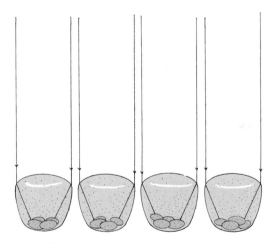

FIGURE 20.3   Lenslike cells of the protonema of the luminous moss. The cells concentrate light on the chloroplasts at their bases.

(After D. von Denffer, W. Schumacher, K. Magdefrau, and F. Ehrendorfer. 1976. Strasburger's *Textbook of Botany,* 30th German ed., translated by P. Bell and D. Coombe. 1976. Redrawn by permission of Gustav Fischer Verlag, Stuttgart, and Longman Group, London and New York.)

a "leafy" gametophyte, usually resembling a tiny, capped urn at the end of a slender stalk.

All bryophytes have similar life cycles, chromosomes, and habitats. However, based on their structure and reproduction, they are divided into three distinct divisions. None of the bryophytes appear closely related to other living plants, and fossils provide little evidence that members of each division are related to those of the other divisions. Botanists speculate that the three lines of bryophytes may have arisen independently from ancestral green algae.

DIVISION HEPATICOPHYTA— LIVERWORTS

The word *wort* simply means "plant" or "herb." In medieval times, when the Doctrine of Signatures held sway, the herbalists of the day thought that some of the bryophytes— specifically those with flattened bodies and liver-shaped lobes (Fig. 20.4)—were useful for treating liver diseases. This belief proved to be without merit, but the name **liverwort** is still universally used today.

Structure and Form

There are about 8,000 known species of liverworts. The most common and widespread liverworts have flattened, lobed, somewhat leaflike bodies called **thalli** (singular: **thallus**). The *thalloid* liverworts, however, constitute only about 20% of the species. The other 80% are "leafy" and superficially resemble mosses.

Liverworts differ from mosses in several details and are considered less complex. Their thalli or "leafy" stages (gametophytes), which develop almost directly from spores, have smoother upper surfaces with various markings and pores; the corners of cell walls of most liverworts are specially thickened. The lower surfaces have many one-celled *rhizoids*. Growth is prostrate instead of upright, and the rhizoids, which look like tiny roots, anchor the plants to the soil and other surfaces.

Thalloid Liverworts

The best-known species of thalloid liverworts are in the genus *Marchantia* (named in honor of a French botanist, N. Marchant) (Fig. 20.5). The most widespread *Marchantia* species is often found on damp soil after a fire. The thallus, which is about 30 cells thick in the center and 10 cells thick at the margin, forks dichotomously as it grows. Each branch has a notch at the apex and a central groove that extends back lengthwise behind the notch. The thalli grow as meristematic cells at the notches divide. Older tissues at the rear decay as the new growth is added. The upper surface of the thallus is divided into diamond-shaped or polygonal segments, the segment lines marking the limits of the chambers below. Each segment has a small bordered pore opening into the interior.

Seen through a microscope, a sectioned liverwort thallus looks a little like a series of covered prickly pear cactus gardens sitting on a decorative rock wall (Fig. 20.5). The "rock wall," which may comprise most of the thallus, con-

A.

B.

FIGURE 20.4 *Marchantia* (a thalloid liverwort). *A.* A thallus with male gametophores. *B.* A thallus with female gametophores.

sists of parenchyma cells that have few, if any, chloroplasts. The tissue apparently stores substances produced in other cells. The bottom layer of cells is an epidermis from which rhizoids and scales arise. The "cactus gardens" consist of

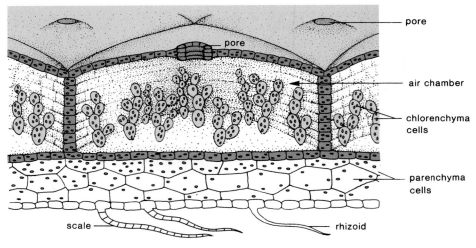

FIGURE 20.5 A section through a portion of a *Marchantia* thallus.

upright branching rows of chlorenchyma cells in an air space. Vertical walls enclose the individual "gardens," which are covered by a slightly dome-shaped layer of epidermal cells. The conspicuous pore, which remains open at all times, is located in the center of each "roof" and looks something like a short, suspended, open-ended barrel.

Marchantia—Asexual Reproduction

Marchantia reproduces asexually by means of **gemmae** (singular: **gemma**). Gemmae are tiny, lens-shaped pieces of tissue that become detached from the thallus. They are produced in small *gemmae cups* scattered over the upper surface of the liverwort gametophyte (Fig. 20.6). Raindrops may splash the gemmae as much as 1 meter (3 feet) away. While gemmae are in the cup, lunularic acid inhibits their further development, but each is capable of growing into a new thallus as soon as it is out of the cup. In addition, parts of an older thallus may die, isolating patches of active tissue, which may then continue to grow independently.

Marchantia—Sexual Reproduction

The gametangia of *Marchantia,* which are produced on separate male and female gametophytes, are more specialized than those of other liverworts. Both types of gametangia are formed on **gametophores,** which are umbrellalike structures borne on slender stalks rising from the central grooves of the thallus (Fig. 20.6). The top of the male gametophore, or **antheridiophore,** is disclike with a scalloped margin, while that of the female gametophore, or **archegoniophore,** looks like the hub and spokes of a wagon wheel.

Antheridia, which are club-shaped male gametangia containing numerous sperms, are produced in rows just beneath the upper surface of the antheridiophore. **Archegonia,** which are flasklike female gametangia each containing a single egg, are also produced in rows and hang neck downward beneath the spokes of the archegoniophore. Raindrops sometimes splash the released sperms, which have numerous flagella, more than 0.5 meter (1.5 feet) away. Fertilization may occur before the stalks of the archegoniophores have finished growing.

After fertilization, the zygote develops into a multicellular **embryo** (an immature *sporophyte*). A knoblike **foot** anchors the sporophyte (the diploid, spore-producing phase) in the tissues of the archegoniophore. The sporophyte hangs suspended by a short, thick stalk called the **seta.** The main part of the sporophyte, in which different types of tissues develop, is called a **capsule.** Liverwort sporophytes typically have no stomata.

Spore mother cells in the capsule undergo meiosis, producing haploid *spores.* Other capsule cells do not undergo meiosis but remain diploid and develop instead into long, pointed **elaters,** which have spiral thickenings and are sensitive to changes in humidity (Fig. 20.7). Spore dispersal in *Marchantia* takes place as the elaters twist and untwist rapidly. In the sporophytes of other liverworts, the elaters may aid spore dispersal with a snapping action or by suddenly expanding.

Until it is mature, the young sporophyte is protected by a caplike tissue, the **calyptra,** which grows out from the gametophyte, and by other membranes covering the capsule. The capsule splits at maturity, and air currents carry the spores away. Under favorable conditions, the spores germinate, producing new gametophytes.

Other thalloid forms, such as the floating or amphibious liverworts, do not produce gametophores. Instead, the archegonia and antheridia develop within the thallus beneath the central grooves, where the sporophytes also are formed. The spores are liberated from the submerged sporophytes as the thallus decays.

"Leafy" Liverworts

"Leafy" liverworts (Fig. 20.8), which are often abundant in tropical jungles and in fog belts, always have two rows of

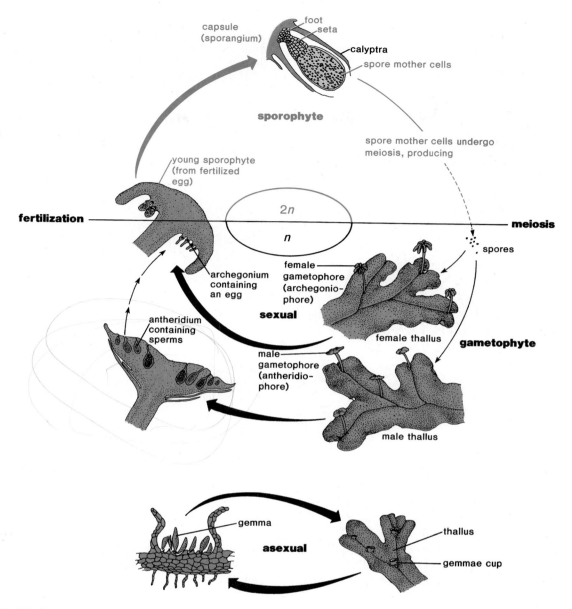

FIGURE 20.6 Life cycle of the thalloid liverwort *Marchantia*.

partially overlapping "leaves" whose cells contain distinctive oil bodies. The "leaves" have no midribs, and unlike the "leaves" of mosses, they often have folds and lobes. In the tropics, the lobes form little water pockets in which tiny animals are nearly always present. It has been suggested that these water pockets may function like the pitchers of pitcher plants. A third row of "underleaves," which are smaller than the other "leaves" and not visible from the top, often is present on the underside of "leafy" liverworts. A few rhizoids, which anchor the plants, develop from the stemlike axis at the base of the "underleaves."

The archegonia and antheridia of the "leafy" liverworts are produced in cuplike structures composed of a few modified "leaves," either in the axils of "leaves" or on separate branches. At maturity, the sporophyte capsule may be pushed out from among the "leaves" as the seta elongates. When a spore germinates, it produces a **protonema,** which consists of a short filament of photosynthetic cells. The protonema soon develops into a new gametophyte plant.

DIVISION ANTHOCEROTOPHYTA— HORNWORTS

Structure and Form

Hornworts, whose mature sporophytes look like miniature, greenish cattle horns, have gametophytes that look like

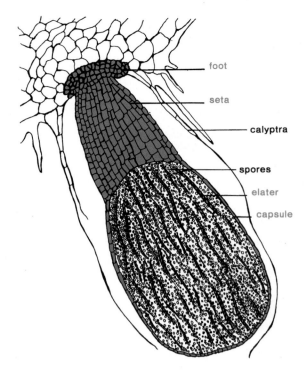

FIGURE 20.7 A longitudinal section through a sporophyte of *Marchantia*.

foot
seta
calyptra
spores
elater
capsule

FIGURE 20.8 *Frullania,* a "leafy" liverwort.

filmy versions of those of thalloid liverworts. They are usually less than 2 centimeters (0.8 inch) in diameter and thrive mostly on moist earth in shaded areas, although some occur on trees. There are about 100 species worldwide, but they are uncommon in arctic regions. They differ from liverworts and mosses in several respects and appear to be only distantly related to them. Hornworts usually have only one large chloroplast in each cell (a few species have up to eight). Each chloroplast has pyrenoids similar to those of green algae. The thalli have pores and cavities filled with mucilage, in contrast with the air-filled pores and cavities of thalloid liverworts. Nitrogen-fixing blue-green bacteria often grow in the mucilage. Rhizoids anchor the plants to the surface.

Asexual Reproduction

Hornworts reproduce asexually primarily by fragmentation or as lobes separate from the main part of the thallus. A few hornworts form tiny tubers that are capable of becoming new gametophytes.

Sexual Reproduction

In sexual reproduction, archegonia and antheridia are produced in rows just beneath the upper surfaces of the ga-

metophytes. Like both mosses and liverworts, some species of hornworts have *unisexual* plants, while other species are *bisexual*.

The distinctive sporophytes of hornworts have numerous stomata. They have no setae (stalks) and look like tiny green broom handles or horns rising through a basal sheath from a foot beneath the surface of the thallus (Fig. 20.9). A meristem above the foot continually increases the length of the sporophyte from the base when conditions are favorable. As growth occurs, spore mother cells surrounding a central rodlike axis in the sporophyte undergo meiosis, producing spores. The tip of the sporophyte horn splits into two or three ribbonlike segments, releasing the spores, and the segments continue to peel back as long as the meristem is producing new tissue at the base.

DIVISION BRYOPHYTA— MOSSES

Structure, Form, and Subclasses

Many different organisms have been called *mosses*. In fact, almost any greenish covering or growth on tree trunks and forest floors has probably been called *moss* at one time or another. Some examples are lichens (e.g., reindeer moss), red algae (e.g., Irish moss), flowering plants (e.g., Spanish moss), and club mosses. Club mosses look somewhat like large true mosses but are vascular plants with xylem and phloem. About 15,000 species of mosses are currently known. These are divided into three different classes, commonly called *peat mosses, true mosses,* and *rock mosses.* Mosses are distinct, both in form and reproduction and possibly in origin, from any other groups of organisms.

out chloroplasts), which adapt them to water absorption and storage. Small green photosynthetic cells are sandwiched between the large cells (Fig. 20.10). The "leaves," which intially are formed in three ranks, usually end up appearing to be arranged in a spiral or alternately on an axis that twists as it grows.

The axis is somewhat stemlike but has no xylem or phloem, although there is often a distinctive central strand of hydroids. At the base, rootlike rhizoids, consisting of several rows of colorless cells, anchor the plant. Some water absorbed by rhizoids rises up the central strand, but most water used by the plant apparently travels up the outside of the plant by means of capillarity. The closely packed habit of many mosses and the fact that they rarely extend more than a few centimeters (2 to 3 inches) into the air favors such outside movements of water, which is absorbed directly through the plant surfaces.

Sexual Reproduction

Sexual reproduction in mosses begins with the formation of multicellular gametangia, usually at the apices of the "leafy" shoots of gametophytes (see Fig. 20.13), although they frequently form on special separate branches. Both male and female gametangia are often produced on the same plant, but in some species they occur on separate plants.

The *archegonia,* which are the female gametangia, are somewhat cylindrical and project upward from the base of the expanded gametophyte tip (Fig. 20.11). When certain cells break down in the swollen base of the archegonium (called the **venter**), a cavity develops in which a single *egg* cell is produced. The part of the archegonium above the venter is called the *neck*. The neck, which may taper toward the tip, contains a narrow *canal*. The canal is at first plugged with cells, but these break down as the archegonium matures, leaving an opening to the outside at the top. Several archegonia usually are produced at the same time, with sterile hairlike, multicellular filaments called *paraphyses* (singular: *paraphysis*) scattered among them.

Male gametangia, which also have paraphyses among them, are sausage-shaped to roundish with walls that are one cell thick. These *antheridia* (Fig. 20.12) are borne on short stalks. A mass of tissue inside each antheridium develops into numerous coiled or comma-shaped *sperm cells*. This mass of sperms is forced out of the top of the antheridium when it absorbs water and swells. After release, the sperm mass breaks up into individual cells, each with a pair of flagella. It is believed that the breakup of the sperm mass is aided, in some cases, by fats produced by the moss, while in other instances rain splash is responsible.

Archegonia release sugars, proteins, acids, or other substances that attract the sperm, and eventually a sperm, after swimming down the neck of an archegonium, unites with the egg, forming a diploid *zygote* (Fig. 20.13). The zygote usually grows rapidly into a spindle-shaped *embryo,* which breaks down the cells at the base of the archegonium

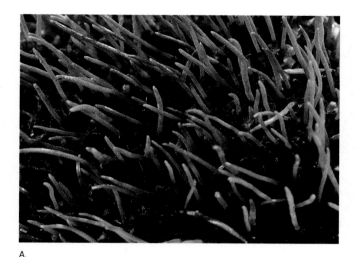

A.

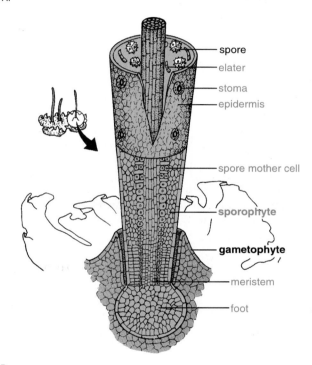

B.

FIGURE 20.9 *A.* Hornworts. *B.* Detail of a hornwort sporophyte.

The "leaves" of moss gametophytes have no mesophyll tissue, stomata, or veins like the leaves of more complex plants. Moss "leaf" cells also are all haploid. The blades are nearly always merely one cell thick, except at the *midrib,* which runs lengthwise down the middle, and they are never lobed or divided nor do they have a petiole. The midrib, which is absent in some genera, occasionally projects beyond the tip in the form of a hair or spine. The "leaf" cells usually contain numerous lens-shaped chloroplasts, except at the midrib. The "leaves" of peat mosses, however, have large transparent water-storage cells (with-

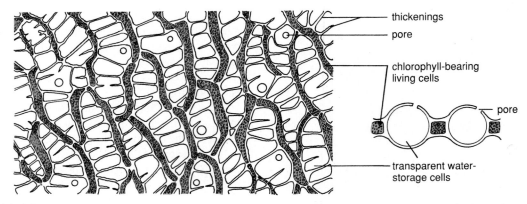

FIGURE 20.10 An enlargement of a portion of a peat moss (*Sphagnum*) leaf. *A*. Surface view. *B*. A further-enlarged cross section of living and dead cells.

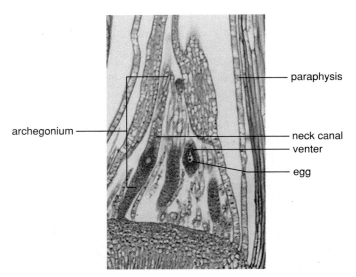

FIGURE 20.11 A longitudinal section through the tip of a female gametophyte of the moss *Mnium*.

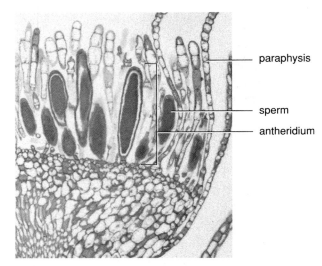

FIGURE 20.12 A longitudinal section through the tip of a male gametophyte of the moss *Mnium*.

and becomes firmly established in the tissues of the stem by means of a swollen knob called a *foot*. As the embryo grows, cells around the venter divide, thereby accommodating its increasing size. The length of the embryo soon exceeds the length of the cavity in the venter. The top of the venter is split off and is left sitting like a pixie cap on top of the embryo. By this time the embryo is a developing *sporophyte*. The pixie cap, called a *calyptra,* remains until the sporophyte is mature. In one genus with the common name of *extinguisher mosses,* the calyptra looks just like a little candlesnuffer, and in the haircap mosses it resembles a miniature, pointed, goatskin cap such as might be worn by a Shakespearean actor.

The cells of the sporophyte become photosynthetic as it develops, remaining so until maturity. The sporophyte, however, depends to varying degrees on the gametophyte for some of its carbohydrate needs as well as for at least a part of its water and minerals, which are absorbed through the foot.

The mature sporophyte, which at first is green and photosynthetic, consists of a *capsule* located at the tip of a slender stalk, the *seta*. Depending on the species, the seta may be less than 1 millimeter (0.04 inch) long, or it may be up to 15 centimeters (6 inches) long. Most, however, are less than 5 centimeters (2 inches) long. The capsule, which may resemble a tiny apple, a pear, an urn, a box, or a wingtip fuel

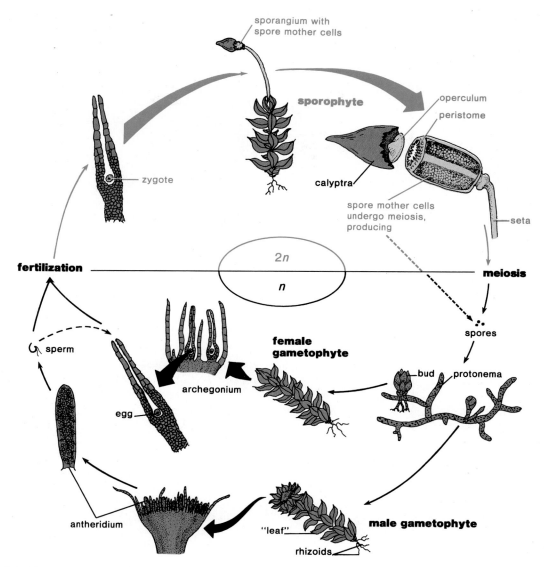

FIGURE 20.13 Life cycle of a moss.

tank of an airplane, usually has from 3 or 4 to over 200 stomata at or near its surface. Unless extremely dry conditions prevail, the stomata normally remain open until the capsule begins to age, and then they close permanently. The free end of the capsule is usually protected by a little rimmed lid, the **operculum,** which falls off at maturity.

As the capsule matures, *spore mother cells* inside it undergo meiosis, producing haploid *spores*. These spores, often numbering in the millions, are released from the free end of the capsule after the operculum has fallen off, usually through a structure called a **peristome.** Most peristomes consist of a circular row, or often two rows, of narrowly triangular and membranous teeth arranged around the rim of the capsule, each row having 16 teeth. The teeth are frequently colored orange or red and are often beautifully sculptured with bars and fringes. They open or close in re-

sponse to changes in humidity. In a few species of mosses, the peristome is a cone-shaped structure with pores through which the spores are released.

Some rock mosses have neither a peristome nor an operculum. The spores in these mosses are released when the capsule splits longitudinally along four lines. In the dung mosses, a putrid odor is given off when the spores are ready for release. Some of the spores adhere to the legs and bodies of flies, which are attracted by the odor, and are disseminated as the insects clean themselves. Most moss spores are, however, simply blown away by the wind, and if they fall in a suitable damp location, they usually germinate relatively quickly.

In most mosses, fine green tubular threads consisting of single rows of cells with chloroplasts first emerge from the spores. These soon branch and grow, forming an algalike *protonema*. The protonema can be distinguished from a

Hibernating Mosses

Mosses are the "amphibians" of the plant world, so-called because they are at home in either semiaquatic environments like moist stream banks or drier habitats such as rock surfaces or the Arctic tundra. Unlike their towering cousins the flowering plants, mosses are dependent on a watery landscape to reproduce. Because moss sperm are flagellated, a film of water is needed for the sperm to swim to the egg. But life on land is not easy, especially when water becomes limited or nonexistent for months. The genetic diversity of mosses is exceptional and they inhabit some of the most inhospitable habitats on earth. Dry heaths, rock faces, tree trunks, and even deserts are home to these remarkable plants. Distinctive adaptations allow moss species this wide range of habitats, including the ability to tolerate drying out, a process known as **desiccation.**

Vascular plants have structural mechanisms that maintain an adequate water supply within plant tissues. These include an impermeable waxy cuticle on leaf surfaces, an internal water transport system (xylem), water-absorbing organs (roots), and leaf pores (stomata) that can close in order to conserve water. Parts of the life cycle such as seeds are especially tolerant to desiccation. Mosses do not possess these adaptations to living in a dry land environment. This means that the internal water balance in mosses is in equilibrium with the atmosphere. When the air is dry mosses are dry. When it rains mosses

Polytrichum commune (hairy cap moss) growing in a deciduous forest of the northeastern United States. (Photo by Daniel Scheirer)

quickly absorb water and become rehydrated. They are opportunists in this regard.

Not all species of mosses are capable of withstanding this desiccation-rehydration cycle, but those that have this ability can "return from the dead" with each cycle. It is as if they have been hibernating. There are several mechanisms that slow water loss from mosses that make the cycle less drastic. Many mosses form dense mats or tufts that create a moist microatmosphere within the tufts and over the plants. *Polytrichum commune* is a common moss that has moderate

---

filamentous green alga by the oblique crosswalls of its cells and by the lens-shaped chloroplasts. If light and other conditions are favorable, tiny "leafy" buds appear at intervals along the protonemal filaments after about two to four weeks of growth. These "leafy" buds develop rhizoids at the base and grow into new "leafy" gametophytes, completing the cycle.

Asexual Reproduction

Some reproduction in mosses does not depend on such a sexual cycle and its Alternation of Generations. It has been demonstrated under laboratory conditions that cells of archegonia and antheridia, paraphyses, leaves, stems, and rhizoids all can develop protonemata. Two American biologists once collected bryophyte gametophyte fragments from a snowbed in the Canadian high arctic and found that 12% of their samples resumed growth in various ways when cul-

tured in the laboratory. They calculated that for each cubic meter (1.3 cubic yards) of snow at their study site there were over 4,000 bryophyte fragments capable of becoming new plants. They suggested that wind dispersal of fragments may be routine in arctic regions. Such dispersal and vegetative reproduction also occur widely in more temperate areas.

HUMAN AND ECOLOGICAL RELEVANCE OF BRYOPHYTES

Some bryophytes and lichens are pioneers on bare rock after volcanic eruptions or other geological upheavals and after the retreat of glaciers. They slowly accumulate mineral and organic matter, which can then be inhabited or utilized by other organisms. This process, called *succession,* is discussed

desiccation tolerance. It lives in habitats such as bogs or temperate, moist forests and can grow up to 40 cm tall. It grows luxuriantly during rainy periods but then twists into rusty red mats upon drying in the sun. These mosses have a thin, waxy cuticle covering their tiny "leaves" (8–10 mm in length) that retards water loss. They also have primitive water conducting cells (hydroids) that move some water through the plant when water is lost to the atmosphere by evaporation. Instead of root hairs, they have an extensive rhizoid system that absorbs some water from the soil. However, if dry conditions persist, all available water is lost to the atmosphere.

Additional mechanisms by which mosses are able to survive these desiccation-rehydration cycles are being discovered. *Tortula ruralis* is one of the most desiccation-tolerant mosses known, and research has centered on its ability to recover after prolonged or repeated desiccation events. Hydrated *T. ruralis* cells are similar to mesophyll cells in the leaves of vascular plants. There are chloroplasts with stacks of grana and a prominent nucleus. Mitochondria are numerous and similar in size and shape to those of higher plants, with internal membranes folded into cristae.

During desiccation, *T. ruralis* dries out, and its "leaves" fold up around the stem. Cells of desiccated plants are damaged. There is extensive plasmolysis as water is lost from the protoplast. Chloroplasts become smaller and more spherical and starch is not present. Internal thylakoid membranes are collapsed and disorganized. Mitochondria in the hydrated cell are elongated with numerous cristae membranes. In the desiccated state, mitochondria are smaller and rounded with few internal cristae membranes. The nucleoplasm of nuclei becomes dense and chromatin is condensed. Nucleoli are prominent with a compact appearance. The plasma membrane is damaged and electrolytes leak out. In all desiccated mosses, photosynthesis stops and respiration slows or ceases.

When the plants are rehydrated, dried "leaves" unfold and return to the normal hydrated state in two minutes. Most of the internal damage is repaired within a few more minutes. Activity of a drought-repair gene increases in the minutes after rehydration, and repair proteins are quickly manufactured and mobilized into action. Respiration resumes after a few minutes. Photosynthesis resumes up to 24 hours later.

Scientists at the USDA's Agricultural Research Service are attempting to locate the genes responsible for drought damage repair in *Tortula* species in the hope of transferring them to crop plants. With crops having more drought-tolerance, rain shortage would be less of a problem than it is now. Arid lands currently unsuited for agriculture could blossom with crop plants genetically engineered with drought-tolerant genes. World population increases of 1.6% yearly means that an additional 78,000 metric tons of grain per day are required just to maintain current consumption levels. A lowly moss might some day be responsible for a revolution in food production.

in Chapter 25. Mosses, in particular, retain moisture, slowly releasing it into the soil. They reduce flooding and erosion and contribute to humus formation. Some mosses grow only in soils that are rich in calcium; the presence of others indicates higher than usual soil salinity or acidity.

When certain mosses are present in a dry area, it is a good indication that running water occurs there during a part of the year. A few mosses are occasionally a problem in water reservoirs, where they may plug entrances to pipes. A few bryophytes are reported to be grazed, along with lichens, by foraging mammals in arctic regions, but bryophytes are not generally edible. Some mosses have been used for packing dishes and stuffing furniture, and Native Americans are reported to have used mosses under splints when setting broken limbs.

By far, the most important bryophytes to humans are the peat mosses. When allowed to absorb water, 1 kilogram (2.2 pounds) of dry peat moss will take up 25 kilograms (55 pounds) of water. Its extraordinary absorptive capacity has made it very useful as a soil conditioner in nurseries and as a component of potting mixtures. Live shellfish and other organisms are shipped in it, and its natural acidity, which inhibits bacterial and fungal growth, gives it antiseptic properties. The antiseptic feature, combined with its absorbency, which is greater than that of cotton, has made it a useful poultice material for application to wounds. It was used for this purpose during the Crimean War of 1854 to 1856 and, as indicated in the chapter introduction, on an emergency basis during World War I. Extensive peat deposits have been formed from the remains of peat mosses that flourished in past eras. Peat, like the undecomposed peat mosses, is used around the world as a soil conditioner and as a fuel. In the manufacture of Scotch whiskey, sprouted barley is dried on a screen over a peat fire. The peat smoke permeates the barley and imparts a smoky flavor to the beverage. See Appendix 1 for the scientific names of all the bryophytes discussed.

Summary

1. Members of the Plant Kingdom have a cuticle and produce their gametes and spores in multicellular organs surrounded by a sterile jacket of protective cells. Their zygotes develop into embryos, and tissues specialized for photosynthesis, conduction, support, anchorage, protection, and reproduction are produced. Sporophyte phases of the life cycles are predominant in more advanced members of the Plant Kingdom.

2. When plant cells divide, cell plates and phragmoplasts, which outside of the Plant Kingdom occur only in certain green algae, are produced. The similarity in pigments, food reserve (starch), and occurrence of cell plates suggests a common ancestor for the green algae and plants. The Plant Kingdom includes three divisions of bryophytes and several divisions of vascular plants.

3. Bryophytes (liverworts, hornworts, mosses) occur in highly varied and also very specific habitats.

4. Water is essential to bryophyte reproduction. Most water is absorbed directly through the plant surfaces.

5. Liverwort gametophytes with flattened, dichotomously forking thalli are common, but about 80% of the liverwort species are "leafy." Liverworts have distinct upper and lower surfaces, with one-celled rhizoids that function in anchorage on the lower surface.

6. A *Marchantia* thallus has a central lengthwise groove along its upper surface and is chambered, each chamber containing chlorenchyma cells and having a surface pore. Rhizoids and scales arise from the thallus base.

7. *Marchantia* reproduces asexually by means of gemmae produced in surface cups and by thallus fragmentation.

8. *Marchantia* reproduces sexually by means of eggs and sperms produced in archegonia and antheridia on archegoniophores and antheridiophores that arise from the thallus.

9. The zygote develops into a sporophyte that is anchored to the archegoniophore by a foot, from which is suspended a capsule connected to the foot by a seta. Spore mother cells in the capsule undergo meiosis, producing spores. Diploid elaters, which aid in spore dispersal, do not undergo meiosis.

10. The calyptra and other membranes protect the spores until they are released as the capsule splits; the spores may then develop into new gametophytes.

11. "Leafy" liverworts have two rows of overlapping "leaves" and, frequently, a third row of "underleaves" not visible from above. The "leaves" often have lobes that retain rainwater.

12. Hornworts, which have one chloroplast with pyrenoids in each cell, resemble liverworts in their gametophytes, but their sporophytes are hornlike and have a meristem above the foot. Hornwort thalli have pores and cavities filled with mucilage in which blue-green bacteria often grow.

13. Asexual reproduction in hornworts is by fragmentation. Sexual reproduction involves archegonia and antheridia produced in rows beneath the upper surface of a thallus. The tip of the hornlike sporophyte splits vertically, releasing the spores.

14. A moss gametophyte consists of an axis to which "leaves" are attached, with rhizoids at the base. The "leaves" are haploid and have no mesophyll, stomata, or veins. Water is absorbed primarily directly through the plant surfaces.

15. Multicellular archegonia and antheridia are produced at the tips of "leafy" shoots. Each archegonium has a cavity, the venter, containing a single egg and a neck through which a sperm gains access to the egg. Sperms are produced in antheridia.

16. After fertilization the zygote grows into an embryo that is attached to the gametophyte by an embedded foot. The sporophyte developing from the embryo consists of a capsule and a seta. A calyptra derived from the gametophyte partially covers the capsule. Spore mother cells in the capsule undergo meiosis, producing spores that are released through the teeth of the peristome, a structure at the tip of the capsule.

17. An operculum that falls off when the spores mature initially protects the peristome. When moss spores germinate, protonemata with "leafy" buds develop. The buds grow into new gametophytes.

18. Mosses may be pioneers, along with lichens, on bare rocks. They are indicators of soil calcium, salinity, and acidity. Mosses are not generally edible, although a few are grazed in arctic regions. Some mosses are used for packing material, but the most significant use is that of peat mosses for soil conditioners. Peat mosses can absorb and retain large amounts of water, and their natural acidity gives them antiseptic properties. Peat deposits, from peat mosses that flourished in past eras, are used for fuel and also as a soil conditioner.

Review Questions

1. What basic features distinguish members of the Plant Kingdom from those of other kingdoms?

2. What features distinguish the bryophytes discussed in this chapter from other plants?

3. How could you tell a hornwort thallus from that of a thalloid liverwort?

4. Contrast the sporophytes of mosses, liverworts, and hornworts.

5. What is a protonema? Do all bryophytes have them? How would you tell a protonema from a green alga?

6. What adaptations do bryophytes have for their particular habitats?

7. Which parts of the life cycles of bryophytes have haploid (*n*) cells? Which parts have diploid (*2n*) cells?

8. Why is a bryophyte "leaf" technically not the same as a flowering plant leaf?

9. Define *calyptra, operculum, capsule, peristome, paraphysis, foot, seta, archegoniophore, thallus, underleaves,* and *elaters.*

10. If you were to single out one bryophyte as being the most important member of its division from a human viewpoint, which would you choose? Why?

Discussion Questions

1. Very few fossils of bryophytes have been found. Suggest reasons for this.

2. Do the multicellular sex structures of plants give them any advantages over other organisms with unicellular sex structures?

3. After reading about the characteristics and uses of peat mosses, can you suggest some possible new uses for these plants?

4. Some bryophytes produce unisexual gametophytes, while others produce bisexual gametophytes. Should one type have any survival or adaptive advantage over the other? Explain.

Additional Reading

Bates, J. W., and A. M. Farmer (Eds.). 1992. *Bryophytes and lichens in a changing environment.* New York: Oxford University Press.

Chopra, R. N., and P. K. Kumra. 1989. *Biology of bryophytes.* New York: Halsted Press.

Conard, H. S., and P. L. Redfearn, Jr. 1979. *How to know the mosses and liverworts,* 2d ed. Pictured Key Nature Series. Dubuque, IA: Wm. C. Brown Publishers.

Crum, H., and L. E. Anderson. 1981. *Mosses of eastern North America.* New York: Columbia University Press.

Dyer, A. F., and J. G. Duckett (Eds.). 1984. *The experimental biology of bryophytes.* San Diego, CA: Academic Press.

Glime, J. M., and D. Saxzena. 1991. *Uses of bryophytes.* Houston, TX: Scholarly Publications.

Scagel, R. F., et al. 1984. *Plants: An evolutionary survey.* Belmont, CA: Wadsworth Publishing Co.

Schultze-Motel, W. (Ed.). 1984. *Advances in bryology,* vol. 2. Forestburgh, NY: Lubrecht & Cramer.

Schuster, R. M. 1966–1974. *The Hepaticae and Anthocerotae of North America,* 4 vols. New York: Columbia University Press.

Zinsmeister, H. D., and R. Mues. 1990. *Bryophytes: Their chemistry and chemical taxonomy.* New York: Oxford University Press.

'Ama'uma'u, (Sadleria cyatheoides) *a small Hawaiian tree fern.*

21

Introduction
to Vascular Plants: Ferns
and Their Relatives

This chapter opens with a brief review of the features that distinguish the major groups of vascular plants without seeds from one another and from the bryophytes, and then it discusses representatives of each division. Included in the discussion are whisk ferns (Psilotum), club mosses (Lycopodium, Selaginella), quillworts (Isoetes), horsetails (Equisetum), and ferns. A digest of the human and ecological relevance of each group is given, and life cycles of representatives are illustrated. The chapter concludes with an examination of fossils, and each type is briefly described. A table showing the geologic time scale is provided.

Some Learning Goals

1. Know the basic structural differences between bryophytes and vascular plants.
2. Distinguish the four divisions of seedless vascular plants from one another.
3. Understand the differences in the life cycles of ground pines (*Lycopodium*) and spike mosses (*Selaginella*).
4. Learn the structural features of horsetail (*Equisetum*) sporophytes.
5. Know how to recognize and explain the functions of all the structures involved in alternation of generations in a fern.
6. Learn 10 important uses of vascular plants that do not produce seeds.
7. Explain what a fossil is and distinguish among the various types of fossils.

O ur survey of all organisms traditionally regarded as plants thus far has taken us from simple one-celled prokaryotic organisms through more specialized eukaryotic protists and fungi and on to the bryophytes. We now take up several divisions of plants whose members have internal conducting tissues but do not produce seeds.

In the introduction to the Plant Kingdom in the preceding chapter, we mentioned that the ancestors of the bryophytes and the vascular plants probably were multicellular green algae, which became established on the land over 400 million years ago. As green organisms moved ashore, they developed several new features, including sterile jackets of cells around gametangia, embryos within protective tissues, and a cuticle. All these features at least partly protected vital parts from drying out. Bryophytes had no significant internal tissues for conducting water, and most water needed internally was absorbed directly through the aboveground parts. External water was required for reproduction.

During the early stages of vascular plant evolution, internal conducting tissues (xylem and phloem) began to develop, true leaves appeared, and roots that function in absorption as well as anchorage developed. At the same time, gametophytes became progressively smaller and more

dependent on and were protected by proportionately larger sporophytes.

Unlike conifers and flowering plants, the primitive vascular plants discussed in this chapter do not produce seeds. They include the ferns and a number of their relatives often referred to as *fern allies.* Four divisions of seedless vascular plants are recognized:

1. **Division Psilotophyta** (*whisk ferns*). Features unique to members of this division of vascular plants include sporophytes that have neither true leaves nor roots and stems and rhizomes that fork evenly.

2. **Division Lycophyta** (*club mosses* and *quillworts*). The stems of these plants are covered with **microphylls** (leaves with a single vein whose trace is not associated with a leaf gap—see Chapter 7). Most microphylls are photosynthetic.

3. **Division Sphenophyta** (*horsetails* and *scouring rushes*). The sporophytes of these plants have ribbed stems containing silica deposits and whorled scale-like microphylls that lack chlorophyll.

4. **Division Pterophyta** (*ferns*). The sporophytes of ferns have **megaphylls** (leaves with more than one vein and a leaf trace that is associated with a leaf gap—see Chapter 7) that are often large and much divided.

Fossils, discussed at the end of the chapter, give us clues to the ancestry of some of these plants and also to the origins of seed plants discussed in the chapters to follow.

DIVISION PSILOTOPHYTA— THE WHISK FERNS

There is nothing very fernlike in the appearance of **whisk ferns,** commonly referred to by their scientific name of *Psilotum,* but they do loosely resemble small, green whisk brooms. Exactly where these plants fit in the Plant Kingdom is not clear. Traditionally, they have been associated with a number of extinct plants called *psilophytes,* which flourished perhaps 400 million years ago, but there isn't much fossil evidence to substantiate the link. Because whisk fern and true fern gametophytes share several features, some botanists classify them with the true ferns. They will be discussed first because, by most present criteria, they are among the simplest of all living vascular plants.

Structure and Form

Psilotum sporophytes consist almost entirely of dichotomously forking (evenly forking) stems and are unique among living vascular plants in having neither leaves nor roots. They usually grow up to 30 centimeters (1 foot) tall but occasionally grow as much as 1 meter (3 feet) or more tall (Fig. 21.1). The visible stems arise from short branching rhizomes just beneath the surface of the ground.

FIGURE 21.1 *Psilotum* sporophytes. The small yellowish objects are clusters of sporangia.

Enations, which are tiny green flaps of tissue without veins, are spirally arranged along the stems. Photosynthesis takes place in the outer cells of the stem (epidermis and cortex), which contains a central cylinder of xylem surrounded by phloem. The xylem is star-shaped in cross section. The rhizomes, which have short rhizoids over their surfaces, function as roots with the aid of mycorrhizal fungi, some of which establish themselves in cortical cells just beneath the epidermis.

Reproduction

Very short stubby branches are produced toward the tips of the angular stems. The branches terminate in small sporangia, which are fused together in threes and resemble miniature yellow pumpkins. After release, spores germinate slowly in the soil, on the "bark" of tree ferns, or in similar habitats. In Hawaii, they also grow in old lava flows.

The gametophytes, which develop from the spores beneath the soil surface, are colorless and easily overlooked. They sometimes resemble tiny, transparent dog bones and are only about 2 millimeters (0.08 inch) wide and seldom more than 6 millimeters (0.25 inch) long. They are cylindrical and may fork dichotomously like the stems. They have no chlorophyll and are *saprobic,* obtaining their nutrients with the aid of fungi. Archegonia and antheridia develop randomly over the surface of the same gametophyte. After a sperm unites with an egg in an archegonium, the zygote develops a foot and a rhizome. As soon as the rhizome establishes itself, upright stems are produced and the rhizome separates from the foot (Fig. 21.2).

Whisk ferns are native to tropical and subtropical regions. In the United States, they are found in Florida, Louisiana, Texas, Arizona, and Hawaii. They are extensively cultivated in Japan and in greenhouses around the world. A close relative, *Tmesipteris,* with leaflike appendages, occurs in Australia and the South Pacific.

Fossil Whisk Fern Look-Alikes

Fossil plants that somewhat resemble whisk ferns have been found in Silurian geological formations (Table 21.1). These formations are estimated to be as much as 400 million years old. One group of these fossil plants, of which *Rhynia* is an example (see Fig. 21.22), had naked stems and spindle-shaped terminal sporangia. A second group of fossils, represented by *Zosterophyllum,* had somewhat rounded sporangia produced along the upper parts of naked stems. *Zosterophyllum* and its relatives first appeared during the Devonian period. They are thought to be ancestral to the club mosses, discussed in the next section. Whisk ferns are of little economic importance. Their spores, which have a slightly oily feel, were once used by Hawaiian men to reduce loincloth irritations of the skin. Hawaiians also made a laxative liquid by boiling whisk ferns in water.

DIVISION LYCOPHYTA— THE CLUB MOSSES AND QUILLWORTS

When I was in high school, my father was interviewed by a newspaper reporter in a hotel room of a large South American city. At the end of the interview, which took place with the rest of the family present, the reporter set up a press camera and said he wanted to take a picture of the family. After posing us, he took a vial from his pocket and poured a little powder along a metal bar attached to one end of a T-shaped device that otherwise looked like a flashlight. He then told us to smile, and a moment later, there was a flash of light followed by a large billow of smoke. I learned later that the flash powder (forerunner of flashbulbs and strobe lights) he had used contained millions of spores of primitive vascular plants called **club mosses.**

The 950 or so known species of club mosses look enough like large true mosses that the great Swedish botanist Linnaeus lumped both together in a single class. Once details of the structure and the form of club mosses were

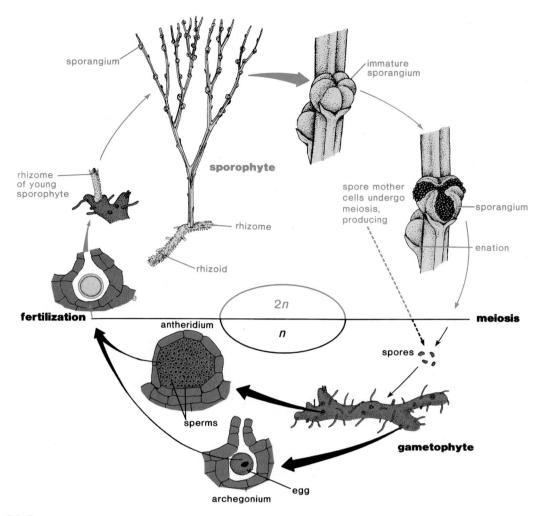

FIGURE 21.2 Life cycle of *Psilotum*.

known, however, it became obvious that they are quite unrelated to true mosses.

Today, there are living representatives of two major and two minor genera of club mosses. Several others became extinct about 270 million years ago. The sporophytes of all have leaves called *microphylls,* which are usually (but not always) quite small; they have true stems and true roots.

The two major genera of club mosses with living members are *Lycopodium,* with about 50 species, and *Selaginella,* with over 700 species. They are distributed throughout the world, but they are more abundant in the tropics and wetter temperate areas.

Lycopodium—Ground Pines

Structure and Form

Lycopodium plants, which often grow on forest floors, are sometimes called *ground pines,* partly because they resemble little Christmas trees, complete with "cones" that are usually

upright or, in a few species, hang down (Fig. 21.3). The stems of ground pine sporophytes are either simple or branched. The plants are mostly less than 30 centimeters (1 foot) tall, although some tropical species grow to heights of 1.5 meters (5 feet) or more. The upright, or sometimes pendent, stems develop from branching rhizomes. The leaves may be whorled or in a tight spiral and are rarely more than 1 centimeter (0.4 inch) long. Adventitious roots, whose epidermal cells often produce root hairs, develop along the rhizomes.

Reproduction

At maturity, some species of ground pines produce kidney bean-shaped sporangia on short stalks in the axils of several of the leaves. Such sporangium-bearing leaves are called **sporophylls.** In other species, the sporophylls have no chlorophyll, are smaller than the other leaves, and are in terminal conelike clusters called **strobili** (singular: **strobilus**). In the sporangia, *spore mother cells* undergo meiosis, producing spores, which are released and carried away by air currents. The spores of some species germinate in a few days

Table 21.1

GEOLOGIC TIME SCALE

ERA	PERIOD	AGE	EARLIEST EVIDENCE OF PLANTS	DURATION (MILLIONS OF YEARS)	MILLION YEARS BEFORE PRESENT
Cenozoic	Quaternary	Age of modern seed plants		63	
	Tertiary				65
Mesozoic	Cretaceous			65	130
	Jurassic		grasses flowering plants	45	175
	Triassic	Age of ancient seed plants		45	220
Paleozoic	Permian		ginkgoes conifers	50	270
	Carboniferous			80	
			seed ferns		350
	Devonian	Age of spore-bearing plants	lycopods horsetails bryophytes	50	
	Silurian			40	400
	Ordovician	Age of bacteria and algae		60	440
	Cambrian		marine algae	100	500
	Precambrian			1,700?	600 / 2,300?

---

if they land in a suitable location, but spores of other species may not germinate for up to several years.

After germination, independent gametophytes develop from the spores. The gametophytes vary in shape, some resembling tiny carrots. The gametophyte body usually develops in the ground in association with mycorrhizal fungi. Some gametophytes, however, develop primarily on the surface where the exposed parts turn green. All types produce both antheridia and archegonia on the same gametophyte, which, in some species, may live for several years.

Zygotes first become embryos with a foot, stem, and leaves and then develop into mature sporophytes (Fig. 21.4). If the gametophyte is underground, chlorophyll does not develop in the young sporophyte until it emerges into the light.

Several sporophytes may be produced from a single gametophyte. A number of ground pines also reproduce asexually by means of small *bulbils* (bulbs produced in the axils of leaves), each of which is capable of developing into a new sporophyte.

Selaginella—Spike Mosses

Structure and Form
The sporophytes of *Selaginella,* the larger of the two major genera of living club mosses, are sometimes called *spike mosses* (Fig. 21.5). The approximately 700 species are widely scattered around the world in wetter areas, but they

A.   B.

FIGURE 21.3   *A. Lycopodium cernum,* a club moss native to the tropics. *B. Lycopodium clavatum,* native to northern and eastern North America.

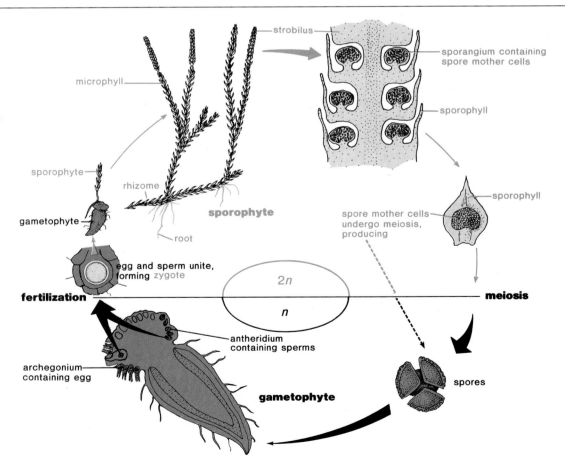

FIGURE 21.4   Life cycle of the ground pine *Lycopodium.*

FIGURE 21.5   *Selaginella,* a spike moss.

are especially abundant in the tropics. A few are common weeds in greenhouses. They tend to branch more freely than ground pines, from which they differ in several respects. The two most obvious differences are (1) the leaves of *Se-laginella* each have a tiny extra appendage, or tongue, called a **ligule,** on the upper surface near the base, and (2) they produce two different kinds of spores and gametophytes—an advanced feature referred to as **heterospory.** The seed bearing coniferous and flowering plants discussed in chapters to follow are all heterosporous.

Reproduction

Sexual reproduction of spike mosses and sexual reproduction of ground pines both involve the production of sporangia, but spike moss sporangia develop on either *microsporophylls* or *megasporophylls* (Fig. 21.6). **Microsporophylls** bear *microsporangia* containing numerous **microspore mother cells,** which undergo meiosis, producing tiny **microspores.** The *megasporangia* of megasporophylls usually contain a **megaspore mother cell,** which, after meiosis, becomes four comparatively large **megaspores.**

Each microspore may become a male gametophyte, which consists simply of a somewhat spherical antheridium surrounded by a sterile jacket of cells within the microspore wall. Either 128 or 256 sperm cells with flagella are produced in each antheridium. A megaspore develops into a female gametophyte, which is also relatively simple in structure. By the time this gametophyte is mature, however, it consists of many cells that have been produced inside the megaspore. As it increases in size, it eventually ruptures its thickened spore wall and produces several archegonia in the exposed seams. The development of both male and female gametophytes often begins before the spores are released from their sporangia. Fertilization and development of new sporophytes are similar to those of ground pines.

Isoetes—Quillworts

Structure and Form

There are about 60 species of *quillworts* (all in the genus *Isoetes*) (Fig. 21.7). Most are found in areas where they are at least partially submerged in water for part of the year. Their leaves (microphylls), which are slightly spoon-shaped at the base, look like green porcupine quills, although they are not stiff and rigid. They are arranged in a tight spiral on a stubby stem, which resembles the corm of a gladiola or a crocus. *Ligules* occur toward the leaf bases. The corms have a vascular cambium and may live for many years. Wading birds and muskrats often eat the corms. The plants are generally less than 10 centimeters (4 inches) tall, but the leaves of one species become 0.6 meter (2 feet) long.

Reproduction

Reproduction is similar to that of spike mosses, with both types of sporangia being produced at the bases of the leaves (Fig. 21.8). Up to 1 million microspores may occur in a single microsporangium.

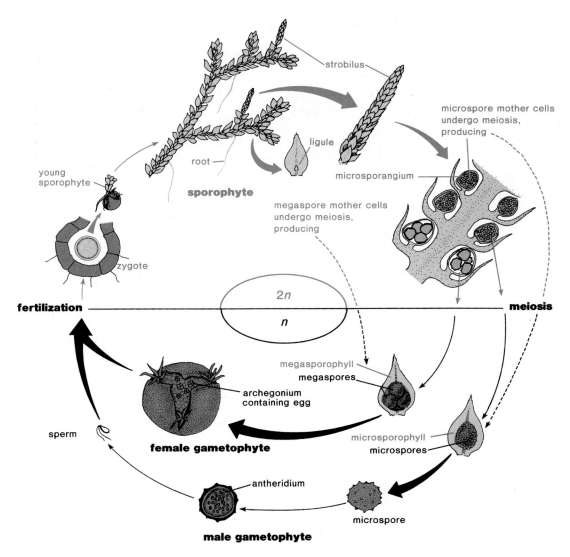

FIGURE 21.6   Life cycle of the spike moss *Selaginella*.

Ancient Relatives of Club Mosses and Quillworts

Some of the ancient and extinct relatives of the club mosses were large and treelike and grew up to 30 meters (100 feet) tall; their trunks were up to 1 meter (3 feet 3 inches) in diameter. They were dominant members of the forests and swamps of the Carboniferous period, which reached its peak some 325 million years ago (see Table 21.1). Quillworts first appeared in the Cretaceous period, about 130 million years ago. Their fossils reveal that even the oldest were remarkably similar to their present-day relatives (Fig. 21.9). A knowledge of the life histories and ancestry of various members of this division enhances our understanding of the development and diversity of the Plant Kingdom as a whole.

Human and Ecological Relevance of Club Mosses and Quillworts

Like the whisk ferns, club mosses and quillworts are of little economic importance today. As mentioned earlier, large numbers of club moss spores produce a flash of light when ignited. This characteristic was exploited at one time in the manufacture of theatrical explosives and photographic flashlight powders. In the past, druggists mixed spore powder with pills and tablets to prevent them from sticking to one another. The spore powder itself has been used for centuries in folk

FIGURE 21.7 Quillwort (*Isoetes*) sporophytes.
(Courtesy Robert A. Schlising)

medicine, particularly for the treatment of urinary disorders and stomach upsets. Some Native Americans used it as a talcum powder for babies, snuffed it to arrest nosebleeds, and applied it following childbirth to staunch hemorrhaging. It was also sometimes used to stop bleeding from wounds.

Club moss extracts have been used in several countries in the past to reduce fevers, but partly because of undesirable side effects, medicinal use has now largely been abandoned. Native Americans of Washington, Oregon, and British Columbia became mildly intoxicated after chewing parts of one local species of club moss. It is reported that they became unconscious if they chewed too much.

Novelty stores sometimes sell a species of spike moss native to Mexico and to the southwestern United States. Known as the *resurrection plant,* it shrivels and rolls up in a ball when dry, appearing to be completely dead, but it quickly unfolds and turns green when sprinkled with water. Other spike mosses have been placed on shelves indoors without water for nearly three years and then have resumed growth when given water. Ground pines and spike mosses have been used ornamentally indoors and outdoors as ground covers. Some ground pines are spray painted and used as Christmas ornaments or in floral wreaths. Several species of *Lycopodium* have been exploited to the extent that they are now on rare and endangered or threatened species lists; they should no longer be collected.

Quillwort corms have been eaten by domestic and wild animals, waterfowl, and humans.

DIVISION SPHENOPHYTA— THE HORSETAILS AND SCOURING RUSHES

Many backpackers and campers have become aware of a unique use of a relatively common and widespread genus of plants (*Equisetum*) known as *horsetails* or *scouring rushes*. Significant deposits of silica accumulate on the inner walls of the epidermal cells of the stems, which make excellent scouring material for dirty metal pots and pans. Native Americans were aware of this scouring property of the plants and used them extensively.

Structure and Form

About 25 species of horsetails (the name usually applied to branching forms, which look a little like a horse's tail) and scouring rushes (unbranched forms) (Fig. 21.10) are scattered throughout all continents including Australia, where they are weeds. They usually grow to less than 1.3 meters (4 feet) tall, but some in the tropics and coastal redwood forests

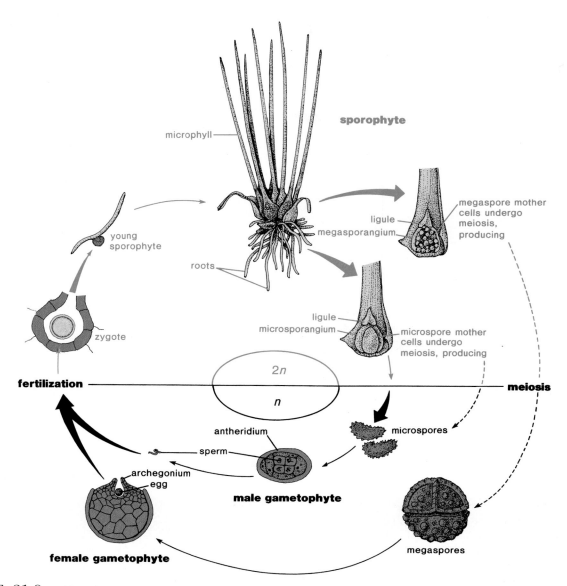

FIGURE 21.8 Life cycle of a quillwort (*Isoetes*).

of California exceed 4.6 meters (15 feet) in height. Where branches occur, they are normally in whorls at regular intervals along the jointed stems. Both branched and unbranched species have tiny scalelike leaves (microphylls) in whorls at the nodes. These leaves are fused together at their bases, forming a collar. They are green when they first appear, but they soon wither and bleach, and virtually all photosynthesis occurs in the stems.

The stems are distinctly ribbed and have obvious nodes and internodes; there are numerous stomata in the grooves between the ribs. A cross section of a stem (Fig. 21.11) reveals that the pith breaks down at maturity, leaving a hollow central canal. There are two cylinders of smaller canals outside of the pith. The inner cylinder consists of water-conducting *carinal canals* that are aligned opposite the ribs of the stem. Each canal has a patch of xylem and phloem to the outside. The canals of the outer cylinder, called *vallecular canals,* contain air. They are larger than the carinal canals and are aligned opposite the "valleys" between the ribs.

The aerial stems develop from horizontal rhizomes, which also have regular nodes, internodes, and ribs. In some species, the rhizomes have adventitious roots and may form extensive branching systems as far as 2 meters (6.5 feet) below the surface of the ground. Both internal stem structure and external features help distinguish the various species of horsetails from one another.

FIGURE 21.9 A small portion of the surface of the fossil lycopod, *Lepidodendron,* showing leaf (microphyll) bases similar to those seen in modern lycopods.

(Specimen courtesy University of Illinois Paleobotany Laboratory)

Reproduction

If stems or rhizomes are broken up by a disturbance, such as a storm or foraging animals, the fragments can grow into new sporophyte plants. In most species, however, reproduction usually involves a sexual process. In the spring, some species produce from rhizomes special cream- to buff-colored nonphotosynthetic stems (Fig. 21.12). Small cone-like **strobili** develop at the tips of these special stems or in other species at the tips of regular photosynthetic stems.

The strobili are usually about 2 to 4 centimeters (0.75 to 1.50 inches) long. Hexagonal, dovetailing plates at the surface of the strobilus give it the appearance of an ellipsoidal honeycomb. Each hexagon marks the top of a sporangiophore, which has five to ten elongate sporangia connected to the rim. The stalks of the sporangiophores are attached to the central axis of the strobilus, and the sporangia, which surround the sporangiophore stalks, point inward. These hidden sporangia are not visible until maturity when the sporangiophores separate slightly. The spores are then released.

When the spore mother cells in the sporangia undergo meiosis, distinctive-appearing green spores are produced. At one pole, the spores have four ribbonlike appendages that are slightly expanded at the tips (Fig. 21.13). The appendages are called **elaters** (not related to the elaters of liverworts). The elaters, which are very sensitive to changes in humidity, aid spores in their dispersal. While spores are being carried by an air current, the elaters are more or less extended like wings. If a spore enters a humid air pocket

A.   B.

FIGURE 21.10 Horsetails (*Equisetum*). *A.* An unbranched species. *B.* A branched species.

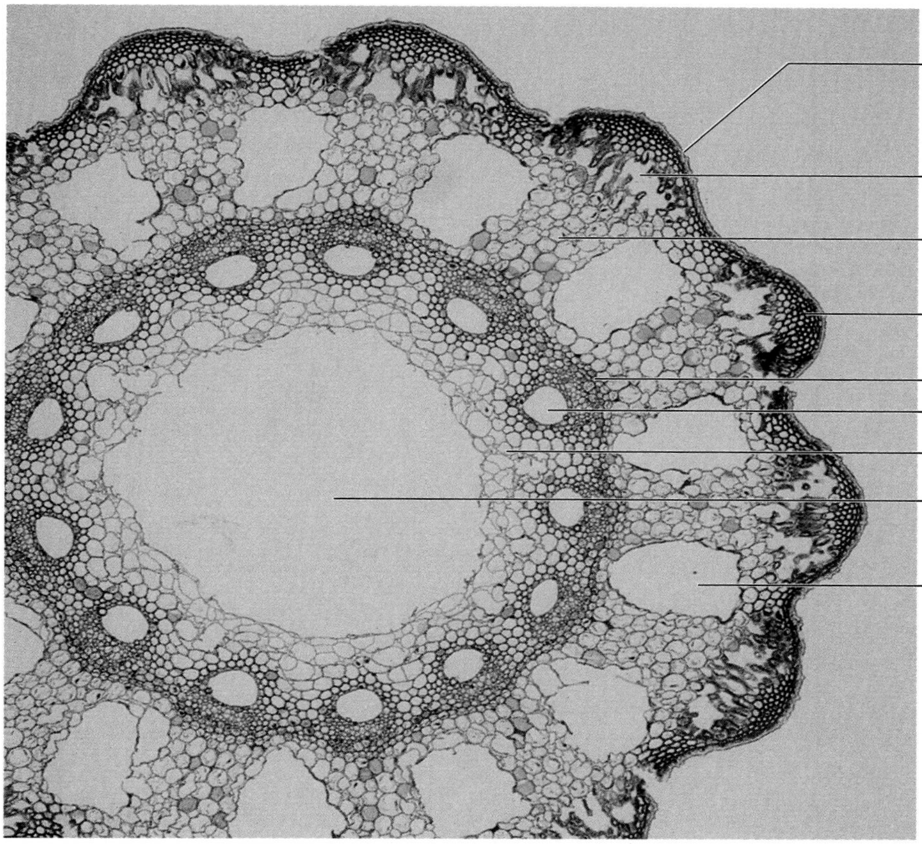
epidermis (silica is deposited mostly on the inner walls of epidermal cells)
chlorenchyma
cortex
sclerenchyma
endodermis
carinal canal
pith
central canal
vallecular canal

FIGURE 21.11 A horsetail (*Equisetum*) stem in cross section. (Photomicrograph by G. S. Ellmore)

above a damp area below, the elaters coil, causing the spore to drop in an area that is more likely to support germination and growth.

After germination, which usually occurs within a week, spores produce lobed, cushionlike, green gametophytes that seldom grow to more than 8 millimeters (0.36 inch) in diameter. Rhizoids anchor them to the surface. At first, about half of the gametophytes are male with antheridia, and the other half are female with archegonia. After a month or two, however, the female gametophytes of most species become bisexual, producing only antheridia from then on. When water contacts mature antheridia, sudden changes in water pressure cause the sperms produced within to be explosively ejected. Several eggs on a female or bisexual gametophyte may be fertilized, and the development of more than one sporophyte is common.

Ancient Relatives of Horsetails

Horsetails and scouring rushes belong to the single remaining genus (*Equisetum*) of a division of several different orders that flourished in the Carboniferous period about 300 million years ago. At that time, some members of this division, like the ancient club mosses, were treelike (Fig. 21.14) and grew over 15 meters (50 feet) tall, while others were vinelike. Some were similar to present-day horsetails in having leaves in whorls (Fig. 21.15), jointed stems with internal canals, and sporangiophores with sporangia hanging down from the rims.

Human and Ecological Relevance of Horsetails and Scouring Rushes

The seventh-century Romans ate the young strobili of field horsetails boiled or fried after mixing them with flour. Native Americans peeled off the tough epidermis of young stems and ate the inner parts either raw or boiled, while others cooked the rhizomes of the giant horsetail for food. Cows, goats, muskrats, bears, and geese also have been known to eat horsetails. Hopi Indians ground dried horsetail stems to flour, which they mixed with cornmeal to make bread or mush. Horsetails have, however, occasionally been reported to be poisonous to horses, and they are not recommended for human consumption.

Native Americans and Asians had several other uses for these plants. One tribe drank the water from the carinal canals of the stem, and another thought the shoots were "good for the blood." It is not known if there was any basis for this latter idea, but there is an old unconfirmed report that field horsetail consumption "produces a decided increase in

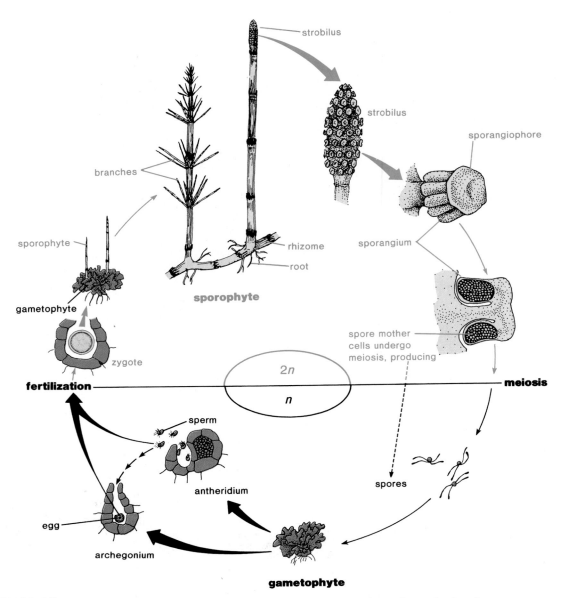

strobilus

strobilus

sporangiophore

branches

sporophyte

rhizome

root

sporophyte

sporangium

gametophyte

spore mother
cells undergo
meiosis, producing

zygote

$2n$

n

fertilization ——————————————————————— **meiosis**

sperm

antheridium

spores

egg

archegonium

gametophyte

FIGURE 21.12   Life cycle of a horsetail (*Equisetum*) that has separate vegetative and reproductive shoots.

blood corpuscles." At least one or two species are known to have a mild diuretic effect (a *diuretic* is a substance that increases the flow of urine), and they have been used in the past in folk medicinal treatment of urinary and bladder disorders. Some have also been used as an antacid or an *astringent* (an astringent is a substance that arrests discharges, particularly of blood). One species was used in the treatment of gonorrhea, and others were used for tuberculosis.

At least two Native American tribes burned the stems and used the ashes to alleviate sore mouths or applied the ashes to severe burns. Members of another tribe ate the strobili of a widespread scouring rush to cure diarrhea, and still others boiled stems in water to make a shampoo for the control of lice, fleas, and mites.

At one time, the use of scouring rush stems for scouring and sharpening was widespread. They were used not only for cleaning pots and pans, but also for polishing brass, hardwood furniture and flooring, and for honing mussel shells to a fine edge. Scouring rushes are still in limited use for these purposes today. Some species of horsetails accumulate certain minerals in addition to silica. Veins of such minerals have been located beneath populations of horsetails by analyzing the plants' mineral contents. This process of analysis involves a chemical treatment of the tissues followed by the use of X-ray equipment.

In the geological past, the giant horsetails and club mosses were a significant part of the vegetation growing in vast swampy areas. In some instances, the swamps were stagnant and slowly sinking, permitting the gradual accumulation of plant remains, which, because of the lack of oxygen in the water, were not readily attacked by decay bacteria. Such circumstances, over aeons of time, were ideal for the

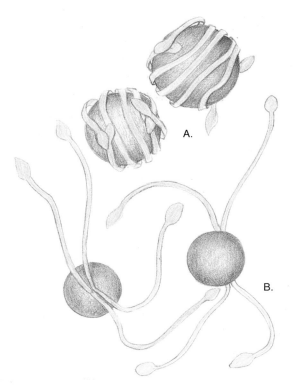

FIGURE 21.15 Reconstruction of the fossil giant horsetail, *Calamites*.

FIGURE 21.13 Horsetail spores. *A*. With elaters coiled. *B*. With elaters spread.

FIGURE 21.14 Reconstruction of a coal age (Carboniferous) forest.

(Photo by Field Museum of Natural History, Chicago)

formation of coal. Today, sectioning a lump of coal thinly enough and examining it under the microscope yields views of bits of tissue and spores of plants that were living hundreds of millions of years ago. One soft coal known as *cannel* consists primarily of the spores of giant horsetails and club mosses that, through the ages, were reduced to carbon.

For the scientific names of species discussed, see Appendix 1.

DIVISION PTEROPHYTA— THE FERNS

When I was a small child, my family had a large potted fern on each side of a fireplace. I became quite attached to the plants and, believing I was removing a "disease," I carefully scraped off the little brownish patches that appeared from time to time on the lower surfaces of the leaves. It wasn't until I went to college that I learned that instead of controlling a disease I had inadvertently been frustrating the sex life of my favorite plants!

I'm sure that if we could take a worldwide opinion poll about ornamental plants, ferns would rank high in popularity. In fact, in some parts of the world it is difficult to find a household without at least one fern either inside or out in the garden. Their leaves display such an infinite and aesthetically pleasing variety of form that Thoreau was once moved to state, "God made ferns to show what He could do with leaves."

Structure and Form

The approximately 11,000 known species of ferns vary in size from tiny floating forms less than 1 centimeter (0.4 inch) in diameter to giant tropical tree ferns up to 25 meters (82 feet) tall. Although fern leaves, commonly referred to as **fronds,** are typically dissected and feathery in appearance (Fig. 21.16), some are undivided, pleated, or tonguelike, and others resemble four-leaf clovers or grow in such a way as to form "nests." In the tropics, the "nest" ferns often accumulate enough humus to provide food and shelter for huge earthworms that are up to 0.6 meter (2 feet) long. Since ferns require external water for their sexual reproduction, they are most abundant in wetter tropical and temperate habitats, but a few are adapted to drier areas.

Reproduction

The sporophyte is the conspicuous phase of the life cycle in all the plants discussed in this chapter (Fig. 21.17). A fern sporophyte consists of the fronds, a stem in the form of a rhizome, and adventitious roots that arise along the rhizome. The fronds, regardless of their ultimate form, usually first appear tightly coiled at their tips. These *croziers,* or *fiddleheads* (Fig. 21.18), then unroll and expand, revealing the blades. At maturity, the blades are often divided into segments called **pinnae** (singular: **pinna**), which are attached to

a midrib, or rachis. A stalk, or petiole, is usually present at the base.

When the fronds have expanded, small, often circular, rust-colored patches of powdery looking material may appear on the lower surfaces of some or all of the blades. Because these patches appear similar to fungal rusts, some fern owners have thought that their plants were diseased and have turned to sprays to deal with the "problem" or have even carefully scraped the patches off. The development of the patches is normal and healthy, however, and examination with a hand lens or dissecting microscope will reveal that they are actually clusters of *sporangia* (Fig. 21.19). The sporangia may be scattered evenly over the lower surfaces of the fronds, but they are often confined to the margins.

The sporangia are mostly found in numerous discrete clusters called **sori** (singular: **sorus**). In many ferns, the sori, while they are developing, are protected by thin individual flaps of colorless tissue called **indusia** (singular: **indusium**). As the sporangia mature, the indusium, which often resembles a tiny semitransparent umbrella attached by its base to the frond surface, shrivels and exposes the sporangia beneath.

Most of the sporangia, which are microscopic and stalked, look something like tiny transparent baby rattles with a conspicuous row of heavy-walled brownish cells along the edge. This row of cells, which looks like a tiny millipede, is called an **annulus.** It functions in catapulting spores out of the sporangium with a distinct snapping action influenced by moisture changes in the cells (Fig. 21.20).

Spore mother cells undergo meiosis in the sporangia, usually producing either 48 or 64 spores per sporangium. Sporangia of some of the primitive adder's tongue ferns may have up to 15,000 spores, however, and the number of sporangia is often so great that it has been estimated that a single beech fern plant will produce a total of 50 million spores.

In certain aquatic or amphibious ferns, two kinds of spores are produced. Single large megaspores are produced in some sporangia, while numerous tiny microspores are produced in others. The vast majority of fern species, however, produce only one kind of spore.

After the spores have been flung out of their sporangia, they are dispersed by wind; relatively few end up in habitats suitable for their survival. Such habitats include shady, wet ledges and rock crevices or moist soil. The spores can also easily be germinated on damp clay flowerpots in the home or greenhouse. Those that germinate in favorable locations produce little Irish valentines, . . . or **prothalli** (singular: **prothallus**). . . as the green heart-shaped gametophytes of these and other seedless vascular plants are more properly called (Fig. 21.21). These structures, which often curl slightly at their edges, may be 5 to 6 millimeters (0.25 inch) in diameter and are visible without a microscope.

Prothalli are only one cell thick, except toward the middle where they are slightly thicker. Antheridia are interspersed among the rhizoids produced on the lower surface of the central area. Archegonia are also produced, usually closer to the notch of the heart-shaped gametophyte. The

FIGURE 21.16 Common ferns with typically dissected fronds. *A.* Cinnamon fern. The cinnamon-colored fertile fronds in the center are nonphotosynthetic and produce large numbers of sporangia. *B.* Holly fern. *C.* A maidenhair fern. *D.* 'Ama'uma'u, a small Hawaiian tree fern.

(*A, B* Courtesy Perry J. Reynolds)

archegonia are somewhat flask-shaped, with curving necks that protrude slightly above the surface, while the antheridia are more spherical and often elevated above the surface on short stalks. A single antheridium may produce from 32 to several hundred sperms, each with few to many flagella.

Only one zygote develops into a young sporophyte on any prothallus, regardless of the number of eggs that may be fertilized. This sporophyte usually has smaller, simpler fronds during its first season of growth, but typical full-sized fronds grow from the persisting rhizomes in succeeding years.

Fossil Relatives of Ferns

Fossil remains of ferns and plants thought to be ancestors of ferns abound in ancient deposits. The ancestors of ferns, sometimes referred to as *preferns* (Fig. 21.22) are found in Devonian formations (see Table 21.1) estimated to be 375 million years old. Most resemble ferns in habit but they have no broad fronds and otherwise are more like whisk ferns than true ferns. Well-preserved fossils of tree ferns related to large tropical ferns of the present day are found in geological strata dating back to between 250 and 320 million years ago (Fig. 21.23).

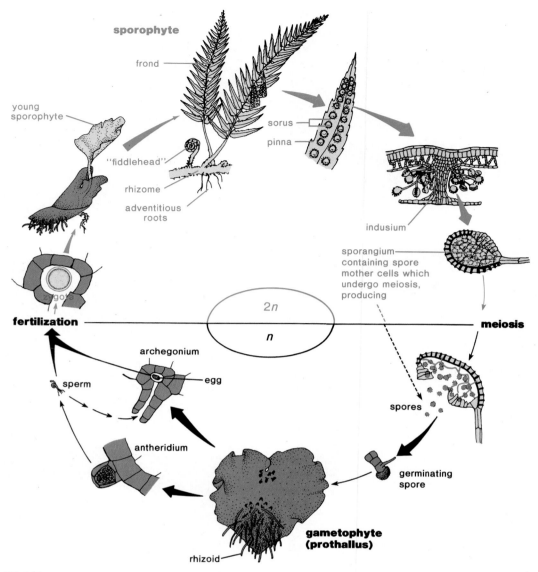

sporophyte

frond

young
sporophyte

sorus

pinna

"fiddlehead"

rhizome

adventitious
roots

indusium

sporangium
containing spore
mother cells which
undergo meiosis,
producing

zygote

fertilization

2*n*

n

meiosis

archegonium

egg

sperm

spores

antheridium

germinating
spore

**gametophyte
(prothallus)**

rhizoid

FIGURE 21.17   Life cycle of a fern.

FIGURE 21.18   A crozier (fiddlehead) of a tropical fern.

Human and Ecological Relevance of Ferns

In some parts of North America, it is unusual to find a house without at least a Boston fern growing within. Ferns make ideal houseplants because many of them are adapted to growing in low light, and most are not as susceptible as other plants to aphids, mites, mealybugs, and similar pests. Outdoors, ferns are equally popular as ornamentals. Some eventually become subjects for an artist's brush or a natural history photographer's camera.

Apart from the pleasing aesthetic aspects of ferns, they function well as air filters. For example, one average-sized Boston fern can, during a single hour, remove about 1,800 micrograms of formaldehyde (a common pollutant

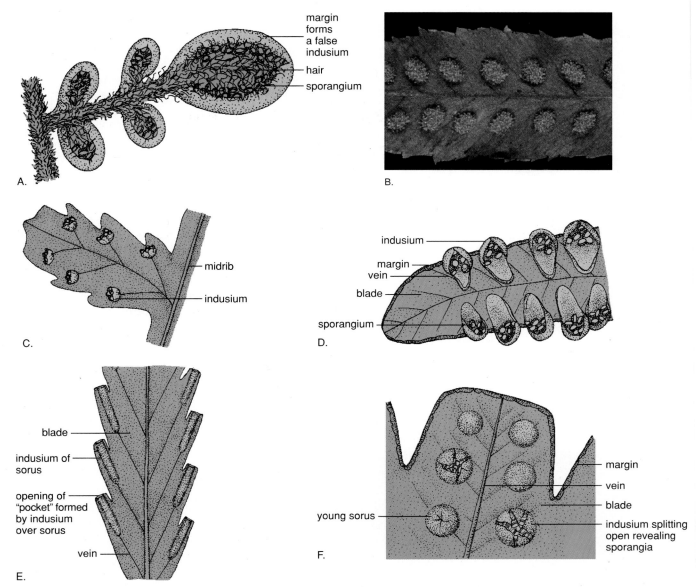

A.

margin
forms
a false
indusium

hair

sporangium

B.

C.

midrib

indusium

D.

indusium

margin
vein

blade

sporangium

E.

blade

indusium of
sorus

opening of
"pocket" formed
by indusium
over sorus

vein

F.

margin

vein

blade

indusium splitting
open revealing
sporangia

young sorus

FIGURE 21.19   Pinnae of fern fronds, showing some types of arrangements of sporangia. *A. Cheilanthes. B. Polypodium. C. Cystopteris. D. Cibotium. E. Davallia. F. Cyathea.*

from carpets) from the air in a typical room measuring 3 meters by 3 meters (10 feet by 10 feet). According to the Environmental Protection Agency, the fern, if properly maintained, can on a daily basis almost completely rid the air of pollutants.

Ferns also have other practical value. Commercial growers of the brilliantly colored anthuriums in Hawaii and elsewhere have found that native tree ferns provide the ideal amount of shade and other environmental conditions needed for bringing their flowers to perfection. Tree fern rhizome or root bark[1] and rhizome bark of certain other species, such as

the royal fern (which produces osmunda bark), are a favorite medium of orchid, bromeliad, and staghorn fern growers. Its texture is well suited to the growth of the orchids' aerial roots, and as the bark slowly decomposes, rainwater trickling over its surface picks up nutrients that are particularly appropriate for these plants. The demand for fern bark for orchids has exceeded the supply for a number of years, and it has become very expensive on the market (Fig. 21.24).

As the young *fiddleheads* of many species of ferns unroll, a dense covering of hairs is visible on the petiole and rachis. In the past, the silky hairs of some of the larger tropical tree ferns (Fig. 21.25) were stripped and used for upholstery, pillow, and mattress stuffing. During the late 1800s, over 1,900 metric tons (2,094 tons) of this material was shipped from Hawaii to the mainland, and if it were

1. Ferns do not have a cork cambium and do not, therefore, produce the bark that is typical of woody seed plants. The cells of the outer layers of fern roots and rhizomes, however, do becomes impregnated with suberin, thus becoming barklike in appearance and function.

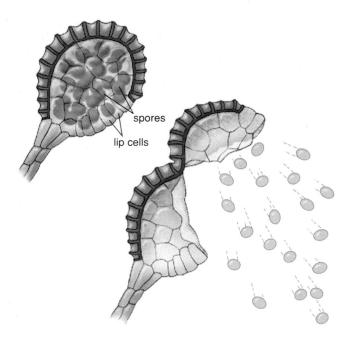

FIGURE 21.20   Release of spores from a fern sporangium. *A.* An intact sporangium. *B.* Spores being ejected as the sporangium splits; the annulus first draws back and then snaps forward.

not for the eventual substitution of alternative materials, these magnificent plants might have been totally destroyed. Some tropical hummingbirds use these hairs along with scales of other ferns to line their nests, and at one time, Polynesians used them in a form of embalming for their dead. The trunks of tree ferns have been used in the construction of small houses in the tropics. Parts of one Hawaiian species of small tree fern and the fronds of an Asian fern yield red pigments used for dyeing cloth.

The bracken fern, which is distributed worldwide, has been used and even cultivated for human food for many years, particularly in Japan and New Zealand, even though it has also long been known to be mildly poisonous to livestock. Recent research in both Europe and Japan has shown conclusively that bracken fronds fed to experimental animals produce intestinal tumors, and because of this the consumption of these fronds for food should be actively discouraged.

Indigenous peoples of many areas where ferns occur have eaten the cooked rhizomes and young fronds of various ferns. Native Americans often baked the rhizomes of sword ferns, lady ferns, and others in stone-lined pits, removed the outer layers, and ate the starchy inner material. Similarly, native Hawaiians ate the starchy core of their tree ferns as emergency food. In Asia, the oriental water fern is still sometimes grown for food in rice paddies and used as a raw or cooked vegetable. In Malaysia, a relative of the lady fern is frequently used as a vegetable.

A.

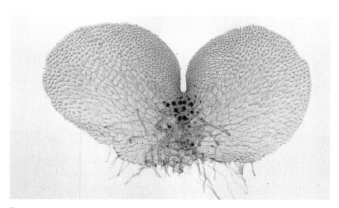

B.

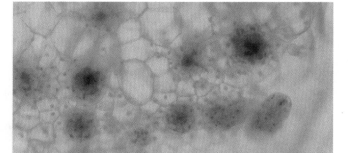

C.

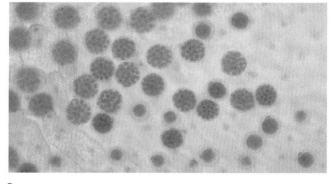

D.

FIGURE 21.21   Fern prothalli. *A.* Surface view, × 10. *B.* A prothallus as seen with the aid of a microscope, × 20. *C.* Archegonia, × 100. *D.* Antheridia × 100.

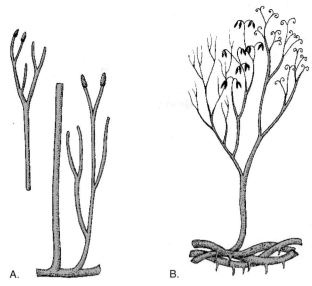

FIGURE 21.22 Ancient preferns. *A*. Rhynia.
B. Psilophyton.

FIGURE 21.23 A fossil fern.

FIGURE 21.24 An orchid plant growing on bark.

Uses of ferns in folk medicine abound. They have been used in the treatment of diarrhea, dysentery, rickets, diabetes, fevers, eye diseases, burns, wounds, eczema and other skin problems, leprosy, coughs, stings and insect bites; as a poison antidote; and for labor pains, constipation, dandruff, and a host of other maladies. The male fern, which is more common in Europe than the United States, contains a drug that is effective in expelling intestinal worms (e.g., tapeworms). Its use for this purpose dates back to ancient times, and it is still occasionally so used, although synthetic medicines have now largely replaced it. The licorice fern, which was used by Native Americans of the Pacific Northwest in the treatment of sore throats and coughs, was also used as a flavoring agent and a sugar substitute. Members of a tribe in California chewed stalks of goldback ferns to quell toothaches and snuffed a liquid made from the fronds of the bird's-foot fern to arrest nosebleeds.

The fronds of bracken and other ferns have been used in the past for thatching houses. Anyone who has placed such fronds in compost piles knows that they break down much more slowly than the leaves of other plants. Bracken fronds are still occasionally used as an overnight bedding base by fishermen and hunters. A substance extracted from these fronds has been used in the preparation of chamois leather, and the rhizome is used in northern Europe in the brewing of ale.

The chain fern, which has large fronds up to 2 meters (6.5 feet) long, has two flexible leathery strands in the petiole and rachis of each frond. Native Americans and others have gathered the fronds for many years to strip these strands for use in basketry and weaving. They do so by gently cracking the long axis with stones to expose the strands, which are then easily removed. The glossy black petioles of the five-finger fern have also been used in intricate basketry patterns by Native Americans. In Southeast Asia, the climbing fern has fronds with a rachis that may grow up to 12 meters (40 feet) long; it is still a favorite material (when available) for the weaving of baskets.

The mosquito fern, which is a floating water fern in the genus *Azolla*, forms tiny plants little bigger than duckweeds. It is found over wide areas where the climate is relatively mild. It sometimes forms such dense floating mats that it is believed to suffocate mosquito larvae, which periodically

A.

B.

FIGURE 21.25 *A*. A Brazilian tree fern. *B*. The growing tip of a tree fern, showing the protective rust-colored hairs.

need to reach the surface for air. This same fern frequently has blue-green bacteria living symbiotically in cavities between cells. The blue-green bacteria fix nitrogen, and it has been shown experimentally that plants without these organisms do not grow as well as those with them.

Many ferns are now commercially propagated through tissue culture (discussed in Chapter 14). The scientific names of ferns mentioned in this chapter are given in Appendix 1.

FOSSILS

Introduction

Several references to fossils have been made in this and preceding chapters, and other references occur in the chapters that follow. We have become increasingly aware of worldwide limitations of energy sources, and we now frequently hear references to *fossil fuels*. Exactly what is a fossil, and how did it come into existence?

Originally, the word *fossil* was applied to anything unusual found in rocks, but its use is now more or less confined to any recognizable prehistoric organic object (or its impres-

sion) that has been preserved from past geological ages in the earth's crust (see Table 21.1). Such a definition includes teeth marks, borings, impressions of footprints and tracks, dung deposits, and deposits of chemicals, which are evidence of the activities of algae and bacteria. Age itself does not make a fossil. Human remains thousands of years old that have been found in caves and tombs are not regarded as fossils, but it is conceivable that other remains of the same age could be preserved in rocks and regarded as fossils.

Fossils are formed in a number of different ways. The conditions for their formation almost always include the accumulation of sediments in an area where plants, animals, or other organisms are present or to which they have been transported. Such areas are found in swamps, oceans, lakes, or other bodies of water. Hard parts such as wood or bones are more likely to be preserved than soft parts. Other environmental factors, such as the presence of salt or antiseptic chemicals or stagnation resulting in low oxygen content of the water, also favor fossil formation.

Quick burial is another factor enhancing the chances of organic material becoming fossilized. Such burial might result when an organism is unable to avoid a rain of ash from

FIGURE 21.26   A compression fossil.

a volcanic eruption or is trapped in a cave or sudden sandstorm. Other forms of quick burial occur when an organism falls into quicksand, a body of water, or asphalt. Descriptions of several common types of fossils follow.

Molds, Casts, Compressions, and Imprints

After silt or other sediment has buried an object and hardened into rock, the organic material may be slowly washed away by water percolating through the pores of the rock. This leaves a space, which may then be filled in with silica deposits. If only a space is left, it is called a *mold;* if it is filled in with silica, it is called a *cast.* Artificial casts of plaster or wax compounds may be made from molds of the original objects. When objects such as leaves are buried by layers of sediment, the sheer weight of the overlying material may compress them to as little as 5% of their original thickness. When this occurs, all that is left is a thin film of organic material and an outline showing some surface details. Virtually no preservation of cells or other internal structure takes place. Such fossils, which are very common, are called *compressions* (Fig. 21.26). The image of a compression, like the details of a foot or a hand pressed into wet cement, is called an *imprint.* Coal is a special type of compression, often involving different plant parts that are thought to have been subjected to enormous pressures after the fallen plants slowly accumulated in a swamp.

Petrifactions

Petrifactions (Fig. 21.27) are uncompressed rocklike materials in which the original cell structure has been preserved. About 20 different mineral substances, including silica and

FIGURE 21.27   Petrified wood.
(Courtesy Sharon Stern)

the salts of several metals, are known to bring about petrifaction. At one time, it was believed that the process occurred through the replacement of the plant parts by minerals in solution, one molecule at a time. It is now believed that chemicals in solution infiltrate the cells and cell walls, where they crystallize and harden, permanently preserving the original material.

Petrifactions can be studied by cutting thin sections with a diamond or carborundum saw and polishing the material with extremely fine grit powder until it is thin enough for light to pass through. Then it can be examined with a compound microscope. Another, simpler method of studying petrifactions involves etching the cut surface with a dilute acid and then applying a plastic or similar film. As soon as the film hardens, it can be peeled off. Such peels display lifelike microscopic details of the surface with which they were in contact. They are commonly used by *paleobotanists* (botanists who study fossil plant materials) in their research.

Coprolites

Coprolites are the fossilized dungs of prehistoric animals and humans. They may contain pollen grains and other plant and animal parts that provide clues to the food and feeding habits of past organisms and cultures.

Unaltered Fossils

When plants or animals fall into bodies of oil or water in which certain substances are present or there is a nearly total lack of oxygen, decay does not occur. Some animals die in snowfields, and their bodies are permanently frozen. In such instances, preservation in the unaltered state occurs. Unaltered fossils, particularly frozen ones, are very rare.

Summary

1. Four divisions of seedless vascular plants are recognized: Psilotophyta (whisk ferns); Lycophyta (club mosses and quillworts); Sphenophyta (horsetails and scouring rushes); and Pterophyta (ferns).

2. Whisk ferns (*Psilotum*) are the simplest of all living vascular plants, consisting of evenly forking green stems that have small protuberances called *enations* but no leaves; roots are also lacking. The stem contains a central cylinder of xylem and phloem.

3. Whisk fern spores germinate into tiny saprophytic gametophytes over the surface of which archegonia and antheridia are scattered. The zygote develops a foot and a rhizome. Upright stems are produced when the rhizome separates from the foot.

4. *Tmesipteris*, an Australian relative of whisk ferns, has leaflike appendages. Fossil plants resembling whisk ferns have been found in Silurian geological formations.

5. There are two genera of club mosses with living members. Ground pines (*Lycopodium*) develop sporangia in the axils of sporophylls. Several sporophytes may be produced from one gametophyte.

6. Spike mosses (*Selaginella*) are heterosporous and have a ligule on each microphyll; microspores develop into male gametophytes with antheridia, and the megaspores develop into female gametophytes with archegonia.

7. Quillworts (*Isoetes*), which are heterosporous, have quill-like microphylls that arise from a cormlike base. The corms have a cambium that remains active for many years.

8. Some fossil relatives of club mosses were large dominant members of the forests and swamps of the Carboniferous period.

9. Club moss spores have been used for flash powder, for medicinal purposes, as talcum powder, and to staunch bleeding. The plants themselves have been used as ornamentals, novelty items, Christmas ornaments, and for intoxicating purposes.

10. Horsetails and scouring rushes (*Equisetum*) accumulate deposits of silica in their epidermal cells and make good scouring material. They occur in both unbranched and branched forms. The stems are jointed and ribbed and have tiny scalelike leaves in whorls at each joint.

11. The stems of *Equisetum* are hollow in the center and contain cylinders of carinal and vallecular canals. The stems arise from rhizomes that branch extensively below the surface of the ground.

12. Some species of *Equisetum* produce nonphotosynthetic stems. Conelike strobili are produced in the spring in all species. *Equisetum* spores have ribbonlike elaters that are sensitive to changes in humidity.

13. Equal numbers of male and female gametophytes are produced, but female gametophytes may become bisexual. The development of more than one sporophyte from a gametophyte is common.

14. Ancient relatives of horsetails were the size of trees when they flourished in the Carboniferous period of 300 million years ago.

15. Horsetails have been used for food after the parts containing silica were removed, but they are not recommended for human consumption. They have also been used medicinally as a diuretic, as an astringent, and in the treatment of venereal disease and tuberculosis. Other uses include as a shampoo, a mineral indicator, and a metal polish. Cannel coal consists primarily of spores of giant horsetails that were reduced to carbon.

16. Fern leaves (fronds) are typically dissected and feathery in appearance but vary greatly in form. They usually first appear as croziers that unroll and expand.

17. Patches of sporangia appear on the lower surfaces of fern fronds. The sporangia commonly occur in sori,

which may be protected by indusia. Each sporangium has an annulus, which functions in catapulting mature spores out of the sporangium.

18. Fern gametophytes (prothalli) develop after spores germinate. The prothalli contain both archegonia and antheridia; only one zygote develops into a sporophyte.

19. Preferns, the ancestors of ferns, are found in Devonian deposits estimated to be 375 million years old. Ferns (especially tree ferns) became so abundant during the latter part of the Carboniferous period that this era was once referred to as the *Age of Ferns,* but the discovery of seeds on some of them raised questions about relationships and the term *Age of Ferns* was dropped.

20. Ferns are used as ornamentals, as air filters, as a source of "bark" for growing orchids and other plants, as a source of stuffing materials for bedding, in tropical construction, as food, and in numerous folk medicinal applications. Other uses include as basketry and weaving material, as an ingredient in brewing ale, and as an ingredient in the preparation of chamois leather. One floating fern, which forms dense mats, is believed to suffocate mosquito larvae.

21. Fossils, which are recognizable prehistoric organic objects, are formed in different ways. Molds, casts, compressions, and imprints are formed when material buried by silt or other sediment has hardened into rock and the organic material has slowly been washed away by water.

22. Petrifactions are uncompressed rocklike materials in which the original cell structure has been preserved. Coprolites are fossilized dungs, which may contain pollen grains and other plant and animal parts. Unaltered fossils are the fossils of plants or animals that fell into bodies of oil or water or snowfields and were thus not subjected to decay.

Review Questions

1. What basic features of the ferns and their relatives distinguish them from any organisms thus far studied?

2. How does the gametophyte of a whisk fern differ from that of a true fern?

3. Which of the fern relatives have significantly functional leaves? In those without conspicuous leaves, how are the carbohydrate needs of the plants met?

4. How does one distinguish among ground pines, spike mosses, and quillworts?

5. How did the ancient ground pines differ from those of the present day?

6. How do the spores and the female gametophytes of horsetails differ from those of any other plants studied thus far?

7. What are the location and function of carinal canals?

8. In your opinion, which have the most human relevance today: club mosses and horsetails of the present or those of the geological past? Why?

9. Define *crozier, rachis, pinna, indusium, prothallus,* and *prefern.*

10. Diagram the life cycle of a typical fern.

11. Summarize present and past human uses of ferns.

12. How are fossils formed, and what different types are recognized?

Discussion Questions

1. Would you assume that there is any significance to the fact that both the sporophytes and the gametophytes of whisk ferns branch in the same manner? If so, what might it be? If not, why not?

2. Do spike mosses, which produce two kinds of spores and gametophytes, have any advantage over ground pines, which produce only one kind of spore and gametophyte?

3. Some gametophytes of fern relatives develop underground, while others develop at the surface. If you were to be a gametophyte, which would you prefer? Why?

4. After looking at the internal structure of a horsetail stem, can you suggest a function for the silica in the ribs?

5. How would we be affected if all ferns were to become extinct in a few years?

Additional Reading

Bir, S. S. 1983. *Pteridophytes: Their morphology, cytology, taxonomy, and phylogeny.* Houston, TX: Scholarly Publications.

Cobb, B. 1975. *A field guide to the ferns and their related families.* Boston: Houghton Mifflin Co.

Grillos, S. J. 1966. *Ferns and fern allies of California.* Berkeley: University of California Press.

Lellinger, D. B. 1985. *A field manual of the ferns and fern allies of the United States and Canada.* Washington, DC: Smithsonian Institution Press.

Lloyd, R. M. 1964. Ethnobotanical uses of California pteridophytes by western American Indians. *American Fern Journal* 54: 76–82.

McHugh, A. 1992. *The cultivation of ferns.* North Pomfret,VT: Trafalgar.

Mickel, J. 1979. *How to know the ferns and fern allies.* Pictured Key Nature Series. Dubuque, IA: Wm. C. Brown Publishers.

Montgomery, J. D., and D. E. Fairbrothers. 1992. *New Jersey ferns and fern allies.* New Brunswick, NJ: Rutgers University Press.

Raghaven, V. 1989. *Developmental biology of fern gametophytes.* New York: Cambridge University Press.

Seward, A. C. (Ed.). 1991. *Fossil plants for students of botany and geology,* 4 vols. Houston, TX: Scholarly Publications.

Stewart, W. N., and G. W. Rothwell. 1993. *Paleobotany and the evolution of plants,* 2d ed. New York: Cambridge University Press.

A female cycad cone (Cycas sp.). Cycads are slow-growing, superficially palmlike gymnosperms. Gymnosperm means naked-seeded, and the developing seeds can be seen here out in the open at the edges of the cone scales.

22

Introduction to Seed Plants: Gymnosperms

O v e r v i e w

The chapter begins by exploring the differences between ferns and seed plants and then gives a brief geological history of gymnosperms. The leaves, roots, and stems of pine trees are discussed, and the life cycle of a typical gymnosperm (pine) is given. Other conifers, such as yews, podocarps, junipers, and redwoods, are mentioned, and a brief discussion of cycads, Ginkgo, Ephedra, Gnetum, and Welwitschia follows. The chapter concludes with a digest of the human and ecological relevance of gymnosperms, with particular emphasis on the conifers.

Some Learning Goals

1. Know the features common to typical male and female conifer strobili and how they differ.
2. Understand what distinguishes the subdivisions and classes of living gymnosperms from one another.
3. Learn the modifications of pine leaves that adapt them to a harsh environment.
4. Indicate where the following structures occur in the life cycle of a pine tree: archegonia, eggs, sperms with flagella, male and female gametophytes, the sporophyte, integument, vessels, spore mother cells, embryo, and pollen grains.
5. Explain the function of each of the following: resin canals, albuminous cells, mycorrhizal fungi, nucellus, generative cell, megaspore.
6. Learn a use for each of at least 10 different conifers.
7. Identify three major representatives of the Subdivision Gneticae.

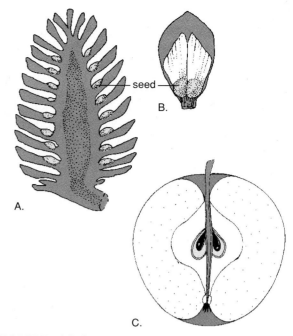

FIGURE 22.1 A comparison between exposed gymnosperm seeds and enclosed angiosperm seeds. *A.* Exposed seeds on a female (woody) cone. *B.* A single female cone scale with two seeds. *C.* A section through an apple, showing the enclosed seeds.

Some of the giant tree ferns, which have large graceful leaves held above an unbranched trunk (see Fig. 21.25), superficially resemble coconut palms. Anyone who has sat at the base of a coconut palm when a coconut has fallen, however, has been reminded of a basic difference between ferns and coconut palms. The palms are flowering plants, which produce seeds like pine trees and other cone-bearing plants, while the ferns produce no seeds at all.

The oldest known seeds were produced by plants that appeared late in the Devonian period, more than 350 million years ago. Seeds provided a significant adaptation for plants that had invaded the land. Unlike spores, seeds have a protective seed coat and a supply of food (usually endosperm) for the embryo, which may be capable of lying dormant through long periods of freezing weather and drought and may, in some instances, even survive fire. This survival value of seeds undoubtedly played a major role in seed plants becoming the dominant vegetation on earth today.

The first seed plants were so fernlike in appearance that they were originally classified as ferns. When fossils with obvious seeds on the fronds were discovered, however, these *pteridosperms* ("seed ferns") were reclassified as *gymnosperms*.

The name **gymnosperm** is derived from two Greek words: *gymnos,* meaning "naked," and *sperma,* "a seed."

The name refers to the exposed nature of the seeds, which are produced on the surface of sporophylls or similar structures instead of being enclosed within a fruit as they generally are in the flowering plants (Fig. 22.1). The seed-bearing sporophylls of the sporophyte are often spirally arranged in female strobili (cones), which develop at the same time as smaller male strobili (cones); the male strobili produce *pollen grains.*

The female gametophyte is produced inside an **ovule** that contains a diploid, fleshy, nutritive **nucellus.** The nucellus is itself enclosed within one or more outer layers of diploid tissue. These outer layers of tissue constitute an **integument,** which becomes a **seed coat** after the fertilization and development of an embryo takes place (see Fig. 22.8).

The sporophytes of gymnosperms are mostly trees and shrubs, although a few are vines. The gametophytes are proportionately more reduced in extent than they are in ferns and their relatives. Unlike the gametophytes discussed so far, they don't grow independently but develop within sporophyte structures.

The Division Pinophyta, which includes all living gymnosperms, is divided into three subdivisions. *Cycadicae* includes the superficially palmlike *cycads; Pinicae* includes the conifers and *Ginkgo;* and *Gneticae* includes three genera of *gnetophytes.*

The largest and most significant group of gymnosperms, the **conifers** (Class Pinatae), includes about 575 species of pines, firs, spruces, hemlocks, redwoods, cedars, and other cone-bearing, woody plants. Fossils of some conifers extend back 290 million years to the late Carboniferous period.

Ginkgo (Class Ginkgoatae) has fan-shaped leaves and seeds enclosed in a fleshy covering; *Gnetophytes* (Subdivision Gneticae) have wood that contains vessels—structural elements unknown in other gymnosperms.

DIVISION PINOPHYTA, CLASS PINATAE—THE CONIFERS

Pines

The largest genus of conifers, *Pinus* (pines) has over 100 living species. They are the dominant trees in the vast coniferous forests of the Northern Hemisphere. They have also been planted extensively in the Southern Hemisphere, but only the Merkus pine occurs there naturally, and its distribution barely extends south of the equator. They include the world's oldest known living organisms, the bristlecone pines, which occur in the White Mountains of eastern central California and in the Snake range on the central Nevada-Utah border. Some trees still standing are about 4,600 years old, and one that was, unfortunately, cut down in 1964 was found to have been about 4,900 years old.

Structure and Form
Pine leaves, which are needlelike, are arranged in clusters or bundles of two to five leaves each (a handful of species have as many as eight or as few as one to a cluster). Regardless of the number of leaves, each cluster (*fascicle*) forms a cylindrical rod if the leaves are held together (Fig. 22.2). Pines often live in areas where the topsoil is frozen for a part of the year, making it difficult for the roots to obtain water. In addition, the leaves may be exposed to high winds and bitterly cold temperatures. Accordingly, they have several modifications that enable them to survive in harsh environments.

Just beneath the epidermis there is a **hypodermis,** consisting of one to several layers of thick-walled cells. The epidermis itself is coated with a thick cuticle. The stomata, instead of being at the surface, are recessed or sunken in small cavities. The veins and their associated tissues are surrounded by an endodermis, and the mesophyll cells do not have the obvious air spaces typical of the spongy mesophyll of the leaves of flowering plants (see figs. 7.4 and 7.8).

Conspicuous **resin canals** develop in the mesophyll. These resin canals, which are found throughout other parts of the plant as well, consist of tubes lined with special cells that secrete *resin*. Resin, which is aromatic and antiseptic, prevents the development of fungi and deters insect attacks. Other conifers apparently produce resin canals in response to injury. Pine fascicles usually *absciss* (i.e., fall off—see Chapter 7) within two to five years of their maturing, but

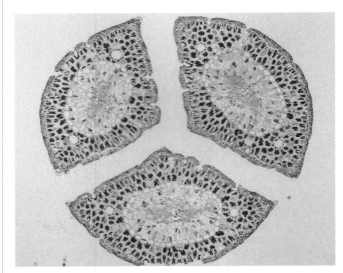

FIGURE 22.2 A cluster (fascicle) of pine needles in cross section, showing how the needles form a cylindrical rod.

those of bristlecone pines persist for up to 30 years. The fascicles are lost a few at a time so that some functional leaves are always present on a healthy tree.

Wood varies considerably in hardness. Most gymnosperm wood, including that of pines, consists primarily of tracheids and differs from the wood of dicots in having no vessel members or fibers. Conifer wood, because of the absence of thick-walled cells, is said to be *soft,* while the wood of broadleaf trees is described as *hard*.

In many conifers, the annual rings of the xylem are often fairly wide as a result of a comparatively rapid growth rate during the growing season. Resin canals are formed both vertically and horizontally throughout various tissues (Fig. 22.3). The bark, which includes the secondary phloem, may be relatively thick. It often becomes 7.5 centimeters (3 inches) or more wide, and in the giant redwood it may become as much as 60 centimeters (2 feet) wide. Companion cells are absent from the phloem, but similar *albuminous cells* apparently perform the same function.

Mycorrhizal fungi are associated with the roots of most conifers. In fact, pine seedlings that germinate in sterilized soil do not grow well at all until the fungi are introduced or allowed to develop. The roots of adjacent pines often interweave. New England pioneers made use of this characteristic in eastern white pines when clearing land. After trees were felled, the settlers would tip over the stumps with the roots still attached and use them for fences. The fences survived for many years as the resin deterred bacteria and other organisms from breaking down the wood.

Reproduction
Pines, like spike mosses, quillworts, and a few of the ferns, produce two kinds of spores (see Fig. 22.8). **Male cones** (*male strobili*—also called *pollen cones*), which consist of papery or membranous scales arranged in a spiral or in whorls around an axis, are usually produced in the spring.

FIGURE 22.3   A portion of a cross section of a pine stem, showing annual rings.

FIGURE 22.4   A cluster of male pine cones.

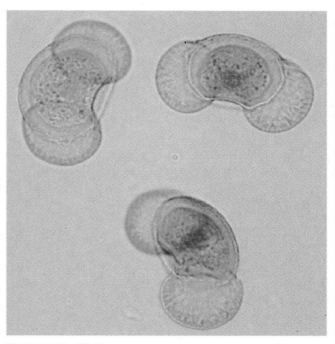

FIGURE 22.5   Pollen grains of a pine, as seen with the aid of a microscope. Each pollen grain has a pair of air sacs that provide added buoyancy.

The male cones develop toward the tips of the lower branches in clusters of up to 50 or more (Fig. 22.4) and are usually less than 4 centimeters (about 1.5 inches) long. **Micro-sporangia** develop in pairs toward the bases of the scales.

Microspore mother cells in the microsporangia each undergo meiosis, producing four haploid **microspores.** These then develop into **pollen grains,** each consisting of four cells and a pair of external air sacs that look something like tiny water wings (Fig. 22.5). These sacs give the pollen

A.

B.

FIGURE 22.6 Female pine cones. *A.* Three immature cones shortly after being produced. *B.* A mature cone.

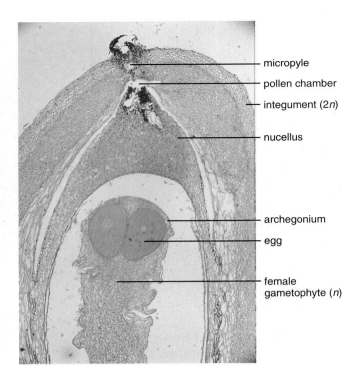

micropyle
pollen chamber
integument (2*n*)
nucellus
archegonium
egg
female gametophyte (*n*)

FIGURE 22.7 A longitudinal section through part of a pine ovule.

produce more than 1 million grains, and there may be hundreds of such clusters on one tree. The grains accumulate as a fine yellow dust on cars, shrubbery, or anything else in the vicinity, and they often form an obvious scum on pools and puddles. Within a few weeks of the pollen's release, the male cones shrivel and fall from the trees.

Megaspores are produced in **megasporangia** located within **ovules** at the bases of the female cone scales. The female cones (*female strobili—*also called *seed cones*) are much larger than the male cones, becoming as much as 60 centimeters (2 feet) long in sugar pines and weighing as much as 2.3 kilograms (5 pounds) in Coulter pines. When mature, they have woody scales with inconspicuous bracts between them, arranged in a spiral around an axis. They are produced on the upper branches of the same tree on which the male cones appear (Fig. 22.6).

The *ovules* (Fig. 22.7), which occur in pairs toward the base of each scale of the immature female cones, are larger and more complex than the microsporangia of male cones. Each ovule contains a *megasporangium* embedded in multicellular nutritive tissue called the **nucellus.** The nucellus, in turn, is surrounded and enclosed by a thick, layered **integument.** The integument has a somewhat tubular channel or pore called a **micropyle** that is pointed toward the cone's central axis. One of the integument layers later becomes the **seed coat** of the seed.

A single **megaspore mother cell** within the megasporangium (nucellus) of each ovule undergoes meiosis,

grains added buoyancy, which may result in the grains being carried great distances by the wind.

Pines produce pollen grains in astronomical numbers. For example, it has been estimated that each of the 50 or more pollen cones commonly found in a single cluster may

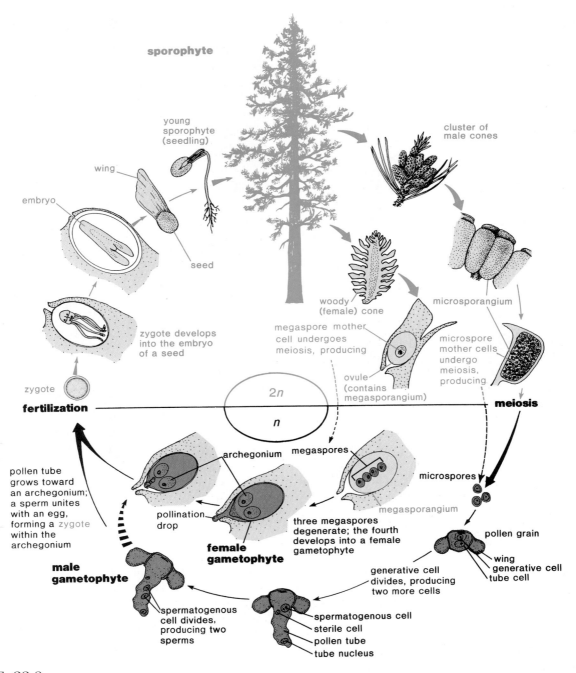

sporophyte

young sporophyte (seedling)

wing

embryo

seed

cluster of male cones

woody (female) cone

microsporangium

megaspore mother cell undergoes meiosis, producing

microspore mother cells undergo meiosis, producing

zygote develops into the embryo of a seed

ovule (contains megasporangium)

2n

n

zygote

fertilization

meiosis

archegonium

megaspores

microspores

pollen tube grows toward an archegonium; a sperm unites with an egg, forming a zygote within the archegonium

pollination drop

female gametophyte

megasporangium

three megaspores degenerate; the fourth develops into a female gametophyte

pollen grain

wing
generative cell
tube cell

male gametophyte

generative cell divides, producing two more cells

spermatogenous cell divides, producing two sperms

spermatogenous cell
sterile cell
pollen tube
tube nucleus

FIGURE 22.8 Life cycle of a pine.

producing a row of four relatively large **megaspores.** Three of the megaspores soon degenerate. Over a period of months, the remaining one slowly develops into a *female gametophyte,* which ultimately may consist of several thousand cells.

As gametophyte development nears completion, two to six *archegonia* differentiate at the end facing the micropyle. Each archegonium contains a single large *egg.* When a stained thin, lengthwise section of a pine ovule is examined with a microscope, only one archegonium or egg may be seen, or the micropyle may appear to be missing.

This is due to the section not having been sliced precisely through the middle of all the structures.

Female cones, which are at first usually reddish or purplish, commonly take two seasons to mature into the green and finally the brownish woody structures with which we are all familiar (Fig. 22.8). During the first spring, the immature cone scales spread apart, and pollen grains carried by the wind sift down between the scales. There they catch in sticky drops of fluid (*pollen drops*) oozing out of the micropyles. As the fluid evaporates, the pollen is drawn down through the micropyle to the top of the nucellus.

After pollination, the scales grow together and close, protecting the developing ovule. Meiosis and megaspore development don't occur until about a month after pollination. After a functional megaspore is produced, the female gametophyte and its archegonia don't mature until more than a year later.

Meanwhile, the pollen grain (immature male gametophyte) produces a **pollen tube,** which slowly grows and digests its way through the nucellus to the area where the archegonia develop. While the pollen tube is growing, two of the original four cells in the pollen grain enter it. One of these, called the **generative cell,** divides and forms two more cells, called the *sterile cell* and the *spermatogenous cell.* The spermatogenous cell divides again, producing two male gametes, or *sperms.* The pine sperms have no flagella, unlike the sperms of other organisms encountered thus far. The germinated pollen grain, with its pollen tube and two sperms, constitutes the mature *male gametophyte.* Notice that no antheridium has been formed.

About 15 months after pollination, the tip of the pollen tube arrives at an archegonium, unites with it, and discharges the contents. One sperm unites with the egg, forming a zygote. The other sperm and remaining cells of the pollen grain degenerate. The sperms of other pollen grains present may unite with the eggs of other archegonia, and each zygote begins to develop into an *embryo.* This is similar to the development of fraternal twins or triplets in animals. At a later stage, an embryo may divide in such a way as to produce the equivalent of identical twins or quadruplets in animals. Normally, however, only one embryo completes development. While this development is occurring, one of the layers of the integument hardens, becoming a *seed coat.* A thin membranous layer of the cone scale becomes a "wing" on each seed. The wing may aid in the seed's dispersal. Squirrels and other animals may also help dispersal by breaking open the cones (see Fig. 22.8).

In other species, such as the lodgepole, jack, and knobcone pines, the cones remain on the tree with the scales closed until they are seared by fire or open with old age. Sometimes, these cones are slowly buried as the cambium adds tissues that increase the girth of the branch. Seeds of such engulfed cones have been reported to germinate after they have been dug out of the stem.

Other Conifers

Some conifers don't produce woody female cones with conspicuous scales, nor do all conifers produce both male and female cones on the same tree. For example, yew (*Taxus*) and California nutmeg (*Torreya*) produce ovules singly at the tips of short axillary shoots. Each ovule is at least partially surrounded by a fleshy, cuplike covering called an **aril.** In yews, this is bright red and open at one end, giving the fleshy seed the appearance of a small red hors d'oeuvre olive with its stuffing removed (Fig. 22.9). The fleshy seeds are produced only on female trees, while the pollen-bearing strobili, or male cones, are produced only on male trees.

FIGURE 22.9 The seeds of yew (*Taxus*), which are not produced in cones, are surrounded at maturity by a red, fleshy, cuplike structure called an *aril.*

Podocarps, which are conifers of the Southern Hemisphere, are widely planted as ornamentals in regions with milder climates. Their fleshy-coated seeds, which are produced singly, are similar to those of yews, but they are not open at one end and have an additional larger appendage at the base (Fig. 22.10). The origin of these fleshy seeds is not clear and has led to speculation that yews and podocarps may have diverged from other conifers very early in the evolution of gymnosperms.

The scales of the female cones of junipers tend to be fleshy at maturity, so they look more like berries than cones. Juniper pollen, as well as that of a number of other conifers, does not have air sacs. Cypress and redwood female cone scales are flattened at the tips and narrow at the base, where they do not overlap one another as do pine cone scales.

The two California species of redwoods are both renowned for their size, height, and longevity. Coastal redwoods occasionally grow to a height of 90 meters (295 feet), and one tree in Humboldt County, California, which is 111.6 meters (366.2 feet) tall, is believed to be the tallest conifer in the world. The other species, usually referred to as *Big Tree*

FIGURE 22.10 Fleshy seeds of a podocarp (*Podocarpus*). Note the large appendage at the base of each seed.

---

or *Giant Redwood,* is confined to the western slopes of California's Sierra Nevada range. It does not grow quite as tall as the coastal redwood but exceeds it in total mass. The General Sherman tree in Sequoia National Park, for example, is 31 meters (101.5 feet) in circumference at the base and over 24 meters (79 feet) in circumference 1.5 meters (5 feet) above the ground. It weighs an estimated 5,594 metric tons (6,167 tons). There are 600,000 board feet of timber in this single tree—enough to build more than 75 five-room houses (although the wood is generally not suitable for construction purposes) or to make 20 billion toothpicks. It is over 3,500 years old.

OTHER GYMNOSPERMS

Living representatives of other gymnosperms are not as numerous or as well-known as the conifers and outwardly don't resemble them at all. In fact, some of them look more like leftover props from a science-fiction movie set. They include the **cycads,** the **ginkgoes,** and the **gnetophytes** (pronounced née-toe-fytes).

Subdivision Pinicae, Class Ginkgoatae—*Ginkgo*

There is only one living species of *Ginkgo* (Fig. 22.11), whose name is derived from Chinese words meaning "silver apricot." The fossil record indicates *Ginkgo* and other members of its family (Ginkgoaceae) were once widely distributed, especially in the Northern Hemisphere. Despite isolated reports to the contrary, there are doubts that ginkgoes now exist anywhere they have not been cultivated, and the plant has often been called a living fossil.

Ginkgoes are often referred to as *maidenhair trees* because their notched, broad, fan-shaped leaves look like larger versions of the individual pinnae of maidenhair ferns. They are widely cultivated in the United States and are popular street trees in some areas. The leaves, which are mostly produced in a spiral on short, slow-growing spurs, have no midrib or prominent veins. Instead, hairlike veins branch dichotomously (fork evenly) and are relatively uniform in their width. They are deciduous and turn a bright golden yellow before abscission in the fall.

Ginkgo is dioecious, with a life cycle similar to that of cycads; the male and female reproductive structures are produced on separate trees, and the sperms have flagella. The mature seeds, which resemble small plums, are enclosed in fleshy seed coats. The flesh is, however, unrelated to that of true fruits. In North America, male trees (propagated from cuttings) are preferred for ornamental purposes because the seed flesh has a nauseating odor and is irritating to the skin of some individuals. However, female trees predominate in China and Korea, where seeds are considered a delicacy, and only enough males are propagated to insure pollination.

Subdivision Cycadicae—The Cycads

Cycads, which look like a cross between a tree fern and a palm, are slow-growing plants of the tropics and subtropics. They have unbranched trunks that grow more than 15 meters (50 feet) tall in a few species and have a crown of large pinnately divided leaves. Extinct members of this division, known as *cycadeoids,* were abundant during the Mesozoic era. Several of the approximately 100 known living species are presently facing extinction. Their life cycles are similar to those of conifers. Each sperm of cycads, however, has from 10,000 to 20,000 spirally arranged flagella. Cycads are dioecious; the male and female strobili, which in some species are huge (e.g., more than a meter [3 feet 3 inches] long with a weight of over 220 kilograms [100 pounds]), are produced on separate plants (Fig. 22.12). The scales of female strobili of some species are covered with feltlike or woolly hairs.

Subdivision Gneticae— The Gnetophytes

The 70 known species of *gnetophytes* are distributed among three distinctive genera. They are unique among the

A.

B.

FIGURE 22.11 *Ginkgo. A.* A mature tree in the fall.
B. Seeds and leaves. *C.* Male strobili.

C.

gymnosperms in having vessels in the xylem. The genus
Ephedra, whose members are sometimes called *joint firs*
(Fig. 22.13), accounts for more than half of the species.
These shrubby plants, which inhabit drier regions, produce
tiny leaves in twos or threes at a node. The leaves turn brown
soon after they appear. The slightly ribbed branches are
often whorled and are photosynthetic when they are young.

In preparation for pollination, a tubular extension, like
the neck of a tiny bottle, extends into the air from the integument of an ovule. Sticky fluid oozes out of this extension,
which constitutes the micropyle, and airborne pollen catches
in it. Male and female strobili may be produced on the same
plant or on different ones, depending on the species.

Most of the remaining species in this division are in
the genus *Gnetum* (Fig. 22.14), which has not been given an
English common name. Its members, which have broad
leaves similar to those of flowering plants, occur in the tropics of South America, Africa, and Southeast Asia. Most are
vinelike, but the best-known species is a tree that grows up
to 10 meters (33 feet) tall.

The third genus, *Welwitschia,* has only one species,
which is confined to the temperate Namib and Mossamedes
deserts of southwestern Africa. Here the average annual
rainfall is only 2.5 centimeters (1 inch), and in some years it
does not rain at all. The plants carry on CAM photosynthesis
and their stomata are open at night. The plants apparently
survive much of the time on dew and condensate from fog
that rolls in off the ocean at night.

A.

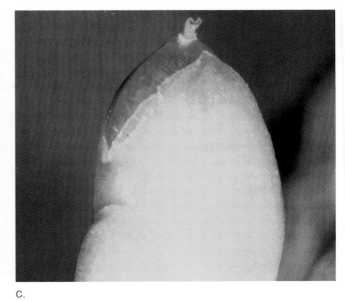

B.

B.

FIGURE 22.12 *A*. A male cycad with a strobilus.
B. A female cycad with a strobilus.

C.

FIGURE 22.13 Joint fir (*Ephedra*). *A*. Part of a single
plant. *B*. Male strobili, ×20. *C*. A female strobilus, ×20.

FIGURE 22.14 A climbing *Gnetum.*

Welwitschia plants are truly extraordinary in appearance. The stem, which rises only a short distance above the surface, is in the form of a large shallow cup that tapers at the base into a long taproot. At maturity, the plants, which may live to be 100 years old, have a crusty, barklike covering on the surface of the stem cup. The stems may be more than 1 meter (3 feet 3 inches) in diameter (Fig. 22.15).

Throughout their lifespan, *Welwitschia* plants usually produce only two leaves. The leaves are wide and straplike, each with a meristem at the base. The meristems constantly add to the length of the leaves, but as the leaves flap about in the wind, they become tattered and split, wearing off at the tips so that they are seldom more than 2 meters (6.5 feet) long. Male and female strobili, which are produced on separate plants, are on axes that emerge from the axils of the leaves so they appear to be growing around the rim of the stem cup.

HUMAN AND ECOLOGICAL RELEVANCE OF GYMNOSPERMS

As a group, the gymnosperms are second only to the flowering plants in their impact on our daily lives. Space does not permit a detailed account of all they contribute, but the following sections provide an overview of some of their uses, past and present.

FIGURE 22.15 A female *Welwitschia* plant in the Namibian Desert.
(Photo by Margaret Marker, in the collection of the National Botanical Institute, Kirstenbosch. Courtesy Fiona Getliffe Norris)

What do we plant when we plant the tree?
We plant the ship, which will cross the sea.
We plant the mast to carry the sails;
We plant the planks to withstand the gales—
The keel, the keelson, the beam, the knee;
We plant the ship when we plant the tree.

What do we plant when we plant the tree?
We plant the houses for you and me.
We plant the rafters, the shingles, the floors,
We plant the studding, the lath, the doors,
The beams, the siding, all parts that be;
We plant the house when we plant the tree.

What do we plant when we plant the tree?
A thousand things that we daily see;
We plant the spire that out-towers the crag,
We plant the staff for our country's flag,
We plant the shade, from the hot sun free;
We plant all these when we plant the tree.

Henry Abbey

Conifers

In the early 1970s, the late author-naturalist Euell Gibbons filmed a series of television commercials in which he mentioned uses of several wild plants for food. In one of the commercials, he rhetorically inquired if his audience had ever eaten a pine tree and added, "many parts are edible."

The edibility of parts of many conifers was known to Native Americans long before Europeans set foot on the North American continent. In fact, early explorers found large numbers of pines stripped of their bark. For centuries, the inner parts (phloem, cambium) had been used for emergency food. The Adirondack Mountains of New York are believed to have received their name from a Mohawk Indian word meaning "tree-eater," in reference to Native American use of the inner bark of eastern white pines. This material (specifically the phloem) contains sugars that make it taste sweet. Some tribes ate the material raw, some dried it and ground it to flour, and others boiled it or stored dried strips for winter food. Early settlers in New England candied strips of eastern white pine inner bark. To prevent scurvy, both they and local Native Americans drank a tea made of the needles, which are rich in vitamin C.

The seeds of nearly all pines are edible, but those of western North America include the larger and better tasting species. The protein content of those analyzed generally ranges between 15% and 30%, with much of the remainder consisting of oils.

California Indians relished gray pine seeds in particular, but even the small seeds of ponderosa pine were eaten raw or made into a meal for soups and bread. Cones of pinyons were collected by tribes of the Southwest and thrown on a fire to loosen the seeds. These were then pounded and made into cakes or soup. The soup was often fed to infants. In Siberia, people crush the seeds of Siberian white pine to obtain a nutritious oil, but its use has declined since corn and cottonseed oils have become available.

Italians and other Europeans cook *pignolias,* the seeds of the stone pine, in stews and soups. The seeds are also used in cakes and cookies, and some are exported to the United States for this purpose. Many of the so-called nuts used by commercial American bakers in cakes and confectionery are really seeds from the east Himalayan chilghoza pine. Other sources include the Mexican stone pine and a few pinyons.

Eastern white pines were often used as masts in sailing vessels. In colonial days, the royal surveyors marked certain trees for the use of the Crown, and severe penalties were imposed on colonists who ignored the ban on the use of any white pine not growing on private land. It was, however, legal for colonists to use white pines that had blown down, which gave rise to the term *windfall.* Eastern white pine wood contains less resin than that of other species and was extensively used for crates, boxes, matchsticks, furniture, flooring, and paneling. By the end of the 19th century, eastern white pines, which originally grew over vast tracts of the northeastern United States and Canada, had been decimated by wholesale logging done with no thought to conservation. Bald cypress trees in the southeastern United States met a similar fate. White pine blister rust also took its toll. Although new growth is now being promoted, most white pine lumber used today comes from large stands of western white pine in the Pacific Northwest.

The trunks of lodgepole pines are used in both the United States and Canada for telephone poles; the straight-grained wood is also used for railroad ties, mine timbers, and pulp.

Smog has severely damaged ponderosa and other native pines in California. For a number of years, the U.S. Forest Service has experimented with Afghanistan pine, a smog- and drought-resistant pine from Russia and adjacent areas, as a replacement for native trees. Growth rates in tests have been very rapid. Rapid growth is a desirable commercial feature, since it permits considerably more timber to be produced than does slow growth, but the wisdom of introducing non-native plants into natural communities is in question, since there are many examples of such activities thoroughly disrupting delicate ecological balances.

The resin produced in the resin canals of conifers is a combination of a liquid solvent called *turpentine* and a waxy substance called *rosin.* When a conifer tree is wounded or damaged by insects, resin usually covers the area, sometimes trapping the insects. Out in the air, the turpentine evaporates quickly, leaving a protective layer of rosin, which deters water loss and fungal attacks. Both turpentine and rosin are very useful products, and a large industry centered in the southern United States and in the south of France is devoted to their extraction and refinement. Turpentine is considered a premier paint and varnish solvent; rosin is used by musicians on violin bows and by baseball pitchers to improve their grip on the ball. Turpentine and rosin are often referred to as *naval stores,* a term that originated when the British Royal Navy used large quantities for caulking and sealing their sailing ships.

Most naval stores and a third or more of the lumber used in the United States today come from a group of southern yellow pines, particularly slash pine. Pitch pine, which was also a source of naval stores before slash and other yellow pines became more profitable, was used in the past for the water wheels of mills. Pitch pine wood was also used as fuel for steam engines, as it produces considerable heat when it burns. Batters apply pine tar (pine tar is a sticky blackish fluid made by heat distillation of pine wood) to the handles of bats to minimize slippage.

In the past, pine pitch (resin) was used by Native Americans for patching canoes, and it has been suggested that Noah's Ark was sealed with pitch (resin) from aleppo pine. Pine resin was used for purifying wine in the first century A.D., and today Greeks still add it to certain wines. *Colofonia* is the Spanish word for a type of resin Monterey pines produced abundantly around the old Spanish capital of Monterey. The early California priest, Padre Arroyo, suggested during the first half of the 19th century that the colony received its name from this Spanish word. *California,* the name ultimately given to the colony, was, however, the name of a mythical paradise in a Spanish novel published in the early 16th century, and no more than an interesting similarity between the two Spanish words may be involved.

The huge kauri pines (*Agathis*) of New Zealand, which are in a family different from that of true pines, are the source of a mixture of resins called *dammar.* Dammar is used in high-quality colorless varnishes and was also the resin originally used in the manufacture of linoleum.

In New Zealand, dammar, also called *amber,* is obtained primarily in fossil form from former or present kauri pine forest areas. Most amber, however, has come from extinct conifers that flourished 60 to 70 million years ago in the Baltic area of the former Soviet Union and from other extinct conifers in what is now the Dominican Republic. It occurs as lumps of translucent material with a deep orange-yellow tint. According to Greek mythology, amber was the congealed tears of Phaëthon's sisters who, while weeping over his death, were turned into trees. Some of the lumps weigh up to 45 kilograms (100 pounds). The supply, which was at its peak at the turn of the century, is now nearing exhaustion. Remarkably lifelike preservations of prehistoric insects millions of years old have occurred in amber (Fig. 22.16).

Other products refined from resin are used in the manufacture of menthol for cigarettes (menthol also occurs naturally in members of the Mint Family), floor waxes, printer's ink, paper coatings, varnishes, and perfumes.

White spruce is the chief source of pulpwood for newsprint (Fig. 22.17) and other paper in North America. Enormous quantities of paper are used every day. A single

FIGURE 22.16 A prehistoric red wolf spider preserved in Dominican Republic amber. The amber is estimated to be between 20 and 30 million years old.
(Photograph by John Yellen)

midweek issue of a large metropolitan newspaper may use an entire year's growth of 50 hectares (123 acres) of these trees, and that amount may double for weekend editions. White spruce is also used to make toilet paper. A large American publishing company, in an attempt to find ways of reducing paper consumption in the United States, tried trimming 2.5 centimeters (1 inch) from the width of all rolls of toilet paper in their building facilities. They found that the employees still used the same number of rolls per month as they had previously. From this, it was calculated that if all rolls of toilet paper were similarly trimmed in width, 1 million trees would be saved each year in the United States alone.

Split roots of the white spruce are quite pliable and were used by Native Americans for lashing canoes and for basketry. Spruce beer, brewed from young twigs and leaves with an added sugar source, such as honey or molasses, was once used as a remedy for scurvy. Resin of white, red, black, and Sitka spruces was used as a type of chewing gum by Native Americans, who sometimes hardened it slightly in cold water. Europeans who have tried it report that it has to be of the right consistency to be enjoyable, since it behaves like unhardened taffy if it is too soft and is bitter if it is too hard. In southeast Alaska, Sitka spruce buds are boiled in sugar water to make spruce bud syrup.

FIGURE 22.17   A modern newsprint factory.
(Courtesy International Paper Company)

The tracheids of spruces have spiral thickenings on the inner walls. These apparently are responsible for giving the wood a resonance that makes it ideal for use as soundboards for musical instruments (Fig. 22.18). Sitka spruce of the Pacific Northwest produces a strong resilient wood that is a favorite material of manufacturers of light aircraft.

Larches, which along with the dawn redwood and bald cypress are exceptions to the rule that conifers are evergreens, have some of the toughest of all conifer woods. Fence posts of larch are known to last 20 years. In the southern and southwestern United States, posts of juniper wood and bald cypress last even longer, some remaining usable for 40 to 50 years or more. The resin of the western larch has been used in the manufacture of baking powder, and the European larch is the source of a special type of turpentine.

There are about 40 species of true firs, which are widely used in the construction, plastic, and paper industries; as ornamentals; and as Christmas trees. The balsam fir produces on its bark blisters containing a clear resin. This resin, known as Canada balsam, was used in the past for cementing optical lenses and is still used on microscope slides for making permanent mounts. It has medicinal properties, too, and was used by New England colonists in sore throat medications.

Douglas fir, found in the mountains of the West, is not a true fir. In the Pacific Northwest, it grows into giant trees

FIGURE 22.18   A violin with a soundboard made from red spruce.

FIGURE 22.19   Bald cypress trees in a southern swamp. Note the "knees" protruding above the surface.

that are second only to the redwoods in size. It is probably the most desired timber tree in the world today. The wood, which is strong and relatively free of knots as a result of rapid growth with less branching than most other conifers, is heavily used in plywoods and is a major source of large beams. A useful wax is extracted from the bark of Douglas firs. Exploitation has nearly eliminated old-growth stands, but large numbers of new trees are being grown in managed forests.

Coastal redwoods are also prized for their wood, which contains substances that inhibit the growth of fungi and bacteria. The wood is light in weight, strong, and soft, but it splits easily. It is used for some types of construction, furniture, posts, and greenhouse benches and for many other purposes. California wineries use it extensively for wine barrels. The Giant Redwoods (Big Trees) are so huge that a double bowling alley was built on the log of the first specimen cut down. They are no longer logged and now serve almost exclusively as tourist attractions in California. They are, however, now being planted as timber trees in Romania and Yugoslavia.

Wood of the bald cypress, found in southern swamps, is like that of the redwood in being very resistant to decay. In the past, it was used for railroad ties, coffins, general construction, guttering, and shingles. The trees are well known for their "knees," which rise above the water as tapering growths from the roots (Fig. 22.19). At one time, it was widely believed that these were a means of admitting oxygen to the roots, but this is now in doubt. The knees are favored for making knickknacks, such as wall ornaments and lamp bases. The leaves of the bald cypress yield a red dye.

A dull red dye can be obtained from the younger bark of the eastern hemlock, which is also a source of tannins for shoe leather. The tanning industry so depleted the native eastern hemlocks that it now has had to use tropical substitutes. The wood of these small trees contains exceptionally hard knots, which can chip an axe blade. It sputters and throws out sparks freely when placed on a fire. Native Americans made a poultice for scrapes and cuts by pounding the inner bark. British Columbia Indians used to scrape out the cambium and phloem of both the western and mountain hemlocks for food, killing the trees in the process.

A Living Fossil?

Just imagine a real-life Jurassic Park, where prehistoric dinosaurs long thought extinct actually are alive. Well, something close to this actually happened in Australia in 1994. David Noble, while hiking on his Christmas holiday in the Wollemi National Park near Sydney, stepped into a stand of trees which until that moment were thought to have been extinct for over 150 million years. Known only as fossils, there they stood—about 40 pine trees growing in an isolated, rugged rain forest gully that had protected them all these years. This is one of the most significant botanical finds of the century—and the trees are among the rarest plants alive.

Now called the Wollemi Pine (*Wollemia nobilis*), these conifer (cone-bearing) trees are so distinctive that they constitute a new genus. While most conifers have dark green leaves, Wollemi pine leaves are brighter and lighter green, almost the color of green apples. The leaves are complex and unusual. The trees are bisexual, with bright green female cones and brown cylindrical male cones on the tips of upper branches. Also distinctive is the corklike knobby bark that is an unusual chocolate brown. The tallest tree is estimated to be 150 years old, towering 130 feet from the ground with a trunk about 10 feet wide. The new trees belong to the plant family Araucariaceae which, while found only in the southern hemisphere today (Norfolk Island Pine is an example), had a worldwide distribution during the Jurassic and Cretaceous periods (208–66 million years ago). This suggests that the Wollemi pines' closest relatives lived when Australia, New Zealand, Africa, South America, and India were all parts of the supercontinent Gondwana. During this period the eastern coast of Australia lay close to the South Pole at a time when worldwide climates were relatively warm to hot and wet.

Sydney's Royal Botanic Gardens and the New South Wales National Parks and Wildlife Service are jointly studying the Wollemi pine with methods ranging from scanning electron microscopy of the leaf and pollen, to DNA extraction and gene-sequencing studies. One priority is to study its propagation methods so that the plant can be established in cultivation. Although the

Drawings of Wollemi pine. *Top:* Leaves and branch; *Middle:* Male (pollen cone) at end of branch; *Bottom:* Female (seed) cone. Scale bar = 2 cm.
Illustration by David MacKay, Royal Botanic Gardens, Sydney.

Wollemi site is within 200 km of Sydney, Australia's largest city, the exact location remains a secret to protect the plant from seed collectors and poachers.

The Wollemi pines, protected in their sheltered spot, not only escaped when their closest relatives died some 50 million years ago, but remained hidden until this exciting, remarkable find. "Wollemi" is an aboriginal word meaning "look around you." What an appropriate name for this real living fossil—it reminds all of us that there remains much to understand and explore in our world.

Eastern white cedar, also known as *arborvitae,* is a favorite ornamental in temperate areas. The wood is pliable, and several Native American tribes used it for canoes. The Atlantic or southern white cedar was the first tree to be used for the construction of pipes for pipe organs in North America. During World War II, old logs of this species found in a swamp in New Jersey were milled and used in the construction of patrol torpedo boats. The fleshy red aril surrounding the seeds of yews is sweet and edible, but the seeds themselves and other parts of the plants are definitely poisonous. The wood, which is tough and resilient, is favored for making bows.

English yew, the wood of longbows, changed history in 1415 and brought an end to the Middle Ages at the Battle of Agincourt, when the English longbow proved superior to heavily armored cavalry. A red or purple dye can be obtained from the bark and roots of yew. Podocarps, two species of which are valuable timber trees in New Zealand, have edible seeds.

In 1989, researchers at the Johns Hopkins Oncology Center reported that 12 women with advanced ovarian cancers that had not responded to traditional therapies (including, in some cases, surgery) had a decrease of 50% or more in the size of their tumors and one woman's tumor disappeared altogether after treatment with *taxol,* a drug obtained from the bark of the Pacific yew tree. Pacific yew trees are small and do not occur in extensive stands. They also grow so slowly it takes more than 70 years to attain their full size, and removal of their bark kills them. Despite the problems of the drug's availability, scientists are excited about the potential of the drug in the fight against ovarian cancer, and the trees are now being mass-produced in nurseries.

The wood of incense cedars has been used in the manufacture of venetian blinds and pencils. Red cedar wood is also used for pencils as well as for cedar chests, closet lining, fence posts, and cigar boxes. It was used at one time in Germany for smoking hams. An aromatic oil used in floor-sweeping compounds is extracted from red cedar wood, and the "berries" of this and related junipers are widely eaten by birds. Many Native American tribes used the berries and inner bark as survival food during bleak winters. Some roasted the dried berries and brewed a beverage from them. Western red cedar was the most important single plant of Native Americans of the Northwest who used it for housing, clothing, nets, canoes, totem poles, medicines, and other purposes. Berries of the dwarf juniper are used to flavor gin. Some authorities indicate that the word *gin* may have been derived from *genievre,* the French word for *juniper berry.*

Other Gymnosperms

Florida arrowroot starch was once obtained from the extensive cortex and pith of a species of cycad whose northern-most distribution occurs in Florida. Before it could be used for human consumption, however, a poisonous substance had to be leached out. Since cycads grow too slowly to make continued preparation of the arrowroot starch profitable, the practice has been abandoned. Today, cycads that are seen outside of their natural habitats are being grown primarily for ornamental purposes. In Louisiana and elsewhere, the large compound leaves are used in Palm Sunday religious services. Some species, which are nearing extinction in the wild, may soon be known only in botanical gardens and conservatories.

Despite the foul odor of the fleshy seed coat of seeds of *Ginkgo,* the starchy food reserves of the seeds themselves have a nutlike flavor punctuated with a hint of shrimp. In the Orient, *Ginkgo* seeds are widely used for food, either boiled or roasted, and are found imported in canned form in Chi-nese food stores of large metropolitan areas in the United States and Canada. Extracts of *Ginkgo* plants have been clinically demonstrated to improve blood circulation in humans, and in 1990 more than 20 million Europeans were routinely taking *Ginkgo* extracts to counteract the effects of aging. Clinical studies have shown that *Ginkgo* leaf extracts do, indeed, improve blood supply to the brain and lungs, thereby improving both short-term memory and breathing.

In the southwestern United States, joint firs (*Ephedra*) are grazed by livestock, and the leaves and stems are still brewed into *Mormon tea.* To offset a slight bitterness, a teaspoon of sugar, honey, or jam is added to each cup of tea. Native Americans and pioneers used a concentrated version of the tea in the treatment of venereal diseases. They also ground the seeds into flour from which a bitter bread was made. The drug *ephedrine,* which is widely used in the treatment of asthma and other respiratory problems, still is extracted from a Chinese species, but most now in use is synthetically produced. The Chinese used ephedrine medicinally more than 4,700 years ago. One species of *Gnetum* is cultivated in Java for its shoots, which are cooked in coconut milk and eaten. Fibers from the bark are made into a rope.

See Appendix 1 for scientific names of species discussed in this chapter.

Summary

1. The exposed seeds of gymnosperms are produced on sporophylls forming a strobilus, or cone. The female (seed) cones and smaller male (pollen) cones are produced on the sporophytes.

2. A female gametophyte develops within a nucellus that is enclosed in an integument inside an ovule. The integument becomes the seed coat of a seed after fertilization, and the female gametophyte nurtures the development of the embryo.

3. One division and three subdivisions of living gymnosperms are recognized. Division Pinophyta includes Subdivision Cycadicae (cycads), Subdivision Pinicae (*Ginkgo* and conifers), and Subdivision Gneticae (gnetophytes).

4. Pines are the most numerous conifers. The needlelike leaves, which are arranged in clusters of two to five, have modifications adapting them to harsh environments.

5. Resin canals that occur throughout the plants secrete resin, which inhibits fungi and certain insect pests.

6. Pine xylem lacks vessel members and fibers and is relatively soft. The phloem lacks companion cells but has albuminous cells, which apparently perform the same function. Pine roots are always associated with mycorrhizal fungi, which are essential to normal development of the plants.

7. Two kinds of spores are produced. Microspores are produced in papery male cones that, in turn, develop in clusters toward the tips of lower branches. The microsporangia develop in pairs toward the base of the male cone scales and give rise to four-celled pollen grains, which occur in huge numbers.

8. Megaspores are formed in ovules at the bases of female cone scales. The integument of the ovule has a pore called the *micropyle*. One megaspore develops into a female gametophyte. The mature female gametophyte contains archegonia.

9. Before the archegonia mature, pollen grains catch in sticky pollination drops between the cone scales. Each pollen grain produces a pollen tube that digests its way down to the developing archegonia, and two of the original four cells in the pollen grain migrate into the tube as it grows. The generative cell divides and produces a sterile cell and a spermatogenous cell that itself divides, producing two sperms.

10. After pollination, one sperm unites with the egg, forming a zygote. The zygote develops into an embryo of a seed that has a membranous wing formed from a layer of the cone scale.

11. Some conifers produce seeds enclosed in fleshy or berrylike coverings. Their evolutionary origin is not clear.

12. *Ginkgo* has small fan-shaped leaves with evenly forking veins. The life cycle of *Ginkgo* is similar to that of cycads. The edible seeds are enclosed in a fleshy covering, which has a rank odor at maturity.

13. Cycads superficially resemble palm trees with unbranched trunks and crowns of large, pinnately divided leaves. They have strobili and life cycles similar to those of conifers, but their sperms, unlike those of pines, have numerous flagella.

14. Gnetophytes all have vessels in their xylem. Half the species are in the genus *Ephedra,* whose members have jointed stems and leaves reduced to scales. *Gnetum* species have broad leaves and occur in the tropics, primarily as vines. *Welwitschia* is confined to southwest African deserts. Its stem is in the form of a shallow cup with straplike leaves that extend from the rim; basal meristems on the leaves constantly add to their length.

15. The seeds and inner bark of pines are edible, and a tea has been made from the leaves. Eastern white pine stems were used as masts for sailing vessels and for crates, furniture, flooring, paneling, and matchsticks. Western white pine is the source of most such lumber today.

16. Resin from pines consists of turpentine and rosin. Turpentine is used as a solvent, and rosin is used by musicians and by baseball players. Dammar from kauri pines is used in colorless varnishes. Amber is fossilized resin. Resin is also used in floor waxes, printer's ink,

paper coatings, perfumes, and in the manufacture of menthol.

17. White spruce is the chief source of newsprint. It was also used for basketry and canoe lashing by Native Americans, with molasses or honey for treating scurvy, and in brewing a beer. Spruce resin was used for a type of chewing gum. Today, the wood is used as soundboards for musical instruments and in the construction of aircraft.

18. Larch and juniper woods are used for fence posts. Firs are used in the construction, paper, ornament, and Christmas tree industries. Douglas fir is probably the most desired timber tree in the world today.

19. Coastal redwoods are also prized for their wood, which is resistant to fungi and insects. Bald cypress wood, used in the past for coffins and shingles, is also resistant to decay.

20. A dye and tannins are obtained from the eastern hemlock. Native Americans used parts of hemlocks for poultices and for food.

21. Eastern white cedar's wood was used for canoes, and that of the Atlantic cedar was used for construction of pipes for pipe organs. Yew wood is used for making bows, and an extract has potential for the treatment of human ovarian cancer.

22. Podocarps of New Zealand have edible seeds. Incense cedar wood is used for cedar chests, cigar boxes, pencils, and fence posts. Juniper berries are used to flavor gin and were used by Native Americans for food and a beverage.

23. Arrowroot starch was once obtained from a cycad. *Ginkgo* seeds are edible, and *Ginkgo* plant extracts are used to improve blood circulation. Mormon tea is brewed from the leaves and stems of joint firs (*Ephedra*), which, in the past, were also a source of the drug ephedrine and a venereal disease treatment. One *Gnetum* species is cultivated in Java for food.

Review Questions

1. What is a gymnosperm? How is it distinguished from any other kind of organism?

2. What is the difference between a seed and a spore?

3. How are the leaves of pines different from those of broadleaf flowering plants?

4. What is a resin canal, and what is its function? Where are resin canals found?

5. How do pines differ in their reproduction from ground pines and ferns?

6. How do pollen grains differ from spores or sperms?

7. Which conifers discussed in this chapter do not have woody female cones?

8. If you had samples of leaves of a pine, a cycad, *Ginkgo,* a joint fir, and *Welwitschia,* construct a key to indicate how you could tell them apart.

9. What parts of a pine are considered edible?

10. What is resin? Discuss some of its uses, past and present.

Discussion Questions

1. *Ginkgo* and cycads have broad leaves, while those of pines are needlelike. Can you suggest any significance of this difference in terms of the climates and habitats involved?

2. If no distinction were made at the level of Kingdom between plants and animals, what would be the equivalent, if any, of sporophyte and gametophyte in humans?

3. Both bristlecone pines and redwoods can live to be thousands of years old. What do you suppose makes this possible?

4. Most of the old-growth stands of conifers in North America are now gone, and others will be gone soon. Much of what has been harvested is being replaced with new growth, sometimes with hybrids and non-native plants. Our forests are essential to our economy as we know it. If you had the power to change the way we manage and exploit our natural forest resources, what would you do differently, assuming you did not want to damage the economy?

5. If money were no object and you wished to landscape your yard primarily with gymnosperms, what would you include, taking into account your particular geographical area? Why?

Additional Reading

Beck, C. B. 1988. *Origin and evolution of gymnosperms.* New York: Columbia University Press.

Bever, D. N. 1981. *Northwest conifers*: *A photographic key.* Portland, OR: Binford and Mort Publishing.

Bold, H. C., C. S. Alexopoulos, and T. Delevoryas. 1987. *Morphology of plants and fungi,* 5th ed. New York: Harper & Row.

Farjon, A. 1990. *A bibliography of conifers.* Champaign, IL: Koeltz Science Books.

Jones, D. L. 1993. *Cycads of the world.* Washington, DC: Smithsonian Institution Press.

Krussman, G. 1985. *Manual of cultivated conifers.* Portland, OR: Timber Press.

Singh, H. 1978. *Embryology of gymnosperms* (*Encyclopedia of plant anatomy,* vol X, no. 2). Forestburgh, NY: Lubrecht & Cramer.

Smith, W. K., and T. M. Hinckley (Eds.). 1994. *Resource physiology of conifers: Acquisition, allocation, and utilization.* San Diego, CA: Academic Press.

Vidaković, M. 1991. *Conifers. Morphology and variation* (M. Soljan, Trans.). Zagreb, Croatia: Graficki Zavod Hrvatske. Distr. by University of Arizona Press.

Weiner, M. A. 1991. *Earth medicine earth food.* New York: Fawcett Book Group.

Chapter Outline

Flowers of the butterfly weed (Asclepias curassavica), *a native of the tropics from Florida to South America. Parts of the plant are used medicinally in tropical areas. In Mexico, for example, root extracts have been used in the treatment of cancers of the intestinal tract, kidneys, and uterus.*

Flowering Plants

Overview————————

This chapter begins with a probe of the differences between angiosperms and gymnosperms and then discusses the theoretical origin of flowering plants. The exceptional diversity of the flowering plants is reviewed next, and the chapter continues with a description of the parallel development of the gametophytes in the anthers and ovules. This leads up to pollination, followed by some details of fertilization and the development of a seed.

A discussion of trends of specialization and classification in flowering plants is followed by a section on pollination ecology. Flower preservation, including simple herbarium techniques and practice, comes next. The chapter closes with a brief survey on the uses of herbaria and a word of caution to plant collectors concerning unnecessary depletion of native floras.

Some Learning Goals

1. Understand the basic differences between angiosperms and gymnosperms.
2. Contrast two principal schools of thought concerning the origin of the flowering plants and the nature of the first flowers.
3. Diagram the life cycle of a flowering plant, indicating shifts from haploid to diploid cells and vice versa.
4. Compare two types of embryo sac development, and learn how a male gametophyte develops.
5. Know the characteristics of flowers associated with specific types of pollinators.
6. Know major trends of specialization in the flowering plants.
7. Know the functions of a herbarium and the techniques of preparing herbarium specimens.

A flowering plant once saved my life in an unconventional way. While in Washington's Olympic National Park, I was conducting field research on a group of relatively small herbaceous plants known as *bleeding hearts* (Fig. 23.1). At one point, I found myself on a slope of loose shale above a steep cliff. Suddenly, the shale started to move and, before I could do anything about it, I was rapidly sliding on my back, feet first, directly toward the edge of the cliff. There were no bushes, trees, or anything stable I could cling to or grab that would at least slow my journey to doom in an avalanche of shale. Just a few feet before I was to hurtle over the cliff, I cannoned into a large clump of bleeding hearts that somehow had rooted securely enough beneath the shale surface to stop my speedy plunge. Obviously, I lived to tell the tale, and it won't surprise you to know that, ever since that experience, I have had a special place in my garden for bleeding hearts!

The Pacific bleeding hearts that saved my life are only one of more than 250,000 known species of flowering plants that comprise, by far, the largest and most diverse of the divisions of the Plant Kingdom. These flowering plants are called **angiosperms.**

FIGURE 23.1   Pacific bleeding heart flowers.

The term *angiosperm* is derived from two Greek words: *angeion,* meaning "vessel," and *sperma,* meaning "seed." The vessel is the **carpel,** which is like an inrolled leaf with seeds along its margins. A seed develops from an *ovule* within a carpel and is part of an *ovary* that becomes a *fruit*. Although the angiosperms generally have organs and tissues similar to those of the gymnosperms, the enclosed ovules and seeds of the angiosperms distinguish them from gymnosperms, which have exposed ovules and seeds.

All angiosperms are presently considered to be in the Division Magnoliophyta (other classifications of flowering plants are discussed in Chapter 16). The Magnoliophyta are divided into two large classes: the Magnoliopsida (dicots) and the Liliopsida (monocots). The features distinguishing members of the two classes from one another are listed in Table 8.1.

Since Darwin's *On the Origin of Species* appeared in 1859, there have been two major theories concerning the origin of angiosperms. One, held by the German botanist Adolph Engler and his followers, suggests that flowering plants evolved from conifers and that primitive flowers are similar in structure to the strobili of conifers. Such flowers are inconspicuous and in clusters, like those of the grasses, oaks, willows, and cattails.

Most botanists now suspect, however, that angiosperms evolved independently from the *pteridosperms* (seed ferns—discussed in Chapter 22). They also think that a flower is really a modified stem bearing modified leaves. The most primitive flower is thought to be one with a long receptacle and many spirally arranged flower parts that are separate and not differentiated into sepals and petals. In addition, the stamens and carpels are flattened. Such flowers are found among relatives of magnolias and buttercups, leading many modern botanists to postulate that all present-day flowering plants are derived from a primitive stock with such characteristics. A further discussion of primitive and advanced characteristics follows later in the chapter.

DIVISION MAGNOLIOPHYTA— THE FLOWERING PLANTS

The plants of members of this division vary greatly in size, shape, texture, form, and longevity. The division includes, for example, the tiny duckweeds, which may be less than 1 millimeter long; all the grasses and palms; many aquatic and epiphytic plants; and most shrubs and trees, including the huge *Eucalyptus regnans* trees of Tasmania, which rival the redwoods in total volume.

Other flowering plants are parasitic. Dodders, for example, occasionally cause serious crop losses as they twine about their hosts and, by means of *haustoria* (shown in Fig. 5.13), intercept food and water in the host xylem and phloem. Broomrapes also parasitize a variety of plants, as do mistletoes. Mistletoes produce some chlorophyll and depend only partially on their hosts for food. Still others, such as the beautiful snowplant (Fig. 23.2) and some of the orchids, are saprophytes. The vast majority of flowering plants, however, produce their food independently through photosynthesis.

Like the gymnosperms, the angiosperms are *heterosporous* (produce two kinds of spores), and their sporophytes are even more dominant than those of the gymnosperms. The female gametophytes are wholly enclosed within sporophyte tissue and reduced to only a few cells. At maturity, the male gametophytes consist of a germinated pollen grain with three nuclei.

Development of Gametophytes

While the flower is developing in the bud, a diploid **megaspore mother cell** differentiates from all the other cells in the ovule (Fig. 23.3). This cell undergoes meiosis, producing four haploid *megaspores*. Soon after they are produced, three of these megaspores degenerate and disappear, but the nucleus of the fourth undergoes mitosis and the cell enlarges. While the cell is growing larger, its two haploid nuclei divide once more. The resulting four nuclei then divide yet another time. Consequently, eight haploid nuclei in all are produced (without walls being formed between them). By the time these three successive divisions have been completed, the cell has grown to many times its original volume. At the same time, two outer layers of cells of the ovule differentiate. These layers, called **integuments**, later become the **seed coat** of the seed. As they develop, they leave a pore, or gap, called the **micropyle** at one end.

At this stage, there are eight haploid nuclei in two groups, four nuclei toward each end of the large cell. One

FIGURE 23.2   Snowplant flowers.
(Courtesy Robert A. Schlising)

nucleus from each group then migrates toward the middle of the cell. These two **polar nuclei** may become a binucleate cell, or they may fuse together, forming a single diploid nucleus. Cell walls also form around the remaining nuclei. In the group closest to the micropyle, one of the cells functions as the female gamete, or *egg*. The other two cells, called *synergids,* either are destroyed or degenerate during or after events that occur later. At the other end, the remaining three cells, called *antipodals,* have no apparent function and later they also degenerate. The large sac, usually containing eight nuclei in seven cells, constitutes the *female gametophyte*. At maturity, it is called an **embryo sac** (Fig. 23.4).

Usually while the female gametophyte is developing, a parallel process that leads to the formation of male gametophytes takes place in the **anthers.** As an anther develops, four patches of tissue differentiate from the main mass of cells. These patches of tissue contain many diploid **microspore mother cells,** each of which undergoes meiosis, producing a quartet (also referred to as a *tetrad*) of **microspores.** Four chambers or cavities (pollen sacs) lined with nutritive *tapetal* cells are visible in an anther cross section by the time the microspores have been produced. As the anther matures, the walls between adjacent pairs of chambers break down so that only two larger sacs remain.

After meiosis, the microspores in the pollen sacs undergo several changes more or less simultaneously. The following three changes are the most important:

1. the nucleus in each microspore divides once;

2. the members of each quartet of microspores separate from one another (in some species the separation does not occur, but this is unusual); and

3. a two-layered wall, whose outer layer is often finely sculptured, develops around each microspore.

When these events are complete, the microspores have become **pollen grains** (Fig. 23.5). The outer layer of the pollen grain wall, called the **exine,** contains chemicals that may later react with other chemicals in the stigma of a

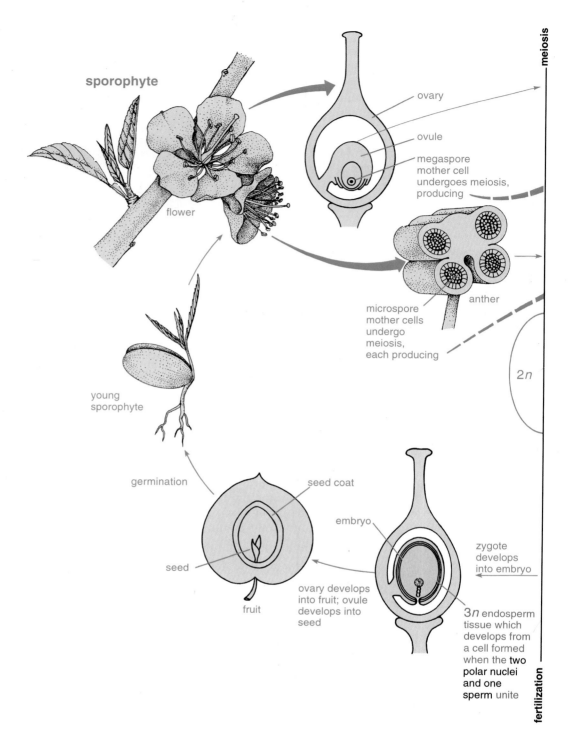

sporophyte

flower

ovary

ovule

megaspore
mother cell
undergoes meiosis,
producing

microspore
mother cells
undergo
meiosis,
each producing

anther

meiosis

2n

young
sporophyte

germination

seed coat

embryo

zygote
develops
into embryo

seed

fruit

ovary develops
into fruit; ovule
develops into
seed

3n endosperm
tissue which
develops from
a cell formed
when the **two
polar nuclei
and one
sperm** unite

fertilization

FIGURE 23.3   Life cycle of a typical flowering plant.

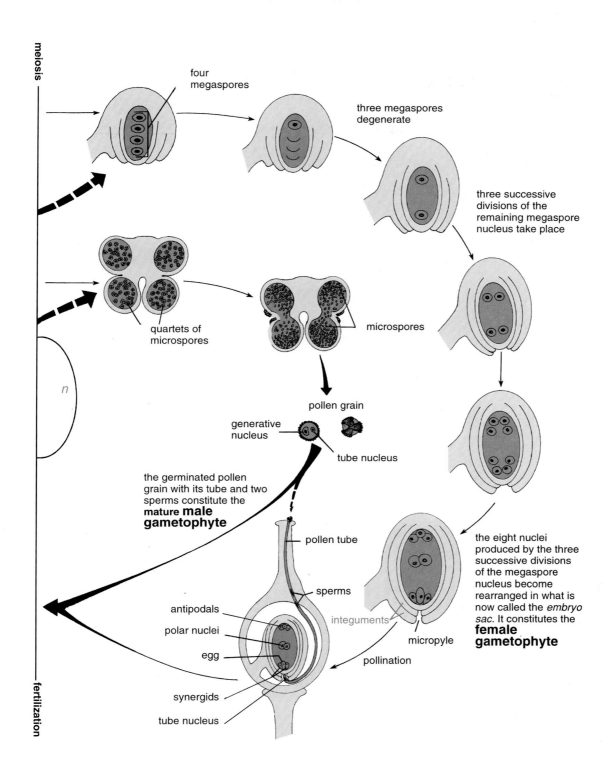

meiosis

four
megaspores

three megaspores
degenerate

three successive
divisions of the
remaining megaspore
nucleus take place

quartets of
microspores

microspores

n

pollen grain

generative
nucleus

tube nucleus

the germinated pollen
grain with its tube and two
sperms constitute the
mature **male
gametophyte**

the eight nuclei
produced by the three
successive divisions
of the megaspore
nucleus become
rearranged in what is
now called the *embryo
sac*. It constitutes the
**female
gametophyte**

pollen tube

sperms

antipodals

polar nuclei

integuments

micropyle

egg

pollination

synergids

tube nucleus

fertilization

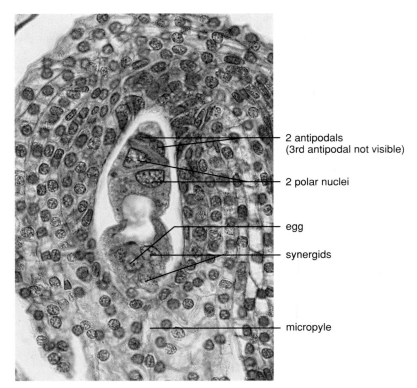

FIGURE 23.4   A mature embryo sac of a lily (*Lilium* sp.).

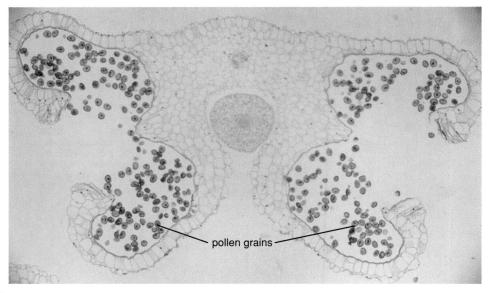

FIGURE 23.5   A cross section of a lily (*Lilium* sp.) anther.

flower. As a result of these reactions, the pollen grain may germinate or its further development may be blocked, depending on whether it originated from the same plant, another plant of the same species, or a plant of a different species.

Many pollen grains have three boat-shaped or porelike thin areas (*apertures*) in the wall, but the number of apertures may range from one to many. The development of the male gametophyte, discussed in the next section, may involve any of the apertures. One of the pollen grain's two nuclei, the **generative nucleus,** will later divide, producing two sperms. The remaining *tube nucleus* is involved in events that take place after the pollen grain has left the anther (Fig. 23.6).

FIGURE 23.6 Scanning electron micrograph of a poppy pollen grain. The furrows are apertures through which a pollen tube may later emerge.

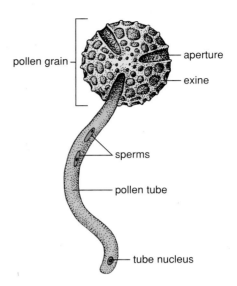

FIGURE 23.7 A mature male gametophyte of a flowering plant.

Pollination

The press and popular magazines often refer to flowers being *pollenized* (a coined word not found in dictionaries), and the implication is made that the dusting of the pollen of one flower upon another is the equivalent of fertilization. Not so! **Pollination** is simply the transfer of pollen grains from an anther to a stigma, nothing more. **Fertilization** involves the union of egg and sperm, and it may not occur until days or weeks or months after pollination has taken place or it may not follow pollination at all.

Most pollination is brought about by insects or wind, but water, birds, bats, other mammals, and gravity act as agents, or pollinators, in many species. The adaptations between the flower and its pollinators can be intricate and precise and may even in certain instances involve force, drugs, or sexual enticement (as shown in Fig. 23.18). A discussion of pollination ecology is given later in this chapter.

Fertilization and Development of the Seed

After pollination has occurred, further development of the male gametophyte may not take place unless the pollen grain is (1) from a different plant of the same species or (2) from a variety different from that of the flower receiving it.

Under suitable conditions, the dense cytoplasm of the pollen grain absorbs substances from the stigma and bulges out through one of the apertures in the form of a tube. This **pollen tube** then grows down between the cells of the stigma and style until it reaches the micropyle. In corn, it may have to grow more than 50 centimeters (20 inches) before it ar-

rives at its destination, but in most plants the distance is considerably less.

The pollen tube's journey may last less than 24 hours and usually does not take more than two days, although there are a few plants in which growth takes over a year. As the tube grows, most of the contents of the pollen grain are discharged into it. The tube nucleus stays at the tip, while the generative nucleus (cell) lags behind and divides by mitosis, usually in the tube, producing two sperms without flagella. Sometimes, the generative nucleus (cell) divides before the pollen tube has formed. The germinated pollen grain with its tube nucleus and two sperms constitutes the *mature male gametophyte* (Fig. 23.7).

When the pollen tube reaches the micropyle, it continues on to the embryo sac, which it enters, destroying a synergid in the process; it then either bursts or forms a pore, discharging its contents. Next, an event unique to angiosperms, called **double fertilization** (or **double fusion**), takes place. One sperm migrates from the synergid to the egg, losing most of its protoplasm along the way. The sperm nucleus then unites with the egg nucleus, forming a **zygote.**

The other sperm also migrates from the synergid toward the polar nuclei and, upon arrival, unites with the polar nuclei, producing a 3*n* (triploid) *endosperm nucleus.* The endosperm nucleus becomes exceptionally active and divides repeatedly by mitosis, with cell cycles being completed every few hours, usually without formation of cell walls. This nutritive 3*n* tissue, called **endosperm,** may eventually have hundreds of thousands of nuclei; it surrounds the *embryo* that develops from the zygote.

In some monocots, such as corn and other grasses, the endosperm tissue becomes an extensive part of the seed, but in most dicots, such as members of the Legume Family (e.g., peas, beans), the endosperm provides nutriment for the embryo that develops from the zygote but disappears by the time the seed is mature. The integuments harden, becoming a *seed coat,* and the remaining haploid nuclei, or cells (antipodals, synergids, and tube nucleus), degenerate. At the conclusion of these various events, the ovule has become a seed, and, at the same time, the ovary matures into a fruit. Seed dispersal and fruits are discussed in Chapter 8.

Other Types of Embryo Sac Development

The process of embryo sac development just described occurs in about 70% of the known flowering plants. The remaining 30% exhibit variations in which the embryo sac has from four to 16 nuclei or cells at maturity, and the endosperm may be 5*n*, 9*n*, or even 15*n*.

One such variation occurs in lilies, which microscope slide manufacturers favor for showing embryo sac development. When the megaspore mother cell undergoes meiosis in a lily, all four of the haploid megaspore nuclei produced remain functional nuclei (not cells). Three of the megaspore nuclei unite, forming one 3*n* nucleus; the fourth megaspore nucleus remains *n*. Both the 3*n* nucleus and the *n* nucleus undergo mitosis twice, resulting in a large cell with four 3*n* nuclei plus four *n* nuclei. The large cell with eight nuclei becomes the embryo sac; one of the polar nuclei is 3*n* and the other polar nucleus is *n*. During fertilization when one sperm (*n*) unites with the two polar nuclei, a 5*n* endosperm nucleus is produced (Fig. 23.8).

Apomixis and Parthenocarpy

Some embryos of seeds can develop *apomictically;* that is, without development or fusion of gametes (sex cells) but with the normal structures (e.g., ovaries) otherwise being involved. An embryo may develop, for example, from a 2*n* nutritive cell or other diploid cell of an ovule, instead of from a zygote. After germination, this makes the plant that develops from such a seed the equivalent of a vegetatively propagated plant. Fruits that develop from ovaries having unfertilized *eggs* (female sex cells) are said to be **parthenocarpic.** Such fruits, which are seedless, are found in navel oranges, supermarket bananas, and certain varieties of figs and grapes. *Apomixis* is discussed in Chapter 15.

To further complicate matters, seedless fruits are not all parthenocarpic. For example, when Thompson seedless grapes are fertilized, the ovules don't develop with the fruit. Also, applying dilute hormone sprays to flowers can bring about artificial parthenocarpy. Seedless tomatoes are often produced in this way. Crossing watermelon varieties with different chromosome numbers results in hybrid watermelons that are seedless because the chromosomes can't pair properly during meiosis, and fertilization and seed formation don't occur.

Trends of Specialization and Classification in Flowering Plants

Ever since Theophrastus grouped plants into trees, shrubs, and herbs in the third century B.C., various classifications

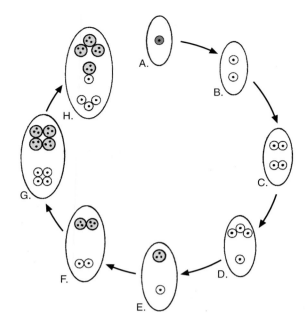

FIGURE 23.8 How the embryo sac of a lily develops. *A.* The process begins with a diploid (2*n*) megaspore mother cell. *B.* and *C.* The megaspore mother cell undergoes meiosis, producing four haploid (*n*) megaspore nuclei. *D.* and *E.* The megaspore nuclei unite, forming a triploid (3*n*) nucleus; the other haploid nucleus remains separate. *F.* and *G.* Both the 3*n* and *n* nucleus undergo two consecutive mitotic divisions, resulting in four 3*n* and four *n* nuclei. *H.* Three of the 3*n* nuclei function as antipodals; the fourth 3*n* nucleus and one *n* nucleus function as polar nuclei. The remaining haploid nuclei function as an egg and two synergids.

have been proposed. The first classifications were merely for convenience and did not necessarily reflect natural relationships. Even today, plants may be lumped together in unnatural groupings in order to make them easier to identify. For example, some wildflower books arrange together all white-flowered species or all yellow-flowered species. There is nothing wrong with such identification arrangements, but because such schemes do not reflect natural relationships (i.e., relationships based on heredity), it is often difficult to recognize in such schemes family characteristics or lineage. We don't infer that all blondes are more closely related to each other than they are to brunettes or that all long-haired dogs are more closely related to each other than they are to short-haired dogs. Modern botanists, therefore, try to group plants according to their natural relationships, which are based on evidence gleaned from breeding experiments, form and structure, chemical components, fossil records, and other features. There are, however, many interpretations of trends in the specialization of flowering and other plants that are based primarily on inference from limited evidence presently available.

Although the information is very incomplete, the fossil record suggests that the flowering plants first appeared about 160 million years ago during the late Jurassic period (see Table 21.1) and that they developed during the

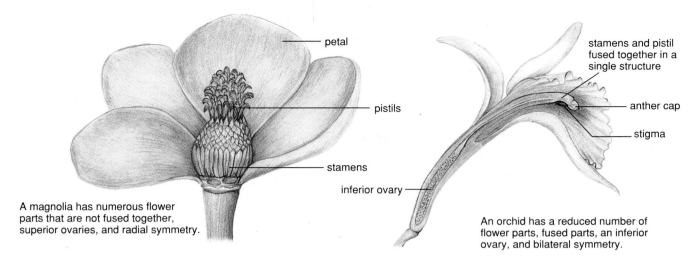

petal

pistils

stamens

inferior ovary

stamens and pistil fused together in a single structure

anther cap

stigma

A magnolia has numerous flower parts that are not fused together, superior ovaries, and radial symmetry.

An orchid has a reduced number of flower parts, fused parts, an inferior ovary, and bilateral symmetry.

FIGURE 23.9   Comparison between a primitive flower, magnolia (*left*) and an advanced flower, orchid (*right*).

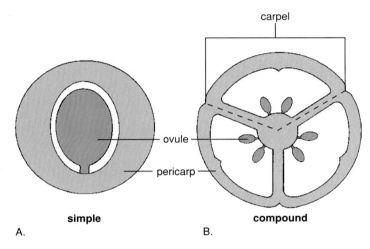

carpel

ovule

pericarp

simple

A.

compound

B.

FIGURE 23.10   Ovaries in cross section. *A.* The ovary of a simple pistil as found in a peach. *B.* The ovary of a compound pistil as found in a lily.

Cretaceous and the ensuing periods of the Cenozoic era into the dominant elements of the flora they are today. Most botanists believe that primitive flowering plants, which they think originated in the Jurassic period, had the following features: their leaves were simple; the flowers had numerous spirally arranged parts that were not fused to each other and were variable in number; the flowers were also *radially symmetrical* or *regular* (i.e., the flowers could be divided into two equal halves along more than one lengthwise plane); and the flowers had both stamens and pistils.

Although today there still are many plants whose flowers have primitive features, various specializations and modifications have occurred since flowers first appeared. Flower parts have become fewer and definite in number. Some parts have fused, and spiral arrangements have been compressed to whorls (Fig. 23.9).

The first pistil is believed to have been formed from a leaflike structure with ovules along its margins. The edges of the blade apparently rolled inward and fused together. Such a fertile blade is called a *carpel*. In due course, the separate carpels of primitive flowers fused together, forming the common *compound pistil* consisting of several carpels, which is found in many of today's angiosperms (Fig. 23.10). Each segment of an orange, for example, represents a single carpel, and three carpels are easily distinguishable in a cross section of a cucumber.

In advanced flowers, the receptacle or other flower parts have fused to the ovary and grown up around it, so that the calyx and the corolla appear to be attached at the top of the ovary. When the ovary is embedded in the receptacle and other parts, it is said to be an **inferior ovary,** in contrast to a more primitive **superior ovary,** which is produced on top of the receptacle with the other flower parts attached around its base. Some flowers have ovaries in an intermediate or *half-inferior* position (Fig. 23.11). Flowers have also tended to become *bilaterally symmetrical* or *irregular* (i.e., capable of

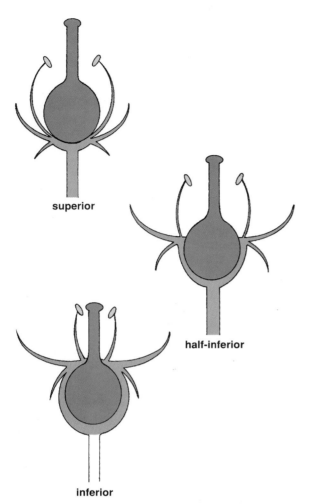

FIGURE 23.11 Ovary positions in flowers: superior (e.g., peach), half-inferior (e.g., cherry), and inferior (e.g., apple).

A.

B.

FIGURE 23.12 Unisexual flowers of a squash. *A.* Male. *B.* Female. Note the inferior ovary beneath the corolla.

being divided into two symmetrical halves only by a single lengthwise plane passing through the axis), as in sweet peas and orchids.

In some families, the flowers have become *imperfect* (*unisexual*). Each imperfect flower has either stamens or a pistil but not both. The Pumpkin Family (e.g., pumpkins, squashes, watermelons, cantaloupes, cucumbers) has imperfect flowers (Fig. 23.12). When both male and female imperfect flowers occur on the same plant, the species is **monoecious.** If male or female flowers occur only on separate plants, however, the species is **dioecious.**

In both the dicots and the monocots, evolutionary specialization has involved reduction and fusion of parts and a shifting of the ovary from a superior to an inferior position, as hypothetical progression is made from primitive families through intermediates to those that are advanced.

Observations on the human and ecological relevance of the flowering plants occupy literally thousands of volumes. A brief overview of several aspects of this subject is given in chapters 24 and 25.

POLLINATION ECOLOGY

When certain *consumers,* such as insects, forage for food among photosynthetic *producers,* such as plants, they often come in contact with flowers. Many insects and other animals become dusted with pollen, and as they feed or collect pollen and nectar, they unknowingly but effectively bring about pollination of the plants they visit. Throughout the evolutionary history of the flowering plants, the pollinators

A.

B.

FIGURE 23.13 Flower markings on coneflowers. *A.* In ordinary light. *B.* In ultraviolet light.

have evidently coevolved with plants. In some instances, the relationship between the two has become highly specialized.

Twenty thousand different species of bees are included among the pollinators of present-day flowering plants. By far, the best known of these are honey bees. Their chief source of food is nectar, but they also gather pollen for their larvae. The flowers that bees visit are generally brightly colored and mostly blue or yellow—rarely pure red. Pure red appears black to bees, and they generally overlook red flowers. Flowers often have lines or other distinctive markings, which may function as honey guides that lead the bees to the nectar. Bees can see ultraviolet light (a part of the spectrum not visible to humans), and some flower markings are visible only in ultraviolet light, making patterns seen by bees sometimes different from those seen by humans (Fig. 23.13).

Many bee-pollinated flowers are delicately sweet and fragrant. In contrast, flowers pollinated by beetles tend to have stronger, yeasty, spicy, or fruity odors. Beetles don't have keen visual senses, and flowers pollinated by them are usually white or dull in color. Some beetle-pollinated flowers don't secrete nectar but either furnish the insects with pollen or have food available on the petals in special storage cells, which the beetles consume.

Some flowers, including the stapelias of South Africa (Fig. 23.14), smell like rotten meat. Short-tongued flies pollinate such flowers, which tend to be dull red or brown. These plants are related to our milkweeds, although superficially they don't resemble milkweeds at all. They are often called *carrion flowers* because of their foul odor and appearance. Flies with longer tongues may also pollinate bee-pollinated flowers.

Moth- and butterfly-pollinated flowers, like bee-pollinated flowers, often have sweet fragrances. Night-flying moths visit flowers that tend to be white or yellow—colors that stand out against dark backgrounds in starlight or moonlight. Some very specialized relationships between moths and flowers occur between certain small moths and

FIGURE 23.14 A *Stapelia* (carrion flower) plant.

members of the genus *Yucca;* these relationships are discussed on the back cover of this book.

Red flowers are sometimes pollinated by butterflies, some of which can detect red colors. The nectaries of these flowers are found at the bases of corolla tubes or spurs, where only moths and butterflies with longer tongues can reach. However, an enterprising bumblebee will occasionally bypass convention and chew through the base of a spur to get at the nectar.

Birds—particularly the hummingbirds of the Americas and the sunbirds of Africa—and the flowers that they pollinate are also adapted to one another (Fig. 23.15). The birds do not have a keen sense of smell, but they have excellent vision. Their flowers are often bright red or yellow and usually have little if any odor. Bird-pollinated flowers also are typically large or are part of a large inflorescence.

Birds are highly active pollinators and tend to burn energy rapidly. They must feed frequently to sustain themselves. Many of the flowers birds prefer produce copious amounts of nectar, thereby assuring repeated visits. The nectar is often produced in long floral tubes, which keep most insects out. Some native California fuchsias and their relatives have long threads extending from each pollen grain (Fig. 23.16). When a hummingbird inserts its long bill into such a flower, the pollen grain threads catch on short stiff hairs located toward the base of its bill, and in this way the bird unknowingly transfers pollen from one flower to another.

Bat-pollinated flowers, found primarily in the tropics, tend to open only at night when the bats are foraging (Fig. 23.17). These flowers are dull in color, and like flowers pollinated by birds, they either are large enough for the animal to insert part of its head or consist of ball-like inflorescences containing large numbers of small flowers that dust the visitor with pollen.

The very large Orchid Family, which has approximately 35,000 species, has pollinators among all the types mentioned. Some of the adaptations between orchid flowers and their pollinators are extraordinary (Fig. 23.18). The pollen grains of most orchids are produced in little sacs called **pollinia** (singular: **pollinium**), which typically have sticky pads at the bases. When a bee visits such a flower, the pollinia are usually deposited on its head. The "glue" of the sticky pads dries almost instantly, causing the pollinia to adhere tightly. In some orchids, the pollinia are forcibly slapped on the pollinator through a trigger mechanism within the flower.

Members of *Ophrys,* a genus of orchids found in Europe and in North Africa, have a modified petal that resembles a female bumblebee or wasp. Male bees or wasps emerge from their pupal stage a week or two before the females and apparently mistake the flowers for potential mates. They try to copulate with the flowers, and while they are doing so, pollinia are deposited on their heads. When they visit other flowers, the pollinia catch in sticky stigma cavities. During a single visit by the pollinators, the pollinia from a previously visited flower and stuck to the insect's head are removed and replaced with fresh pollinia.

The pollinia of orchids pollinated by moths and butterflies are attached to their long tongues with sticky clamps instead of pads. Some bog orchids attach their pollinia to the eyes of their pollinators, which happen to be female mosquitoes. After a few visits, the mosquitoes are blinded and

FIGURE 23.15 A hummingbird visiting a flower.

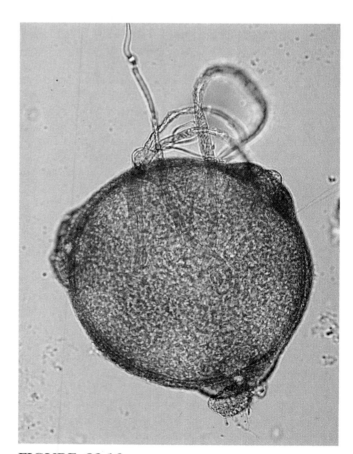

FIGURE 23.16 A pollen grain of a California fuchsia, showing the threads that catch in the short bristle hairs at the base of a hummingbird's bill, ×1,000.

FIGURE 23.17   A bat visiting a flower of an organ-pipe cactus. The bat's head is covered with pollen.

(Photo by Donna J. Howell)

HERBARIA AND PLANT PRESERVATION

The botanical resources of many universities and other institutions include **herbaria** (singular: **herbarium**), which are something like libraries of dried and pressed plants arranged so that specific plants are easily retrieved. Properly prepared and maintained specimens should remain in excellent condition for 300 or more years. Formal training and experience aren't needed to make one's own herbarium. The materials, some simple equipment (which may be homemade), and the ability to follow a few relatively elementary procedures are all that is necessary.

Methods

The moisture content of flowers and other plant parts to be preserved should be reduced as quickly as possible, with a minimum of distortion. This is usually done with the aid of a *plant press* (Fig. 23.19). This simple device consists of two pieces of plywood (or other wood materials or thin metal plates) with dimensions of approximately 30 × 46 centimeters (12 × 18 inches) and a pair of webbing or leather straps to go around the boards. Place a number of *felts* (sheets of heavy blotting paper) of similar dimensions between the boards. Place a folded page of newspaper between each pair of blotters and intersperse a few sheets of stiff cardboard in the stack of blotters.

Wash off any soil clinging to the roots of a specimen to be pressed, and lay the plant out on one of the newspaper sheets. Carefully straighten out leaves, as well as petals and other plant parts, so that they are not folded during pressing. Pencil notes on where, when, and by whom the specimen was collected on the newspaper or use a number corresponding to such notes in a separate field notebook. Then fold the newspaper over the specimen and place it between two blotters (felts).

Many specimens may be placed in a press at one time or, if there is space, between the same sheets of newspaper. Place only one species in a single fold of newspaper, however, as mixing species invariably leads to some being pressed better than others, due mostly to the varying amounts of woody tissues within the stems. After returning the newspaper and specimen to the press, tighten the straps around it as much as possible and leave it to dry in the sun or near a heater (but not close enough to scorch!) for three or four days. Unless the leaves were succulent or wet at the time they were placed in the press, they should be dry enough to mount on paper at this time. If a press is not available, plants may be pressed between newspapers and blotters by placing heavy weights on top of them.

If possible, 100% rag-content paper should be used for mounting the specimen, as pulp-content papers deteriorate with age. The paper should also be of heavy weight to lend some support to the now brittle specimen. The pressed plant(s) may be attached to the paper in one of several ways.

unable to continue their normal activities—a striking example of biological controls within an ecosystem.

Among the most bizarre of the orchid pollination mechanisms are those in which the pollinator is dunked in a pool of watery fluid secreted by the orchid itself; the pollinator escapes under water through a trapdoor. The route of the insect ensures contact with pollinia and stigma surfaces. In other orchids with powerful narcotic fragrances, pollinia are slowly attached to the drugged pollinator. When the transfer of pollinia has been completed, the fragrance abruptly disappears. The temporarily stupefied insect then recovers and flies away.

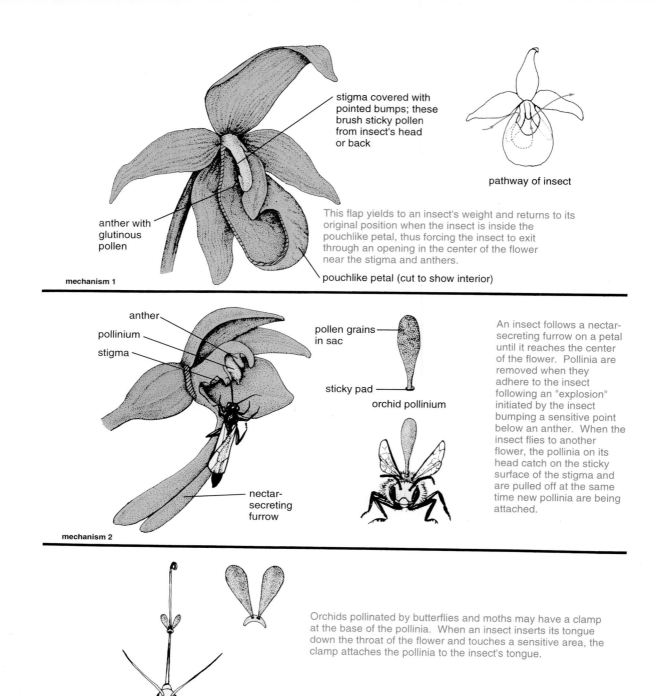

stigma covered with pointed bumps; these brush sticky pollen from insect's head or back

anther with glutinous pollen

pathway of insect

This flap yields to an insect's weight and returns to its original position when the insect is inside the pouchlike petal, thus forcing the insect to exit through an opening in the center of the flower near the stigma and anthers.

pouchlike petal (cut to show interior)

mechanism 1

anther

pollinium

stigma

pollen grains in sac

sticky pad

orchid pollinium

An insect follows a nectar-secreting furrow on a petal until it reaches the center of the flower. Pollinia are removed when they adhere to the insect following an "explosion" initiated by the insect bumping a sensitive point below an anther. When the insect flies to another flower, the pollinia on its head catch on the sticky surface of the stigma and are pulled off at the same time new pollinia are being attached.

nectar-secreting furrow

mechanism 2

Orchids pollinated by butterflies and moths may have a clamp at the base of the pollinia. When an insect inserts its tongue down the throat of the flower and touches a sensitive area, the clamp attaches the pollinia to the insect's tongue.

mechanism 3

FIGURE 23.18   Some pollinating mechanisms in orchids.

Some workers spread a good white library glue on a glass plate and then place the specimen on the wet glue so that one side is covered with it. They then transfer the specimen to herbarium paper, which normally measures 29 × 42 centimeters (11.5 × 16.5 inches). Others place the specimen on the paper first and add glue or liquid plastic at strategic points. The bottom right-hand corner of the paper should be kept clear for a label indicating the scientific name of the plant, collection information, the collector's name, and the collection date (Fig. 23.20).

Professional botanists also number their individual collections, giving a collection number after the collector's

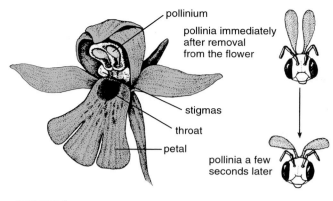

pollinium

pollinia immediately after removal from the flower

stigmas

throat

petal

pollinia a few seconds later

mechanism 4

In the showy orchids, there are two separate stigma patches. When an insect visits a flower, the pollinia are attached to its head; then they twist outward and forward so that, when the insect visits another flower, the pollinia are simultaneously deposited in the separate stigmas while new pollinia are attached to the insect's head.

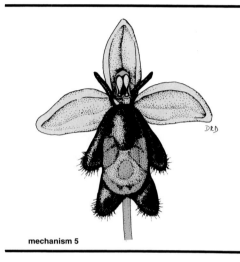

mechanism 5

One genus of orchids native to Europe and North Africa has a petal modified to resemble a female wasp. Male wasps try to copulate with the "female" and in doing so bump a sensitive area containing pollinia. The pollinia are "glued" to a wasp's head in the process and carried to another flower. The wasps thus effectively bring about cross-pollination.

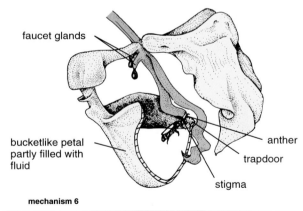

faucet glands

bucketlike petal partly filled with fluid

anther

trapdoor

stigma

mechanism 6

The bucket orchids of South America have faucet glands that partially fill a bucketlike petal with fluid. In visiting the flower, the insects may fall into the fluid. They escape by a trapdoor. The pathway of an insect assures contact with both the stigma and anthers containing pollinia, since they are located just in front of the trapdoor.

name. The label should measure about 7.5 × 12.5 centimeters (3 × 5 inches), and its paper should also be of 100% rag content. Collectors then place specimens in manila folders and store them in some systematic fashion so that retrieval of individual specimens is easily accomplished. Major institutional herbaria usually arrange the plants by families in a presumed evolutionary sequence and then arrange the genera and species alphabetically within the families. Mothballs or similar material should be added to the cabinet or storage area to prevent insect damage during storage.

Some people are interested in pressing flowers for use in dry arrangements on cards, placemats, or other decorative

FIGURE 23.19   A plant press.

items. In such cases, the pressing is done in the same manner as for the herbarium specimens, but the material is then further manipulated in one of several ways. For wall art, mount the specimens on a piece of smooth cardboard, covered with clear plastic or glass, mat them, and frame them (Fig. 23.21). To make decorative notepaper, place pressed flowers on the paper and cover them with rice paper or facial tissue. Then brush a mixture of one part white glue to three parts water over the tissue, causing it to become a permanent mount, with the specimen clearly visible through the thin paper film. Clear contact paper cut to the appropriate size makes a good mount for pressed flowers placed on cards or placemats. Pressed flowers can also be embedded in clear plastic poured into molds.

Flowers can also be dried without pressing. With a little patience, relatively lifelike three-dimensional preservations can be made, although flowers with petals that are not easily detached lend themselves more readily to three-dimensional drying than do others. Most such drying is done in a shoebox, but almost any type of container can be used. Cover the bottom of the box with about 2 centimeters (0.8 inch) of sand, silica gel, or borax mixture. Then gently lay the fresh flower, with a little of the peduncle or stem still attached, on the surface. After this, slowly drizzle more sand, silica gel, or borax mixture by hand into the box until the entire flower is buried, being careful not to create air pockets around any parts (Fig. 23.22).

Each of the three drying agents mentioned has certain advantages and drawbacks. Sand must be thoroughly washed several times to be certain it is perfectly clean before use. The sand also needs to be completely dry, uniformly fine, and if possible, have individual grains that are relatively rounded rather than angular. Such sand with rounded grains is found around the Great Salt Lake in Utah and is available in arts and crafts stores. It takes about two weeks to dry most flowers with sand. Silica gel is also available commercially and is the quickest drying agent of the three mentioned, usually completing the job in four to

five days (if the box is placed in a microwave oven, the drying process may be completed in as little as five minutes). Its granules tend to be of different sizes, however, giving the flower surface a slightly irregular texture. It also tends to darken certain colors, and it is expensive. When borax is used, two parts of it should be mixed with one part sand or cornmeal. It dries flowers in about three weeks, but is sometimes difficult to remove completely from the dried flower surfaces.

When drying has been completed, tilt the container and slowly and gently pour out the sand or other drying material in an uninterrupted motion. If any petals have come loose, glue them in place later. Wire may be inserted in the stem to add rigidity, and after any clinging granules have been carefully removed, soft, powdered colored chalk may be dusted on to restore any fading of color. It is recommended that beginners use fairly large flowers whose petals are not easily detached (e.g., zinnias, roses) for their first attempts at three-dimensional flower drying.

A Word of Caution to Collectors

Dried plant collections of herbaria, in particular, have proved invaluable in botanical research in the past and will continue to do so in the future. They have facilitated quick identifications of plants in emergency situations in which children have eaten plants or plant parts suspected of being poisonous and have helped pin down specific plants that have caused allergic reactions. Herbaria also have been involved in archaeological research, helping determine the uses of plants by past cultures and have been used for teaching purposes at various educational levels. Herbarium specimens have been useful in criminal litigation for both the prosecution and the defense and have been the primary source of information on the distribution of plants with potential for new agricultural and horticultural crops or with possible medicinal values. Most of the unraveling of problems pertaining to natural relationships of plants begins in a herbarium, and, without these plant libraries, our ability to increase our knowledge along many practical and theoretical lines would be severely restricted.

Literally hundreds of plants native to North America are now on rare and endangered species lists, and thousands more are in similar predicaments on other continents. *The day has come when both professional and amateur persons interested in plants must discipline themselves to exercise extreme caution in collecting native plants. Collectors should first know what they are collecting or otherwise refrain, and collecting for private collections without serious purpose should be strictly limited. Except for certain types of research, a good photograph of a plant may actually be preferable to a dried specimen and aesthetically more pleasing. It is sincerely hoped that, as much as possible, each reader will confine a collection to photographs of native plants.*

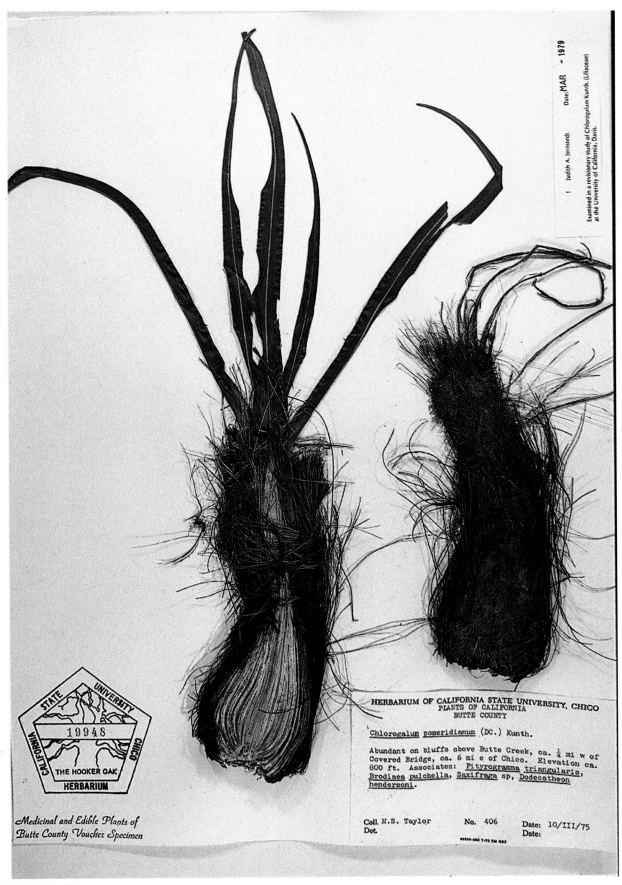

The following text appears within the herbarium specimen label in the image:

Date MAR – 1979

1. Judith A. Jernsted

Examined in a revisionary study of Chlorogalum Kunth. (Liliaceae)
at the University of California, Davis.

STATE UNIVERSITY
CALIFORNIA   CHICO
19948
THE HOOKER OAK
HERBARIUM

Medicinal and Edible Plants of
Butte County Voucher Specimen

HERBARIUM OF CALIFORNIA STATE UNIVERSITY, CHICO
PLANTS OF CALIFORNIA
BUTTE COUNTY

Chlorogalum pomeridianum (DC.) Kunth.

Abundant on bluffs above Butte Creek, ca. ¼ mi w of
Covered Bridge, ca. 6 mi e of Chico. Elevation ca.
800 ft. Associates: Pityrogramma triangularis,
Brodiaea pulchella, Saxifraga sp, Dodecatheon
hendersoni.

Coll. M.S. Taylor   No. 406   Date: 10/III/75
Det.   Date:

FIGURE 23.20   A mounted herbarium specimen.

FIGURE 23.21   Pressed flowers framed as wall art.

FIGURE 23.22   Three-dimensional drying. A flower placed in a box is embedded in sand, silica gel, or a borax mixture.

Summary

1. Flowering plants (angiosperms) have ovules and seeds completely enclosed within carpels, which comprise ovaries that become fruits. There is one division of flowering plants (Magnoliophyta); it is divided into two classes (dicots and monocots).

2. Englerian botanists believe flowering plants evolved from conifers. Most contemporary botanists believe they evolved separately from pteridosperms and that the first flowers had many separate, flattened parts spirally arranged on an elongate receptacle.

3. Flowering plants are heterosporous.

4. Flowering plant gametophytes develop in separate structures. The female gametophyte (embryo sac) develops in the ovule. Integuments, which later become a seed coat, surround the embryo sac. Pollen grains developed in anthers become male gametophytes.

5. Pollination is the transfer of pollen grains from an anther to a stigma, brought about by insects, wind, and other agents.

6. After pollination, a pollen tube may grow from a pollen grain to the embryo sac; the tube nucleus remains at its tip, and the generative nucleus divides, producing two sperms.

7. Following the discharge of the contents of the pollen tube into the embryo sac, a sperm unites with the egg, forming a zygote; the other sperm simultaneously unites with the polar nuclei, forming a $3n$ endosperm nucleus.

8. The endosperm nucleus becomes nutritive endosperm tissue that may become part of the seed or be used up by the seed's embryo.

9. Due to variations in how an embryo sac develops, some flowering plants produce $5n$, $9n$, or $15n$ endosperm tissue.

10. Some artificial groupings of flowers to aid identification do not reflect natural relationships. Sources of evidence used to try to group plants naturally include fossils that suggest the flowering plants first appeared about 160 million years ago.

11. Primitive flowering plants had simple leaves and numerous spirally arranged flower parts that were not fused to each other; they possessed both stamens and pistils and were radially symmetrical (regular).

12. Specializations include a reduction in the number of parts; fusion of parts; appearance of compound pistils composed of several individual carpels; inferior ovaries; bilateral symmetry (irregular flowers); and unisexual flowers.

13. Monoecious species have both male and female flowers on the same plants; dioecious species have male and female flowers on separate plants.

14. Bee-pollinated flowers are delicately sweet and fragrant and tend to be blue or yellow in color.

15. Beetle-pollinated flowers tend to have stronger odors and are usually white or dull in color.

16. Some fly-pollinated flowers emit foul odors.

17. Moth-pollinated flowers tend to be white or yellow.

18. Bird-pollinated flowers are usually bright red or yellow and have much nectar but little odor.

19. Most orchids produce pollen grains in pollinia that adhere to or clamp onto parts of visiting insects. The flowers of some orchid species have developed bizarre pollination mechanisms.

20. Herbaria are libraries of dried and pressed plants arranged so that specific plants may be readily located. Properly pressed plants may last for hundreds of years.

21. A plant to be pressed is placed in a plant press between sheets of newspaper and absorbent material. Dry specimens are affixed to sheets of high-quality paper with a label giving collection information.

22. Plant parts may be pressed for art, placemats, and so on. Flowers may be preserved three-dimensionally.

23. Because so many plants are now on rare and endangered species lists, collectors should try to confine their future collecting to photographs as much as possible.

Review Questions

1. What are the basic differences between gymnosperms and angiosperms?

2. What is the function of the integuments?

3. How are the pollen grains of flowering plants different from those of pine trees?

4. What is the functional equivalent of an archegonium in a flower?

5. What is the function of endosperm, and how does it originate?

6. What differences are there between the embryo sacs of peaches and lilies?

7. Distinguish between radial and bilateral symmetry.

8. What distinguishes primitive flowers from those that are advanced?

9. In general, how do flowers pollinated by bees differ from those pollinated by moths, flies, beetles, and birds?

10. What is a herbarium?

Discussion Questions

1. The world's largest flowers are on inconspicuous parasitic plants that resemble a fungus mycelium. If the flowers were not present, would it be possible to tell that the plant is not a fungus? How?

2. In most embryo sacs with eight nuclei, only the egg and the polar nuclei do not degenerate. What would happen if the megaspore itself functioned as an egg and did not go on to produce other nuclei in an embryo sac?

3. It takes only one pollen grain to initiate the development of an ovule into a seed, yet a single flower may produce many thousands of pollen grains. Do you suppose such huge numbers of pollen grains are really necessary? Why?

4. Are such items as the name of the collector and the date significant on a herbarium specimen? Explain.

5. Is it really important that plants be classified? What would be the consequences if they were not?

Additional Reading

Bold, H. C., C. S. Alexopoulos, and T. Delevoryas. 1987. *Morphology of plants and fungi,* 5th ed. New York: Harper & Row.

Cronquist, A. 1979. *How to know the seed plants.* Dubuque, IA: Wm. C. Brown Publishers.

Cronquist, A. 1988. *The evolution and classification of the flowering plants,* 2d ed. New York: New York Botanical Garden.

Faegri, K., and L. van der Pijl. 1972. *The principles of pollination ecology,* 2d ed. Elmsford, NY: Pergamon Press.

Geesink, R. 1991. *Thommer's analytical key to the families of flowering plants.* New York: State Mutual Book and Periodical Service.

Hughes, N. F. 1994. *The enigma of angiosperm origins.* New York: Cambridge University Press.

Johri, B. M. (Ed.). 1984. *Embryology of angiosperms.* New York: Springer-Verlag.

Joosten, T. 1988. *Flower drying with a microwave.* Asheville, NC: Altamont Press.

Kung, S. D. 1994. *Discoveries in plant biology,* 2 vols. Riveredge, NJ: World Scientific Publications.

Raven, P. H., R. F. Evert, and S. E. Eichhorn. 1992. *Biology of plants,* 5th ed. New York: Worth Publishers.

Stace, C. 1992. *Plant taxonomy and biosystematics,* 2d ed. New York: Cambridge University Press.

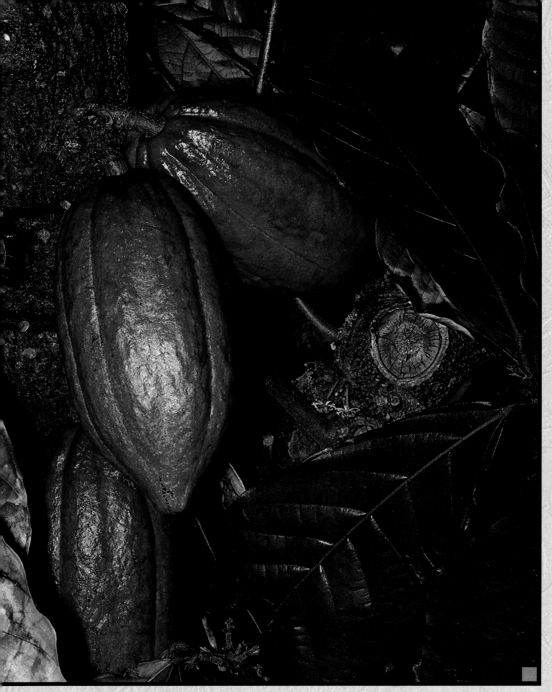

Pods of a tropical cacao tree (Theobroma cacao). *The seeds are the source of chocolate.*

24

Flowering Plants and Civilization

P*uccoons* are herbaceous plants that grow on dry plains and slopes throughout the western United States and British Columbia. One puccoon, which has greenish-yellow flowers, has seeds that are so hard it was given the common name of *stoneseed* in Nevada and California.

Native American women of the Shoshoni tribe of Nevada reportedly drank a cold water infusion of the roots of stoneseed every day for six months to ensure permanent sterility. Biologist Clellan Ford became curious about these reports and gave extracts of stoneseed plants to mice. He found the extracts effectively eliminated the estrous cycle of the mice and decreased the weights of the ovaries, the thymus, and the pituitary glands.

Although this reported Native American use of a plant was demonstrated experimentally to have a basis in fact, distinguishing between fact and fantasy in recorded past uses of plants is often difficult, particularly if the plants have become rare or extinct. But this type of research is essential today if we're going to save potential sources of medicinal drugs and other useful plant products before they are eliminated by clearing of land and other symbols of "progress."

Primitive peoples, despite occasional misguided superstition and folklore, did cure some of the diseases they treated with plants, even though they didn't know the scientific reasons for the results. As indicated earlier, botanists who have recognized this have teamed with anthropologists, medical doctors, and interpreters to interview tribal and other primitive medical practitioners in the tropics of the Americas and Africa. They are sifting through as much of this information as possible before it is too late. Their work is already leading to useful new discoveries.

ORIGIN OF CULTIVATED PLANTS

Alphonse de Candolle, a Swiss botanist, published in 1822 a book entitled *Origin of Cultivated Plants,* based on data he gathered from many sources. He deduced that cultivated plants probably originated in areas where their wild relatives grow.

Ninety-four years later, in 1916, N. I. Vavilov, a Russian botanist, began a follow-up of de Candolle's work. During the next 20 years, he expanded on de Candolle's work and modified his conclusions. Vavilov became persuaded, as a result of his research, that most cultivated plants differ appreciably from their wild relatives. He also concluded that dispersal centers of cultivated plants are characterized by the presence of dominant genes in plant populations, with recessive genes becoming apparent toward the margins of a plant's distribution. Vavilov recognized eight centers of origin of cultivated plants, with some plants originating in more than one center. The centers (Fig. 24.1), some of which are subdivided, are as follows.

Chinese Center

The Chinese Center (1), the earliest independent center, consists of mountainous areas and adjacent lowlands of western and central China. Cultivated plants believed to have originated here include millets, soybean and several other legumes, bamboo, radish, eggplant, cucumber, some citrus, peach, apricot, walnut, persimmon, tea, some sugar canes, and hemp.

Indian Center

The Indian Center (2A), second in importance only to the Chinese Center, consists of Burma and India (exclusive of the northwest portion). Included with the cultivated plants believed by Vavilov to have originated here are rice, sorghum, mung and several other beans, gourds, yam, mango, orange and other citrus fruits, sugar cane, bowstring hemp, black pepper, betel nut, cardamon, gum arabic, henna, senna, strychnine, and rubber plant.

Indo-Malayan Center

The Indo-Malayan Center (2B) consists of Indochina, Malaysia, Java, Borneo, Sumatra, and the Philippines. Vavilov believed giant bamboo, ginger, banana, breadfruit, candlenut, coconut, clove, nutmeg, Manila hemp, and many tropical fruits originated here.

Central Asiatic Center

The Central Asiatic Center (3) consists of northwest India, Afghanistan, and adjacent Soviet provinces. Wheat, garden pea, lentil, mustard, safflower, cotton, garlic, carrot, onion, basil, pear, almond, grape, apple, and other fruits and nuts are believed to have originated here.

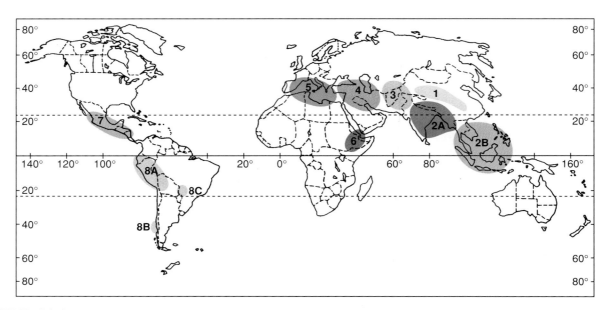

FIGURE 24.1 Major centers of origin of cultivated plants.

(After Vavilov, N. I. 1951. *The origin, variation, immunity and breeding of cultivated plants.* (K. Starr Chester, Trans.). New York: The Ronald Press Company, © 1951.)

Near-Eastern Center

The Near-Eastern Center (4) consists of the interior of Asia Minor, Iran, Transcaucasia, and the highlands of Turkmenistan. Cultivated plants believed to have originated here include several types of wheat, rye, barley and oats, alfalfa, vetch, anise, poppy, cantaloupe, cabbage, lettuce, fig, pomegranate, cherry, and hazelnut.

Mediterranean Center

The Mediterranean Center (5) consists of areas bordering the Mediterranean Sea. Cultivated plants from this region include additional varieties of wheat, fava beans, clover, flax, black mustard, olive, carob, beet, parsley, leek, chive, savory, celery, parsnip, rhubarb, caraway, thyme, hyssop, lavender, peppermint, sage, rosemary, hop, and chufa.

Abyssinian Center

The Abyssinian Center (6) consists of Ethiopia and Somaliland. Among the cultivated plants believed to have originated here are additional varieties of wheat, sesame, garden cress, coffee, okra, indigo, and fenugreek.

South Mexican and Central American Center

The South Mexican and Central American Center (7) consists of southern Mexico, Guatemala, El Salvador, Honduras, Nicaragua, Belize, and Costa Rica. Maize (corn), common bean, lima bean, sweet potato, pepper, some cotton varieties, prickly pear, papaya, cashew, cacao (source of chocolate), cherry tomato, and annatto (a dye and spice plant) are believed to have originated here.

South American Center

The South American Center (8A) consists of the highlands of Peru, Ecuador, and Bolivia. Several species of potatoes (other than the common white or Irish potato), tomato, pumpkin, marigold, cocaine plant, Egyptian cotton, guava, tobacco, and quinine tree are believed to have originated here.

Chiloe Center

The Chiloe Center (8B) is confined to an island off the coast of southern Chile. The common white or Irish potato and the strawberry are believed to have originated here.

Brazilian-Paraguayan Center

The Brazilian-Paraguayan Center (8C) consists of Brazil and Paraguay. Despite the large diversity of species in this huge area, few cultivated plants are believed to have originated here, but some have become major crop plants. Manioc (source of cassava), peanut, passion fruit, pineapple, Brazil nut, and Pará rubber tree are among the cultivated plants from this area.

Each of these centers is isolated by mountain ranges, deserts, or oceans, and together they constitute less than 3% of the total land area of the earth. Agricultural development appears to have arisen independently in each center as primitive peoples investigated various members of the diverse local floras and in time came to make use of the plants available to them. As of 1990, considerably less than 1% of the more than 250,000 species of flowering plants known had become significant crop plants.

SELECTED FAMILIES OF FLOWERING PLANTS

The following is a survey of a few of the well-known plant families, indicating not only past uses of some of their members but also possible future uses in a few instances, along with biological notes and miscellaneous observations.

The reader is cautioned against assuming that past medicinal uses were always effective or without harmful side effects and is urged to refrain from experimenting with such plants.

More than 300 families of flowering plants are recognized today. Flowers and fruits are considerably more reliable and stable indicators of heredity than are leaves or other vegetative plant parts, and so flowering plant families are distinguished from one another mostly on the basis of flower and fruit parts and structure. Some families include a mere handful of species, while others are very large, their members numbering in the thousands. Space permits only brief discussion of a few of the larger, better-known families here, and the reader is referred to the additional readings for other sources of information.

The families are taken up in a more or less phylogenetic sequence, beginning with those that have more generalized flower structure and progressing to those whose flowers are considered specialized. Generalized flowers tend to have numerous parts that are not fused together and ovaries that are superior. In specialized flowers, the parts are reduced in number and often fused together; the ovary is inferior. (See Fig. 23.9 and the amplification of this subject that begins on page 410.) The following is an abbreviated key to the families discussed:

1. Flowers with parts in fours or fives or multiples thereof; seeds with two cotyledons (DICOTS)
 2. Petals separate from one another or lacking
  3. Petals present
   4. Stamens more than twice as many as the petals
    5. Stamens, petals, and sepals attached to the rim of a cup surrounding the one to many pistils . . . . .Rose Family (Rosaceae)
    5. Stamens, petals, and sepals not attached to the rim of a cup
     6. Pistils several to many in each flower . . . . .Buttercup Family (Ranunculaceae)
     6. Pistil one
      7. Ovary superior . . . . .Poppy Family (Papaveraceae)
      7. Ovary inferior . . . . .Cactus Family (Cactaceae)
   4. Stamens not more than twice as many as the petals
    8. Herbaceous vines; fruit a pepo . . . . .Pumpkin Family (Cucurbitaceae)
    8. Primarily herbs, shrubs, and trees; fruit not a pepo
     9. Fruit a legume . . . . .Legume Family (Fabaceae)
     9. Fruit not a legume
      10. Fruit a silique or silicle . . . . .Mustard Family (Brassicaceae)
      10. Fruit not a silique or silicle
       11. Ovary superior; stems square in cross section; leaves opposite; fruit of four nutlets . . . . .Mint Family (Lamiaceae)
       11. Ovary inferior; stems rounded in cross section; leaves alternate; fruit a schizocarp . . . . .Carrot Family (Apiaceae)
  3. Petals lacking; calyx sometimes petal-like
       12. Ovary of three carpels and usually elevated on a gynophore; anthers splitting lengthwise . . . . .Spurge Family (Euphorbiaceae)
       12. Ovary of one carpel; gynophore lacking; anthers splitting by raised flaps . . . . .Laurel Family (Lauraceae)
 2. Petals fused together
       13. Flowers not in a head; each flower with its own receptacle; ovary superior . . . . .Nightshade Family (Solanaceae)
       13. Flowers in a head, several to many florets on a common receptacle, comprising a flowerlike inflorescence; ovary inferior . . . . .Sunflower Family (Asteraceae)
1. Flowers with parts in threes or multiples thereof; seed with one cotyledon (MONOCOTS)
        14. Flowers inconspicuous; without petals or sepals . . . . .Grass Family (Poaceae)
        14. Flowers conspicuous; the petals and sepals mostly similar in coloration
         15. Ovary superior; petals all alike . . . . .Lily Family (Liliaceae)
         15. Ovary inferior; one petal different in form from the other two . . . . .Orchid Family (Orchidaceae)

DICOTS

The Buttercup Family (Ranunculaceae)

Nearly all the 1,500 members of the Buttercup Family are herbaceous. The flowers, whose petals often vary in number, have numerous stamens and several to many pistils with superior ovaries (Fig. 24.2). Most have dissected leaves with no stipules and with petioles that are slightly expanded at the base. Well-known representatives include ornamental plants such as buttercup, columbine, larkspur, anemone, monkshood, and *Clematis* (Fig. 24.3). Most members of the Buttercup Family are concentrated in north temperate and arctic regions.

Columbine flowers, which have five spurred petals, resemble a circle of doves, and their name comes from *columba*, the Latin word for "dove." A blue and white species of columbine is the state flower of Colorado. Native Americans, for control of diarrhea, made a tea from boiled columbine roots, and members of at least two tribes believed columbine seeds to have aphrodisiac properties. A man would pulverize the seeds, and after rubbing them in the palms of his hands, he would try to shake hands with the woman of his choice, believing the woman would then succumb to his advances. Others crushed and moistened the seeds and applied them to the scalp to repel lice.

Most members of the family are at least slightly poisonous, but the cooked leaves of cowslips have been used for food, and the well-cooked roots of the European bulbous buttercup are considered edible. The European buttercup, in its natural state, causes blistering on the skin of sensitive individuals. East Indian fakirs are reported to deliberately blister their skin with buttercup juice in order to appear more pitiful when begging. Native Americans of the West gathered buttercup achenes, which they parched and ground into meal for bread. Others made a yellow dye from buttercup flower petals. Karok Indians made a blue stain for the shafts of their arrows from blue larkspurs and Oregon grape berries.

Goldenseal, which is still sold in health food stores, is a plant that was once abundant in the woods of temperate eastern North America. It has become virtually extinct in the wild because of relentless collecting by herb dealers. They sold the root for various medicinal uses, including remedying inflamed throats, skin diseases, and sore eyes. At least one Native American tribe mixed the pounded root in animal fat and smeared it on the skin as an insect repellent.

Monkshood yields a drug complex called *aconite*, which was once used in the treatment of rheumatism and neuralgia. Although popular as garden flowers, monkshoods are very poisonous. Death may follow within a few hours of ingestion of any part of the plant. Most species have purplish to bluish or greenish flowers, but one Asian monkshood, called *wolfsbane*, has yellow flowers. Wolf hunters in the past used to poison the animals with a juice obtained from wolfsbane roots.

FIGURE 24.2   A buttercup flower.

The Laurel Family (Lauraceae)

The Laurel Family is a primitive family whose flowers have no petals but whose six sepals are sometimes petal-like. The stamens, which occur in three or four whorls of three each, are a curiosity because the anthers open by flaps that lift up. The ovary is superior. Most of the approximately 1,000 species in this family are tropical evergreen shrubs and trees, many with aromatic leaves. The family received its name from the famous laurel cultivated for centuries in Europe. Its foliage was used by the ancient Greeks to crown victors in athletic events and later was used in the conferring of academic honors.

Several important spices come from members of this family. Powdered cinnamon is the pulverized bark of a small tree that is native to India and Sri Lanka, although it is also grown commercially elsewhere. Cassia, which is very similar, is today often sold interchangeably with cinnamon. Cinnamon oil is distilled from young leaves of the trees. Use of cinnamon and cassia dates back thousands of years. They were used in perfumes and anointing oils at the time of Moses, and other records reveal their use in Egypt at least 3,500 years ago.

Camphor has been used since ancient times. This evergreen tree, which is native to China, Japan, and Taiwan, is the main source of camphor essence still used in cold remedies and inhalants, insecticides, and perfumes. The essence is distilled from wood chips. Some American cities and towns with milder climates have been using camphor trees as street trees, because of their capacity to withstand smog.

Sassafras trees, which are native to the eastern United States and eastern Asia, also have spicy-aromatic wood. A flavoring widely used in toothpaste, chewing gum,

A.

B.

C.

D.

FIGURE 24.3　Representatives of the Buttercup Family. *A.* Columbine. *B. Hepatica. C.* Monkshood. *D. Isopyrum.* (*C.* Courtesy Donald E. Brink, Jr.)

mouthwashes, and soft drinks used to be obtained by distillation of wood chips and bark. Sassafras tea still is considered a refreshing beverage. Sassafras is also an ingredient of some homemade root beers, and in the southern states an alcoholic beer has been made by adding molasses to boiled

sassafras shoots and allowing the mixture to ferment. In Louisiana, powdered sassafras leaves (called *filé*) have been used as a thickening and flavoring agent for gumbo. It has in the past been used by country physicians for treating hypertension and for inducing a sweat in those with respiratory infections. Reports indicate that in large doses it has a narcotic-stimulant effect, and it is also reported to be carcinogenic. Most sassafras flavorings now in use are artificial.

The sweet bay, used as a flavoring agent in gravies, sauces, soups, and meat dishes, comes from the leaves of the laurel. Leaves of the related California bay (Fig. 24.4) are sometimes used as a substitute for sweet bay and for making Christmas wreaths. This tree, which is native to California,

FIGURE 24.4   A fruit and leaves of a California bay tree.

FIGURE 24.5   A prickly poppy flower.

also occurs in southwestern Oregon, where it is known as *myrtle* (true myrtles, however, belong to a different family). Its wood, which is hard, can be polished to a high luster and is used for making a variety of bowls, ornaments, and other smaller, wooden articles.

Early settlers in the West and Native Americans of the region used California bay for the relief of rheumatism, bathing in hot water to which a quantity of leaves had been added. The nutlike fruits (drupes) were roasted and used for winter food. A leaf was placed under a hat on the head to cure a headache (but even a small piece of leaf placed near the nostrils can produce an almost instant headache!). A few leaves placed on top of flour or grains in a canister will keep weevils away, and small branches have been used as chicken roosts to repel bird lice and fleas. A leaf or two rubbed on exposed skin functions as a mosquito repellent.

Avocados are also members of the Laurel Family. The fruits, which have more energy value by weight than do red meats, are rich in vitamins and iron.

The Poppy Family (Papaveraceae)

Most members of the Poppy Family are herbs distributed throughout temperate and subtropical regions north of the equator, but several poppies occur in the Southern Hemisphere, and a number are widely planted as ornamentals. Poppies, like buttercups, tend to have numerous stamens, but most have a single pistil (Fig. 24.5). Most also have milky or colored sap, and their sepals usually fall off as the flowers open. All members produce alkaloidal drugs.

Bloodroot is a pretty early flowering spring plant of eastern North American deciduous forests. A bright reddish sap, which is produced in its rhizomes, was used by some Native Americans as a facial dye, an insect repellent, and a cure for ringworm. Children today still paint their nails with it. It has a very bitter taste, except when ingested in minute amounts. The bitterness made it effective in inducing vomiting, but members of one tribe use it to treat sore throats after compensating for the bitterness by squeezing a few drops on a lump of maple sugar so that it could be held in the mouth.

Opium poppies have had a significant impact on societies of both the past and the present. Opium itself was described by Dioscorides in the first century A.D., and ancient Assyrian medical texts refer to both opium and opium poppies. Opium smoking, which does not extend back nearly as far as the use of the drug in other ways, became a major problem in China in the 1600s. Smoking opium has given way to other forms of use in recent years. The substance is obtained primarily by making small gashes in the green capsules of the poppies (Fig. 24.6). The crude opium appears as a thick, whitish fluid oozing out of the gashes and is scraped off. It contains two groups of drugs. One group contains the narcotic and addicting drugs morphine and codeine, which are best known for their widespread medicinal use as painkillers and cough suppressants. Members of the other group are neither narcotic nor addictive. They include papaverine, which is used in the treatment of circulatory diseases, and noscapine, which is used as a codeine substitute because it functions like codeine in suppressing coughs but does not have its side effects.

Heroin, a scourge of modern societies, is a derivative of morphine. It is from four to eight times more powerful than morphine as a painkiller. Less than 100 years ago, it was advertised and marketed in the United States as a cough suppressant. It is estimated that 75% of American drug

FIGURE 24.6   Immature opium poppy capsules that were gashed with a razor blade. Note the opium-containing latex oozing from the gashes.

FIGURE 24.7   Shepherd's purse.

addicts used heroin up until the early 1980s, but cocaine has now largely replaced it. The loss to society in terms of its economic, physical, and moral impact is enormous, since addicts frequently commit violent crimes to obtain the funds needed to support their habits, which often cost well over $200 per day.

The seeds of opium poppies contain virtually no opium and are widely used in the baking industry as a garnish. They also contain up to 50% edible oils, which are used in the manufacture of margarines and shortenings. Another type of oil obtained from the seeds after the edible oils have been extracted is used in soaps and paints.

The Mustard Family (Brassicaceae)

The original Latin name for the Mustard Family, still in widespread use today, was *Cruciferae*. The name describes the four petals of the flowers, which are arranged in the form of a cross. The flowers also have four sepals, usually four nectar glands, and six stamens, two of which are shorter than the other four. All members produce siliques or silicles (shown in Fig. 8.17), which are unique to the family. All 2,500 species of the family produce a pungent, watery juice, and nearly all are herbs distributed primarily throughout the temperate and cooler regions of the Northern Hemisphere.

Among the widely cultivated edible plants of the Mustard Family are cabbage, cauliflower, brussels sprouts, broccoli, radish, kohlrabi, turnip, horseradish, watercress, and rutabaga. Some edible members are also widespread weeds. The leaves of shepherd's purse (Fig. 24.7), for example, can be cooked and eaten, and the seeds can be used for bread meal. Other wild edible members are several cresses, peppergrass, sea rocket, toothwort, and wild mustard. Wild mustards are often weeds in row crops. Their leaves are sometimes sold as vegetable greens in markets.

The seeds of wild mustard, shepherd's purse, and several other members of this family produce a sticky mucilage when wet. Biologists at the University of California at Riverside discovered a potential new use for these seeds. They fed pelleted alfalfa rabbit food to mosquito larvae in water tanks, which they were using for experiments on mosquito control. They noticed that the larvae, which had to come to the surface at frequent intervals for air, often stuck to the pellets and suffocated. Curious, the workers examined the pellets under a microscope and found that they contained mustard seeds. Evidently, the field where the alfalfa had been harvested had also contained mustard plants. The scientists then tried heating the mustard seeds to kill them and found that this did not affect production of mucilage by wet seeds. It was calculated that 0.45 kilogram (1 pound) of such seeds could kill about 25,000 mosquito larvae. A few mosquito abatement districts have used the seeds effectively, but

FIGURE 24.8   A Sitka rose.

FIGURE 24.9   A raspberry.

experiments are needed to determine if there is a practical way to harvest many more seeds and control mosquitoes by such nonpolluting means on a much larger scale.

Native Americans mixed the tiny seeds of several members of this family with other seeds and grains for bread meal and gruel. To prevent or reduce sunburn, Zuni Indians applied a water mixture of ground western wallflower plants to the skin. Watercress, which is widely known as a salad plant, has had many medicinal uses ascribed to it. During the first century A.D., for example, Pliny listed more than 40 medicinal uses. Native Americans of the West Coast of the United States treated liver ailments with a diet consisting exclusively of large quantities of watercress for breakfast, abstinence from any further food until noon, and then resumption of an alcohol-free but otherwise normal diet for the remainder of the day. This was repeated until the disease, if curable, disappeared.

Dyer's woad, a European plant that has become naturalized and established in parts of North America, is the source of a blue dye that was used for body markings by the ancient Anglo-Saxons. Another member of the family, camelina, has been grown in the Netherlands for the oil that is obtained from its seeds. Camelina oil has been used in soaps and was once used as an illuminant for lamps.

The Rose Family (Rosaceae)

The Rose Family includes more than 3,000 species of trees, shrubs, and herbs distributed throughout much of the world. The flowers characteristically have the basal parts fused into a cup, with petals, sepals, and numerous stamens attached to the cup's rim (Fig. 24.8). The family is divided into subfamilies on the basis of flower structure and fruits. The flowers of one group have inferior ovaries and produce pomes for fruits. Flowers of other groups have ovaries that are superior or partly inferior and produce follicles, achenes, or drupes or clusters of drupelets.

The economic impact of members of the Rose Family is enormous, with large tonnages of stone fruits (e.g., cherries, apricots, peaches, plums), pome fruits (e.g., apples, pears), and aggregate fruits such as strawberries, blackberries, loganberries, and raspberries being grown annually in temperate regions of the world (Fig. 24.9).

Members of this family have been relevant to humans in many other ways in the past and still continue to be so. Roses themselves, for example, have for centuries been favorite garden ornamentals of countless numbers of gardeners, and the elegant fragrance of some roses delights many people. In Bulgaria and neighboring countries, a major perfume industry has grown up around the production from damask roses of a perfume oil known as *attar* (or *otto*) *of roses*. In a valley near Sofia, more than 200,000 persons are involved in the industry, whose product brought more than $2,200 per kilogram ($1,000 per pound) during the 1970s. A considerable quantity of the oil is blended with less expensive substances in the perfume industry. Perfume workers are reported rarely to develop respiratory disorders, suggesting that the plant extracts may have medicinal properties.

The fruits of wild roses, called *hips* (Fig. 24.10), are exceptionally rich in vitamin C. In fact, they may contain as much as 60 times the vitamin C of citrus fruit. Native Americans from coast to coast included rose hips in their diets (except for members of a British Columbia tribe, who believed they gave one an "itchy seat"), and it is believed that this practice contributed to scurvy being unknown among them. During World War II when food supplies became scarce in some European countries, children in particular were kept healthy on diets that included wild rose hips. The hips also contain, in addition to vitamin C, significant amounts of iron, calcium, and phosphorus. Today, many Europeans eat *Nyppon Sopa,* a sweet, thick pureé of rose hips, whenever they have a cold or influenza.

FIGURE 24.10 Mature rose hips.

After giving birth, the women of one western Native American tribe drank western black chokecherry juice to staunch the bleeding. Other tribes frequently made a tea from blackberry roots to control diarrhea. Five hundred Oneida Indians once cured themselves of dysentery with blackberry root tea, while many nearby white settlers, who refused to use "Indian cures," died from the disease. Men of certain tribes used older canes of roses for arrow shafts (presumably after removing the prickles!). Wild blackberries, raspberries, salmonberries, thimbleberries, dewberries, juneberries, and strawberries all provided food for Native Americans and early settlers, and they are still eaten today, either fresh or in pies, jams, and jellies. A spiced blackberry cordial is still a favorite for "summer complaints" in southern Louisiana. Wild strawberries are considered by many to be distinctly superior in flavor to cultivated varieties.

The Legume Family (Fabaceae)

The Legume Family is the third largest of the approximately 300 families of flowering plants, with only the Sunflower and Orchid Families having more species. Its 13,000 members, which are cosmopolitan in distribution, include many important plants. The flowers range in symmetry from radial (regular) to bilateral (irregular). The irregular flowers have a characteristic *keel* (which is a boat-shaped fusion of two petals enclosing the pistil), two *wing petals,* and a larger *banner* petal (Fig. 24.11). The stamens in such flowers are generally fused in the form of a tube around the ovary. The common feature that keeps the members together in one family is the fruit, which is a legume (shown in Fig. 8.16).

Important crop plants include peas, many kinds of beans (e.g., kidney, lima, garbanzo, broad, mung, tepary), soybeans, lentils, peanuts, alfalfa, sweet clover, licorice, and wattle. Wattle is an Australian tree that is grown commercially as a source of tannins for leather tanning. Carob, which is widely used as a chocolate substitute, is also a member of this family. Several copals (hard resins used in varnishes and lacquers) are obtained from certain legume plants, as are gum arabic and gum tragacanth, which are used in mucilages, pastes, paints, and cloth printing.

Important dyes, such as indigo, logwood (used in staining tissues for microscope slides and now scarce), and woadwaxen (a yellow dye), come from different legume plants. Locoweeds, which have killed many horses, cattle, and sheep, particularly in the southwestern United States, belong to a large genus (*Astragalus*) of about 1,600 species. The poisonous principle in those species affecting livestock seems to vary in concentration according to the soil type in which the plants are growing. Other poisonous legumes are lupines, jequirity beans, black locusts, and mescal beans.

About 90% of the members of the Legume Family exhibit leaf movements, but few are as rapid as those of the sensitive plant (*Mimosa pudica*), whose leaves fold within seconds in response to a disturbance. Sensitive plants, which grow as weeds in the tropics and the deep South of the United States, are discussed in Chapter 11 and are shown in Figure 11.17. Many of the movements of other legume plant leaves are correlated primarily with day length.

Clovers were widely used in the past by gatherers of wild food plants. The leaves are difficult to digest in quantity, but the rhizomes were gathered and usually roasted or steamed in salt water and then dipped in grease before being eaten. The seeds of both clovers and vetches also were gathered and either ground for meal or cooked in a little water and eaten as a vegetable. Today, seeds of several legumes, including alfalfa and mung beans, are popular for their sprouts, which are widely used in salads and Oriental dishes. A tropical bean called *winged bean* (Fig. 24.12) has unusually high levels of protein, and all parts of the plant are edible. It is presently being grown in several widely scattered tropical and subtropical regions and also is being marketed on a limited scale in some temperate zones. It is believed to have great potential for improving the diet of undernourished peoples throughout the tropics.

The Spurge Family (Euphorbiaceae)

Although many of the members of the Spurge Family are tropical, they are widespread in temperate regions both north and south of the equator. The stamens and pistils are produced in separate flowers, which often lack a corolla and are inconspicuous. In true spurges (*Euphorbia*), the female flower is elevated on a stalk called a *gynophore* and is surrounded by several male flowers that each consist of little more than an anther. Both the female and male flowers are inserted on a cup composed of fused bracts, the cup usually having distinctive glands on the rim. This type of inflorescence is called a *cyathium* (Fig. 24.13). Sometimes, the inconspicuous flowers are surrounded by brightly colored

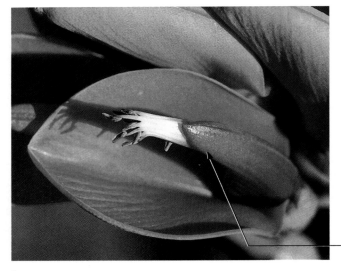

A.

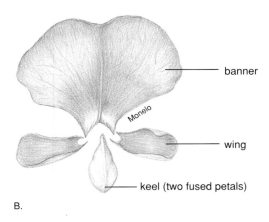

banner

Monelo

wing

keel (two fused petals)

B.

stamens (nine stamens are fused in a tube that surrounds the pistil; the tenth stamen is separate)

C.

FIGURE 24.11 *A.* Irregular (bilaterally symmetrical) flowers of a coral tree. *B.* Parts of a sweet pea flower. *C.* An inflorescence composed of regular (radially symmetrical) legume flowers.

FIGURE 24.12 A winged bean. Winged beans are highly nutritious, are easy to grow in the tropics, and hold promise for improving diets in developing countries.

bracts (e.g., *poinsettia,* shown in Fig. 7.18) that give the inflorescence the appearance of a single large flower. Most members of this large family produce a milky latex, and a number of species are poisonous.

Sooner or later, many gardeners experience "urges to purge spurges," as some members of the family are exceptionally aggressive weeds that reproduce very rapidly. Other large tropical spurges closely resemble columnar cacti. Several economically important plants are cultivated, particularly in frost-free regions. For example, an estimated 90 million metric tons (100 million tons) of cassava are harvested annually from plants cultivated in the jungles of South America, Africa, and eastern Asia. The roots, which develop thickened storage areas that resemble large sweet potatoes (shown in Fig. 5.16), are a diet staple of the tropics, as much as white potatoes and cereals are of temperate areas. Poisonous principles are removed by boiling, fermenting, or squeezing out the juice. In dried and powdered form, the cassava is known as *farina.* In Western countries, tapioca is prepared by forcing heated cassava pellets through a mesh while it is being agitated. Cassava starch is also used as a base for the production of alcohol, acetone, and other industrial chemicals.

Another cultivated spurge of the tropics is the Pará rubber tree, the source of the crude rubber from which most rubber products are made today. Although wild South American trees were the original commercial source of rubber, rubber trees have been widely planted in Indonesia, Africa, and adjacent areas. The trees, which vary in height from less than 5 to over 50 meters (16 to 164 feet), produce most of the latex in the inner bark.

The laticifers in which latex is secreted spiral around inside the trunk at an angle of about 30 degrees. Accordingly, cuts are made at the same angle into the inner bark to obtain maximum yields of latex, which trickles down into collecting cups that are attached to the tree. After collection, the latex is coagulated by chemicals or smoke and then shipped in sheet or crumbled form to processing plants. Sometimes, an anticoagulant is mixed with it, and the liquid

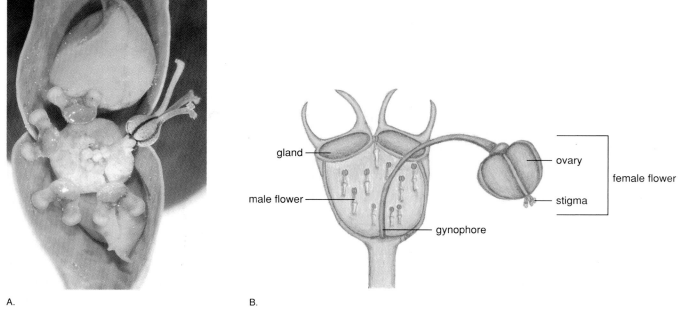

FIGURE 24.13   A spurge. *A.* Inflorescence. *B.* An individual cyathium.

is transferred in tankers. Much of the world's rubber goes into the manufacture of automobile and aircraft tires, but other products made from rubber are legion.

The Pará rubber tree should not be confused with the broad-leaved ornamental known as a *rubber plant*. The rubber plant, which is popular as a houseplant, also produces latex but is a member of the Fig Family (Moraceae).

The latex of other spurges may hold a key to future sources of fuel and lubricating oils. In 1976, Melvin Calvin, a University of California Nobel Prize winner, proposed the use of latex of gopher plants as a source of materials for oil. He estimated that such plants, which can grow in semidesert areas, would produce 10 to 50 barrels of oil per year on 0.4 hectare (1 acre) of land at a cost of $3 to $10 per barrel.[1]

A spurge called *candelilla* occurs in remote areas of Mexico. On its stems it produces a wax that is used in the making of candles and other wax products. Still another spurge produces seeds with a special oil used in plasticizers, and castor oil (from castor beans) is used in the manufacture

of nylon, plastics, and soaps. Castor beans themselves are very poisonous—as few as one to three are sufficient to kill a child. The plants, which grow very rapidly, are popular as ornamentals but are one of the leading natural causes of poisoning among American children.

A Mexican jumping bean is the seed of a certain spurge in which a small moth has laid an egg. When the egg hatches, the grub periodically changes position with a jerk, causing the seed to jump. The crown-of-thorns is an ornamental plant with somewhat flexible twisting stems bearing vicious-looking spines. Some believe it to have been the plant from which the crown of thorns for the head of Christ was made. Poinsettias, or Christmas flowers, are favorite yuletide plants in various parts of the world. Tung oil, used in oil paints and varnishes, and Chinese vegetable tallow, a substance used in the manufacture of soap and candles, are two more commercially important products obtained from the seeds of cultivated members of the Spurge Family.

The Cactus Family (Cactaceae)

Cacti, native only to the Americas, include many highly regarded ornamentals that have been exported around the world. The flowers are usually showy (Fig. 24.14), with numerous stamens, petals, and sepals. The sepals are often colored like the petals, and the inferior ovary develops into a berry. There are possibly more than 1,500 species, most occurring in drier subtropical regions. The leaves of many are reduced in size or missing, with the fleshy, flattened or cylindrical, often fluted stems carrying on the photosynthesis of the plants (Fig. 24.15). Many cacti can tolerate high temperatures, and some can withstand up to several years without moisture. They vary in size from pinheadlike forms to the

---

1. *Jojoba,* a member of the Box Family (Buxaceae), is a desert shrub with an acorn-sized capsule containing a large oily seed with about 50% liquid wax content. This high-grade wax and another found in the seeds of meadow foams, which are members of the Meadow Foam Family (Limnanthaceae), are the only known natural substitutes for sperm whale oil, a vital ingredient in engine lubricants. The importing of sperm oil into the United States was banned in 1970 to protect the nearly extinct large ocean mammals, and an expensive synthetic substitute is being used. Experiments and research are now in progress to find improved strains of jojoba and to test the feasibility of its being grown on a large scale. Two California counties have been using a mixture of petroleum oil and jojoba oil since 1981 in the transmissions of public transportation buses to see if the mixture will keep them from overheating. The results are promising and may reduce the need for transmission oil changes from once every 50,000 miles to once every 100,000 miles.

A.

B.

FIGURE 24.14   Cactus flowers.

American supermarkets, taste a little like pears. Prickly pear fruits also have seeds that Native Americans of the Southwest dried and ground for flour they used in *atole,* a staple food. A good syrup is obtained from boiling the fruits of prickly pears and also those of the giant saguaros. In the past, cactus candy was made by partly drying strips of barrel cactus and boiling them in saguaro fruit syrup, but the cactus is now usually boiled in cane sugar syrup.

Native Americans of the Southwest used to scoop out barrel cacti, dry them, and use them for pots. They also mixed the sticky juice of prickly pear cacti in the mortar used in constructing their adobe huts. In Texas, a poultice of prickly pear stem was applied to spider bites. Hopi Indians chewed raw cholla cactus as a treatment for diarrhea, and the skeletons of these cacti were used for flower arrangements.

In the middle of the 19th century, Australians planted a few imported prickly pear cacti in the dry interior. These cacti found no natural enemies in their new environment and multiplied rapidly, infesting more than 24 million hectares (60 million acres) within 75 years. In 1925, in an effort to control them, Australia introduced an Argentine moth among the cacti. The moth's caterpillars, which feed on prickly pear cacti, gradually brought the plants under control, and the land was made usable again.

Another cactus parasite, the cochineal insect (related to the mealybug, a common houseplant pest), feeds on prickly pear cacti in Mexico. At one time, the insects were collected for a crimson dye they produce, which was used in lipstick and rouge before aniline dyes were introduced.

Peyote cacti are small buttonlike plants that have no spines, with roots resembling those of carrots. They contain several drugs, the best known of which is *mescaline,* a powerful hallucinogen. Dried slices of peyote have been used in native religious ceremonies in Mexico for centuries and more recently by at least 30 tribes of Native Americans. The drug gives the user a variety of hallucinations in vivid colors.

The Mint Family (Lamiaceae)

The 3,000 members of the Mint Family are relatively easy to distinguish since they have a unique combination of angular stems (which are square in cross section), opposite leaves, and bilaterally symmetrical (irregular) flowers (Fig. 24.16). Most also produce aromatic oils in the leaves and stems. The superior ovary has four parts, with each of the four divisions developing into a nutlet. Included in the family are such well-known plants as rosemary, thyme, sage (not to be confused with sagebrush, which is in the Sunflower Family), oregano, marjoram, basil, lavender, catnip, peppermint, and spearmint.

Mint oils can be distilled at home with ordinary canning equipment. Whole plants (or at least the foliage) are loosely packed to a depth of about 10 centimeters (4 inches) or more in the bottom of a large canning pot. Then a wire rack or other support is also put in the pot, and a bowl is placed in the middle on the rack. Enough water is added to

giant saguaro (shown in Fig. 25.19), which can attain heights of 15 meters (50 feet) and weigh more than 4.5 metric tons (5 tons). They generally grow exceptionally slowly and, because they need so little care, make good houseplants for sunny windows.

In 1944, a marine pilot was forced to bail out of his aircraft over the desert near Yuma, Arizona. Until he was rescued five days later, he survived the intense heat and low humidity of the area by chewing the juicy pulp of barrel cacti in the vicinity. Since then, the use of cacti for emergency fluids and food has been recommended in most desert survival manuals.

Most cacti have edible fruits, and only three cacti (peyote, living rock, and hedgehog cactus) are known to be poisonous. Prickly pear fruits, which are occasionally sold in

A.

B.

C.

D.

FIGURE 24.15 Cacti. *A.* Prickly pear cacti. *B.* Peyote. *C.* A barrel cactus. *D.* An organ-pipe cactus.

cover the vegetation, the pot is placed on a range, and the lid is inverted over it. The water is brought to a boil, and as it boils, ice is placed on the inverted lid. The oils vaporize, and condense when they contact the cold lid, dripping then from the low point into the bowl (Fig. 24.17). Of course, some moisture also condenses, but the oil, being lighter, floats on top. Peppermint oil is easy to collect this way and will keep for a year or two in a refrigerator.

Mint oils have been used medicinally and as an antiseptic in different parts of the world. Mohegan Indians used catnip tea for colds, and dairy farmers in parts of the midwestern United States used local mint oils to wash their milking equipment. As a result, mastitis, a common disease of dairy cattle, was seldom encountered in their herds. Horehound, a common mint weed of Europe, has become naturalized on other continents and is cultivated in France. A leaf

FIGURE 24.16　Flowers of lamb's ear mint.

FIGURE 24.17　A simple apparatus for distilling mint oil at home.

FIGURE 24.18　A peppermint plant in flower. Oil from the leaves is a source of menthol.

extract is still used in horehound candy and cough medicines. In England, it is a basic ingredient of horehound beer. Vinegar weed, also known as *blue curls,* is a common fall-flowering plant of western North America. Native Americans of the area used it in cold remedies, for the relief of toothaches, and in a bath for the treatment of smallpox. It was also used to stupefy fish.

Menthol, the most abundant ingredient of peppermint oil (Fig. 24.18), is widely used today in toothpaste, candies, chewing gum, liqueurs, and cigarettes. Most American mint is grown commercially in the Columbia River basin of Oregon and Washington. Geese are sometimes used in the mint fields to control both insects and weeds, since they do not interfere with the growth of the mint plants themselves.

Ornamental mints include salvias and the popular variegated-leaf *Coleus* plants, neither of which has typical mint oils in the foliage. *Chia* (Fig. 24.19), another relatively odorless mint, is confined to the drier areas of western North America. Native Americans parched chia seeds and used them in gruel. The seeds, which become mucilaginous when wet, were also ground into a paste that was placed in the eye to aid in the removal of dirt particles. The paste was also used as a poultice for gunshot wounds, and Spanish Californians made a refreshing drink from ground chia seeds, lemon juice, and sugar. Chia seeds reportedly contain an unidentified substance that has effects similar to those of caffeine. Before the turn of the century, one physician reported that a tablespoon of chia seeds was sufficient to sustain a man on a 24-hour endurance hike. Since that time,

FIGURE 24.19 Chia.

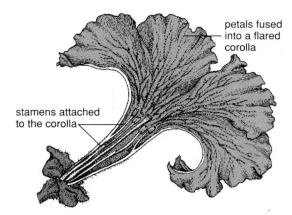

petals fused into a flared corolla

stamens attached to the corolla

FIGURE 24.20 A sectioned petunia flower.

backpackers have experimented with the seeds, and results tend to support the earlier claim. A thorough scientific investigation of the matter is needed. Chia seeds are presently sold commercially for making into a paste that is spread on clay models and then watered. The seeds sprout and resemble green hair.

The Nightshade Family (Solanaceae)

Flowers of the Nightshade Family, which is concentrated in the tropics of Central and South America, have fused petals, with the stamen filaments fused to the corolla so they appear to arise from it (Fig. 24.20). The superior ovary develops into a berry or a capsule. The more than 3,000 species of the family have alternate leaves and occur as herbs, shrubs, trees, or vines. Well-known representatives include tomato, white potato, eggplant, pepper, tobacco, and petunia.

Many nightshades produce poisonous drugs, some of which have medicinal uses. One of the best-known medicinal drug producers is the deadly nightshade of Europe. A drug complex called *belladonna* is extracted from its leaves. Belladonna, which was used in the "magic potions" of the past and also for dilating human pupils for cosmetic purposes, is now the source of several widely used drugs, including atropine, scopolamine, and hyoscyamine. Atropine is used in shock treatment, for relief of pain, to dilate eyes, and to counteract muscle spasms. Scopolamine is used as a tranquilizer, and hyoscyamine has effects similar to those of atropine. Capsicum, obtained from a pepper, is used as a gastric stimulant and is also a principal ingredient of *mace,* which is used to repel human or animal assailants. Capsaicin, derived from peppers, is used in ointments for the relief of arthritic and neuropathic pain.

Jimson weed (shown in Fig. 8.6) is also a source of medicinal drugs that have been used in the treatment of asthma and other ailments. The drugs can be fatal if ingested in sufficient quantities but have been much used in controlled amounts in Native American rituals of the past. Records indicate that users became temporarily insane but had no recollection of their activities when the effects of the drug wore off. The drug solanine is present in most if not all members of the family. Many arthritis sufferers apparently are sensitive to solanine, and a number of arthritics have reported partial relief through total avoidance of consuming members of this family (including potatoes, tomatoes, peppers, and eggplant).

Tobacco cultivation occupies more than 800,000 hectares (2 million acres) of American farmland. In its dried form, tobacco contains 1% to 3% of the drug nicotine. Nicotine is used in certain insecticides, and it is also used for killing intestinal worms in farm livestock. It is, however, an addictive drug. The evidence that human tobacco use is a primary cause of heart and respiratory diseases including lung cancer and other cancers such as those of the mouth and throat mounts almost daily. The only "benefit" it may have to humans appears to be as a killer of leeches. It is said that leeches attaching themselves to heavy smokers will drop off dead within five minutes from nicotine poisoning—a very dubious justification for continued human use!

Tomatoes are among the most popular of all "vegetables." About 18 million metric tons (20 million tons) are grown annually around the world. The plants are

FIGURE 24.21   A mechanical tomato harvester harvesting a field.

day-neutral (day lengths are discussed in Chapter 11), and even though they require warm night temperatures (16°C or 60°F) to set fruit well, they are easily cultivated in greenhouses when natural conditions are unfavorable. Most commercially grown American tomatoes are processed into juice, tomato paste, and catsup. In Italy, a small amount of edible oil is extracted from the seeds after the pulp has been removed. Most American tomatoes are grown in California, where they are harvested with special machinery developed during the 1960s when inexpensive labor became unavailable (Fig. 24.21).

The white or Irish potato is one of the most important foods grown in temperate regions of the world, with annual production estimated at well over 270 million metric tons (300 million tons). The leading producers are China, Poland, the United States, and countries of the former Soviet Union, which account for about 30% of the total. It is believed that white potatoes originated on an island off the coast of Chile and were sent back to Europe by Spanish invaders of South America in the 16th century. In the 1840s, late blight infested and destroyed the potato crop of Ireland, causing severe famine. Irish settlers subsequently emigrated to the United States, Canada, Australia, and other parts of the world.

When potato tubers are exposed to the sun, they turn green at the surface. Poisonous drugs are produced in the green areas. These have proved fatal to both animals and humans and should never be eaten.

The Carrot Family (Apiaceae)

Many members of the Carrot Family, which is widespread in the Northern Hemisphere, have savory-aromatic herbage. The flowers tend to be small and numerous and are arranged in umbels. The ovary is inferior, and the stigma is two-lobed. The petioles of the leaves, which are generally dissected, usually form sheaths around the stem at their bases. Included in the 2,000 members of the family are dill, celery, carrot, jicama, parsley, caraway, coriander, fennel, anise, and parsnip. Anise is one of the earliest aromatics mentioned in literature. It is used for flavoring cakes, curries, pastries, and candy. Pocket gophers apparently are attracted by its aroma, and some poison baits are enhanced with anise. A liqueur known as *anisette* is flavored with it.

Another liqueur, called *kümmel,* is flavored with caraway seeds, which are well-known for their use in rye and pumpernickel breads.

Some members of the Carrot Family are poisonous. Water hemlock (Fig. 24.22) and poison hemlock, which are common weeds in ditches and along streams, are deadly and have often been fatal to unwary wild-food lovers. Socrates is believed to have died as a result of ingesting poison hemlock, which should not be confused with cone-bearing hemlock trees.

Several members of the Carrot Family, such as cow parsnip, squawroot, and hog fennel, have edible roots and were used for food by Native Americans. The reader is

FIGURE 24.22   A water hemlock in flower.

A.

B.

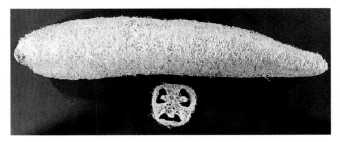

C.

FIGURE 24.23   Fruits and items associated with members of the Pumpkin Family (Cucurbitaceae). *A.* A Hawaiian ceremonial gourd with feathers attached and gourds used in South America for drinking maté. A metal straw that strains out the maté leaves is resting in one gourd. *B.* Cantaloupes. *C.* A luffa (vegetable sponge).

advised, however, to be absolutely certain of the identity of such plants before experimenting with them.

The Pumpkin Family (Cucurbitaceae)

Although most species in the Pumpkin Family are tropical or subtropical, many occur in temperate areas of both the Northern and Southern Hemispheres. Plants are prostrate or climbing herbaceous vines with tendrils. The flowers have fused petals, and female flowers have an inferior ovary with three carpels. All are unisexual. Some species have both male and female flowers on the same plant, while others have only male or only female flowers on one plant. In male flowers, the stamens cohere to varying degrees, depending on the species. The family has about 700 members, several of which have numerous horticultural varieties.

This family includes many important edible plants, and some have been cultivated for so long that they are unknown in the wild state. Well-known members of the family include pumpkins, squashes, cucumbers, cantaloupes (Fig. 24.23 B), and watermelons. The vegetable sponge (Fig. 24.23 C), which resembles a large cucumber when it is growing, has a highly netted fibrous skeleton, which serves as a bath sponge after the soft tissues have been removed.

Gourds found in Mexican caves have been dated back to 7000 B.C. Various types of gourds (Fig. 24.23 A) serving many purposes are still grown today. Some are scooped out and used for carrying liquids or for storing food, particularly grains. South Americans drink maté, a tea, from gourds, which are also used for several types of musical instruments. In parts of Africa, gourds are used to catch monkeys. A type with a narrow neck is scooped out and partly filled with corn or other grains. One end of a

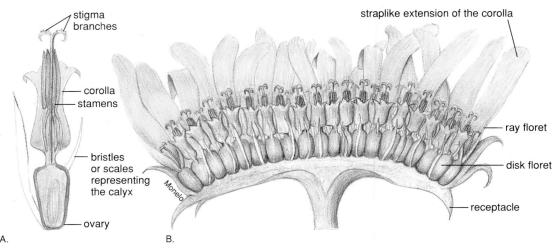

FIGURE 24.24   Parts of a sunflower. *A.* A section through a single floret. *B.* A section through an inflorescence.

rope is then tied to the gourd and the other to a stake driven into the ground. When a monkey tries to grab a fistful of grain, it finds that the neck will not allow its bulging hand to be removed. Most monkeys do not realize that letting go of the grain would allow them to escape, and they stubbornly hang on until they are captured.

Melonette, a small cucumberlike vine of the southeastern United States, has seeds that can be purgative (drastically laxative). Other cucumberlike plants of the western states, manroots (shown in Fig. 5.8), produce huge water-storage roots, some weighing as much as 90 kilograms (200 pounds). These roots were crushed by Native Americans and thrown into dammed streams to stupefy fish. An oil from the seeds was applied to the scalp as a remedy for infections that caused hair loss.

The Sunflower Family (Asteraceae)

The Sunflower Family, with approximately 20,000 species, is the second largest of the flowering plant families in terms of number of species. The individual flowers are called **florets.** They are usually tiny and numerous but are arranged in a compact inflorescence so that they resemble a single flower. A sunflower or daisy, for example, consists of dozens if not hundreds of tiny flowers crowded together, with those around the margin having greatly developed corollas that extend out like straps, forming what appear to be the "petals" of the inflorescence (Fig. 24.24 gives details of a sunflower inflorescence). In dandelions, all the individual florets of the inflorescence have narrow straplike extensions.

Well-known members of this family include lettuce, endive, chicory, Jerusalem artichoke, globe artichoke, dahlia, chrysanthemum, marigold, sunflower, and thistle.

Santonin, obtained from flower buds of a relative of sagebrush that is native to the Middle East, is used as an intestinal worm remedy. Tarragon, used as a spice in meat dishes and pickles, comes from another relative of sagebrush. Pyrethrum is a natural insecticide obtained from certain chrysanthemum flowers. Fructose, a sugar, is obtained from the tubers of Jerusalem artichokes and dahlias. Dahlias are also renowned for their huge showy flowers, while Jerusalem artichokes are often eaten as a vegetable.

Marigolds are favorite plants of organic gardeners. Their roots are said to release a substance that repels nematodes, and the odor of the leaves repels whiteflies and other insects. Unfortunately, snails seem to be immune and voraciously consume the foliage.

Many members of this family were widely used by Native Americans. The seeds of balsamroot and mule ears were used for food. Balsamroot plants as a whole were eaten raw or cooked, and in the West, extracts of both mule ears and tarweeds were used to treat poison oak inflammations. Salsify and thistle roots were also used for food.

Young leaves and roots of dandelions have been eaten for centuries, and the flowers have been used to make wine. Roasted dandelion and chicory roots have been used as a coffee substitute. During World War II, chicory was grown as a crop specifically for use as a coffee adulterant.

The dried and boiled leaves of American yarrow are said to make a nourishing broth. The rhizomes of European yarrow, which has become naturalized in North America, contains an anaesthetic that when chewed numbs the tongue and gums, and it has been used to ease the discomfort of teething toddlers; the substance has also been used in suppression of menstruation. This plant is believed to have been used by Achilles in treating the wounds of his soldiers.

Sunflowers themselves were widely used by Native Americans for their seeds, which were ground into a meal for bread. They are grown commercially today primarily for the edible oil extracted from the seeds and for birdseed, but the seeds are increasingly being eaten by modern Americans.

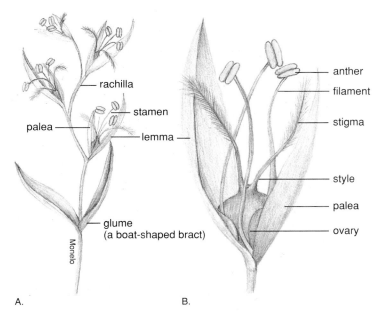

FIGURE 24.25   Grass flowers. *A*. An expanded grass spikelet. *B*. An enlargement of a single flower (floret).

Monocots

The Grass Family (Poaceae)

Although there are possibly only one-fourth as many species of grasses as there are members of the Sunflower Family, the Grass Family is the largest family of flowering plants in terms of numbers of individuals and is the most widely distributed. The flowers of grasses are highly specialized in structure and have a terminology all their own (Fig. 24.25). The calyx and corolla are represented by tiny, inconspicuous scales, and the flowers are protected by boat-shaped bracts. The stigmas, when they are exposed, are feathery, and the leaves sheathe the stem at their bases.

All of the cereals, including wheat, barley, rye, oats, rice, and corn, belong in this family, which includes nine of the ten most important crop plants in the world. Indeed, civilization as we know it would be vastly different without them. More than 900 million metric tons (1 billion tons) of cereals, feeding more than half of the world's population, are harvested each year, primarily in the Orient, North America, and Europe.

Sugar cane (Fig. 24.26), from which about 55 million metric tons (60 million tons) of sugar is extracted annually, is grown at lower elevations throughout humid tropical areas. It is a large grass, often growing to heights of 6 meters (20 feet). After the cane is harvested, the liquid portion is squeezed out. The solid waste is sometimes made into paper or particle board. It has also been converted to gasoline in South Africa and is now used in the production of electric power in Hawaii. Sugar is crystallized out of the liquid portion and separated by centrifuging. The dark remnant, molasses, may be used to produce rum or alcohol.

Grasses have been widely used by primitive peoples for making mats and baskets and for thatching huts. One variety

FIGURE 24.26   Sugar cane plants.

of sorghum is grown for its fibers, which are made into brooms, although natural broom fiber has now been largely replaced by synthetic materials. Some varieties are grown for their seeds, which are processed into cereal flours. Others are

FIGURE 24.27   A tiger lily flower.

FIGURE 24.28   *Sansevieria* plants.

a source of silage and a carnaubalike wax. Citronella oil, once a common ingredient of mosquito repellents, is obtained from a grass grown in Indonesia. It is now used in cheaper soaps, cosmetics, and perfumes. Related grasses are the source of lemon grass oil, which is used like citronella oil.

Juice squeezed from fresh young grass mowings has a high protein content and is being investigated as a source of protein for future human consumption.

The Lily Family (Liliaceae)

Although the 4,550 members of this family are particularly abundant in the tropics and subtropics, they occur in almost any area that supports vegetation. The flowers are often large, and their parts are all in multiples of three, with the sepals and petals frequently resembling each other in color and form (Fig. 24.27).

In addition to numerous types of lilies and daffodils used widely as ornamentals, the family includes asparagus, sarsaparilla, squill, meadow saffron, bowstring hemp, and *Aloe*. Sarsaparilla, which was at one time widely used for flavoring soft drinks and medicines, is obtained from the roots of a genus of woody vines whose stems are often covered with prickles. The bulbs of squills are the source of a rodent poison and also of a drug used as a heart stimulant.

Meadow saffron is the source of *colchicine,* a drug once used to treat rheumatism and gout but now much more widely used in experimental agriculture to interfere with spindle formation in cells so that the chromosome number of plants may be artificially increased. This increasing of the chromosome number can result in larger and more vigorous varieties of plants. Meadow saffron should not be

confused with true saffron, a member of the Iris Family and the source of the world's most expensive spice and a powerful yellow dye.

Bowstring hemps are related to the familiar, seemingly indestructible houseplants called *sansevierias* (Fig. 24.28), which have long, narrow, stiff leaves that stand upright from the base. The plants are cultivated in tropical Africa for their long fibers, which are used for string, rope, bowstrings, mats, and cloth. New Zealand flax, a larger plant, is grown in South America and New Zealand for similar purposes but is also widely used in ornamental plantings.

Several *Aloe* species produce juices used in shampoos, cosmetics, and sunburn lotions and in the treatment of X-ray and other burns. African *Aloe* species are prized as ornamentals in areas with milder climates. Their thick, fleshy leaves have short spines along the margins. The spines were once used for phonograph needles.

Many lily bulbs are edible and were widely used for food by Native Americans, but they should no longer be eaten, as such use may endanger the survival of wild species. *(Caution: Lily bulbs should not be confused with those of daffodils and their relatives. Daffodil and related bulbs are highly poisonous.)*

FIGURE 24.29   An uprooted California soaproot plant.

FIGURE 24.30   Bamboo orchids growing wild in Hawaii.

The California soaproot (Fig. 24.29), which is confined to California and southern Oregon, was important as food for Native Americans of the region and had several additional uses. The large bulbs are covered with coarse fibers, which were removed and tied to sticks to make small brooms. The bulbs themselves produce a lather in water and were used for soap. Numbers of bulbs were crushed and thrown in small streams that had been dammed. Fish would be stupefied and float to the surface. The bulbs were generally eaten after being roasted in a stone-lined fire pit. While they were roasting, a sticky juice would ooze out. This was used for gluing feathers to arrow shafts.

A resin used in stains and varnishes exudes from the stem of dragon's blood plants. Grass trees of Australia yield resins used in sealing waxes and varnishes.

The Orchid Family (Orchidaceae)

Members of this very large family, which, according to some authorities, has more than 35,000 species, are very widely distributed although, like members of the Lily Family, they are especially abundant in the tropics. In many genera, the number of individual plants at any one location may be quite small—sometimes limited to a single plant.

The flowers are exceptionally varied in size and form, and the habitats of the plants are equally diverse. The flowers of one Venezuelan species have a diameter of less than 1 millimeter (one 25th of an inch), while those of a species native to Madagascar may be more than 45 centimeters (about 18 inches) long. One species of *Dendrobium* orchid from Java has flowers that are so delicate they perish within five or six minutes of opening. Many orchids are epiphytic on the bark of trees. During its five-month flowering season, one epiphytic species of Malaysia and the Philippine Islands produces 10,000 flowers on plants that weigh more than 1 metric ton. Others are aquatic or terrestrial, and a saprophytic species (shown in Fig. 8.1) native to western Australia grows and flowers entirely underground.

Orchids have three sepals and three petals, with one of the petals (the lip petal) differing in form from the other two (Fig. 24.30). The stamens and pistil are united in a unique single structure, the *column* (shown in Fig. 23.9). The stigma usually consists of a sticky depression on the column, and the anthers, which contain sacs of pollen called *pollinia,* are covered with a cap until they are removed by an insect or other pollinator. The specific adaptations between orchid flowers and their pollinators is extraordinary and sometimes bizarre (as illustrated in Fig. 23.18).

Coffee and Caffeine

Of all the foods, beverages, and products provided by flowering plants, the one plant that millions of people have contact with every day is the plant known as *Coffea arabica*. Americans take their coffee break seriously. We drink some 350 million cups of coffee daily, which is enough coffee to turn the Metrodome into a deep caffeine pool. Here are some frequently asked questions concerning coffee and caffeine. So grab a cup of coffee, pull up your favorite chair, and sit down and relax while you read about the world's favorite beverage.

Where does coffee grow?

Coffee is a small tree or shrub native to the mountains of Ethiopia but, because of its popularity as a drink, was spread by traders to many tropical regions of the world. Today, Brazil and Colombia are the world's leading coffee producers. Coffee is derived from the seeds (beans) of *Coffea arabica* or *Coffea robusta*, two species with seventy or more recognized varieties between them. The trees require a tropical or subtropical climate and need about 60–100 inches of rain per year. They grow best at mountain elevations of 3,000–6,000 feet where the temperature is cool and relatively stable (68°F) because they are unable to tolerate frost. Trees start producing fruit around 3–5 years of age.

Botanically speaking, what is a coffee "bean"?

In Costa Rica, the crop season begins with the first "flowering rains" that come in March or April. Flowering buds that were produced at the end of the dry season are now provoked to flower by the gentle rains. The white delicate flowers have little aroma and bloom for three days before falling off the trees. Each pollinated flower will produce one fruit, called a "cherry" because of its red color. The coffee cherry contains two seeds that are tightly packed against each other, which results in each seed having a flat side. The coffee "beans" are actually the two seeds of each fruit. Coffee fruits change in color from dark green to deep red and are hand-picked when ripe (red color). Each tree is usually harvested three or four times because of the uneven maturation of fruits. A single tree can yield up to 5–6 pounds of fruit per year.

How much caffeine does a cup of coffee have?

A 7-ounce cup of coffee prepared by the drip method contains 115–175 mg of caffeine, while instant coffee contains 65–100 mg. Decaffeinated instant coffee contains 2–3 mg of caffeine.

How much caffeine is in soft drinks?

The following table compares the caffeine content of several brands of soft drinks.

Soft Drink (12-oz. can)	Caffeine Content (mg)
Jolt	72.0
Mountain Dew	54.0
Coca-Cola	45.6
Dr. Pepper	39.6
Pepsi-Cola	38.4
RC Cola	36.0

Orchids have minute seeds that are often produced in prodigious numbers (e.g., a single fruit of certain orchid species may contain up to 1 million seeds). Each seed consists of only a few cells, and, in order for a seed to germinate, it must become associated with a specific mycorrhizal fungus that produces substances necessary for its development. Once a seed has germinated, it may take from six to twelve or more years for the first flower to appear.

Contrary to popular belief, some orchids can be grown relatively easily on a windowsill that has bright light but not direct sunlight (see Appendix 4). Because orchids are among the most beautiful and prized of flowers, a large industry has grown up around their culture and propagation, which is discussed in the section on mericloning and tissue culture in Chapter 14. One species, the vanilla orchid, is grown commercially in the tropics for its fruits, which are the source of true vanilla flavoring.

See Appendix 1 for the scientific names of the plants mentioned in this chapter.

Summary

1. Distinguishing between fact and fantasy in reported past uses of plants is often difficult. Teams of specialists are interviewing tribal medicine men and women of the tropics to try to save potentially useful plants from extinction.

2. N. I. Vavilov determined that there were eight major centers of distribution of cultivated plants. The centers are located in China, India and Indo-Malaysia, Central Asia, the Near East, the Mediterranean, the Abyssinian area of Africa, Central America and adjacent South Mexico, and South America (with independent centers in Peru-Ecuador-Bolivia, Chiloe Island, and Brazil-Paraguay).

Chemically speaking, what is caffeine?

Caffeine is a plant alkaloid found not only in coffee beans but also in tea leaves and cola nuts. It works as a stimulant of the central nervous system and the respiratory system, as well as increasing heart rate and constricting blood vessels. As a stimulant, caffeine makes us more aware and alert and able to work without drowsiness. It is reported that coffee was discovered in the Near East (perhaps in Yemen) by goats who became frisky and sleepless after eating the wild fruit. A shepherd is said to have located the shrub that the goats had chewed and described it as "a shrub with leaves like laurel, flowers like jasmine and little dark berries the chief object of the animals' nibbling."

What are the effects of caffeine on the body?

Caffeine stimulates or increases heart rate, respiration, and basal metabolic rate, and increases the production of stomach acid and urine. These effects translate into a stimulating "lift" that a person may feel. Generally, one feels less drowsy, less fatigued, and more alert and focused after consuming caffeine.

Where does the word "coffee" come from?

It is probably derived from the Arabic word "kahveh" that means "stimulating," and is thought to have been used to describe the effects of eating coffee seeds (beans).

What does caffeine taste like?

Caffeine is very bitter to the taste as are other plant alkaloids. Caffeine is added as a flavoring agent to beverages such as root beer because it produces the sharp bitterness characteristic of the drink.

How is coffee decaffeinated?

If green (unroasted) coffee beans are rinsed with hot water above 175°F caffeine is extracted from the beans. However, in the process of removing caffeine, hot water alone is not used because it strips away too many of the essential flavors and aromatic elements contained in the beans. Along with water, a decaffeinating chemical is generally used such as methylene chloride or ethyl acetate. One common method is to place the green beans in a rotating drum and soften them with steam for 30 minutes. They are then rinsed for approximately 10 hours with methylene chloride that removes the caffeine from the beans. Methylene chloride containing the caffeine is drained off and the beans are steamed a second time for 8–12 hours to remove any remaining methylene chloride solvent. Finally, the beans are air or vacuum dried to remove excess moisture before the roasting process. Most decaffeinated coffee contains less than 0.1 parts per million (ppm) of residual methylene chloride, 100 times less than the maximum level of 10 ppm allowed by the U.S. Food and Drug Administration.

---

3. Flowering plant families are surveyed in phylogenetic sequence.

4. The Buttercup Family (Ranunculaceae) includes buttercup, columbine, larkspur, anemone, monkshood, *Clematis,* goldenseal, and wolfsbane.

5. The Laurel Family (Lauraceae) includes cinnamon, camphor, sassafras, sweet bay, California bay (or myrtle), avocado, and laurel.

6. The Poppy Family (Papaveraceae) includes bloodroot and opium poppy (the source of medicinal drugs and heroin).

7. The Mustard Family (Brassicaceae) includes cabbage, cauliflower, brussels sprouts, broccoli, radish, kohlrabi, turnip, horseradish, watercress, rutabaga, shepherd's purse, western wallflower, dyer's woad, and camelina.

8. The Rose Family (Rosaceae) includes stone fruits (e.g., cherry, apricot, peach, plum), strawberry, raspberry, rose, and related wild species.

9. The Legume Family (Fabaceae) includes pea, bean, lentil, peanut, soybean, alfalfa, clover, licorice, wattle, indigo, logwood, locoweed, sensitive plant (*Mimosa pudica*), and winged bean.

10. The Spurge Family (Euphorbiaceae) includes spurge, poinsettia, cassava, Pará rubber, gopher plant, candelilla, castor bean, crown-of-thorns, Mexican jumping bean, and tung oil tree.

11. The Cactus Family (Cactaceae) includes all cacti (e.g., prickly pear, cholla, barrel cactus, peyote).

12. The Mint Family (Lamiaceae) includes rosemary, thyme, sage, oregano, marjoram, basil, lavender,

catnip, peppermint, spearmint, horehound, salvia, *Coleus,* and chia.

13. The Nightshade Family (Solanaceae) includes tomato, white potato, eggplant, pepper, tobacco, belladonna, petunia, and jimson weed (a source of hallucinogenic drugs).

14. The Carrot Family (Apiaceae) includes dill, caraway, celery, carrot, parsley, jicama, coriander, fennel, anise, parsnip, water hemlock, poison hemlock, cow parsnip, squawroot, and hog fennel.

15. The Pumpkin Family (Cucurbitaceae) includes pumpkin, squash, cucumber, cantaloupe, watermelon, vegetable sponge, gourd, melonette, and manroot.

16. The Sunflower Family (Asteraceae) includes sunflower, dandelion, lettuce, endive, chicory, Jerusalem and globe artichokes, dahlia, chrysanthemum, marigold, thistle, sagebrush, pyrethrum, balsamroot, tarweed, and yarrow.

17. The Grass Family (Poaceae) includes all cereals (e.g., wheat, barley, rye, oats, rice, corn), sugar cane, sorghum, citronella, and lemon grass.

18. The Lily Family (Liliaceae) includes lilies, asparagus, sarsaparilla, squill, meadow saffron, bowstring hemp, *Aloe,* and New Zealand flax.

19. The Orchid Family (Orchidaceae) has highly specialized flowers. It includes a species that is the source of vanilla flavoring.

Review Questions

1. To which flowering plant family does each of the following belong: Poinsettia, lupine, columbine, peach, pear, cinnamon, sarsaparilla, belladonna, peyote, horehound, rubber, gourd, jimsonweed, parsley, sorghum, asparagus, broccoli, lettuce, tomato, opium?

2. Make a list of the poisonous plants in the families discussed.

3. Which plants mentioned are or have been used for medicines?

4. Which plants mentioned have been used for tools or utensils?

5. Native Americans made extensive use of plants for a wide variety of purposes in the past. List such uses for as many plants as possible.

Discussion Questions

1. Scientific investigations often take a great deal of time and money. Is it worth the effort to check out scientifically the past uses of plants?

2. A return to exclusively herbal medicines is being advocated in some quarters. Is this a good idea? What are the pros and cons?

3. Would you expect drugs produced naturally by plants to be more effective or better than drugs produced synthetically? Why?

4. If you were asked to single out the three most important families of the 16 discussed, which would you choose? Why?

5. A number of wild edible plants were mentioned in this chapter. What would happen if a large portion of the population were to gather these wild plants as a major source of food?

Additional Reading

Advisory Committee on Technology Innovation. 1975. *Underexploited tropical plants with promising economic value.* Washington, DC: National Academy of Sciences.

Akerle, O., et al. (Eds.). 1991. *Conservation of medicinal plants.* New York: Cambridge University Press.

Brill, S., and E. Dean. 1994. *Identifying and harvesting edible and medicinal plants.* New York: Hearst Books.

Gibbons, E. 1987. *Stalking the wild asparagus.* New York: David McKay Co.

Harrington, H. D. 1974. *Edible native plants of the Rocky Mountains.* Albuquerque NM: University of New Mexico Press.

Hornsok, L. (Ed.). 1993. *Cultivation and processing of medicinal plants.* New York: Wiley & Sons.

Kirk, D. R. 1975. *Wild edible plants of the western United States,* color ed. Healdsburg, CA: Naturegraph Publishers.

Krochmal, A., and C. Krochmal. 1984. *A field guide to the medicinal plants.* New York: Times Books.

Kunkel, G. 1984. *Plants for human consumption: Annotated checklist of edible phanerogams and ferns.* Forestburgh, NY: Lubrecht & Cramer.

Lewis, W. H., and M. P. F. Elvin-Lewis. 1982. *Medical botany: Plants affecting man's health.* New York: John Wiley & Sons, Inc.

Simpson, B. B., and M. Conner-Orgarzaly. 1986. *Economic botany.* Hightstown, NJ: McGraw-Hill Book Co.

Sweet, M. 1976. *Common edible and useful plants of the West,* rev. ed. Healdsburg, CA: Naturegraph Publishers.

Turner, N. J., and A. F. Szczawinski. 1991. *Common poisonous plants and mushrooms of North America,* 2d ed. Portland, OR: Timber Press.

Van Allen Murphey, E. 1987. *Indian uses of native plants,* 3d ed. Glenwood, IL: Meyer Books.

Wijesekera, R. O. 1991. *The medicinal plant industry.* Boca Raton, FL: CRC Press.

Chapter Outline

A pond succession scene.

Ecology

25

Overview ─────────────────────────────■

This chapter explores some of the ecological topics not already discussed elsewhere in the text.

Regional issues, including acid rain, water contamination, the destruction of wetlands, and hazardous waste are discussed first. Then global issues are introduced with a discussion of populations, communities, and ecosystems. This discussion is followed by a brief look at producers, primary and secondary consumers, decomposers, and food chains. Then energy flow through an ecosystem is considered, and the nitrogen and carbon cycles are explored. Discussions of succession follow, and the subject of fire ecology is broached. Human disruption of ecosystems, including the greenhouse effect, ozone depletion, and loss of biodiversity, are covered next. Descriptions of the major biomes of North America and telescopic views of their flora and fauna conclude the chapter.

Some Learning Goals

1. Know at least 10 ways in which humans have disrupted ecosystems.
2. Know the functions of producers, primary consumers, secondary consumers, and decomposers in an ecosystem.
3. Understand the cycling of energy in an ecosystem.
4. Learn how nitrogen and carbon are cycled.
5. Define *succession* and know a succession that begins with bare rocks or one that begins with water.
6. Define *ecotype, secondary succession, eutrophication,* and *climax vegetation.*
7. Learn the major biomes of North America and describe the principal living members of each.

It has been estimated that the total human population of the world was less than 20 million in 6000 B.C. During the next 7,750 years, it rose to 500 million; by 1850, it had doubled to 1 billion, and 70 years later, it had doubled again to 2 billion. The 4.48 billion mark was reached in 1980 and, within five years, it had grown to 4.89 billion. It is presently increasing by nearly 100 million annually, and estimates for the year 2000 are 6.25 billion. The earth remains constant in size, but humans obviously have occupied a great deal more of it over the past few centuries or at least have greatly increased in density of population.

In feeding, clothing, and housing ourselves, we humans have had a major impact on our environment. We have cleared natural vegetation from vast areas of land and drained wetlands. We have dumped wastes and other pollutants into rivers, oceans, and lakes and added pollutants to the atmosphere, and we have killed pests and plant disease organisms with poisonous substances. These poisonous substances have also killed natural predators and other useful organisms and, in general, have thoroughly disrupted the delicately balanced ecosystems that existed before humans began degrading their natural surroundings.

If we are to survive on this planet, it is incumbent upon us to stop increasing in numbers and to supplant the many unwise agricultural and industrial practices that have accompanied the burgeoning of human populations with practices more in tune with restoring some ecological balance. Agricultural practices of the future will have to include the return of organic material to the soil after each harvest, instead of consisting solely of adding inorganic fertilizers. Harvesting of timber and other crops will have to be done in a manner that prevents topsoil erosion, and the practice of clearing brush with chemicals will have to be abolished. Industrial pollutants will have to be captured, dismantled, and recycled whenever possible.

Many substances that now are still largely discarded (e.g., garbage, paper products, glass, metal cans) will also have to be recycled on a much larger scale. Biological controls (discussed in Appendix 2) will have to replace the use of poisonous controls whenever possible. Water and energy conservation will have to be universally practiced, and rare plant species, with their largely unknown gene potential for future crop plants, will need to be saved from extinction by preservation of their habitats and by other means. The general public will have to be made even more aware of the urgency for wise land management and conservation—which will be especially needed when pressures are exerted by influential forces promoting unwise measures in the name of "progress"—before additional large segments of our natural resources are irreparably damaged or lost forever. Alternatives appear to be nothing less than death from starvation, respiratory diseases, poisoning of our food and drink, and other catastrophic events that could ensure the premature demise of large segments of the world's population.

In recent years, scientists, and increasingly the general public, have become alarmed about these effects of human carelessness on our environment. It wasn't until the 1980s, however, that damage to forests and lakes caused by acid rain, the "greenhouse effect," contamination of ground water by nitrates and pesticides, reduction of ozone shield, major global climatic changes, loss of biodiversity in general, loss of tropical rain forests in particular, and other aspects of *ecology* gained widespread publicity.

Ecology is the broad discipline of study involving the relationships of plants and animals to each other and to their environment. The term *ecology,* which has now become a household word, was first proposed by Ernst Haeckel in 1869, but the discipline traces its origins back to early civilizations when humans learned to modify their surroundings through the use of tools and fire.

Various aspects of ecology were touched on in earlier chapters. For example, soils and organisms associated with them were discussed in Chapter 5. Some of the adaptations and modifications of leaves associated with specific environments were discussed in Chapter 7. The effects of light and temperature on plants were covered in chapters 10 and 11. Other important ecological topics comprise the bulk of this chapter, but the reader is referred to the Additional Reading section for a sampling of works covering a number of these themes in greater depth.

Since it is possible to review only a few of the major aspects of ecological study and environmental considerations in a text of this nature, this chapter will briefly explore some of the issues at two broad levels: (1) **regional issues,** and (2) **global issues.**

REGIONAL ISSUES

Although funding for solving many types of ecological problems tends to be woefully inadequate, the allocations to regional issues usually gain better public support because they are highly visible, and progress in dealing with them can be followed.

In the past two decades alone, concern about the effects of pollution on lands, waters, and peoples has been widely reported in the media and in scientific journals. During this time, attempts to correct or stem environmental damage have resulted in requirements that many construction projects file environmental impact reports before proceeding. These reports provide information that helps various agencies evaluate the possible effects a proposed project may have on the flora, fauna, and physical environment and determine what mitigations may be necessary before the project can or should be approved. The process has sometimes resulted in a great deal of controversy and emotional debate between landowners who believe their rights are being usurped and those who are convinced that preservation of the environment should supersede material considerations. Brief discussions of some regional issues follow.

Acid Rain

Acid rain occurs after fossil fuels are burned. The burned fuels release sulphur and nitrogen compounds into the atmosphere. There, chemical reactions brought about by sunlight and rain convert the compounds to nitric acid and sulphur dioxide. These combine with moisture droplets and become acid rain, which has a pH of less than 5.5, as compared with a pH of about 6.0 for unpolluted rainwater. The acid rain adversely affects or even kills many living organisms on which it falls. Mycorrhizal fungi, which are beneficially associated with the roots of many trees, are particularly susceptible to acid rain. Higher acidity of soil water also reduces the capacity of plant roots to obtain needed mineral nutrients.

Uncut trees have been dying prematurely in the world's forests. Some have died because of a lack of water during successive years of low rainfall or from other causes, such as insect infestations or salt scattered to melt ice and snow on roads. There seems to be little question, however, that acid rain has had a major role in stunting or destroying trees downwind from industrial sites. In recent years, there has been increasing international attention given to and cooperation in reducing industrial emissions that play a role in the production of acid rain.

Acid rain also affects nonliving materials. In 1989, Dr. Merle Robertson, an authority on pre-Columbian art, reported that the natural weathering of ancient Mayan ruins in southern Mexico has been accelerated by acid rain over the past decade or two. The acid rain apparently originates when moisture-laden winds blow over major oil refineries and other industrial sites located downwind from the ruins.

Water Contamination

Water becomes contaminated in different ways. Much of the pollution in our lakes and streams comes from the dumping of toxic industrial wastes and from runoff over polluted land. Included in other sources of pollution are pesticides, the exhaust and other emissions of aircraft and ships, and airborne pollutants originating with the combustion of fossil fuels. Groundwater supplies become polluted from pesticides and wastes trickling through the soil, from septic tanks, and from garden and farm fertilizers. Even deep wells, which seldom became polluted in the past, increasingly are becoming contaminated with unacceptably high levels of nitrates and other substances that harm the health of water-dependent humans and animals. In response to the problem, many communities are improving their water treatment plants, and many families are installing water filtration systems in their homes.

Long-range goals for curbing water pollution include a tightening of restrictions on the dumping of wastes and greater improvements in municipal water purification plants and systems. Genetic engineering and bacteria also probably will play a major role in the future. For example, a bacterium that can remove more than 99% of 2,4,5-T (a major component of Agent Orange—discussed in Chapter 11) from a contaminated environment has already been bred, and several other bacteria are being genetically engineered to enhance their capacities for breaking down other toxic wastes.

Destruction of Wetlands

Swamps, marshes, lagoons, floodplains, and other wetland areas have historically been regarded as wastelands and have routinely been drained and converted to drier, more "productive" agricultural and industrial sites. Only 6% of California's original wetlands, for example, remain today, and more than 90% of the wetlands once along the shores of Lake Ontario are now gone. Wetlands are, however, far from the wastelands they were once thought to be. Consider that the plants and other organisms in just 1 hectare (2.47 acres) of tidal wetland can perform the same recycling functions as about $150,000 of the latest wastewater treatment equipment—at no cost to taxpayers. At the same time, the wetland provides habitat and shelter for a wide variety of wildlife, many species of which are now threatened with extinction (Fig. 25.1).

Some attitudes toward wetlands have begun to change as the public has slowly become aware of the folly of eliminating them. City and regional planners in many areas now either ban new developments on wetland or require a proposed development that would encroach on a wetland to create new wet areas equal in extent and quality to those that may be lost. In addition, projects now in progress in many areas involve the restoration of damaged or lost wetlands.

FIGURE 25.1   Butte Sink, California. A typical wetland, as seen from the air.

Hazardous Waste

Too many members of earlier generations routinely drained on the ground or into storm drains the oil from their vehicle crankcases or took the used oil and other hazardous wastes to the dump. Highly toxic industrial wastes were also disposed of both in and at the outskirts of cities and towns. In 1996, it was estimated that some 12 million children in the United States were living less than 4 miles from hazardous waste sites. With the advent of atomic energy, radioactive wastes were sometimes insufficiently isolated from living organisms (including humans), with catastrophic results. Even when hazardous wastes aren't unceremoniously dumped, serious accidents and spills take place, with the effects sometimes lingering indefinitely, as, for example, in and around the former Soviet Union's Chernobyl atomic meltdown site.

Today, there are concerted efforts to curb the disposal of hazardous wastes and to greatly reduce the probabilities of accidents and spills. At most solid waste dumps, it is illegal to dispose of even empty latex paint cans, let alone more toxic materials, and heavy fines are levied on those found disposing of industrial wastes in an improper manner. Moneys from a U.S. government superfund are being used, with some success, to clean up selected old hazardous waste sites. The process thus far is slow and inadequate, but the increased restrictions on disposal methods and, as previously noted, the

genetic engineering of bacteria that can dismantle and render harmless many types of wastes hold promise for the future.

GLOBAL ISSUES

There are many issues, such as ozone depletion, worldwide changes in climate, and loss of biodiversity, that are global in scope and long lasting in impact. Strategies for attacking the problems emphasize the need for multinational dialogue and cooperation between specialists in fields as diverse as biology, chemistry, economics, and political science.

To comprehend the big picture, we'll first examine some basic principles and then apply the principles to global matters.

Populations, Communities, and Ecosystems

We recognize that plants, animals, and other organisms tend to be associated in various ways with one another and also with their physical environment. For example, the term *forest* is applied to **populations** (groups of individuals of the same species) of trees or other plants that form a **plant community** (unit composed of all the populations of plants occurring in a given area). The lichen and moss flora on a rock also constitute a community, as do the various seaweeds in a tidepool (Fig. 25.2). But these communities also invariably have

FIGURE 25.2   A tidepool community at low tide.

animals and other living organisms associated with them. It is preferable, therefore, to refer to the unit composed of all the populations of living organisms in a given area as a **biotic community.** Considered together, the communities and their physical environments, which interact with each other and are interconnected by physical, chemical, and biological processes, constitute **ecosystems.** Some populations, communities, and ecosystems may be microscopic in extent while others are much larger or even global.

Populations

Populations may vary in number, in density, and in the total mass of individuals. Depending on circumstances, a field biologist may investigate a population in various ways. If, for example, a conservation organization is concerned about the preservation of a rare or threatened species, the organization may simply count the **number** of individuals. If a count is not feasible, the organization may estimate population **density** (number of individuals per unit volume—e.g., five blueberry bushes per square meter). If the individuals in a population vary greatly in size or are unevenly scattered, a better estimate of the population's importance to the ecosystem may be calculated by determining the **biomass** (total mass of the individuals present). Population biology studies also include, among many components, physiological and reproductive ecology with the examination and tabulation of factors, such as pollination, seed dispersal, germination, and seedling survival and establishment in plant populations.

Communities

Communities are composed of populations of one to many species of organisms living together in the same location. Similar communities occur under similar environmental conditions, although actual species composition can vary considerably from one location to another. A community is difficult to define precisely, because species of one community may also occur in other communities. Furthermore, species of one com-

munity may have specific genetic adaptations to that community. If individuals are transplanted to a second, different community where the same species occurs, the transplanted individuals may not necessarily be able to survive alongside their counterparts, which are themselves adapted to this second community. Individuals adapted to specific communities within their overall distribution are called *ecotypes,* and areas transitional between communities are called *ecotones*.

Analysis and classification of communities is important in the preparation of maps that form the basis of activities such as land-use planning, forestry, natural resource management, and military maneuvers.

Ecosystems

Living organisms interacting with one another and with factors of the nonliving environment constitute an *ecosystem*. The nonliving factors of the environment include light, temperature, oxygen level, air circulation, fire, precipitation, energy, and soil type. The distribution of a plant species in an ecosystem is controlled mostly by temperature, precipitation, soil type, and the effects of other living organisms (biotic factors). In Mediterranean climates, such as those that occur in parts of California and Chile, nearly all precipitation occurs during the winter months and the summers are dry. This type of climate favors spring annuals that complete their life cycles by summer and evergreen shrubs that can tolerate long periods of drought. Forests may occur in areas where heavy winter snowfall soaks deeply into the soil, compensating to a certain extent for the lack of summer moisture.

The leaves and other parts of plant species that occur naturally in areas of low precipitation and high temperatures (*xerophytes*) generally are adapted to their particular environment through modifications that reduce transpiration. These modifications are discussed under "Specialized Leaves" in Chapter 7 and "Regulation of Transpiration" in Chapter 9. Plants of arid areas may also have specialized forms of photosynthesis, such as CAM photosynthesis

FIGURE 25.3 A food web.

(discussed in Chapter 10). Similarly, plants that grow in water (*hydrophytes*) are modified for aquatic environments.

The distribution of plant species is influenced by the mineral content of soils. For example, serpentine soils—which contain relatively high amounts of magnesium, iron, usually nickel, and chromium and low amounts of calcium and nitrogen—often support species that are not found on nearby nonserpentine soils. Biotic factors, such as competition for light, the mineral nutrients and water available, and grazing by the animal members of the biotic community, also influence the distribution of plant species.

Ideally, ecosystems sustain themselves entirely through photosynthetic activity, energy flow through food chains, and the recycling of nutrients. Organisms that are capable of carrying on photosynthesis capture light energy and convert it, along with carbon dioxide and water, to energy-storing molecules. Such organisms are called **producers.** Animals, such as cows, caribou, and caterpillars, and other organisms that feed directly on producers are called **primary consumers. Secondary consumers,** such as tigers, toads, and tsetse flies, feed on primary consumers. **Decomposers** break down organic materials to forms that can be reassimilated by the producers. The foremost decomposers in most ecosystems are bacteria and fungi.

In any ecosystem, the producers and consumers interact, forming **food chains** or interlocking *food webs,* which determine the flow of energy through the different levels (Fig. 25.3). Since most organisms have more than a single source of food and are themselves often consumed by a variety of consumers, there are considerable differences in the length and intricacy of food chains or webs.

Light energy itself, which enters at the producer level, cannot be recycled in an ecosystem. Only about 1% of the light energy falling on a temperate zone community is actually converted to organic material. As the organisms at each level respire, energy gradually dissipates as heat into the atmosphere. Additionally, parts of organisms are not consumed at subsequent levels in the food chain. For example, the energy in leaves that fall from a plant before a herbivore grazes on it is released when decomposers (bacteria and fungi) break down the leaves. It has been estimated that when cattle graze, only about 10% of the energy stored by the green plants they consume is converted to animal tissue, with most of the remaining energy dissipating as heat into the atmosphere. When we eat beef, only roughly 10% of its stored energy is utilized by our bodies to manufacture new blood cells and otherwise sustain life. The remaining energy is converted to heat. Obviously, then, in a long food chain, the final consumer gains only a tiny fraction of the energy originally captured by the producer at the bottom of the chain.

Conversely, there is proportionately much less loss of energy between levels in short food chains. Let's assume, for example, that for every 100 calories of light energy falling on lettuce plants each day, 10 calories are converted to lettuce tissue. Then suppose that the lettuce is fed to hogs. Again, only 10%, or 1 calorie, of the original energy may be converted to animal tissue. If as secondary consumers we eat

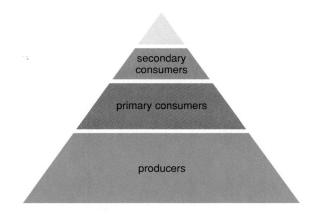

FIGURE 25.4 An energy pyramid of an ecosystem. There is much more energy at the bottom than at the top.

such pork, our bodies, in turn, may utilize 10% of the energy available in the pork, or only 0.1% of the original energy available. If, however, we eat the lettuce directly, we end up with the hog's percentage, which is 10 times more of the original solar energy than we would get if we ate the hog that ate the lettuce. (The actual amount of energy per gram of fatty hog meat is considerably higher than that found in a gram of lettuce tissue.)

From this, it follows that a vegetarian diet makes much more efficient use of solar energy than one that relies heavily on meats, and where food is scarce or humans are very abundant (as in India or Ethiopia), humans become virtual vegetarians. It also follows that, in terms of the numbers of individuals and the total mass, there is a sharp reduction at each level of the food chain. In a given portion of ocean, for example, there may be billions of microscopic algal producers supporting millions of tiny crustacean consumers, which, in turn, support thousands of small fish, which meet the food needs of scores of medium-sized fish, which are finally consumed by one or two large fish (Fig. 25.4). In other words, one large fish may very well depend on a billion tiny algae to meet its energy needs every day.

The interrelationships and interactions among the components of an ecosystem can be quite complex, but all function together in a regulatory fashion. An increase in food made available by producers can result in an increase in consumers, but the increased number of consumers reduces the available food, which then inevitably leads to a reduction of consumers to earlier levels. While cyclical in nature, the net result is sustained self-maintenance of the ecosystem. This is the basis for the so-called *Balance of Nature.*

Interactions Between Plants, Herbivores, and Other Organisms

While it is easy to see that the total mass of consumers is largely determined by the total mass of food made available by the producers, the interactions among producers themselves, between predators and prey, and between the decomposers and the other members of the ecosystem are usually more subtle. Many flowering plants produce

A.

B.

FIGURE 25.5 A species of *Acacia* that is host to ants that live in its hollow thorns. The ants attack any organisms, large or small, that come in contact with the plant. The plant provides food for the ants through nectaries at the bases of the thorns (*A*) and nitrogenous bodies at the tips of the leaflets (*B*).

substances that either inhibit or promote the growth of other flowering plants. Black walnut trees, for example, produce a substance that wilts tomatoes and potatoes and injures apple trees that come in contact with black walnut

roots. Many other plants produce *phytoalexins* (chemicals that kill or inhibit disease fungi or bacteria), making them resistant to various diseases (see discussion under "Use of Resistant Varieties" in Appendix 2).

Conversely, some bacteria and fungi limit higher plant growth by producing various inhibitory chemical compounds. Parasitism limits population size in other bacteria, fungi, and flowering plants. The degree of parasitism varies considerably with the organisms involved. Some of the species of the Figwort Family, which includes snapdragons and similar plants, have no chlorophyll and depend entirely on their flowering plant hosts for their energy and other nutritional needs. Other related species do have chlorophyll but apparently also require supplemental food from their hosts. Still other species often parasitize the roots of certain plants but are also capable of existing independently.

Mycorrhizal fungi are intimately associated with the roots of most woody and many other plants in such a way that both organisms derive benefit (such associations, called *mutualisms,* are discussed in Chapter 5). The fungi greatly increase the absorptive surface of the root, usually playing a major role in the absorption of phosphorus and other nutrients, while obtaining energy from root cells.

Thomas Belt, a naturalist of more than 100 years ago, first called attention to an association between tropical ants and thorny, rapidly growing species of *Acacia* (Fig. 25.5). *Acacia,* which has large hollow thorns at the base of each leaf, is host to ants that feed on sugars, fats, and proteins produced by petiolar nectaries and special bodies at the tip of each leaflet. The ants live within the hollow thorns and vigorously attack any other organisms, from insects to large animals, that come in contact with the plant. They also kill, by *girdling,* any plant that touches the *Acacia* (girdling is discussed under "bridge grafting" in Appendix 4). Experiments have shown that, when ants are removed from these *Acacia* species, the plants grow very slowly and usually soon die from insect attacks or from shading by other plants.

Large herbivorous animals, such as deer and moose, feed on a wide variety of plants, each differing in nutritional value. Each plant species also produces different combinations, types, and amounts of chemical compounds in addition to proteins, fats, and carbohydrates. Many of the chemical compounds are poisonous to the consumers, but the animals do not display symptoms of poisoning because their digestive systems are capable of breaking down the compounds and eliminating or excreting them to a limited degree. The limitations imposed by such compounds result in the consumers varying their diet, seeking familiar foods and being wary of new ones. If a plant species did not have some natural defense, such as chemical compounds or structural modifications (for example, spines), primary consumers of all kinds, from insects to elephants, would soon render that species extinct. In an ecosystem, the defenses that both producers and consumers have against each other have been developed through a process of coevolution resulting from natural selection and are maintained in delicate balance.

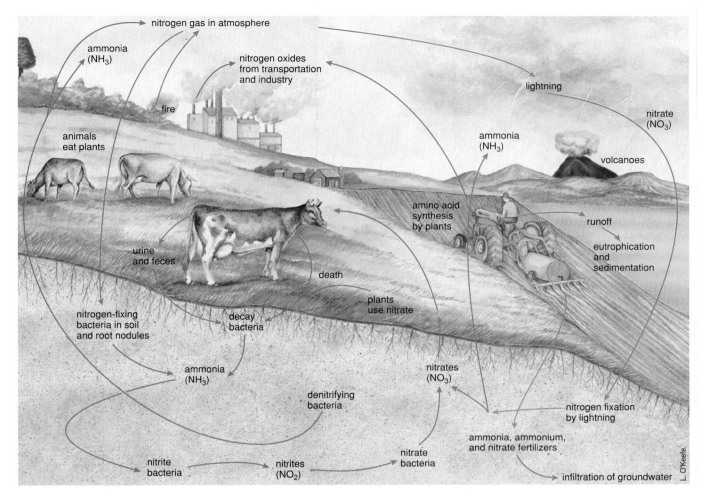

FIGURE 25.6 The nitrogen cycle.

NATURAL CYCLES

Water and elements such as nitrogen, carbon, and phosphorus are constantly recycling throughout nature. We'll use the nitrogen and carbon cycles as examples.

The Nitrogen Cycle

Much of the protoplasm of living cells consists of protein. Nitrogen, the most abundant element in our atmosphere, constitutes about 18% of the protein. There are nearly 69,000 metric tons of nitrogen in the air over each hectare of land (35,000 tons per acre), but the total amount of nitrogen in the soil seldom exceeds 3.9 metric tons per hectare (2 tons per acre) and is usually considerably less. This discrepancy results from the nitrogen of the atmosphere being chemically inert, which is another way of saying that it will not combine readily with other molecules. It is, therefore, largely unavailable to plants and animals for their use in building proteins and other substances containing nitrogen.

Most of the nitrogen supply of plants (and indirectly of animals) is derived from the soil in the form of inorganic compounds and ions taken in by the roots. These compounds and ions include those that contain nitrogen chemically combined with oxygen or hydrogen. They are produced by bacteria and fungi as these break down the more complex molecules of dead plant and animal tissues to simpler ones. Some nitrogen from the air is also *fixed*—that is, converted to ammonia or other nitrogenous compounds by various *nitrogen-fixing bacteria*. Some of these organisms gain access to various plants, particularly legumes (e.g., peas, beans, clover, alfalfa), through the root hairs, with the plant producing root nodules in which the bacteria multiply (root nodules are shown in Fig. 5.15). Others live free in the soil.

Figure 25.6 shows that there is a constant flow of nitrogen from dead plant and animal tissues into the soil and from the soil back to the plants. Decay bacteria and fungi can break down enormous quantities of dead leaves and other tissues to tiny fractions of their original volumes within a few days to a few months. If they were abruptly to cease their activities, the available nitrogen compounds would be completely exhausted and the carbon dioxide supply needed for photosynthesis seriously depleted within a few decades. Forests and prairies would die as the accumulations of shed leaves, bodies, and debris buried the living plants and shielded their leaves from the light essential to

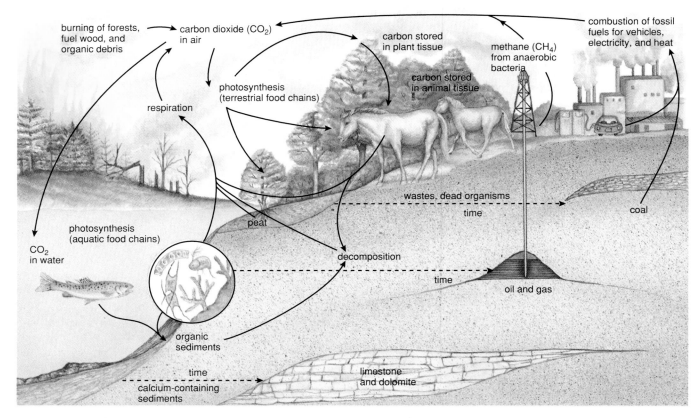

FIGURE 25.7   The carbon cycle.

photosynthetic activity. At present, even with the various bacteria involved in the nitrogen cycle functioning normally, the total amount of nitrogen in the soil is not being increased by their activities but is merely being recycled.

Significant amounts of nitrogen are continually being lost as water leaches it out or carries it away through erosion of topsoil. More is lost with each harvest, the average crop removing about 25 kilograms per hectare (25 pounds per acre) per year. This nitrogen loss from the harvesting of crops can be sharply reduced if vegetable and animal wastes are recycled and returned to the soil each year. While bacteria are decomposing tissues, they use nitrogen, and little is available until they die and release their accumulations into the soil. Accordingly, crops should not be planted in soils to which only partially decomposed materials have been added. Likewise, when sawdust, straw, or other organic mulches are spread around plants in a garden to control weeds and conserve soil moisture, less nitrogen will be available to the growing plants until the mulches have been decomposed.

Weeds and stubble are often controlled by burning. Fire, however, causes serious loss of nitrogen, which has to be replaced. It has been estimated that the annual combined loss of nitrogen from the soil in the United States from fire, harvesting, and other causes exceeds 21 million metric tons (23 million tons), and only 15.5 million metric tons (17 million tons) are replaced by natural means. To offset the net

loss, some 32 million metric tons (35 million tons) of inorganic fertilizers are applied to the soils each year. If organic matter is not added at the same time, however, this application of inorganic fertilizers, combined with the annual burning of stubble, may eventually result in the creation of a *hardpan* soil.

Hardpan develops through the gradual accumulation of salt residues, which dissolve humus and disrupt the structure of the soil, causing the clay particles to clump and also producing colloids that are impervious to moisture. In hardpan soils and others low in oxygen (e.g., flooded areas), *denitrifying bacteria* use nitrates instead of oxygen in their respiration, depleting the remaining soil nitrogen.

Precipitation returns a little nitrogen to the soil from the atmosphere, where it has accumulated as a result of the action of light on industrial pollutants, fixation by flashes of lightning, and diffusion of ammonia released through decay. The activities of nitrogen-fixing bacteria and volcanoes also contribute to the natural replenishment of nitrogen by converting it to forms that can be utilized by plants.

The Carbon Cycle

In the process of recycling nitrogen, bacteria also play a major role in recycling carbon and many other substances (Fig. 25.7). One of the two raw materials of photosynthesis, carbon dioxide, constitutes 0.03% of our atmosphere. The combined plant life of the oceans and the land masses is estimated to use about 14.5 billion metric tons (16 billion tons) of carbon obtained from carbon dioxide every year. Respiration by all living organisms constantly replaces carbon dioxide, with perhaps as much as 90% or more of it being produced by the incredible numbers of decay bacteria and fungi as they decompose tissues.

The burning of fossil fuels by the internal combustion engines of industry and transportation releases lesser but significantly increasing amounts of carbon dioxide into the air, and a small amount originates with fires and volcanic activity. At the present rate of use by photosynthetic plants, it has been calculated that all of the carbon dioxide of our atmosphere would be used up in about 22 years if it were not constantly being replenished.

Scientists have found that plant growth can be accelerated by increasing levels of carbon dioxide in the air around them, and some commercial nurseries are now pumping carbon dioxide along rows of bedding plants as a "fertilizer." Recent studies indicate that while C_3 plants respond to increased carbon dioxide levels, C_4 plants do not, suggesting that in natural communities the response of C_3 plants to higher levels of carbon dioxide may give them a competitive edge over C_4 plants.

Anaerobic bacteria produce large volumes of carbon-containing methane gas, which is discussed on page 459.

In the web of life, nutrients constantly cycle and are recycled. Carbon, nitrogen, water, phosphorus, and other molecules have for aeons been passing through cycles. Some molecules that were a part of a primeval forest that became compressed and turned to coal may have become part of another plant after the coal burned. Then the new plant may have been eaten by an animal, which, in turn, contributed molecules to a part of yet another living organism. Just think of where the molecules in your own body may have been in the past few billion years. They may well have been a part of some prehistoric seaweed, a saber-toothed tiger, a mighty dinosaur, or even all three!

SUCCESSION

After a volcano spews lava over a landscape, or after an earthquake or a landslide exposes rocks for the first time, there is initially no sign of life on the lava or rock surfaces. Within a few years or sometimes within a few months or even weeks, living organisms begin to appear, and a sequence of events known as *succession* takes place. During succession, the species of plants and other organisms that first appear gradually alter their environment as they carry on their normal activities, such as metabolism and reproduction. In time, the accumulation of wastes, dead organic material, and inorganic debris and other changes (such as of shade and water content in the habitat) favor different species, which may replace the original ones. These, in turn, modify the environment further so that yet other species become established.

Succession occurs whenever and wherever there has been a disturbance of natural areas on land or in water. It proceeds at varying rates, depending on the climate, the soils, and the animals or other organisms in the vicinity. Ecologists recognize a number of variations of two basic types of succession. *Primary succession* involves the actual formation of soil in the beginning stages, while *secondary succession* takes place in areas that previously were covered with soil and vegetation.

Primary Succession

Xerosere

One of the most universal types of primary succession, called a **xerosere,** begins with bare rocks and lava that have been exposed through glacial or volcanic activity or through landslides. Initially, the rocks are sometimes subjected to alternate thawing and freezing, at least in temperate to colder areas. Tiny cracks or flaking may occur on the surface as a result. Lichens often become established on such surfaces (Fig. 25.8). They produce acids that very slowly etch the rocks, and as they die and contribute organic matter, they are replaced by other, larger lichens. Certain rock mosses adapted to long periods of desiccation also may become established, and a small amount of soil begins to build up. This is augmented by dust and debris blown in by the wind. Eventually, enough of a mat of lichen and moss material is present to permit some ferns or even seed plants to become established, and the pace of soil buildup and rock breakdown accelerates.

If deep cracks appear in the rocks, the larger seeds may widen them further as they germinate and the roots expand in girth. It has been calculated that germinating seedlings can exert a force of up to 31.635 kilograms per square centimeter (450 pounds per square inch). Indeed, there are known instances of seedlings splitting rocks that weigh several tons (Fig. 25.9).

As soil buildup continues, larger plants take over, and eventually the vegetation reaches an equilibrium in which the associations of plants and other organisms remain the same until another disturbance takes place or climatic changes occur. Such relatively stable plant associations are referred to as **climax vegetation.**

The climax vegetation of deciduous forests in eastern North America is dominated by maples and beeches, oaks, hickories, and hemlocks or other combinations of trees. In desert regions, various cacti form a conspicuous part of the climax vegetation, while in the Pacific Northwest, large coniferous trees predominate. In parts of the Midwest, prairie grasses and other herbaceous plants form the climax vegetation, and in wet tropical regions, a complex association of trees and herbs constitutes the climax.

Occasionally, when a volcano produces ash instead of lava that buries existing landscape and associated vegetation, some of the successional stages involving lichens and mosses may be bypassed, with larger plants becoming the

FIGURE 25.8   Early stages of succession on a Hawaiian road after an earthquake produced cracks and a hole. Ferns have become established in the cracks and hole less than three years after the disruption.

FIGURE 25.9   A blue oak that grew from an acorn lodged in a small crack of the rock. Over the years, the oak has split the rock apart.

successional pioneers. This occurred following the series of ash eruptions, accumulation of debris, and mud flows of Mount St. Helens in the state of Washington during the early 1980s.

Hydrosere

Succession takes place in wet habitats as well as in drier ones, and such succession is called a **hydrosere.** In the northern parts of midwestern states, such as Michigan, Wisconsin, and Minnesota, ponds and lakes of various sizes abound. Many were left behind by retreating glaciers and often have no streams draining them. The water that evaporates from them is replaced annually by precipitation runoff. They also grow a tiny bit smaller each year as a result of succession (Fig. 25.10).

This succession often begins with algae either carried in by the wind or transported on the muddy feet of waterfowl and wading birds. The algae tend to multiply in shallow water near the margin, and with each reproductive cycle, the dead parts sink to the bottom. Floating plants, such as duckweeds, may then appear, often forming a band around the body of water just offshore (Fig. 25.11). When nutrients, oxygen, pH, and temperatures are low, peat mosses become the dominant floating plants. There are presently about 4 million square kilometers (1.5 million square miles) of peat bogs throughout the world. Next, water lilies and other rooted aquatic plants with floating leaves become established, each group of plants contributing to the organic material on the bottom, which slowly turns to muck. Cattails and other flowering plants that produce their inflorescences above the water often take root in the muck around the edges, and the accumulation of organic material accelerates.

Meanwhile, the algae, duckweeds, peat mosses, and other plants move farther out, and the surface area of

FIGURE 25.10 Succession in a pond in northern Wisconsin.
(Courtesy Robert A. Schlising)

exposed water gradually diminishes. Grasslike sedges become established along the damp margins and sometimes form floating mats as their roots interweave with one another. Dead organic material accumulates and fills in the area under the sedge mats, and herbaceous and shrubby plants then move in. As the margins become less marshy, coniferous trees whose roots can tolerate considerable moisture (e.g., tamaracks or eastern white cedars) gain a foothold, eventually growing across the entire site as the pond or lake disappears. The trees continue to aid in the formation of true soil, and in due course, the climax vegetation takes over. No visible trace of the pond or lake now remains, and the only evidence of its having been there lies beneath the surface, where fossil pollen grains, bits of wood, fossilized fish skeletons, and other material reveal the past history. Such succession may take thousands of years and has never been witnessed from start to finish. The evidence that it does occur, however, is extensive and compelling.

Under natural conditions, some stream-fed lakes and ponds eventually become filled with silt and debris, although this, too, may take thousands of years to occur. The streams that feed these lakes bring in silt, and the nutrient content of the water rises as dissolved organic and inorganic materials (particularly nitrogen and phosphorus) are

FIGURE 25.11 Duckweeds floating on a pond in early stages of succession.

brought in. This gradual to relatively rapid enrichment, called **eutrophication,** facilitates the growth of algae and other organisms, which also add their debris to the bottom of the lake. When sewage and other pollutants are permitted to enter the lake, the process of eutrophication may be greatly accelerated through stimulation of the growth of aquatic organisms. Eutrophication may also be accelerated when trees are cleared from land surrounding lakes prior to the construction of summer homes and resorts. The cleared land erodes more readily, with precipitation runoff carrying soil into the water. Regardless of size, all bodies of water, including rivers, are subjected to these processes.

Secondary Succession

Secondary succession, which occurs more rapidly than primary succession, may take place if soil is already present and there are surviving species in the vicinity. In fact, survivors strongly condition subsequent succession. Many secondary successions follow human disturbances (e.g., land being cleared when timber is harvested or land is converted to farmland). Other secondary successions follow fires. Grasses and other herbaceous plants become established on burned or logged land. These usually are followed by trees and shrubs that have widely dispersed seeds (e.g., aspen and sumac in the Midwest and East and chaparral plants, such as chamise and gooseberries, in the West). After going through fewer stages than are typical of primary succession, the climax vegetation takes over, often within 100 years.

Fire Ecology

Natural fires started primarily by lightning and the activities of prehistoric humans have occurred for thousands of years in North America and on other continents. Trees, such as the giant redwood and ponderosa pines, although scarred by certain types of fires, often survive, and the dates of fires can be determined by the proximity of the scars to specific annual rings (as shown in Fig. 6.7). In the West, growth rings of ponderosa pines show that in the past, forests of such pines used to burn on an average of every six to seven years. These fires and the climate, topography, and soil combined to have a profound effect on various biomes.

As humans became civilized, major and largely successful efforts to control fires were made and this, in turn, significantly altered vegetational patterns. As knowledge of the role of fires in the maintenance of ecosystems has accumulated, it has become apparent that trying to eliminate fires, at least in certain areas, disrupts natural habitats more in the long run than does allowing them to occur, and agencies such as the U.S. National Park Service now may allow fires at higher elevations to run their natural course.

Fires actually benefit grasslands, chaparral, and forests by converting accumulated dead organic material to mineral-rich ash, whose nutrients are recycled within the ecosystem. If the soil has been subjected to fire, some of its nutrients and organic matter may have been lost and the composition of

FIGURE 25.12   Secondary succession on a burned area. California bay trees are resprouting the first season after a burn.

microorganisms originally present is likely to have been altered. Losses are offset, however, by the fact that soil bacteria, including blue-green bacteria, which are capable of fixing nitrogen from the air, increase in numbers after a fire, and there is a decrease in fungi that cause plant diseases.

In some areas, such as the prairie states of the Midwest, grasses are better adapted to fire than are woody plants. They produce seeds within a year or two after germination, and perennial grass buds, which are at the tips of rhizomes close to or beneath the surface where they escape the most intense heat of fires, usually survive, producing new growth the first season after a fire. Accordingly, a fire destroys only one season's growth of grass, often after growth has been completed. Shrubs, however, have much of their living tissue above ground, and a fire may destroy several years' growth. In addition, woody plants often do not produce seeds until several years after a seed germinates. Many shrubs do sprout from burned stumps, particularly in chaparral areas (Fig. 25.12), but repeated burning keeps them small, thereby favoring grasses. There is evidence that at least some of the North American grasslands originated and were maintained because of fire. Since grassland fires have largely been controlled, many of the areas have now been invaded by shrubs that were once confined to watercourses.

Fires also play a role in the composition of forests. In the mountains of east-central California, gooseberry and deerbrush appear in abundance after a fire, but their numbers stabilize within 15 to 30 years when larger trees return to the area. Ponderosa, jack and southern longleaf pines, and Douglas firs (which do not tolerate shade) are among the species that repeatedly replace themselves after fires, and seeds of some species rarely germinate until they have been exposed to fire. The majority of chaparral species, both woody and herbaceous, are so adapted to fire that their seeds also will not germinate, as a rule, until fires remove accumulated litter and toxic wastes produced by the plants during growth.

The Greenhouse Effect

The *greenhouse effect* refers to a global rise in temperature due to the accumulation in the atmosphere of gases that permit radiation from the sun to reach the earth's surface but prevent the heat from escaping back into space. Carbon dioxide and methane, the two most common gases involved in the greenhouse effect, have been part of our atmosphere for millions of years, but others, such as chlorofluorocarbons, are relatively recently produced by-products of the manufacture of refrigerants, plastics, and aerosol cans.

Carbon Dioxide

D. L. Lindstrom of the University of Illinois at Chicago and D. R. MacAyeal of the University of Chicago recently examined records of cores of ancient ice from Siberia, Scandinavia, and the Arctic Ocean. Using computers to simulate the status of ice and atmosphere going back 30,000 years, they found that levels of carbon dioxide had increased enough at the end of the most recent ice age to melt the ice and raise the earth's temperature. Their findings suggest that cycles of ice ages followed by shorter warm periods may have been caused solely by rising and falling levels of carbon dioxide.

In 1986, worldwide carbon dioxide emissions from transportation and industrial sources totaled somewhat less than 5 billion tons, but in 1987, the total rose to more than 5.5 billion tons and has continued to rise. The burning of fossil fuels and deforestation (which destroys the major recyclers of carbon dioxide) have been largely responsible for a 25% increase of carbon dioxide in the atmosphere since 1850. In the last 25 years alone, the earth's atmosphere has become 0.4°C (0.7°F) warmer, and between 1983 and 1990, the surface temperature of the ocean rose about 0.8°C (1.5°F).

Those may not seem to be significant amounts, but during the last ice age in North America when ice covered the northern United States and Canada, the average temperature of the earth at sea level was only 4°C (7°F) colder than it is now. In 1989, Mostafa Tolba, head of the United Nations Environment Program, estimated that if the current levels of gas released into the atmosphere continue, the earth's temperature is likely to rise between 1.4 and 4.3°C (2.5 and 8.0°F) in the next 50 years.

Higher temperatures melt ice at the poles and cause water to expand. Higher ocean levels cause inundation of low-lying, often densely populated, coastal areas. The Environmental Protection Agency estimates that for each 30 centimeters (11.8 inches) the sea rises, it encroaches 100 feet inland, and already during the past century, worldwide ocean levels have risen 12.7 centimeters (5 inches). It is estimated that 7,000 square miles of land in the United States alone will be flooded if the temperatures rise as predicted. Major shifts in population to higher latitudes could follow, with the grain belts of the midwestern United States shifting into Ontario and the fertile crescents of Asia shifting north into Russia. Higher temperatures can also have major effects on winds, currents, and weather patterns, causing droughts and creating deserts in some areas, while bringing about heavy rainfall in others.

Methane

Swamps and wetlands have long been known to be sources of methane produced by anaerobic bacteria. Many animals produce methane during digestive processes, and large amounts of this gas are produced by the wood-digesting organisms within the guts of termites. Anaerobic bacteria in rice paddies produce significant quantities of methane. The total annual production of methane in the atmosphere has been slowly increasing in recent years. A small part of this increase may stem from the prodigious mushrooming in numbers of termites in cleared tropical rain forest areas, which, in 1995, were being destroyed at the rate of more than 100 acres per minute, 24 hours a day.

Ozone

Methane gas and chlorofluorocarbons (CFCs), which are inert chemicals used for refrigeration and other industrial purposes, are broken down into active compounds by sunlight at high altitudes. The breakdown products destroy *ozone,* a form of oxygen that in the stratosphere provides a natural shield for living organisms against intense ultraviolet radiation. Increased ultraviolet radiation correspondingly increases skin cancers. The accelerating weakening of the ozone shield has been recognized as a serious global problem by both the United States and the European Economic Community. In 1987, the United States proposed a 50% reduction of production and uses of chlorofluorocarbons by the year 2000, and in 1989, the European Economic Community proposed a total ban on uses of chlorofluorocarbons, also by the year 2000. In 1990, however, developing nations, such as India, China, and Brazil, had plans to expand the production of chlorofluorocarbons and contended that a ban would place them at an economic disadvantage. Since global cooperation is urgently needed, the major industrial nations have been seeking ways to allay the economic concerns of third-world countries. An international meeting, with mixed results, was convened in 1992 to try to foster global cooperation on this issue.

Chlorofluorocarbons apparently are not the only force actively destroying the protective ozone layer. Bromine-based compounds called *halons,* which are commonly found in electronic equipment, such as computer protection

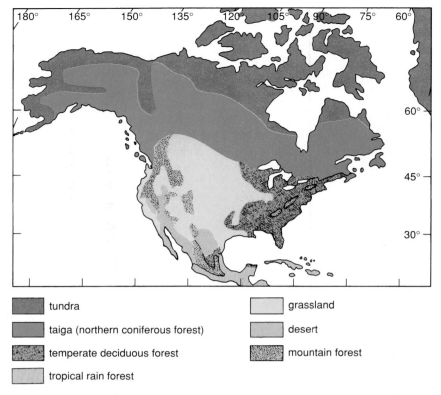

FIGURE 25.13   Major biomes of North America.

Legend

- tundra
- taiga (northern coniferous forest)
- temperate deciduous forest
- tropical rain forest
- grassland
- desert
- mountain forest

systems, and in portable fire extinguishers, are reported to be as much as three to ten times more destructive of ozone than are chlorofluorocarbons. Halon concentrations in the atmosphere increased about 20% a year between 1980 and 1986, according to the Environmental Protection Agency. Some scientists believe the actual concentrations to be as much as 50% higher than that and are strongly recommending that powders and other inert gases be substituted for halons in fire extinguishers of the future.

Loss of Biodiversity

Most plants and animals are adapted, sometimes in subtle ways, to the habitats in which they occur. When their natural habitats are destroyed, a few species may be able to adapt to new habitats, but most are not capable of doing so and ultimately perish. Ever since living organisms first appeared on this planet, various species have become extinct as climates and other environmental factors have changed. However, the rates of extinction have accelerated enormously in just the last half-century as many types of habitats, from tropical rain forests and swamps to deserts, have been damaged or destroyed. Literally thousands of species are now facing extinction, many within the next 10 years.

If food and fiber plants are not particularly endangered by loss of natural habitats, why should we be concerned if seemingly unimportant wild species disappear? One of the many consequences of such losses pertains to the origins of our crop plants. Virtually all food, fiber, medicinal, and other useful plants have been developed from wild species, and keeping crops from succumbing to diseases often depends on our ability to breed new disease-resistant strains by tapping the gene pools of the plants' wild cousins. The gene pools usually have developed over thousands or even millions of years, and if they disappear, the most skillful plant breeders may be no match for rapidly mutating fungal and other diseases that attack our crops.

MAJOR BIOMES OF NORTH AMERICA

Biotic communities considered on a global or at least on a continental scale are referred to as *biomes*. Several important biomes occupy the North American continent (Fig. 25.13), and most are also represented on other continents.

Tundra

Tundra occupies vast areas of the earth's land surfaces (about 20% in all), primarily above the Arctic Circle (Fig. 25.14). It has the appearance of being treeless, although miniature willows and birches only 2.5 to 50.0 centimeters (1 inch to 2 feet) tall do survive in some areas. Another characteristic of arctic tundra is the presence of *permafrost* (permanently frozen soil) at a depth of from 10 to 20 centimeters

FIGURE 25.14   Alaskan tundra in the summer.

(from 4 to 8 inches) to about 1 meter (3 feet 3 inches) below the surface. The level of the permafrost determines the depth to which plant roots can penetrate. Precipitation averages less than 25 centimeters (10 inches) a year, but because of the permafrost, it is largely held at or near the surface of the land. There is a very short growing season of only two to three months, with frost possible on any day of the year. However, temperatures can soar to 27°C (81°F) or higher during a long midsummer day. The vegetation of arctic tundra is dominated by dwarf shrubs, sedges, grasses, lichens, and mosses. During the brief growing season, tiny perennial plants produce brightly colored flowers and form brilliant mats over the topsoil, which is largely organic and generally only 5.0 to 7.5 centimeters (2 to 3 inches) deep.

Patches of alpine tundra occur above timberline in mountains below the Arctic Circle. Permafrost is frequently absent from alpine tundra areas, and grasses, sedges, and herbs tend to predominate.

The biome is exceptionally fragile. A car driven across the tundra compresses the soil sufficiently to kill plant roots, and the tracks are evident many years later. Occasionally, sheep grazing tundra have been observed to pull up patches of the matted vegetation, leaving exposed edges. High winds then can catch the exposed edges and rip away larger segments of mat, leaving barren patches called *blowouts*.

Animals of the tundra include lemmings, the little rodents that reproduce prodigiously every three or four years and then decline in numbers. The arctic fox and snowy owl also fluctuate in numbers, the fluctuations being correlated with the size of the populations of lemmings, which constitute their principal food. Caribou, polar bears (near the coast), shrews, marmots, ptarmigans, loons, plovers, jaegers, and arctic terns also are included in the tundra fauna.

Taiga

The *taiga,* also referred to as *northern coniferous* or *boreal forest,* is found adjacent to and south of the arctic tundra (Fig. 25.15). The vegetation is dominated by evergreen trees, such as spruces, firs, and pines. Tamaracks (larches) also occur in the taiga. Birches, aspens, and willows may be found in some of the wetter areas. Many perennials and a few shrubs occur, but there are few annuals. Snow blankets the region during the winters, which are long and severe, with temperatures dropping to –50°C (–58°F) or lower during the coldest months. In summer, the temperatures often reach 20°C to 30°C (68°F to 86°F). Most precipitation occurs in the summer, ranging from about 25 centimeters (10 inches) to more than 100 centimeters (39 inches) in parts of western North America. Many lakes, ponds, and

FIGURE 25.15   Taiga in Alaska.

FIGURE 25.16   A scene within the eastern deciduous forest.
(Courtesy Robert A. Schlising)

marshes dot the region, which is inhabited by a variety of birds, including jays, warblers, and nuthatches. Many rodents, such as shrews and jumping mice, and larger mammals, notably moose and deer, occur. Ermine and wolverines also make their home in the taiga, and caribou overwinter in this biome.

Temperate Deciduous Forest

Most deciduous trees are broad-leaved species that shed their leaves annually during the fall and remain dormant during the shorter days and colder temperatures of winter. Most *temperate deciduous forests* (Fig. 25.16) are found, like the taiga, on large continental masses in the Northern Hemisphere. In North America, an example of this type of forest occurs from the Great Lakes region south to the Gulf of Mexico and roughly extends from the Mississippi River to the eastern seaboard, although some do not consider the eastern coastal plain to be part of the temperate deciduous forest biome. Temperatures within the area vary a great deal but normally fall below 4°C (39°F) in midwinter and rise above 20°C (68°F) in summer. The climax trees of the forest are well adapted to subfreezing temperatures as long as the cold is accompanied by precipitation or snow cover. Most of the annual precipitation, which totals between 50 and 165 centimeters (20 to 65 inches), occurs in the summer.

Some of the most beautiful of all the broad-leaved trees are found in a variety of associations in this biome. In the upper Midwest, sugar maple and American basswood predominate. Sugar maple, birch, and oak are also found to the Northeast, where they tend to be associated with the stately American beech. In the west-central part of the forest, oaks and hickories abound. Oaks also are abundant along the eastern slopes of the Appalachian Mountains, where American chestnuts were once a conspicuous part of the flora. The chestnuts have now virtually disappeared, having succumbed to chestnut blight disease, which was introduced and began killing the trees during the early 1900s. Oaks and hickories extend into the southeastern United States, where they become associated with pines and other tree species, such as the bald cypress.

Before the arrival of European immigrants, a mixture of large deciduous trees that included maples, ashes, basswood, beeches, buckeyes, hickories, oaks, tulip trees, and magnolias was found on the eastern slopes and valleys of the Appalachian Mountains. Some of the trees grew over 30 meters (100 feet) tall and had trunks up to 3 meters (10 feet) in diameter. Except for in a few protected pockets in the Great Smoky Mountains National Park, the largest trees of this rich forest have been all but eliminated through logging. A smaller-treed extension of this forest is found in an area northeast of Baton Rouge, Louisiana, western Tennessee, and Kentucky and in southern Illinois. American elms, also once a part of the forest, are rapidly disappearing as Dutch elm disease fells both wild trees and those planted along city streets. One small midwestern town, which had some 600 elms planted along its streets, found itself left with only six live trees within a year or two after Dutch elm disease struck.

A mixture of deciduous trees and evergreens occurs on the northern and southeastern borders of the forest. Hemlocks and eastern white pine are found from New England west to Minnesota and south to the Appalachians. The once vast stands of eastern white pine are now largely gone, their valuable lumber having been used for construction and other purposes. Some have been lost to still another tree disease, white pine blister rust, but scattered trees remain, particularly where they have been protected. Various pines dominate the eastern coastal plain from New Jersey to Florida and west to east Texas. Pitch pine is common in New Jersey, while some of the southern pines (e.g., long-leaf, slash,

loblolly) are now cultivated in the southeastern United States for wood pulp, turpentine, lumber, and other commercially valuable products.

During the summer, the trees of the deciduous forests form a relatively solid canopy that keeps most direct sunlight from reaching the floor. Many of the showiest spring flowers of the region (such as bloodroot, hepatica, trilliums, and violets) flower before the trees have leafed out fully and complete most of their growth within a few weeks. Other plants that can tolerate more shade, principally members of the Sunflower Family (e.g., asters and goldenrods), flower in succession in forest openings from midsummer through fall.

Animal life in the eastern deciduous forest includes red and gray foxes, raccoons, opossums, many rodents such as gray squirrels, snakes such as copperheads and black rat snakes, salamanders, and a wide variety of birds, including hawks, flickers, and mourning doves.

Grassland

Naturally occurring *grasslands* are found toward the interiors of continental masses (Fig. 25.17). They tend to intergrade with forests, woodlands, or deserts at their margins, depending on precipitation patterns and amounts. A grassland may receive as little as 25 centimeters (10 inches) of rainfall or as much as 100 centimeters (39 inches) annually. Temperatures can range from 45°C (113°F) in midsummer to −45°C (−50°F) in midwinter.

Large areas of grassland are now used for growing cereal crops (particularly corn and wheat) and for grazing cattle. Before it was destroyed, the American prairie was a remarkable sight. In Illinois and Iowa, the grasses grew over 2 meters (6 feet 6 inches) tall during an average season and another meter taller during a wet one. A dazzling display of wildflowers began before the young perennial grasses emerged in the spring and continued throughout the growing season. I have counted over 50 species of flowering plants in bloom at the same time on 1 hectare (2.47 acres) of protected grassland in the middle of spring.

Grasslands occurring in areas with a Mediterranean climate (e.g., the Great Central Valley of California), where most of the precipitation takes place during the winters, usually include *vernal pools,* some of which may be more than 50,000 years old. These are temporary accumulations of rainwater that evaporate after the rains come to an end. Their unique floras include an orderly sequence of flowering plants, some appearing initially at the pool margins, with each species forming a distinct zone or band until the water is gone (Fig. 25.18). Some species flower only in the damp soil and drying mud that remains. The seeds of many species germinate under water.

In North America, huge herds of buffalo once grazed the prairies, which grew on fertile soils, but these disappeared as the settlers cultivated more and more of the land and hunters slaughtered more and more of the large animals. By 1889, there were only 551 left, but breeding and conservation has since increased their numbers.

FIGURE 25.17 Undisturbed western grassland in the spring.

FIGURE 25.18 Zonation in a vernal pool in northern California.
(Courtesy Robert A. Schlising)

FIGURE 25.19   A desert community in the southwestern United States.

Grassland animals include cottontails, jackrabbits, gophers, mice, and pronghorns. As indicated, buffalo were once abundant. Some 15,000 of these large animals now are protected in national parks, game preserves, and private ranches. Various sparrows (e.g., vesper and savannah sparrows) and other birds still find homes in uncultivated tracts of this biome.

Desert

If you ask the average person to describe a *desert,* the response will probably include words such as *sand, heat, mirage, oasis,* and *camels* (Fig. 25.19). These features are indeed found in some of the world's large deserts that are located in the vicinity of 30 degrees latitude both north and south of the equator, but deserts occur wherever precipitation is consistently low or the soil is too porous to retain water. Most receive less than 12.5 centimeters (5 inches) of rain per year. The low humidity results in wide daily temperature ranges. On a summer day when the temperature has reached over 35°C (95°F), it will generally fall below 15°C (59°F) the same night. The light intensities reach higher peaks in the dry air than they do in areas where atmospheric water vapor filters out some of the sun's rays. Some desert plants have become adapted to these higher light intensities through the evolution of crassulacean acid metabolism (CAM) photosynthesis (discussed in Chapter 10), a process by which certain organic acids accumulate in the chlorophyll-containing parts of the plants during the night and are converted to carbon dioxide during the day. This permits much more photosynthesis to take place than would otherwise be possible, since most of the carbon dioxide of the atmosphere is excluded from the plants during the day by the stomata, which remain closed (thereby also retarding water loss).

Other adaptations of desert plants include thick cuticles, water-storage tissues in stems and leaves, leaves with a leathery texture and/or reduced in size, and even a total absence of leaves. In cacti and other succulents without functional leaves, the stems take over the photosynthetic activities of the plants. Desert perennials are adapted to the biome in various ways. Cacti and similar succulents have widespread shallow root systems that can rapidly absorb water from the infrequent rains. The water can then be stored for long periods in the interior of the stems, which are modified for such storage. Other perennials grow from bulbs that are dormant for much of the year. Annuals provide a spectacular display of color and variety, particularly during an occasional season when above-average precipitation has occurred. The seeds of the annuals often germinate after a fall or winter rain, and the plants then grow slowly or remain in a basal, circular cluster of leaves (rosette) for several months before producing flowers in the spring. Literally hundreds of different species of desert annuals may occur within a few square kilometers (1 or 2 square miles) of typical desert in the southwestern United States.

FIGURE 25.20   Coastal redwoods in northern California.

biome also occur in other parts of the West, particularly toward the southern limits of the mountains.

The trees tend to be very large, particularly in and to the west of the Cascade Mountains of Oregon and Washington and on the western slopes of the Sierra Nevada. Part of the reason for the huge size of trees such as Douglas fir is the high annual rainfall, which exceeds 250 centimeters (100 inches) in some areas. The world's tallest conifers, the coastal redwoods of California (Fig. 25.20), are at low elevations between the Pacific Ocean and the California outer coast ranges. Here they apparently depend more on fog, which reduces transpiration rates, than on large amounts of rain for their size and longevity.

One of the characteristics of mountain forests is a fairly conspicuous altitudinal zonation of species. In other words, one encounters different associations of plants when proceeding from sea level up the mountainsides. At lower elevations in both the Rocky Mountains and the Sierra Nevada, the predominant conifer is ponderosa pine. At lower elevations in the northern part of the Cascades, Douglas firs, western red cedars, and western hemlocks are more common. At intermediate elevations in the Sierra Nevada, sugar pine, white fir, and Jeffrey pine take over, while at higher elevations, other species of pines, firs, and hemlocks predominate.

Most of the mountain forest biome has comparatively dry summers. This has led to frequent forest fires, even before human carelessness became a factor. Several tree species are well adapted to fires. The Douglas fir, for example, has a very thick protective bark that can be charred without transmitting sufficient heat to the interior to damage more delicate tissues, and its seedlings thrive in open areas after a fire. When the bark of the giant redwoods of the Sierra Nevada is burned, the trees are rarely killed. This has undoubtedly contributed to the great age and size of many of the trees. The cones of some of the pine trees (e.g., knobcone pine) remain closed and do not release their seeds until a fire opens them, while seeds of several other species germinate best after they have been exposed to fire. These attributes of members of the mountain forest biome, as mentioned in the section on fire ecology, have led to the practice in some of our national parks of occasionally allowing fires at higher elevations to run their natural course. The higher-than-normal incidence of fires occurring since humans came in large numbers to the forest has made it necessary, however, to control fires in most instances, even though in doing so we may be interfering with natural cycles that would otherwise occur.

Animal life in the mountain forests includes many different rodents (especially chipmunks and voles), bears, mountain lions, bobcats, mountain beavers (not related to true beavers), mule deer, and elk. Large birds such as the golden eagle and many small birds, including mountain chickadees, warblers, and juncos, are an integral part of this biome.

Desert animals are adapted, in many instances, to foraging at night when it is cooler. These animals include various mice and kangaroo rats, snakes (notably rattlesnakes and king snakes), chuckwallas, and lizards (e.g., gila monsters). Various thrashers, doves, and flycatchers are included in the bird life.

Mountain Forests

In the geologic past, deciduous forests extended to western North America. As the climate changed and summer rainfall was reduced, conifers largely replaced the deciduous trees, although some (e.g., maples, birches, aspens, oaks) still remain, particularly at the lower elevations. Today, coniferous forests occupy vast areas of the Pacific Northwest and extend south along the Rocky Mountains and the Sierra Nevada and California coast ranges. Isolated pockets of this

FIGURE 25.21 A tropical rain forest scene.
(Courtesy Mary Lane Powell)

Tropical Rain Forest

About 5% of the earth's surface, representing nearly half of the forested areas of the earth, are included in the *tropical rain forest* biome (Fig. 25.21). The biome is distributed throughout those areas of the tropics where annual rainfall amounts normally range between 200 and 400 centimeters (79 and 157 inches) and where temperatures range between 25°C and 35°C (77°F and 95°F). Although monthly rainfall amounts vary, there is no dry season and some precipitation occurs throughout all 12 months of the year, frequently in the form of afternoon cloudbursts. The humidity seldom drops below 80%.

Such climatic conditions support a diversity of flora and fauna so great that the number of species exceeds those of all the other biomes combined. The forests are dominated by broadleaf evergreen trees, whose trunks are often un-branched for as much as 40 or more meters (160 feet), with luxuriant crowns that form a beautiful dark green and sev-eral-layered canopy. The root systems are shallow and the tree trunks often buttressed (buttress roots are shown in Fig. 5.12). There are literally hundreds of species of such trees, each usually represented by widely scattered specimens.

Most of the plants of the rain forests are woody, al-though not all of them are evergreen. Several of the deciduous tree species shed their leaves from some branches, retain the leaves on others, and flower on yet other branches all at the same time, while branches of adjacent trees of the same species are losing their leaves or flowering at different times. Numerous hanging woody vines and even more numerous *epiphytes*—especially orchids and bromeliads—can be seen on or attached to tree branches. The epiphyte roots, which are not parasitic, have no contact with the ground. The plants are sustained entirely by rainwater that accumulates in their leaf bases and by their own photosynthesis. Traces of minerals, also necessary to the growth of the epiphytes, accumulate in the rainwater as it trickles over decaying bark and dust.

The multilayered canopy is so dense that very little light penetrates to the floor, and the few herbaceous plants that survive are generally confined to openings in the forest. Despite the lush growth, there is little accumulation of litter or humus, and the soil is relatively poor. Decomposers rapidly break down any leaves or other organic material on the forest floor, and the nutrients released by decomposition are quickly recycled or leached by the heavy rains.

A few larger animals, with adaptations for moving through the mesh of branches in the rain forest, are found on the forest floor. Such animals include peccaries, tapirs, and anteaters. Most of the great numbers of animals, however, live out their lives in the canopies. Tree frogs, with adhesive pads on their toes, are common in tropical rain forests, as are various monkeys, sloths, opossums, tree snakes, and lizards. Ants abound, and many of the extraordinary variety of other insects are adapted to their environment through excellent camouflages. Large flocks of parrots and other birds feed on the abundance of insects as well as the available fruits.

The Future of the Tropical Rain Forest Biome

In the 1960s, major plans were developed to convert the Amazon rain forest into large farms, hydroelectric plants, and mines. At the beginning of the 1990s, gold mining activities were filling the rivers with silt, and, as indicated in the discussion of the greenhouse effect, the tropical rain forests were being destroyed or damaged for commercial purposes at the rate of more than 100 acres per minute. In one study of Amazon rain forest birds conducted in 1989, it was found that a population of low-flying species in a 25-acre section of the forest fell by 75% within just six weeks after adjacent land had been cleared, and 10 of the 48 species apparently disappeared completely.

This situation has been repeated on numerous occasions, and confirmed by a project cosponsored by the Smithsonian Institution and Brazil's Institute for Research in the Amazon, which during 10 years of study demonstrated that cutting large expanses of forest into smaller pieces separated by as little as 10 meters (33 feet) of cleared land can have disastrous effects on the ecology of the entire forest, with most of the bird and other animal species originally present disappearing permanently. Very little of the biome, which is the home of more than 50% of all living species of organisms, is presently protected from commercial development. Many of the organisms are doomed to extinction, often before they have been described for the first time, and the biome will essentially have vanished within 20 years if governments and individuals do not take definitive action to halt the large-scale destruction.

John Muir, Father of America's National Park System

Today America's National Parks are overcrowded as some 270 million people enjoy their beauty each year, relaxing in the great outdoors, hiking in fields and mountains, or navigating their rivers. Whether it's Yosemite, the Grand Canyon, Acadia, or one of the many others, each park has a unique and special appeal. While these magnificent parks are the result of many people's vision and foresight, perhaps none more so than John Muir who lived from 1838 to 1914. Often called the "Father of the National Park System," he influenced presidents, members of Congress, and "just plain folks" by his love of nature and his writings about it. Always traveling and exploring various ecosystems, he visited Alaska five times, walked 1,000 miles from Indianapolis to the Gulf of Mexico, as well as walking across California's San Joaquin Valley to explore the Sierra Nevada. Later he wrote of viewing the Sierra Nevada range for the first time, "Then it seemed to me the Sierra should be called not the Nevada, or 'Snowy Range,' but the Range of Light, the most divinely beautiful of all the mountain chains I have ever seen."

Wherever he roamed in the wilderness, he noticed increasing waste and destruction from various sources, whether overgrazing by cattle and sheep or overcutting by loggers. It was his love for the American forest that

Shafts of light penetrate the coastal redwoods of Muir Woods.

Summary

1. Human populations have increased dramatically in the past few centuries, and the disruption of ecosystems by the activities directly or indirectly associated with the feeding, clothing, and housing of billions of people threatens the survival of not only humans but many other living organisms.

2. Ecology is the study of the relationships between organisms and their environment. It is a vast field with many facets. Regional ecological issues include acid rain, water contamination, destruction of wetlands, and hazardous waste.

3. Global ecological issues include populations, communities, ecosystems, interactions between organisms, and loss of biodiversity.

4. Acid rain, which damages or kills living organisms, is produced when sulphur and nitrogen compounds released by the burning of fossil fuels are converted to nitric acid and sulphur dioxide by sunlight and rain.

5. Water contamination occurs when toxic wastes, pesticides, septic tanks, and fertilizers wash or leach into surface and groundwater. Bacteria may in the future be used to break down various contaminants.

6. Wetlands, which were once considered wastelands, have been drained, but are now being increasingly protected.

7. Hazardous waste sites are slowly being detoxified, but progress is slow.

8. Populations vary in numbers, in density, and in the total mass of individuals. Populations of one community may also occur in another community, and ecotypes may be specifically adapted to a single community.

9. Precipitation, temperature, soils, and biotic factors play roles in determining the distributions of plant species in an ecosystem.

drove him to work tirelessly for its preservation. He wrote some 300 articles and 10 books that told of his travels and contained his naturalist philosophy. He called on everyone to "climb the mountains and get their good tidings." He wrote that the wilderness should be preserved, viewed as a treasure rather than a resource to be exploited. Through his writing, he drew attention to the destruction forests were facing and worked to find remedies.

One solution was the formation of national parks. Largely through Muir's efforts, Congress, in 1890, created Yosemite National Park. He was also personally involved in the creation of Sequoia, Grand Canyon, Mount Rainier, and Petrified Forest National Parks. He went one step further by founding the Sierra Club in 1892 and served as its first president until his death in 1914. He and his friends conceived the Sierra Club with the mission to "do something for wildness and make the mountains glad." His idea was to form an association that would work to protect Yosemite National Park.

This brought him to the attention of President Theodore Roosevelt who visited Muir in Yosemite in 1903. Together with Muir, under the trees of the park, Roosevelt's innovative conservation programs began to take form.

John Muir's words and life inspire environmental activism today; they also teach the importance of enjoying nature and protecting it. Muir wrote a warning that is just as timely today as it was 100 years ago.

The axe and saw are insanely busy, chips are flying thick as snowflakes and every summer thousands of acres of priceless forests, with their underbrush, soil, springs, climate, scenery and religion, are vanishing away in clouds of smoke, while, except in the national parks, not one forest guard is employed. Any fool can destroy trees. They cannot run away; and if they could, they would still be destroyed—chased and hunted down as long as fun or a dollar could be got out of their bark hides, branching horns or magnificent bole backbones. Few that fell trees plant them; or would planting avail much towards getting back anything like the noble primeval forests.[1]

A living memorial to John Muir is located 12 miles north of San Francisco, Muir's home in his later years. Muir Woods is a 560-acre national monument consisting of a grove of majestic coastal redwoods along a canyon floor. The towering redwoods and lush canyon ferns growing along Redwood Creek make this a forest of tranquility. Only a 30-minute drive from San Francisco, Muir Woods is visited by over 1.5 million people each year. He declared that this grove of coastal redwoods was, "the best tree-lovers monument that could possibly be found in all the forests of the world."

John Muir—lover of forests, roamer of the wilderness, inspired writer, and fighter for conservation.

1. *Atlantic Monthly,* no. 80, 1897.

10. Xerophytes and hydrophytes have modifications of leaves and other organs.

11. Species distribution is influenced by soil mineral content and biotic factors, such as competition for light, nutrients, and water.

12. Ecosystems ideally sustain themselves entirely through photosynthetic activity, energy flow through food chains, and the recycling of nutrients. Producers carry on photosynthesis; consumers feed on producers. Primary consumers feed directly on producers, and secondary consumers feed on primary consumers.

13. Decomposers break down organic materials to forms that can be reassimilated by the producers. In any ecosystem, the producers and consumers comprise food chains, which determine the flow of energy through the different levels. Food chains vary in length and intricacy and, because of their interconnections, form food webs.

14. Energy itself is not recycled in an ecosystem; it gradually escapes in the form of heat as it passes from one level to another. In a long food chain, the final consumer gains only a tiny fraction of the energy originally captured by the producer at the bottom of the chain.

15. When producers increase the amount of food available, consumers increase correspondingly; this reduces the available food, resulting in a self-maintaining ecosystem. The composition of an ecosystem may be influenced by its living components through the secretion of growth-inhibiting substances, the secretion of growth-promoting substances, and parasitism.

16. Some nitrogen from the air is fixed by the nitrogen-fixing bacteria found in legumes and other plants. In the nitrogen cycle, there is a constant flow of nitrogen from dead plant and animal tissues into the soil and from the soil back to the plants.

17. Water leaches nitrogen from the soil and carries it away when erosion occurs. Other nitrogen is lost from harvesting crops, but the loss can be reduced if wastes

are decomposed and annually returned to the soil. Fire also causes nitrogen loss.

18. Replacement of nitrogen loss by the application of chemical fertilizers can eventually create hardpan by altering the soil structure. Bacteria also recycle carbon and other substances, such as water and phosphorus.

19. Succession occurs whenever there has been a disturbance of natural areas on land or in water.

20. Primary succession involves the formation of soil in the beginning stages, while secondary succession takes place in areas previously covered with vegetation. A xerosere is a primary succession in which bare rock or lava is converted to soil through the activities of lichens, plants, and physical forces in an orderly progression of events over a period of time.

21. Climax vegetation becomes established at the conclusion of succession and remains until or unless a disturbance disrupts it.

22. A hydrosere is initiated in a wet habitat and culminates in a climax vegetation. As a lake or other body of water is filled in with silt and debris, eutrophication facilitating the growth of algae and other organisms occurs.

23. Secondary succession, which proceeds more rapidly than does primary succession, may take place if soil is present. It may occur after fires.

24. The greenhouse effect refers to a global rise in temperature due to carbon dioxide, methane, and other gases preventing the sun's radiant heat from escaping back into space.

25. The carbon dioxide and methane levels in the earth's atmosphere and the earth's temperature have been rising. It is predicted that polar ice will, as a result, melt and flooding of low-lying coastal areas will occur.

26. In the stratosphere, sunlight converts methane, chlorofluorocarbons, and halons into active compounds that destroy the ozone shield that protects us from intense ultraviolet radiation.

27. Loss of biodiversity has serious consequences, including the potential loss of means with which to combat crop diseases.

28. Biotic communities considered on a global or at least a continental scale are called *biomes*. Major biomes of North America include tundra, taiga, temperate deciduous forests, grasslands, deserts, mountain forests, and tropical rain forests.

29. Tundra is found primarily above the Arctic Circle. It includes few trees and many lichens and grasses and is characterized by the presence of permafrost (permanently frozen soil) below the surface. Alpine tundra occurs above timberline in mountains below the Arctic Circle.

30. Taiga is dominated by coniferous trees, with birches, aspens, and willows in the wetter areas. Many perennials and few annuals occur in taiga.

31. Temperate deciduous forests are dominated by trees such as sugar maples, American basswoods, beeches, oaks, and hickories, with evergreens such as hemlocks and eastern white pines toward the northern and southeastern borders.

32. Grasslands are found toward the interiors of continental masses. Those located in Mediterranean climatic zones usually include vernal pools with unique annual floras. Many grasslands have been converted to agricultural use.

33. Deserts are characterized by low annual precipitation and wide daily temperature ranges, with plants adapted both in form and metabolism to the environment.

34. Mountain forests occupy much of the Pacific Northwest and extend south along the Rocky Mountains and California mountain ranges. The primarily coniferous tree species tend to be in zones determined by altitude. The forests have mostly dry summers, and some of the species associated with it have thick bark that protects the trees from fires, which occur frequently in this type of climate. Other species (e.g., knobcone pine) depend on fires for normal distribution and germination of seeds.

35. The tropical rain forests constitute nearly half of all forest land and contain more species of plants and animals than all the other biomes combined. Numerous woody plants and vines form multilayered canopies, which permit very little light to reach the floor. The soils are poor, with nutrients released during decomposition being rapidly recycled. The biome is being destroyed so rapidly that it will disappear within 20 years if the destruction is not halted.

Review Questions

1. How is acid rain formed?

2. What are the principal threats to global drinking water sources?

3. Distinguish among a plant community, a biotic community, an ecosystem, and a biome.

4. What is the difference between a primary consumer and a secondary consumer?

5. How are short food chains more efficient in the use of solar energy than are long ones?

6. Why is the diet of larger herbivorous animals varied?

7. What types of checks and balances exist in an ecosystem?

8. How are nitrogen and carbon recycled in nature?

9. What is primary succession as opposed to secondary succession?

10. What are general characteristics of climax vegetation as compared with nonclimax vegetation?

11. What is the greenhouse effect?

12. What is the ozone shield, and what significant role does it play? What threats are there to its existence?

13. Characterize desert and tundra biomes.

14. What is the function of fire in grassland and mountain forest biomes?

15. What kinds of plants predominate in tropical rain forests?

16. Why does the loss of the tropical forest biome have very serious implications for the future?

Discussion Questions

1. Humans have disrupted ecosystems almost everywhere they have established themselves, at least in industrialized countries. Do you believe that humans could also improve an ecosystem? Explain.

2. If a vegetarian diet makes more efficient use of solar energy, should we all strive to become vegetarians for this reason? Explain.

3. Besides the hypothetical example given (in which it was observed that eating lettuce directly was a more efficient use of solar energy than feeding it to hogs and then eating the pork), can you think of other ways in which food chains might be shortened?

4. Peanut butter has many of the nutrients needed in human nutrition. On the basis of what you have learned about the diet of animals in an ecosystem, do you think it would be a good idea to live on peanut butter and water as a means of saving money? Explain.

5. Could succession take place in an abandoned swimming pool? Explain.

6. Fire has been a natural phenomenon in several biomes for thousands of years, and most plant species are adapted to it in various ways in the ecosystem where it occurs regularly. Should we then not extinguish forest and grassland fires when they occur? Explain.

7. Much of the tropical rain forest biome is located in third world countries, which have precarious economies. Bearing that in mind, what could be done to halt the destruction of this priceless and vital part of our ecosystem?

Additional Reading

Barbour, M., et al. 1987. *Terrestrial plant ecology,* 2d ed. Menlo Park, CA: Benjamin/Cummings Publishing Co., Inc.

Barbour, M. G., and W. D. Billings. 1990. *North American terrestrial vegetation.* New York: Cambridge University Press.

Dafni, A. 1993. *Pollution ecology: A practical approach.* New York: Oxford University Press.

Daubenmire, R. 1978. *Plant geography.* San Diego, CA: Academic Press, Inc.

Fowden, L., et al. 1993. *Plant adaptation to environmental stress.* New York: Chapman and Hall.

Glenn-Lewin, D. C., et al. (Eds.). 1992. *Plant succession: Theory and prediction.* New York: Chapman and Hall.

Koopowitz, H., and H. Kaye. 1983. *Plant extinction: A global crisis.* Washington, DC: Stone Wall Press.

Kormondy, E. J. 1984. *Concepts of ecology,* 3d ed. Englewood Cliffs, NJ: Prentice-Hall.

Kozlowski, T. T., and C. E. Ahlgren (Eds.). 1974. *Fire and ecosystems.* San Diego, CA: Academic Press, Inc.

Rice, E. L. 1983. *Allelopathy,* 2d ed. San Diego, CA: Academic Press, Inc.

Silvertown, J. 1987. *Introduction to plant population ecology,* 2d ed. New York: Halsted Press.

Whitmore, T. C. 1990. *An introduction to tropical rain forests.* Fair Lawn, NY: Oxford University Press.

Appendix 1

Scientific Names of Organisms Mentioned in the Text

This is an alphabetical list of the organisms whose scientific names may not be mentioned in the text. Common and scientific names of organisms mentioned in Appendices 2 through 4 are provided within the respective appendices.

COMMON NAMES AND SCIENTIFIC NAMES OF ORGANISMS

COMMON NAME	SCIENTIFIC NAME	COMMON NAME	SCIENTIFIC NAME
Aardvark	*Orycteropus* spp.	**Algae, yellow-green**	members of Class Xanthophyceae, Division Chrysophyta, Kingdom Protoctista
Abrasives, horsetail source of	*Equisetum* spp.		
Absinthe liqueur, source of ingredients	*Pimpinella anisum, Artemisia absinthium,* and others	**Almond**	*Prunus amygdalus*
Aconite, source of	*Aconitum* spp.	**Aloe juice, source of**	*Aloe barbadensis, A. ferox, A. vera,* and others
Afghanistan pine	*Pinus eldarica*	**'Ama'uma'u**	*Sadleria cyatheoides*
Agar, source of	*Gelidium* spp. and others	**Amaryllis**	*Amaryllis* spp.
Agave	*Agave* spp.	**Amoeba**	*Amoeba proteus* and others
Air plant	*Kalanchoë* spp.	**Amoeba, fungal internal parasites of**	*Cochlonema verrucosum* and others
Alder	*Alnus* spp.		
Alfalfa	*Medicago sativa*	**Amoeba, fungal trappers of**	*Dactylella* spp. and others
Alfalfa caterpillar	*Colias philodice*	**Anabaena**	*Anabaena* spp.
Algae	members of Kingdom Protoctista—all divisions	**Anemone**	*Anemone* spp.
Algae, brown	members of Division Phaeophyta, Kingdom Protoctista	**Angelica**	*Angelica archangelica*
		Anise	*Pimpinella anisum*
		Annatto	*Bixa orellana*
Algae, coralline	*Bossiella* spp., *Corallina* spp., *Lithothamnion* spp., and others	**Ant**	*Formica* spp. and others
		Anteater	*Myrmecophaga jubata*
		Ants, bullhorn *Acacia*	*Pseudomyrmex ferruginea*
Algae, flatworm	*Platymonas* spp.	**Aphid**	*Anuraphis* spp., *Aphis* spp., and others
Algae, golden-brown	members of Class Chrysophyceae, Division Chrysophyta, Kingdom Protoctista		
		Apple	*Pyrus malus*
		Apricot	*Prunus armeniaca*
Algae, green	members of Division Chlorophyta, Kingdom Protoctista	**Arbor-vitae**	*Thuja occidentalis*
		Archaebacteria	members of Division Archaebacteria, Subkingdom Archaebacteriobionta, Kingdom Monera
Algae, platelike colonial green	*Pediastrum* spp., *Chaetopeltis* spp., and others		
		Arrowroot, Florida, source of	*Zamia floridana*
Algae, red	members of Division Rhodophyta, Kingdom Protoctista	**Artichoke, globe**	*Cynaria scolymus*
		Artichoke, Jerusalem	*Helianthus tuberosus*
		Arum Lily Family	Araceae
Algae, snowbank	*Chlamydomonas nivale* and others	**Ash, Oregon**	*Fraxinus latifolia*
		Ash, blue	*Fraxinus quadrangulata*
Algae, sponge	*Chlorella* spp., *Zoochlorella* spp.	**Ash, white**	*Fraxinus americana*

COMMON NAMES AND SCIENTIFIC NAMES OF ORGANISMS

COMMON NAME	SCIENTIFIC NAME	COMMON NAME	SCIENTIFIC NAME
Aspen	*Populus tremuloides* and others	Bacteria, green sulphur	*Chlorobium* spp., *Chloropseudomonas* spp., *Prosthecochloris* spp., and others
Aster	*Aster* spp.		
Astringent, horsetail source of	*Equisetum arvense, E. debile,* and others	Bacteria, hydrogen	*Hydrogenomonas* spp.
Autograph tree (Fig. 8.18 C)	*Clusia rosea*	Bacteria, ice-minus	*Pseudomonas syringiae*
Avocado	*Persea americana* and others	Bacteria, iron	*Gallionella* spp., *Sphaerotilus* spp.
Azalea	*Rhododendron* spp.	Bacteria, kefir	*Lactobacillus bulgaricus,* *Streptococcus lactis*
Baby blue eyes	*Nemophila menziesii*		
Baby powder, ground pine source of	*Lycopodium clavatum*	Bacteria, lactic acid	*Lactobacillus delbrueckii* and others
Bacteria, acetone-producing	*Clostridium acetobutylicum* and others	Bacteria, Legionnaire's disease	*Legionella pneumophilia*
Bacteria, acidophilus	*Lactobacillus acidophilus*	Bacteria, luminescent	*Achromobacter* spp., *Flavobacterium* spp., *Photobacterium* spp., *Pseudomonas* spp., *Vibrio* spp., and others
Bacteria, ammonifying	*Clostridium* spp., *Micrococcus* spp., *Proteus* spp., *Pseudomonas* spp., and others		
Bacteria, anthrax	*Bacillus anthracis*	Bacteria, meningitis	*Neisseria meningitidis* and others
Bacteria, blue-green	members of Class Cyanobacteriae, Division Eubacteria, Subkingdom Eubacteriobionta, Kingdom Monera	Bacteria, methane	*Methanobacterium* spp., *Methanococcus* spp., *Methanosarcina* spp., and others
Bacteria, botulism	*Clostridium botulinum*	Bacteria, milky spore disease	*Bacillus popilliae*
Bacteria, brucellosis	*Brucella abortus, B. suis, B. melitensis*	Bacteria, mosquito-killing	*Bacillus thuringiensis* var. *israelensis*
Bacteria, BT	*Bacillus thuringiensis*		
Bacteria, bubonic plague	*Yersinia pestis*	Bacteria, nitrate (nitrifying)	*Nitrobacter* spp.
Bacteria, buttermilk	*Streptococcus lactis,* *S. cremoris, Leuconostoc citrovorum,* and others	Bacteria, nitrite (nitrosifying)	*Nitrosomonas* spp.
Bacteria, butyl alcohol	*Clostridium acetobutylicum* and others	Bacteria, nitrogen-fixing	*Azotobacter* spp., *Clostridium pasteurinum, Rhizobium* spp.
Bacteria, cheese—*see* Bacteria, buttermilk		Bacteria, paratyphoid fever	*Salmonella paratyphi*
Bacteria, cholera	*Vibrio cholerae*	Bacteria, pneumonia (some forms)	*Streptococcus pneumoniae* and others
Bacteria, decay	*Clostridium* spp., *Micrococcus* spp., *Proteus* spp., *Pseudomonas* spp., and others	Bacteria, PPLO	*Mycoplasma pneumoniae*
		Bacteria, pseudomonad	*Pseudomonas* spp.
Bacteria, dentrifying	*Micrococcus denitrificans,* *Thiobacillus denitrificans,* and others	Bacteria, purple nonsulphur	*Rhodomicrobium* spp., *Rhodopseudomonas* spp., *Rhodospirillum* spp.
Bacteria, dextran	*Leuconostoc mesenteroides*	Bacteria, purple suphur	*Amoebobacter* spp., *Lamprocystis* spp., *Rhodothece* spp., and others
Bacteria, diphtheria	*Corynebacterium diphtheriae*		
Bacteria, ensilage	*Lactobacillus delbrueckii,* *L. plantarum,* and others	Bacteria, salmonella, ("food poisoning")	*Salmonella* spp.
Bacteria, frost-damage preventing	*Pseudomonas syringiae*	Bacteria, salt	*Halococcus* spp., *Halobacterium* spp.
Bacteria, gas gangrene	*Clostridium novyi,* *C. perfringens, C. septicum*	Bacteria, sauerkraut	*Leuconostoc* spp. and others
Bacteria, glutamic acid producing	*Arthrobacter* spp., *Brevibacterium* spp., *Micrococcus* spp.	Bacteria, sorbose	*Acetobacter suboxydans*
		Bacteria, spotted fever	*Rickettsia rickettsii*
Bacteria, gonorrhea	*Neisseria gonorrhoeae*	Bacteria, "strep" throat	*Streptococcus* spp.
Bacteria, grease- and oil-dissolving	*Pseudomonas aeruginosa*	Bacteria, sulfolobus	*Sulfolobus* spp., *Thermoplasma* spp., *Thermoproteus* spp.,

COMMON NAMES AND SCIENTIFIC NAMES OF ORGANISMS

COMMON NAME	SCIENTIFIC NAME	COMMON NAME	SCIENTIFIC NAME
Bacteria, sulphur	*Desulfovibrio* spp., *Thiobacillus* spp., and others	Beaver, mountain	*Aplodontia rufa*
		Bedstraw	*Galium* spp.
		Bee, honey	*Apis mellifera*
Bacteria, syphilis	*Treponema pallidum*	Beech	*Fagus grandifolia*
Bacteria, tetanus	*Clostridium tetani*	Beefsteak morel	*Helvella* spp.
Bacteria, tularemia	*Francisella tularensis*	Beet, garden	*Beta vulgaris*
Bacteria, typhoid fever	*Salmonella typhi*	Beet, sugar	*Beta vulgaris* (horticulturally selected strains)
Bacteria, typhus fever	*Rickettsia prowazeki* and others	Beetle	member of Order Coleoptera, Class Insecta, Phylum Arthropoda, Kingdom Animalia
Bacteria, vinegar	*Acetobacter* spp.		
Bacteria, whooping cough	*Bordetella pertussis*		
Bacteria, yogurt	*Streptococcus thermophilus*		
Balsa	*Ochroma lagopus*	Beetle, elm bark	*Hylurgopinus rufipes, Scolytus multistriatus*
Balsamroot	*Balsamorhiza* spp.		
Bamboo	*Bambusa* spp.	Beetle, fungi used for food by ("ambrosia")	*Ambrosiella* spp., *Monilia*, spp.
Banana	*Musa paradisiaca* and others		
Banyan tree	*Ficus* spp.	Beetle, scarab	member of Family Scarabaeidae—*see* Beetle
Baobab	*Adansonia digitata, A. gregorii*		
		Begonia	*Begonia* spp.
Barbasco	*Lonchocarpus nicou* var. *utilis* and others	Belladonna, source of	*Atropa belladonna*
		Betony, wood	*Pedicularis canadensis*
Barberry	*Berberis* spp.	Big tree	*Sequoiadendron giganteum*
Barberry, common	*Berberis vulgaris*	Bigleaf maple (Fig. 8.20)	*Acer macrophyllum*
Bark, green algae that inhabit	*Protococcus* spp.[1]	Birch	*Betula papyrifera* and others
		Bird's-nest fungus (Fig. 19.21)	*Crucibulum levis*
Barley	*Hordeum* spp.		
Barn swallow	*Hirundo rustica erythrogaster*	Birth control pills, fungi used in manufacture of	*Rhizopus nigricans, R. arrhizus*
		Bittersweet	*Celastrus scandens*
Basil	*Ocimum basilicum*	Black bread mold	*Rhizopus stolonifer* and others
Basswood	*Tilia* spp.		
Bat	*Eidolon* spp., *Epomophorus* spp., and others	Black stem rust of wheat	*Puccinia graminis*
		Blackberry	*Rubus argutus, R. laciniatus, R. procerus,* and others
Bat (Fig. 23.17)	*Leptonycteris sanbornii*		
Bay, California	*Umbellularia californica*	Blackbird	*Euphagus* spp. and others
Bay, sweet	*Laurus nobilis*	Bladderwort	*Utricularia* spp.
Bean, broad	*Vicia faba*	Blazing star	*Liatris ligulistylis*
Bean, castor	*Ricinus communis*	Bleeding, ground pine used to arrest	*Lycopodium clavatum*
Bean, garbanzo	*Cicer arietinum*		
Bean, garden	*Phaseolus vulgaris*	Bleeding heart	*Dicentra* spp.
Bean, jequirity	*Abrus precatorius*	Bleeding heart, eastern	*Dicentra eximia*
Bean, kidney	*Phaseolus vulgaris*	Bleeding heart, Pacific	*Dicentra formosa*
Bean, lima	*Phaseolus lunatus*	Bloodroot	*Sanguinaria canadensis*
Bean, Mexican jumping	*Sebastiana* spp., *Sapium* spp.	Blue curls	*Trichostema* spp.
Bean, mung	*Phaseolus aureus*	Blue-green bacteria, Lake Chad edible	*Spirulina* sp.
Bean, pinto	*Phaseolus vulgaris*		
Bean, scarlet runner	*Phaseolus coccineus*	Blue-green bacteria, Red Sea	*Trichodesmium erythraeum*
Bean, tepary	*Phaseolus acutifolius* var. *latifolius*		
		Blue-green bacteria, thermal	*Bacillosiphon induratus, Synechococcus* spp., and others
Bean, winged	*Psophorcarpus tetragonolobus*		
Bear	*Ursus* spp. and others	Blue jay	*Cyanocitta cristata*
Bear, polar	*Thalarctos maritimus*	Blueberry	*Vaccinium* spp.
Bearberry	*Arctostaphylos uva-ursi*	Bobcat	*Lynx rufus*
		Bolete	*Boletus* spp. and others
		Bollworm	*Pectinophora gossypiella*
		Bowstring fibers, source of	*Sansevieria metalaea*
		Bowstring hemp, source of	*Sansevieria* spp.
		Box elder	*Acer negundo*

[1]These algae are known under several names (*Desmococcus, Phytoconis, Pleurococcus, Protococcus*), and uncertainty exists as to which name has priority. The green algal component of certain lichens, *Trebouxia*, also occurs independently on bark.

COMMON NAMES AND SCIENTIFIC NAMES OF ORGANISMS

COMMON NAME	SCIENTIFIC NAME	COMMON NAME	SCIENTIFIC NAME
Boysenberry	*Rubus* hybrids, with *R. ursinus* as one parent	**Camelina**	*Camelina sativa*
		Camellia	*Camellia* spp.
Brazil nut	*Bertholettia excelsa*	**Camphor, source of**	*Cinnamomum camphora*
Breadfruit	*Artocarpus* spp.	**Candelilla**	*Euphorbia antisyphilitca*
Bridalwreath	*Spiraea vanhouttei* hybrids and others	**Candlenut**	*Aleurites moluccana*
		Cankerworm	*Alsophila pometaria* and others
Broccoli	*Brassica oleracea* var. *botrytis*	**Cantaloupe**	*Cucumis melo*
Bromeliad Family	Bromeliaceae	**Caraway**	*Carum carvi*
Broomrape	*Orobanche* spp.	**Cardamon**	*Elettaria cardamomum*
Brussels sprouts	*Brassica oleracea* var. *gemmifera*	**Caribou**	*Rangifer* spp.
		Carnation	*Dianthus caryophyllus* and others
Bryophyte (*see also* individual listings)	member of Division Anthocerotophyta, Division Hepaticophyta, or Division Bryophyta, Kingdom Plantae	**Carnauba wax, source of**	*Copernicia cerifera*
		Carnaubalike wax, source of	*Stipa tenacissima*
		Carob	*Ceratonia siliqua*
Bryopsid	member of Class Chlorophyceae, Division Chlorophyta, Kingdom Protoctista	**Carpetweed Family**	Aizoaceae
		Carrot	*Daucus carota*
		Cashew	*Anacardium occidentale*
		Cassava	*Manihot esculenta*
Buckeye	*Aesculus* spp.	**Cassia**[2]	*Cinnamomum cassia*
Buckwheat	*Fagopyrum esculentum*	**Catalpa**	*Catalpa* spp.
Buffalo	*Bison bison*	**Caterpillar**	larval stage of member of Order Lepidoptera, Phylum Arthropoda, Kingdom Animalia
Bullhorn *Acacia*	*Acacia cornigera*		
Burn treatment, (ashes) horsetail source of	*Equisetum hyemale* and others		
Butcher's-broom	*Ruscus* spp.	**Catnip**	*Nepeta cataria*
Buttercup	*Ranunculus* spp.	**Cattail**	*Typha* spp.
Buttercup, European bulbous	*Ranunculus bulbosa*	**Cattle**—*see* **Cow**	
		Cauliflower	*Brassica oleracea* var. *botrytis*[3]
Buttercup Family	Ranunculaceae		
Butterfly	member of Superfamily Papilionoidea, Order Lepidoptera, Phylum Arthropoda, Kingdom Animalia	**Caussu wax, source of**	*Calathea lutea*
		Cedar, Atlantic white	*Chamaecyparis thyoides*
		Cedar, eastern red	*Juniperus virginiana*
		Cedar, eastern white	*Thuja occidentalis*
		Cedar, incense	*Libocedrus decurrens*
Butterwort	*Pinguicula* spp.	**Cedar, southern white**	*Chamaecyparis thyoides*
Button snakeroot	*Eryngium* spp.	**Cedar, western red**	*Thuja plicata*
Cabbage	*Brassica oleracea*	**Celery**	*Apium graveolens*
Cabbage Family	Brassicaceae (Cruciferae)	**Cellular slime mold**	member of Division Acrasiomycota, Kingdom Protoctista
Cabbage looper	*Trichoplusia ni*		
Cabbage worm	*Pieris rapae*	**Century plant**	*Agave* spp.
Cacao	*Theobroma cacao*	**Chamise**	*Adenostoma fasciculatum*
Cactus (Fig. 24.14 A)	*Hamatocactus setispinus*	**Chard**	*Beta vulgaris* var. *cicla*
Cactus, barrel	*Mamillaria* spp. and others	**Cheese bacteria**—*see* **Bacteria, buttermilk**	
Cactus, cholla	*Opuntia* spp. (cylindrical forms)		
		Cherry	*Prunus avium* and others
Cactus Family	Cactaceae	**Chestnut**	*Castanea* spp.
Cactus, giant saguaro	*Carnegia gigantea*	**Chestnut, American**	*Castanea americana*
Cactus, hedgehog	*Echinocereus* spp. and others	**Chestnut blight**	*Endothia parasitica*
Cactus, living rock	*Ariocarpus fissuratus* and others	**Chia**	*Salvia columbariae*
Cactus, organ-pipe	*Lemaireocereus* spp.		
Cactus, prickly pear	*Opuntia* spp.		
Cajeput, source of	*Melaleuca leucodendron*		
Calabazilla	*Cucurbita foetidissima*		
Calendula	*Calendula* spp.		
Camel	*Camelus* spp.		

[2]This should not be confused with the genus *Cassia,* the source of senna in the Legume Family, or cassie, a perfume oil whose source is another member of the Legume Family, *Acacia farnesiana.*

[3]Broccoli and cauliflower are two different forms of the same variety.

COMMON NAMES AND SCIENTIFIC NAMES OF ORGANISMS

COMMON NAME	SCIENTIFIC NAME	COMMON NAME	SCIENTIFIC NAME
Chickadee, mountain	*Parus gambeli*	**Coral tree**	*Erythrina crista-galli*
Chickweed (Himalayan)	*Stellaria decumbens*	**Cordage fibers, source of**	*Agave sisalina, A.*
Chicle, source of	*Achras sapota*		*heterocantha, A.*
Chicory	*Cichorium intybus*		*lophantha, Phormium*
Chinese vegetable tallow	*Sapium sebiferum*		*tenax,* and others
Chipmunk	*Eutamias* spp. and others	**Coriander**	*Coriandrum sativum*
Chlamydomonas	*Chlamydomonas* spp.	**Corn**	*Zea mays*
Chloroxybacteria	*Chloroxybacteriae* Division	**Corn borer, European**	*Pyrausta nubialis*
	Eubacteria etc.	**Cotton**	*Gossypium* spp.
Chlorella	*Chlorella* spp.	**Cottonwood**	*Populus deltoides* and others
Chocolate, source of	*Theobroma cacao*	**Cow**	*Bos* spp.
Chokecherry	*Prunus virginiana* var.	**Cow parsnip**	*Heracleum lanatum*
	melanocarpa	**Cowslip**	*Caltha palustris*
Cholla	*Opuntia* spp. (cylindrical	**Crabapple**	*Crataegus* spp.
	forms)	**Cranberry**	*Vaccinium macrocarpus*
Christmas flower	*Euphorbia pulcherrima*	**Cress, garden**	*Lepidium sativum, Barbarea*
Chrysanthemum	*Chrysanthemum* spp.		*verna,* and others
Chuckwalla	*Sauromalus obesus*	**Cress, rock**	*Arabis* spp.
Chufa	*Cyperus esculentus*	**Crocus**	*Crocus* spp.
Chytrids	member of Division	**Crocus, autumn**	*Colchicum autumnale*
	Chytridiomycota,	**Crown-of-thorns**	*Euphorbia milii* var.
	Kingdom Protoctista		*splendens* and others
Cilantro	*Coriandrum* sp.	**Crozier, tropical tree fern**	*Sadleria cyatheoides,*
Cinnamon	*Cinnamomum zeylanicum*		*Dicranopteris* sp.
Citric acid, fungal	*Aspergillus niger* and others	**Crustacean**	member of Class Crustacea,
producers of			Phylum Arthropoda,
Citronella oil, source of	*Cymbopogon nardus*		Kingdom Animalia
Citrus	*Citrus* spp.	**Cryptophyte**	member of Class
Citrus Family	Rutaceae		Cryptophyceae, Division
Clematis	*Clematis* spp.		Chrysophyta, Kingdom
Clover	*Trifolium* spp.		Protoctista
Clover, bur	*Medicago hispida*	**Cucumber**	*Cucumis sativus*
Clover, sweet	*Melilotus* spp.	**Cucumber, squirting**	*Ecballium elaterium*
Cloves	*Eugenia aromatica*	**Cycad**	*Dioon edule, Encephalartos*
Club fungus	member of Class		*altensteinii*
	Basidiomycetes, Division	**Cyclamen**	*Cyclamen* spp.
	Eumycota, Kingdom Fungi	**Cypress, bald**	*Taxodium distichum*
Club moss	member of Division	**Daffodil**	*Narcissus* spp.
	Lycophyta, Kingdom	**Dahlia**	*Dahlia* spp.
	Plantae	**Daisy**	*Dimorphotheca* spp., *Layia*
Cobra plant	*Darlingtonia californica*		spp., and others
Cocaine, source of	*Erythroxylon coca*	**Dandelion**	*Taraxacum officinale*
Cochineal insect	*Dactylopius coccus*	**Dandruff, fern(s) used in**	*Adiantum capillus-veneris,*
Cocklebur	*Xanthium* spp.	**treatment of**	*Polystichum munitum*
Cockroach	*Blatta orientalis, Blatella*	**DDT-like compound, algal**	*Laurencia* spp. and others
	germanica, and others	**producers of**	
Cockroach plant	*Haplophyton cimicidum*	**Death angel**	*Amanita phalloides*
Cockscomb	*Celosia* spp.	**Deer**	*Odocoileus* spp. and others
Coffee	*Coffea arabica*	**Deer, mule**	*Odocoileus hemionus*
Coleus	*Coleus blumei* and others	**Deerbrush**	*Ceanothus integerrimus*
Columbine	*Aquilegia* spp.	**Desmids**	*Closterium* spp., *Cosmarium*
Columbine (Fig. 24.3 A)	*Aquilegia formosa*		spp., and others
Compass plant	*Lactuca serriola*	**Dewberry**	*Rubus* hybrids with *R.*
Coneflower	*Rudbeckia* sp.		*ursinus* as one parent
Coneflower, Asian	*Strobilanthes* spp.	**Diatom**	*Biddulphia* spp., *Cymbella*
Copal, sources of	*Agathis alba, Copaifera*		spp., *Navicula* spp., and
	demeussei, Hymenea		others
	coubaril, Trachylobium	**Diatom (Fig. 18.4)**	*Cymatopleura solea*
	verrucosum, and others		
Copperhead	*Ancistrodon contortrix*		

COMMON NAMES AND SCIENTIFIC NAMES OF ORGANISMS

COMMON NAME	SCIENTIFIC NAME	COMMON NAME	SCIENTIFIC NAME
Dicot	member of Class Magnoliopsida, Division Magnoliophyta, Kingdom Plantae	**Eucalyptus oil, source of**	*Eucalyptus* spp.
		Euglenoids	members of Division Euglenophyta, Kingdom Protoctista
Digitalis, source of	*Digitalis purpurea*	**Fennel**	*Foeniculum vulgare*
Dill	*Anethum graveolens*	**Fenugreek**	*Trigonella foenumgraecum*
Dinoflagellate	member of Division Pyrrophyta, Kingdom Protoctista	**Fern(s), adder's tongue**	*Ophioglossum* spp.
		Fern(s), amphibious	*Marsilea* spp. and others
		Fern(s), aquatic (floating)	*Azolla* spp., *Salvinia* spp.
Dinoflagellate, midnight-bioluminescent	*Gonyaulax polyedra*	**Fern(s)—source of astringent**	*Actiniopteris radiata, Drynaria quercifolia, Pteridium aquilinum,* and others
Dischidia	*Dischidia rafflesiana*		
Dodder	*Cuscuta* spp.		
Dogbane	*Apocynum* spp.	**Fern(s), beech**	*Thelypteris* spp.
Dogwood	*Cornus* spp.	**Fern, bird's foot**	*Pellaea mucronata*
Dove	member of Family Columbidae, Class Aves, Phylum Vertebrata, Kingdom Animalia	**Fern, Boston**	*Nephrolepis exaltata*
		Fern, bracken	*Pteridium aquilinum*
		Fern, Brazilian tree (Fig. 21.25)	*Cyathea* sp.
Dove, mourning	*Zenaidura macroura*	**Fern—used in treating burns**	*Polystichum munitum*
Downy mildew of grape	*Plasmopora viticola*		
Dragon tree	*Dracaena draco*	**Fern, chain**	*Woodwardia fimbriata*
Dragon's blood	*Dracaena* spp., *Daemonorops* spp.	**Fern, cinnamon**	*Osmunda cinnamomea*
		Fern, climbing (Asian)	*Lygodium salicifolium*
Duckweed	*Lemna* spp., *Wolffia* spp., and others	**Fern(s)—used in treating coughs**	*Adiantum aethiopicum, A. lunulatum, Polypodium glycyrrhiza*
Dulse	*Rhodymenia* spp.		
Dung mosses (on dung of carnivores)	*Tayloria* spp.	**Fern(s)—used in treating dandruff**	*Adiantum capillus-veneris, Polystichum munitum*
Dung mossess (on dung of herbivores)	*Splachnum* spp.		
		Fern—used in treating diabetes	*Adiantum caudatum*
Dutch elm disease	*Ceratocystis ulmi*		
Dutchman's breeches	*Dicentra cucullaria*	**Fern(s)—used in treating diarrhea**	*Botrychium lunaria, B. ternatum, Pteridium aquilinum,* and others
Dyer's woad	*Isatis tinctoria*		
Dyes, sources of	see listing in Appendix 3		
Eagle, golden	*Aguila chrysaëtos*	**Fern(s)—used as diuretics**	*Adiantum venustum, Lygodium japonicum*
Earth star	*Geaster* spp. and others		
Earthworm	*Lumbricus* spp. and others	**Fern(s)—sources of dyes**	*Sadleria cyatheoides* (trunk), *Sphenomeris chusana* (fronds)
Ebony	*Diospyros ebenum*		
Eelworm (nematode)	member of Class Nernatoda, Phylum Aschelminthes, Kingdom Animalia		
		Fern(s)—used in treating dysentery	*Botrychium lunaria, B. ternatum, Pteridium aquilinum,* and others
Eelworms (nematodes), fungi that trap with constricting rings	*Dactylaria* spp., *Arthrobotrys dactyloides*		
		Fern—used in treating eczema	*Lygodium flexuosum*
Eelworms (nematodes), fungi that trap with passive rings	*Dactylella* spp.	**Fern—used in treating eye diseases**	*Asplenium adiantum-nigrum*
		Fern—used to reduce fevers	*Marsilea quadrifolia*
Eggplant	*Solanum melongena*	**Fern, five-finger**	*Adiantum pedatum*
Elephant	*Elephas* spp., *Loxodonta* spp.	**Fern(s)—used as food**	*Athyrium filix-femina, Dryopteris austriaca, D. filix-mas, Polystichum munitum,* and others
Elephant ears	*Colocasia* spp.		
Elk	*Cervus canadensis*		
Elm	*Ulmus* spp.		
Elm bark beetle	*Hylurgopinus rufipes, Scolytus multistriatus*	**Fern(s), fossil**	*Psaronius* spp., *Thamnopteris* spp., and others
Endive	*Cichorium endivia*	**Fern, goldback**	*Pentagramma triangularis*
Ergot	*Claviceps purpurea*	**Fern(s), Hawaiian tree**	*Cibotium* spp., *Sadleria cyatheoides*
Ermine	*Mustela erminea*		
Eucalyptus, Tasmanian giant	*Eucalyptus regnans*	**Fern, holly**	*Polystichum lonchitis*

COMMON NAMES AND SCIENTIFIC NAMES OF ORGANISMS

COMMON NAME	SCIENTIFIC NAME	COMMON NAME	SCIENTIFIC NAME
Fern(s)—used by hummingbirds	*Cyathea arborea, Lophosoria quadripinnata, Nephelea mexicana*	Fish	member of Class Pisces, Phylum Vertebrata, Kingdom Animalia
Fern—used for treating insect stings and bites	*Adiantum capillus-veneris*	Fish, flashlight	*Anomalops katoptron, Photoblepharon palpebratus*
Fern—used for easing labor pains	*Athyrium filix-femina*	Fish molds	*Saprolegnia* spp. and others
Fern, lady	*Athyrium filix-femina*		
Fern(s)—used as laxatives	*Asplenium trichomanes, Polypodium vulgare*	Flashlight powder, ground pine source of	*Lycopodium* spp.
Fern—used in treating leprosy	*Marsilea quadrifolia*	Flatworm	*Convoluta roscoffensis*
		Flax	*Linum* spp.
Fern, licorice	*Polypodium glycyrrhiza*	Flax, New Zealand	*Phormium tenax*
Fern(s)—poisonous to livestock	*Onoclea sensibilis, Pteridium aquilinum*	Flea	member of Order Siphonaptera, Phylum Arthropoda, Kingdom Animalia
Fern, maidenhair	*Adiantum* spp.		
Fern—edible Malaysian (relative of lady fern)	*Athyrium esculentum*	Flicker	*Colaptes* spp.
Fern, male	*Dryopteris filix-mas*	Flour, Hopi Indian horsetail source of	*Equisetum laevigatum*
Fern, mosquito	*Azolla caroliniana*	Flowerpot leaf plant	*Dischidia rafflesiana*
Fern, nest	*Asplenium nidus*	Fly	member of Order Diptera, Phylum Arthropoda, Kingdom Animalia
Fern—used to arrest nosebleeds	*Pellaea mucronata*		
Fern(s)—used for orchid bark	*Cibotium* spp., *Osmunda* spp.	Fly agaric	*Amanita muscaria*
Fern, Oriental water	*Ceratopteris thalictroides*	Fly, tsetse	*Glossina morsitans, G. palpalis*
Fern—used as poison antidote	*Polystichum squarrosum*	Fly, white	*Aleurocanthus woglumi* and others
Fern(s)—used in treating rickets	*Asplenium ruta-muraria, Osmunda regalis*	Flycatcher	*Empidonax* spp., *Myiarchus* spp., and others
Fern(s), seed (Pteridosperms)	*Lyginopteris* spp., *Medullosa* spp., and others	Fossil, compression (Fig. 21.26)	*Annularia radiata*
Fern—used for shampoo	*Dryopteris dilatata*	Four-o'clock Family	Nyctaginaceae
Fern(s)—used for stuffing mattresses, pillows, upholstery	*Cibotium* spp., *Sadleria* spp.	Fox, arctic	*Alopex lagopus*
		Fox, gray	*Urocyon cinereoargentus*
Fern, sword	*Polystichum munitum*	Fox, red	*Vulpes fulva*
Fern—used in treating toothache	*Pentagramma triangularis*	Foxglove	*Digitalis* spp.
		Frangipanni	*Plumeria* spp.
Fern(s), tree	*Cyathea* spp., *Ctenitis* spp., *Dicksonia* spp., *Sphaeropteris* spp., and others	Frog	*Rana* spp. and others
		Fruit fly	*Drosophila melanogaster* and others
Fern, tropical climbing	*Dicranopteris linearis*	Fuchsia, California	*Epilobium* spp.
Fern—used for expelling worms	*Dryopteris filix-mas*	Fumitory, Himalayan	*Corydalis gerdae*
Fern(s)—used for treating wounds	*Lygodium circinatum, Ophioglossum vulgatum*	Fungi, ant and termite nest	A *Leucoagaricus* sp. has been identified, but vast majority are unknown
Fevers, fern used to reduce	*Marsilea quadrifolia*	Fungi that produce antibiotics	*Penicillium* spp., *Cephalosporium* spp., and others
Fevers, ground pine used to reduce	*Lycopodium clavatum*	Fungi that cause aspergilloses	*Aspergillus fumigatus, Candida albicans, Coccidiodes immitis,* and others
Fig, common	*Ficus carica*		
Fig, tropical	*Ficus* spp.		
Fig, tropical (Fig. 5.12)	*Ficus macrophyllus*	Fungi that cause athlete's foot	*Trichophyton* spp.
Figwort Family	Scrophulariaceae		
Filaree	*Erodium* spp.	Fungi—used by beetles for food	*Ambrosiella* spp., *Monilia* spp.
Fir, balsam	*Abies balsamea*		
Fir, Douglas	*Pseudotsuga menziesii*	Fungi, bird's-nest	*Nidularia* spp., *Crucibulum levis*
Fir, white	*Abies concolor*		

COMMON NAMES AND SCIENTIFIC NAMES OF ORGANISMS

COMMON NAME	SCIENTIFIC NAME	COMMON NAME	SCIENTIFIC NAME
Fungi—used in manufacturing birth control pills	*Rhizopus nigricans,* *R. arrhizus*	Fungus, white piedra	*Trichosporon beigeli*
		Fungus—used in manufacturing yellow food-coloring agent	*Blakeslea trispora*
Fungi, bracket	*Fomes* spp., *Daedalea* spp., and others	Funori, source of	*Gloiopeltis* spp.
Fungi, brown fruit rot	*Monolinia fruticola* and others	Fur, green algae that inhabit animal	*Trentepohlia* spp.
Fungi, cap-thrower	*Pilobolus* spp.	Gentian, source of	*Gentiana* spp.
Fungi, cheese	*Penicillium camembertii* (for Camembert cheese), *P. roquefortii* (for blue, Gorgonzola, Roquefort, and Stilton cheeses)	Geranium	*Geranium* spp., *Pelargonium* spp.
		Geranium Family	Geraniaceae
		Gila monster	*Heloderma suspectum*
		Ginger	*Zingiber officinale* and others
Fungi, citric acid-producing	*Aspergillus niger* and others	Ginseng, source of	*Panax* spp.
Fungi, flavor-producing	*Aspergillus* spp.	Giraffe	*Giraffa camelopardalis*
Fungi, hallucinogenic	*Amanita muscaria, Conocybe* spp., *Panaeolus* spp., *Psilocybe* spp., and others	Gladiolia	*Gladiolus* spp.
		Gloeocapsa	*Gloeocapsa* spp.
		Goat	*Capra* spp.
Fungi, horse dung	*Pilobolus* spp.	Golden-brown algae	members of Class Chrysophyceae, Division Chrysophyta, Kingdom Protoctista
Fungi, industrial alcohol-producing	*Aspergillus* spp.		
Fungi, insect-parasitizing	members of Order Laboulbeniales, Class Ascomycetes, Division Eumycota, Kingdom Fungi, and others	Golden chain tree	*Laburnum anagyroides*
		Goldenrod	*Solidago* spp.
		Goldenseal	*Hydrastis canadensis*
		Goldenweed	*Haplopappus gracilis*[4]
Fungi, jelly	*Auricularia* spp., *Exidia* spp., *Tremella* spp., and others	Goose	*Branta* spp. and others
		Gooseberry	*Ribes* spp.
Fungi, meat-tenderizing	*Thamnidium* spp.	Goosefoot Family	Chenopodiaceae
Fungi, ringworm	*Epidermophyton* spp., *Mircosporium* spp., *Trichophyton* spp.	Gopher	*Geomys* spp., *Thomomys* spp.
		Gopher plant	*Euphorbia lathyrus*
		Gopher, pocket	*Geomys bursarius* and others
Fungi, shelf—*see* Fungi, bracket		Gourd	*Lagenaria siceraria* and others
Fugi, *shoyu*	*Aspergillus oryzae, A. soyae*	Grape	*Vitis* spp.
Fungi—used in silvering of mirrors	*Aspergillus* spp.	Grapefruit	*Citrus paradisi*
		Grass	*Bromus* spp. and others[5]
Fungi—used in manufacturing soap	*Penicillium* spp.	Grass, Bermuda	*Cynodon dactylon*
		Grass, crested wheat	*Agropyron cristatum*
Fungi, soil	*Fusarium* spp. and others	Grass Family	Poaceae (Gramineae)
Fungi, soy sauce	*Aspergillus oryzae, A. soyae*	Grass, Indian	*Sorghastrum nutans*
Fungi, *sufu*	*Actinomucor elegans, Mucor* spp.	Grass, pampas (Fig. 7.4)	*Cortaderia selloana*
		Grass tree	*Xanthorrhea* spp.
Fungi, *teonanacatl* (sacred)	*Conocybe* spp., *Panaeolus* spp., *Psilocybe* spp., and others	Greenbrier	*Smilax* spp.
		Ground pine	*Lycopodium* spp.
		Ground pine—used for baby powder	*Lycopodium clavatum*
Fungus, bracket (Fig. 19.13 C)	*Polyporus sulphureus*	Ground pine—used to arrest bleeding	*Lycopodium clavatum*
Fungus, bracket (Fig. 19.19)	*Phacolus* sp.	Ground pine, fossil relative of	*Baragwanathia* spp., *Drephanophycus* spp., *Protolepidodendron* spp., and others
Fungus, chlorine-assimilating	*Aspergillus terreus*		
Fungus, cup (Fig. 19.7)	*Caloscypha fulgens*		
Fungus, "foolish seedling"	*Gibberella fujikuroi*		
Fungus, *miso*	*Aspergillus oryzae*	Ground pine—used as intoxicant	*Lycopodium selago*
Fungus—used in producing plastics	*Aspergillus terreus*		
Fungus, *tempeh*	*Rhizopus oligosporus*		
Fungus—used in manufacturing toothpaste	*Aspergillus niger*		

[4]Species with four chromosomes per cell.

[5]About 4,500 species of the Grass Family, Poaceae (Gramineae), in all.

COMMON NAMES AND SCIENTIFIC NAMES OF ORGANISMS

COMMON NAME	SCIENTIFIC NAME	COMMON NAME	SCIENTIFIC NAME
Ground pine—used for ornaments	*Lycopodium clavatum, L. complanatum,* and others	Horsetail, giant	*Equisetum telmateia*
Ground pine—used to reduce fevers	*Lycopodium clavatum*	Horsetail, Hopi Indian flour source	*Equisetum laevigatum*
Guava	*Psidium guajava*	Horsetail—used as shampoo	*Equisetum hyemale*
Gum arabic, source of	*Acacia senegal* and others	Horsetail, treelike fossil	*Calamites* spp.
Gum tragacanth, source of	*Astragalus echidenaeformis, A. gossypinus, A. gummifer,* and others	Horsetail—used as water source	*Equisetum telmateia*
Guppy	*Lebistes reticulatus*	Hot springs, blue-green bacteria of	*Bacillosiphon induratus, Synechococcus* spp., and others
Hawk	*Buteo* spp., *Falco* spp., and others	"Human hair" slime mold	*Stemonitis* spp.
Hazelnut	*Corylus* spp.	Hummingbird	*Archilocus* spp. and others
Heath	*Erica* spp. and others	Hummingbird, Oasis (Fig. 23.15)	*Rhodopis vesper*
Heath Family	Ericaceae	Hummingbirds, ferns used by (for nest material)	*Cyathea arborea, Lophosoria quadripinnata, Nephelea mexicana*
Hemlock, eastern	*Tsuga canadensis*		
Hemlock, mountain	*Tsuga mertensiana*	Hummingbirds, tropical	*Chlorostilbon maugaeus* and others
Hemlock, poison	*Conium maculatum*		
Hemlock, water	*Cicuta* spp.	Hyacinth	*Hyacinthus* spp.
Hemlock, western	*Tsuga heterophylla*	Hyacinth, grape	*Muscari* spp.
Hemp	*Cannabis sativa*	Hyacinth, water	*Eichhornia crassipes*
Hemp, Manila	*Musa textilis*	Hyssop	*Hyssopus officinalis*
Hemp, Mauritius	*Furcraea gigantea*	Ice plant	*Mesembryanthemum* spp.
Henbit	*Lamium amplexicaule*	India, toxic blue-green bacteria of	*Lyngbya majuscula*
Henna	*Lawsonia inermis*		
Hepatica	*Hepatica* spp.	Indian pipe	*Monotropa* spp.
Hepatica (Fig. 1.8; Fig. 24.3 C)	*Hepatica americana*	Indian warrior	*Pedicularis densiflora*
		Indigo	*Indigofera* spp.
Hickory	*Carya* spp.	Insects—*see individual entries*	
Hog	*Sus scrofa* and others		
Hog fennel	*Lomatium* spp.	Insects, fern used for treating stings and bites of	*Adiantum capillus-veneris*
Holly, American	*Ilex opaca*		
Honeybee	*Apis mellifera*		
Hoopoe	*Upupa africana*	Ipecac, source of	*Cephaelis ipecacuanha*
Hop hornbeam	*Ostrya virginiana*	Iris	*Iris* spp.
Hops	*Humulus lupulus*	Iris, butterfly	*Moraea* sp.
Horehound	*Marrubium vulgare*	Iris Family	Iridaceae
Hornwort	*Anthoceros* spp.	Ironwood, South American	*Krugiodendron ferreum*
Horse	*Equus caballus*	Ivy, Boston	*Parthenocissus tricuspidata*
Horseradish	*Rorippa armoracia*	Ivy, English	*Hedera helix*
Horsetail	*Equisetum* spp.	Ivy, poison	*Toxicodendron radicans*
Horsetail (Fig. 21.10 A)	*Equisetum hyemale*	Jacaranda	*Jacaranda* spp.
Horsetail (Fig. 21.10 B)	*Equisetum telmateia*	Jaeger	*Stercorarius* spp.
Horsetail(s)—used as abrasive	*Equisetum* (all spp.)	Jicama	*Pachyrhizus* spp.
		Jimson weed	*Datura* spp.
Horsetail(s)—used as astringent	*Equisetum arvense, E. debile,* and others	Jimson weed (Fig. 8.6)	*Datura stramonium*
		Jojoba	*Simmondsia californica*
Horsetail(s)—used for treating burns	*Equisetum hyemale* and others	Joshua tree	*Yucca brevifolia*
		Jumping mouse	*Zapus hudsonius, Napaeozapus insignis*
Horsetail—used for treating diarrhea	*Equisetum hyemale*		
		Junco	*Junco* spp.
Horsetail(s)—used as diuretic	*Equisetum arvense, E. debile,* and others	Junco, slate-colored	*Junco hyemalis*
		Juneberry	*Amelanchier* spp.
Horsetail—used for treating dysentery	*Equisetum hyemale*	Juniper	*Juniperus* spp.
		Juniper, dwarf	*Juniperus communis* and others
Horsetail, field	*Equisetum arvense*		
Horsetail(s), fossil	*Equisetites* spp., *Hyenia* spp., *Sphenophyllum* spp., and others	Jute	*Corchorus* spp.

COMMON NAMES AND SCIENTIFIC NAMES OF ORGANISMS

COMMON NAME	SCIENTIFIC NAME	COMMON NAME	SCIENTIFIC NAME
Kauri pine	*Agathis australis, A. robusta*	**Lichen, natural dye**	*Parmelia* spp., *Usnea* spp., and others
Kelp	*Alaria* spp., *Dictyoneurum* spp., *Egregia* spp., *Laminaria* spp., *Lessoniopsis* spp., *Nereocystis* spp., and others	**Lichen, perfume stabilizer**	*Evernia* spp.
		Lichen, reindeer ("reindeer moss")	*Cladonia* spp., *Cetraria islandica*
		Lichens, crustose (Fig. 19.33 A)	
		black	*Rinodina* sp.
Kelp, giant	*Macrocystis pyrifera*	chartreuse	*Acarospora citrina*
Knotweed	*Polygonum arenastrum*	gray	*Psora* sp.
Kohlrabi	*Brassica oleracea* var. *caulorapa*	orange-red	*Caloplaca elegans*
		yellow	*Candelariella vitellina*
Kudzu	*Pueraria lobata*	**Lichens used as miniature trees and shrubs**	*Cladonia* spp. and others
Kumquat	*Fortunella japonica*	**Licorice, source of**	*Glycyrrhiza glabra*
Labor pain, fern used to ease	*Athyrium filix-femina*	**Lignum vitae**	*Guaiacum officinale*
Lamb's ears	*Stachys byzantina*	**Lilac**	*Syringa vulgaris*
Larch, eastern	*Larix laricina*	**Lily**	*Lilium* spp. and others
Larch, European	*Larix decidua*	**Lily Family**	Liliaceae
Larch, western	*Larix occidentalis*	**Lily, giant water**	*Victoria amazonica*
Larkspur, blue	*Delphinium* spp.	**Lily, kaffir**	*Clivia* sp.
Larkspur, red	*Delphinium nudicaule*	**Lily, tiger**	*Lilium pardalinum*
Late blight of potato	*Phytophthora infestans*	**Lily, water**	*Nymphaea* spp. and others
Laurel	*Laurus nobilis*	**Lily, wood**	*Lilium superbum*
Laurel Family	Lauraceae	**Lime**	*Citrus aurantifolia*
Lavender	*Lavandula officinalis*	**Litmus indicator dye, source of**	*Rocella* spp.
Laxative, ferns used as	*Asplenium trichomanes, Polypodium vulgare*		
Leaf curl, peach	*Taphrina deformans*	**Liverwort**	member of Division Hepaticophyta, Kingdom Plantae
Leaf miner	*Agromyza* spp. and others		
Leaf roller	*Archips argyrospila* and others	**Liverwort, leafy (Fig. 20.8)**	*Frullania* sp.
		Liverworts, leafy	*Frullania* spp., *Jungermannia* spp., *Porella* spp., and others[7]
Leafy liverwort—*see* Liverwort, leafy			
Legume Family	Leguminosae (Fabaceae)	**Liverworts, thalloid**	*Marchantia* spp. and others
Lemming	*Lemmus* spp., *Dicrostonyx groenlandicus*	**Livestock, ferns poisonous to**	*Onoclea sensibilis, Pteridium aquilinum*
Lemon	*Citrus limon*	**Lizard**	*Sceloporus* spp. and others
Lemongrass oil, source of	*Cymbopogon citratus, C. flexuosus*	**Lobeline sulphate, source of**	*Lobelia inflata*
		Locoweed	*Astragalus* spp.
Lentil	*Lens esculenta*	**Locust, black**	*Robinia pseudo-acacia*
Leprosy, fern used for treating	*Marsilea quadrifolia*	**Locust, honey**	*Gleditsia triacanthos*
		Loganberry	*Rubus hybrids*, with *R. ursinus* as one parent
Lettuce	*Lactuca sativa*		
Lichen (symbiotic association of an alga and a fungus)	Class Lichens, Kingdom Fungi[6]	**Loon**	*Gavia* spp.
		Louse	Order Mallophaga, Order Anaplura, Class Insecta, Phylum Arthopoda, Kingdom Animalia
Lichen, foliose (Fig. 19.32)	*Physcia* sp.		
Lichen, foliose (Fig. 19.33 B)	*Parmelia* sp.		
Lichen, fruticose (Fig. 19.33 C)	*Usnea* sp.	**Luffa**	*Luffa cylindrica*
		Lupine	*Lupinus* spp.
Lichen, grazed by North African sheep	*Lecanora* spp.	**Lupine, tree (with seed valves)**	*Lupinus arboreus*
Lichen, litmus	*Rocella* spp.	**Madder Family**	Rubiaceae
		Magnolia	*Magnolia* spp.
		Mallow	*Malva* spp.

[6]The lichens are arbitrarily treated as a class within Kingdom Fungi because the fungal component of each species of lichen is unique to the species, while the algal component may be common to more than one species of lichen.

[7]There are thousands of leafy liverwort species assigned to about 200 genera.

COMMON NAMES AND SCIENTIFIC NAMES OF ORGANISMS

COMMON NAME	SCIENTIFIC NAME	COMMON NAME	SCIENTIFIC NAME
Mango	*Mangifera indica*	**Moss rose**	*Portulaca grandliflora*
Mangrove	*Rhizophora mangle* and others	**Mosses, annual (bare soil)**	*Acaulon* spp., *Ephemerum* spp., and others
Mangrove, black	*Avicennia nitida*	**Mosses, antler and bone**	*Tetraplodon* spp.
Manila hemp	*Musa textilis*	**Mosses, dung (on dung of carnivores)**	*Tayloria* spp.
Manioc—*see* **Cassava**		**Mosses, dung (on dung of herbivores)**	*Splachnum* spp.
Manroot	*Marah* spp.	**Mosses, extinguisher**	*Encalypta* spp.
Maple	*Acer* spp.	**Mosses, haircap**	*Polytrichum* spp.
Maple, bigleaf (Fig. 8.20)	*Acer macrophyllum*	**Mosses—indicator for absence of calcium**	*Andreaea* spp., *Rhacomitrium lanuginosum*
Maple, hard	*Acer saccharum*		
Maple, silver	*Acer saccharinum*	**Mosses—indicator for presence of calcium**	*Didymodon* spp., *Desmatodon* spp., and others
Maple, sugar	*Acer saccharum*		
Marigold	*Tagetes* spp.		
Marijuana	*Cannabis sativa*	**Mosses—indicator for saline (salty) soil**	*Pottia* spp.
Marjoram	*Majorana hortensis*		
Maté	*Ilex paraguariensis* ·	**Mosses—indicator for seasonal running water**	*Fontinalis* spp.
Meadowfoam	*Limnanthes* spp.		
Mealy bugs	*Pseudococcus* spp.	**Mosses, peat**	*Sphagnum* spp.
Melon	*Cucumis melo*	**Mosses, pygmy**—*see* **Mosses, annual**	
Melon, honeydew	*Cucumis melo* (variety)		
Melonette	*Melothria pendula*	**Mosses, rock**	*Grimmia* spp. and others
Mermaid's wineglass	*Acetabularia* spp.	**Mosses, sphagnum**	*Sphagnum* spp.
Mesquite	*Prosopis glandulosa*	**Mosses—used with splints**	*Philonotis* spp., *Fontinalis* spp., and others
Mildew, powdery	*Erysiphe* spp. and others		
Milkweed	*Asclepias* spp.	**Moth**	member of Order Lepidoptera, Class Insecta, Phylum Arthropoda, Kingdom Animalia
Milkweed, swamp	*Asclepias incarnata*		
Millet	*Pennisetum glaucum, Setaria italica,* and others		
Millipede	member of Class Diplopoda, Phylum Arthropoda, Kingdom Animalia	**Moth, Argentine**	*Cactoblastis cactorum*
		Moth, codling	*Carpocapsa pomonella*
		Moth, gypsy	*Porthetria dispar*
Mint—*see* **Peppermint, Spearmint,** *etc.*		**Moth, Mexican jumping bean**	*Carpocaps asaltitans*
Mint Family	Lamiaceae (Labiatae)	**Moth mullein**	*Verbascum blattaria*
Mistletoe	*Phoradendron* spp.	**Moth, Yucca**	*Pronuba* spp., *Tegeticula* spp.
Mistletoe, dwarf	*Arceuthobium* spp.	**Mountain beaver**	*Aplodontia rufa*
Mite	member of Order Acarina, Phylum Arthropoda, Kingdom Animalia	**Mouse**	*Mus musculus, Peromyscus* spp., and others
Mollusc	member of Phylum Mollusca, Kingdom Animalia	**Mouse, jumping**	*Zapus hudsonius, Napaeozapus insignis*
Monkey	*Ateles dariensis* and others	**Mulberry**	*Morus* spp.
Monkey flower	*Mimulus* spp.	**Mulberry, red**	*Morus rubra*
Monkshood	*Aconitum* spp.	**Mule ears**	*Wyethia* spp.
Monocot	member of Class Liliopsida, Division Magnoliophyta, Kingdom Plantae	**Mullein**	*Verbascum thapsus*
		Mushroom[8]	*Agaricus* spp. and others
		Mushroom, Black Forest	*Lentinus edodes*
Moose	*Alces americana*	**Mushroom, pore (Fig. 19.18)**	*Serillus pungens*
Morel	*Morchella esculenta* and others	**Mushroom, shaggy mane**	*Coprinus comatus*
Morel, false	*Helvella* spp.	**Mushrooms, common woods (Fig. 19.13 A)**	*Russula* sp.
Morning glory	*Ipomoea violacea* and others		
Mosquito	*Anopheles* spp., *Culex* spp., and others	**Mushrooms, inky cap**	*Coprinus* spp.
Moss	member of Division Bryophyta, Kingdom Plantae		
Moss, Irish	*Chondrus crispus*		
Moss, luminous	*Schistostega pennata*		

[8]*Mushroom* is a term generally applied to the fruiting bodies with stalked, caplike structures produced by members of the Class Basidiomycetes, Division Eumycota, Kingdom Fungi; the term is also loosely applied to some of the fruiting bodies of members of other classes of true fungi. There are thousands of known species.

COMMON NAMES AND SCIENTIFIC NAMES OF ORGANISMS

COMMON NAME	SCIENTIFIC NAME	COMMON NAME	SCIENTIFIC NAME
Mushrooms, *teonanacatl* (sacred)	*Conocybe* spp., *Panaeolus* spp., *Psilocybe* spp., and others	**Orchids, vanilla**	*Vanilla* spp.
		Oregano	*Origanum vulgare* and others
Muskrat	*Ondatra zibesthicus*	**Oregon grape**	*Berberis aquifolium* and others
Mustard	*Brassica campestris,* *B. nigra,* and others	**Organpipe cactus**	*Lemaireocereus* spp.
Mustard, cultivated	*Brassica alba, B. juncea,* and others	**Ornaments, ground pines used for**	*Lycopodium complanatum,* *L. clavatum,* and others
Mustard Family	Brassicaceae (Cruciferae)	**Osage orange**	*Maclura pomifera*
Mustard, tumble	*Sisymbrium altissimum*	**Oscillatoria**	*Oscillatoria* spp.
Myrrh tree	*Commiphora* spp.	**Our Lord's Candle**	*Yucca whipplei*
Myrtle[9]	*Umbellularia californica*[9]	**Owl, snowy**	*Nyctea scandiaca*
Nasturtium (garden)	*Tropaeolum majus*	**Painted lady**	*Echeveria derenbergii*
Neem tree	*Melia* sp.	**Palm, coconut**	*Cocos nucifera*
Nematode	member of Class Nematoda, Phylum Aschelminthes, Kingdom Animalia	**Palm, date**	*Phoenix dactylifera*
		Palm Family	Palmaceae
		Palm, panama hat	*Carludovica palmata*
Nettle	*Urtica* spp.	**Palm, Seychelles Island**	*Lodoicea maldivica*
Nicotine relative (nornicotine), source of	*Duboisia hopwoodii,* *Nicotiana tabacum*	**Palm, wax**	*Copernicia cerifera*
		Pansy	*Viola tricolor*
Nightshade, deadly	*Atropa belladonna*	**Papaya**	*Carica papaya*
Nightshade Family	Solanaceae	**Parsley**	*Petroselinum sativum*
Nosebleeds, fern used to arrest	*Pellaea mucronata*	**Parsley Family**	Apiaceae (Umbelliferae)
		Parsnip	*Pastinaca sativa*
Nostoc	*Nostoc* spp.	**Passion fruit**	*Passiflora* spp.
Nutmeg	*Myristica fragrans*	**Patchouli oil, source of**	*Pogostemon cablin* and others
Nutmeg, California	*Torreya californica*		
Oak	*Quercus* spp.	**Pea**	*Pisum sativum*
Oak, blue	*Quercus douglasii*	**Pea, sweet**	*Lathyrus odoratus*
Oak, cork	*Quercus suber*	**Peach**	*Prunus persica*
Oak, Hooker	*Quercus lobata*	**Peach leaf curl**	*Taphrina deformans*
Oak, live (Fig. 9.7)	*Quercus wislizenii*	**Peanut**	*Arachis hypogaea*
Oak, poison	*Toxicodendron diversilobum*	**Pear**	*Pyrus communis*
Oak, white	*Quercus alba*	**Pecan**	*Carya illinoensis*
Oats	*Avena* spp.	**Peccary**	*Pecari angulatus, Tayassus pecari*
Oedogonium	*Oedogonium* spp.		
Olibanum tree	*Boswellia* spp.	**Penicillin mold**[11]	*Penicillium* spp.[11]
Olive	*Olea europaea*	**Pennyroyal**	*Hedeoma pulegioides*
Onion	*Allium cepa*	**Peony**	*Paeonia* spp.
Orange	*Citrus sinensis*	**Peperomia**	*Peperomia* spp.
Orchid	*Cattleya* spp. and others[10]	**Pepper**	*Capsicum anuum,* *C. frutescens*[12]
Orchid, bamboo	*Arundina graminifolia*		
Orchid "bark," fern sources of	*Cibotium* spp., *Osmunda* spp.	**Peppergrass**	*Lepidium* spp.
		Peppermint	*Mentha piperita*
Orchid, *Bletilla* (Fig. 8.18 B)	*Bletilla* sp.	**Persimmon**	*Diospyros* spp.
Orchid, bucket	*Coryanthes* sp.	**Petitgrain oil, source of**	*Citrus aurantium* var. *amara*
Orchid with cladophylls	*Epidendrum* spp.	**Petunia**	*Petunia* spp.
Orchid Family	Orchidaceae	**Peyote**	*Lophophora williamsii*
Orchid, showy	*Orchis* spp.	**Phoebe**	*Sayornis phoebe*
Orchid, underground-flowering	*Rhizanthella gardneri*	**Pillbugs**	*Cylisticus convexus* and others
Orchids, saprophytic	*Corallorhiza* spp., *Eburophyton austinae,* and others		

[9]This plant, also known as the California bay, is in the Laurel Family. True myrtles are in the Myrtle Family (Myrtaceae).

[10]Depending on which authorities are followed, the number of orchid species (all in the family Orchidaceae) may exceed 30,000.

[11]The original penicillin producer discovered by Sir Alexander Fleming was *Penicillium notatum;* current commercially used penicillin producers are strains of *Penicillium chrysogenum.*

[12]The drug capsicum, whose active ingredient is the oleoresin capsaicin, is derived from these species, and garden peppers include these and other species of *Capsicum.* Condiment or black pepper is derived from the unrelated *Piper nigrum.*

COMMON NAMES AND SCIENTIFIC NAMES OF ORGANISMS

COMMON NAME	SCIENTIFIC NAME	COMMON NAME	SCIENTIFIC NAME
Pine	*Pinus* spp.	Polyanthus	*Primula polyanthus* and
Pine, Afghanistan	*Pinus eldarica*		hybrids
Pine, Aleppo	*Pinus halepensis*	Pomegranate	*Punica granatum*
Pine, bristlecone	*Pinus longaeva*	Poor man's pepper	*Lepidium virginicum*
Pine, Chilghoza	*Pinus gerardiana*	Popcorn	*Zea mays* (horicultural
Pine, Colorado bristlecone	*Pinus aristata*		variety)
Pine, Coulter	*Pinus coulteri*	Poplar	*Populus* spp.
Pine, digger—*see* Pine,		Poppy	*Papaver* spp. and others
gray		Poppy, bush	*Dendromecon rigida*
Pine, eastern white	*Pinus strobus*	Poppy, California	*Eschscholzia californica*
Pine, European stone	*Pinus pinea*	Poppy Family	Papaveraceae
Pine, gray	*Pinus sabiniana*	Poppy, Mexican	*Hunnemannia* spp.
Pine, jack	*Pinus banksiana*	Poppy, opium	*Papaver somniferum*
Pine, Jeffrey	*Pinus jeffreyi*	Poppy, Oriental	*Papaver orientale*
Pine, kauri	*Agathis australis, A. robusta*	Poppy, prickly (Fig. 24.5)	*Argemone glauca*
Pine, knobcone	*Pinus attenuata*	Porcupine	*Erethizon* spp., *Hystrix* spp.[13]
Pine, loblolly	*Pinus taeda*	Portulaca Family	Portulacaceae
Pine, longleaf	*Pinus palustris*	Potato, Irish	*Solanum tuberosum*
Pine, Merkus	*Pinus merkusii*	Potato, sweet	*Ipomea batatas*
Pine, Mexican pinyon	*Pinus cembroides*	Potato vine	*Solanum jasminoides*
Pine, Mexican stone	*Pinus cembroides*	Potato, white—*see* Potato,	
Pine, Monterey	*Pinus radiata*	Irish	
Pine, pinyon	*Pinus edulis, P. monophylla,*	Powder puff flower (Fig. 24.11c)	*Calliandra inaequilatera*
	P. quadrifolia	Powdery mildew	*Erysiphe* spp. and others
Pine, pitch	*Pinus rigida*	Prayer plant	*Maranta* spp.
Pine, ponderosa	*Pinus ponderosa*	Preferns	*Cladoxylon* spp.,
Pine, shortleaf	*Pinus echinata*		*Protopteridium* spp., and
Pine, Siberian white	*Pinus sibirica*		others
Pine, slash	*Pinus caribaea, P. elliottii*	Primrose	*Primula* spp.
Pine, southern yellow—*see*		Primrose	
Pine, loblolly; Pine,		Chloroxybacteria	member of Class
longleaf; Pine, shortleaf;			Chloroxybacteriae,
and Pine, slash			Division Eubacteria,
Pine, stone—*see* pine,			Subkingdom
European stone *and* Pine,			Eubacteriobionta,
Mexican stone			Kingdom Monera
Pine, sugar	*Pinus lambertiana*	Pronghorn	*Antilocarpa americana*
Pine, western white	*Pinus monticola*	Ptarmigan	*Lagopus* spp.
Pine, western yellow	*Pinus ponderosa*	Pteridosperms	*Lyginopteris* spp., *Medullosa*
Pineapple	*Ananas comosus*		spp., and others
Pinedrops	*Pterospora* spp.	Puffballs	*Calvatia* spp., *Lycoperdon*
Pistachio	*Pistacia vera*		spp.
Pitcher plants	*Sarracenia* spp. and others	Pulque, source of	*Agave* spp.
Pitcher plants, Asian	*Nepenthes* spp. and others	Pumpkin	*Cucurbita pepo*
Plantain	*Plantago* spp.	Pumpkin Family	Cucurbitaceae
Plastic, fungus used in	*Aspergillus terreus*	Puncture vine	*Tribulus terrestris*
production of		Purple laver	*Porphyra* spp.
Plasticizers, source of oil for	*Euphorbia agascae*	Purple laver (Fig. 18.27)	*Porphyra tenera*
Plover	*Charadrius* spp. and others	Puya (rare)	*Puya raimondii*
Plum	*Prunus domestica* and others	Quillwort	*Isoetes* spp.
Podocarps, New Zealand	*Podocarpus dacrydoides,*	Quillwort, fossil relatives of	*Isoetites* spp.
timber	*P. totara*	Quince	*Cydonia oblonga*
Podocarps, ornamental	*Podocarpus macrophylla,*	Quinine, source of	*Cinchona* spp.
	P. nagi, and others	Rabbit	*Oryctolagus cuniculus*
Poinsettia	*Euphorbia pulcherrima*	Rabbit, cottontail	*Sylvilagus* spp.
Poison antidote, fern used	*Polystichum squarrosum*	Rabbit, jack	*Lepus* spp.
for		Raccoon	*Procyon lotor*
Poison ivy	*Toxicodendron radicans*	Radish	*Raphanus sativus*
Poison oak	*Toxicodendron diversilobum*		

[13]*Hystrix* is also a name for a genus of grasses.

COMMON NAMES AND SCIENTIFIC NAMES OF ORGANISMS

COMMON NAME	SCIENTIFIC NAME	COMMON NAME	SCIENTIFIC NAME
Rafflesia	*Rafflesia* spp.	**Salmon**	*Oncorhynchus* spp., *salmo salar*, and others
Rafflesia (Fig. 8.2)	*Rafflesia micropylora*	**Salmonberry**	*Rubus spectabilis*
Ragweed	*Ambrosia* spp.	**Salsify**	*Tragopogon* spp.
Raspberry	*Rubus* spp.	**Saltbush**	*Atriplex* spp.
Raspberry, red	*Rubus idaeus, R. strigosus,* and their hybrids	**Salvia**	*Salvia* spp.
		Sansevieria	*Sansevieria* spp.
Rat	*Rattus norvegicus, R. rattus,* and others	**Santonin, source of**	*Artemisia cina*
		Sargassum (Fig. 18.21)	*Sargassum* sp.
Rat, kangaroo	*Dipodomys* spp.	**Sarsaparilla, source of**	*Smilax* spp.
Rat snake, black	*Elaphe obsoleta*	**Sassafras**	*Sassafras albidum*
Rattlesnake	*Crotalus* spp.	**Sausage tree, African**	*Kigelia pinnata*
Redbud, eastern	*Cercis canadensis*	**Savory**	*Satureia hortensis*
Redbud, western	*Cercis, occidentalis*	**Saxifrage**	*Saxifraga* spp.
Redwood, coastal	*Sequoia sempervirens*	**Screw pine**	*Pandanus* spp.
Redwood, dawn	*Metasequoia glyptostroboides*	**Sea anemone**	*Stephanauge* spp. and others
		Sea hare	*Aplysia californica*
Redwood, giant	*Sequoiadendron giganteum*	**Sea lettuce**	*Ulva* spp.
Reindeer	*Rangifer* spp.	**Sea palm**	*Postelsia palmaeformis*
Reserpine, source of	*Rauwolfia serpentina*	**Sea rocket**	*Cakile edentula*
Resurrection plant	*Selaginella lepidophylla*	**Sedge**	*Carex* spp. and others
Rhododendron	*Rhododendron* spp.	**Seed ferns (Pteridosperms)**	*Lyginopteris* spp., *Medullosa* spp., and others
Rhubarb	*Rheum rhaponticum*		
Rice	*Oryza sativa*	**Senna**	*Cassia senna* and others
Rice-paper plant	*Tetrapanax papyrifera*	**Sensitive plant**	*Mimosa pudica*
Robin	*Turdus migratorius*	**Sesame**	*Sesamum indicum*
Rock cress	*Arabis* sp.	**Seychelles Island palm**	*Lodoicea maldivica*
Rock-rose, European	*Helianthemum vulgare*	**Shaggy mane mushroom**	*Coprinus comatus*
Rock tripe	*Umbilicaria* spp.	**Shampoo, fern used as**	*Dryopteris dilatata*
Rockweeds	*Fucus* spp., *Pelvetia* spp., and others	**Shampoo, horsetail used as**	*Equisetum hyemale*
		Sheep	*Ovis* spp.
Rose	*Rosa* spp.	**Shepherd's purse**	*Capsella bursa-pastoris*
Rose, damask	*Rosa damascena*	**Shrimp**	*Crago* spp. and others
Rose Family	Rosaceae	**Sisal**	*Agave sisalina*
Rose, Sitka (Fig. 24.8)	*Rosa rugosa*	**Skunk**	*Mephitis* spp.
Rosemary	*Rosmarinus officinalis*	**Slime mold**	member of Divisions Myxomycota, and Acrasiomycota, Subkingdom Myxobionta, Kingdom Protoctista
Rotenone relative, source of	*Tephrosia vogelii*		
Rubber, Pará	*Hevea brasiliensis*		
Rubber plant	*Ficus elastica*		
Ruellia	*Ruellia portellae* and others		
Rust, apple	*Gymnosporangium juniperi-virginianum*	**Slime mold (Fig. 18.30 A)**	*Lamproderma* sp.
		Slime mold (Fig. 18.30 B)	*Lycogala epidendrum*
Rust, black stem of wheat	*Puccinia graminis*	**Slime mold, cellular**	member of Division Acrasiomycota, Kingdom Protoctista
Rust, corn	*Puccinia sorghi*		
Rust, poplar leaf spot	*Melampsora medusae*		
Rust, rock cress	*Puccinia monoica*	**Sloth**	*Bradypus* spp., *Choleopus* spp.
Rust, white pine blister	*Cronartium ribicola*		
Rutabaga	*Brassica campestris* var. *napobrassica*	**Smut**	*Ustilago* spp. and others
		Smut, corn	*Ustilago maydis*
Rye	*Secale cereale*	**Snail**	*Haplotrema concava* and others
Ryegrass	*Lolium* spp.		
Safflower	*Carthamus tinctorius*	**Snapdragon**	*Antirrhinum majus*
Saffron, meadow	*Colchicum autumnale*	**Snowplant**	*Sarcodes sanguinea*
Saffron (true)	*Crocus sativus*	**Snowy owl**	*Nyctea scandiaca*
Sage[14]	*Salvia officinalis*[14]	**Soaproot, California**	*Chlorogalum pomeridianum*
Sage, Jerusalem	*Phlomis fruticosa*	**Sorghum**	*Sorghum* spp.
Saguaro	*Carnegia gigantea*	**Sorrel**	*Oxalis* spp.

[14]This sage, which is in the Mint Family (Lamiaceae), should not be confused with sagebrush, which is in the Sunflower Family (Asteraceae).

COMMON NAMES AND SCIENTIFIC NAMES OF ORGANISMS

COMMON NAME	SCIENTIFIC NAME	COMMON NAME	SCIENTIFIC NAME
Sorrel, redwood	*Oxalis oregana*	Sundew relative used for flypaper	*Drosophyllum lusitanicum*
Southern yellow pine—*see* Pine, loblolly; Pine, longleaf; Pine, shortleaf; and Pine slash		Sunflower	*Helianthus annuus*
		Sunflower Family	Asteraceae (Compositae)
		Sweet pea	*Lathyrus odoratus*
Soybean	*Glycine max*	Sword fern	*Polystichum munitum*
Spanish moss	*Tillandsia* spp.	Sycamore	*Platanus* spp.
Sparrow, savannah	*Passerculus sandwichensis*	Tamarack	*Larix* spp.
Sparrow, vesper	*Pooecetes gramineus*	Tamarisk	*Tamarix* spp.
Spearmint	*Mentha spicata*	Tangerine	*Citrus reticulata*
Spiderwort	*Tradescantia* spp.	Tapir	*Tapirus* spp.
Spiderwort, European	*Tradescantia paludosa*	Taro	*Colocasia* spp.
Spike moss	*Selaginella* spp.	Tarragon	*Artemisia dracunculus*
Spike moss, fossil relatives of	*Lepidodendron* spp., *Sigillaria* spp., and others	Tarweed	*Grindelia* spp.
		Tarweed, western (Fig. 4.12 A)	*Calycadenia* sp.
Spinach	*Spinacia oleracea*		
Spirogyra	*Spirogyra* spp.	Tea	*Camellia sinensis*
Sponge	*Spongilla* spp. and others	Tent caterpillar	*Malacosoma americanum* and others
Sponge, vegetable	*Luffa cylindrica*		
Spring beauty	*Claytonia virginica*	*Teonanacatl* (sacred) mushrooms	*Conocybe* spp., *Panaeolous* spp., *Psilocybe* spp., and others
Spruce	*Picea* spp.		
Spruce, black	*Picea mariana*		
Spruce, red	*Picea rubens*	Tequila, source of	*Agave* spp.
Spruce, Sitka	*Picea sitchensis*	Termite	*Odontotermes* spp., *Reticulitermes* spp., and others
Spruce, white	*Picea glauca*		
Spurge	*Euphorbia* spp.		
Spurge (Fig. 24.13)	*Euphorbia peplus*	Thalloid liverworts	*Marchantia* spp. and others
Spurge Family	Euphorbiaceae	Thimbleberry	*Rubus parviflorus*
Squash	*Cucurbita mixta, C. pepo,* and others	Thistle	*Cirsium* spp. and others
		Thistle, Canada	*Cirsium arvense*
Squawroot	*Perideridia* spp.	Thrasher	*Toxostoma* spp.
Squill	*Scilla* spp.	Thyme	*Thymus vulgaris* and others
Squills	*Urginea maritima*	Ti plant	*Cordyline* spp.
Squirrel	*Citellus* spp., *Sciuris* spp., and others	Tiger	*Panthera tigris*
		Toad	*Bufo americanus*
Squirrel corn	*Dicentra canadensis*	Tobacco	*Nicotiana tabaccum*
Squirrel, gray	*Sciurus carolinensis*	Tomato	*Lycopersicon esculentum*
Squirting cucumber	*Ecballium elaterium*	Tomato fruitworm	*Heliothis armigera*
Stapelia (Fig. 23.14)	*Stapelia similis*	Tomato, Galápagos	*Lycopersicon esculentum* var. *minor, L. pimpinellifolium*
Stinkhorn	*Mutinus* spp., *Phallus* spp., and others		
		Tomato hornworm	*Protoparce quinquemaculata*
Stinkhorn, common (Fig. 19.12)	*Mutinus caninus*	Toothwort	*Dentaria* spp.
		Tortoise, giant Galápagos	*Testudo elephantopus porteri*
Stonecrop	*Sedum* spp. and others		
Stoneseed	*Lithospermum ruderale*	Touch-me-not	*Impatiens* spp.
Stonewort	*Chara* spp., *Nitella* spp.	Tree fern	*Cyathea* spp., *Ctenitis* spp., *Dicksonia* spp., *Sphaeropteris* spp., and others
Strawberry	*Fragaria* spp.		
String-of-pearls	*Senecio rowellianus*		
Strychnine, source of	*Strychnos* spp.		
Stuffing, ferns used for mattresses, etc.	*Cibotium* spp., *Sadleria* spp., and others	Tree-of-heaven	*Ailanthus altissima*
		Trillium	*Trillium* spp.
Sugar cane	*Saccharum officinarum* and others	Truffles	*Tuber* spp.
		Tulip	*Tulipa* spp.
Sumac	*Rhus* spp.	Tulip tree	*Liriodendron tulipifera*
Sunbird	*Anthodiaeta* spp., *Notiocinnyris* spp., and others	Tumbleweeds	*Amaranthus albus, Salsola pestifera,* and others
		Tung oil, source of	*Aleurites* spp.
Sundew	*Drosera* spp.	Turmeric, source of	*Curcuma longa*
		Turnip	*Brassica rapa*

COMMON NAMES AND SCIENTIFIC NAMES OF ORGANISMS

COMMON NAME	SCIENTIFIC NAME	COMMON NAME	SCIENTIFIC NAME
Turtle	*Chelydra* spp., *Chrysemys* spp., and others	**Wattle**	*Acacia decurrens* and others
		Weaver birds	*Anaplectes* spp., *Hyphantoris* spp., and others
Twinflower	*Linnaea borealis*		
Ulothrix	*Ulothrix* spp.	**Webworm, fall**	*Hyphantria cunea*
Ultraviolet light, flowers seen in (Fig. 23.13)	*Rudbeckia* sp.	**Welwitschia**	*Welwitschia mirabilis*
		Whale, sperm	*Physeter catodon*
Unicorn plant	*Proboscoidea* spp.	**Wheat**	*Triticum* spp. and their hybrids
Venus flytrap	*Dionaea muscipula*		
Vetch	*Vicia* spp.	**Wheel tree**	*Trochodendron aralioides*
Vetchling, yellow	*Lathyrus aphaca*	**Whisk fern**	*Psilotum* spp.
Vinegar weed	*Trichostema* spp.	**Whisk fern, fossil relatives of**	*Asteroxylon* spp., *Psilophyton* spp., *Rhynia* spp., and others
Violet	*Viola* spp.		
Violet, African	*Saintpaulia* spp.		
Violet, gold	*Viola douglasii*	**Whisk fern, living relatives of**	*Tmesipteris* spp.
Virginia creeper	*Parthenocissus quinquefolia*		
Virus[15]		**White pine blister rust**	*Cronartium ribicola*
Vole	*Microtus* spp. and others	**Willow**	*Salix* spp.
Wahoo	*Euonymus alata* and others	**Willow Family**	Salicaceae
Wake-robin	*Trillium* spp.	**Window leaves, plants with**	*Fenestraria* spp. and others
Wallflower, western	*Erysimum capitatum*	**Wintergreen oil, sources of**	*Gaultheria procumbens* and others
Walnut	*Juglans* spp.		
Walnut, black	*Juglans nigra, J. hindsii*	**Wisteria**	*Wisteria sinensis* and others
Warbler	*Dendroica* spp. and others	**Witch hazel**	*Hamamelis virginiana*
Watercress	*Nasturtium officinale*	**Woad, dyer's**	*Isatis tinctoria*
Water fern, oriental	*Ceratopteris thalictroides*	**Woadwaxen**	*Genista tinctoria*
Water molds	member of Division Oomycota, Subkingdom Mastigobionta, Kingdom Protoctista	**Wolfsbane**	*Aconitum vulparia*
		Wolverine	*Gulo luscus*
		Woodpecker	*Dendrocopus* spp. and others
		Wormwood	*Artemisia absinthium*
Water net	*Hydrodictyon* spp.	**Wounds, ferns used for treating**	*Lygodium circinatum, Ophioglossum vulgatum*
Water silk	*Spirogyra* spp.		
Water weed	*Elodea* spp.	**Yam**	*Dioscorea* spp.
Water weed, yellow	*Ludwigia repens*	**Yareta**	*Azorella yareta*
Watermelon	*Citrullus vulgaris*	**Yarrow, American**	*Achillea lanulosum*
		Yarrow, European	*Achillea millefolium*
		Yeast	*Saccharomyces* spp.
		Yew	*Taxus* spp.
		Yucca	*Yucca* spp.
		Zebra	*Equus zebra* and others
		Zinnia	*Zinnia elegans* and others

[15]Depending on the classification used, viruses may not have a scientific name. Many are named after the disease they cause; e.g., tobacco mosaic virus causes tobacco mosaic disease. One classification attempts to give them at least a Latin prefix, so the virus for warts is *Papavovirus;* for smallpox, *Poxvirus;* for polio, *Picornavirus;* for measles and mumps, *Paramyxovirus.*

Appendix 2

Biological Controls

If you were to ask the average farmer or backyard gardener how to control a particular insect or plant pest, you might be given the name of some poisonous spray or bait that has proved "effective" in the past. Evidence that spraying with such substances yields only temporary results, however, has been mounting for many years, and the spraying is frequently followed by even larger invasions of pests. Also, the residues of poisonous sprays often accumulate in the soil and disrupt the microscopic living flora and fauna essential to the soil's health. The problem is compounded and the ecology further upset when large amounts of inorganic fertilizers are added. As increasing numbers of people become aware of the devastating effects of pesticides and herbicides on the environment, they have turned to **biological controls** as an alternative to the use of poisonous sprays. To the surprise of some, such controls are often more effective than traditional controls.

Poisonous sprays often promote pest invasions because the sprays usually kill beneficial insects along with the undesirable ones. In addition, the pests, through mutations, often become resistant to the sprays. In undisturbed natural areas, weeds are never a problem, and even though pests may be present they seldom destroy the ecological community. Why is this so? You'll recall from Chapter 25 that all members of a community are in ecological balance with one another. The plants produce a variety of substances that may either repel or attract insects, inhibit or promote the growth of other plants, and generally contribute to the health of the community as a whole.

Virtually all insects have their own pests and diseases, as do most other living organisms. Each pest ensures, at least indirectly, that the various species of a community are perpetuated. This principle of nature can be applied, to a certain extent, to farming and gardening. The following are some general and specific biological controls that either are now in widespread use or are in various tests showing promise for the future.

GENERAL CONTROLS

Establishment of Beneficial Insects

Ladybugs (Family Coccinellidae)
The small and often colorful beetles called *ladybugs,* and particularly their larval stages, consume large numbers of aphids, thrips, insect eggs, weevils, and other pests. They are obtainable from various commercial sources (e.g., Bozeman Bio-Tech, P.O. Box 3146, Bozeman, MT 59772; Arbico, P.O. Box 4247, Tucson, AZ 85738), but, if given a chance, they probably will establish themselves without being imported. When obtained from outside of the local area, they should be placed in groups at the bases of plants on which pests are present, preferably in the early evening after watering.

Lacewings (Families Chrysopidae and Hemerobiidae)
Lacewings are slow-flying, delicate-winged insects that consume large numbers of aphids, mealybugs, and other pests. They lay their eggs on the undersides of leaves, each egg being borne at the tip of a slender stalk. The larvae consume the immature stages of leafhoppers, bollworms, caterpillar eggs, mites, scale insects, thrips, aphids, and other destructive pests. Commercial sources include Arbico, P.O. Box 4247, Tucson, AZ 85738, and All Pest Control, 6030 Grenville Lane, Lansing, MI 48910.

Praying Mantis (Family Mantidae)

About 20 species of *praying mantis* are now established in the United States. These are voracious feeders that prey somewhat indiscriminately on flying insects and sometimes even on other mantises. They can be established by tying their egg cases to tree branches or at other locations above the ground. The egg cases, which form compact masses about 2.5 to 5.0 centimeters (1 to 2 inches) long, are obtainable from various commercial sources, including Bob Bauer, 311 Ford Rd., Howell, NJ 07731, and Bozeman Bio-Tech, P.O. Box 3146, Bozeman, MT 59772.

Trichogramma Wasps (Family Trichogrammatidae)

Trichogramma wasps are minute insects, mostly less than 1 millimeter (1/25 inch) long; they parasitize insect eggs and are known to have significantly reduced populations of well over 100 different insect pests, including alfalfa caterpillars, armyworms, cabbage loopers, cutworms, hornworms, tent caterpillars, and the larvae of many species of moths. As with other insects used as biological controls, trichogramma wasps should not be released unless there are pest eggs in the vicinity, as the wasps may otherwise parasitize eggs of beneficial butterflies and other useful insects. They are available from commercial sources such as Unique Insect Control, P.O. Box 15376, Sacramento, CA 95851, and New Earth, 4422 East Hwy. 44, Shepherdsville, KY 40165.

Ichneumon Wasps (Family Ichneumonidae)

The *ichneumon wasps* belong to a very large family of wasps that are mostly stingless. These tiny wasps tend to be slender and have long ovipositors that are sometimes longer than the body. Most insects are parasitized by at least one species of ichneumon. Many species parasitize the larval stages of insects, consuming the host internally after hatching from eggs deposited on the body; alternatively, they may complete development in a later stage. Ichneumons will usually appear naturally in a backyard or farm population of pests if toxic sprays and other unnatural conditions have not interfered with their normal activities.

Tachinid Flies (Family Tachinidae)

Many members of the large family of *tachinid flies* resemble houseflies or bumblebees. All parasitize other insects, including a large variety of caterpillars, Japanese beetles, European earwigs, grasshoppers, gypsy moths, tomato worms, sawflies, and various beetles. Contact Unique Insect Control, P.O. Box 15376, Sacramento, CA 95851, for further information.

Use of Pathogenic Bacteria

Bacillus thuringiensis (BT) is one of several pathogenic bacteria registered for use on edible plants in the United States. It reproduces only in the digestive tracts of caterpillars and is harmless to humans and all other wildlife, including earthworms, birds, and mammals. It is exceptionally effective against a wide range of caterpillars, such as tomato hornworms and fruitworms, cabbage worms and loopers, grape leaf rollers, corn borers, cutworms, fall webworms, and tent caterpillars. It is mass-produced and sold in a powdered spore form at nurseries and garden supply stores under the trade names of *Dipel, Biotrol,* and *Thuricide.* The powder is mixed with water and applied as a spray.

Establishment of Toads and Frogs

It has been estimated that a single adult toad will consume about 10,000 insects and slugs in one growing season. Toads and frogs feed at night when snails, slugs, sowbugs, earwigs, and other common pests are active.

Use of Beneficial Nematodes

Several species of these abundant microscopic roundworms are notorious for damaging economically important crops when they invade plant roots and other underground organs. Most species, however, are either harmless or beneficial to plants. They have been used successfully in parasitizing cabbage worm caterpillars, codling moth larvae, Japanese beetle grubs, and tobacco budworms, and have shown considerable potential against other pests. One species that has been particularly effective in controlling ants, beetles, bugs, flies, wasps, and many other insects is the caterpillar nematode (*Neoaplectana carpocapsae*). It carries a symbiotic bacterium (*Xenorhabdus nematophilus*), which multiplies rapidly in the host, killing most insects within 24 hours of initial contact. It may be obtained from Nematode Farm, Inc., 2617 San Pablo Avenue, Berkeley, CA 94702.

Use of Limonoid Sprays

Limonoids are bitter substances found in the rinds, seeds, and juice of citrus fruits (especially grapefruit). If the rinds and seeds of two or three fruits are ground up and soaked overnight in a pint of water, and the solid material is strained out, the liquid may then be sprayed on plants. The bitter principle apparently stops or reduces the feeding of larvae on the foliage. In experiments, limonoid sprays have proved effective against corn earworm, fall armyworm, tobacco budworm, and pink bollworm but undoubtedly will deter many other pests as well.

Use of Liquefied Pest Sprays

Jeff Cox, an editor of Rodale's *Organic Gardening* magazine, called attention to this method of pest control in the magazine in October 1976, and again in May 1977. Insect pests or slugs are gathered in small quantities and liquefied with a little water in a blender. The material is then further diluted with water and sprayed throughout the infested area. It is not known why spraying with "bug juice" is effective against pests. It is known, however, that virtually all organisms harbor viruses. It has been theorized that even the inactive viruses carried by healthy insects and slugs may somehow be activated in the process of liquefaction. The viruses would be spread throughout an entire yard or farm if all parts of the area were sprayed. Most viruses are highly specific, generally attacking a single species of organism.

M. Sipe, a Florida entomologist who has recommended the "bug juice" technique, has also suggested that the odor of the liquefied insects possibly attracts their predators and parasites or that the insects' distress **pheromones** (naturally produced insect chemicals that influence sexual or other behavior) are released by the blender, with the pheromones acting as an insect repellent. Possibly the observed effects of spraying "bug juice" are the result of a combination of viruses, predator attraction, and repellent pheromones. Sipe warns that if one tries this method of pest control, care should be taken to use only pest species and only those that are doing significant damage. Failure to heed this warning could disrupt the activities of natural predators and other natural controls present.

This approach still needs further testing and investigation of its safety for use by humans, but preliminary results in various areas of North America have thus far yielded impressive results with no evidence of harm to humans or beneficial organisms.

Use of Resistant Varieties

Many plants may kill or inhibit disease fungi or bacteria with chemicals known as *phytoalexins*. Phytoalexins are synthesized at the point of attack or invasion by the pathogen and are toxic to the fungus or bacterium. In selecting for improved fruit quality, vigor of growth, or other desirable characteristics, horticulturists in the past have sometimes unknowingly bred out a plant's capacity to produce certain phytoalexins, although general vigor is usually accompanied by disease resistance. Now that this aspect of a plant's defense mechanisms is known, breeders are concentrating on developing varieties capable of producing phytoalexins against various fungi, bacteria, and even nematodes. Several tomato varieties, for example, are listed as being *VFN*. The letters *V* and *F* indicate a resistance to *Verticillium* and *Fusarium* (common pathogenic fungi), while the letter *N* denotes a resistance to *root-knot nematodes*.

Other aspects of plant disease resistance include thick cuticles, the secretion of gums, resins, and other metabolic products that may interfere with fungal and bacterial spore germination, and the presence within all the cells of the plant of chemical compounds toxic to pathogens.

Interplanting with Plants That Produce Natural Insecticides or Substances Offensive to Pests

Many plant species produce substances that repel a significant number of pests, but none produce anything that repels all pests. Among the best-known plant producers of insect repellents are marigolds, garlic, and members of the Mint Family such as pennyroyal, peppermint, basil, and lavender. An expanded discussion of this subject is given in Appendix 3.

Specific Controls

Weeds

In 1974, the Weed Science Society of America published a special committee report (*Weed Science* 22:490–95) on the biological control of weeds, summarizing the status of projects on the biological control of weeds with insects and plant pathogens in the United States and Canada. Table A2.1 is condensed from that report and supplemented with additional information. Many other biological controls for these and other weeds are currently under investigation.

Insects

The maintenance of ecological balance in nature includes a vast array of predator-prey relationships between animals, birds, insects, and other organisms. Specific biological controls for several types of insect pests, in addition to the general controls previously discussed, are given in Table A2.2.

Companion planting

The *Additional Reading* list reveals that the literature on the chemical interactions between plants and also between plants and their consumers is already extensive. Despite the scientific evidence on the subject to date, however, a significant amount of the "backyard biological control" that is practiced today is based primarily on empirical information. Such information has been obtained from thousands of gardeners and farmers who have tried various techniques with their plantings and pest controls. As a result, they have come to conclusions that certain strategies work while others do not, but they have not deliberately set up controlled experiments, nor have they necessarily understood the scientific basis for what they have observed. This does not mean that their observations are not useful or that they are invalid. In fact, such empirical observations have often been the inspiration for investigations and experiments by scientists. The scientific investigations have sometimes revealed that the empirical observations were biased or not carefully made or that erroneous conclusions had been drawn, but, frequently, sound scientific bases for these observations have been uncovered.

Further insights into how plants inhibit or enhance the growth of others and into the nature of their resistance to disease or their insect-repelling mechanisms continue to occur. Observations of such phenomena in the past have led organic gardeners and others to the practice of *companion planting*, which involves the interplanting of various crops and certain other plants in such a way that each species derives some benefit from the arrangement. The companion planting list shown in Table A2.3, based primarily on empirical information, appeared in the February 1977 issue of *Organic Gardening and Farming* magazine. It is included here with the permission of Rodale Press, Inc.

Table A2.1

SPECIFIC WEEDS AND AGENTS INVOLVED IN THEIR BIOLOGICAL CONTROL

WEED	AGENT(S) OF BIOLOGICAL CONTROL
Alligator weed (*Alternanthera philoxeroides*)	Flea beetles (*Agasicles hygrophila*)
Bladder campion (*Silene cucubalus*)	Tortoise beetle (*Cassida hemisphaerica*)
Brazil peppertree (*Schinus terebinthifolius*)	Weevil (*Bruchus atronotatus*) and others
Brushweed (*Cassia surattensis*)	Imperfect fungus (*Cephalosporium* sp.)
Curly dock (*Rumex crispus*)	Rust (*Uromyces rumicis*)
Curse (*Clidemia hirta*)	Thrip (*Liothrips urichi*) and others
Cypress spurge (*Euphorbia cyparissias*)	Sphinx moth (*Hyles euphorbiae*)
Dalmatian toadflax (*Linaria dalmatica*)	Leaf miner (*Stagmatophora serratella*) and others
Emex (*Emex australis*)	Seed weevils (*Apion antiquum*) and others
Gorse (*Ulex europaeus*)	Seed weevils (*Apion ulicis*) and others
Halogeton (*Halogeton glomeratus*)	Casebearer (*Coleophora parthenica*) and others
Hawaiian blackberry (*Rubus penetrans*)	Sawflies (*Pamphilius sitkensis, Priophorus morio*) and others
Jamaica feverplant (*Tribulus terrestris*)	Weevils (*Microlarinus* spp.)
Joint vetch (*Aeschynomene virginica*)	Imperfect fungus (*Colletotrichum gloeosporioides*)
Klamath weed (*Hypericum perforatum*)	Leaf beetles (*Chrysolina* spp.); buprestid beetle (*Agrilus hyperici*)
Lantana (*Lantana camara*)	Seed weevil (*Apion* sp.); ghost moth (*Hepialus* sp.); plume moth (*Platyptilia pusillidactyla*); hairstreaks (*Strymon* spp.); and others
Leafy spurge (*Euphorbia esula*)	Wood-boring beetle (*Oberea* sp.) and others
Mediterranean sage (*Salvia aethiopis*)	Snout beetles (*Phrydiuchus* spp.)
Milkweed vine (*Morrenia odorata*)	Oomycete fungus (*Phytophthora citrophthora*); rust (*Aecidium asclepiadinum*)
Prickly pear (*Opuntia* spp.)	Moth (*Cactoblastis cactorum*); cochineal insects (*Dactylopius* spp.); and others
Puncture vine (*Tribulus terrestris*)	Weevils (*Microlarinus* spp.)
Scotch broom (*Cytisus scoparius*)	Seed weevil (*Apion fuscirostre*) and others
Skeleton weed (*Chondrilla juncea*)	Gall mite (*Aceria chondrillae*); root moth (*Bradyrrhoa gilveolella*); rust (*Puccinia chondrillina*); powdery mildews (*Erysiphe cichoracearum, Leveillula taurica*)
Spiny emex (*Emex spinosa*)	Seed weevil (*Apion antiquum*)
Tansy ragwort (*Senecio jacobaea*)	Seed fly (*Hylemya seneciella*); cinnabar moth (*Tyria jacobaeae*); leaf beetle (*Longitarsus jacobaeae*)
Thistles:	
Bull thistle (*Cirsium vulgare*)	Tortoise beetle (*Cassida rubiginosa*)
	Weevil (*Ceuthorrhynchidius horridus*);
Canada thistle (*Cirsium arvense*)	Weevil (*Ceutorhynchus litura*); flea beetle (*Altica carduorum*); stem gall fly (*Urophora cardui*)
Diffuse knapweed (*Centaurea diffusa*)	Seed fly (*Urophora affinis*)
Italian thistle (*Carduus pycnocephalus*)	Flea beetles (*Rhinocyllus conicus, Psylliodes chalcomera*); weevil (*Ceutorhynchus trimaculatus*)
Milk thistle (*Silybum marianum*)	Flea beetle (*Rhinocyllus conicus*)
Musk thistle (*Carduus nutans*)	Weevils (*Ceutorhynchus trimaculatus, Ceuthorrhynchidius horridus; Rhinocyllus conicus*); flea beetle (*Psylliodes chalcomera*)
Perennial sowthistle (*Sonchus arvensis*)	Peacock fly (*Tephritis dilacerata*)
Plumeless thistle (*Carduus acanthoides*)	Tortoise beetle (*Cassida rubiginosa*); seed weevil (*Rhinocyllus conicus*); weevil (*Ceuthorrhynchidius horridus*)
Russian thistle (*Salsola kali* var. *tenuifolia*)	Casebearer (*Coleophora parthenica*) and others
Slenderflower thistle (*Carduus tenuiflorus*)	Weevil (*Ceutorhynchus trimaculatus*)
Spotted knapweed (*Centaurea maculosa*)	Seed fly (*Urophora affinis*) and others
Star thistle (*Centaurea nigrescens*)	Weevil (*Ceuthorrhynchidius horridus*)
Yellow star thistle (*Centaurea solstitialis*)	Seed fly (*Urophora siruna-seva*)
Water hyacinth (*Eichhornia crassipes*)	Weevils (*Neochetina bruchi, N. eichhorniae*); moth (*Sameodes albiguttalis*)
Water purslane (*Ludwigia palustris*)	Snout beetle (*Nanophyes* sp.)

Table A2.2

SPECIFIC BIOLOGICAL CONTROLS FOR SEVERAL TYPES OF INSECT PESTS

INSECT	CONTROL
Ants (about 8,000 spp. within the Superfamily Formicoidea)	Ants that carry aphids into trees and consume ripening fruits can be prevented from getting farther than the trunk by applying a band of sticky material around the trunk. A commercial preparation sold under the trade name of Tanglefoot is particularly effective. A water suspension of ground hot peppers (*Capsicum* spp.) used as a spray can act as an ant deterrent. *Caution:* Many ants are beneficial to a balanced ecology; they should not be decimated indiscriminately.
Grasshoppers (there are several families of grasshoppers, but the insects that usually constitute the most serious pests are species of *Melanoplus,* **Family Acrididae)**	In 1980, the Environmental Protection Agency permitted private companies to begin the mass culture of a protozoan, *Nosema locustae,* for use in controlling rangeland grasshoppers. Tests have shown that properly timed applications of spores mixed with wheat bran can reduce grasshopper populations by up to 50%.
Gypsy moths (*Porthetria dispar*)	Parasitic wasps (*Apanteles flavicoxis, A. indiensis*) imported from India lay their eggs in gypsy moth caterpillars and kill large numbers.
Japanese beetles (*Popillia japonica*)	The pathogenic bacterium *Bacillus popillae,* which is sold commercially, is specific for Japanese beetle larvae. It causes what is known as *milky spore disease* in the grubs while they are still in the soil, and it is very destructive.
Mealybugs (*Pseudococcus* spp.)	The small brown beetles called *crypts* (*Cryptolaemus montrouzieri*) effectively control mealybugs in greenhouses and also outdoors on apple, pear, peach, and citrus trees. Order from Rincon-Vitova Insectaries, Inc., P.O. Box 95, Oak View, CA 93022.
Mosquitoes (*Culex* spp., *Anopheles* spp., and others)	The bacterium *Bacillus thuringiensis* var. *israelensis* has proved to be very effective in destroying mosquito larvae. A fungus (*Lagenidium giganteum*) has also proved highly effective against mosquito larvae if the temperature is above 20°C (68°F). The bacterium is available from several sources, including Abbott Laboratories, Dept. 95-M, 1400 Sheridan Rd., N. Chicago, IL 60064, and Sandoz, Inc., 480 Camino del Rio S., San Diego, CA 92108.
Red spider mites (*Tetranychus telarius*)	Predatory mites (*Phytolesius persimilis*, which works best when weather is not hot, and *Amblyseius californicus,* which is more effective in hot weather) effectively control populations of red spider mites.
White flies (*Trialeurodes vaporariorum*)	A minute wasp, *Encarsia formosa,* parasitizes white flies exclusively. The wasps have been known to be very effective in greenhouses. They are obtainable from White Fly Control Co., Box 986, Milpitas, CA 95035, and Rincon-Vitova Insectaries, Inc., P.O. Box 95, Oak View, CA 93022. White flies are attracted to the color yellow. Large numbers of white flies are trapped when a yellow board is sprayed or painted with any sticky substance and placed in the vicinity of the pests.

SOME SOURCES OF HERB PLANTS AND SEEDS

China Herb Co., 428 Soledad, Salinas, CA 93901

Cottage Herbs, P.O. Box 100, Troy, ID 83871

De Giorgi Co., 6011 N St., Omaha, NE 68117

Fragrant Fields, Dongola, IL 62926

Herbs-Liscious, 1702 S. Sixth St., Marshalltown, IA 50158

Jude Herbs, Box 56360, Huntington Station, NY 11746

Otto Richter and Sons, Box 260, Goodwood, Ontario, LOC 1A0

PG Nursery, R18, Box 470, Bedford, IN 47421

Putney Nursery, Putney, VT 05346

Rawlinson Garden Seed, 269 College Rd., Truro, Nova Scotia B2N 2P6

Sanctuary Seeds, 2388 West Fourth Avenue, Vancouver, British Columbia V6K 1P1

Sea Island Savory Herbs, 5920 Chisolm, John's Island, SC 29455

Shoestring Seeds, P.O. Box 2261, Martinsville, VA 24113

Story House Herb Farm, Route 7, Box 246, Murray, KY 42071

Sunnybrook Farms Nursery, Box 6, Chesterland, OH 44026

Sunshine Herbs and Flowers, Rt. 1, Box 234, Comer, GA 30629

The Thyme Garden, 20546-0 Alsea Hwy., Alsea, OR 97324

Thompson and Morgan, Inc., P.O. Box 1308, Jackson, NJ 08527

Wildwood Herbal, P.O. Box 746, Albemarle, NC 28002

Willhite Seed Company, Box 23, Poolville, TX 76076

Table A2.3

COMPANION PLANTS

PLANT	COMPANIONS AND EFFECTS
Asparagus	Tomatoes, parsley, basil.
Basil	Tomatoes (improves growth and flavor); said to dislike rue; repels flies and mosquitoes.
Beans	Potatoes, carrots, cucumbers, cauliflower, cabbage, summer savory, most other vegetables and herbs; around houseplants when set outside.
Beans (bush)	Sunflowers (beans like partial shade, sunflowers attract birds and bees), cucumbers (combination of heavy and light feeders), potatoes, corn, celery, summer savory.
Beets	Onions, kohlrabi.
Borage	Tomatoes (attracts bees, deters tomato worm, improves growth and flavor), squash, strawberries.
Cabbage Family	Potatoes, celery, dill, chamomile, sage, thyme, mint, pennyroyal, rosemary, lavender, beets, onions. Aromatic plants deter cabbage worms.
Carrots	Peas, lettuce, chives, onions, leeks, rosemary, sage, tomatoes.
Catnip	Plant in borders; protects against flea beetles.
Celery	Leeks, tomatoes, bush beans, cauliflower, cabbage.
Chamomile	Cabbage, onions.
Chervil	Radishes (improves growth and flavor).
Chives	Carrots; plant around base of fruit trees to discourage insects from climbing trunk.
Corn	Potatoes, peas, beans, cucumbers, pumpkin, squash.
Cucumbers	Beans, corn, peas, radishes, sunflowers.
Dill	Cabbage (improves growth and health), carrots.
Eggplant	Beans.
Fennel	Most plants are supposed to dislike it.
Flax	Carrots, potatoes.
Garlic	Roses and raspberries (deters Japanese beetle); with herbs to enhance their production of essential oils; plant liberally throughout garden to deter pests.
Horseradish	Potatoes (deters potato beetles); around plum trees to discourage curculios.
Lamb's quarters	Nutritious edible weed; allow to grow in modest amounts in corn.
Leek	Onions, celery, carrots.
Lettuce	Carrots and radishes (lettuce, carrots, and radishes make a strong companion team), strawberries, cucumbers.
Lovage	Plant here and there in garden.
Marigolds	The workhorse of pest deterrents. Keeps soil free of nematodes; discourages many insects. Plant freely throughout garden.

ADDITIONAL READING

Bosch, R. van den, et al. 1981. *An introduction to biological control.* New York: Plenum Publishing Company.

Burges, H. D., and N. W. Hussey (Eds). 1981. *Microbial control of pests and plant diseases, 1970–1980.* New York: Academic Press, Inc.

Carson, R. 1994. *Silent spring.* Boston: Houghton Mifflin Co.

Cook, R. J., and K. F. Baker. 1983. *Nature and practice of biological control of plant pathogens.* St. Paul, MN: American Phytopathological Society.

Debach, P., and D. Rosen. 1991. *Biological control by natural enemies,* 2d ed. New York: Cambridge University Press.

Goeden, R. D., L. A. Andres, T. E. Freeman, P. Harris, R. L. Pienkowski, and C. R. Walker. 1974. Present status of projects on the biological control of weeds with insects and plant pathogens in the United States and Canada. *Weed Science* 22:490–95.

Henry, J. E. 1981. Natural and applied control of insects by protozoa. *Annual Review of Entomology* 26:49–73.

Hoy, M., and G. L. Cunningham. 1983. *Biological control of pests by mites: Proceedings of a conference.* Oakland, CA: Agricultural and Natural Resources, University of California.

Huffaker, C. B., and R. L. Rabb (Eds). 1984. *Ecological entomology.* San Diego, CA: Academic Press, Inc.

Jutsum, A. R., and R. F. S. Gordon (Eds). 1989. *Insect pheromones in plant protection.* New York: John Wiley and Sons, Inc.

Metcalf, R. L., and E. R. Metcalf. 1991. *Plant kairomones in insect control.* New York: Chapman and Hall.

Mukerji, K. G., and K. L. Garg (Eds). 1988. *Biocontrol of plant diseases.* (2 vols.) Boca Raton, FL: CRC Press.

Rice, E. L. 1984. *Allelopathy,* 2d ed. San Diego, CA: Academic Press, Inc.

Rice, E. L. 1983. *Pest control with nature's chemicals: Allelochemics and pheromones in gardening and agriculture.* Norman, OK: University of Oklahoma Press.

Whittaker, R. H., and P. P. Feeny. 1971. Allelochemics: Chemical interactions between species. *Science* 171:757–70.

Table A2.3

COMPANION PLANTS

PLANT	COMPANIONS AND EFFECTS
Marjoram	Here and there in garden.
Mint	Cabbage family; tomatoes; deters cabbage moth.
Mole plant	Deters moles and mice if planted here and there throughout the garden.
Nasturtium	Tomatoes, radishes, cabbage, cucumbers; plant under fruit trees. Deters aphids and pests of cucurbits.
Onion	Beets, strawberries, tomato, lettuce (protects against slugs), beans (protects against ants), summer savory.
Parsley	Tomato, asparagus.
Peas	Squash (when squash follows peas up trellis), plus grows well with almost any vegetable; adds nitrogen to the soil.
Petunia	Protects beans; beneficial throughout garden.
Pigweed	Brings nutrients to topsoil; beneficial growing with potatoes, onions, and corn; keep well thinned.
Potato	Horseradish, beans, corn, cabbage, marigold, limas, eggplant (as trap crop for potato beetle).
Pot marigold	Helps tomato; plant throughout garden as deterrent to asparagus beetle, tomato worm, and many other garden pests.
Pumpkin	Corn.
Radish	Peas, nasturtium, lettuce, cucumbers; a general aid in repelling insects.
Rosemary	Carrots, beans, cabbage, sage; deters cabbage moth, bean beetles, and carrot fly.
Rue	Roses and raspberries; deters Japanese beetle. Keep it away from basil.
Sage	Rosemary, carrots, cabbage, peas, beans; deters some insects.
Southernwood	Cabbage; plant here and there in garden.
Soybeans	Grows with anything, helps everything.
Spinach	Strawberries.
Squash	Nasturtium, corn.
Strawberries	Bush beans, spinach, borage, lettuce (as a border).
Summer savory	Beans, onions. Deters bean beetles.
Sunflower	Cucumbers.
Tansy	Plant under fruit trees; deters pests of roses and raspberries; deters flying insects, Japanese beetles, striped cucumber beetles, squash bugs; deters ants.
Tarragon	Good throughout garden.
Thyme	Here and there in garden; deters cabbage worm.
Tomato	Chives, onion, parsley, asparagus, marigold, nasturtium, carrot, limas.
Turnip	Peas.
Valerian	Good anywhere in garden.
Wormwood	As a border; keeps animals from the garden.
Yarrow	Plant along borders, near paths, near aromatic herbs; enhances essential oil production of herbs.

Appendix 3

Useful Plants and Poisonous Plants

WILD EDIBLE PLANTS

Words of Caution

Literally thousands of native and naturalized plants, or at least parts of them, have been used for food and other purposes by Native Americans and the immigrants who came later from other quarters of the globe. Table A3.1 has been compiled from a variety of sources; I have had opportunities to sample only a fraction of these plants myself and thus cannot confirm the edibility of all of the plants listed. *The reader is cautioned to be certain of the identity of a plant before consuming any part of it.* Cow parsnip (*Heracleum lanatum*) and water hemlocks (*Cicuta* spp.), for example, resemble each other in general appearance, but although cooked roots of cow parsnip have been used for food for perhaps many centuries, those of water hemlocks are very poisonous and have caused many human fatalities.

As was indicated in Chapter 21, many species of organisms are now on rare and endangered species lists, and a number of them are doomed to extinction within the next few years. Although the wild edible plants discussed here are not included on such lists, it might not take much indiscriminate gathering to endanger their existence as well. Because of this, one should exercise the following rule of thumb: *Never reduce a population of plants by more than 10% when collecting them for any purpose!*

Table A3.1

WILD EDIBLE PLANTS

PLANT	SCIENTIFIC NAME	USES
Amaranth	*Amaranthus* spp.	Young leaves used like spinach; seeds ground with others for flour.
Arrow grass	*Triglochin maritima*	Seeds parched or roasted. (***Caution***: *The plant is otherwise poisonous.*)
Arrowhead	*Sagittaria latifolia*	Tubers used like potatoes.
Balsamroot	*Balsamorhiza* spp.	Whole plant edible, especially when young, either raw or cooked.
Basswood	*Tilia* spp.	Fruits and flowers ground together to make a paste that can serve as a chocolate substitute; winter buds edible raw; dried flowers used for tea.
Bedstraw	*Galium aparine*	Roasted and ground seeds make a coffee substitute.
Beechnuts	*Fagus grandifolia*	Seeds used like nuts; oil extracted from seeds for table use.
Biscuit root	*Lomatium* spp.	Roots eaten raw or dried and ground into flour; seeds edible raw or roasted.
Bitterroot	*Lewisia rediviva*	Outer coat of the bulbs should be removed to eliminate the bitter principle; bulbs are then boiled or roasted.
Blackberry (wild)	*Rubus* spp.	Fruits edible raw and in pies, jams, and jellies.
Black walnut	*Juglans nigra*	Nut meats highly edible.
Bladder campion	*Silene cucubalus*	Young shoots (less than 5 cm tall) cooked as a vegetable.
Blueberry	*Vaccinium* spp.	Fruits edible raw, frozen, and in pies, jams, and jellies.
Bracken fern	*Pteridium aquilinum*	Young uncoiling leaves (fiddleheads) cooked like asparagus; rhizomes also edible but usually tough. (***Caution***: *Recent evidence indicates that frequent consumption of bracken fern can cause cancer of the intestinal tract.*)

Table A3.1

WILD EDIBLE PLANTS

PLANT	SCIENTIFIC NAME	USES
Broomrape	*Orobanche* spp.	Entire plant eaten raw or roasted.
Bulrush (Tule)	*Scirpus* spp.	Roots and young shoot tips edible raw or cooked; pollen and seeds also edible.
Butternut	*Juglans cinerea*	Nut meats edible.
Caraway	*Carum carvi*	Young leaves in salads; seeds for flavoring baked goods and cheeses.
Cattail	*Typha* spp.	Copious pollen produced by flowers in early summer is rich in vitamins and can be gathered and mixed with flour for baking; rhizomes can be cooked and eaten like potatoes.
Chicory	*Cichorium intybus*	Leaves edible raw or cooked; dried, ground roots make a coffee substitute.
Chokecherry	*Prunus virginiana*	Fruits make excellent jelly or can be cooked with sugar for pies and cobblers.
"Coffee" (wild)	*Triosteum* spp.	Berries dried and roasted make a coffee substitute.
Common chickweed	*Stellaria media*	Plants cooked as a vegetable.
Corn lily	*Clintonia borealis*	Youngest leaves can be used as a cooked vegetable.
Cow parsnip	*Heracleum lanatum*	Roots and young stems cooked.
Cowpea	*Vigna sinensis*	"Peas" and young pods cooked as a vegetable (plant naturalized in southern United States).
Crab apple	*Pyrus* spp.	Jelly made from fruits.
Crowberry	*Empetrum* spp.	Fruits should first be frozen then cooked with sugar.
Dandelion	*Taraxacum officinale*	Leaves rich in vitamin A; dried roots make a coffee substitute; wine made from young flowers.
Dock	*Rumex* spp.	Leaves cooked like spinach; tartness of leaves varies from species to species and sometimes from plant to plant—tart forms should be cooked in two or three changes of water.
Douglas fir	*Pseudotsuga menziesii*	Cambium and young phloem edible; tea made from fresh leaves.
Elderberry	*Sambucus* spp.	Fresh flowers used to flavor batters; fruits used in pies, jellies, wine. (**Caution**: *Other parts of the plant are poisonous.*)
Evening primrose	*Oenothera hookeri, O. biennis,* and others	Young roots cooked.
Fairy bells	*Disporum trachycarpum*	Berries edible raw.
Fennel	*Foeniculum vulgare*	Leaf petioles edible raw or cooked.
Fireweed	*Epilobium angustifolium*	Young shoots and leaves boiled as a vegetable.
Ginger (wild)	*Asarum* spp.	Rhizomes can be used as substitute for true ginger.
Gooseberry	*Ribes* spp.	Berries edible cooked, dried, or raw; make excellent jelly.
Grape (wild)	*Vitis* spp.	Berries usually tart but can be eaten raw; make good jams and jellies.
Grass	Many genera and species	Seeds of most can be made into flour; rhizomes of many perennial species can be dried and ground for flour.
Greenbrier	*Smilax* spp.	Roots dried and ground; refreshing drink made with ground roots, sugar, and water.
Groundnut	*Apios americana*	Tubers cooked like potatoes.
Hawthorn	*Crataegus* spp.	Fruits edible raw and in jams and jellies.
Hazelnut	*Corylus* spp.	Nuts edible raw or roasted.
Hickory	*Carya* spp.	Nuts edible.
Highbush cranberry	*Viburnum trilobum*	Fruits make excellent jellies and jams.
Huckleberry	*Vaccinium* spp.	Berries eaten raw or in jams and jellies.
Indian paintbrush	*Castilleja* spp.	Flowers of many species edible. (**Caution**: *In certain soils, plants absorb toxic quantities of selenium.*)
Indian pipe	*Monotropa* spp.	Whole plant edible raw or cooked.
Juniper	*Juniperus* spp.	"Berries" dried, ground, and made into cakes.
Labrador tea	*Ledum* spp.	Tea made from young leaves.
Lamb's quarters	*Chenopodium album*	Leaves and young stems cooked as a vegetable.
Licorice	*Glycyrrhiza lepidota*	Roots edible raw or cooked.
Mallow	*Malva* spp.	Leaves and young stems used as vegetable (use only small amounts at one time).
Manzanita	*Arctostaphylos* spp.	Berries eaten raw, used in jellies or pies, or made into "cider."

Table A3.1

WILD EDIBLE PLANTS

PLANT	SCIENTIFIC NAME	USES
Maple	*Acer* spp.	Sugar maples (*Acer saccharum*) well known for the sugar content of the early spring sap; other species (e.g., box elder, *A. negundo,* and bigleaf maple, *A. macrophyllum*) also contain usable sugars in their early spring sap.
Mariposa lily	*Calochortus* spp.	Bulbs edible raw or cooked.
Mayapple	*Podophyllum peltatum*	Fruit good raw or cooked. (**Caution**: *Other parts of the plant are poisonous.*)
Maypops	*Passiflora incarnata*	Fruits edible raw or cooked.
Miner's lettuce	*Claytonia perfoliata*	Leaves eaten raw as a salad green.
Mint	*Mentha arvensis* and others	Leaves of several mints used for teas.
Mormon tea	*Ephedra* spp.	Tea from fresh or dried leaves (add sugar to offset bitterness); seeds for bitter meal.
Mulberry	*Morus* spp.	Fruits of the red mulberry (*M. rubra*) used raw and in pies and jellies; fruits of white mulberry (*M. alba*) edible but insipid.
Mushrooms	Many genera and species	***Utmost caution*** *should be exercised in identifying mushrooms before consuming them. Although poisonous species are in the minority, they are common enough. Edible forms that are relatively easy to identify include morels (Morchella esculenta), most puffballs (Lycoperdon spp.), and inky cap mushrooms (Coprinus spp.).*
Mustard	*Brassica* spp.	Leaves used as vegetable; condiment made from ground seeds.
Nettles	*Urtica* spp.	Leaves and young stems boiled like spinach.
New Jersey tea	*Ceanothus americanus*	Tea from leaves.
Nutgrass	*Cyperus esculentus* and others	Tubers edible raw.
Oak	*Quercus* spp.	Acorns were widely used ground for flour by native North Americans; all contain bitter tannins that must be leached out before use.
Onion (wild)	*Allium* spp.	Bulbs edible raw or cooked.
Orach	*Atriplex patula* and others	Leaves and young stems cooked as a vegetable.
Pawpaw	*Asimina triloba*	Fruit edible raw or cooked.
Pennycress	*Thlaspi arvense*	Young leaves edible raw.
Peppergrass	*Lepidium* spp.	Immature fruits add zest to salads; seeds spice up meat dressings.
Persimmon	*Diospyros virginiana*	Fully ripened fruits edible raw or cooked.
Pickerel weed	*Pontederia cordata*	Fruits edible raw or dried.
Pigweed (*see* **Amaranth**)		
Pines	*Pinus* spp.	Cambium and young phloem edible; tea from fresh needles rich in vitamin C.
Pipissewa	*Chimaphila umbellata*	Drink made from boiled roots and leaves (cool after boiling).
Plantain	*Plantago* spp.	Young leaves eaten in salads or as cooked vegetable.
Prairie turnip	*Psoralea esculenta*	Turniplike roots cooked like potatoes.
Prickly pear	*Opuntia* spp.	Fruits and young stems peeled edible raw or cooked.
Purple avens	*Geum rivale*	Liquid from boiled root has flavor similar to chocolate.
Purslane	*Portulaca oleracea*	Leaves and stems cooked like spinach.
Raspberry (wild)	*Rubus* spp.	Fruits edible raw or in pies, jams, and jellies.
Redbud	*Cercis* spp.	Flowers used in salads; cooked young pods edible.
Rose (wild)	*Rosa* spp.	Fruits (hips) exceptionally rich in vitamin C; hips edible raw, pureed, or candied.

Table A3.1

WILD EDIBLE PLANTS

PLANT	SCIENTIFIC NAME	USES
Salsify	*Tragopogon* spp.	Roots edible raw or cooked.
Saltbush	*Atriplex* spp.	Seeds nutritious. (***Caution***: *In certain soils, plants can absorb toxic amounts of selenium.*)
Sassafras	*Sassafras albidum*	Tea from roots. (***Caution***: *Large quantities have a narcotic effect.*) Leaves and pith used for Louisiana filé.
Serviceberry	*Amelanchier* spp.	All fruits edible (mostly bland).
Shepherd's purse	*Capsella bursa-pastoris*	Leaves cooked as vegetable; seeds edible parched or ground for flour.
Showy milkweed	*Asclepias speciosa*	Flowers edible raw or cooked; young shoots cooked.
Soap plant	*Chlorogalum pomeridianum*	Bulbs slow-baked and eaten like potatoes after fibrous outer coats removed.
Solomon's seal	*Polygonatum* spp.	Rootstocks dried and ground for bread flour.
Sorrel	*Oxalis* spp.	Leaves mixed in salads.
Spatterdock	*Nuphar polysepalum*	Seeds placed on hot stove burst like popcorn and are edible as such; peeled tubers edible boiled or roasted.
Speedwell	*Veronica americana* and others	Leaves and stems used in salads.
Spring beauty	*Claytonia* spp.	Bulbs edible raw or roasted.
Strawberry (wild)	*Fragaria* spp.	Fruits superior in flavor to cultivated varieties.
Sunflower	*Helianthus annuus*	Seeds edible raw or roasted; seeds yield cooking oil.
Sweet cicely	*Osmorhiza* spp.	Roots have aniselike flavor.
Sweet flag	*Acorus calamus*	Young shoots used in salads; roots candied.
Thistle	*Cirsium* spp.	Peeled stems edible; roots edible raw or roasted.
Vetch	*Vicia* spp.	Tender green pods edible baked or boiled.
Watercress	*Nasturtium officinale*	Leaves edible raw in salads or cooked as a vegetable.
Waterleaf	*Hydrophyllum* spp.	Young shoots edible raw in salads; shoots and roots cooked as vegetable.
Water plantain	*Alisma* spp.	Bulblike base of the plant dried and then cooked.
Water shield	*Brasenia schreberi*	Tuberlike roots peeled and then dried to be ground for flour or boiled.
Winter cress	*Barbarea* spp.	Leaves and young stem edible as cooked vegetable.
Yarrow	*Achillea lanulosa*	Plant dried and made into broth. (***Caution***: *The closely related and widespread European yarrow—A. millefolium—is somewhat poisonous.*)
Yellow pond lily (*see* **Spatterdock**)		
Yew	*Taxus* spp.	Bright red pulpy part of berries edible. (***Caution***: *Seeds and leaves poisonous.*)

POISONOUS PLANTS

Literally thousands of plants contain varying amounts of poisonous substances. In many instances, the poisons are not present in sufficient quantities to cause adverse effects in humans when only moderate contact or consumption is involved and cooking may destroy or dissipate the substance. Some plants have substances that produce toxic effects in some organisms but not in others. Ordinary onions (*Allium cepa*), for example, occasionally poison horses or cattle yet are widely used for human food, and poison ivy (*Toxicodendron radicans*) or poison oak (*Toxicodendron diversilobum*) produce dermatitis in some individuals but not in others. Table A3.2 and Table A3.3 include plants that are native to, or cultivated in, the United States and Canada.

Table A3.2

PLANTS KNOWN TO HAVE CAUSED HUMAN FATALITIES

PLANT	SCIENTIFIC NAME	POISONOUS PARTS
Angel's trumpet	*Datura suaveolens*	All parts, especially seeds and leaves.
Azalea	*Rhododendron* spp.	Leaves and flowers (however, poisoning is rare).
Baneberry	*Actaea* spp.	Berries and roots.
Belladonna	*Atropa belladonna*	All parts, especially fruits and roots.
Black cherry	*Prunus serotina*	Bark, seeds, leaves. (***Caution***: *Seeds of most cherries, plums, and peaches contain a poisonous principle.*)
Black locust	*Robinia pseudo-acacia*	Seeds, leaves, inner bark.
Black snakeroot	*Zigadenus* spp.	Bulbs.
Buckeye	*Aesculus* spp.	Seeds, leaves, shoots, flowers.
Caladium	*Caladium* spp.	All parts.
Carolina jessamine	*Gelsemium sempervirens*	All parts.
Castor bean	*Ricinus communis*	Seeds.
Chinaberry	*Melia azedarach*	Fruits and leaves.
Daphne	*Daphne mezereum*	All parts.
Death camas (*see* Black snakeroot)		
Dieffenbachia	*Dieffenbachia* spp.	All parts.
Duranta	*Duranta repens*	Berries.
English ivy	*Hedera helix*	Berries and leaves.
False hellebore	*Veratrum* spp.	All parts.
Foxglove	*Digitalis purpurea*	All parts.
Golden chain	*Laburnum anagyroides*	Seeds and flowers.
Jequirity bean	*Abrus precatorius*	Seeds.
Jimson weed	*Datura stramonium* and other *Datura* spp.	All parts, especially seeds.
Lantana	*Lantana camara*	Unripe fruits.
Lily of the valley	*Convallaria majalis*	All parts.
Lobelia	*Lobelia* spp.	All parts.
Mistletoe	*Phoradendron* spp.	Berries.
Monkshood	*Aconitum* spp.	All parts.
Moonseed	*Menispermum canadense*	Fruits.
Mountain laurel	*Kalmia latifolia*	Leaves, shoots, flowers.
Mushrooms	Many genera and species, especially *Amanita* spp.	All parts.
Nightshade	*Solanum* spp.	Unripened fruits. (***Caution***: Common potatoes are Solanum tuberosum; *a poisonous principle is produced in tubers exposed to light.*)
Oleander	*Nerium oleander*	All parts.
Poke	*Phytolacca americana*	Roots and mature stems.
Rhododendron (*see* Azalea)		
Rhubarb	*Rheum rhaponticum*	Leaf blades. (***Caution***: *Although young petioles are widely eaten, dangerous accumulations of a poisonous substance can occur in leaf blades.*)
Rubber vine	*Cryptostegia grandiflora*	All parts.
Sandbox tree	*Hura crepitans*	Milky sap and seeds.
Tansy	*Tanacetum vulgare*	Leaves, flowers.
Tung tree	*Aleurites fordii*	All parts, especially seeds.
Water hemlock	*Cicuta* spp.	Roots.
White snakeroot	*Eupatorium rugosum*	All parts.
Yellow oleander	*Thevetia peruviana*	All parts, especially fruits.
Yew	*Taxus* spp.	All parts except "berry" pulp.

Table A3.3

OTHER PLANTS PRODUCING SIGNIFICANT QUANTITIES OF POISONOUS SUBSTANCES

PLANT	SCIENTIFIC NAME	POISONOUS PARTS
Amaryllis	*Amaryllis* spp.	Bulbs.
Autumn crocus	*Colchicum autumnale*	All parts.
Bittersweet	*Celastrus scandens*	Seeds.
Bleeding hearts	*Dicentra* spp.	All parts.
Bloodroot	*Sanguinaria canadensis*	All parts.
Blue cohosh	*Caulophyllum thalictroides*	Fruits, leaves.
Boxwood	*Buxus sempervirens*	Leaves.
Buckthorn	*Rhamnus* spp.	Fruits.
Bushman's poison	*Acokanthera* spp.	All parts.
Buttercup	*Ranunculus* spp.	All parts; toxicity varies from species to species; mostly cause blistering.
Buttonbush	*Cephalanthus occidentalis*	Leaves.
Chincherinchee	*Ornithogalum thyrsoides*	All parts.
Crown of thorns	*Euphorbia milii*	Milky latex.
Culver's root	*Veronicastrum virginicum*	Root.
Daffodil	*Narcissus* spp.	Bulbs.
Desert marigold	*Baileya radiata*	All parts.
Dutchman's breeches	*Dicentra cucullaria*	All parts.
Fly poison	*Amianthium muscaetoxicum*	Leaves, roots.
Four-o'clock	*Mirabilis jalapa*	Seeds, roots
Gloriosa lily	*Gloriosa* spp.	All parts.
Goldenseal	*Hydrastis canadensis*	Rhizomes, leaves.
Holly	*Ilex aquifolium*	Berries.
Horse chestnut	*Aesculus hippocastanum*	Seeds, flowers, leaves.
Hyacinth	*Hyacinthus* spp.	Bulbs.
Hydrangea	*Hydrangea* spp.	Buds, leaves.
Jack-in-the-pulpit	*Arisaema triphyllum*	Roots, leaves.
Jessamine	*Cestrum* spp.	Leaves, young stems.
Jonquil (*see* Daffodil)		
Karaka nut	*Corynocarpus laevigata*	Seeds.
Kentucky coffee tree	*Gymnocladus dioica*	Fruits.
Larkspur	*Delphinium* spp.	Young plants, seeds.
Lignum vitae	*Guaiacum officinale*	Fruits.
Locoweed	*Astragalus* spp.	Location of poisonous principles varies from species to species; plants more of a problem for livestock than for humans.
Lupine	*Lupinus* spp.	Location of poisonous principles varies from species to species, but primarily in pods and seeds.
Marijuana	*Cannabis sativa*	Resins secreted by glandular hairs among flowers.
Mayapple	*Podophyllum peltatum*	All parts except ripe fruits.
Mescal bean	*Sophora secundiflora*	Seeds.
Milkweed	*Asclepias* spp.	All parts.
Narcissus (*see* Daffodil)		
Ngaio	*Myoporum laetum*	Leaves.
Opium poppy	*Papaver somniferum*	Unripe fruits.
Philodendron	*Philodendron* spp.	Leaves, stems.
Pittosporum	*Pittosporum* spp.	Fruits, leaves, stems.
Poinsettia	*Euphorbia pulcherrima*	Milky latex.
Poison ivy	*Toxicodendron radicans*	Leaves.
Poison oak	*Toxicodendron diversilobum*	Leaves.
Poison sumac	*Toxicodendron vernix*	Leaves.
Prickly poppy	*Argemone* spp.	Seeds, leaves.
Privet	*Ligustrum vulgare*	Fruits.
Sneezeweed	*Helenium* spp.	All parts.
Snow-on-the-mountain	*Euphorbia marginata*	Milky latex.
Squirrel corn	*Dicentra canadensis*	All parts.
Star-of-Bethlehem	*Ornithogalum umbellatum*	All parts.
Sweet pea	*Lathyrus* spp.	Seeds.
Tobacco	*Nicotiana tabacum*	Leaves (when eaten).

MEDICINAL PLANTS

Although a significant number of modern medicines include synthetic substances, many still contain drugs naturally produced by plants, and as recently as 50 years ago the vast majority of medicines used in the treatment of human diseases and ailments were plant-produced. Table A3.4 includes a sampling of plants associated with past and some present medicinal uses. Some of the drugs concerned are prescribed for specific ailments by modern medical practitioners, while others are a part of folk medicine still practiced in rural areas. **Caution:** *Do not use any of the plants listed here for medicinal purposes without consulting a physician.*

Table A3.4

PLANTS ASSOCIATED WITH MEDICINAL USES

PLANT	SCIENTIFIC NAME	USES
Aloe	*Aloe* spp. (especially *Aloe vera*)	Juice from leaves contains chrysophanic acid, which promotes healing of burns.
American mountain ash	*Pyrus americana*	Liquid made from steeping inner bark in water used as astringent; tea of berries used as wash for piles; berries eaten to prevent or cure scurvy.
Anemone	*Anemone canadensis*	Pounded boiled root applied to wounds as antiseptic.
Arnica	*Arnica* spp.	Plants applied as poultice to bruises and sprains.
Astragalus	*Astragalus* spp.	Plant extracts said to boost immune system; also lower blood pressure.
Balm of Gilead	*Populus X gileadensis*	Buds used as ingredient in cough syrups.
Balsam poplar	*Populus balsamifera*	Buds made into ointment, placed in nostrils by Native Americans for relief of congestion.
Bilberry/blueberry	*Vaccinium* spp.	Some evidence that anthocyanosides in fruit when ingested increase oxygen flow to eyes, reducing progression of cataracts, glaucoma, and macular degeneration.
Blackberry	*Rubus* spp.	Tea of roots used by northern California Native Americans to cure dysentery.
Black cohosh	*Cimicifuga racemosa*	Dried roots used in cough medicines and for rheumatism.
Black currant	*Ribes nigrum*	Oil from seeds used to improve suppleness of skin and to reduce skin dryness.
Black haw	*Viburnum prunifolium*	Bark used in treatment of asthma and for relieving menstrual irregularities.
Bloodroot	*Sanguinaria canadensis*	Native Americans used rhizome for ringworm, as insect repellent, and for sore throat.
Blue cohosh	*Caulophyllum thalictroides*	Tea of root drunk by Native Americans and early settlers a week or two before giving birth to promote rapid parturition.
Boneset	*Eupatorium perfoliatum*	Water infusion of dried plant tops widely used to treat fevers and colds.
Borage	*Borago officinalis*	Oil from seeds contains GLA and other oils beneficial in human nutrition.
Broom snakeweed	*Gutierrezia sarothrae*	Navajo Indians applied chewed plant to insect stings and bites of all kinds.
Burdock	*Arctium lappa*	Used as insulin substitute in folklore; root extract used in 17th century for veneral diseases.
Button snakeroot	*Eryngium* spp.	Natchez Indians inserted chewed stem in nostrils to arrest nosebleed.
California bay	*Umbellularia californica*	Yuki Indians put leaves in bath of hot water for relief of rheumatism; leaves used as insect repellent.
Camphor	*Cinnamomum camphora*	Oil from leaves and wood used in cold remedies and liniments.
Camptotheca	*Camptotheca acuminata*	Extracts from flowers and immature fruits yield camptothecin, which has given evidence of being effective against certain forms of cancer.
Cascara	*Rhamnus purshiana*	Bark extract widely used in laxatives.
Catnip	*Nepeta cataria*	Leaf tea used for treatment of colds and to relieve infant colic.
Cat's claw	*Uncaria tomentosa*	Plant extracts used in treatment of intestinal problems, including diverticulosis and Crohn's disease.

Table A3.4

PLANTS ASSOCIATED WITH MEDICINAL USES

PLANT	SCIENTIFIC NAME	USES
Cayenne pepper	*Capsicum frutescens*	Used to reduce mucus drainage (recent evidence suggests it may be carcinogenic if ingested); capsaicin extracts used in ointments to relieve pains of arthritis and neuropathy.
Cherry (wild)	*Prunus serotina*	Tea brewed from bark used for coughs and colds.
Chia	*Salvia columbariae*	Mucilaginous seeds used by Spanish Californians to make refreshing drink. Seeds contain caffeinelike principle that enabled Native Americans to perform unusual feats of endurance; seed paste used in eyes irritated by foreign matter.
Cinchona	*Cinchona* spp.	Bark yields quinine drugs used in treating malaria.
Club moss	*Lycopodium clavatum*	Spores dusted on wounds or inhaled by Native Americans to arrest nosebleeds.
Coca	*Erythroxylon coca*	Cocaine from leaves used as local anesthetic; South American laborers use it as a stimulant.
Coleus	*Coleus forskolii*	Plant extracts used in treatment of hypertension, allergies, glaucoma, and psoriasis.
Cotton	*Gossypium* spp.	Cotton root bark used by black slaves and Native Americans to induce abortions.
Cranberry	*Vaccinium oxycoccum*	Fruit juice drunk to treat female yeast infections.
Creosote bush	*Larrea divaricata*	Decoction from leaves used by Native Americans as a cure-all but especially for respiratory problems.
Cubebs	*Piper cubeba*	Dried fruit best known as a condiment but also used in treatment of asthma.
Damiana	*Turnera diffusa*	Dried leaves used for minor pain and as laxative, stimulant, and flavoring for a liqueur.
Deadly nightshade	*Atropa belladonna*	Belladonna, drug complex extracted from leaves, contains drugs atropine, hyoscyamine, and scopolamine; these used as opium antidote, in shock treatments, and for dilation of pupils. Scopolamine also used as tranquilizer for "twilight sleep" in childbirth.
Dogbane	*Apocynum androsaemifolium*	Roots boiled in water and resulting liquid used as heart medication (contains drug similar in action to digitalis).
Dogwood	*Cornus* spp.	Inner bark boiled in water, and resulting liquid drunk to reduce fevers.
Elderberry	*Sambucus* spp.	Source of *sambucol,* which is reported to have antiviral properties.
Ephedra	*Ephedra* spp.	Drug ephedrine, widely used to relieve nasal congestion and low blood pressure, obtained from stems (most ephedrine now in use is synthetic).
Ergot	Source: *Claviceps purpurea* on cereal grains	Used to treat migraine headaches and to control bleeding after childbirth.
Eyebright	*Euphrasia officinalis*	Once used in treatment of eye diseases.
Evening primrose	*Oenothera* spp.	Seeds are source of GLA oils beneficial in human nutrition.
Feverfew	*Chrysanthemum parthenium*	Dried flowers used in treatment of migraine headaches; to induce abortion and menstruation; and as insecticide.
Flax	*Linum usitatissimum*	Cold-processed seed oils are rich source of GLA, beneficial in human nutrition.
Flowering ash	*Chionanthus virginicus*	Bark used as laxative and in treatment of liver ailments.
Foxglove	*Digitalis purpurea*	Drug digitalis, widely used as a heart stimulant, obtained from leaves.
Garlic	*Allium sativum*	Evidence that *allicin* extracted from bulbs inhibits common cold and other viruses; regular consumption appears to lower risk of stomach cancer.
Gentian	*Gentiana catesbaei*	Catawba Indians applied hot water extract of roots to sore backs; liquid drunk as remedy for stomachaches.
Geranium (wild)	*Geranium maculatum*	Dried roots used for dysentary, diarrhea, and hemorrhoids.
Ginger	*Zingiber officinale*	Said to aid digestion and reduce nausea.
Ginger (wild)	*Asarum* spp.	Extract of rhizome used as a broad-spectrum antibiotic.

Table A3.4

PLANTS ASSOCIATED WITH MEDICINAL USES

PLANT	SCIENTIFIC NAME	USES
Ginkgo	*Ginkgo biloba*	Evidence that concentrated leaf extract improves oxygen-carrying capacity of capillaries, especially those of the brain, and improves memory.
Ginseng	*Panax* spp.	Considered a general panacea, especially in the Orient.
Goldenseal	*Hydrastis canadensis*	Rhizome source of alkaloidal drugs used in treatment of inflamed mucous membranes; also used as tonic.
Goldthread	*Coptis groenlandica*	Native Americans boiled plant and gargled the liquid for sore or ulcerated mouths.
Gotu kola	*Cola nitida*	Seeds contain up to 3.5% caffeine and 1% theobromine which may lessen fatigue; related to kola nuts of cola fame.
Green hellebore	*Helleborus viridis*	Plant extract used to treat hypertension (drug now synthesized); Thompson Indians used in small amounts to treat syphilis.
Green tea	*Camellia sinensis*	Unfermented leaves source of polyphenols, which appear to reduce incidence of cancers in regular users.
Hemlock	*Tsuga* spp.	Native Americans made tea of inner bark to treat colds and fevers.
Horehound	*Marrubium vulgare*	Extract from dried tops of plants used in lozenges for relief of sore throats and colds.
Horsetail	*Equisetum* spp.	Plants boiled in water; liquid used as a delousing shampoo or as a gargle for mouth ulcers.
Indigo (wild)	*Baptisia tinctoria*	Native Americans boiled plant and used liquid as an antiseptic for skin sores.
Ipecac	*Cephaelis ipecacuana*	Drug from roots and rhizome used to treat ameobic dysentery; also used as an emetic.
Jimson weed	*Datura* spp.	Drugs atropine, hyoscyamine, stramonium, and scopolamine obtained from seeds, flowers, and leaves. Stramonium used for knockout drops and in treatment of asthma (*see* Deadly nightshade).
Joe-pye weed	*Eupatorium purpureum*	Dried root said to prevent formation of gallstones.
Joshua tree	*Yucca brevifolia*	Cortisone and estrogenic hormones made from sapogenins produced in the roots.
Juniper	*Juniperus* spp.	Tea of "berries" drunk by Zuni Indian women to relax muscles following childbirth.
Kansas snakeroot	*Echinacea angustifolia*	Dried roots used as antiseptic in treatment of sores and boils, periodontal disease, and sinus drainage problems.
Kava	*Piper methysticum*	Leaf tea used as sedative, muscle relaxant, and pain reliever.
Licorice	*Glycyrrhiza glabra*	Rhizomes source of licorice used in cough drops and for the soothing of inflamed mucous membranes.
Lily of the valley	*Convallaria majalis*	All parts of plant contain a heart stimulant similar to digitalis.
Lobelia	*Lobelia inflata*	Drug lobeline sulphate, obtained from dried leaves, used in antismoking preparations and in treatment of respiratory disorders.
Mandrake	*Mandragora officinarum*	Extracts of plant used in folk medicine as painkiller (drugs hyoscyamine, podophyllin, and mandragorin have been isolated; podophyllin used experimentally in treatment of paralysis).
Manroot	*Marah* spp.	Native Americans used oil from seeds to treat scalp problems and crushed roots for relief from saddle sores.
Marginal fern	*Dryopteris marginalis*	Rhizomes contain oleoresin, used in expulsion of tapeworms from intestinal tract.
Marijuana	*Cannabis sativa*	Tetrahydrocannabinol obtained from resinous hairs in inflorescences; ancient medicinal drug of China.
Mayapple	*Podophyllum peltatum*	Podophyllin, obtained from roots, used experimentally in treatment of paralysis; dried root power used on warts.
Maypop	*Passiflora incarnata*	Dried leaves used as sedative; Native Americans used juice as treatment for sore eyes.
Mesquite	*Prosopis glandulosa*	Native Americans mixed dry leaf powder with water and used liquid to treat sore eyes.

Table A3.4

PLANTS ASSOCIATED WITH MEDICINAL USES

PLANT	SCIENTIFIC NAME	USES
Mexican yam	*Dioscorea floribunda*	Tuberous roots produce up to 10% diosgenin, a precursor of progesterone and cortisone, and are source of DHEA (dihydroepiandrosterone), a complex hormone naturally produced by humans. DHEA levels decline with aging; there is some evidence that controlled DHEA supplementation in older persons retards some aspects of aging.
Milk thistle	*Silybum marianum*	Silymarin extracted from plants has antioxidant properties that appear to be especially beneficial to liver.
Milkweed	*Asclepias syriaca*	Quebec Indians promoted temporary sterility by drinking infusion of pounded roots.
Mistletoe	*Phoradendron flavescens*	Mistletoe.
Monkshood	*Aconitum napellus*	Source of aconite once used in treatment of rheumatism and neuralgia.
Mulberry	*Morus rubra*	Rappahannock Indians applied milky latex of leaf petioles to scalp for ringworm.
Mullein	*Verbascum thapsus*	Native Americans smoked leaves for respiratory ailments and asthma; flowers once widely used in cough medicines.
Onion (wild)	*Allium* spp.	Bulbs eaten to prevent scurvy; Cheyenne Indians applied bulbs in poultice to boils.
Opium poppy	*Papaver sominferum*	Morphine and codeine obtained from latex of immature fruits.
Oregon grape	*Berberis aquifolium*	Bark tea drunk by Native Americans to settle upset stomach; used in strong doses for treatment of venereal diseases.
Pacific yew	*Taxus brevifolia*	*Taxol*, promising anticancer agent, extracted from bark.
Pansy (wild)	*Viola* spp.	Plants ground up and applied to skin sores.
Pennyroyal	*Mentha pulegium*	Native Americans used leaf tea in small amounts for relief of headaches and flatulence and to repel chiggers. (Toxic in larger amounts.)
Persimmon	*Diospyros virginiana*	Liquid from boiled fruit used as astringent. Fruits have high beta carotene content; leaves said to be high in vitamin C.
Peyote	*Lophophora williamsii*	Alcoholic extract of plant used as antibiotic.
Pine	*Pinus* spp.	Pycnogenols extracted from bark have powerful antioxidant properties.
Pinkroot	*Spigelia marilandica*	Powdery root very effective in expulsion of roundworms from intestinal tract.
Pipssisewa	*Chimaphila umbellata*	Native Americans steeped plant in water and used liquid to draw out blisters.
Pitcher plant	*Sarracenia purpurea*	Native Americans used root widely as smallpox cure (records indicate it was effective).
Plantain	*Plantago ovata* and other spp.	Seed husks (known as *psyllium*) widely used in bulking laxatives.
Pleurisy root	*Asclepias tuberosa*	Liquid from roots boiled in water used in treatment of respiratory problems.
Prickly ash	*Zanthoxylum americanum*	Bark and berries widely used by Native Americans for toothache (pieces inserted in cavities); liquid infusion drunk for venereal diseases.
Purple coneflower	*Echinacea purpurea*	Plant extracts used to boost the immune system.
Quassia	*Picraea excelsa; Quassia amara*	Wood extract used as pinworm remedy and as insecticide.
Rauwolfia	*Rauwolfia serpentina*	Reserpine obtained from roots; drug used in treatment of mental illness and in counteracting effects of LSD.
Saffron (meadow)	*Colchicum autumnale*	Drug colchicine from corms used in past for treatment of gout and back disc problems but now used mostly for experimental doubling of chromosome numbers in plants.
Sarsaparilla	*Aralia nudicaulis*	Cough medicines made from roots.
Sassafras	*Sassafras albidum*	Tea of root bark used to induce sweating; used externally as liniment.
Self-heal	*Prunella vulgaris*	Native Americans applied plants in poultices to boils.
Seneca snakeroot	*Polygala senega*	Bark boiled in water and then liquid applied to snakebites. Taken internally as abortifacient; used in cough remedy.

Table A3.4

PLANTS ASSOCIATED WITH MEDICINAL USES

PLANT	SCIENTIFIC NAME	USES
Senna	*Cassia senna* and other spp.	Leaf extract used as purgative.
Skeleton weed	*Lygodesmia juncea*	Widely used by Native American women to increase milk flow.
Skullcap	*Scutellaria laterifolia*	Dried plant used as anticonvulsive in treatment of epilepsy and as sedative.
Slippery elm	*Ulmus fulva*	Dried inner bark, which contains an aspirinlike substance, used to soothe inflamed membranes.
Spicebush	*Lindera benzoin*	Berries, buds, and bark brewed for tea used to reduce fevers.
Spruce	*Picea* spp.	Cree Indians ate small immature female cones for treatment of sore throat.
Squills	*Urginea maritima*	Bulbs of red variety are source of heart stimulant; bulbs of white variety are widely used as rodent killer.
Stoneseed	*Lithospermum ruderale*	Shoshoni women reported to have drunk cold water infusion of roots daily for six months to ensure permanent sterility (experiments with mice suggest substance to the reports).
Strophanthus	*Strophanthus* spp.	Seeds are major source of cortisone and also source of heart stimulant.
Strychnine plant	*Strychnos nox-vomica*	Strychnine extracted from seeds widely used as insect and animal poison; principal ingredient in blowgun darts used by South American aborigines. Minute amounts stimulate the central nervous systems and relieve paralysis.
Sumac	*Rhus* spp. (especially *R. glabra*)	Native Americans applied leaf decoction as remedy for frostbite; fruits and liquid made from leaves applied to poison ivy rash and gonorrhea sores. Root chewed for treatment of mouth ulcers.
Sweet flag	*Acorus calamus*	Boiled root applied to burns; root chewed for relief of colds and toothache.
Sweet gum	*Liquidambar styraciflua*	Bud balsam used to treat chigger bites; balsam also used in insect fumigating powders.
Sword fern	*Polystichum munitum*	Boiled rhizome used by Native Americans to treat dandruff; sporangia applied to burns.
Tamarind	*Tamarindus indica*	Fruit pulp used as laxative.
Turmeric	*Curcuma longa*	Rhizome extracts used to lower cholesterol levels and to prevent blood clots.
Valerian	*Valeriana septentrionalis*	Pulverized plant applied to wounds.
Velvet bean	*Mucuna* spp.	Seeds contain L-dopa used in treatment of Parkinson's disease.
Virginia snakeroot	*Aristolochia serpentaria*	Native Americans used tea of plant for reducing high fevers.
Wahoo	*Euonymus atropurpureus*	Bark steeped in water; liquid has digitalislike effect on heart.
Watercress	*Rorippa nasturtium-aquaticum*	Some evidence that daily consumption reduces development of lung cancer in smokers.
Western wallflower	*Erysimum capitatum*	Zuni Indians ground plant with water and applied it to skin to prevent sunburn.
Willow	*Salix* spp.	Chickasaw Indians snuffed infusion of roots as remedy for nosebleed; Pomo Indians boiled bark in water and applied liquid for relief of skin itches. Fresh inner bark contains salicin, an aspirinlike compound used to reduce fevers.
Wintergreen	*Gaultheria procumbens*	Oil from leaves used as folk remedy for body aches and pains.
Witch hazel	*Hamamelis virginiana*	Oil distilled from twigs used as external medicine.
Wormseed	*Chenopodium ambrosioides*	Oil from seeds used to expel intestinal worms.
Wormwood	*Artemisia* spp.	Yokia Indians made tea from leaves to treat bronchitis; other Native Americans used tea as cold remedy.
Yarrow	*Achillea millefolium*	Native Americans used plant infusion for treating wounds, earaches, and burns.
Yellow lady's slipper	*Cypripedium calceolus*	Dried root used for relief of insomina and as sedative.
Yellow nut grass	*Cyperus esculentus*	Paiute Indians pounded tubers with tobacco leaves and applied mass in wet dressing for treatment of athlete's foot.
Yerba santa	*Eriodictyon californicum*	Native Americans smoked leaves or drank leaf tea for treatment of colds or asthma.

HALLUCINOGENIC PLANTS

Although a few hallucinogenic substances produced by animals have been isolated and some have been synthesized, the majority of known hallucinogens are produced by plants. Table A3.5 is not a complete list, but it includes the better-known sources. Refer to *Additional Reading* for further information.

SPICE PLANTS

The word *spice* describes any aromatic plant or part of a plant used to flavor or season food; spices are also used to add scent or flavor to manufactured products (Table A3.6). Although spices have no nutritional value, they add a pleasurable zest to meals, and before food preservation was possible, they helped make unappealing but still edible food palatable.

The value placed on spice plants was responsible for changing the course of Western civilization as a principal motive behind the voyages of discovery.

Table A3.5

HALLUCINOGENIC SUBSTANCES PRODUCED BY PLANTS

PLANT	SCIENTIFIC NAME	PART USED	PRINCIPAL ACTIVE SUBSTANCE
Ajuca	*Mimosa hostilis*	Roots	Nigerine
Belladonna	*Atropa belladonna*	Leaves	Hyoscyamine, scopolamine
Caapi	*Banisteriopsis caapi*	Wood	Harmine
Canary broom	*Cytisus canariensis*	Seeds	Cytisine
Catnip	*Nepeta cataria*	Leaves	Unknown
Cohoba	*Piptadena peregrina*	Seeds (snuff)	Tryptamines
Coral bean	*Erythrina* spp.	Seeds	Unknown
Cubbra borrachera	*Methysticodendron amnesianum*	Leaves	Scopolamine
Ergot fungus	*Claviceps purpurea*	Rhizomorph	Ergine (LSD)
Fly agaric	*Amanita muscaria*	Mushroom cap	Ibotenic acid, muscimol
Henbane	*Hyoscyamus* spp.	Leaves	Hyoscyamine, scopolamine
Iboga	*Tabernanthe iboga*	Root bark	Ibogaine
Jimson weed	*Datura* spp.	All parts	Scopolamine
Kava kava	*Piper methysticum*	Root (large amounts of beverage produce hallucinations)	Myristicinlike compound
Mace	*Myristica fragrans*	Aril of seed	Myristicin
Mescal bean	*Sophora secundiflora*	Seeds	Cytisine
Morning glory	*Ipomoea violacea*	Seeds	Ergine
Nutmeg	*Myristica fragrans*	Seeds	Myristicin
Ololiuqui	*Rivea corymbosa*	Seeds	Tubicoryn
Peyote	*Lophophora williamsii*	Stems	Mescaline
Psilocybe mushrooms	*Psilocybe* spp, *Concocybe* spp., *Panaeolus* spp., and others	All parts	Psilocybin, psilocin
Rape dos Indios	*Maquira sclerophylla*	Dried plant (snuff)	Unknown
San Pedro	*Trichocereus pachanoi*	Stems	Mescaline
Sassafras	*Sassafras albidum*	Root bark (large amounts of tea)	Safrole
Sweet flag	*Acorus calamus*	Dried root	Asarone, β-asarone
Syrian rue	*Peganum harmala*	Seeds	Harmine
Vygie	*Mesembryanthemum expansum*	All parts	Mesebrine
Wood rose	*Argyreia nervosa*	Seeds	Ergoline alkaloids
Yakee (Parica)	*Virola* spp.	Resin from inner surface of freshly removed bark (snuff)	Tryptamine
Yohimbehe	*Corynanthe* spp.	Bark	Yohimbine

Table A3.6

PLANTS USED TO SEASON OR FLAVOR

SPICE	SCIENTIFIC NAME OF PLANT	PARTS USED; REMARKS	PRINCIPAL SOURCE
Allspice	*Pimenta dioica*	Powdered dried fruit.	Jamaica
Almond	*Prunus amygdalus*	Oil from seed used for flavoring baked goods.	Mediterranean; U.S.
Angelica	*Angelica archangelica*	Stems candied; oil from seeds and roots used in liqueurs.	Europe; Asia
Anise	*Pimpinella anisum*	Oil distilled from fruits used for flavoring.	Widely cultivated
Arrowroot	*Maranta arundinacea*	Powdered root used in milk puddings, baked goods.	South America
Asafoetida	*Ferula asafoetida*	Powdered gum from stems and roots used in minute quantities with fish.	Middle East
Balm (Melissa)	*Melissa officinalis*	Oil from leaves used in beverages; leaves used as food flavoring.	U.S.; Mediterranean
Basil	*Ocimum basilicum*	Leaves used in meat dishes, soups, sauces.	Mediterranean
Bay	*Laurus nobilis*	Leaves used in soups, sauces.	Europe
Bell pepper	*Capsicum frutescens*	Dried diced fruit used in chip dips, salad dressings.	Widely cultivated
Bergamot	*Monarda didyma*	Leaves used with pork. (*Note*: A perfume oil obtained from a variety of orange—*Citrus aurantium* var. *bergamia*—is also called *bergamot*.)	North America (*Monarda*); Italy (*Citrus*)
Black pepper	*Piper nigrum*	Dried fruits used as condiment.	India; Indonesia
Borage	*Borago officinalis*	Leaves used as beverage flavoring.	England
Burnet	*Sanguisorba minor*	Used in soups and casseroles.	Eurasia
Calamus	*Acorus calamus*	Powdered rhizome used for flavoring.	Europe; Asia; North America
Capers	*Capparis spinosa*	Flower buds used for flavoring relishes, pickles, sauces.	Mediterranean
Caraway	*Carum carvi*	Seeds used in breads, cheeses; seed oil used in the liqueur kümmel.	North America; Europe
Cardamom	*Elletaria cardamomum*	Dried fruit and seeds used for flavoring baked goods. (*Note*: Several false cardamoms—*Amomum* spp.—are sold commercially.)	India; Sri Lanka; Central America
Cassia	*Cinnamomum cassia*	Powdered bark used as cinnamon substitute.	Southeast Asia
Cayenne pepper	*Capsicum* spp.	Powdered dried fruits used in chili powder, Tabasco sauce.	American tropics
Celery	*Apium graveolens*	Seeds used in celery salt, soups.	Europe; U.S.
Chervil	*Anthriscus cerefolium*	Used as parsley substitute.	Europe; Near East
Chives	*Allium schoenoprasum*	Leaves and bulbs used with sour cream, butter.	Widely cultivated
Chocolate	*Theobroma cacao*	Ground seeds used for flavoring.	Africa; South America
Cilantro	*Coriandrum sativum*	Leaves used in avocado dip and with poultry.	Europe
Cinnamon	*Cinnamomum zeylanicum*	Ground bark used for flavoring baked goods; oil from leaves used as flavoring, clearing agent.	Seychelles; Sri Lanka
Citrus	*Citrus* spp.	Fruits, especially rinds, source of flavoring oil.	Mediterranean; South Africa; U.S.
Cloves	*Syzgium aromaticum*	Dried flower buds used as flavoring.	Zanzibar

Table A3.6

PLANTS USED TO SEASON OR FLAVOR

SPICE	SCIENTIFIC NAME OF PLANT	PARTS USED; REMARKS	PRINCIPAL SOURCE
Coffee	*Coffea arabica*	Roasted seeds source of mocha-coffee flavoring.	Tropics
Coriander	*Coriandrum sativum*	Ground seed used in German frankfurters, curry powders.	Mediterranean
Cubebs	*Piper cubeba*	Dried fruits used as seasoning.	East Indies
Cumin	*Cuminum cyminum*	Ground seeds used with meats, pickles, cheeses, curry.	Mediterranean
Curry		Spicy condiment containing several ingredients, such as turmeric, cumin, fenugreek, and zedoary.	India
Dill	*Anethum graveolens*	Seeds used in pickling brines; leaves used for seasoning meat loaves, sauces.	Europe; Asia
Dittany	*Origanum dictamnus*	Leaves used as seasoning for poultry, meats.	Crete
Eucalyptus	*Eucalyptus* spp.	Oil from leaves used in toothpastes, flavoring agents.	Australia
Fennel	*Foeniculum vulgare*	Seeds used in baked goods.	Europe
Fenugreek	*Trigonella foenumgraecum*	Oil distilled from seeds used in pickle, chutney, curry powders, imitation maple flavoring.	Widely cultivated
Filé (*see* Sassafras)			
Garlic	*Allium sativum*	Fresh or dry bulbs used for meat seasonings.	Widely cultivated
Ginger	*Zingiber officinale* and others	Dried rhizomes used for flavoring many foods and drinks.	India; Taiwan
Grains of paradise	*Afromomum melegueta*	Seeds used to flavor beverages and medicines.	West Africa
Hops	*Humulus lupulus*	Dried inflorescences of female plants used in brewing beer.	Europe; North America
Horseradish	*Rorippa armoracia*	Grated fresh root used as condiment.	Europe; North America
Juniper	*Juniperus* spp.	"Berries" used to season beef roasts, poultry, sauces.	North America
Licorice	*Glycyrrhiza glabra*	Dried rhizome and root used to flavor pontefract cakes, candies.	Middle East
Lovage	*Ligusticum scoticum*	Stems candied; seeds used in pickling sauces; celery substitute.	Europe
Mace	*Myristica fragrans*	Aril of seed used for flavoring beverages, foods.	Grenada; Indonesia; Sri Lanka
Marigold	*Tagetes* spp.	Petals substituted for saffron in rice dishes, stews.	Widely cultivated
Marjoram	*Marjorana hortensis*	Leaves used in stews, dressings, sauces.	Mediterranean
Mustard	*Brassica* spp.	Ground seeds used in meat condiment.	Europe; China
Nasturtium	*Tropaeolum majus*	Flowers, seeds, leaves used in salads.	Widely cultivated
Nutmeg	*Myristica fragrans*	Seeds used for flavoring foods, beverages.	Grenada; Indonesia; Sri Lanka
Oregano	*Origanum vulgare* and others	Leaves used as seasoning with poultry, meats.	Europe
Paprika (*see* Cayenne pepper)			
Parsley	*Petroselinum sativum*	Leaves used as meat garnish and flavoring in sauces.	Widely cultivated
Pepper, black	*Piper nigrum*	Dried drupes ground for condiment.	Madagascar; India

Table A3.6

PLANTS USED TO SEASON OR FLAVOR

SPICE	SCIENTIFIC NAME OF PLANT	PARTS USED; REMARKS	PRINCIPAL SOURCE
Peppermint	*Mentha piperita*	Oil from leaves used for food, drink, dentifrice flavoring (much commercial menthol is derived from *Mentha arvensis* grown in Japan).	U.S.; countries of the former Soviet Union
Pimiento	*Capsicum* spp.	Bright red fruits of a cultivated variety of pepper used in stuffing olives and in cold meats, cheeses.	Central and South America
Poppy	*Papaver somniferum*	Seeds used in baking.	Widely cultivated
Rosemary	*Rosmarinus officinalis*	Oil from leaves used in perfumes, soaps.	Mediterranean
Rue	*Ruta graveolens*	Flavoring for fruit cups, salads.	Europe
Saffron	*Crocus sativus*	Dried stigmas used to flavor oriental-style dishes.	Spain; India
Sage	*Salvia officinalis*	Leaves used in poultry and meat dressings.	Yugoslavia
Sarsaparilla	*Smilax* spp.	Roots source of flavoring for beverages, medicines.	American Tropics
Sassafras	*Sassafras albidum*	Bark and wood yield flavoring for beverages, toothpaste, gumbo.	U.S.
Savory (summer)	*Satureia hortensis*	Leaves used in green bean and bean salads, in lentil soup, with fish.	Mediterranean
Savory (winter)	*Satureia montana*	Leaves used as seasoning in stuffings, meat loaf, stews.	Europe
Scallion	*Allium fistulosum*	Leaves used in wine cookery, soups.	Widely cultivated
Sesame	*Sesamum indicum*	Seeds used in baking.	Asia
Shallot	*Allium ascalonicum*	Bulbs and leaves used in Colbert butter, wine cookery.	Widely cultivated
Southernwood	*Artemisia abrotanum*	Leaves used to flavor cakes.	Europe
Star anise	*Illicium verum*	Fruits used in candy and cough drops.	China
Stonecrop	*Sedum acre*	Dried leaves (ground) used as pepper substitute.	Europe
Tarragon	*Artemisia dracunculus*	Leaves and flowering tops used in pickling sauces.	Europe
Thyme	*Thymus vulgaris*	Leaves used in meat and poultry dishes, soups, sauces.	Widely cultivated
Tonka bean	*Dipteryx* spp.	Seeds source of flavoring for tobacco; vanilla substitute (now largely synthesized).	American tropics
Turmeric	*Curcuma longa*	Rhizomes powdered and used in curry powders, meat flavoring.	India; China
Vanilla	*Vanilla planifolia*	Flavoring extracted from fruits; used in foods, drinks.	Malagasay Republic
Wintergreen	*Gaultheria procumbens*	Oil from leaves and bark used as flavoring for confections, toothpaste.	U.S.
Zedoary	*Curcuma zedoaria*	Dried rhizome used in liqueurs, curry powders.	India

DYE PLANTS

In the recent and ancient past, dyes from many different plants were used to color cotton, linen, and other fabrics. Since the middle of the 19th century, however, natural dyes have been almost completely replaced in industry by synthetic dyes, and today the use of natural dyes is largely confined to the work of individual hobbyists.

Any reader interested in experimenting with natural dyes is encouraged not only to choose those plant materials included in Table A3.7 but to try any local plants available. The experimenter will soon find that quite unexpected colors may be derived from plants, as the colors of fresh flowers, bark, or leaves often bear little relationship to the colors of the dyes. For methods of dyeing, see the footnote given under the heading of Lichens in Chapter 19 and references in *Additional Reading*.

Table A3.7

PLANT SOURCES OF NATURAL DYES

PLANT OR DYE	SCIENTIFIC NAME OF PLANT SOURCE	REMARKS
Acacia	*Acacia* spp.	Brown dyes from bark and fruits.
Alder	*Alnus* spp.	Brownish dyes from bark.
Alkanet	*Alkanna tinctoria*	Red dye from roots.
Annatto	*Bixa orellana*	Yellow or red dye from pulp surrounding seeds.
Bamboo	*Bambusa* spp.	Light green dye from leaves.
Barberry	*Berberis vulgaris*	Grayish dye from leaves.
Barwood	*Baphia nitida*	Purplish dyes from wood.
Bearberry	*Arctostaphylos uva-ursi*	Yellowish dye from leaves.
Bedstraw	*Galium* spp.	Light reddish brown dyes from roots.
Birch	*Betula* spp.	Light brown to black dyes from bark.
Black cherry	*Prunus serotina*	Red dye from bark; gray to green dyes from leaves.
Black walnut	*Juglans nigra*	Rich brown dye from bark; brown dye from walnut hulls.
Bloodroot	*Sanguinaria canadensis*	Red dye from rhizomes.
Blueberry	*Vaccinium* spp.	Blue to gray dye from mature fruits (tends to fade).
Bougainvillea	*Bougainvillea* spp.	Light brownish dyes from floral bracts.
Brazilwood	*Caesalpinia* spp.	Reddish dyes from wood.
Buckthorn	*Rhamnus* spp.	Green dyes from fruits.
Buckwheat	*Fagopyrum esculentum*	Blue dye from stems.
Buckwheat (wild)	*Eriogonum* spp.	Dark gold, pale yellow, and beige dyes from stems and flowers.
Buffaloberry	*Shepherdia argentea*	Red dye from fruit.
Butternut	*Juglans cinerea*	Yellow to grayish brown dyes from fruit hulls.
Cocklebur	*Xanthium strumarium*	Dark green dye from stems and leaves.
Coffee	*Coffee arabica*	Light brown dye from ground, roasted seeds.
Cudbear (Archil)	*Rocella* spp. (lichen)	Red dye obtained by fermentation of thallus.
Cutch	*Acacia* spp.; *Uncaria gambir*	Brown to drab green dyes from stem gums.
Dock	*Rumex* spp.	Light brown dyes from stems and leaves.
Dogwood	*Cornus florida*	Red dye from bark; purplish dye from root.
Doveweed	*Eremocarpus setigerus*	Light to olive green dye from entire plant.
Dyer's rocket	*Reseda luteola*	Orangish dye from all parts.
Elderberry	*Sambucus* spp.	Blackish dye from bark; purple, blue, or dark brown dyes from fruits.
Eucalyptus	*Eucalyptus* spp.	Beige dyes from bark.
Fennel	*Foeniculum vulgare*	Yellow dyes from shoots.
Fig	*Ficus carica*	Green dyes from leaves and fruits.
Fustic	*Chlorophora tinctoria*	Yellow, bright orange, and greenish dyes from heartwood.
Gamboge	*Garcinia* spp.	Yellow dye from resins that ooze from cuts made on stems.
Giant reed	*Arundo donax*	Pale yellow dye from leaves.
Grape	*Vitis* spp.	Bright yellow to olive green dyes from leaves.
Hawthorn	*Crataegus* spp.	Pink dye from ripe fruits.
Hemlock	*Tsuga* spp.	Reddish brown dye from bark.
Henna	*Lawsonia inermis*	Orange dye from shoots and leaves.
Hickory	*Carya tomentosa*	Yellow dye from bark.
Hollyhock	*Althaea rosea*	Purplish black dye from flower petals.
Horsetail	*Equisetum* spp.	Tan dyes from all green parts.
Indigo	*Indigofera tinctoria*	Bright blue dyes from leaves.
Kendall green (*see* Woadwaxen)		
Larkspur	*Delphinium* spp.	Blue dyes from petals.
Lichens	Many genera and species	Many lichens yield brilliant shades of yellows, golds, and browns with various mordants.
Litmus	*Rocella tinctoria*	Famous pink-to-blue pH indicator dye from thallus.
Logwood	*Haematoxylon campechianum*	Dark blue purple dye from heartwood.
Lokao	*Rhamnus* spp.	Green dye from wood.

Table A3.7

PLANT SOURCES OF NATURAL DYES

PLANT OR DYE	SCIENTIFIC NAME OF PLANT SOURCE	REMARKS
Lupine	*Lupinus* spp.	Greenish dyes from flowers.
Madder	*Rubia tinctorium*	Bright red dye from roots.
Madrone	*Arbutus menziesii*	Brown dye from bark.
Manzanita	*Arctostaphylos* spp.	Beige to dull yellow dyes from dried fruits.
Maple	*Acer* spp.	Pink dye from bark.
Marsh marigold	*Caltha palustris*	Yellow dye from petals.
Milkweed	*Asclepias speciosa*	Pale yellow dyes from leaves.
Morning glory	*Ipomoea violacea*	Gray green dye from blue flowers.
Mullein	*Verbascum thapsus*	Gold dyes from leaves.
Oak	*Quercus* spp.	Yellow dye from bark.
Onion	*Allium cepa*	Reddish brown dyes from dry outer bulb scales of red onions; yellow dyes from similar parts of yellow onions.
Oregon grape	*Berberis aquifolium*	Yellow dyes from roots.
Osage orange	*Maclura pomifera*	Yellow, gray, and green dyes from fruits; yellow orange dye from wood.
Peach	*Prunus persica*	Green dyes from leaves.
Poke	*Phytolacca americana*	Red dyes from mature fruits.
Pomegranate	*Punica granatum*	Dark gold dye from fruit rinds.
Prickly lettuce	*Lactuca serriola*	Green dye from leaves.
Privet	*Ligustrum vulgare*	Yellow green dye from leaves; deep gray dye from berries.
Quercitron	*Quercus velutina*	Bright yellow dye from bark.
Rhododendron	*Rhododendron* spp.	Tan dyes from leaves.
Safflower	*Carthamus tinctorius*	Reddish dye from flower heads.
Saffron	*Crocus sativus*	Powerful yellow dye from stigmas.
Sage	*Salvia officinalis*	Yellow dye from shoots.
Sandalwood	*Pterocarpus santalinus*	Red dye from wood.
Sappanwood	*Caesalpinia sappan*	Red dye from heartwood.
Sassafras	*Sassafras albidum*	Orange brown dye from bark.
Scotch broom	*Cytisus scoparius*	Yellow dye from all parts of plant.
Smoke tree	*Cotinus coggyria*	Orange yellow dye from wood (dye sometimes called *young fustic*).
Smooth sumac	*Rhus glabra*	Grayish brown dye from bark.
St. John's wort	*Hypericum* spp.	Light brownish dyes from leaves.
Tansy	*Tanacetum* spp.	Yellow, green dyes from leaves.
Toyon	*Heteromeles arbutifolia*	Reddish brown dyes from leaves.
Turmeric	*Curcuma longa*	Orangish dye from rhizome.
Woad	*Isatis tinctoria*	Blue dye from leaves.
Woadwaxen	*Genista tinctoria*	Yellow dye from all parts.
Yerba santa	*Eriodictyon californicum*	Rich dark brown dyes from leaves.

ADDITIONAL READING

Adrosko, R. J. 1971. *Natural dyes and home dyeing.* New York: Dover Publications.

Bliss, A. 1986. *North American dye plants,* rev. ed. New York: Boulder, Co: Juniper House.

Furst, P. E. 1992. *Mushrooms: Psychedelic fungi,* rev. ed. Edgemont, PA: Chelsea House Publications.

Graedon, J., and T. Graedon. 1995. *The people's guide to deadly drug interactions.* New York: St. Martin's Press.

Lewis, W. H., and M. P. F. Elvin-Lewis. 1977. *Medical botany.* New York: John Wiley & Sons, Inc.

Loewenfeld, C. 1989. *Herb gardening: Why and how to grow herbs.* Winchester, MA: Faber and Faber.

Merory, J. 1968. *Food flavorings: Composition, manufacture and use,* 2d ed. New York: Chemical Publishing Co.

Schultes, R. E., and A. Hoffman. 1980. *The botany and chemistry of hallucinogens,* 2d ed. Springfield, IL: Charles C. Thomas, Publishers.

Spoerke, D. G., Jr. 1992. *Herbal medications,* 2d ed. Santa Barbara, CA: Woodbridge Press Publishing Company.

Spoerke, D. G., Jr., and S. Smolinske (Eds.). 1990. *Toxicity of houseplants.* Boca Raton, FL: CRC Press.

Vogel, V. J. 1990. *American Indian medicine.* Norman, OK: University of Oklahoma Press.

See also the *Additional Reading* entries in Chapter 24.

Appendix 4

Houseplants and Home Gardening

GROWING HOUSEPLANTS

If sales volume is an indication, houseplants have never been more popular in the United States than they are now. Many are easy to grow and will brighten windowsills, planters, and other indoor spots for years if a few simple steps are followed to ensure their health and vigor.

Water

Houseplants are commonly overwatered, resulting in the unnecessary development of rots and diseases (see Table A4.1). As a rule, the surface of the potting medium should be dry to the touch before watering, but the medium should not be allowed to dry out completely unless the plant is dormant. Care should be taken, particularly during the winter, that the water is at room temperature. Rainwater, if available, is preferred to tap water, particularly if the water has a high mineral content or is chlorinated. Broad-leaved plants should periodically have house dust removed with a damp sponge (never use detergents to clean surfaces—they remove protective waxes). Many plants benefit from a daily misting with water, particularly in heated rooms.

Containers

In time, plants may develop too extensive a root system for the pots in which they are growing (becoming what is commonly called *root bound*). Nutrients in the potting medium may become exhausted, salts and other residues from fertilizers and water may build up to the point of inhibiting growth, or the plants themselves may produce substances that accumulate until they interfere with the plant's growth. To resolve these problems, periodically repot the plants and, if necessary, divide them at the time of repotting.

Temperatures

Most houseplants don't thrive where the temperatures are either too high or too low (see Table A4.1). In general, they tend to do best with minimum temperatures of about 13°C (55°F) and maximum temperatures of about 29°C (84°F). Many houseplants that prefer

Table A4.1

COMMON AILMENTS OF HOUSEPLANTS

PROBLEM SYMPTOMS	POSSIBLE CAUSES
Wilting or collapse of whole plant	Lack of water; too much heat; too much water or poor drainage resulting in root rot.
Yellowish or pale leaves	Insufficient light; too much light; microscopic pests (especially spider mites); too much or too little fertilizer.
Brown, dry leaves	Humidity too low; too much heat; poor air circulation; lack of water.
Tips and margins of leaves brown	Mineral content of water too high; drafts; too much sun or heat; too much or too little water.
Ringed spots on leaves	Water too cold.
Leaves falling off	Improper watering or water too cold; excessive use of fertilizer or wrong fertilizer; too much sun or, if lower leaves drop only, too little light.
Stringy growth	Insufficient light; too much fertilizer.
Base of plant soft or rotting	Overwatering.
No flowers or flower buds drop	Too much or too little light; night temperatures too high.
Water does not drain	Drain hole plugged; potting mixture has too much clay.
Mildew present	Fungi present—arrest with sulphur dust.

COMMON PESTS	CONTROLS*
Aphids	Wash off under faucet or spray with soapy water (not detergent); pyrethrum or rotenone sprays also effective.
Mealybugs	Remove with cotton swabs dipped in alcohol; spray with Volck oil.
Scale insects	Remove by hand; spray with Volck oil.
Spider mites	Use sprays containing small amounts of xylene (as soon as possible—spider mites multiply very rapidly).
Thrips	Spray with pyrethrum/rotenone or Volck oil sprays.
White flies	Spray with soapy water or pyrethrum/rotenone sprays every 4 days for 2 weeks (only the adults are susceptible to the sprays).

*For additional controls, see the biological controls in Appendix 2.

warmer temperatures while actively growing also benefit from a rest period at lower temperatures after flowering.

Light

Next to overwatering, insufficient light is the most common contributor to the decline or death of houseplants (see Table A4.1). This does not mean that houseplants do better in direct sunlight—such light frequently damages them—but filtered sunlight (as, for example, through a muslin curtain) is usually better for the plant than the light available in the middle of the room. Plants can also thrive in artificial light of appropriate quality. Ordinary incandescent bulbs have too little light of blue wavelengths, and ordinary fluorescent tubes emit too little red light. A combination of the two, however, works very well. Generally, the wattage of the incandescent bulbs should be only one-fourth that of the fluorescent tubes in such a combination. Several types of fluorescent tubes specially balanced to imitate sunlight are also available.

Humidity

Dry air is hard on most houseplants. The level of humidity around the plants can be raised by standing the pots in dishes containing gravel or crushed rock to which water has been added. The humidity level can also be raised through the use of humidifiers, which come in a variety of sizes and capacities. Daily misting, as mentioned previously, can also help.

COMMON HOUSEPLANTS

Here is an explanation of the symbols given with each plant:

Water

 = Give little water (applies primarily to cacti and succulents; these plants store water in such a way that the soil can be completely dry for a week or two without their being adversely affected).

 = Water regularly but not excessively; wait until the potting medium surface is dry to the touch before watering.

 = Immerse pot in water for a few minutes each week and water frequently, never allowing the potting medium to become dry; do not, however, leave the base of the pot standing in water.

 = Give little to regular water.

= Give regular to frequent water.

Minerals in hard water are taxing on houseplants, and commercial water softeners do not improve water for the plants. Use rainwater or filtered water if at all possible; otherwise, repot more often.

Temperature

= cool; maximum 13°C–16°C (55°F–61°F); minimum 5°C–7°C (41°F–45°F)

= cool to medium; maximum 18°C–21°C (65°F–70°F); minimum 10°C–13°C (50°F–55°F)

= medium; maximum 30°C (86°F)

= medium to warm

= warm

Many houseplants are native to the tropics, where they thrive under year-round warm temperatures, while cacti and succulents prefer cool winters. The closer a houseplant's environment is to its natural environment, the better the plant will grow (see Table A4.2).

Light

= Needs shading or indirect daylight.

= Needs bright light but needs to be screened from direct sunlight.

= Needs direct sunlight.

= Needs shading from bright light.

= Needs bright to direct light.

As mentioned previously, improper lighting is second only to over-watering as a cause of problems for houseplants; generally, they are given too little light and, occasionally, too much. A south-facing windowsill may be ideal for certain plants in midwinter but excessively bright in midsummer; conversely, a north-facing windowsill may have enough light for certain plants in midsummer but not in midwinter. Adjustable screens permit manipulation of daylight to suit the plants involved.

Humidity

= Will tolerate dry air.

= Needs dry to regular air.

= Will tolerate the air in most houses provided it is mist-sprayed occasionally.

= Needs regular to humid air.

= Needs high humidity; use a humidifier if possible.

Virtually all plants with or symbols benefit from having a pan of gravel with water beneath the pot.

Potting Medium

= Requires a porous, slightly acid medium that drains immediately.

= Requires a loam that is slightly alkaline (e.g., a mixture of sand and standard commercial potting medium).

= Requires a peaty potting mixture and acid fertilizer.

Table A4.2

ENVIRONMENTS SUITABLE FOR COMMON HOUSEPLANTS

PLANT	SCIENTIFIC NAME	ENVIRONMENTAL REQUIREMENTS	REMARKS
Aechmea	*Aechmea fasciata*		*See* Bromeliad; produces side shoots that should be propagated as main plant dies after flowering.
African lily	*Agapanthus* spp.		Do not repot until pot is full; keep cool in winter.
African violet	*Saintpaulia* spp.		Let rest under cooler conditions after flowering; dislikes cold water.
Agave	*Agave* spp.		Keep cool and dry in winter.
Algerian ivy	*Hedera canariensis*		Resembles variegated English ivy.
Aloe	*Aloe* spp.		Keep cool and dry in winter.

Table A4.2

ENVIRONMENTS SUITABLE FOR COMMON HOUSEPLANTS

PLANT	SCIENTIFIC NAME	ENVIRONMENTAL REQUIREMENTS	REMARKS
Aluminum plant	*Pilea cadierei*		Plants do not usually survive long in houses.
Amaryllis	*Amaryllis* spp.		Let leaves die back in fall; put bulb in cool, dark place until early spring; then repot, water, and fertilize weekly.
Anthurium	*Anthurium* spp.		If it grows without flowering, try putting it in cooler location for a few weeks.
Aphelandra	*Aphelandra squarrosa*		Mist-spray frequently; fertilize regularly.
Aralia (*see* **Fatsia**)			
Asparagus fern	*Asparagus plumosus*		Not a true fern; repot annually; fertilize weekly.
Aspidistra	*Aspidistra* spp.		Sometimes called "cast iron plant" because it can stand neglect.
Aucuba	*Aucuba japonica*		Must be kept cool in winter.
Avocado	*Persea* spp.		Easily propagated from seed; provides good greenery but will not produce fruit indoors.
Azalea	*Rhododenron* spp.		Needs acid fertilizer; avoid warm locations.
Bamboo (dwarf)	*Bambusa angulata*		Needs good air circulation, bright light.
Begonia	*Begonia* spp.		Easily propagated from leaves; repot regularly.
Bilbergia	*Bilbergia* spp.		*See* Bromeliad; tough plant that can tolerate some neglect.
Birdcatcher plant	*Pisonia umbellifera*		Sticky exudate on fruits attracts birds in the plant's native habitat of New Zealand; strictly a foliage plant in houses.
Bird of paradise plant	*Strelitzia reginae*		Can be grown outdoors in milder climates.
Bird's nest fern	*Asplenium nidus*		Produces a spongelike mass of roots at base; requires much water and regular fertilizing.
Black-eyed Susan	*Thunbergia alata*		Annual climbing vine; grow from seed.

ENVIRONMENTS SUITABLE FOR COMMON HOUSEPLANTS

PLANT	SCIENTIFIC NAME	ENVIRONMENTAL REQUIREMENTS	REMARKS
Bloodleaf	*Iresine herbstii*		Easily propagated.
Boston fern	*Nephrolepis exaltata*		Needs regular watering and fertilizing.
Bromeliads	many species		These plants absorb virtually all their water and nutrients through their leaves; they should not be placed in regular potting soil, nor should they be watered with high calcium content water. They produce offshoots that should be propagated as main plant dies after flowering.
Cacti	many species		Contrary to popular belief, these slow-growing plants should not be grown in pure sand. Add some humus to potting mixture and withhold water in winter; keep cool in winter.
Caladium	*Caladium* spp.		Must have high humidity; keep root at 18°C (65°F) in pot during winter.
Calceolaria	*Calceolaria herbeohybrida*		Discard after flowering.
Calla lily	*Zantedeschia aethiopica*		After flowering, allow plant to dry up; repot in fall and start over.
Cape jasmine	*Gardenia jasminoides*		Needs night temperatures below 22°C (72°F) to initiate flowering; needs cool temperatures in winter.
Carrion flower	*Stapelia* spp.		Cactuslike plants with foul-smelling flowers.
Century plant (*see* **Agave**)			
Chinese evergreen	*Aglaonema costatum*		Needs warm temperatures and much water all year.
Chrysanthemum	*Chrysanthemum* spp.		Plants may be artificially dwarfed through use of chemicals; flowering initiated by short days.
Cineraria	*Senecio cruentus*		Needs cool temperatures; discard after flowering.

Table A4.2

ENVIRONMENTS SUITABLE FOR COMMON HOUSEPLANTS

PLANT	SCIENTIFIC NAME	ENVIRONMENTAL REQUIREMENTS	REMARKS
Cliff brake	*Pellaea rotundifolia*		Hanging basket fern; needs minimum temperature above 10°C (50°F).
Coffee	*Coffea arabica*		Handsome foliage plant; self-pollinating variety will produce fruit.
Coleus	*Coleus blumei*		To control size, restart plants from cuttings annually.
Copperleaf	*Acalypha wilkesiana*		Seldom survives average house environment for long.
Corn plant	*Dracaena massangeana*		Uses much water when large; easy to grow.
Croton	*Codiaeum* spp.		Needs constant high humidity and bright light.
Crown of thorns	*Euphorbia milii* and *E. splendens*		Deviation from watering routine may result in loss of leaves, but plant generally recovers.
Cyclamen	*Cyclamen* spp.		Fertilize weekly; keep cool; withhold water after flowering for few weeks, then start over.
Donkey tail	*Sedum morganianum*		Keep cool and dry in winter.
Dracaena	*Dracaena* spp.		Many kinds—all easy to grow and tolerant of some neglect.
Dumbcane	*Dieffenbachia* spp.		Needs regular fertilizing; keep away from small children (poisonous).
Dwarf banana	*Musa cavendishii*		Will produce small edible bananas if given enough light, water, and humidity.
Dwarf cocos palm	*Microcoleum weddelianum*		Keep temperature above 18°C (65°F) at all times.
Echeveria	*Echeveria* spp.		Keep cool in winter.
English ivy	*Hedera helix*		Needs cool temperatures to grow at its best.
False aralia	*Dizygotheca elegantissima*		Benefits from frequent mist-spraying.
Fatshedera	*Fatshedera lizei*		Climbing plant.
Fatsia	*Fatsia japonica*		Also called *Aralia*.

Table A4.2

ENVIRONMENTS SUITABLE FOR COMMON HOUSEPLANTS

PLANT	SCIENTIFIC NAME	ENVIRONMENTAL REQUIREMENTS	REMARKS
Ferns	many species		Water regularly; propagate from spores or runners.
Figs:			
Climbing fig	*Ficus pumila*		Damp-sponge leaves regularly.
Fiddleleaf fig	*Ficus lyrata*		Damp-sponge leaves regularly.
Weeping fig	*Ficus benjamina*		Damp-sponge leaves regularly.
Fingernail plant	*Neoregelia* spp.		*See* Bromeliad; name from red tips of leaves.
Fittonia	*Fittonia* spp.		Strictly terrarium plants—humidity too low elsewhere.
Flame violet	*Episcia cupreata*		Add charcoal and peat to potting medium.
Flowering maple	*Abutilon striatum*		Needs bright light to flower.
Fuchsia	*Fuchsia* spp.		Soil must be alkaline.
Gardenia (*see* **Cape jasmine**)			
Geranium	*Pelargonium* spp.		Make cuttings annually and discard parent plants each fall; keep cool through winter. Available with scents of orange, rose, or coconut.
Gloxinia	*Sinningia speciosa*		Fertilize heavily and water frequently; after flowering, withhold water and keep bulb cold for a few weeks.
Goldfish plant	*Columnea* spp.		Pot in mixture of leaf mold, fern bark, peat moss, and charcoal; use only rainwater or filtered water.
Grape ivy	*Rhoicissus rhomboidea*		Tolerates low light better than most plants.
Haworthia	*Haworthia* spp.		Aloelike plants that need minimum temperatures above 10°C (50°F) in winter.
Hen and chickens	*Sempervivum tectorum*		Keep cool in winter.
Hibiscus	*Hibiscus rosa-sinensis*		Fertilize weekly.

Table A4.2

ENVIRONMENTS SUITABLE FOR COMMON HOUSEPLANTS

PLANT	SCIENTIFIC NAME	ENVIRONMENTAL REQUIREMENTS	REMARKS
Hippeastrum (*see* **Amaryllis**)			
Holly fern	*Cyrtomium falcatum*		Relatively tough fern; keep cool in winter.
Houseleek	*Sempervivum* spp.		Keep cool in winter.
Hydrangea	*Hydrangea* spp.		Prune after flowering; keep cool in winter. Pink-flowering plant can be converted to blue by changing the soil to acid and vice versa.
Impatiens	*Impatiens* spp.		Exceptionally easy to propagate from cuttings.
Ivy-arum	*Rhaphidophora aurea*		Can tolerate some neglect.
Jade plant	*Crassula argenta* and others		Keep cool in winter.
Kaffir lily	*Clivia miniata*		Save the plant's energy by removing flowers as they wither.
Kalanchoë	*Kalanchoë* spp.		Withhold water and fertilizer for a few weeks after flowering, then repot and start over.
Lantana	*Lantana camara*		Fertilize twice a month; can be espaliered.
Madagascar jasmine (*see* **Stephanotis**)			
Maidenhair fern	*Adiantum* spp.		Mist-spray regularly.
Meyer fern	*Asparagus densiflora* var. *meyeri*		Fertilize regularly; repot annually.
Moneywort	*Lysimachia nummularia*		Hanging pot plant; needs bright light.
Moonstones	*Pachyphytum* spp.		Keep cool in winter.
Moses in the cradle	*Rhoeo* spp.		Can tolerate some neglect.
Mother-in-law's tongue (*see* **Sansevieria**)			
Mother of thousands	*Saxifraga* spp.		Also called *Saxifrage;* hanging plant. Plantlets formed on runners can be removed and grown separately.

Table A4.2

ENVIRONMENTS SUITABLE FOR COMMON HOUSEPLANTS

PLANT	SCIENTIFIC NAME	ENVIRONMENTAL REQUIREMENTS	REMARKS
Neanthe palm	*Chamaedorea elegans*		Sometimes called *Parlor palm;* stays less than 1 meter tall.
Norfolk Island pine	*Araucaria* spp.		Needs cold temperatures 2°C–3°C (36°F–38°F) in winter.
Octopus tree	*Schefflera arboricola*		Does best under cool conditions.
Oleander	*Nerium oleander*		Keep pot cool in winter for better flowering; keep away from small children (poisonous).
Orchid	thousands of species	no single set of environmental conditions applies	Contrary to popular belief, the common *Cattleya* and related orchids do not need high temperatures and humidity; most can get along with a minimum temperature of 13°C (56°F) at night and a minimum humidity of 40%. Most need bright light. They should never be placed in soil; pot them in sterilized pots with chips of fir bark or shreds of tree fern bark. *See Additional Reading* for culture references.
Oxtongue	*Gasteria* spp.		Can tolerate some neglect; needs a cool and relatively dry winter to flower.
Palms	many species		Use deep pots; fertilize regularly.
Parlor palm	*Howea fosteriana*		One of the easiest palms to grow.
Peperomia	*Peperomia* spp.		Many kinds; keep warm, humid; fertilize regularly.
Persian violet	*Exacum affine*		Needs good air circulation.
Philodendron	*Philodendron* spp.		Relatively tough plants; repot each spring.
Piggyback plant	*Tolmiea menziesii*		Plantlets formed on leaves can be separated and propagated.

Table A4.2

ENVIRONMENTS SUITABLE FOR COMMON HOUSEPLANTS

PLANT	SCIENTIFIC NAME	ENVIRONMENTAL REQUIREMENTS	REMARKS
Pineapple	*Ananas comosus*		*See* Bromeliad; easily grown from the top of a pineapple. If plant has not flowered after one year, enclose it in a plastic bag with a ripe apple for a few days (ethylene from apple should initiate flowering); no temperature below 15°C (59°F).
Pink polka dot plant	*Hypoestes sanguinolenta*		Susceptible to diseases and pests.
Pittosporum	*Pittosporum* spp.		Put several cuttings in one pot for bushy appearance.
Poinsettia	*Euphorbia pulcherrima*		After flowering, let plant dry under cool conditions until it loses its leaves; then restart.
Prayer plant	*Maranta leuconeura*		Name derived from fact that leaves fold together in evening.
Primrose	*Primula* spp.		Needs much water; does well outside in cool weather.
Purple tiger	*Calathea amabilis*		Use pots that are broader than they are deep.
Rosary plant	*Ceropegia woodii*		Hanging pot plant whose potting medium must drain well or plant will not survive.
Rubber plant	*Ficus elastica*		Do not overwater!
Sago palm	*Cycas revoluta*		Very slow growing (not a palm but a gymnosperm); never allow to dry out, but be certain water drains.
Sansevieria	*Sansevieria* spp.		Perhaps the toughest of all houseplants—nearly indestructible.
Satin pothos	*Scindapsus pictus*		Basket or pot plant.
Screw pine	*Pandanus* spp.		If given space, can become large; mist-spray often.
Selaginella (*see* Spike moss)			

Table A4.2

ENVIRONMENTS SUITABLE FOR COMMON HOUSEPLANTS

PLANT	SCIENTIFIC NAME	ENVIRONMENTAL REQUIREMENTS	REMARKS
Sensitive plant	*Mimosa pudica*		Leaves fold when touched; does not usually last more than a few months in most houses.
Shrimp plant	*Beloperone guttata*		Winter temperatures should be above 15°C (59°F).
Spathe flower	*Spathiphyllum wallisii*		Prefers warm winters and even warmer summers.
Spider plant	*Chlorophytum comosum*		Plantlets formed at tips of stems can be propagated separately.
Spiderwort	*Tradescantia* spp.		Easy to grow; do not overwater.
Spike moss	*Selaginella* spp.		Can become a weed in greenhouses.
Splitleaf philodendron	*Monstera deliciosa*		Plant adapts to various indoor locations quite well.
Sprenger fern	*Asparagus densiflora* var. *sprengeri*		Fertilize weekly; repot annually.
Staghorn fern	*Platycerium* spp.		Tough plant; immerse in water weekly.
Stephanotis	*Stephanotis floribunda*		Use very little fertilizer; keep cool in winter but water sparingly.
Stonecrop	*Sedum* spp.; *Crassula* spp.		Keep cool in winter.
Stove fern	*Pteris cretica*		Water and fertilize regularly.
String-of-pearls	*Senecio rowellianus*		Keep cool in winter.
Sundew	*Drosera* spp.		Sterilize pots; grow only on sphagnum moss.
Syngonium	*Syngonium podophyllum*		Repot annually in spring.
Tillandsia	*Tillandsia* spp.		*See* Bromeliad; best-known species is called *Spanish moss*.
Ti plant	*Cordyline terminalis*		Needs high humidity; seems to do better with other plants in pot.
Treebine	*Cissus antarctica*		Plant dislikes acid potting medium.

ENVIRONMENTS SUITABLE FOR COMMON HOUSEPLANTS

PLANT	SCIENTIFIC NAME	ENVIRONMENTAL REQUIREMENTS	REMARKS
Umbrella plant	*Cyperus* spp.		One of very few plants that needs to stand in water.
Velvetleaf	*Gynura sarmentosa*		Gets "stringy" but is easily restarted from cuttings.
Venus flytrap	*Dionaea muscipula*		Sterilize pots; grow only in sphagnum moss; repot annually.
Vriesia	*Vriesia* spp.		*See* Bromeliad.
Wandering Jew	*Zebrina pendula*		Easy to grow; do not overwater.
Wax plant	*Hoya* spp.		Climber; leave pot in one place—does not like to be moved.

GROWING VEGETABLES

Seed Germination

Many gardeners germinate larger seeds (e.g., squash, pumpkin) in damp newspaper. Soak a few sheets of newspaper in water for a minute and then hang them over a support for about 15 minutes or until the water stops dripping. Then line up the seeds in a row on the newspaper, wrap them, and place the damp mass in a plastic bag. Tie off the bag or seal it and place it in a warm (not hot!), shaded location (the floor beneath a running refrigerator is an example). Depending on the species, germination should occur within 2 to several days.

Tiny seeds (e.g., carrots, lettuce) may be mixed with clean sand before sowing to bring about a more even distribution of the seed in the rows.

Transplanting

Roots should be disturbed as little as possible when seedlings or larger plants are transplanted. Even a few seconds' exposure to air will kill root hairs and smaller roots. They should be shaded (e.g., with newspaper) from the sun and transplanted late in the day or on a cool, cloudy day if at all possible. To minimize the effects of transplanting, immediately water the seedlings or plants in their new location with a dilute solution of vitamin B/hormone preparation (e.g., Superthrive).

Bulb or Other Plants with Food-Storage Organs

All plants with food-storage organs (e.g., beets, carrots, onions) develop much better in soil that is free of lumps and rocks. If possible, the areas where these plants are to be grown should be dug to a minimum depth of 30 centimeters (12 inches) and the soil sifted through a 0.7 centimeter (approximately 0.25 inch) mesh before planting. Obviously, such a procedure is not always practical, but it can yield dramatic results.

Cutworms

Cutworms forage at or just beneath the surface of the soil. Their damage to young seedlings can be minimized by placing a collar around each plant. Tuna cans with both ends removed make effective collars when pressed into the ground to a depth of about 1 to 2 centimeters (0.4 to 0.8 inch).

Protection Against Cold

Some seedlings can be given an earlier start outdoors if plastic-topped coffee cans with the bottoms removed are placed over them. The plastic lids can be taken off during the day and replaced at night during cool weather. Conical paper frost caps can serve the same purpose.

Watering

Proper watering promotes healthy growth. It is much better to water an area thoroughly (e.g., for 20 to 30 minutes) every few days than to wet the surface for a minute or two daily. Shallow daily watering promotes root development near the surface, where midsummer heat can damage the root system. Conversely, too much watering can leach minerals out of the topsoil.

Fertilizers

Manures, bone and blood meals, and other organic fertilizers, which release the nutrients slowly and do not "burn" young plants, are preferred. Plants will utilize minerals from any available source, but in the long run the plants will be healthier and subject to fewer problems if they are not given sudden boosts with liquid chemical fertilizers.

Pests and Diseases

Biological controls are listed in Appendix 2. If sprays must be used, biodegradable substances, such as rotenone and pyrethrum, should be used.

COMMON VEGETABLES AND THEIR NUTRITIONAL VALUES

The nutritional values (NV) given are per 100 grams (3.5 ounces), edible portion, as determined by the U.S. Department of Agriculture.

Asparagus

NV (spears cooked in water): 20 calories; protein 2.2 gm; fats 0.2 gm; vit. A 900 I.U.; vit. B1 0.16 mg; vit. B2 0.18 mg; niacin 1.5 mg; vit. C 26 mg; fiber 0.7 gm; calcium 21 mg; phosphorus 50 mg; iron 0.6 mg; sodium 1 mg; potassium 208 mg.

Asparagus can be started from seed, but time until the first harvest can be reduced by a year or two if planting begins with one-year-old root clusters of healthy, disease-resistant varieties (e.g., "Mary Washington"). Asparagus requires little care if appropriate preparations are made before planting in the permanent location. Seeds should be sown sparsely and the seedlings thinned to about 7.5 centimeters (3 inches) apart. Before transplantation the following spring, dig a trench 30 to 60 centimeters (12 to 24 inches) deep and about 50 centimeters (20 inches) wide in an area that receives full sun, usually along one edge of the garden. If the soil is heavy, place crushed rock or gravel on the bottom of the trench to provide good drainage. Add a layer of steer manure about 10 centimeters (4 inches) thick, followed by about 7.5 centimeters (3 inches) of rich soil that has been prepared by thorough mixing with generous quantities of steer manure and bone meal. Place root clusters about 45 centimeters (18 inches) apart in the trench and cover with about 15 centimeters (6 inches) of prepared soil (be sure not to allow root clusters to dry out). As the plants grow, gradually fill in the trench. If one-year-old roots are planted, wait for 2 years before harvesting tips; if 2-year-old roots are planted, some asparagus may be harvested the following year. In all cases, no harvesting should be done after June, so that the plants may build up reserves for the following season. Cut shoots below the surface but well above the crown before the buds begin to expand. Cut all stems to the ground after they have turned yellow later in the season.

Beans

String or Snap Beans

NV (young pods cooked in water): 25 calories; protein 1.6 gm; fats 0.2 gm; vit. A 540 I.U.; vit. B1 0.07 mg; vit. B2 0.09 mg; niacin 0.5 mg; vit. C 12 mg; fiber 1 gm; calcium 50 mg; phosphorus 37 mg; iron 0.6 mg; sodium 4 mg; potassium 151 mg.

String or snap beans are warm-weather plants, although they can be grown almost anywhere in the United States. Wait until all danger of frost has passed and the soil is warm. Prepare the soil, preferably the previous winter, by digging to a depth of 30 centimeters (12 inches) and mixing in aged manure and bone meal. Pulverize the soil just before sowing; if soil has a low pH, add lime. Plant seeds thinly in rows about 40 to 50 centimeters (16 to 20 inches) apart; thin plants to 10 centimeters (4 inches) apart when the first true leaves have developed. Beans respond unfavorably to a very wet soil—do not overwater! In areas with hot summers, beans also prefer some light shade, particularly in midafternoon. As the bean plants grow, nitrogen-fixing bacteria invade the roots and supplement the nitrogen supply. Early vigorous growth can be enhanced by inoculating the seeds with such bacteria, which are available commercially in a powdered form. To maintain a continuous supply of green beans, plant a new row every two to three weeks during the growing season until two months before the first predicted frost. Cultivate regularly to control weeds, taking care not to damage root systems. Do not harvest or work with beans while the plants are wet, as this may invite disease problems.

Pole Beans

NV similar to those of string beans.

Soil preparation and cultivation are the same as for string beans. Plant beans in hills around poles that are not less than 5 centimeters (2 inches) in diameter and are at least 2 meters (6.5 feet) tall. As beans twine around their supports, it helps to tie them to the support with plastic tape as they grow. If harvested before the pods are mature, pole beans will produce over a longer period of time than will bush varieties.

Lima Beans

NV (immature seeds cooked in water): 111 calories; protein 7.6 gm; fats 0.5 gm; vit. A 280 I.U.; vit. B1 0.18 mg; vit. B2 0.10 mg; niacin 1.3 mg; vit. C 17 mg; fiber 1.8 gm; calcium 47 mg; phosphorus 121 mg; iron 2.5 mg; sodium 1 mg; potassium 422 mg.

Lima beans take longer to mature than other beans and are more sensitive to wet or cool weather. They definitely need warm weather to do well. Prepare soil and cultivate as for string beans.

Soybeans

NV (dry, mature seeds): 403 calories; protein 34.1 gm; fats 17.7 gm; vit. A 80 I.U.; vit. B1 1.10 mg; vit. B2 0.31 mg; niacin 2.2 mg; vit. C[1]; fiber 4.9 gm; calcium 226 mg; phosphorus 554 mg; iron 8.4 mg; sodium 5 mg; potassium 1,677 mg.

Broad (Fava) Beans

NV (dry, mature seeds): 338 calories; protein 25.1 gm; fats 1.7 gm; vit. A 70 I.U.; vit. B1 0.5 mg; vit. B2 0.3 mg; niacin 2.5 mg; vit. C[1]; fiber 6.7 gm; calcium 47 mg; phosphorus 121 mg; iron 2.5 mg; sodium 1 mg; potassium 422 mg.

Unlike other beans, broad beans need cool weather for their development. Sow as early as possible (in mild climates they may be sown in the fall, as they can withstand light frosts). Since the plants occupy a little more space than do bush beans, plant in rows about 0.9 to 1.0 meter (3 feet or more) apart and thin to about 20 centimeters (8 inches) apart in the rows. After the first pods mature, pinch out the tips to promote bushier development. To most palates, broad beans do not taste as good as other types of beans.

Beets

NV (cooked in water): 32 calories; protein 1.1 gm; fats 0.1 gm; vit. A 20 I.U.; vit. B1 0.03 mg; vit. B2 0.04 mg; vit. C 6 mg; fiber

[1]Values not available. *See also* Broad (Fava) Beans.

0.8 gm; calcium 14 mg; phosphorus 23 mg; iron 0.5 mg; sodium 43 mg; potassium 208 mg.

Beets will grow in a variety of climates but do best in cooler weather. They can tolerate light frosts and can be grown on a variety of soil types, although they prefer a sandy loam supplemented with well-aged organic matter. As with any bulb or root crop, they develop best in soil that is free of rocks and lumps. Beet "seeds" are really fruits containing several tiny seeds. Plant them in rows 40 to 60 centimeters (16 to 24 inches) or more apart and thin to about 10 centimeters (4 inches) apart in the rows after germination. After harvesting, the beets will keep in cold storage for up to several months. The leaves, if used when first picked, make a good substitute for spinach.

Broccoli

NV (spears cooked in water): 26 calories; protein 3.1 gm; fats 0.3 gm; vit. A 2,500 I.U.; vit. B1 0.09 mg; vit. B2 0.2 mg; niacin 0.8 mg; fiber 1.5 gm; calcium 88 mg; phosphorus 62 mg; iron 0.8 mg; sodium 10 mg; potassium 267 mg.

Broccoli is a cool-weather plant that will thrive in any good prepared soil, providing that it has not been heavily fertilized just prior to planting (fresh fertilizer promotes rank growth). The plants can stand light frosts and are planted in both the spring and fall in areas with mild climate. Although broccoli may continue to produce during the summer, most growers prefer not to keep the plants going during warm seasons because of the large numbers of pest insects they may attract.

Sow seeds indoors and transplant outdoors after danger of killing frosts has passed. Place plants about 0.9 to 1.0 meter (3 feet or more) apart and keep well watered. Keep area weeded and pests under control. Harvest heads (bundles of spears) while they are still compact. Smaller heads will develop very shortly after the first harvest; if these are removed regularly, the plants will continue to produce for some time, although the heads become smaller as the plants age.

Cabbage

NV (raw): 24 calories; protein 1.3 gm; fats 0.2 gm; vit. A 130 I.U.; vit. B1 0.05 mg; vit. B2 0.05 mg; niacin 0.3 mg; vit. C 47 mg; fiber 1 gm; calcium 49 mg; phosphorus 29 mg; iron 0.4 mg; sodium 20 mg; potassium 233 mg.

Growth requirements of cabbage are similar to those of broccoli.

Carrots

NV (raw): 42 calories; protein 1.1 gm; fats 0.2 gm; vit. A 11,000 I.U.; vit. B1 0.6 mg; vit. B2 0.5 mg; niacin 0.6 mg; vit. C 8 mg; fiber 1 gm; calcium 37 mg; phosphorus 36 mg; iron 0.7 mg; sodium 47 mg; potassium 341 mg.

Carrots are hardy plants that can tolerate a wide range of climate and soils, but the soil must be well prepared: free of rocks and lumps and preferably not too acid. The seeds are slow to germinate. Plant in rows a little more than 30 centimeters (12 inches) apart and thin seedlings to about 5 centimeters (2 inches) apart in the rows. Weed the rows regularly until harvest. Carrots keep well in belowground storage containers when freezing weather arrives.

Cauliflower

NV (cooked in water): 22 calories; protein 2.3 gm; fats 0.2 gm; vit. A 60 I.U.; vit. B1 0.09 mg; vit. B2 0.08 mg; niacin 0.6 mg; vit. C 55 mg; fiber 1 gm; calcium 21 mg; phosphorus 42 mg; iron 0.7 mg; sodium 9 mg; potassium 206 mg.

Growth requirements of cauliflower are similar to those of broccoli except that heavier fertilizing is required. As cauliflower heads develop, protect them from the sun by tying the larger leaves over the tender heads. Harvest while the heads are still solid.

Corn

NV (cooked sweet corn kernels): 83 calories; protein 3.2 gm; fats 1 gm; vit. A 400 I.U. (yellow varieties; white varieties have negligible vit. A content); vit. B1 0.11 mg; vit. B2 0.10 mg; niacin 1.3 mg; vit. C 7 mg; fiber 0.7 gm; calcium 3 mg; phosphorus 89 mg; iron 0.6 mg; sodium trace; potassium 165 mg.

There are several types of corn (e.g., popcorn, flint corn, dent corn), but sweet corn is the only type grown to any extent by home gardeners. It can be grown in any location where there is at least a 10-week growing season and warm summer weather.

Corn does best in a fertile soil, which should be prepared by mixing with compost and liberal amounts of chicken manure or fish meal. Since corn is wind-pollinated, it can be helpful to grow the plants in several short rows at right angles to the prevailing winds rather than in a single long row. For best results, use only fresh seeds and plant in rows 60 centimeters (24 inches) apart for dwarf varieties and 90 centimeters (36 inches) apart for standard varieties. Thin to 20 to 30 centimeters (8 to 12 inches) apart in the rows after the plants have produced three to four leaves. Cultivate frequently to control weeds. The corn is ready to harvest when the silks begin to wither.

Cucumber

NV (raw, with skin): 15 calories; protein 0.9 gm; fats 0.1 gm; vit. A 250 I.U.; vit. B1 0.03 mg; vit. B2 0.04 mg; niacin 0.2 mg; vit. C 11 mg; fiber 0.6 mg; calcium 25 mg; phosphorus 27 mg; iron 1.1 mg; sodium 6 mg; potassium 160 mg.

Until drought- and disease-resistant varieties were developed in recent years, cucumbers were considered rather temperamental plants to grow. The newer varieties are no more difficult to raise than are most other common vegetables.

The soil should be a light loam—neither too heavy nor too sandy. It should be mixed with well-aged manure and compost and heaped into small mounds about 2 meters (6.5 feet) apart. Five to six seeds should be planted in each mound about 2.5 centimeters (1 inch) below the surface in the middle of the spring. When the plants are about 1 decimeter (4 inches) tall, thin to three plants per mound. Cultivate regularly and, to promote continued production, pick all cucumbers as soon as they attain eating size.

Eggplant

NV (cooked in water): 19 calories; protein 1 gm; fats 0.2 gm; vit. A 10 I.U.; vit. B1 0.05 mg; vit. B2 0.04 mg; niacin 0.5 mg; vit. C 3 mg; fiber 0.9 gm; calcium 11 mg; phosphorus 21 mg; iron 0.6 mg; sodium 1 mg; potassium 150 mg.

Eggplant is strictly a hot-weather plant that is sensitive to cold weather or dry periods and needs heavy fertilizing. Since seedling development is initially slow, plant the seeds indoors about 2 months before the plants will be set out, which should be about 5 to 6 weeks after the last average date of frost.

Eggplants do best in enriched sandy soils that are supplemented with additional fertilizer once a month. Never permit them to dry out. Place the seedlings about 70 to 80 centimeters (28 to 32 inches) apart in rows 0.9 to 1.0 meter (3 feet or more) apart. Some staking of the plants may be desirable. The fruits are ready to harvest when they have a high gloss. They are still edible after greenish streaks appear and the gloss diminishes, but they are not as tender at this stage.

Lettuce

NV (crisp, cabbage-head varieties): 13 calories; protein 0.9 gm; fats 0.1 gm; vit. A 330 I.U.; vit. B1 0.06 mg; vit. B2 0.06 mg; niacin 0.3 mg; vit. C 6 mg; fiber 0.5 gm; calcium 20 mg; phosphorus 22 mg; iron 0.5 mg; sodium 9 mg; potassium 175 mg.

NV (leaf varieties): 18 calories; protein 1.3 gm; fats 0.3 gm; vit. A 1,900 I.U.; vit. B1 0.05 mg; vit. B2 0.08 mg; niacin 0.4 mg; vit. C 18 mg; calcium 68 mg; phosphorus 25 mg; iron 1.4 mg; sodium 9 mg; potassium 264 mg.

This favorite salad plant comes in a wide variety of types and forms, all of which do better in cooler weather, although a few of the leaf types (e.g., oak leaf) can tolerate some hot periods. As long as the individual plants are given room to develop and the soil is not too acid, most varieties can be grown on a wide range of soil types.

Since lettuce can stand some frost, sow the seeds outdoors as early in the spring as the ground can be cultivated. Do not cover the seeds with more than a millimeter or two of soil—they need light to germinate. Mix the soil with a well-aged manure and a general-purpose fertilizer a week or two before sowing. Plant seedlings about 30 centimeters (12 inches) apart in rows 30 to 40 centimeters (12 to 16 inches) apart. For best results, do not allow the soil to dry out and plant only varieties suited to local conditions. The most common crisp, cabbage-head varieties found in produce markets will not form heads in hot weather, and many others will bolt (begin to flower) during hot weather and longer days. Cultivate weekly between rows to promote rapid growth and to control weeds.

Onion

NV (raw): 38 calories; protein 1.5 gm; fats 0.1 gm; vit. A 40 I.U. (yellow varieties only); vit. B1 0.03 mg; vit. B2 0.04 mg; niacin 0.2 mg; vit. C 10 mg; fiber 0.6 gm; calcium 27 mg; phosphorus 36 mg; iron 0.5 mg; sodium 10 mg; potassium 157 mg.

These easy-to-grow vegetables do best in fertile soils that are free of rocks and lumps, are well drained, and are not too acid or sandy.

Onions may take several months to mature from seed. The viability of the seed decreases rapidly after the first year. Bulb formation is determined by day length rather than by the total number of hours in the ground. Because of these characteristics of onions, most gardeners prefer to purchase *sets* (young plants that already have a small bulb) from commercial growers, although green or bunching onions are still easily grown from seed.

Plant the sets upright 6.0 to 7.5 centimeters (2.5 to 3.0 inches) apart in rows and firm in place. They will need little care except for weeding, watering, and occasional shallow cultivation until harvest about 14 weeks later. The onions are mature when the

tops fall over. After pulling them from the ground, allow them to dry in the shade for 2 days. Then remove the tops 2 to 3 centimeters (about 1 inch) above the bulbs and spread out the bulbs to continue curing for 2 to 3 more weeks. After this, store them in sacks or other containers that permit air circulation until needed.

Peas

NV (green, cooked in water): 71 calories, protein 5.4 gm; fats 0.4 gm; vit. A 540 I.U.; vit. B1 0.28 mg; vit. B2 0.11 mg; niacin 2.3 mg; vit. C 20 mg; fiber 2 gm; calcium 23 mg; phosphorus 99 mg; iron 1.8 mg; sodium 1 mg; potassium 196 mg.

Peas are strictly cool-weather plants that generally produce poorly when the soil becomes too warm. Seeds should be planted in the fall or very early spring. As is the case with beans, peas receive a better start if the seeds are inoculated with nitrogen-fixing bacteria (see discussion of beans) at planting time. Prepare the soil by mixing thoroughly with liberal amounts of aged manure and bone meal. Plant the seeds about 2.5 centimeters (1 inch) deep in heavy soil or 5 centimeters (2 inches) deep in light, sandy soil, about 2.5 centimeters (1 inch) apart in single rows for dwarf bush varieties or 15 centimeters (6 inches) apart in double files for standard varieties, with intervals of 80 to 90 centimeters (32 to 36 inches) between the rows. After germination, thin the plants to 10 centimeters apart. Place support wires, strings, or chicken wire between the rows at the time of planting; peas will not do well without such support.

Green peas should be picked while still young and cooked or frozen immediately, as the sugars that make them sweet are converted to starch within 2 to 3 hours after harvest.

Peppers

NV (raw sweet or bell peppers): 22 calories; protein 1.2 gm; fats 0.2 gm; vit. A 420 I.U.; vit. B1 0.08 mg; vit. B2 0.08 mg; niacin 0.5 mg; vit. C 128 mg; fiber 1.4 gm; calcium 9 mg; phosphorus 22 mg; iron 0.7 mg; sodium 13 mg; potassium 213 mg.

Peppers, like eggplants, are strictly hot-weather plants for most of their growing season, but unlike eggplants they actually do better toward the end of their season if temperatures have moderated somewhat. Sweet or bell peppers are closely related and have similar cultural requirements.

Plant seeds indoors 8 to 10 weeks before the outdoor planting date, which is generally after the soil has become thoroughly warm. They will grow in almost any sunny location in a wide variety of soils, but to obtain the large fruits seen in produce markets fertilize the plants heavily and water regularly. Plant seedlings 50 to 60 centimeters (20 to 24 inches) apart in rows that are 60 to 90 centimeters (24 to 36 inches) apart. Sweet peppers can be harvested at almost any stage and are still perfectly edible after they have turned red.

Potatoes

NV (baked in skin): 93 calories; protein 2.6 gm; fats 0.1 gm; vit. A trace; vit. B1 0.10 mg; vit. B2 0.04 mg; niacin 1.7 mg; vit. C 20 mg; fiber 0.6 gm; calcium 9 mg; phosphorus 65 mg; iron 0.7 mg; sodium 4 mg; potassium 503 mg.

Potatoes grow best in a rich, somewhat acid, well-drained soil that has had compost or well-aged manure added to it. They are subject to several diseases, and it is advisable to use disease-free seed potatoes purchased from a reputable dealer. Two to 3 weeks before the average date of the last spring frost, plant the seed potatoes whole or cut into several pieces, making sure that each piece has at least one eye. Place the potato pieces about 30 centimeters (12 inches) or more apart at a depth of about 12 to 15 centimeters (5 to 6 inches) in rows 0.9 to 1.0 meter (about 3 feet) apart. Later plantings are feasible. Spread a thick mulch (e.g., straw) over the area after planting to keep soil temperatures down and to retain soil moisture.

Potatoes are ready for harvest when the tops start turning yellow, but they may be left in the ground for several weeks after that if the soil is not too wet. After harvest, wash the potatoes immediately and place them in a dry, cool, dark place until needed. If left exposed to light, the outer parts of the potato turn green; poisonous substances are produced in these tissues, and such potatoes should be discarded.

Spinach

NV (raw): 26 calories; protein 3.2 gm; fats 0.3 gm; vit. A 8,100 I.U.; vit. B1 0.10 mg; vit. B2 0.20 mg; niacin 0.6 mg; vit. C 51 mg; fiber 0.6 gm; calcium 93 mg; phosphorus 51 mg; iron 3.1 mg; sodium 71 mg; potassium 470 mg.

Spinach is a cool-season crop that goes to seed as soon as the days become long and warm. It should be planted in the fall or early spring. If protected by straw or other mulches, it will overwinter in the ground in most areas and be ready for use early in the spring. Spinach has a high nitrogen requirement and reacts negatively to acid soils. It is otherwise easy to grow. Mix the soil thoroughly with aged manure and bone meal before planting. Plant seedlings 3 to 5 centimeters (1 to 2 inches) apart in rows 40 to 50 centimeters (16 to 20 inches) apart. Keep the plants supplied with adequate moisture and their growing area free of weeds. Harvest the whole plant when a healthy crown of leaves develops.

Squash

NV (cooked zucchini): 12 calories; protein 1 gm; fats 0.1 gm; vit. A 300 I.U.; vit. B1 0.05 mg; vit. B2 0.08 mg; niacin 0.8 mg; vit. C 9 mg; fiber 0.6 gm; calcium 25 mg; phosphorus 25 mg; iron 0.4 mg; sodium 1 mg; potassium 141 mg.

All varieties of squash are warm-weather plants, and all are targets of a variety of pests. Thorough preparation of the soil before planting pays dividends in production and in the health of the plants. Mix compost and aged manure with the soil and heap the soil in small hills about 1.2 meters (4 feet) apart from one another. Plant four to five seeds in each hill and thin the seedlings to three after they are about 10 centimeters (4 inches) tall. Summer squashes (e.g., zucchini) mature in about 2 months, while winter squashes (e.g., acorn) can take twice as long to mature. Summer squashes should be harvested while very young—they can balloon, seemingly overnight, into huge fruits. Winter squashes should be harvested before the first frost; only clean, undamaged fruits will store well. Keep such squashes laid out in a cool, dry place and not piled on top of one another. Check them occasionally for the development of surface fungi.

Tomatoes

NV (raw, ripe): 22 calories; protein 1.1 gm; fats 0.2 gm; vit. A 900 I.U.; vit. B1 0.06 mg; vit. B2 0.04 mg; niacin 0.7 mg; vit. C 23 mg;

fiber 0.5 gm; calcium 13 mg; phosphorus 27 mg; iron 0.5 mg; sodium 3 mg; potassium 244 mg.

These almost universally used fruits are easy to grow providing one understands a few basic aspects of their cultural requirements:

1. Many tomato plants normally will not initiate fruit development from their flowers when night temperatures drop below 14°C (57°F) or day temperatures climb above 40°C (104°F). For the earliest yields, seeds may be germinated indoors several weeks before the plants are to be placed outside, but little is accomplished by transplanting before the night temperatures begin to remain above 14°C (57°F); in addition, some growers insist that plants given an early start indoors do not always do as well later as those germinated outdoors.

2. Tomatoes require considerably more phosphorus in proportion to nitrogen from any fertilizers added to the soil where they are to be grown. Many inexperienced gardeners make the mistake of giving the plants lawn or general-purpose fertilizers that are proportionately high in nitrogen. As a result, the plants may grow vigorously but produce very few tomatoes. Give tomatoes bone meal, tomato food (Magamp is an excellent commercial slow-release preparation), or steer manure mixed with superphosphate.

3. Tomato plants seem more susceptible than most to soil fungi and to root-knot nematodes. The damage caused by these organisms may not become evident until the plants begin to bear. Then the lower leaves begin to wither, and yellowing progresses up the plant or there seems to be a general loss of vigor and productivity. Using disease- and nematode-resistant varieties (usually indicated by the letters V, F, and N on seed packets) is by far the simplest method of controlling these problems. Another effective control involves dipping the seedling roots in an emulsion of 0.25% corn oil in water when transplanting; experiments have shown that the corn oil greatly reduces root-knot nematode infestation.

4. Many garden varieties of tomatoes need to be staked to keep fruits off the ground where snails and other organisms can gain easy access to them. When using wooden stakes, be sure they have a diameter of 5 centimeters (2 inches) or more, and tie the plants securely to the stakes with plastic tape. Thinner stakes are likely to break or collapse when the plants grow to a height of 2 meters (6.5 feet) or more. Some growers prefer to use heavy wire tomato towers instead of stakes.

5. Hornworms and tomato worms almost invariably appear on tomato plants during the growing season. They can virtually strip a plant and ruin the fruits if not controlled. Fortunately control is simple and highly effective with the use of *Bacillus thuringiensis,* which was discussed in Appendix 2.

6. The eating season for garden tomatoes can be extended for about 2 months past the first frost if all the green tomatoes are picked before frost occurs. Place the tomatoes on sheets of newspaper on a flat surface in a cool, dry place, where they will ripen slowly a few at a time. Generally, the taste of tomatoes ripened in this way is superior to that of hothouse tomatoes sold in produce markets. Be sure when picking the green tomatoes to handle them very gently, as they bruise very easily, and molds quickly develop in the bruised areas.

Pruning

A good gardener or orchardist makes a habit of pruning trees, shrubs, and other plants regularly for a variety of reasons. He or she may wish to improve the quality and size of the fruits and flowers, restrict the size of the plants, keep the plants healthy, shape the shrubbery, or generally get more from the plants.

Except for spring-flowering ornamental shrubs, which should be pruned right after flowering, most maintenance pruning is done in the winter when active growth is not taking place. It usually involves removal of portions of stems, but it can also involve roots. When a terminal bud is removed, the axillary or lateral bud just below the cut will usually develop into a branch, and a bushier growth will result. Some gardeners pinch off terminal buds routinely to encourage such growth. The following sections provide a few generalities and specifics pertaining to several types of plants.

Fruit Trees

When young fruit trees are first planted, all except four or five stems and any damaged roots should be pruned off. The remaining stems should be cut back so that there is one central leader about 1 meter (3 feet) tall, with shorter side branches facing in different directions (Fig. A4.1). When cutting the stems, be careful to cut in such a way that the axillary or lateral bud just below the cut is facing outward. Each succeeding year, prune back new growth to within a few centimeters of the previous year's cut. Remove any dead or diseased branches, and prune stems that have grown so that they are rubbing against each other. Cut out the central leader the second year in peach trees so that the interior of the tree is left relatively open. Regularly remove any sucker shoots that develop from the base or along the trunk of the tree.

Grapevines

There are several methods of pruning grapevines, depending on the type of vine and the circumstances under which the vines are being grown. In general, grapevines should be pruned heavily in late winter for best fruit production. After allowing a central trunk to develop, cut back each shoot, regardless of its length, so that no more than three axillary buds remain. Exceptions to this rule involve situations in which the vines are trained on arbors or wires, when the shoots initially may be allowed to grow longer. Even then, however, after the desired training has taken place (Fig. A4.2), pruning should be heavy for best results.

Roses

Rose bushes should be pruned heavily—they will recover! In general, new stems should be cut back to within 10 to 20 centimeters (4 to 8 inches) of their point of origin, with care taken that the top remaining axillary bud of each stem is pointing outward. This promotes growth that leaves the center open for better air circulation. Any dead or diseased canes should be removed and the number of remaining canes limited to three or four per plant.

Raspberries, Blackberries, and Their Relatives

Berry canes are biennial. They are produced from the base the first year, branch during the summer, and usually produce fruit on the branches the second year, although in milder climates they may also produce fruit on the first year's growth. The canes die after the second year.

Old, dead canes should be removed and all but three or four canes developing from each crown should be pulled out when the

FIGURE A4.1   A young peach tree. *A*. Before pruning. *B*. After pruning.

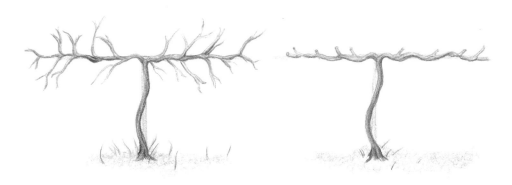

A.   B.

FIGURE A4.2   A grapevine. *A.* Before pruning. *B.* After pruning.

FIGURE A4.3   A bonsai plant. This Sitka spruce tree is little more than 30 centimeters (1 foot) tall and is over 40 years old.
(Courtesy Guy Downing)

ground is soft. New canes should be cut back to lengths of 1 meter (3 feet) or less in the spring. Branches of 1-year-old canes should be cut back in early spring to lengths of about 30 centimeters (12 inches) for larger fruits.

Bonsai

Container-grown trees that are dwarfed through the constant careful pruning of both roots and stems, the manipulation of soil mixtures, and the weighting of branches are called **bonsai** (Fig. A4.3). Some of these dwarfed trees attain ages of 50 to 75 years or more and may be less than 1 meter (3 feet) tall. Bonsai is an art that requires knowledge of the environmental requirements and tolerances of individual species.

In general, bonsai growers pinch out new growth above a bud every few days during a growing season but never prune when a plant is dormant. Refer to *Additional Reading* for more information on the subject.

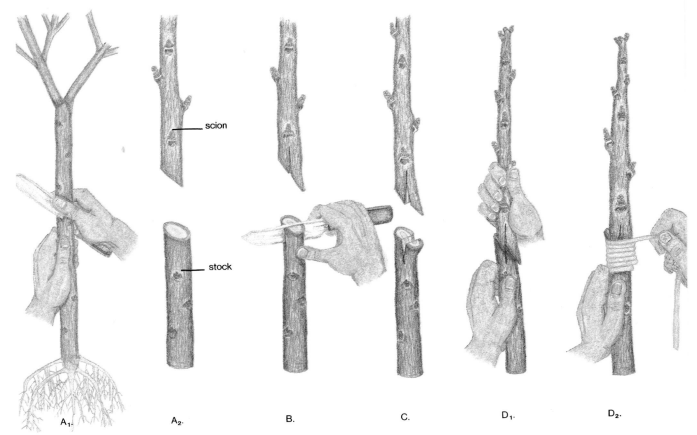

FIGURE A4.4 Stages in whip, or tongue, grafting. *A.* A smooth tangential cut is made at the bottom of the scion and at the top of the stock. *B.* Vertical cuts are made back into the centers of the stock and scion. *C.* The cuts are slightly widened to form a little tongue on each portion. *D.* The scion is inserted into the stock as tightly as possible without forcing a split. The graft then is bound with rubber strips and sealed with grafting wax.

Major Types of Grafting

Whip or Tongue Grafting

Whip, or *tongue grafting,* is widely used for relatively small material—that is, wood between 0.70 and 1.25 centimeters (0.25 and 0.50 inch) in diameter. The stock and scion (rooted portion and aerial portion, respectively) should be nearly identical in diameter to bring about maximum contact between the cambia. The scion should contain two or three buds, and the cuts on both the stock and scion should be made in an internode. As shown in Figure A4.4, a smooth tangential cut about 5 centimeters (2 inches) long is made with a sharp sterilized knife at the bottom of the scion and at the top of the stock. The angles of both cuts should also be as nearly identical as possible, and there should be no irregularities or undulations in the surfaces (such as those caused by a dull cutting instrument). A second cut is then made in both the stock and scion about one-third of the distance from the tip of the cut surfaces. This cut is made back into the wood, nearly parallel to the first cut so that it forms a little tongue. The scion is then inserted into the stock as tightly as possible, taking care not to force a split. In addition, the bottom edge of the scion should not protrude past the bottom of the cut of the stock. The process is completed by binding the materials with plastic tape and adding grafting wax.

If the stock and scion are not identical in diameter, it is still possible to obtain a graft if care is taken to bring the cambia in close contact along one edge of the cuts (Fig. A4.5).

Splice Grafting

Splice grafting is sometimes used with plants in which the pith is extensive. It is essentially the same as whip grafting, except a second cut isn't made and no tongue is formed.

Cleft Grafting

Cleft grafting is used routinely when the diameter of the stock is considerably greater than that of the scion or scions. First, cut the stock branch or trunk at right angles, making sure the bark is not torn. If some bark is pulled loose by the saw, make a new cut. Commercial growers often minimize detachment of the bark by making a cut one-third of the way in on one side and then making a cut slightly lower on the opposite side. This usually leaves a surface with clean edges. Next, hammer a meat cleaver or heavy knife 5.0

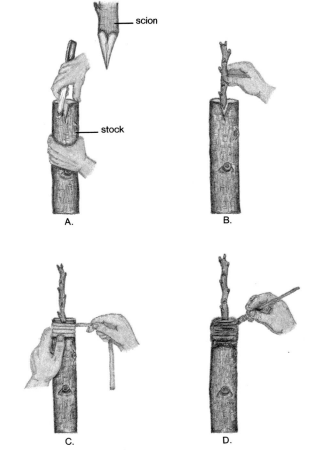

FIGURE A4.5 A whip or tongue graft in which the stock and scion are of different diameters.

to 7.5 centimeters (2 to 3 inches) into the wood to make a vertical cut or shallow split. Insert a wedge temporarily into the cut to keep it open. Insert scions, usually 7.5 to 10.0 centimeters (3 to 4 inches) long, on each side of the cut toward the outer parts so that as much of the cambia as possible is in contact. When the wedge is removed, the fit should be tight enough to prevent the scions from easily being pulled out by hand. Finally, seal any exposed surface with grafting wax (Fig. A4.6).

Side Grafting

Side grafting is often used with stocks that are about 2.5 centimeters (1 inch) in diameter. Make a cut about 2.5 centimeters (1 inch) deep with a heavy knife or chisel at an angle of 20 to 30 degrees to the surface of the stock. Cut the bottom end of the scion, which should be 7.5 to 10.0 centimeters (3 to 4 inches) long and about 0.75 centimeter (0.25 inch) in diameter, into a smooth wedge about 2.5 centimeters (1 inch) long. Then bend the stock slightly to open up the cut and insert the scion, making certain that the maximum contact between each cambium is obtained. Then release the pressure to ensure a tight fit (Fig. A4.7). Follow this by sealing with grafting wax. Side grafting may be used to replace limbs lost through storm or other damage or for cosmetic purposes, such as improving the symmetry of the plant.

A variation of this graft, the *side tongue graft,* is often used with small broad-leaved evergreen plants. This graft involves slicing about 2.50 to 3.25 centimeters (1.0 to 1.5 inches) of stem out of the side toward the base to a depth of 0.75 centimeter (0.25 inch) and then making a second, smaller cut to form a tongue within the original cut. A scion, prepared in similar fashion to the side-grafted scion, is then inserted, and the graft is tied and sealed.

Approach Grafting

If two related plants tend not to form grafts very well by other means, *approach grafting* can be tried. Two independently growing plants, at least one of which is usually in a container, are prepared by making smooth cuts identical in length and depth at the same height on both stems. A tongue sometimes also is cut in both exposed areas, and then the two parts are fitted together, tied, and sealed (Fig. A4.8). This can be done at any time of the year but is most likely to be successful during periods of active growth. After union is achieved, the top of one plant may be cut off above the graft and the bottom of the other removed below the graft.

Inarching is a variation of approach grafting sometimes used to save a valuable tree whose root system has been damaged. Plant young seedlings or trees of the same kind around the base of the tree. When they have become established, cut the tip of each seedling, which should be 0.75 to 1.25 centimeters (0.25 to 0.50 inch) thick, vertically for about 15 centimeters (6 inches) on the side nearest the main tree. Then make vertical cuts of similar length on the tree to the exact width of the prepared seedling tip and deep enough to expose the cambium; leave a small flap of bark at the top to cover the tip of the seedling. Next, fit the prepared seedling tips into the slots and nail them in with four to six flat-headed nails and seal the entire area with grafting wax. If any side shoots appear from the seedlings after the grafting union has developed, prune them off. Because the larger tree will be producing a considerable amount of food, the seedlings often grow very rapidly after successful inarching (Fig. A4.9).

Bridge Grafting

Sometimes in temperate and colder regions, a particularly deep snowpack may prevent rodents and other animals from reaching their usual winter food. When this occurs, they may turn to the bark of trees and gnaw off a band of tissue, sometimes a decimeter or two wide. The damage usually extends through the phloem and cambium, as these tissues are the most palatable to the animals. This stripping of tissues is frequently referred to as **girdling** a tree and, if left untreated, will probably result in the tree's death through starvation of the roots, since the phloem cannot conduct food past the damaged area.

The tree can often be saved, however, through *bridge grafting,* particularly if the grafting is done in the early spring just as new growth is beginning. Cut the scions from dormant twigs of the same tree and keep them in a refrigerator until needed. Clean and prepare the damaged area by cutting out any remaining dead or tattered tissues. Then insert the scions above and below the girdle, about 5.0 to 7.5 centimeters (2 to 3 inches) apart, with the natural bottom ends facing down and the tip ends up (Fig. A4.10). The graft will not succeed if a scion is put in upside down.

Bud Grafting (Budding)

Bud grafting, or **budding,** is a form of grafting that utilizes a single bud. The method is widely used in commercial nurseries, partly

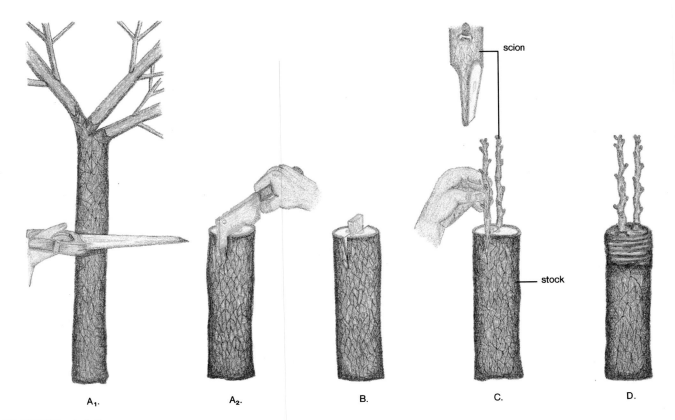

FIGURE A4.6   Stages in cleft grafting. *A.* The stock is cut transversely and a meat cleaver or heavy knife is driven into the wood to make a vertical cut or split. *B.* A wedge is temporarily inserted into the vertical cut to keep it open. *C.* Scions are inserted into the cut in the vicinity of the cambium, and the wedge is removed. *D.* The exposed surfaces are sealed with grafting wax.

because a single team of two or three workers can produce over a thousand such grafts a day and also because frequently more than 95% of the grafts are successful.

Budding is usually done in the summer when the season's axillary buds are mature and while the sap of the stocks is flowing freely. Budding is generally most successful when plump leaf buds (not flower or mixed buds) of the current season's growth are grafted to healthy stocks that are 2 to 3 years old.

Prepare the stock by removing all the leaves and side branches below and in the vicinity of the point at which the graft is to be made. Then make a T-shaped incision through the bark with the aid of a sterilized, razor-sharp knife. The transverse cut should be roughly 1.25 centimeters (0.5 inch) wide and the vertical cut about 2.5 to 3.0 centimeters (1.0 to 1.2 inches) long. Both cuts should be no deeper than the cambium, and the bark should peel back easily at the junction of the two cuts.

Prepare a *bud stick* by removing all leaf material except for 1.25 centimeters (0.5 inch) of the petioles, which are left to serve as handles. Then carefully cut a bud from the stick so that an oval piece of tissue about 1.75 centimeters (0.75 inch) in diameter surrounds it. If the bud separates from the tissue, discard it and cut another one. Next, insert the bud and its oval shield into the T-shaped cut of the stock and fold the flaps of bark over the shield, leaving the bud exposed. Use flat rubber strips to tie the T shut so that only the bud remains visible. After growth begins, cut off the stock just

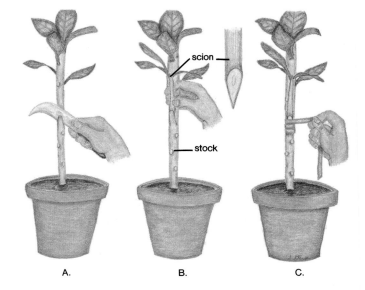

FIGURE A4.7   How a side graft is made. A tangential cut is made on the side of the stock and a prepared scion is inserted.

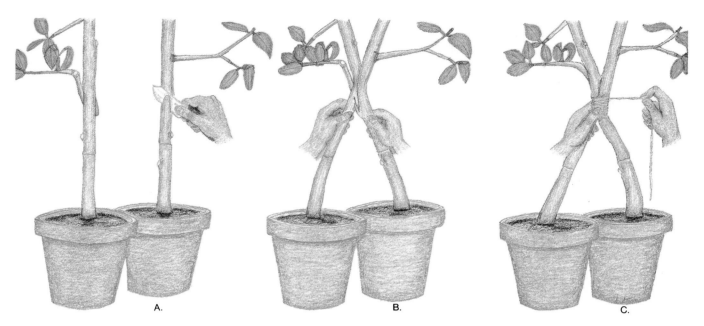

FIGURE A4.8   An approach graft. Two independently growing plants are prepared and grafted together as shown. *A.* Smooth slanting vertical cuts identical in length and depth and at the same height are made on both stems; the cuts, however, go in opposite directions. *B.* The two parts are wedged together. *C.* The united area is tied and sealed.

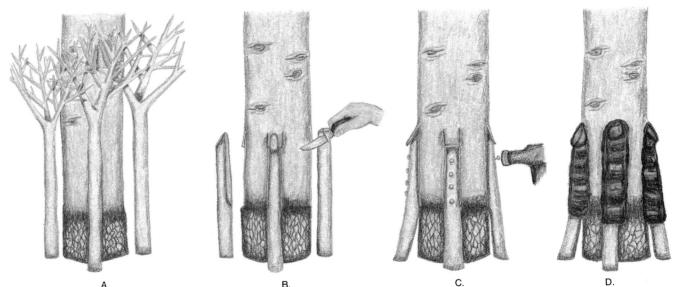

FIGURE A4.9   Inarching. *A.* Established seedlings, which had previously been planted around the base of the tree, are cut vertically at their tips. Vertical slots of similar length are made in the tree adjacent to the seedling tips. *B.* The seedling tips are nailed into the slots. *C.* Growth of the seedling bases may be very rapid if the inarching is successful.

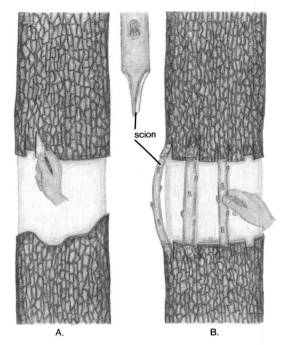

FIGURE A4.10   Bridge grafting. Scions of small diameter are cut at an angle at each end and inserted into prepared slots cut above and below the girdled area.

above the bud. The stock should not be cut off any earlier because the bud derives benefit from the transport of substances up and down the stock (Fig. A4.11).

Bud grafting can also be done in the spring, as soon as possible after growth of the stock begins but before the bud sticks become active. The bud material is frequently cut and stored in a refrigerator before growth begins. In areas with long growing seasons, bud grafting may also be done in early summer if the bark still peels back easily, but it should not be done any later because a young tree needs to produce sufficient growth before fall to be healthy and vigorous the following season.

In thick-barked trees, such as the Pará rubber tree and some of the nut trees, a rectangular patch of bark is cut out of the stock and replaced with a similar patch containing a bud. Normally, the patch is not more than 5 centimeters (2 inches) in diameter. This method is slower than the other budding methods described but generally gives much better results in species with thick bark.

Root Grafting

Whip, or tongue, grafts with roots or pieces of roots used for stocks are sometimes used in the propagation of apple, pear, and other fruit trees. After the grafts have been made, they are usually stored in a cool, damp place for about 2 months and then refrigerated until early spring, when they are planted before growth starts. After growth begins, they are checked to make sure the scion is not producing its own roots, or the advantages of the original rootstock will be lost. These advantages may include disease resistance or dwarfing.

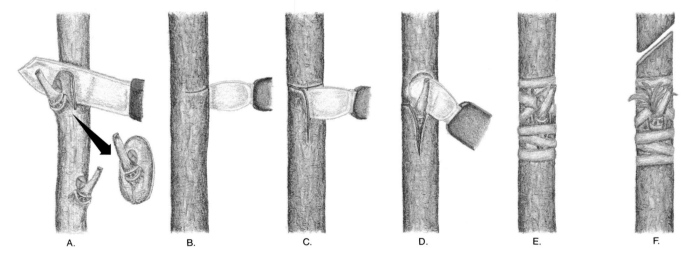

FIGURE A4.11   Budding (bud grafting). Leaves and side branches are removed from the stock below the point at which the graft is to be made. A *bud stick* is prepared by removing all leaf material except a short portion of each petiole. *A*. A bud, with an oval patch of tissue surrounding it, is cut from the bud stick. *B*. A T-shaped incision is made in the bark of the stock. *C*. The bark is pried up slightly at the corners. *D*. The bud and its surrounding tissue are inserted into the T-shaped incision on the stock. *E*. The flaps of bark are folded over the tissue, leaving the bud exposed, and the T is tied shut with flat rubber strips. *F*. After growth begins, the stock is cut off just above the bud.

Double-Working

In *double-working* grafts, an *interstock* consisting of a stem segment varying in length from 2.5 centimeters to 30.0 centimeters (1 to 12 inches) or more is grafted between the stock and scion (i.e., three sections of stem are used for two grafts). Double-working grafts are used for special purposes. One such purpose is dwarfing, usually achieved by using special combinations of materials but sometimes by this method of grafting. A young tree is cut off above the ground, an additional segment is cut to serve as the interstock, and the stock, interstock, and top (scion) are grafted together with the interstock inverted. This method will work only with certain varieties, as inversion normally effectively blocks the flow of materials up and down the stem. Other purposes of double-working grafts—in which the interstock is not inverted—include the influencing of growth so as to promote greater flower production than would otherwise occur, providing a disease- or cold-resistant trunk, and circumventing graft incompatibility (the failure of grafts to form permanent unions). With regard to this last purpose, if scions of certain varieties will not produce permanent unions when grafted to stocks of another variety but will form good grafts with a third variety, it may be possible to graft the third variety to the stock so that it can function as an interstock, thereby circumventing the problem.

ADDITIONAL READING

Adams, C. F. 1981. *Nutritional value of American foods in common units.* Washington, DC: Government Printing Office.

Baker, J. 1991. *Jerry Baker's happy, healthy house plants.* New York: New American Library/Dutton.

Bonar, A., and D. MacCarthy. 1991. *Practical gardening: How to grow and use herbs.* New York: Sterling Publishing Co., Inc.

Davidson, W. 1988. *Houseplants.* New York: State Mutual Book and Periodical Service.

Hartmann, H. T., D. E. Kester, and F. T. Davies Jr. 1990. *Plant propagation: Principles and practices,* 5th ed. Englewood Cliffs, NJ: Prentice-Hall.

Hill, L. 1986. *Pruning simplified* (updated ed.). Pownal, VT: Storey Communications, Inc.

Hill, L. 1992. *Fruits and berries for the home garden* (rev. ed.). Pownal, VT: Garden Way Publishing.

Kramer, J. 1992. *Know your houseplants.* New York: Lyons and Burford Pubs., Inc.

Lesniewicz, P. 1984. *Bonsai: The complete guide to art and technique.* Englewood Cliffs, NJ: Sterling Publishing Co., Inc.

Sunset Editors. 1977. *How to grow orchids.* Menlo Park, CA: Lane Publishing Co.

Sunset Editors. 1983. *House plants,* 4th ed. Menlo Park, CA: Lane Publishing Co.

Wyman, D. 1987. *The gardening encyclopedia* (updated). New York: Macmillan.

Appendix 5

Metric Equivalents and Conversion Tables

METRIC SYSTEM OF MEASUREMENT

APPLICATION	INTERNATIONAL SYSTEM OF UNITS	ENGLISH SYSTEM EQUIVALENTS
Length	Kilometer	0.62137 miles
	Meter	39.37 inches
	Centimeter	0.3937 inch
	Millimeter	0.03937 inch
	Micrometer (Micron)	0.00003937 inch
	Nanometer	0.00000003937 inch
	Angstrom	0.0000000003937 inch
Mass (Weight)	Metric ton	2,200 pounds
	Kilogram	2.2 pounds
	Gram	0.03527 ounce
	Milligram	0.00003527 ounce
Volume	Liter	1.06 quart
	Milliliter	0.00106 quart
	Cubic meter	35.314 cubic feet
	Cubic centimeter	0.061 cubic inch
Area	Hectare	2.471 acres
Temperature	To convert Celsius to Fahrenheit, multiply the Celsius figure by 9, divide the total by 5, and add 32 (see next page).	

CONVERSION TABLE

TO CONVERT	TO	MULTIPLY BY
Millimeters	Inches	0.039
Centimeters	Inches	0.39
Meters	Yards	1.094
Kilometers	Miles	0.6214
Inches	Centimeters	2.54
Feet	Centimeters	30.48
Yards	Meters	0.9144
Miles	Kilometers	1.609
Grams	Ounces	0.035
Kilograms	Pounds	2.205
Ounces	Grams	28.35
Pounds	Grams	453.6
Pounds	Kilograms	0.4536

(Table continues)

TO CONVERT	TO	MULTIPLY BY
Milliliters	Fluid ounces	0.03
Liters	Quarts	1.057
Liters	Gallons	0.2642
Fluid ounces	Milliliters	29.57
Quarts	Liters	0.9463
Gallons	Liters	3.785
Hectares	Acres	2.471
Square kilometers	Square miles	0.3861
Square inches	Square feet	6.944×10^{-3}
Square feet	Square inches	144.0
Acres	Hectares	0.4047
Square miles	Square kilometers	2.590

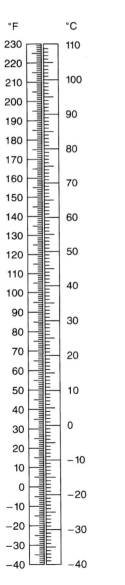

Temperature Conversion Scale

To convert Fahrenheit to Celsius use the following formula:

$$C = {}^5/_9 \left({}^\circ F - 32\right)$$

To convert Celsius to Fahrenheit use the following formula:

$$C = {}^9/_5 \, {}^\circ C + 32$$

Glossary

A

abscisic acid (ab-siz′ik as′id) **(ABA)** a growth-inhibiting hormone of plants; involved with other hormones in dormancy (p. 186)

abscission (ab-sizh′un) the separation of leaves, flowers, and fruits from plants after the formation of an abscission zone at the base of their petioles, peduncles, and pedicels (p. 116)

achene (uh-keen′) a single-seeded fruit in which the seed is attached to the pericarp only at its base (p. 130)

acid (as′id) a substance that dissociates in water, releasing hydrogen ions (p. 20)

active transport (ak′tiv trans′port) the expenditure of energy by a cell in moving a substance across a plasma membrane against a diffusion gradient (p. 148)

adventitious (ad-ven-tish′uss) said of buds developing in internodes or on roots or of roots developing along stems or on leaves (pp. 66, 88)

aerobic respiration (air-oh′bik res-puh-ray′shun) respiration that requires free oxygen (pp. 171, 172)

agar (ah′gur) a gelatinous substance produced by certain red algae and also a few brown algae; often used as a culture medium, particularly for bacteria (pp. 308, 310)

aggregate fruit (ag′gruh-git froot) a fruit derived from a single flower having several to many pistils (p. 127)

algin (al′jin) a gelatinous substance produced by certain brown algae; used in a wide variety of food substances and in pharmaceutical, industrial, and household products (pp. 304, 309)

allele (uh-leel′) one of a pair of genes at the identical location (*locus*) on a pair of homologous chromosomes (p. 214)

Alternation of Generations (ol-tur-nay′shun uv jen-ur-ay′shunz) alternation between a haploid gametophyte phase and a diploid sporophyte phase in the life cycle of sexually reproducing organisms (p. 209)

amino acid (ah-mee′noh as′id) one of the organic, nitrogen-containing units from which proteins are synthesized; there are about 20 in all proteins (p. 22)

anaerobic respiration (an-air-oh′bik res-puh-ray′shun) respiration in which the hydrogen removed from the glucose during glycolysis is combined with an organic ion (instead of oxygen) (p. 171)

angiosperm (an′jee-oh-spurm) a plant whose seeds develop within ovaries that mature into fruits (p. 402)

annual (an′you-ul) a plant that completes its entire life cycle in a single growing season (pp. 83, 122)

annual ring (an′you-ul ring) a single season′s production of xylem (wood) by the vascular cambium (p. 84)

annulus (an′yu-luss) a specialized layer of cells around a fern sporangium; aids in spore dispersal through a springlike action; also a membranous ring around the stipe of a mushroom (p. 372)

anther (an′thur) the pollen-bearing part of a stamen (p. 403)

antheridiophore (an-thur-id′ee-oh-for) a stalk that bears an antheridium (p. 348)

antheridium (pl. **antheridia**) (an-thur-id′ee-um; pl. an-thur-id′ee-ah) the male gametangium of certain algae, fungi, bryophytes, and vascular plants other than gymnosperms and angiosperms (pp. 301, 315, 348)

anthocyanin (an-thoh-sy′ah-nin) a water-soluble pigment found in cell sap; anthocyanins vary in color from red to blue (p. 39)

antibiotic (an-tee-by-ot′ik) a substance produced by a living organism that interferes with the normal metabolism of another living organism (p. 335)

apical dominance (ay′pi-kul dom′i-nunts) suppression of growth of lateral buds by hormones (p. 188)

apical meristem (ay′pi-kul mair′i-stem) a meristem at the tip of a shoot or root (pp. 50, 63)

apomixis (ap-uh-mik′sis) reproduction without fusion of gametes or meiosis in otherwise normal sexual structures (p. 248)

archegoniophore (ahr-kuh-goh′nee-oh-for) a stalk bearing an archegonium (p. 348)

archegonium (pl. **archegonia**) (ahr-kuh-goh′nee-um; pl. ahr-kuh-goh′nee-ah) the multicellular female gametangium of bryophytes and most vascular plants other than angiosperms (p. 348)

aril (air'il) an often brightly colored appendage surrounding the seed of certain plants (e.g., yew) (p. 388)

ascus (pl. **asci**) (as'kus; pl. as'eye) one of often numerous, frequently fingerlike hollow structures in which the fusion of two haploid nuclei is followed by meiosis; a row of ascospores (usually eight) is ultimately produced in each ascus on or within the sexually initiated reproductive bodies of cup (sac) fungi (p. 323)

asexual reproduction (ay-seksh'yule ree-proh-duk'shun) any form of reproduction not involving the union of gametes (p. 204)

assimilation (uh-sim-i-lay'shun) cellular conversion of raw materials into protoplasm and cell walls (p. 16)

atom (at'um) the smallest individual unit of an element that retains the properties of the element (p. 16)

ATP (ay-tee-pee) adenosine triphosphate, a molecule with three phosphate groups found in all living cells; the principal vehicle for energy storage and exchange in cell metabolism (p. 164)

autotrophic (aw-toh-troh'fik) descriptive of an organism capable of sustaining itself through conversion of inorganic substances to organic material (p. 273)

auxin (awk'sin) a growth-regulating substance produced either naturally by plants or synthetically (p. 182)

axil (ak'sil) the angle formed between a twig and the petiole of a leaf; normally the site of an *axillary bud* (also called *lateral bud*) (p. 80)

B

backcross (bak'kross) a cross involving a hybrid and one of its parents (p. 218)

bacteriophage (bak-teer'ee-oh-fayj) a virus whose host is a bacterium (p. 285)

bark (bahrk) tissues of a woody stem between the vascular cambium and the exterior (p. 85)

base (bayss) a substance that dissociates in water, releasing hydroxyl (OH⁻) ions (p. 20)

basidiospore (bah-sidd'ee-oh-spor) a spore produced on a basidium (p. 329)

basidium (pl. **basidia**) (buh-sid'ee-um; pl. buh-sid'ee-ah) one of usually numerous, frequently club-shaped hollow structures in which the fusion of two haploid nuclei is followed by meiosis, the four resulting nuclei becoming externally borne basidiospores; basidia are produced on or within sexually initiated reproductive bodies of the club fungi (e.g., mushrooms, puffballs) (pp. 328, 329)

berry (bair'ee) a thin-skinned fruit that usually develops from a compound ovary and commonly contains more than one seed (p. 126)

biennial (by-en'ee-ul) a plant that normally requires two seasons to complete its life cycle, the first season's growth being strictly vegetative (p. 122)

biological controls (by-oh-loj'i-kull kun-trohlz') the use of natural enemies and inhibitors in combating insect pests and other destructive organisms (pp. 67, 487)

biome (by'ohm) similar biotic communities considered on a worldwide scale (e.g., desert biome, grassland biome) (p. 460)

biotechnology (by-oh-tek-nol'-oh-jee) the manipulation of organisms, tissues, cells, or molecules for specific applications primarily intended for human benefit (p. 228)

biotic community (by-ot'ik kuh-myu'nit-ee) an association of plants, animals, and other organisms (e.g., woodland) (p. 449)

blade (blayd) the conspicuous, flattened part of a leaf (also called *lamina*) or seaweed (pp. 103, 303)

bond (bond) a force that holds atoms together (p. 19)

bonsai (bon-sy') container-grown plants (usually trees) that have been dwarfed artificially through skillful pruning and manipulation of the growing medium (p. 528)

botany (bot'an-ee) science involving the study of plants (p. 6)

botulism (bot'yu-lizm) poisoning from consumption of food infected by botulism bacteria (p. 275)

bract (brakt) a structure that is usually leaflike and modified in size, shape, or color (p. 112)

bryophyte (bry'oh-fyt) a photosynthetic, terrestrial, aquatic, or epiphytic embryo-producing plant without xylem and phloem (e.g., mosses, liverworts, hornworts) (p. 344)

budding (bud'ing) a form of grafting in which the scion is a bud surrounded by a small shield of tissue; also a form of asexual reproduction in which a new cell develops to full size from a protuberance arising from a mature cell, as in yeasts (pp. 323, 530)

bulb (buhlb) an underground food-storage organ that is essentially a modified bud consisting of fleshy leaves that surround and are attached to a small stem (p. 91)

bundle scar (bun'dul skahr) a small scar left by a vascular bundle within a leaf scar when the leaf separates from its stem through abscission (p. 80)

bundle sheath (bun'dul sheeth) the parenchyma and/or sclerenchyma cells surrounding a vascular bundle (p. 107)

C

callose (kal'ohs) a complex carbohydrate that develops in sieve tubes following an injury; commonly associated with the sieve areas of sieve-tube members (p. 16)

callus (kal'uss) undifferentiated tissue that develops around injured areas of stems and roots; also the undifferentiated tissue that develops during tissue culture (pp. 16, 234)

Calvin cycle (kal'vin sy'kuhl) see *carbon-fixing* reactions (p. 164)

calyptra (kuh-lip'truh) tissue from the enlarged archegonial wall of many mosses that forms a partial or complete cap over the capsule (p. 348)

calyx (kay'liks) collective term for the sepals of a flower (p. 124)

cambium (kam'bee-um) a meristem producing secondary tissues; see *vascular cambium, cork cambium* (p. 41, 50)

capillary water (kap'i-lair-ee waw'tur) water held in the soil against the force of gravity; available to plants (p. 75)

capsule (kapp'sool) a dry fruit that splits in various ways at maturity, often along or between carpel margins; also the main part of a sporophyte, in which different types of tissues develop (pp. 128, 348)

carbohydrate (kahr-boh-hy'drayt) an organic compound containing carbon, hydrogen, and oxygen, with twice as many hydrogen as oxygen atoms per molecule (p. 21)

carbon-fixing reactions (kahr′bon fixing ree-ak′shunz) a cyclical series of chemical reactions that utilizes carbon dioxide and energy generated during the light reactions of photosynthesis. Sugars are produced, some of which are stored as insoluble carbohydrates while others are recycled; the reactions are independent of light and occur in the stroma of chloroplasts (pp. 163, 164)

carpel (kahr′pul) an ovule-bearing unit that is a part of a pistil (pp. 126, 402)

caryopsis (kare-ee-op′siss) a dry fruit in which the pericarp is tightly fused to the seed; does not split at maturity (p. 131)

Casparian strip (kass-pair′ee-un strip) a band of suberin around the radial and transverse walls of an endodermal cell (p. 65)

cell (sel) the basic structural and functional unit of living organisms; in plants it consists of protoplasm surrounded by a cell wall (p. 28)

cell biology (sel by-ol′uh-jee) the biological discipline involving the study of cells and their functions (p. 10)

cell cycle (sel sy′kul) a sequence of events involved in the division of a cell (p. 40)

cell membrane (sel mem′brayn) see *plasma membrane*

cell plate (sel playt) the precursor of the middle lamella; forms at the equator during telophase (p. 43)

cell sap (sel sap) the liquid contents of a vacuole (p. 39)

cell wall (sel wawl) the relatively rigid boundary of cells of plants and certain other organisms (p. 31)

centromere (sen′truh-meer) the dense constricted portion of a chromosome to which a spindle fiber is attached (also called *kinetochore*) (p. 41)

chemiosmosis (kem-ee-oz-moh′siss) a theory that energy is provided for phosphorylation by protons being "pumped" across inner mitochondrial and thylakoid membranes (p. 175)

chiasma (pl. **chiasmata**) (kyaz′mah; pl. ky-az′mah-tah) the X-shaped configuration formed by two chromatids of homologous chromosomes as they remain attached to each other during prophase I of meiosis (p. 205)

chlorenchyma (klor-en′kuh-mah) tissue composed of parenchyma cells that contain chloroplasts (p. 51)

chlorophyll (klor′uh-fil) green pigments essential to photosynthesis (p. 36)

chloroplast (klor′uh-plast) an organelle containing chlorophyll, found in cells of most photosynthetic organisms (p. 36)

chromatid (kroh′muh-tid) one of the two strands of a chromosome; united by a centromere (p. 205)

chromatin (kroh′muh-tin) a readily staining complex of DNA and proteins found in chromosomes (p. 39)

chromoplast (kroh′muh-plast) a plastid containing pigments other than chlorophyll; the pigments are usually yellow to orange (p. 36)

chromosome (kroh′muh-sohm) a body consisting of a linear sequence of genes and composed of DNA and proteins; chromosomes are found in cell nuclei and appear in contracted form during mitosis and meiosis (p. 39)

cilium (pl. **cilia**) (sil′ee-um; pl. sil′ee-uh) a short hairlike structure usually found on the cells of unicellular aquatic organisms, normally in large numbers and arranged in rows; the most common function of cilia is propulsion of the cell (p. 196)

circadian rhythm (sur-kay′dee-an rith′um) a mostly daily rhythm of growth and activity found in living organisms (p. 193)

cladistics (kluh-diss′tiks) a classification system based on analysis of shared features (p. 263)

cladophyll (klad′uh-fil) a flattened stem that resembles a leaf; also called *phylloclade* (p. 93)

class (klas) a category of classification between a division and an order (pp. 257, 258)

climax vegetation (kly′maks vej-uh-tay′shun) vegetational association that perpetuates itself indefinitely at the culmination of ecological succession (p. 455)

coenocytic (see-nuh-sit′ik) having many nuclei not separated from one another by crosswalls, as in the hyphae of water molds (p. 315)

cohesion-tension theory (koh-hee′zhun ten′shun thee′uh-ree) theory that explains the rise of water in plants as resulting from a combination of cohesion of water molecules in vessels and tracheids and tension on the water columns brought about by transpiration (p. 150)

coleoptile (koh-lee-op′tul) a protective sheath surrounding the emerging shoot of seedlings of the Grass Family (Poaceae) (e.g., corn, wheat) (p. 136)

coleorhiza (koh-lee-uh-ry′zuh) a protective sheath surrounding the emerging radicle (immature root) of members of the Grass Family (Poaceae) (e.g., corn, wheat) (p. 136)

collenchyma (kuh-len′kuh-muh) tissue composed of cells with unevenly thickened walls (p. 52)

colloid (kol′oyd) a substance consisting of a medium in which fine particles are permanently dispersed (p. 74)

community (kuh-myu′nit-ee) a collective term for all the living organisms sharing a common environment and interacting with one another (p. 449)

companion cell (kum-pan′yun sel) a specialized cell derived from the same parent cell as the closely associated sieve-tube member immediately adjacent to it (in angiosperm phloem) (p. 54)

compound (kom′pownd) a substance whose molecules are composed of two or more elements (p. 18)

compound leaf (kom′pownd leef) a leaf whose blade is divided into distinct leaflets (p. 104)

conidium (pl. **conidia**) (kuh-nid′ee-um; pl. kuh-nid′ee-uh) an asexually produced fungal spore formed outside of a sporangium (p. 323)

conifer (kon′i-fur) a cone-bearing tree or shrub (p. 384)

conjugation (kon-juh-gay′shun) a process leading to the fusion of isogametes in algae, fungi, and protozoa; also the means by which certain bacteria exchange DNA (p. 300)

conjugation tube (kon-juh-gay′shun t(y)oob) a tube permitting transfer of a gamete or gametes between adjacent cells, as in *Spirogyra* or desmids (p. 299)

cork (kork) tissue composed of cells whose walls are impregnated with suberin at maturity; the outer layer of tissue of an older woody stem; produced by the cork cambium (p. 57)

cork cambium (kork kam′bee-um) a narrow cylindrical sheath of cells between the exterior of a woody root or stem and the central vascular tissue; produces *cork* to its exterior and *phelloderm* to its interior; also called *phellogen* (pp. 41, 50, 82)

corm (korm) a vertically oriented, thickened food-storage stem that is usually enveloped by a few papery nonfunctional leaves (p. 91)

corolla (kuh-rahl′uh) collective term for the petals of a flower (p. 124)

cortex (kor'teks) a primary tissue composed mainly of parenchyma; the tissue usually extends between the epidermis and the vascular tissue (pp. 64, 81)

cotyledon (kot-uh-lee'dun) an embryo leaf ("seed leaf") that usually either stores or absorbs food (pp. 83, 136)

covalent bond (koh-vay'luhnt bond) a force provided by pairs of electrons that travel between two or more atomic nuclei; the force holds atoms together and keeps them at a stable distance from each other (p. 19)

crossing-over (kross'ing oh'vur) the exchange of corresponding segments of chromatids between homologous chromosomes during prophase I of meiosis (p. 205)

cuticle (kyut'i-kul) a waxy or fatty layer of varying thickness on the outer walls of epidermal cells (p. 55)

cutin (kyu'tin) the waxy or fatty substance of which a cuticle is composed (pp. 55, 105)

cyclosis (sy-kloh'sis) the flowing or streaming of cytoplasm within a cell (p. 39)

cytochrome (sy'toh-krohm) iron-containing protein involved in molecule transfer in an electron transport system (p. 167)

cytokinesis (sy-toh-kuh-nee'sis) cell division (p. 41)

cytokinin (syt-uh-ky'nin) a growth hormone involved in cell division and several other metabolic activities of cells (p. 186)

cytology (sy-tol'uh-jee) see *cell biology*

cytoplasm (sy'tuh-plazm) the protoplasm of a cell exclusive of the nucleus (p. 31)

cytoplasmic streaming (sy-tuh-plaz'mik streem'ing) see *cyclosis*

D

dark reactions (dahrk ree-ak'shunz) see *carbon-fixing reactions*

day-neutral plant (day new'trul plant) a plant that is not dependent on specific day lengths for the initiation of flowering (p. 197)

deciduous (duh-sij'yu-wuss) shedding leaves annually (pp. 80, 116)

decomposer (dee-kuhm-poh'zur) organism (e.g., bacterium, fungus) that breaks down organic material to forms capable of being recycled (p. 451)

development (duh-vel'up-ment) changes in the form of a plant resulting from growth and differentiation of its cells into tissues and organs (p. 181)

dicotyledon (dy-kot-uh-lee'dun) a class of angiosperms whose seeds commonly have two cotyledons; frequently abbreviated to *dicot* (pp. 83, 122)

dictyosome (dik'tee-uh-sohm) see *Golgi body*

differentially permeable membrane (dif-uh-rensh'uh-lee pur'mee-uh-bul mem'brayn) a membrane through which different substances diffuse at different rates (p. 145)

differentiation (dif-uh-ren-shee-ay'shun) the change of a relatively unspecialized cell to a more specialized one (e.g., the change of a cell just produced by a meristem to a vessel member or fiber) (p. 181)

diffusion (dif-fyu'zhin) the random movement of molecules or particles from a region of higher concentration to a region of lower concentration, ultimately resulting in uniform distribution (p. 144)

digestion (duh-jes'jin) an enzyme-controlled conversion of complex, usually insoluble substances to simpler, usually soluble substances (pp. 16, 177)

dihybrid cross (dy-hy'brid kross) a cross involving two different pairs of chromosomes in the parents (p. 217)

dikaryotic (dy-kair-ee-ot'ik) having a pair of nuclei in each cell or a type of the mycelium in club fungi (p. 329)

dioecious (dy-ee'shuss) having unisexual flowers or cones, with the male flowers or cones confined to certain plants and the female flowers or cones of the same species confined to other different plants (p. 410)

diploid (dip'loyd) having two sets of chromosomes in each cell; the 2*n* chromosome number characteristic of the sporophyte generation (p. 209)

diuretic (dy-yu-ret'ik) a substance tending to increase the flow of urine (p. 370)

division (duh-vizh'un) the largest undivided category of classification of organisms within a kingdom; equivalent of phylum in animals (p. 258)

DNA (dee-en-ay) standard abbreviation of deoxyribonucleic acid, the carrier of genetic information in cells and viruses (pp. 24, 221)

dominant (dom'uh-nint) the member of a pair of genes that masks or suppresses the phenotypic expression of the recessive gene (p. 214)

dormancy (dor'man-see) a period of growth inactivity in seeds, buds, bulbs, and other plant organs even when environmental conditions normally required for growth are met (pp. 136, 200)

double fusion (dub'ul fu'shun) the more or less simultaneous union of one sperm and egg (forming a zygote) and union of another sperm and polar nuclei (forming a primary endosperm nucleus) that occur in the embryo sac of flowering plants (p. 407)

drupe (droop) a simple fleshy fruit whose single seed is enclosed within a hard endocarp (p. 126)

E

ecology (ee-kol'uh-jee) the biological discipline involving the study of the relationships of organisms to each other and to their environment (p. 446)

ecosystem (ee'koh-sis-tim) a system involving interactions of living organisms with one another and with their nonliving environment (p. 449)

egg (eg) a nonmotile female gamete (pp. 204, 301)

elater (el'uh-tur) a straplike appendage (usually occurring in pairs) attached to a horsetail (*Equisetum*) spore; also, a somewhat spindle-shaped sterile cell occurring in large numbers in liverwort sporangia; both types of elaters facilitate spore dispersal (pp. 348, 368)

electron (ee-lek'tron) a negatively charged particle of an atom (p. 17)

element (el'uh-mint) one of more than 1 types of matter, most existing naturally but some human-made, each of which is composed of one kind of atom (p. 16)

embryo (em'bree-oh) immature sporophyte that develops from a zygote within an ovule or archegonium after fertilization (pp. 61, 344, 348)

embryo sac (em'bree-oh sak) the female gametophyte of angiosperms, which, in approximately 70% of the species investigated, contains eight nuclei (p. 403)

enation (ee-nay'shun) one of the tiny, green leaflike outgrowths on the stems of whisk ferns (*Psilotum*) (p. 360)

endocarp (en'doh-kahrp) the innermost layer of a fruit wall (p. 126)

endodermis (en-doh-dur'mis) a single layer of cells surrounding the vascular tissue (stele) in roots and some stems; the cells have Casparian strips (p. 65)

endoplasmic reticulum (en-doh-plaz′mik ruh-tik′yu-lum) a complex system of interlinked double membrane channels subdividing the cytoplasm of a cell into compartments; parts of it are lined with ribosomes (p. 34)

endosperm (en′doh-spurm) a food-storage tissue that develops through divisions of the primary endosperm nucleus; digested by the sporophyte after germination in some species (e.g., corn) or before maturation of the seed in other species (e.g., beans) (p. 407)

energy (en′ur-jee) the capacity to do work; some forms of energy are heat, light, and kinetic (p. 20)

enzyme (en′zym) one of numerous complex proteins that speeds up a chemical reaction in living cells without being used up in the reaction (i.e., it catalyzes the reaction) (pp. 18, 181)

epicotyl (ep′uh-kaht-ul) the part of an embryo or seedling above the attachment point of the cotyledon(s) (p. 136)

epidermis (ep-uh-dur′mis) the exterior tissue, usually one cell thick, of leaves, young stems and roots, and other parts of plants (p. 55)

epiphyte (ep′uh-fyt) an organism that is attached to and grows on another organism without parasitizing it (p. 291)

ergotism (ur′got-izm) a disease resulting from ingestion of foods made with flour containing ergot fungus (p. 326)

essential element (uh-sen′shul el′uh-mint) one of 18 elements generally considered essential to the normal growth, development, and reproduction of most plants (p. 154)

ethylene (eth′uh-leen) a simple, naturally produced, gaseous hormone that inhibits plant growth and promotes the ripening of fruit (p. 186)

etiolation (ee-tee-oh-lay′shun) a condition characterized by long internodes, poor leaf development, and pale, weak appearance due to a plant's having been deprived of light (p. 198)

eukaryotic (yu-kair-ee-ot′ik) pertaining to cells having distinct membrane-bound organelles, including a nucleus with chromosomes (p. 30)

eutrophication (yu-troh-fuh-kay′shun) the gradual enrichment of a body of water through the accumulation of nutrients, resulting in a corresponding increase in algae and other organisms (p. 458)

exine (ek′seen *or* ek′syne) the outer layer of the wall of a pollen grain or spore (p. 403)

exocarp (ek′soh-kahrp) the outermost layer of a fruit wall (p. 126)

eyespot (eye′spot) a small, often reddish structure within a motile unicellular organism; appears to be sensitive to light (also called *stigma*) (p. 295)

F

F₁ (eff wun) first filial generation; the offspring of a cross between two parent plants (p. 214)

F₂ (eff too) second filial generation; the offspring of F₁ plants (p. 214)

FAD (eff-ay-dee) flavin adenine dinucleotide, a hydrogen acceptor molecule involved in the Krebs cycle of respiration and in photosynthesis (pp. 168, 172)

family (famm′uh-lee) a classification category between genus and order (p. 258)

fat (fat) an organic compound containing carbon, hydrogen, and oxygen but with proportionately much less oxygen than is present in a carbohydrate molecule (p. 22)

fermentation (fur-men-tay′shun) respiration in which the hydrogen removed from the glucose during glycolysis is transferred back to pyruvic acid, creating substances such as ethyl alcohol or lactic acid (p. 171)

fertilization (fur-til-i-zay′shun) formation of a zygote through the fusion of two gametes (pp. 210, 407)

fiber (fy′bur) a long thick-walled cell whose protoplasm often is dead at maturity (p. 53)

filament (fil′uh-mint) threadlike body of certain algae and fungi; also the stalk portion of a stamen (pp. 124, 268)

fission (fish′un) the division of cells of bacteria and related organisms into two new cells (pp. 268)

flagellum (pl. **flagella**) (fluh-jel′um; pl. fluh-jel′uh) a fine threadlike structure protruding from a motile unicellular organism or the motile cells produced by multicellular organisms; functions primarily in locomotion (pp. 196, 268)

floret (flor′et) a small flower that is a part of the inflorescence of members of the Sunflower Family (Asteraceae) and the Grass Family (Poaceae) (p. 438)

florigen (flor′uh-jen) one or more hormones once thought from circumstantial evidence to initiate flowering but which have never been isolated or proved to exist (p. 198)

follicle (foll′uh-kuhl) a dry fruit that splits along one side only (p. 128)

food chain (food chayn) a natural chain of organisms of a community wherein each member of the chain feeds on members below it and is consumed by members above it, with autotrophic organisms (producers) being at the bottom; interconnected food chains are referred to as *food webs* (p. 451)

foot (foot) the basal part of the embryo of bryophytes and other plants; attached to and absorbs food from the gametophyte (p. 348)

fossil (fos′ul) the remains or impressions of any natural object that has been preserved in the earth's crust (p. 373)

frond (frond) a fern leaf; term occasionally also applied to palm leaves (p. 372)

fruit (froot) a mature ovary usually containing seeds; term also somewhat loosely applied to the reproductive structures of groups of plants other than angiosperms (p. 125)

fucoxanthin (fyu-koh-zan′thin) a brownish pigment occurring in brown and other algae (p. 293)

G

gametangium (pl. **gametangia**) (gam-uh-tan′jee-um; pl. gam-uh-tan′jee-ah) any cell or structure in which gametes are produced (p. 305)

gamete (gam′eet) a sex cell; one of two cells that unite, forming a zygote (p. 204)

gametophore (guh-me′toh-for) a stalk on which a gametangium is borne (p. 348)

gametophyte (guh-me′toh-fyte) the haploid (*n*) gamete-producing phase of the life cycle of an organism that exhibits Alternation of Generations (pp. 209, 210)

gemma (pl. **gemmae**) (jem′uh; pl. jem′ee) a small outgrowth of tissue that becomes detached from the parent body and is capable of developing into a complete new plant or other organism; gemmae are produced in cuplike structures on liverwort thalli and are also produced by certain fungi (p. 348)

gene (jeen) a unit of heredity; part of a linear sequence of such units occurring in the DNA of chromosomes (pp. 24, 214)

generative cell (jen'uh-ray-tiv sel) the cell of the male gametophyte of angiosperms that divides, producing two *sperms;* also, the cell of the male gametophyte of gymnosperms that divides, producing a *sterile cell* and a *spermatogenous cell* (p. 388)

genetic engineering (juh'net'ik en-juh-neer'ing) the introduction, by artificial means, of genes from one form of DNA into another form of DNA (p. 228)

genetics (juh-net'iks) the biological discipline involving the study of heredity (pp. 10, 213)

genotype (jeen'oh-typ) the genetic constitution of an organism; may or may not be visibly expressed, as contrasted with *phenotype* (p. 214)

genus (pl. **genera**) (jee'nus; pl. jen'er-ah) a category of classification between a family and a species (p. 256)

gibberellin (jib-uh-rel'in) one of a group of plant hormones that have a variety of effects on growth; they are particularly known for promoting elongation of stems (p. 185)

gill (gil) one of the flattened plates of compact mycelium that radiate out from the stalk on the underside of the caps of most mushrooms (p. 329)

girdling (gurd'ling) the removal of a band of tissues extending inward to the vascular cambium on the stem of a woody plant (p. 530)

gland (gland) a small body of variable shape and size that may secrete certain substances but that also may be functionless (pp. 57, 106)

glycolysis (gly-kol'uh-sis) the initial phase of all types of respiration in which glucose is converted to pyruvic acid without involving free oxygen (p. 171)

Golgi body (gohl'jee bod'ee) an organelle consisting of disc-shaped, often branching hollow tubules that apparently function in accumulating and packaging substances used in the synthesis of materials by the cell; also called *dictyosome;* collectively, the Golgi bodies of a cell may be referred to as the *Golgi apparatus* (p. 35)

graft (graft) the union of a segment of a plant, the *scion,* with a rooted portion, the *stock* (pp. 242, 529)

grain (grayn) see *caryopsis*

granum (pl. **grana**) (gra'num; pl. gra'nuh) a series of stacked thylakoids within a chloroplast (p. 36)

gravitational water (grav-uh-tay'shun-ul waw'tur) water that drains out of the pore spaces of a soil after a rain (p. 75)

gravitropism (grav-uh-troh'pism) growth response to gravity (p. 191)

ground meristem (grownd mair'i-stem) meristem that produces all the primary tissues other than the epidermis and stele (e.g., cortex, pith) (pp. 50, 63, 81)

growth (grohth) progressive increase in size and volume through natural development (pp. 15, 181)

guard cell (gahrd sel) one of a pair of specialized cells surrounding a stoma (pp. 57, 106)

guttation (guh-tay'shun) the exudation of water in liquid form from leaves due to root pressure (p. 153)

gymnosperm (jim'noh-spurm) a plant whose seeds are not enclosed within an ovary during their development (e.g., pine tree) (p. 383)

H

haploid (hap'loyd) having one set of chromosomes per cell, as in gametophytes; also referred to as having *n* chromosomes (as contrasted with 2*n* chromosomes in the *diploid* cells of sporophytes) (p. 209)

haustorium (pl. **haustoria**) (haw-stor'ee-um; pl. haw-stor'ee-uh) a protuberance of a fungal hypha or plant organ such as a root that functions as a penetrating and absorbing structure (p. 69)

heartwood (hahrt'wood) nonliving, usually darker-colored wood whose cells have ceased to function in water conduction (p. 85)

herbaceous (hur-bay'shuss *or* ur-bay'shuss) referring to nonwoody plants (p. 83)

herbal (hur'bul *or* ur'bul) a 16th- and 17th-century botany book that emphasized medicinal uses, edibility, and other utilitarian functions of plants (p. 7)

herbarium (pl. **herbaria**) (hur-bair'ee-um *or* ur-bair'ee-um; pl. hur-bair'ee-uh) a collection of dried pressed specimens, usually mounted on paper and provided with a label that gives collection information and an identification (p. 413)

heterocyst (het'uh-roh-sist) a transparent, thick-walled, slightly enlarged cell occurring in the filaments of certain blue-green bacteria (p. 279)

heterospory (het-uh-ross'por-ee) the production of both microspores and megaspores (p. 364)

heterotrophic (het-ur-oh-troh'fick) incapable of synthesizing food and therefore dependent on other organisms for it (p. 272)

heterozygous (het-uh-roh-zy'guss) having a pair of genes with contrasting characters at the same location on homologous chromosomes (p. 215)

holdfast (hold'fast) attachment organ or cell at the base of the thallus or filament of certain algae or blue-green bacteria (pp. 298, 303)

homologous chromosomes (hoh-mol'uh-guss kroh'muh-sohmz) pairs of chromosomes that associate together in prophase I of meiosis; each member of a pair is derived from a different parent (p. 205)

homozygous (hoh-moh-zy'guss) having a pair of genes with identical characters at the same location on a pair of homologous chromosomes (p. 214)

hormone (hor'mohn) an organic substance generally produced in minute amounts in one part of an organism and transported to another part of the organism where it controls or affects growth and development (p. 181)

hybrid (hy'brid) offspring of two parents that differ in one or more genes (pp. 216, 233)

hydathode (hy'duh-thohde) structure at the tip of a leaf vein through which water is forced by root pressures (p. 153)

hydrolysis (hy-drol'uh-sis) the breakdown of complex molecules to simpler ones as a result of the union of water with the compound; usually controlled by enzymes (p. 177)

hydrosere (hy'droh-sear) a primary succession that is initiated in a wet habitat (p. 456)

hygroscopic water (hy-gruh-skop'ik waw'tur) water that is chemically bound to soil particles and therefore unavailable to plants (p. 75)

hypha (pl. **hyphae**) (hy'fuh; pl. hy'fee) a single, usually tubular, threadlike filament of a fungus; *mycelium* is a collective term for hyphae (p. 320)

hypocotyl (hy-poh-kot'ul) the portion of an embryo or seedling between the radicle and the cotyledon(s) (p. 136)

hypodermis (hy-poh-dur'mis) a layer of cells immediately beneath the epidermis and distinct from the parenchyma cells of the cortex in certain plants (pp. 109, 384)

hypothesis (hy-poth'uh-sis) a postulated explanation for some observed facts that must be tested experimentally before it can be accepted as valid or discarded as incorrect (p. 6)

I

imbibition (im-buh-bish'un) adsorption of water and subsequent swelling of organic materials because of the adhesion of the water molecules to the internal surfaces (p. 146)

indusium (pl. **indusia**) (in-dew'zee-um; pl. in-dew'zee-uh) the small, membranous, sometimes umbrellalike covering of a developing fern *sorus* (p. 372)

inferior ovary (in-feer'ee-or oh'vuh-ree) an ovary to which parts of the calyx, corolla, and stamens have become more or less united so they appear to be attached at the top of it (pp. 124, 409)

inflorescence (in-fluh-res'ints) a collective term for a group of flowers attached to a common axis in a specific arrangement (p. 124)

inorganic (in-or-gan'ik) descriptive of compounds having no carbon atoms (p. 20)

integument (in-teg'yu-mint) the outermost layer of an ovule; usually develops into a seed coat; a gymnosperm ovule usually has a single integument, and an angiosperm ovule usually has two integuments (pp. 383, 386, 403)

intermediate-day plant (in-tur-me'dee-ut day plant) a plant that has two critical photoperiods; it will not flower if the days are either too short or too long (p. 197)

internode (in'tur-nohd) a stem region between nodes (p. 80)

ion (eye'on) a molecule or atom that has become electrically charged through the loss or gain of one or more electrons (p. 19)

isogamy (eye-sog'uh-me) sexual reproduction in certain algae and fungi having gametes that are alike in size (p. 299)

isotope (eye'suh-tohp) one of two or more forms of an element that have the same chemical properties but differ in the number of neutrons in the nuclei of their atoms (p. 17)

K

kinetochore (kuh-net'uh-kor) see *centromere*

kingdom (king'dum) the highest category of classification (e.g., Plant Kingdom, Animal Kingdom) (p. 258)

knot (not) a portion of the base of a branch enclosed within wood (p. 95)

Krebs cycle (krebz' sy'kul) a complex series of reactions following glycolysis in aerobic respiration that involves ATP, mitochondria, and enzymes and that results in the combining of free oxygen with protons and electrons from pyruvic acid to make water (p. 172)

L

lamina (lam'uh-nuh) see *blade*

lateral bud (lat'uh-rul bud) see *axil*

laticifer (luh-tis'uh-fur) specialized cells or ducts resembling vessels; they form branched networks of *latex*-secreting cells in the phloem and other parts of plants (p. 86)

leaf (leef) a flattened, usually photosynthetic structure arranged in various ways on a stem (pp. 50, 102)

leaf gap (leef gap) a parenchyma-filled interruption in a stem's cylinder of vascular tissue immediately above the point at which a branch of vascular tissue (*leaf trace*) leading to a leaf occurs (p. 81)

leaflet (leef'lit) one of the subdivisions of a compound leaf (p. 104)

leaf scar (leef skahr) the suberin-covered scar left on a twig when a leaf separates from it through abscission (p. 80)

leaf trace (leef trays) see *leaf gap*

legume (leg'yoom) a dry fruit that splits along two "seams," the seeds being attached along the edges (p. 128)

lenticel (lent'uh-sel) one of usually numerous, slightly raised, somewhat spongy groups of cells in the bark of woody plants; lenticels permit gas exchange between the interior of a plant and the external atmosphere (pp. 58, 83)

leucoplast (loo'kuh-plast) a colorless plastid commonly associated with starch accumulation (p. 37)

light reactions (lyt ree-ak'shunz) a series of chemical and physical reactions through which light energy is converted to chemical energy with the aid of chlorophyll molecules; in the process, water molecules are split, with hydrogen ions and electrons being produced and oxygen gas being released; ATP and $NADPH_2$ also are created (pp. 163, 164)

lignin (lig'nin) a polymer with which certain cell walls (e.g., those of wood) become impregnated (pp. 33, 52)

ligule (lig'yool) the tiny tonguelike appendage at the base of a spike moss (*Selaginella*) or quillwort (*Isoetes*) leaf; also, the outgrowth from the upper and inner side of a grass leaf at the point where it joins the sheath; also, the conspicuous straplike portion of the corolla of an outer floret in the flower head of a member of the Sunflower Family (Asteraceae) (p. 364)

linkage (link'ij) the tendency of two or more genes located on the same chromosome to be inherited together (p. 218)

lipid (lip'id) a general term for fats, fatty substances, and oils (p. 21)

locule (lok'yool) a cavity within an ovary or a sporangium (p. 128)

long-day plant (long day plant) a plant in which flowering is not initiated unless exposure to more than a critical day length occurs (p. 197)

M

mass flow hypothesis (mass flo hypoth'uh-sus) see *pressure-flow hypothesis*

megaphyll (meg'uh-fill) a leaf having branching veins; it is associated with a leaf gap (p. 359)

megaspore (meg'uh-spor) a spore that develops into a female gametophyte (pp. 364, 386, 387)

megaspore mother cell (meg'uh-spor muth'ur sel) a diploid cell that produces megaspores upon undergoing meiosis (pp. 364, 386, 403)

meiocyte (my'oh-syt) see *spore mother cell*

meiosis (my-oh'sis) the process of two successive nuclear divisions through which segregation of genes occurs and a single diploid ($2n$) cell becomes four haploid (n) cells (p. 204)

mericloning (mair′i-kloh-ning) multiplication of plants through culturing and artificial dividing of shoot meristems (p. 235)

meristem (mair′i-stem) a region in which undifferentiated cells divide (pp. 41, 50)

mesocarp (mez′uh-karp) the middle region of the fruit wall that lies between the exocarp and the endocarp (p. 126)

mesophyll (mez′uh-fil) parenchyma (chlorenchyma) tissue between the upper and lower epidermis of a leaf (p. 107)

metabolism (muh-tab′uh-lizm) the sum of all the interrelated chemical processes occurring in a living organism (p. 16)

microfilament (my′kroh-fil′uh-mint) a protein filament involved with cytoplasmic streaming and with contraction and movement in animal cells (p. 38)

microphyll (my′kroh-fil) a leaf having a single unbranched vein not associated with a leaf gap (p. 359)

micropyle (my′kroh-pyl) a pore or opening in the integuments of an ovule through which a pollen tube gains access to an embryo sac or archegonium of a seed plant (pp. 386, 403)

microspore (my′kroh-spor) a spore that develops into a male gametophyte (pp. 364, 385, 403)

microspore mother cell (my′kroh-spor muth′ur sel) a diploid cell that produces microspores upon undergoing meiosis (pp. 364, 385, 403)

microsporophyll (my-kroh-spor′uh-fil) a leaf, usually reduced in size, on or within which microspores are produced (p. 364)

microtubule (my′kroh-t(y)oo-byul) an unbranched tubelike proteinaceous structure commonly found inside the plasma membrane where it apparently regulates the addition of cellulose to the cell wall (p. 38)

middle lamella (mid′ul luh-mel′uh) a layer of material, rich in pectin, that cements two adjacent cell walls together (p. 31)

midrib (mid′rib) the central (main) vein of a pinnately veined leaf or leaflet (p. 105)

mitochondrion (pl. **mitochondria**) (my-toh-kon′dree-un; pl. my-toh-kon′dree-uh) an organelle containing enzymes that function in the Krebs cycle and the electron transport chain of aerobic respiration (p. 36)

mitosis (my-toh′sis) nuclear division, usually accompanied by cytokinesis, during which the chromatids of the chromosomes separate and two genetically identical daughter nuclei are produced (p. 40)

mixture (miks′chur) a substance containing two or more ingredients, the atoms and molecules of which retain a separate identity and are not in a fixed proportion to one another (p. 19)

molecule (mol′uh-kyul) the smallest unit of an element or compound retaining its own identity; consists of two or more atoms (pp. 16, 18)

monocotyledon (mon-oh-kot-uh-lee′dun) a class of angiosperms whose seeds have a single cotyledon; the term is commonly abbreviated to *monocot* (pp. 83, 122)

monoecious (moh-nee′shuss) having unisexual male flowers or cones and unisexual female flowers or cones both on the same plant (p. 410)

monohybrid cross (mon-oh-hy′brid kross) a cross involving a single pair of genes with contrasting characters in the parents (p. 217)

monokaryotic (mon-oh-kair-ee-ot′ik) having a single nucleus in each cell or unit of the mycelium in club fungi (p. 328)

monomer (mon′oh-mur) a simple individual molecular unit of a polymer (p. 21)

motile (moh′tul) capable of independent movement (p. 268)

multiple fruit (mul′tuh-pul froot) a fruit derived from several to many individual flowers in a single inflorescence (p. 128)

mushroom (mush′room) a sexually initiated phase in the life cycle of a club fungus, usually consisting of an expanded *cap* and a *stalk* (*stipe*) (p. 329)

mutation (myu-tay′shun) an inheritable change in a gene or chromosome (pp. 234, 247)

mycelium (my-see′lee-um) a mass of fungal hyphae (p. 320)

mycorrhiza (pl. **mycorrhizae**) (my-kuh-ry′zuh; pl. my-kuh-ry′zee) a symbiotic association between fungal hyphae and a plant root (p. 69)

N

n (en) having one set of chromosomes per cell (*haploid*) (p. 209)

NAD (en-ay-dee) nicotinamide adenine dinucleotide, a molecule that during respiration temporarily accepts electrons whose negative charges are balanced by also accepting protons and thereby hydrogen atoms (pp. 172, 174)

NADP (en-ay-dee-pee) nicotinamide adenine dinucleotide phosphate, a high-energy storage molecule that temporarily accepts electrons from Photosystem I in the light reactions of photosynthesis (p. 167)

nastic movement (nass′tik moov′mint) a nondirected movement of a flat organ (e.g., petal, leaf) in which the organ alternately bends up and down (p. 190)

neutron (new′tron) an uncharged particle in the nucleus of an atom (p. 17)

node (nohd) region of a stem where one or more leaves are attached (pp. 50, 80)

nonmeristematic tissue (non-mair′i-stem-atic tish′yu) a tissue composed of cells that have assumed various shapes and sizes related to their functions as they matured following their production by a meristem (p. 51)

nucellus (new-sel′us) ovule tissue within which an embryo sac develops (pp. 383, 386)

nuclear envelope (new′klee-ur en′vuh-lohp) a porous double membrane enclosing a nucleus (p. 38)

nucleic acid (new-klay′ik as′id) see *DNA, RNA*

nucleolus (pl. **nucleoli**) (new-klee′oh-luss; pl. new-klee′oh-ly) a somewhat spherical body within a nucleus; contains primarily RNA and protein; there may be more than one nucleolus per nucleus (p. 38)

nucleotide (new′klee-oh-tyd) the structural unit of DNA and RNA (p. 24)

nucleus (new′klee-uss) the organelle of a living cell that contains chromosomes and is essential to the regulation and control of all the cell's functions; also, the core of an atom (pp. 17, 38)

nut (nutt) one-seeded dry fruit with a hard, thick pericarp; develops with a cup or cluster of bracts at the base (p. 131)

O

oil (oyl) a fat in a liquid state (p. 22)

oogamy (oh-og′uh-mee) sexual reproduction in which the female gamete or egg is nonmotile and larger than the male gamete or sperm, which is motile (p. 301)

oogonium (pl. **oogonia**) (oh-oh-goh′nee-um; pl. oh-oh-goh′nee-ah) a female sex organ of certain algae and fungi; consists of a single cell that contains one to several eggs (pp. 301, 315)

operculum (oh-per′kyu-lum) the lid or cap that protects the peristome of a moss sporangium (p. 353)

orbital (or′buh-till) a volume of space in which a given electron occurs 90% of the time (p. 17)

order (or′dur) a category of classification between a class and a family (p. 258)

organelle (or-guh-nel′) a membrane-bound body in the cytoplasm of a cell; there are several kinds, each with a specific function (e.g., mitochondrion, chloroplast)[1] (p. 31)

organic (or-gan′ik) pertaining to or derived from living organisms and pertaining to the chemistry of carbon-containing compounds (p. 20)

osmosis (oz-moh′sis) the diffusion of water or other solvents through a differentially permeable membrane from a region of higher concentration to a region of lower concentration (p. 145)

osmotic potential (oz-mot′ik puh-ten′shil) potential pressure that can be developed by a solution separated from pure water by a differentially permeable membrane (the pressure required to prevent osmosis from taking place) (p. 145)

osmotic pressure (oz-mot′ik presh′ur) see *osmotic potential*

ovary (oh′vuh-ree) the enlarged basal portion of a pistil that contains an ovule or ovules and usually develops into a fruit (p. 124)

ovule (oh′vyool) a structure of seed plants that contains a female gametophyte and has the potential to develop into a seed (pp. 124, 383, 386)

P

palisade mesophyll (pal-uh-sayd′ mez′uh-fil) mesophyll having one or more relatively uniform rows of tightly packed, elongate, columnar parenchyma (chlorenchyma) cells beneath the upper epidermis of a leaf (p. 107)

palmately compound (pahl′mayt-lee kom′pownd) having leaflets or principal veins radiating out from a common point (p. 104)

1. Ribosomes, which are considered organelles, are an exception in that they are not bounded by a membrane.

palmately veined (pahl′mayt-lee vaynd) see *palmately compound*

papilla (pl. **papillae**) (puh-pil′uh; pl. puh-pill′ay) a small, usually rounded or conical protuberance (p. 299)

parenchyma (puh-ren′kuh-muh) thin-walled cells varying in size, shape, and function; the most common type of plant cell (p. 51)

parthenocarpic (par-thuh-noh-kar′pik) developing fruits from unfertilized ovaries; the resulting fruit is, therefore, usually seedless (p. 408)

passage cell (pas′ij sel) a thin-walled cell of an endodermis (p. 65)

pectin (pek′tin) a water-soluble organic compound occurring primarily in the middle lamella; becomes a jelly when combined with organic acids and sugar (p. 31)

pedicel (ped′i-sel) the individual stalk of a flower that is part of an inflorescence (p. 124)

peduncle (pee′dun-kul) the stalk of a solitary flower or the main stalk of an inflorescence (p. 124)

peptide bond (pep′tyd bond) the type of chemical bond formed when two amino acids link together in the synthesis of proteins (p. 22)

perennial (puh-ren′ee-ul) a plant that continues to live indefinitely after flowering (p. 122)

pericarp (per′uh-karp) collective term for all the layers of a fruit wall (p. 126)

pericycle (per′uh-sy-kul) tissue sandwiched between the endodermis and phloem of a root; often only one or two cells wide in transverse section; the site of origin of lateral roots (p. 65)

periderm (pair′uh-durm) outer bark; composed primarily of cork cells (p. 57)

peristome (per′uh-stohm) one or two series of flattened, often ornamented structures (teeth) arranged around the margin of the open end of a moss sporangium; the teeth are sensitive to changes in humidity and facilitate the release of spores (p. 353)

petal (pet′ul) a unit of a corolla; usually both flattened and colored (p. 124)

petiole (pet′ee-ohl) the stalk of a leaf (pp. 80, 103)

$P_{far-red}$ or P_{fr} (pee-far-red *or* pee-ef-ahr) a form of phytochrome (which see) (p. 198)

pH scale (pee-aitch) a symbol of hydrogen ion concentration indicating the degree of acidity or alkalinity (p. 20)

phage (fayj) see *bacteriophage*

phellogen (fel′uh-jun) see *cork cambium*

phenotype (fee′noh-typ) the physical appearance of an organism (p. 214)

pheromone (fer′uh-mohn) something produced by an organism that facilitates chemical communication with another organism (p. 489)

phloem (flohm) the food-conducting tissue of a vascular plant (p. 54)

photon (foh′ton) a unit of light energy (p. 165)

photoperiodism (foh-toh-pir′ee-ud-izm) the initiation of flowering and certain vegetative activities of plants in response to relative lengths of day and night (p. 197)

photosynthesis (foh-toh-sin′thuh-sis) the conversion of light energy to chemical energy; water, carbon dioxide, and chlorophyll are all essential to the process, which ultimately produces carbohydrate, with oxygen being released as a by-product (pp. 16, 161)

photosynthetic unit (foh-toh-sin-thet′ik yew′nit) one of two groups of about 250 to 4 pigment molecules each that function together in chloroplasts in the light reactions of photosynthesis; the units are exceedingly numerous in each chloroplast (p. 163)

photosystem (foh′toh-sis-tum) collective term for a specific functional aggregation of photosynthetic units (p. 165)

phytochrome (fy′tuh-krohm) protein pigment associated with the absorption of light; it is found in the cytoplasm of cells of green plants and occurs in interconvertible active and inactive forms ($P_{far\ red}$ and P_{red}); facilitates a plant's capacity to detect the presence (or absence) and duration of light (p. 198)

pilus (pl. **pili**) (py′lis; pl. py′lee) the equivalent of a conjugation tube in bacteria (p. 269)

pinna (pl. **pinnae**) (pin′uh; pl. pin′ee) a primary subdivision of a fern frond; the term is also applied to a leaflet of a compound leaf (p. 372)

pinnately compound (pin′ayt-lee kom′pownd) having leaflets or veins attached on both sides of a common axis (e.g., rachis, midrib) (p. 104)

pinnately veined (pin′ayt-lee vaynd) see *pinnately compound*

pistil (pis′tul) a female reproductive structure of a flower; composed of one or more carpels and consisting of an ovary, style, and stigma (p. 124)

pit (pit) a more or less round or elliptical thin area in a cell wall; pits occur in pairs opposite each other, with or without shallow, domelike borders (pp. 46, 54)

pith (pith) central tissue of a dicot stem and certain roots; consists of parenchyma cells that become proportionately less of the volume of woody plants as cambial activity increases the organ's girth (p. 81)

plankton (plank'ton) free-floating aquatic organisms that are mostly microscopic (p. 296)

plant anatomy (plant uh-nat'uh-mee) the botanical discipline that pertains to the internal structure of plants (p. 8)

plant community (plant kuh-myu'nuh-tee) an association of plants inhabiting a common environment and interacting with one another (p. 448)

plant ecology (plant ee-koll'uh-jee) the science that deals with the relationships and interactions between plants and their environment (p. 9)

plant geography (plant jee-og'ruh-fee) the botanical discipline that pertains to the broader aspects of the space relations of plants and their distribution over the surface of the earth (p. 9)

plant morphology (plant mor-fol'uh-jee) the botanical discipline that pertains to plant form and development (p. 9)

plant physiology (plant fiz-ee-ol'uh-jee) the botanical discipline that pertains to the metabolic activities and processes of plants (p. 8)

plant taxonomy (plant tak-son'uh-mee) the botanical discipline that pertains to the classification, naming, and identification of plants (p. 9)

plasma membrane (plaz'muh mem'brayn) the outer boundary of the protoplasm of a cell; also called *cell membrane,* particularly in animal cells (p. 33)

plasmid (plaz'mid) one of up to 30 or 40 small, circular DNA molecules usually present in a bacterial cell (p. 228)

plasmodesma (pl. **plasmodesmata**) (plaz-muh-dez'muh; pl. plaz-muh-dez'muh-tah) minute strands of cytoplasm that extend between adjacent cells through pores in the walls (p. 33)

plasmodium (pl. **plasmodia**) (plaz-moh'dee-um; pl. plaz-moh'dee-ah) the multinucleate, semiviscous liquid, active form of slime mold; moves in a "crawling-flowing" motion (p. 312)

plasmolysis (plaz-mol'uh-sis) the shrinking in volume of the protoplasm of a cell and the separation of the protoplasm from the cell wall due to loss of water via osmosis (p. 146)

plastid (plas'tid) an organelle associated primarily with the storage or manufacture of carbohydrates (e.g., *leucoplast, chloroplast*) (p. 36)

plumule (ploo'myool) the terminal bud of the embryo of a seed plant (p. 136)

polar nuclei (poh'lur new'klee-eye) nuclei, frequently two in number, that unite with a sperm in an embryo sac, forming a primary endosperm nucleus (p. 403)

pollen grain (pahl'un grayn) a structure derived from the microspore of seed plants that develops into a male gametophyte (pp. 124, 385, 403)

pollen tube (pahl'un t(y)oob) a tube that develops from a pollen grain and conveys the sperms to the female gametophyte (pp. 388, 407)

pollination (pahl-uh-nay'shun) the transfer of pollen from an anther to a stigma (p. 407)

pollinium (pl. **pollinia**) (pah-lin'ee-um; pl. pah-lin'ee-ah) a cohesive mass of pollen grains commonly found in members of the Orchid Family (Orchidaceae) and the Milkweed Family (Asclepiadaceae) (p. 412)

polymer (pahl'i-mur) a large molecule composed of many monomers (p. 21)

polyploidy (pahl'i-ploy-dee) having more than two complete sets of chromosomes per cell (p. 234)

pome (pohm) a simple fleshy fruit whose flesh is derived primarily from the receptacle (p. 127)

population (pop-yew-lay'shun) a group of organisms, usually of the same species, occupying a given area at the same time (pp. 448, 449)

P$_{red}$ or **P**$_r$ (pee-red *or* pee-ahr) a form of phytochrome (which see) (p. 198)

pressure-flow hypothesis (presh'ur floh hy-poth'uh-sis) the theory that food substances in solution in plants flow along concentration gradients between the sources of the food and sinks (places where the food is utilized) (p. 151)

primary consumer (pry'mer-ree kon-soo'mur) organism that feeds directly on producers (p. 451)

primary tissue (pry'mer-ee tish'yu) a tissue produced by an apical meristem (e.g., epidermis, cortex, primary xylem and phloem, pith) (p. 50)

primordium (pry-mord'ee-um) an organ or structure (e.g., leaf, bud) at its earliest stage of development (p. 81)

procambium (proh-kam'bee-um) a tissue produced by the primary meristem that differentiates into primary xylem and phloem (pp. 50, 63, 81)

producer (pruh-dew'sur) an organism that manufactures food through the process of photosynthesis (p. 451)

prokaryotic (proh-kair-ee-ot'ik) having a cell or cells that lack a distinct nucleus and other membrane-bound organelles (e.g., bacteria) (pp. 30, 267)

proplastid (proh-plas'tid) a tiny, undifferentiated organelle that can duplicate itself and that may develop into a chloroplast, leucoplast, or other type of plastid (p. 37)

protein (proht'ee-in *or* proh'teen) a polymer composed of many amino acids linked together by peptide bonds (p. 22)

prothallus (pl. **prothalli**) (proh-thal'us; pl. proh-thal'eye) the gametophyte of ferns and their relatives; also called *prothallium* (p. 372)

protoderm (proh'tuh-durm) the primary meristem that gives rise to the epidermis (pp. 50, 63, 81)

proton (proh'ton) a positively charged particle in the nucleus of an atom (p. 17)

protonema (proh-tuh-nee'muh) a green, usually branched, threadlike or sometimes platelike growth from a bryophyte spore; gives rise to "leafy" gametophytes (p. 349)

protoplasm (proh'tuh-plazm) the living part of a cell (includes the cytoplasm and nucleus) (p. 20)

pruning (proon'ing) removal of portions of plants for aesthetic purposes, for improving quality and size of fruits or flowers, or for elimination of diseased tissues (p. 527)

pyrenoid (py'ruh-noyd) a small body found on the chloroplasts of certain green algae and hornworts; associated with starch accumulation; may occur singly on a chloroplast or may be numerous (p. 297)

pyruvic acid (py-roo'vik as'id) the organic compound that is the end product of the glycolysis phase of respiration (p. 172)

Q

quiescence (kwy′ess-ens) a state in which a seed or other plant part will not germinate or grow unless environmental conditions normally required for growth are present (p. 200)

R

rachis (ray′kiss) the axis of a pinnately compound leaf or frond extending between the lowermost leaflets or pinnae and the terminal leaflet or pinna (corresponds with the midrib of a simple leaf) (p. 104)

radicle (rad′i-kuhl) the part of an embryo in a seed that develops into a root (pp. 61, 136)

ray (ray) radially oriented tiers of parenchyma cells that conduct food, water, and other materials laterally in the stems and roots of woody plants; generally continuous across the vascular cambium between the xylem and the phloem; the portion within the wood is called a *xylem ray,* while the extension of the same ray in the phloem is called a *phloem ray* (pp. 54, 85)

receptacle (ree-sep′tuh-kul) the commonly expanded tip of a peduncle or pedicel to which the various parts of a flower (e.g., calyx, corolla) are attached (p. 124)

recessive (ree-ses′iv) descriptive of a member of a pair of genes whose phenotypic expression is masked or suppressed by the dominant gene (p. 214)

red tide (red tyd) the marine phenomenon that results in the water becoming temporarily tinged with red due to the sudden proliferation of certain dino-flagellates that produce substances poisonous to animal life and humans (p. 294)

reproduction (ree-proh-duk′shun) the development of new individual organisms through either sexual or asexual means (p. 15)

resin canal (rez′in kuh-nal′) a tubular duct of many conifers and some angio-sperms that is lined with resin-secreting cells (pp. 85, 384)

respiration (res-puh-ray′shun) the cellular breakdown of sugar and other foods, accompanied by release of energy; in aerobic respiration, oxygen is utilized (pp. 16, 171, 174)

rhizoid (ry′zoyd) delicate root- or root-hairlike structures of algae, fungi, the gametophytes of bryophytes, and certain structures of a few vascular plants; function in anchorage and absorption but have no xylem or phloem (p. 315)

rhizome (ry′zohm) an underground stem, usually horizontally oriented, that may be superficially rootlike in appearance but that has definite nodes and internodes (p. 88)

ribosome (ry′boh-sohm) a granular particle composed of two subunits consisting of RNA and proteins; ribosomes lack membranes, are the sites of protein synthesis, and are very numerous in living cells (p. 35)

RNA (ar-en-ay) the standard abbreviation for *ribonucleic acid,* an important cellular molecule that occurs in three forms, all involved in communication between the nucleus and the cytoplasm and in the synthesis of proteins (pp. 24, 222)

root (root) a plant organ that functions in anchorage and absorption; most roots are produced below ground (p. 50)

root cap (root kap) a thimble-shaped mass of cells at the tip of a growing root; functions primarily in protection (p. 62)

root hair (root hair) a delicate protu-berance that is part of an epidermal cell of a root; root hairs occur in a zone behind the growing tip (p. 64)

runner (run′ur) a stem that grows horizontally along the surface of the ground; typically has long internodes; see also *stolon* (p. 88)

S

salt (salt) a substance produced by the bonding of ions that remain after hydrogen and hydroxyl ions of an acid and a base combine to form water (p. 20)

samara (sah-mair′uh) a dry fruit whose pericarp extends around the seed in the form of a wing (p. 132)

saprobe (sap′rohb) an organism that obtains its food directly from nonliving organic matter (p. 272)

sapwood (sap′wood) outer layers of wood that transport water and minerals in a tree trunk; sapwood is usually lighter in color than heartwood (p. 85)

science (sy′ints) a branch of study involved with the systematic observation, recording, organization, and classification of facts from which natural laws are derived and used predictively (p. 6)

scion (sy′un) a segment of plant that is grafted onto a stock (p. 529)

sclereid (sklair′id) a sclerenchyma cell that usually has one axis not conspicuously longer than the other; may vary in shape and is heavily lignified (p. 52)

sclerenchyma (skluh-ren′kuh-muh) tissue composed of lignified cells with thick walls; functions primarily in strengthening and support (p. 52)

secondary consumer (sek′on-dair-ee kon-soo′mer) an organism that feeds on other consumers (p. 451)

secondary tissue (sek′un-der-ee tish′yu) a tissue produced by the vascular cambium or the cork cambium (e.g., virtually all the xylem and phloem in a tree trunk) (p. 64)

secretory cell (see′kruh-tor-ee sel) cell (or tissue) producing a substance or sub-stances that are moved outside the cells (p. 53)

secretory tissue (see′kruh-tor-ee tish′yu) see *secretory cell*

seed (seed) a mature ovule containing an embryo and bound by a protective seed coat (pp. 124, 204)

seed coat (seed′ koht) the outer boundary layer of a seed; developed from the integument(s) (pp. 383, 386, 403)

semipermeable membrane (sem-ee-pur-me-uh-bil mem-brayn) see *differentially permeable membrane*

senescence (suh-ness′ints) the breakdown of cell components and membranes that leads to the death of the cell (p. 188)

sepal (see′puhl) a unit of the calyx that frequently resembles a reduced leaf; sepals often function in protecting the unopened flower bud (p. 124)

sessile (sess′uhl) without petiole or pedicel; attached directly by the base (p. 103)

seta (see′tuh) the stalk of a bryophyte sporophyte (p. 348)

sexual reproduction (seksh′yule ree-proh-duk′shun) reproduction involving the union of gametes (p. 204)

short-day plant (short day plant) a plant in which flowering is initiated when the days are shorter than its critical photoperiod (p. 197)

sieve plate (siv playt) an area of the wall of a sieve-tube member that contains several to many perforations that permit cytoplasmic connections between similar adjacent cells, the cytoplasmic strands being larger than plasmodesmata (p. 54)

sieve tube (siv t(y)oob) a column of sieve-tube members arranged end to end; food is conducted from cell to cell through sieve plates (p. 54)

sieve-tube member (siv t(y)oob mem'bur) a single cell of a sieve tube (p. 54)

silique (suh-leek') a dry fruit that splits along two "seams," with the seeds borne on a central partition (p. 128)

simple leaf (sim'pul leef) a leaf with the blade undivided into leaflets (p. 104)

solvent (sol'vent) a substance (usually liquid) capable of dissolving another substance (p. 145)

sorus (pl. **sori**) (sor'uss; pl. sor'eye) a cluster of sporangia; the term is most frequently applied to clusters of fern sporangia (p. 372)

species (spee'seez; *species* is spelled and pronounced the same way in either singular or plural form; there is no such thing as a *specie*) the basic unit of classification; a population of individuals capable of interbreeding freely with one another but, which, because of geographic, reproductive, or other barriers, do not in nature interbreed with members of other species (p. 256)

sperm (spurm) a male gamete; except for those of red algae and angiosperms, sperms are frequently motile and are usually smaller than the corresponding female gametes (pp. 204, 301)

spice (spyss) an aromatic organic plant product used to season or flavor food or drink (p. 506)

spindle (spin'dul) an aggregation of fiberlike threads (microtubules) that appears in cells during mitosis and meiosis; some threads are attached to the centromeres of chromosomes, whereas other threads extend directly or in arcs between two invisible points designated as poles (p. 42)

spine (spyn) a relatively strong, sharp-pointed, woody structure usually located on a stem; usually a modified leaf or stipule (p. 110)

spongy mesophyll (spun'jee mez'uh-fil) mesophyll having loosely arranged cells and numerous air spaces; generally confined to the lower part of the interior of a leaf just above the lower epidermis (p. 107)

sporangiophore (spuh-ran'jee-uh-for) the stalk on which a sporangium is produced (p. 321)

sporangium (pl. **sporangia**) (spuh-ran'jee-um; pl. spuh-ran'jee-uh) a structure in which spores are produced; may be either unicellular or multicellular (pp. 313, 321)

spore (spor) a reproductive cell or aggregation of cells capable of developing directly into a gametophyte or other body without uniting with another cell (*note:* a bacterial spore is not a reproductive cell but is an inactive phase that enables the cell to survive under adverse conditions); *sexual spores* formed as a result of meiosis are often called *meiospores;* spores produced by mitosis may be referred to as *vegetative spores* (pp. 210, 313)

spore mother cell (spor muth'ur sel) a diploid cell that becomes four haploid spores or nuclei as a result of undergoing meiosis (pp. 210, 348)

sporophyll (spor'uh-fil) a modified leaf that bears a sporangium or sporangia (p. 361)

sporophyte (spor'uh-fyt) the diploid (2*n*) spore-producing phase of the life cycle of an organism exhibiting Alternation of Generations (pp. 209, 210)

stamen (stay'min) a pollen-producing structure of a flower; consists of an anther and usually also a filament (p. 124)

stele (steel) the central cylinder of tissues in a stem or root; usually consists primarily of xylem and phloem (p. 83)

stem (stem) a plant axis with leaves or enations (p. 50)

stigma (stig'muh) the pollen receptive area of a pistil; also, the eyespot of certain motile algae (p. 124)

stipe (styp) the supporting stalk of seaweeds, mushrooms, and certain other stationary organisms (p. 329)

stipule (stip'yool) one of a pair of appendages of varying size, shape, and texture present at the base of the leaves of some plants (pp. 80, 103)

stock (stok) the rooted portion of a plant to which a scion is grafted (p. 242)

stolon (stoh'lun) a stem that grows vertically below the surface of the ground; typically has relatively long internodes; see also *runner* (p. 90)

stoma (pl. **stomata**) (stoh'muh; pl. stoh'mah-tuh) a minute pore or opening in the epidermis of leaves, herbaceous stems, and the sporophytes of hornworts (*Anthoceros*); flanked by two guard cells that regulate its opening and closing and thus regulate gas exchange and transpiration; the guard cells and pore are also collectively referred to as a *stoma* (pp. 36, 57, 106)

strobilus (pl. **strobili**) (stroh'buh-luss; pl. stroh'buh-leye) an aggregation of sporophylls on a common axis; usually resembles a cone or is somewhat conelike in appearance (pp. 361, 368)

stroma (stroh'muh) a region constituting the bulk of the volume of a chloroplast or other plastid; contains enzymes that in chloroplasts play a key role in carbon fixation, carbohydrate synthesis, and other photosynthetic reactions (p. 36)

style (styl) the structure that connects a stigma and an ovary (p. 124)

suberin (soo'buh-rin) a fatty substance found primarily in the cell walls of cork and the Casparian strips of endodermal cells (pp. 58, 82)

succession (suk-sesh'un) an orderly progression of changes in the composition of a community from the initial development of vegetation to the establishment of a climax community (p. 455)

sucrose (soo'krohs) a disaccharide composed of glucose and fructose; the primary form in which sugar produced by photosynthesis is transported throughout a plant (p. 21)

superior ovary (soo-peer'ee-or oh'vuh-ree) an ovary that is free from the calyx, corolla, and other floral parts, so the sepals and petals appear to be attached at its base (pp. 124, 409)

symbiosis (sim-by-oh'siss) an intimate association between two dissimilar organisms that benefits both of them (mutualism) or is harmful to one of them (parasitism) (p. 267)

syngamy (sin'gam-mee) a union of gametes; fertilization (p. 210)

T

2*n* (too-en) having two sets of chromosomes; diploid (p. 209)

3*n* (three-en) having three sets of chromosomes; triploid (p. 408)

tendril (ten′dril) a slender structure that coils on contact with a support of suitable diameter; usually is a modified leaf or leaflet, and aids the plant in climbing (p. 109)

thallus (pl. **thalli**) (thal′uss; pl. thal′eye) a multicellular plant body that is usually flattened and not organized into roots, stems, or leaves (pp. 304, 338, 347)

thylakoid (thy′luh-koyd) coin-shaped membranes whose contents include chlorophyll and that are arranged in stacks that form the *grana* of chloroplasts (p. 36)

tissue (tish′yu) an aggregation of cells having a common function (p. 50)

tissue culture (tish′yu kult′yur) the culture of isolated living tissue on an artificial medium (p. 234)

tracheid (tray′kee-id) a xylem cell that is tapered at the ends and has thick walls containing pits (p. 54)

transpiration (trans-puh-ray′shun) loss of water in vapor form; most transpiration takes place through the stomata (p. 103)

tropism (troh′pizm) the response of a plant organ or part to an external stimulus, usually in the direction of the stimulus (p. 190)

tuber (t(y)oo′bur) a swollen, fleshy underground stem (e.g., white potato) (p. 90)

turgid (tur′jid) firm or swollen because of internal water pressures resulting from osmosis (p. 146)

turgor movement (turr′gor moov′mint) the movement that results from changes in internal water pressures in a plant part (p. 192)

turgor pressure (tur′gur presh′ur) pressure within a cell resulting from the uptake of water (p. 146)

U

unisexual (yu-nih-seksh′yu-ul) a term usually applied to a flower lacking either stamens or a pistil (p. 410)

V

vacuolar membrane (vak-yu-oh′lur mem′brayn) the delimiting membrane of a cell vacuole; also called *tonoplast* (p. 39)

vacuole (vak′yu-ohl) a pocket of fluid that is separated from the cytoplasm of a cell by a membrane; may occupy more than 99% of a cell's volume in plants; also, food-storage or contractile pockets within the cytoplasm of unicellular organisms (p. 39)

vascular bundle (vas′kyu-lur bun′dul) a strand of tissue composed mostly of xylem and phloem and usually enveloped by a bundle sheath (p. 83)

vascular cambium (vas′kyu-lur kam′bee-um) a narrow cylindrical sheath of cells that produces secondary xylem and phloem in stems and roots (pp. 41, 50)

vascular plant (vas′kyu-lur plant) a plant having xylem and phloem (p. 344)

vein (vayn) a term applied to any of the vascular bundles that form a branching network within leaves (p. 107)

venter (ven′tur) the site of the egg in the enlarged basal portion of an archegonium (p. 351)

vessel (ves′uhl) one of usually very numerous cylindrical "tubes" whose cells have lost their cytoplasm; occur in the xylem of most angiosperms and a few other vascular plants; each vessel is composed of *vessel members* laid end to end; the perforated or open-ended walls of the vessel members permit water to pass through freely (p. 54)

vessel member (ves′uhl mem′bur) a single cell of a vessel (p. 54)

virus (vy′riss) a minute particle consisting of a core of nucleic acid, usually surrounded by a protein coat; incapable of growth alone and can reproduce only within, and at the expense of, a living cell (p. 283)

vitamin (vyt′uh-min) a complex organic compound produced primarily by photosynthetic organisms; various vitamins are essential in minute amounts to facilitate enzyme reactions in living cells (p. 182)

W

whorled (wirld) having three or more leaves or other structures at a node (p. 103)

X

xerosere (zer′roh-sear) a primary succession that initiates with bare rock (p. 455)

xylem (zy′lim) the tissue through which most of the water and dissolved minerals utilized by a plant are conducted; consists of several types of cells (p. 54)

Z

zoospore (zoh′uh-spor) a motile spore occurring in algae and fungi (p. 297)

zygote (zy′goht) the product of the union of two gametes (pp. 204, 407)

Index

C

water, *5*, 446, 447
Polychlorinated biphenyl, 161
Polygalacturonase, 239
Polyhedrosis virus, 288
Polymer, 21, 22, 24, 55
Polypeptide chain, 22, 34, 76, 223, *224, 285*
Polyploidy, **233–34**, 248
Polypodium, 375
Polysaccharide, 21, 31, 35, 160, 164, 278
Polysiphonia, 307–8
Polytrichum, 354
Pome, 127, 428
Pomegranate, 5, 126, 422
Ponderosa pine, 393, 458, 459, 466
Poor man's pepper, 200
Popcorn, 136
Poplar, 332, 334
Poplar leaf spot rust, 332
Poppy, 6, *62*, 128, 134, 138, *406*, 422, 426. *See also* California poppy; Mexican poppy; etc.
Poppy Family (Papaveraceae), 423, **426–27**
Population, 220, 246, 247–51, **448–49**
Populus, 256
Pore fungi, *330*
Portulaca Family, 116
Postelsia, 303
Potassium, potassium ion, 74, 148, 152, 154–56, 186, 195, 274, 309
Potato beetle, 229
Potato blight. *See* Late blight of potato.
Potato, Irish, white, 2, 3, 9, 90, *92,* 125, 187, 197, 213, 231, 274, 422, 430, 435, 436, 452, 525–26
Potato, sweet, *66,* 72, 125, 242, 422, 430
Potato vine, 109
Potential energy, 20
Powder puff flower, *430*
Powdery mildew, 323
Powdery scab of potato, 313
PPLO pneumonia, 276
PPLOs. *See* Pleuropneumonialike organisms.
Pq, 167
Prairie, 453, 455, 463
Prayer plant, 190, *194*
Praying mantis, 488
Precursor, 182, 184, 185
Prefern, 455
Preprophase band, 41
Preservation of plants, **413–18**
Pressure-flow hypothesis, **151–52**
Pressure potential, 146
Prickle, 93, 110, *111,* 134, 440
Prickly pear cactus, *93,* 347, 422, 432, *433*
Prickly poppy, *426*
Priestly, Joseph, 165
Primary
    consumer, *450–52*
    growth, 50
    meristem, 50, *62,* 63

phloem, 63, 65, 81, 83, *86–87*
    root, 61, *137*
    succession, 455–58
    tissue, 31, 50, 81, 83
    wall, 31, *33, 44, 52*
    xylem, 63, 65, 66, 81, 83, *86–87*
Primitive flower, 402, 409
Primordium, 81, *82,* 103, 124, 186
Primrose, 6
Procambium, 50, *51, 62,* 63, 81, *82,* 84, 86
Prochlorobacteria. *See* Chloroxybacteria.
Prochloron, 282, *283*
Prochlorophyta, 282
Prochlorothrix, 282
Producer, 410, *450, 451*
Progametangium, 321, *322*
Prokaryotic cell, 30, 167, 258, **266–74,** 279, 321
Proline, 148
Pronghorn, 464
Propagation. *See* Vegetative propagation.
Propagative roots, 66
Prophase, **41–42, 205, 207,** *208*
Proplastid, 37
Prop root, 67, *68, 137*
Protective layer (abscission), 116
Protein, 16, **22–24,** *34,* 38, 39, 76, 86, 115, 118, 139, 148, 160, 165, 169, 177, 187, 228, 229, 231, 284, 326, 393, 429, 440
    sequencer, 229
    structure, 22–*24,* 222–24
    synthesis, 24, 34, **222–24,** 229
Protein Z, 167
Prothallus, 372–74
Protist, 258, 292, 315, 321
Protoctista, 258–62, 279, **290–316**
    distinctions between Kingdomes Protoctista and Fungi, **321**
Protoderm, 50, *51, 62,* 63, 81, *82*
Protomere, *285*
Proton, 17–19, 165, 175
Protonema, *346, 349, 353,* 354
Proton pump, 148, 175
Protoplasm, 20, 31, **33–39,** 46, 58, 85, 150, 196, 235, 292
Protoplast fusion, **235–36**
Protostele, 83
Protozoa, 279, 292, 311, 315, 335
Pruning, **527–28**
Pseudomonad bacteria, 278, 326
Pseudoplasmodium, 313
Psilocybe mushrooms. *See* Teonanacatl mushrooms.
Psilophyton, 377
Psilotophyta, psilophytes, 262, **359–61**
Psilotum, **359–61**
Psittacosis, 275
Ptarmigan, 461
Pteridosperm, 383, 402
Pterophyta, 263
Puccoon, 421
Puffball, 320, 328, 330, *331*

Pulp (wood), 95–97, 394
Pulque, 117
Pulvinus, 190, 192, 194, *195,* 196
Pump, ion, 148
Pump, proton, 148, 175
Pumpkin, 127, 181, 422, 437
Pumpkin Family (Cucurbitaceae), 66, 109, 127, 136, 140, 188, 423, **437–38**
Punctuated equilibrium theory, 251
Puncture vine, 134, *135*
Punnett, R. C., 218
Punnett square, 218
Pun-tsao, 5
Purine, 221–24
Purple laver, 309, 310
Purple nonsulphur bacteria, 273
Purple sulphur bacteria, 273
Pygmy moss, 345
Pyrenoid, *297, 299,* 300, 350
Pyrethrum, 438
Pyrimidine, 221–24
Pyrrophyta, 262, **294–95.** *See also* Dinoflagellates.
Pyruvic acid, 169, 172–76

Q

Quantum, 165
Quiescence, **200**
Quillworts, 359, **364–67,** 384
    human and ecological relevance of, 365–66
Quince, 127
Quinine, 422

R

R group, 22–24
Rabbit, 248, 249
    cottontail, 465
    jack, 465
Rabies, 283, 284
Raccoon, 463
Raceme, *125*
Rachilla, *439*
Rachis, 104, 109, 372, 377
Radial symmetry (of flowers), 409
Radicle, 61, 136, *137,* 201
Radioactive tracer, 151
Radioactivity, 17, 76, 151, 213, 234, 235, 340, 448
Radish, 39, 66, 128, 181, 213, 421, 427
Rafflesia, 122
Ragweed, 197
Rain, acid, 446
Rain forest, tropical, 320, 446, 459, 460
Rambler, 93
Ranunculaceae. *See* Buttercup Family.
Raphe, 293
Raspberry, 93, *106,* 110, *111,* 126, 127, 240, *428,* 429, 527
Rat, 276, *277*
Rat snake, black, 463

Rattlesnake, 466
Ray, 54, 85, *90,* 95
    initial, 54
    parenchyma, 54
    phloem, 85, *86–87*
    xylem, 85, *86–87*
Ray floret, *438*
Reaction center molecule, 165, 167
Receptacle, *124,* 127, 128, 304, *305,* 409, *438*
Receptive hyphae, *333*
Recessive, **214**
Recombinant DNA technology, **228–32**
Recycling, 3, 272, 320, 447, 451–55, 459
Red algae, 76, **305–8,** 350
Red cedar, 398
Red cedar, western, 398, 466
Red fox, 463
Red light, 198
Red mangrove, *140*
Red mulberry, 95
Red pepper, 37
Red Sea, 279
Red spruce, *95,* 394, 395
Red tide, 294–95
Redbud
    eastern, 249, *250*
    western, 249, *250*
Reduction, 165
Reduction division, 205
Red wolf spider, *394*
Redwood, 31, 181, 384, 388, 396
    coastal, 366, 388, 389, 396, *466*
    dawn, 395
    giant, 384, 388, 389, 396, 458, 466
Redwood sorrel, 196
Regional (ecological) issues, **447–48**
Refrigerant, 459
Rehydration, 354, 355
Reindeer, 334, 340, 345
Reindeer moss, 350
Release factor, 224
Repair enzyme, 228
Reproduction, 15, 204. *See also* Asexual reproduction; Sexual reproduction.
Reproductive isolation, **248–51**
Reproductive leaves, *112*
Reserpine, 72
Reservoir (of *Euglena*), 295
Resin, 53, 85, 93, 117, 177, 384, 393, 429, 441
Resin canal, duct, 53, 85, *91, 109,* 384, *385*
Resolution, 29
Respiration, 16, 86, 138, 155, 160, **171–77,** 454
    aerobic, 138, **171–77**
    anaerobic, 138, **171–73**
    comparison between photosynthesis and, 177
    details of, **171–77**
    electron transport chain, 167–68, 172–73, **175–76**